LF

Hazardous Materials Spills Handbook

Hazardous Materials Spills Handbook

Gary F. Bennett
Professor of Biochemical Engineering
The University of Toledo

Frank S. Feates
Director, Nuclear Waste Management
U.K. Department of the Environment

Ira Wilder
Chief, Oil and Hazardous Materials Spills Branch
U.S. Environmental Protection Agency

McGraw-Hill Book Company
New York · St. Louis · San Francisco · Auckland
Bogotá · Hamburg · Johannesburg · London · Madrid
Mexico · Montreal · New Delhi · Panama · Paris
São Paulo · Singapore · Sydney · Tokyo · Toronto

Library of Congress Cataloging in Publication Data
Main entry under title:

Hazardous materials spills handbook.
    Includes index.
    1.Hazardous substances—Accidents—Handbooks,
manuals, etc.   I.Bennett, Gary F.   II.Feates,
Frank S.   III.Wilder, Ira, 1945—   .
T55.3.H3H43   363.1′79   82-123
ISBN 0-07-004680-8   AACR2

1 2 3 4 5 6 7 8 9 0   HDHD   8 9 8 7 6 5 4 3 2

ISBN 0-07-004680-8

The editors for this book were Diane D. Heiberg and Beatrice E. Eckes,
the designer was Mark E. Safran, and the production supervisor was Ter-
esa F. Leaden. It was set in Baskerville by University Graphics, Inc.

Printed and bound by Halliday Lithograph.

To our wives: Judy, Gwenda, and Jo

Contents

Contributors / xi

Preface / xv

Introduction / *Gary F. Bennett, Frank S. Feates, and Ira Wilder* / I-1

CHAPTER 1 International Laws and Regulations / 1-1

PART 1 Intergovernmental Maritime Consultative Organization / *C. Hugh Thompson* / 1-2

PART 2 Convention on the Prevention of Marine Pollution by Dumping Wastes and Other Matter / *John E. Portmann* / 1-13

PART 3 Transport of Dangerous Goods by Air / *D. W. Dines* / 1-21

CHAPTER 2 National Laws and Regulations / 2-1

PART 1 United States / *H. D. Van Cleave* / 2-2

PART 2 Canada / *John E. MacLatchy* / 2-10

PART 3 Europe / *J. Bentley* /   2-14

PART 4 U.S.S.R. / *William J. Lacy* /   2-20

CHAPTER 3 Information Systems and Reporting /   3-1

PART 1 U.S. Coast Guard Systems / *Mark E. Ives, Richard V. Harding, Michael C. Parnarouskis, Robert J. Embrie, Richard G. Potts, and Kirk R. Karwan* /   3-2

PART 2 Other United States Systems / *George J. Moein* /   3-9

PART 3 European Systems / *R. F. Cumberland* /   3-25

PART 4 Canadian Systems / *Robert A. Beach* /   3-45

CHAPTER 4 Impact /   4-1

PART 1 Municipal Facilities / *C. Joseph Touhill* /   4-2

PART 2 Environmental Effects / *John Cairns, Jr.* /   4-15

PART 3 Economic Effects / *Jonathan E. Amson* /   4-22

CHAPTER 5 Assessment of Hazard and Risk / *V. C. Marshall* /   5-1

CHAPTER 6 Prevention /   6-1

PART 1 Plant Operations / *William B. Katz* /   6-2

PART 2 Maritime Transportation / *William A. Creelman* /   6-25

PART 3 Rail Transportation / *Deborah K. Shaver and Robert M. Graziano* /   6-43

PART 4 Tank Truck Transportation / *Clifford J. Harvison* /   6-51

CHAPTER 7 Response Plans /   7-1

PART 1 United States Federal Governmental Plans / *Kenneth E. Biglane and James J. Yezzi, Jr.* /   7-3

PART 2 United States Local Governmental Plans / *Larry R. Froebe* /   7-15

PART 3 European Governmental Plans / *A. J. Fairclough* / 7-29

PART 4 United States Industrial Plans / *Randolph A. Jensen* / 7-43

PART 5 European Industrial Plans / *A. H. Smith* / 7-56

PART 6 Land Transportation Plans / *G. Stapleton* / 7-64

PART 7 British Maritime Transportation Plans / *H. P. Lunn* / 7-72

PART 8 Coast Guard Organization for United States Coastal Areas / *Gregory N. Yaroch* / 7-77

PART 9 Fire Service Role / *Charles L. Page* / 7-86

CHAPTER 8 Spills into Watercourses / 8-1

PART 1 Sampling, Analysis, and Detection / *Joseph P. Lafornara* / 8-2

PART 2 Bioassay / *Royal J. Nadeau* / 8-14

PART 3 Dispersion Modeling / *Michael T. Kontaxis and Joseph A. Nusser* / 8-22

CHAPTER 9 Spill Cleanup / 9-1

PART 1 Field-Implemented Measures / *Robert C. Scholz* / 9-2

PART 2 Technology Development / *Frank J. Freestone and John E. Brugger* / 9-24

PART 3 Biological Measures / *Neal E. Armstrong* / 9-40

PART 4 Chemical and Physical Measures / *W. W. Eckenfelder, Jr.* / 9-50

CHAPTER 10 Volatile Materials / 10-1

PART 1 Commercially Available Monitors for Airborne Hazardous Materials / *Sheridan J. Rodgers* / 10-3

PART 2 Atmospheric Dispersion / *Richard H. Schulze* / 10-13

PART 3 Vapor Hazard Control / *Ralph Hiltz* / 10-21

PART 4 Ammonia / *Phani Raj* /   10-34

PART 5 Chlorine / *Robert L. Mitchell, Jr.* /   10-53

PART 6 Flammable Liquid Gases / *L. Edward Brown and Larry M. Romine* /   10-63

CHAPTER 11 Case Histories: Volatile Materials /   11-1

PART 1 Pesticide Fires / *Russell E. Diefenbach* /   11-2

PART 2 Bulk Terminals: Silicon Tetrachloride Incident / *William C. Hoyle* /   11-11

PART 3 Seveso Accident: Dioxin / *Alex P. Rice* /   11-18

CHAPTER 12 Case Histories: Land and Water Spills /   12-1

PART 1 Polychlorinated Biphenyls / *Al J. Smith* /   12-2

PART 2 Phenol / *Austin Shepherd* /   12-12

PART 3 Pesticides / *Lee Frisbie* /   12-18

CHAPTER 13 Personnel Safety Equipment / *Dennis Rome* /   13-1

CHAPTER 14 Ultimate Disposal /   14-1

PART 1 Transportation / *Ronald J. Buchanan* /   14-2

PART 2 Landfill / *A. Parker* /   14-11

PART 3 Incineration on Land / *Rudy G. Novak and Charles Pfrommer, Jr.* /   14-20

PART 4 Incineration at Sea / *E. E. Finnecy* /   14-37

CHAPTER 15 The Future / *Dennis M. Stainken* /   15-1

APPENDIX Conversion Factors / A-1

Index follows the Appendix.

Contributors

Jonathan E. Amson, Biological Science Administrator, Office of Water Regulations and Standards, U.S. Environmental Protection Agency (Chap. 4, Part 3, "Economic Effects")

Neal E. Armstrong, Ph.D., Professor of Civil Engineering, University of Texas at Austin (Chap. 9, Part 3, "Biological Measures")

Robert A. Beach, Manager, National Environmental Emergency Centre, Environmental Protection Service, Environment Canada (Chap. 3, Part 4, "Canadian Systems")

Gary F. Bennett, Ph.D., Professor of Biochemical Engineering, The University of Toledo, Toledo, Ohio (Introduction)

J. Bentley, Land Waste Division, U.K. Department of the Environment (Chap. 2, Part 3, "Europe")

Kenneth E. Biglane, Director, Hazardous Response Support Division, U.S. Environmental Protection Agency (Chap. 7, Part 1, "United States Federal Governmental Plans")

L. Edward Brown, President, Energy Analysts, Inc., Norman, Oklahoma (Chap. 10, Part 6, "Flammable Liquid Gases")

John E. Brugger, Ph.D., Physical Scientist, Oil and Hazardous Materials Spills Branch, U.S. Enivronmental Protection Agency (Chap. 9, Part 2, "Technology Development")

Ronald J. Buchanan, Ph.D., Manager, Environmental Affairs, Conversion Systems, Inc., Horsham, Pennsylvania (Chap. 14, Part 1, "Transportation")

John Cairns, Jr., Ph.D., University Distinguished Professor and Director, Center for Environmental Studies, Virginia Polytechnic Institute and State University, Blacksburg, Virginia (Chap. 4, Part 2, "Environmental Effects")

William A. Creelman, President, Transport Division, National Marine Service, Inc., St. Louis, Missouri (Chap. 6, Part 2, "Maritime Transportation")

R. F. Cumberland, Manager, National Chemical Emergency Centre, Harwell Laboratory, Oxfordshire, England (Chap. 3, Part 3, "European Systems")

Russell E. Diefenbach, Environmental Emergency Section, Region V, U.S. Environmental Protection Agency (Chap. 11, Part 1, "Pesticide Fires")

D. W. Dines, IATA Restrictive Lists Manager, British Airways, Hounslow, England (Chap. 1, Part 3, "Transport of Dangerous Goods by Air")

W. W. Eckenfelder, Jr., Distinguished Professor of Environmental and Water Resources Engineering, Vanderbilt University, Nashville, Tennessee (Chap. 9, Part 4, "Chemical and Physical Measures")

Robert J. Embrie, Marine Environmental Protection Division, U.S. Coast Guard (Chap. 3, Part 1, "U.S. Coast Guard Systems")

A. J. Fairclough, U.K. Department of Transport (Chap. 7, Part 3, "European Governmental Plans")

Frank S. Feates, Ph.D., Director, Nuclear Waste Management, U.K. Department of the Environment (Introduction)

E. E. Finnecy, Hazardous Materials Service, Harwell Laboratory, Oxfordshire, England (Chap. 14, Part 4, "Incineration at Sea")

Frank J. Freestone, Chief, Hazardous Spills Staff, Oil and Hazardous Materials Spills Branch, U.S. Environmental Protection Agency (Chap. 9, Part 2, "Technology Development")

Lee Frisbie, President, Environment Plus, Olathe, Kansas (Chap. 12, Part 3, "Pesticides")

Larry R. Froebe, Ph.D., Regional Response Team Manager, Ecology and Environment, Inc., Dallas, Texas (Chap. 7, Part 2, "United States Local Governmental Plans")

Robert M. Graziano, Vice President, Marketing, O. H. Materials, Inc., Findlay, Ohio (Chap. 6, Part 3, "Rail Transportation")

Richard V. Harding, Marine Environmental Protection Division, U.S. Coast Guard (Chap. 3, Part 1, "U.S. Coast Guard Systems")

Clifford J. Harvison, Managing Director, National Tank Truck Carriers, Inc., Washington, D.C. (Chap. 6, Part 4, "Tank Truck Transportation")

Ralph Hiltz, Senior Scientist, MSA Research Corporation, Evans City, Pennsylvania (Chap. 10, Part 3, "Vapor Hazard Control")

William C. Hoyle, Ph.D., Corporate Analytical Services, The Continental Group, Inc., Downers Grove, Illinois (Chap. 11, Part 2, "Bulk Terminals: Silicon Tetrachloride Incident")

Mark E. Ives, Marine Environmental Protection Division, U.S. Coast Guard (Chap. 3, Part 1, "U.S. Coast Guard Systems")

Randolph A. Jensen, P.E., President, Jensen Consultants, Middletown, Kentucky (Chap. 7, Part 4, "United States Industrial Plans")

Kirk R. Karwan, Marine Environmental Protection Service, U.S. Coast Guard (Chap. 3, Part 1, "U.S. Coast Guard Systems")

William B. Katz, President, Illinois Chemical Corporation, Highland Park, Illinois (Chap. 6, Part 1, "Plant Operations")

Michael T. Kontaxis, Project Engineer, Hydroqual, Inc., Mahwah, New Jersey (Chap. 8, Part 3, "Dispersion Modeling")

William J. Lacy, Director of Research and Development, Water and Hazardous Waste Monitoring, U.S. Environmental Protection Agency (Chap. 2, Part 4, "U.S.S.R.")

Joseph P. Lafornara, Ph.D., Acting Chief, Chemical Evaluation and Safety Section, Environmental Response Team, U.S. Environmental Protection Agency (Chap. 8, Part 1, "Sampling, Analysis, and Detection")

H. P. Lunn, General Manager, Cargo and Operational Services Division, Overseas Containers Ltd., London, England (Chap. 7, Part 7, "British Maritime Transportation Plans")

John E. MacLatchy, Legislative Adviser, Environmental Protection Service, Environment Canada (Chap. 2, Part 2, "Canada")

V. C. Marshall, Director of Safety Services, University of Bradford, Bradford, England (Chap. 5, "Assessment of Hazard and Risk")

Robert L. Mitchell, Jr., Executive Director, The Chlorine Institute, New York (Chap. 10, Part 5, "Chlorine")

George J. Moein, Chief, Emergency Response and Control Section, Region IV, U.S. Environmental Protection Agency (Chap. 3, Part 2, "Other United States Systems")

Royal J. Nadeau, Ph.D., Acting Chief, Environmental Impact Section, Environmental Response Team, U.S. Environmental Protection Agency (Chap. 8, Part 2, "Bioassay")

Rudy G. Novak, IT Enviroscience, Inc., Knoxville, Tennessee (Chap. 14, Part 3, "Incineration on Land")

Joseph A. Nusser, President, Nusser and Associates, Ho-Ho-Kus, New Jersey (Chap. 8, Part 3, "Dispersion Modeling")

Charles L. Page, Training Specialist, Fire Protection Training Division, Texas Engineering Extension Service, The Texas A&M University System, College Station, Texas (Chap. 7, Part 9, "Fire Service Role")

A. Parker, Hazardous Materials Service, Harwell Laboratory, Oxfordshire, England (Chap. 14, Part 2, "Landfill")

Michael C. Parnarouskis, Marine Environmental Protection Division, U.S. Coast Guard (Chap. 3, Part 1, "U.S. Coast Guard Systems")

Charles Pfrommer, Jr., IT Enviroscience, Inc., Knoxville, Tennessee (Chap. 14, Part 3, "Incineration on Land")

John E. Portmann, Ph.D., Fisheries Laboratory, Burnham on Crouch, Essex, England, U.K. Ministry of Agriculture, Fisheries, and Food (Chap. 1, Part 2, "Convention on the Prevention of Marine Pollution by Dumping Wastes and Other Matter")

Richard G. Potts, Marine Environmental Protection Division, U.S. Coast Guard (Chap. 3, Part 1, "U.S. Coast Guard Systems")

Phani Raj, President, Technology and Management Systems, Burlington, Massachusetts (Chap. 10, Part 4, "Ammonia")

Alex P. Rice, Director of Client Services, Cremer and Warner, Ltd., London, England (Chap. 11, Part 3, "Seveso Accident: Dioxin")

Sheridan J. Rodgers, MSA Research Corporation, Evans City, Pennsylvania (Chap. 10, Part 1, "Commercially Available Monitors for Airborne Hazardous Materials")

Dennis Rome, Lieutenant Commander, U.S. Coast Guard (Chap. 13, "Personnel Safety Equipment")

Larry M. Romine, Senior Engineer, Energy Analysts, Inc., Norman, Oklahoma (Chap. 10, Part 6, "Flammable Liquid Gases")

Robert C. Scholz, Environmental Research Center, Rexnord Inc., Milwaukee, Wisconsin (Chap. 9, Part 1, "Field-Implemented Measures")

Richard H. Schulze, President, Trinity Consultants, Dallas, Texas (Chap. 10, Part 2, Atmospheric Dispersion")

Deborah K. Shaver, Manager of Hazardous Materials Programs, Systems Technology Laboratory, Inc., Arlington, Virginia (Chap. 6, Part 3, "Rail Transportation")

Austin Shepherd, Manager of Environmental Control, Singer Company, Stamford, Connecticut (Chap. 12, Part 2, "Pheonol")

A. H. Smith, M.C.I.T., Distribution Manager, Laporte Industries, Ltd., Widnes, Cheshire, England (Chap. 7, Part 5, "European Industrial Plans")

Al J. Smith, P.E., Chief, Environmental Emergency Branch, Air and Waste Management Division, U.S. Environmental Protection Agency (Chap. 12, Part 1, "Polychlorinated Biphenyls")

Dennis M. Stainken, Allied Corporation, Morristown, New Jersey (Chap. 15, "The Future")

G. Stapleton, Motorway Maintenance Department, Hereford and Worcester County Council, Warndon, Worcestershire, England (Chap. 7, Part 6, "Land Transportation Plans")

C. Hugh Thompson, Ph.D., Director, Environmental Affairs, Aerojet-General Corporation, Sacramento, California (Chap. 1, Part 1, "Intergovernmental Maritime Consultative Organization")

C. Joseph Touhill, Ph.D., President, Baker/TSA, Inc., Pittsburgh, Pennsylvania (Chap. 4, Part 1, "Municipal Facilities")

H. D. Van Cleave, Acting Director, Emergency Response Division, Office of Emergency and Remedial Response, U.S. Environmental Protection Agency (Chap. 2, Part 1, "United States")

Ira Wilder, Chief, Oil and Hazardous Materials Spills Branch, U.S. Environmental Protection Agency (Introduction)

Gregory N. Yaroch, Lieutenant Commander, U.S. Coast Guard (Chap. 7, Part 8, "Coast Guard Organization for United States Coastal Areas")

James J. Yezzi, Jr., Physical Scientist, Oil and Hazardous Materials Spills Branch, U.S. Environmental Protection Agency (Chap. 7, Part 1, "United States Federal Governmental Plans")

Preface

Hazardous materials have been and will continue to be articles of commerce which, when properly used and converted, provide some of the materials that are an everyday part of our standard of living. When properly handled, they pose no threat or danger. Unfortunately, no matter how careful the planning and how great the care, accidents can and do occur. It is then critical that we negate or minimize the effects of such accidents on the population and the environment.

Work to contain spilled hazardous materials and minimize damage has been carried out for many years on an individual, seat-of-the-pants basis. It is only since 1970 that work in developing methodologies for containment and mitigation has been carried out on a planned basis. At the same time, the public has demanded stricter rules for shipping and handling hazardous materials. This effort is now providing systems and methodologies for the sound handling of hazardous materials and the unfortunate accidents that may occur in their transport. These then serve as a basis for developing regulations to guide the transporters.

The editors of the *Hazardous Materials Spills Handbook* have been leaders in this field since 1970. They have been active in the field and have seen it grow and mature. They bring to the *Handbook* a unique insight into the advances and an appreciation of the problems because of their long leadership in the area.

The material presented in the *Handbook* provides a basis for anyone to manage hazardous materials movement and develop plans for the prevention of accidents and for the mitigation of spills and other incidents that may occur. The authors of the various chapters and parts are all in the forefront of their respective fields and are all intimately acquainted with the management of hazardous materials.

As an early worker in this exciting field, I consider it a pleasure to see this compendium, whose time and need is now, available for use by those who are now active and will be active in the future in ensuring that we can continue to handle, as articles of commerce, those hazardous materials which are so necessary to support the lifestyle which we have and aspire to.

Peter B. Lederman, Ph.D., P.E.
Vice President,
Roy F. Weston

Hazardous Materials Spills Handbook

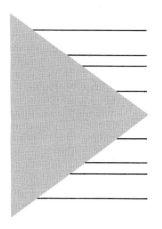

Introduction

Gary F. Bennett, Ph.D.
Professor of Biochemical Engineering
The University of Toledo

Frank S. Feates, Ph.D.
Director, Nuclear Waste Management
U.K. Department of the Environment

Ira Wilder
Chief, Oil and Hazardous Materials Branch
U.S. Environmental Protection Agency

Magnitude of the Problem / 3
Severity of the Problem / 5
Transportation / 7
Effective Response / 9
Assistance / 9
Prevention / 10
Safety / 11
Disposal / 12
Conclusion / 12

The wrecking of the *Torrey Canyon* in 1967 on the Seven Stones Rocks in the English Channel, with the loss of 117,000 metric tons of Kuwait crude oil, focused worldwide attention on the increasing problem faced by even major water bodies because of oil spills. As a result of widespread awareness of the magnitude of the oil spill problem, governmental activities began in the 1970s in Britain, with research on oil removal techniques at the Warren Spring Laboratories, while in the United States, Congress expressed concern in 1970 in Public Law 91-224, the Water Quality Improvement Act. As a result, the U.S. Environmental Protection Agency (U.S. EPA) published its requirements for oil spill prevention control and countermeasure (SPCC) plans on December 11, 1973.

The problems of hazardous materials, though known, were not addressed until later in the 1970s. In the United States, one of the first steps taken by the U.S. EPA was to form a research laboratory at Edison, New Jersey, to address the urgent problems of prevention, control, and cleanup of hazardous materials spills. In the United Kingdom, the Hazardous Materials Service at Harwell was set up with similar objectives.

Little guidance was available to these early workers, and indeed there was not even widespread realization or acceptance that the problem was severe. An initial action was to sponsor a National Conference on Control of Hazardous Material Spills in Houston in March 1972[1] to define the problem and suggest solutions. The papers presented at this conference clearly demonstrated the need for development of hazardous material spills control technology.

The U.S. Congress addressed the problem of hazardous material spills in Public Law 92-500, the Water Quality Improvement Act of 1972. In the preamble to Sec. 311, Congress stated its intention that no hazardous materials be spilled into any navigable body

Fig. 1 Fire in 1974 at the Bulk Terminal Co. in Calumet Harbor. [*Photograph courtesy Illinois Department of Transportation.*]

of water of the United States. To this end, it directed the U.S. EPA to develop a list of hazardous chemicals, specifying penalties for spilling and determining reportable quantities and removability. However, it was not until 1979 that these regulations were finally in place, covering almost 300 chemicals. The tortuous path taken by the U.S. EPA to promulgate its regulations is described in detail by Bennett and Wilder.[2] The Superfund law, the Comprehensive Environmental Response, Compensation, and Liability Act of 1980 (see Chap. 2, Part 1, and Chap. 15), will result in new regulations designating additional hazardous chemicals and establishing reportable quantities for their release not only into navigable waters but also into groundwaters, soils, sediments, and the atmosphere.

In the United Kingdom, the Deposit of Poisonous Wastes Act of 1972 and the Control of Pollution Act of 1976 had the same goals as Sec. 311 of Public Law 92-500.

MAGNITUDE OF THE PROBLEM

Worldwide, billions of kilograms of hazardous chemicals are produced daily. According to data published in *Chemical and Engineering News*[3] for 1979, production of the 50 most prolific chemicals totals almost 266 billion kg a year (Table 1). Almost one-half of the 50 most common chemicals are deemed hazardous (Table 2) according to the U.S. EPA. In 1979 their production totaled almost 138 billion kg (about 377 million kg a day) and was growing at an annual rate of 7.6 percent.

Of course, not all chemicals are hazardous, nor do all hazardous chemicals have the same degree of toxicity. The U.S. EPA has categorized chemicals according to toxicity criteria: X is the most toxic and D the least, with A, B, and C being intermediate. If 1 lb (0.45 kg) of a chemical in Category X is spilled into a watercourse, federal authorities must be notified, but it is not until 5000 lb (2268 kg) of Category D chemicals reaches a watercourse that notification is required.[4]

It is inevitable that when chemicals are produced, stored, or shipped, they will be

TABLE 1 1979 Production Rate of the 50 Top Chemicals in the United States

Rank, 1979		Production, 1979 (billion kg)	Average annual change, 1978–1979 (percent)	Hazardous chemical category list*
1	Sulfuric acid	38.07	2.2	C
2	Lime	17.59	0.0	
3	Ammonia	16.44	5.4	B
4	Oxygen	16.03	−0.2	
5	Nitrogen	13.57	5.4	
6	Ethylene	13.24	12.5	
7	Sodium hydroxide	11.24	15.7	C
8	Chlorine	10.99	9.6	A
9	Phosphoric acid	9.19	8.3	D
10	Nitric acid	7.77	7.9	C
11	Sodium carbonate	7.49	−0.4	
12	Ammonium nitrate	7.08	8.2	
13	Propylene	6.49	9.9	
14	Urea	6.14	24.2	
15	Benzene	5.77	16.3	C
16	Toluene	5.38	55.2	C
17	Ethylene dichloride	5.36	7.5	D
18	Ethyl benzene	3.87	1.7	C
19	Vinyl chloride	3.42	8.6	
20	Styrene	3.39	4.0	C
21	Methanol	3.36	15.1	
22	Terephthalic acid	3.29	22.0	
23	Carbon dioxide	3.21	5.7	
24	Xylene	3.13	12.4	C
25	Formaldehyde	2.93	1.1	C
26	Hydrochloric acid	2.70	6.6	D
27	Ethylene oxide	2.39	5.4	
28	Ethylene glycol	2.09	17.9	
29	p-Xylene	1.90	18.8	C
30	Cumene	1.81	18.3	
31	Ammonium sulfate	1.79	1.8	
32	1,3-Butadiene	1.61	0.9	
33	Acetic acid	1.51	19.8	C
34	Carbon black	1.51	−0.6	
35	Phenol	1.34	10.1	C
36	Acetone	1.13	−0.8	
37	Aluminum sulfate	1.12	−5.9	D
38	Cyclohexane	1.09	3.4	C
39	Sodium sulfate	1.06	0.4	
40	Propylene oxide	1.02	9.8	D
41	Calcium chloride	0.92	−1.5	
42	Acrylonitrile	0.92	15.4	B
43	Vinyl acetate	0.90	18.6	C
44	Isopropyl alcohol	0.89	13.9	
45	Adipic acid	0.82	11.1	D
46	Sodium silicate	0.70	−3.1	
47	Sodium tripolyphosphate	0.68	2.3	
48	Acetic anhydride	0.68	0.7	C
49	Titanium dioxide	0.67	5.7	
50	Ethanol	0.59	3.1	
	Total	265.94	7.6	

*Hazardous material category defined by the U.S. Environmental Protection Agency under Sec. 311 of the Water Quality Improvement Act (Public Law 92-500) of 1972. See G. F. Bennett and I. Wilder, "Evolution of Hazardous Material Spills Regulations in the United States," *Journal of Hazardous Materials*, vol. 4, January 1981, p. 257; U.S. Environmental Protection Agency, "Hazardous Substances: Definition, Designation, Revocation of Regulations, Proposed Expansion of Criteria of Designation and Proposed Determination of Reportable Quantities," *Federal Register*, vol. 44, Feb. 16, 1979, pp. 10266–10284.
    SOURCE: "Top 50 Chemicals," *Chemical and Engineering News*, vol. 58, June 11, 1980, p. 36.

TABLE 2   Hazardous Materials Found in the Top 50 United States–Produced Chemicals

Hazardous materials category*	Harmful quantity, kg (lb)		Number of chemicals in category	1979 production (billion kg)
X	0.45	(1.0)	0	0.00
A	4.5	(10)	1	10.99
B	45	(100)	2	17.36
C	454	(1000)	15	88.97
D	2270	(5000)	6	20.21
Total			24	137.53

*Hazardous material category defined by the U.S. Environmental Protection Agency under Sec. 311 of the Water Quality Improvement Act (Public Law 92-500) of 1972. See G. F. Bennett and I. Wilder, "Evolution of Hazardous Material Spills Regulations in the United States," *Journal of Hazardous Materials,* vol. 4, January 1981, p. 257; U.S. Environmental Protection Agency, "Hazardous Substances: Definition, Designation, Revocation of Regulations, Proposed Expansion of Criteria of Designation and Proposed Determination of Reportable Quantities," *Federal Register,* vol. 44, Feb. 16, 1979, pp. 10266–10284.

SOURCE: "Top 50 Chemicals," *Chemical and Engineering News,* vol. 58, June 11, 1980, p. 36.

spilled. Indeed, Dawson and Stradley[5] estimated that 0.025 percent of the sulfuric acid produced would be spilled and that approximately 60 percent of the spillage would be caused by transportation accidents. They also estimated that almost one-half of the spilled acid would reach water bodies, causing substantial harm to aquatic systems.

One of the first comprehensive attempts to define the magnitude of the spill problem on a national scale was made by Buckley and Wiener,[6] looking back at 15,000 spills in the United States over a 2½-year period. The five materials spilled most often and the average spilled volume are shown in Table 3.

A hazard potential analysis based on the size of spill and toxicity was also devised[6] to rank the danger posed by spilled material (Table 4). From this analysis, one can see that it is neither the most prevalent chemical (e.g., sulfuric acid) nor the most toxic (e.g., ethyl parathion) that represents the greatest environmental threat.

SEVERITY OF THE PROBLEM

"Destruction and death struck Texas City this morning at 9:15 when the French ship Grandcamp, loading nitrate at the docks exploded throwing a fiery column into the land, killing an untold number and destroying almost the entire water front area, including the Texas City Terminal and the Monsanto Chemical Plant" (*Texas City Press,* Sunday, April 1947).[7]

TABLE 3   Most Frequently Spilled Materials as Identified in an EPA Study

Chemical spilled	Average spill	Maximum spill
Sulfuric acid	30 m^3	1,100 m^3
Ammonia nitrate	16,000 kg	910,000 kg
Caustic soda	15 m^3	190 m^3
Hydrochloric acid	13 m^3	360 m^3
Ethyl parathion	0.62 m^3	14 m^3

SOURCE: J. L. Buckley and S. A. Wiener, *Hazardous Material Spills: A Documentation and Analysis of Historical Data,* EPA-600/2-78-066, U.S. Environmental Protection Agency, April 1978.

TABLE 4   Hazard Potential for
Hazardous Chemical Spills

Chemical spilled	Hazard potential
Anhydrous ammonia	6.00
Toluene	5.86
Nitric acid	4.73
Phenol	4.53
Methanol	3.57
Xylene	3.40

SOURCE: J. L. Buckley and S. A. Wiener, *Hazardous Material Spills: A Documentation and Analysis of Historical Data,* EPA-600/2-78-066, U.S. Environmental Protection Agency, April 1978.

This explosion made clear the dangerous aspect of hazardous chemicals, but it was not the first, nor was it the last, of the hazardous chemical explosions. To appreciate the danger posed by hydrocarbons, one has only to look at data from a list of the 100 worst chemical explosions resulting from unconfined (organic) vapor clouds (Table 5). This list was compiled by Gugan[8] in his comprehensive study that evolved from the Flixborough disaster in England, in which cyclohexane exploded, killing 28 people.

Even small quantities of chemicals present a potential danger according to an article in *The Wall Street Journal:* "After 50 years some fear Edison's vials now might be vile; officials close historic site in New Jersey to check 10,000 items for danger."[9]

Transportation accidents have had tragic results. Two of note in 1978 were a freight train derailment in Florida, in which a chlorine release killed 8 persons; and the killing of 150 persons at a camp site in Spain by a runaway propylene tanker.

The environment too has suffered. One incident was described in 1972 by William Nye,[10] then director of natural resources for the state of Ohio. Some demented person threw 1 kg of endrin-contaminated strychnine into a small Ohio lake. Virtually all living

Fig. 2   Petroleum refinery tank fire in Toledo, Ohio. [*Photograph courtesy Associated Chemical and Environmental Services, Oregon, Ohio.*]

TABLE 5   Unconfined Vapor Cloud Explosions: Selected Data from the Work of Gugan

Year	Place	Chemical	Amount (kg)	Blast	Number killed
1944	Cleveland, Ohio	Methane	182,000	No	213
1948	Ludwigshafen, West Germany	Dimethyl ether	30,000	Yes	209
1959	Meldrin, Georgia	Liquefied petroleum gas	......	Yes	23
1961	Texas	Cyclohexane	3,860	Yes	1
1966	Feyzin, France	Propane	432,000	No	17
1971	Longview, Texas	Ethylene	......	Yes	3
1972	Brazil	Butane	......	Yes	37
1974	Flixborough, England	Cyclohexane	36,000	Yes	28
1975	Beek, Netherlands	Propylene	5,450	Yes	14

SOURCE: K. Gugan, *Unconfined Vapor Cloud Explosions,* Gulf Publishing Co., Houston, 1978.

things died; the entire fish population was destroyed, the only surviving aquatic invertebrates being tadpoles. In those early days of spill response, there was no mobile equipment, proven methods, or successful examples of past cleanup. As a result, cleanup of the lake was difficult and expensive. But lessons were learned and needs highlighted, pointing the way to future development of cleanup technology.

TRANSPORTATION

Although much larger quantities of chemicals are stored at fixed-base facilities than are in transport at any one time, the transportation system presents unique spill problems for these reasons:

1. Accidents may happen at any place or any time, and some occur at the most inopportune times and places.
2. The exact nature of the chemical involved in a spill may not be known to responding personnel, as placards may have been destroyed, the driver injured, or shipping papers lost.
3. Knowledgeable response personnel and equipment may not be available (i.e., response may be made by a fire department that lacks a trained hazardous material response team; most departments in the United States do not have such teams).

And when a transportation accident happens, the result can be dramatic, as in the propylene tanker explosion in Spain or the chlorine release in Florida.

The problem faced at the scene of a railroad derailment–caused chlorine spill can be contrasted with procedures and policies in force at the Dow Chemical Company's plant in Louisiana, where contingency planning for a spill includes computerized models of a dispersing chlorine plume and the training of both in-house and local emergency response personnel in what to do, whom to evacuate, and even how to recognize the odor of chlorine.[11]

Hence transportation-related spills present different and often very serious problems.

Fig. 3   Spilled solid chemicals in a railroad transportation accident. [*Photograph courtesy Associated Chemical and Environmental Services, Oregon, Ohio.*]

Moreover, they are not rare. In 1977, according to the U.S. Coast Guard,[12] there were 12,638 reported spills totaling 46,000 m³ into the navigable waters of the United States. Of these spills, 10,556 were oil-related, 234 involved hazardous materials, and the remaining 1848 involved materials other than those classified as oil or hazardous materials. Volumes spilled in the three categories were 34,000, 1600, and 10,000 m³ respectively. In 1978 there were 14,495 incidents in which a total of 66,000 m³ of liquid was spilled.

The U.S. National Transportation Safety Board (NTSB) reported[13] that during the period 1975–1977 a total of 10,481 railroad cars containing hazardous materials were involved in accidents, 27,720 people were evacuated, and $70 million in railroad equipment was damaged. In 1977 alone, 32 were killed and 744 injured, with 30 of the deaths and 503 of the injuries resulting from highway accidents (Table 6).

The most dangerous chemicals are shown in Table 7. Explosive hydrocarbons (e.g., gasoline, 17 deaths), caused the most numerous deaths, whereas various gases and corrosives (bromine, 101 injuries; sulfuric acid, 83; ammonia, 81; chlorine, 44; and nitric acid, 43) caused the most numerous injuries.

TABLE 6   U.S. Transportation Accidents Resulting in Injury or Death in 1977

Transport mode	Deaths	Injuries
Highway	30	503
Railroad	2	231
Air	0	10
Water	0	0

SOURCE: J. M. Winton, "Chemical Cargo Safety Gets Them through Track," *Chemical Week*, vol. 126, July 5, 1978, p. 29.

TABLE 7   Injuries and Deaths by Hazardous
Commodity Spilled in Transportation Accidents

Hazardous commodity	Deaths	Injuries
Acetaldehyde	0	1
Ammonia	5	81
Bromine	0	101
Chlorine	0	44
Gasoline	17	31
Methyl methacrylate	1	7
Nitric acid	0	43
Liquefied petroleum gas	1	22
Sodium hydroxide	0	17
Sulfuric acid	0	83

SOURCE: J. M. Winton, "Chemical Cargo Safety Gets The
Through Track," *ChemicalWeek*, vol. 126, July 5, 1978, p. 29.

EFFECTIVE RESPONSE

The *Proceedings of the 1974 National Conference on Control of Hazardous Material Spills*[14] gives quite a different picture from that described at the 1972 conference. In 1974, papers were presented which described the beginnings of effective response and cleanup technology development. The U.S. EPA presented the first of a series of papers on its mobile activated-carbon system, which had been used effectively to clean a creosote spill in the Little Menomonee River in 1972. Since that auspicious beginning, this unit has been used more than 40 times to remove pesticides and other organics such as PCBs from contaminated water bodies. Moreover, the basic premise, that a mobile carbon adsorption system is needed in spill response, has been proved, and the unit developed by the U.S. EPA has been duplicated by several commercial firms.

The spillage of oil and hazardous materials has created a whole new industry, with firms specializing in spill response and cleanup. These contractors are normally called upon to assist the spiller in cleanup operations, and in that capacity they can work under the direction of either the spiller or the government. Details of spill response, management, scenarios, and cleanup coordination have been described by Smith in his book *Managing Hazardous Substances Accidents*.[15]

ASSISTANCE

Concomitant with the development of equipment for spill cleanup, information systems to supply emergency personnel with needed data on the hazards of the chemicals involved were being developed. In 1975, the U.S. Coast Guard produced the four-volume *CHRIS Manual*, which in 1978 underwent a revision (mainly an expansion that doubled the size of the chemical data base) and computerization.[16] The U.S. EPA proceeded on a parallel path, developing its OHM-TADS system,[17] which is totally computerized, although hard copy is available by portable terminal. Both computer systems are accessible by telephone at 1-800-424-8802, which is also the number to call for the legally required reporting of spills in the United States.

Assistance to spill responders is also available in the United States on a 24-h basis through the CHEMTREC system, operated by the Chemical Manufacturers Association: 1-800-424-9300.

A development reported at the 1980 National Conference on Control of Hazardous Material Spills[18] revealed plans to link by telephone and computer CHEMTREC and the National Response Center.[19] Similar developments in providing assistance have occurred at Harwell[20] in the United Kingdom, where the emphasis is on professional chemists as well as on computers. The Canadians have also developed a computer-assisted system comparable with that in the United States.[21]

Modern mathematical techniques are now being employed for dispersion prediction from plant facilities when the potential problems and terrain are known. Gaussian air-dispersion models are being used to evaluate the potential downwind concentration of volatile chemicals such as chlorine that would result from a spill. In water, mixing models are being used to predict time-distance concentrations of soluble toxic materials in order to determine the passage of dangerous levels of contaminants past a water intake.

At the federal level, the U.S. EPA and the U.S. Coast Guard offer assistance when a major spill occurs; these agencies have the ability to utilize the National Oil and Hazardous Substances Pollution Contingency Plan,[15] which gives access to the resources of a number of governmental agencies (see also Chap. 7, Part 1). On the operating level, they can activate EPA's environmental response team (ERT), a group of scientists who offer on-scene technical advice; they can also call for sophisticated equipment such as the mobile activated-carbon system, mobile laboratories, and stream diversion systems to assist in spill evaluation and cleanup.

The U.S. Coast Guard also has its own strike teams to respond to spills of oil and hazardous materials.[22] In the United States there are three such teams of 18 persons each, ready to leave on short notice, supplied with air-transportable equipment, and trained to go into action on arrival.

The transportation industry too has been responsive to national needs in spill response. The United States railroads announced at a conference in April 1980[23] a new information system involving a seven-digit code to identify chemicals, backed by a response system to give information on the hazards of those chemicals. To improve future decision-making procedures, the NTSB is producing maps that feature a time-sequenced display of dispersion patterns for spills along with information on the weather at the time the spill is reported, the number of injuries and fatalities, the spill location, and a synopsis of the accident scenario.[24]

The U.S. EPA's R&D group in Edison, New Jersey, has produced unique equipment to improve spill response and cleanup. It has developed units as small as a hand-carried spill-diking system and as large as mobile physical-chemical treatment systems, mobile laboratories, and mobile incineration systems. One company has produced a commercial model of EPA's developed portable detection kit for hazardous materials. These developments have been described in numerous papers at the national conferences and are also discussed in Chap. 9.[1,14,18,25,26]

Additional developments made through industrial research and development (independently of EPA) include:

1. A universal adsorbent that picks up virtually all hazardous substances[27]
2. Mutant bacterial cultures that will degrade spilled materials such as phenol and oil[28]

PREVENTION

Spill response and cleanup are generally viewed as negative, nonproductive ventures and expensive ones at that. The key to avoiding the cost of a spill is prevention, and prevention of spills was the thrust of the successful SPCC regulations issued by the U.S. EPA in 1973.[29]

Fig. 4 Vacuum truck is operating to retrieve spilled chemicals. [*Photograph courtesy Associated Chemical and Environmental Services, Oregon, Ohio.*]

These regulations require that all United States facilities storing oil in amounts greater than 5 m³ above ground and 76 m³ below ground prepare a thoroughly thought-out and preengineered plan to prevent spills, a scenario to describe where a spilled liquid will flow if a spill does occur, and a plan to clean up the spilled material. Aspects of this prevention plan should include, according to the regulations, needed physical changes in storage facilities such as dikes to contain the contents of a tank if it fails, high-level alarms to guard against overfilling, and administrative controls such as supervision of tank filling, tanker unloading, etc. Although not yet fully promulgated, similar proposed regulations were published by the EPA to control hazardous materials in addition to oil.[30] Chapter 6 of the *Handbook* describes both of the above aspects of prevention: (1) the physical facility (one should examine one's facility with a "new pair of critical eyes" to spot and eliminate spill potentials) and (2) an administrative response plan for an industrial complex which suggests protocols for responding to spills.

SAFETY

When oil spills are being cleaned up, the safety of response teams is rarely a major concern, provided precautions are taken to prevent a fire. But hazardous material cleanup is totally different owing to the toxic nature of many chemicals. For instance, each chemical has specific properties and can present radically different hazards from another "equally hazardous chemical"; chlorine is dangerous when inhaled, acids cause skin burns, and lead compounds are toxic when absorbed dermally. Exposure to some chemicals can result in death in a very short time when a person is exposed to a sufficiently high concentration; other chemicals, benzene, for example, may cause cancer when a person is exposed to low doses over a long period of time.

Safety equipment that should be used by cleanup personnel can be as simple as a face mask and a fire fighter's turnout gear. For more dangerous situations, personnel may be required to wear approved masks to remove organic vapors (at ambient concentrations not to exceed 10 times the approved threshold-limit value). For the more hazardous chemicals, personnel may be required to wear acid suits and self-contained breathing apparatus. Finally and ultimately, as the case may demand, complete body protection by impenetrable sets of protective clothing and self-contained breathing apparatus may be required.

What to wear depends upon a knowledge of the hazards involved, and that decision depends upon an accurate identification of the chemicals. Both of the foregoing are assisted by reference to one of the many hazardous material data banks such as OHM-TADS or CHRIS.

But data require interpretation, especially in the field. Hence, needed consultants to a safety-conscious response and cleanup team are the industrial hygienist and the industrial toxicologist. The complexities of field exposure and interpretation of human impact require these trained professionals.

Rome addresses the problem of personnel safety equipment in Chap. 13. He notes the unique hazards that a responder may face: chemical toxicity, fire, oxygen deficiency, and explosion. Having identified the problem, he then discusses various types of protective clothing and breathing apparatus and their suitability in various situations.

DISPOSAL

The last but perhaps one of the most difficult aspects of spill cleanup is disposal, which is now subject to extensive regulations, e.g., those issued on May 19, 1980,[31] under the U.S. Resource Conservation and Recovery Act. The law requires each generator to have an EPA identification number, each shipment to be manifested, and each disposer to be licensed; cleanup contractors are not exempt.

In spill cleanup, careful consideration should be given to the ultimate fate of the recovered material. Can it be recycled, incinerated, landfilled, stabilized, or detoxified? And since the cost of disposal can be significant, care should be taken in the cleanup process to accumulate as little hazardous waste as possible; i.e., if one uses 1000 kg of sand to adsorb 10 kg of spilled toxic chemicals, one has 1010 kg of toxic hazardous waste; alternatively one might have used 20 kg of a commercial adsorbent to perform the same task. The initial cost of the adsorbent might have been higher than that of sand, but the overall cost, considering both disposal and adsorbent purchase, might be lower.

CONCLUSION

The field of spill cleanup and response is new, vibrant, and dramatically changing. A need not even recognized 20 years ago has been addressed and partially met. New technology, new equipment, and recognition of hazards have taken cleanup response from simplistically applying straw to adsorb spilled oil through complex activated-carbon adsorption of pesticides. Now the same techniques are being applied to the leakage problem of hazardous waste sites. The latter can be considered the third generation of spill response, at least in the United States, where the Love Canal problem and the chronic leakage of chemicals buried long ago have moved the cleanup industry into the forefront of public attention.[32]

REFERENCES

1. "Control of Hazardous Material Spills," *Proceedings of the 1972 National Conference,* U.S. Environmental Protection Agency, Houston, March 1972.

2. G. F. Bennett and I. Wilder, "Evolution of Hazardous Material Spills Regulations in the United States," *Journal of Hazardous Materials,* vol. 4, January 1981, p. 257.

3. "Top 50 Chemicals," *Chemical and Engineering News,* vol. 58, June 11, 1980, p. 36.

4. U.S. Environmental Protection Agency, "Hazardous Substances: Definition, Designation, Revocation of Regulations, Proposed Expansion of Criteria of Designation and Proposed Determination of Reportable Quantities," *Federal Register,* vol. 44, Feb. 16, 1979, pp. 10266–10284.

5. G. W. Dawson and M. W. Stradley, "A Methodology for Quantifying the Environmental Risks from Spills of Hazardous Materials," in G. F. Bennett (ed.), *Water—1975,* American Society of Chemical Engineers Symposium Series, no. 151, vol. 71, 1975, p. 349.

6. J. L. Buckley and S. A. Wiener, *Hazardous Material Spills: A Documentation and Analysis of Historical Data,* EPA-600/2-78-066, U.S. Environmental Protection Agency, April 1978.

7. P. C. Ditzel, "Texas City, 1947. Death on the Docks. Tanker Explosion Kills 561," *Firehouse,* February 1980, p. 46.

8. K. Gugan, *Unconfined Vapor Cloud Explosions,* Gulf Publishing Co., Houston, 1978.

9. "After 50 Years, Some Fear Edison's Vials Now Might Be Vile." *The Wall Street Journal,* May 7, 1980.

10. W. B. Nye, "The Hazardous Material Experience in Shawnee Lake, Ohio—A Case History," *Proceedings of the First National Conference on Control of Hazardous Material Spills,* Houston, March 1972, p. 217.

11. "They Learn How to Face Emergencies. Dow Plant Tightens Procedures, Prepares Outsiders Too," *Chemical Week,* vol. 126, July 19, 1978, p. 45.

12. *Pollution Incidents in and around U.S. Waters,"* COMDTINST, M1645012, U.S. Coast Guard, Department of Transportation, July 1980.

13. J. M. Winton, "Chemical Cargo: Safety Gets the Through Track," *Chemical Week,* vol. 126, July 5, 1978, p. 29.

14. G. F. Bennett (ed.), "Control of Hazardous Material Spills," *Proceedings of the 1974 National Conference on Control of Hazardous Material Spills,* American Society of Chemical Engineers and U.S. Environmental Protection Agency, San Francisco, August 1974.

15. A. J. Smith, *Managing Hazardous Substances Accidents,* McGraw-Hill Book Company, New York, 1981.

16. *CHRIS,* vol. 1: *Condensed Guide to Chemical Hazards,* vol. 2: *Hazardous Chemical Data,* vol. 3: *Hazards Assessment Handbook,* vol. 4: *Response Method Handbook,* U.S. Coast Guard, Washington, 1978.

17. J. Wright and C. R. Gentry, "Oil and Hazardous Materials Technical Assistance Data System," in G. F. Bennett (ed.), *Proceedings of the 1976 National Conference on Control of Hazardous Material Spills,* New Orleans, April 1976, p. 101.

18. "Control of Hazardous Material Spills," *Proceedings of the 1980 National Conference on Control of Hazardous Material Spills,* Louisville, Ky., May 1980, U.S. Environmental Protection Agency, U.S. Coast Guard, and Vanderbilt University.

19. J. C. Clow and J. C. Zercher, "The Coast Guard's National Response Center and CHEMTREC of the Chemical Manufacturers Association," *Proceedings of the 1980 National Conference on Control of Hazardous Material Spills,* Louisville, Ky., May 1980, p. 358.

20. R. F. Cumberland, "Role of the National Chemical Emergency Centre at Harwell, U.K.," in G. F. Bennett (ed.), *Proceedings of the 1978 National Conference on Control of Hazardous Material Spills,* Miami Beach, Fla., April 1978, p. 60.

21. C. S. L. McNeil, "Computer Support for Environmental Emergency Management at Environment Canada," in G. F. Bennett (ed.), *Proceedings of the 1974 National Conference on Control of Hazardous Material Spills,* San Francisco, August 1978, p. 75.

22. H. D. Williams, "The Coast Guard's National Strike Force and Hazardous Substances," in G. F. Bennett (ed.), *Proceedings of the 1974 National Conference on Control of Hazardous Material Spills,* San Francisco, August 1974, p. 38.

23. D. K. Guinan, "The Railroad Industry Hazard Information and Response System," *Proceedings of the 1980 National Conference on Control of Hazardous Material Spills,* Louisville, Ky., May 1980, p. 350.

24. L. Benner and R. Rote, "NTSB Hazardous Material Spill Maps: A New Safety Information Resource," presented at American Institute of Chemical Engineers meeting, Philadelphia, June 1980.

25. G. F. Bennett (ed.), "Control of Hazardous Material Spills," *Proceedings of the 1974 National Conference on Control of Hazardous Material Spills,* New Orleans, April 1976, Information Transfer, Washington.

26. G. F. Bennett (ed.), "Control of Hazardous Materials Spills," *Proceedings of the 1978 National Conference on Control of Hazardous Material Spills,* Miami Beach, Fla., April 1978, Information Transfer, Washington.

27. R. E. Temple, R. J. Esterhay, W. T. Gooding, and G. F. Bennett, "A New Universal Sorbent for Hazardous Spills," *Journal of Hazardous Materials,* vol. 4, 1980, p. 185.

28. C. S. McDowell, H. H. Bourgeois, Jr., and T. G. Zitrides, "Biological Methods for the *in Situ* Cleanup of Oil Spill Residues," presented at the Coastal and Off-Shore Oil Pollution Conference, New Orleans, September 1980.

29. U.S. Environmental Protection Agency, "Oil Prevention—Non-Transportation Related On-Shore and Offshore Facilities," *Federal Register,* vol. 38, Dec. 11, 1973, pp. 34164–34170.

30. U.S. Environmental Protection Agency, "National Pollutant Discharge Elimination System—Proposed Requirements for Spill Prevention Control and Countermeasure Plans to Prevent Discharges of Hazardous Materials from Certain Facilities," *Federal Register,* vol. 43, Sept. 1, 1978, pp. 39276–39290.

31. U.S. Environmental Protection Agency, "Hazardous Waste Management," *Federal Register,* vol. 45, May 19, 1980, pp. 33063–33588.

32. G. F. Bennett and H. Bernard (eds.), *Management of Uncontrolled Hazardous Waste Sites,* Hazardous Materials Control Research Institute, Silver Spring, Md., 1980.

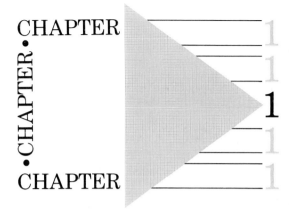

CHAPTER
•CHAPTER•
CHAPTER

1
1
1
1
1

International Laws and Regulations

PART 1
Intergovernmental Maritime Consultative
Organization / 1-2

PART 2
Convention on the Prevention of Marine
Pollution by Dumping Wastes and Other
Matter / 1-13

PART 3
Transport of Dangerous Goods by Air / 1-21

PART 1
Intergovernmental Maritime Consultative Organization

C. Hugh Thompson, Ph.D.
Director, Environmental Affairs
*Aerojet-General Corporation**

International Convention for the Prevention of Pollution from Ships / 1-5
Intervention Protocol / 1-11
Other Actions / 1-11

Hazardous materials, dangerous goods, harmful substances, toxic pollutants, and other such items are familiar terms in the regulations and activities of the Intergovernmental Maritime Consultative Organization (IMCO). IMCO was created by the UN Maritime Conference in 1948 and was established in 1959 after acceptance by the required number of member states. It is composed primarily of maritime nations working on technical matters affecting shipping. The cooperation and exchange of information among governments emphasize safety at sea and have more recently been establishing standards for control of pollution of the sea from ships. IMCO responsibilities include convening international conferences on shipping matters and drafting maritime conventions and agreements. This part of Chap. 1 provides insight into some of IMCO's activities pertaining to control of hazardous materials in the marine environment.

IMCO is authorized under the United Nations Charter, which states that "various specialized agencies, established by intergovernmental agreement and having wide international responsibilities, as defined in their basic instruments, in economic, social, cultural, educational, health, and related fields, shall be brought into relationship with the United Nations." The Economic and Social Council (ECOSOC), one of six principal organs of the United Nations, serves as the establishing body to negotiate the individual relationships of specialized agencies such as IMCO with the United Nations. With the concurrence of the UN Council, the agreements are then approved by the UN General Assembly. These specialized agencies then function as separate and autonomous organizations. They control their own membership and legislative and executive bodies, establish their own budgets and secretariats, and use the coordinating functions and resources of ECOSOC.

IMCO annual reports are submitted in accordance with ECOSOC resolutions, such as Resolutions 497 (XVI), 1090E and 1090F (XXIX), and 1172 (XLI). These reports document activities of the IMCO Assembly, the IMCO Council, and special IMCO committees. The IMCO Assembly is the policymaking body and includes all member states. It approves work programs and budgets, elects the IMCO Council, and approves the appointment of the IMCO secretary general. The Council governs IMCO between ses-

---

*When this part was written, the author was employed by the Pacific Northwest Laboratories of Battelle Memorial Institute.

sions of the Assembly and is composed of at least 16 member states. The Maritime Safety Committee (MSC) has traditionally been the third part of IMCO and is composed of at least 14 member states.

A fourth part of IMCO is the Marine Environment Protection Committee (MEPC), which was created during the eighth IMCO Assembly meeting (November 1973) and was designated as the appropriate IMCO body to deal with amendments under the new conventions pertaining to marine pollution. MEPC meetings began in the spring of 1974.

IMCO has had an interest in hazardous materials for many years. The International Maritime Dangerous Goods Code provides guidance to aid compliance with the provisions of the 1960 International Convention for the Safety of Life at Sea (SOLAS). Prior to this guidance, efforts were dedicated to hazardous materials at the 1948 Safety of Life at Sea Conference and the 1929 International Conference of Safety of Life at Sea. The UN Committee of Experts on the Transport of Dangerous Goods in 1956 provided a general framework to ECOSOC for all modes of hazardous materials transportation including classification activities and the Dangerous Goods Code. These actions all placed primary emphasis upon safety with little attention to pollution control and were, therefore, conducted under the purview of the MSC.

Prior to 1971, IMCO had recognized the need for more specialized examination of marine environmental issues. An MSC Subcommittee on Marine Pollution was formed and began, in coordination with a Subcommittee on Ship Design and Equipment, to provide a forum for member states to delineate the nature of pollution problems and alternative solutions. These subcommittees met repeatedly over the next 2½ years to conduct the technical work necessary to meet IMCO Assembly Resolution A.176VI. This resolution had established the goal of convening an international conference in 1973 for the purpose of preparing a suitable international agreement for placing restraints on the contamination of the sea, land, and air by ships, vessels, and other equipment operating in the marine environment. Other factors focusing attention upon marine pollution were the preparations necessary for the 1972 UN Conference on the Human Environment (Stockholm). IMCO efforts to control pollution of the sea were summarized for the MSC and its Subcommittee on Marine Pollution in February and April 1971 in *An Appreciation of the Present Situation with Regard to Marine Pollution from Ships, Vessels, and Other Equipment Operating in the Marine Environment,* highlights of which are presented in Table 1-1.

Since 1954 measures concerning marine pollution resulting from the transport of oil and other hazardous or toxic materials by seagoing vessels have been adopted in the following areas:

1. Working toward the prohibition of deliberate discharge

2. Working toward the prevention of accidental discharge

3. Developing appropriate national governmental powers for dealing with pollution

4. Providing redress (liability and compensation) for damage

5. Recommending restrictive measures for the discharge of certain pollutants: domestic sewage and food-processing waste, pesticides, inorganic and organic wastes, oil and oil dispersants, solid objects, and dredging spoil and inert wastes

6. Recommending methods for dealing with spillages

7. Working toward prevention of oil pollution from offshore drilling rigs

8. Categorizing hazardous and toxic substances and compounds and developing control measures for certain categories

IMCO functions through committees and subcommittees, which in turn form ad hoc or standing working groups composed of member government delegations and industrial expert observers. Reports are prepared and work flows through these groups with the assistance of full-time IMCO technical secretariats. One method of work directly applicable to IMCO's activities in hazardous materials is that of providing the secretariat function to the Joint Group of Experts on the Scientific Aspects of Marine Pollution (GESAMP).

GESAMP is an international, interdisciplinary group of scientists and engineers who

TABLE 1-1   Highlights of IMCO Pollution Control–Related Activities
as of April 1971

1. IMCO was charged in Art. 1 of the basic convention to provide machinery to facilitate cooperation and regulation in all technical matters affecting shipping.

Substantial interest in oil pollution

---

2. The 1954 Convention for Prevention of Pollution of the Sea by Oil was given to IMCO in 1959 to address the issue of deliberate, nonaccidental discharges of oil from ships.

3. The 1954 convention was amended in 1962.

4. The 1954 convention was further amended in 1969 to limit the total amount of oil ($\frac{1}{15,000}$ of cargo) discharged at 32.4 L/km and to prohibit discharges of oil within 92.6 km of land.

5. Plans were made for the 1973 MARPOL Convention.

6. Preparations were made for the 1972 UN Conference on the Human Environment in Stockholm.

7. Technical advances were made in tank-cleaning operations, shore reception facilities, oil-content meters, and oil-water separators as well as in enforcement advantages.

8. Provisions of the 1960 SOLAS Convention such as navigation, port advisory services, pilots, maneuvering data, and traffic separation were used to reduce accidental discharges.

9. The International Convention Relating to Intervention on the High Seas in Cases of Oil Pollution Casualties of 1969 gave threatened coastal states actions to be taken.

10. The Civil Liability Convention for Oil Pollution Damage of 1969 gave the basis for owners or operators of ships to pay third-party damages.

Interest in pollution by agents other than oil

---

11. Safety to crew and ship from handling and storage of chemical packaged cargoes was enhanced by the IMCO Dangerous Goods Code, but pollution was not considered.

12. GESAMP was requested to identify cargoes other than oil, including bulk cargoes, that required pollution prevention. Materials were divided into two categories, those that caused major hazards and those that caused localized hazards, distinguished mainly by the size of the cargo released.

13. Speculations were offered about the range of controls and their relationship to existing legal instruments.

14. IMCO was faced with the following issues, for which it had sparse data and few methods of evaluating importance and establishing control: ocean dumping of sewage, food-processing wastes, pesticides, organic and inorganic wastes, oil and oil dispersants, solid objects, dredging spoil and inert wastes, sewage from ships, and air pollution from ships.

15. Discharge of pollutants arising from shore-based industry, shipboard pollutants, deliberate or accidental discharge of tank washings, and bilge for cargoes other than oil; prevention of pollution by offshore operational equipment; and legal responsibilities all needed resolution to clarify IMCO's mission to protect the marine environment.

SOURCE: MSC XXII/10(6)/1/Add.1.

are sponsored by the United Nations and seven specialized agencies including the UN Environment Program, Food and Agriculture Organization, World Health Organization, World Meteorological Organization, IMCO, and International Atomic Energy Agency. These agencies pool some resources dedicated to environmental problems, present issues of broad significance, and work in close association with experts in special working groups and at annual sessions of GESAMP. For several years one of these working groups has been preparing evaluations of the hazards of harmful substances carried by ships. This work, once approved by GESAMP, is used by subcommittee delegations to IMCO's MEPC meetings to categorize these substances under Annex II of the International Convention for Prevention of Pollution from Ships. GESAMP scientific reports and studies as well as session reports are published in more than one language by each of the sponsoring agencies. Copies of reports are often available from the GESAMP Administrative Secretary, Intergovernmental Maritime Consultative Organization, 101-104 Picadilly, London W1V OAE, United Kingdom.

INTERNATIONAL CONVENTION FOR THE
PREVENTION OF POLLUTION FROM SHIPS

The International Convention for the Prevention of Pollution from Ships (MARPOL) is an agreement reached between 78 countries in London on November 2, 1973. It consists of 20 articles which generally administer five technical annexes designed to control discharges of oil, noxious liquid substances, harmful packaged goods, sewage, and garbage from ships. The target date for the MARPOL Convention to come into force is 1984. Ratification by 15 countries is required for the convention to come into force within 1 year thereafter. These 15 countries together must have commercial merchant fleets constituting at least 50 percent of the gross tonnage of the world's merchant shipping.

The term "harmful substance" means any substance which if introduced into the sea is liable to create hazards to human health, to harm living resources and marine life, to damage amenities, or to interfere with legitimate uses of the sea and includes any substances subject to control by the present convention. This generalized term includes oil and other materials as noted in the several annexes.

While the various articles of the convention generally apply to hazardous materials (liquid noxious substances), Art. 8 is of special significance. This article requires reporting, by the master of the ship to the party to the convention (the ship's flag state), of incidents involving harmful substances when a discharge or a potential discharge other than that permitted under the convention occurs or when a discharge occurs to save ship or crew, to combat pollution, or to conduct scientific research. The intent of Art. 8 is similar to governmental notification requirements under United States law for incidents or releases of hazardous materials, e.g., hazardous materials transportation as found in 49 CFR 171.15 and 171.16 for railway and highway, 14 CFR 103.28 for air, 46 CFR 2.20-65 and 2.20-70 for water, and various U.S. Environmental Protection Agency (U.S. EPA) rules promulgated under the Clean Water Act (40 CFR 120 et seq.), as well as in 33 CFR 153.203 for incidents that are hazardous to navigation.

Enforcement of the convention is primarily the responsibility of the flag state (the country of registry of the vessel, as opposed to coastal nations which have a maritime interest). Article 4 defines violations, while Art. 6 outlines provisions for detection and enforcement. Certificates of inspection, delay of ships, settlement of disputes, and communication of information to IMCO are provided for in Arts. 5, 7, 10, and 11.

The mechanics of the convention are contained in the remaining articles, which specify general obligations, definitions, applicability, procedures for ratification, amendment, denunciation, and deposition, and languages. A specific effort has been made in Art. 17

to deal with the need for technical cooperation by providing a role for the UN Environment Program in training, equipping, and encouraging research to further the aims of the convention.

While Annex I of MARPOL, pertaining to the control of oil discharges, was rather fully developed with 25 regulations and 3 appendixes, Annex II (described below) was designed in 13 regulations and 5 appendixes to address initially control of noxious liquid substances carried in bulk. The concerns addressed under Annex II are closely associated with United States concerns addressed under Sec. 311 of the Federal Water Pollution Control Act as amended and regulations promulgated under 40 CFR 116 and following. Other IMCO work, such as the Bulk Chemicals Code, is closely related to Annex II and therefore requires careful consideration of the environmental factors in a safety context.

Annex III pertains to prevention of pollution by harmful substances carried by sea in packaged forms or in freight containers, portable tanks, or road and rail tank wagons. These concerns are similar to those being addressed under United States law, e.g., in Title 49 of the *Code of Federal Regulations* and in parts of Title 46. Annex III consists of eight brief regulations, but these must be recognized as extensions of work conducted for years under the SOLAS Convention and are directly related to the IMCO Dangerous Goods Code and the work of the ECOSOC Committee of Experts on the Transport of Dangerous Goods.

Pollution potential and control of hazardous materials in transportation became an issue when the MARPOL Convention was being drafted. The question of what problems must be caused, and over what period of time, to be regarded as susceptible to regulation is still an issue. Some of the environmental, health, and transportation statutes and regulations of the United States have continued along the same confusing lines, and there are more than 20 overlapping authorities and requirements. Internationally a similar overlap created a gap. An ECOSOC group of experts identified materials as dangerous under circumstances which it determined were of priority concern. These transportation priority classes are explosives, gases, flammable liquids, oxidizers, poisons and infectious substances, radioactive substances, corrosives, and miscellaneous dangerous substances. IMCO followed this lead and produced the Dangerous Goods Code, which specifies packaging, labeling, storage, and other information concerning the identification and properties of materials shipped by vessel. It is the last-named miscellaneous class which creates concern. Neither of the groups (ECOSOC- or IMCO-oriented) was able, prior to MARPOL, to determine that materials which posed an environmental hazard from acute exposure or materials which posed a hazard to humans from prolonged exposure merited attention for the purpose of material carriage by rail, road, or sea. Within the United States, the same phenomenon is evident, and it is an issue of regulatory consideration by the U.S. Department of Transportation.

Annex III of MARPOL is quite an important bridge which the convention drafters attempted to establish between transportation and environmental authorities. Resolution 19 of the 1973 IMCO International Conference on Marine Pollution recommends that IMCO examine the need to revise the Dangerous Goods Code to include harmful substances. The MEPC actively began considering this issue, and progress was noted in 1977 (MEPC Circular 50 of July 14, 1977). The IMCO Subcommittee on the Carriage of Dangerous Goods has evaluated the pollution potential issue and formed an opinion (CDG XXIX/16, Annex 7, November 16, 1978) that polluting substances should be included in the Dangerous Goods Code and that substances which only present a hazard to the marine environment and are not mentioned in the code should be included in Class 9 (miscellaneous dangerous substances). This opinion was expressed to the MSC with a suggestion that it be forwarded to the MEPC. The wisdom of this suggested classification remains to be seen.

Class 9 carries the lowest hazards warning and involves the minimum of care in shipment. The concern of the transportation community has been to avoid diluting the existing hazard classes with the environmentally hazardous materials. It was recognized, however, that no final IMCO decision would be appropriate until information which would provide a fully satisfactory response to Resolution 19 as noted above had been gathered.

Annexes IV and V pertain to regulations for the prevention of pollution by sewage and garbage discharged from ships. These annexes are optional to the extent that a country may become a party to the convention without accepting either or both of them.

Hazardous materials (noxious liquid substances in bulk) were uniquely considered in Annex II of MARPOL, and the system developed by the participating delegations is explained below. The systems of designation and classification used were considered methodologies for implementing United States hazardous substance regulations under Sec. 311 of the Water Pollution Control Act but were regarded as not meeting all the requirements as interpreted by the U.S. EPA. The provisions of Annex II will require accommodation on the part of the United States government such as new implementing legislation, modifying U.S. EPA regulations, or using the Ports and Waterways Safety Act (administered by the U.S. Coast Guard).

Annex II provides a mechanism for identifying liquid substances carried in bulk which the parties to the convention regard as sufficiently harmful to merit control at sea. The emphasis on control at sea is important when the scope of the initiating IMCO Assembly Resolution A.176VI and the final convention definitions of harmful substances, noxious liquid substances, and clauses of convention applicability are considered. The latter considerations moved the emphasis of the convention from protection of the total marine environment, which would include shorelines, beaches, and air sheds, to addressing offshore releases from vessels as a priority. A careful examination of *Guidelines for Categorization of Noxious Liquid Substances* (App. I of Annex II) illustrates the artifacts left over from earlier marine environmental control efforts (e.g., concern with BOD_5, deposits on the seafloor, etc.) which were proposed as most significant in the nearshore and in harbors. These concerns were given lower priority in the regulatory considerations of Annex II because it was recognized that member states would address the nearshore discharge problems for which many of these lesser effects posed the greatest hazard potential.

Operation of Annex II involves:

1. Listing materials carried by ships
2. Developing hazard profiles for each material and categorizing them
3. Establishing special areas
4. Stringently limiting discharges of Category A materials with less control over remaining categories, using Annex II regulations
5. Establishing discharge reception facilities at ports and maintaining cargo record books
6. Surveys and certification of ships

The system of material or substance listing has evolved from the initial list (developed by ad hoc GESAMP–member state delegations to the Subcommittee on Marine Pollution of the MSC) to lists and data supplied by member states, noting current or new materials to be shipped, and to other IMCO group lists such as those of the Subcommittee on Bulk Chemicals and the Subcommittee on the Carriage of Dangerous Goods.

Listed materials are rated by GESAMP, and hazard profiles are developed according to the criteria noted in Table 1-2. The hazard profiles are taken by IMCO and made available to the MEPC, in which they are classified by the delegations into five categories by using the guidelines noted in Table 1-3. Substances categorized as A are more stringently controlled than those categorized as D. Regulation 3 (given in part in Table 1-4) explains the emphasis which member states wish to place upon these rated materials.

TABLE 1-2 Criteria for Hazard Profiles of Selected Substances Submitted to IMCO by GESAMP

Column A: Bioaccumulation

+ Bioaccumulated and liable to produce a hazard to aquatic life or human health
0 Not known to be significantly bioaccumulated
Z Short retention, of the order of 1 week or less
T Liable to produce tainting of seafood
B Bioaccumulated but the hazard to aquatic life or human health unknown (low acute aquatic toxicity)

Column B: Damage to Living Resources

Ratings	TL_M 96* (mg/L)
4 Highly toxic	< 1
3 Moderately toxic	1–10
2 Slightly toxic	10–100
1 Practically nontoxic	100–1000
0 Nonhazardous	> 1000

*Toxicity limit of 96 h, which is roughly equivalent to the toxic lethal value.

BOD = problem caused primarily by high oxygen demand.
D = deposits liable to blanket seafloor.

Column C: Hazard to Human Health—Oral Intake

Ratings	LD_{50}* (mg/kg)
4 Highly hazardous	< 5
3 Moderately hazardous	5–50
2 Slightly hazardous	50–500
1 Practically nonhazardous	500–5000
0 Nonhazardous	> 5000

*Lethal dose 50.

Column D: Hazard to Human Health—Skin Contact and Inhalation (Solution)

II Hazardous (solution)
I Slightly hazardous (solution)
0 Nonhazardous (solution)

Column E: Reduction of Amenities—Ratings

xxx Highly objectionable because of persistence, smell, or poisonous or irritant characteristics: beaches liable to be closed
xx Moderately objectionable because of the above characteristics, but with short-term effects leading to temporary interference with use of beaches
x Slightly objectionable; noninterference with use of beaches
0 Posing no problem

All Columns

Ratings in parentheses, ( ), indicate insufficient data available to the panel or the GESAMP working group on specific substances; hence extrapolation was required.
NA = not applicable.
— = data not available to the panel or to the GESAMP working group.

SOURCE: Annex IV, as amended, *Report of the Fourth Session of Joint Group of Experts on the Scientific Aspects of Marine Pollution,* World Meteorological Organization, Geneva, September 1972.

This system illustrates how worldwide scientific data can be offered to a standard-making group to produce practical determinations with technically justifiable bases. With some effort, therefore, toxicological information for a given material can be traced through the GESAMP hazard profiles and IMCO categorization to the regulatory provisions contained in Annex II. The system allows shippers to anticipate a rating and prepare a material accordingly before that material is formally included.

Regulatory control of Annex II is designed around "prohibited" discharges, i.e., discharges limited by concentration, quantity, and location. The "special area" concept was developed to accommodate unique concerns in the Baltic and Black Sea areas. Substances for shipment are intended to be evaluated and either regulated or clearly set aside (App. III, Annex II). Regulation 4 includes provisions for acceptable discharges of bilge or ballast water.

Prohibited discharges include Category A materials and tank washings, which must go to reception facilities until the concentration of the material is below the level given in

TABLE 1-3   Guidelines for the Categorization of Noxious Liquid Substances

Category A:	Substances which are bioaccumulated and are liable to produce a hazard to aquatic life or human health; or which are highly toxic to aquatic life (as expressed by a hazard rating of 4, defined by a TL_M* of less than 1 mg/L); and additionally certain substances which are moderately toxic to aquatic life (as expressed by a hazard rating of 4, defined by a TL_M of 1 mg/L or more, but less than 10 mg/L) when particular weight is given to additional factors in the hazard profile or to special characteristics of the substance.
Category B:	Substances which are bioaccumulated with a short retention of the order of 1 week or less, or which are liable to produce tainting of seafood, or which are moderately toxic to aquatic life (as expressed by a hazard rating of 3, defined by a TL_M of 1 mg/L or more, but less than 10 mg/L); and additionally certain substances which are slightly toxic to aquatic life (as expressed by a hazard rating of 2, defined by a TL_M of 10 mg/L or more, but less than 100 mg/L) when particular weight is given to additional factors in the hazard profile or to special characteristics of the substance.
Category C:	Substances which are slightly toxic to aquatic life (as expressed by a hazard rating of 2, defined by a TL_M of 10 mg/L or more, but less than 100 mg/L); and additionally certain substances which are practically nontoxic to aquatic life (as expressed by a hazard rating of 1, defined by a TL_M of 100 mg/L or more, but less than 1000 mg/L) when particular weight is given to additional factors in the hazard profile or to special characteristics of the substance.
Category D:	Substances which are practically nontoxic to aquatic life (as expressed by a hazard rating of 1, defined by a TL_M of 100 mg/L or more, but less than 1000 mg/L, or causing deposits blanketing the seafloor with a high biochemical oxygen demand (BOD), or highly hazardous to human health, with an LD_{50}† of less than 5 mg/kg, or producing moderate reduction of amenities because of persistence, smell, or poisonous or irritant characteristics, possibly interfering with use of beaches, or moderately hazardous to human health, with an LD_{50} of 5 mg/kg or more, but less than 50 mg/kg, and producing a slight reduction of amenities.

Other Liquid Substances (for the Purposes of Regulation A of This Annex):   Substances other than those categorized in Categories A, B, C, and D.

---

*TL_M = toxicity limit of 96 h, which is roughly equivalent to the toxic lethal value.
†Lethal dose 50.
SOURCE: 1973 MARPOL Convention, App. I, Annex I.

TABLE 1-4   Regulation 3: Categorization and Listing of Noxious Liquid Substances*

(1) For the purpose of the Regulations of this Annex, except Regulation 13, noxious liquid substances shall be divided into four categories as follows:

   (a) Category A   Noxious liquid substances which if discharged into the sea from tank cleaning or deballasting operations would present a major hazard to either marine resources or human health or cause serious harm to amenities or other legitimate uses of the sea and therefore justify the application of stringent anti-pollution measures.

   (b) Category B   Noxious liquid substances which if discharged into the sea from tank cleaning or deballasting operations would present a hazard to either marine resources or human health or cause harm to amenities or other legitimate uses of the sea and therefore justify the application of special anti-pollution measures.

   (c) Category C   Noxious liquid substances which if discharged into the sea from tank cleaning or deballasting operations would present a minor hazard to either marine resources or human health or cause minor harm to amenities or other legitimate uses of the sea and therefore require special operational conditions.

   (d) Category D   Noxious liquid substances which if discharged into the sea from tank cleaning or deballasting operations would present a recognizable hazard to either marine resources or human health or cause minimal harm to amenities or other legitimate uses of the sea and therefore require some attention in operational conditions.

*In part.
SOURCE: 1973 MARPOL Convention, Annex II.

App. II, Annex II (this level nominally appears to be below the data for acute marine effect). Any discharges of residual material remaining in the tank must be diluted with water of not less than 5 percent of the tank volume, released below the waterline, and operated not less than 22.2 km (12 nautical mi) from the nearest land in water not less than 25 m deep and while the vessel is en route at a speed of at least 3.6 m/s (7 knots), or 2.1 m/s (4 knots) for non-self-propelled vessels.* Discharge into special areas is further controlled by Regulation 7 requirements pertaining to reception facilities and their adequacy.

Materials in Category B may be discharged during a tank washing only when the vessel is en route as noted in Category A, the concentration of the material in the wake of the vessel is not more than 1 mg/L, and the total quantity does not exceed 1 m³ or $\frac{1}{3000}$ of tank capacity. These materials may be discharged into special areas only if the residue of the tank washing has been diluted with at least 0.5 percent of the tank volume and discharged into the required reception facility.

Category C materials may be discharged from a tank washing when the vessel is en route as noted for Category A, the concentration of the material in the wake of the vessel is not greater than 10 mg/L, and the total quantity does not exceed 3 m³ or $\frac{1}{1000}$ of the tank volume. Control of discharges of these materials into special areas is more stringent, with concentration in the wake of the vessel not to exceed 1 mg/L.

Category D materials may be discharged from a tank washing in any waters when the vessel is en route, as noted above, except that the depth of the water can be less than 25 m; the washing mixture must be at least 10 parts water to 1 part of substance.

Procedures for the implementation of these operations are specified in Regulation 8; they use surveyors and cargo record books. Provisions for the ventilation of tanks and other alternatives for reducing the concentration of tank-washing residues are provided along with alternatives for determining potential concentrations of materials in the wake

*"Non-self-propelled" is intended to apply to rigs being towed out to be put in place, as for oil drilling.

of each specific vessel. The cargo record book format and elements of the surveys are provided in Regulations 9 and 10. Heavy reliance upon the governments of the parties to the convention was considered the most viable manner of control, especially with respect to discharges to shore reception facilities and associated computational alternatives. IMCO plays a coordinating and monitoring role between the vessel flag state and the member state responsible for the reception facility, using IMCO standards and regulations as a basis of operation.

An International Pollution Prevention Certificate for the Carriage of Noxious Liquid Substances in Bulk is to be issued to any ship engaged in such transport. The certificate is to be issued, for a period not to exceed 5 years, by the flag state which is a party to the convention. These precautions establish a check on both the equipment and the operational knowledge of the ship's master, thus enhancing the potential for compliance at sea. Accidental pollution is recognized in Regulation 13 as a special problem, requiring each party to be more active in the design, construction, and operation of ships in order to minimize the release of polluting substances during an accident.

The Subcommittee on Bulk Chemicals is actively working on detailed procedures and arrangements to facilitate the implementation of Annex II. Engineering considerations are being drafted into guideline formats that provide parties to the convention with precise instructions on such topics as reception facilities in ports, standards and procedures for the discharge of liquid noxious substances, and evaluations of hazards of materials shipped in bulk on the basis of reexamination of GESAMP profiles. This type of reexamination is expected to continue as more data are made available for existing materials and as new materials require bulk shipment.

INTERVENTION PROTOCOL

The 1973 International Conference on Marine Pollution produced another significant agreement on the shipment of hazardous materials, the Protocol Relating to the Intervention on the High Seas in Case of Marine Pollution by Substances Other Than Oil. This agreement outlines, in 11 articles, steps which a party to the convention may take when a vessel's cargo on the high seas poses a reasonably grave or imminent danger of pollution to its coastline or related interests.

The protocol is designed to operate on a list of substances other than oil (IMCO began preparing this list through the MEPC) with the aid of experts qualified to give advice regarding the substances' potential threat. The parties are thus enabled to take whatever measures are necessary to mitigate or eliminate pollution on the high seas. This protocol follows and refers to the 1969 International Convention Relating to Intervention on the High Seas in Cases of Oil Pollution Casualties. The United States has passed legislation on this convention (Public Law 93-248) and is therefore prepared to take intervention steps concerning oil pollution.

It should be emphasized that "intervention" is a serious matter and that the full burden of proof on the necessity and propriety of any initiated actions rests with the intervening party. This protocol is related to the MARPOL Convention through several ties; a clear tie consists of the notification requirements found in Art. 8 of the convention.

OTHER ACTIONS

Other considerations pertaining to IMCO and control of hazardous materials include ocean dumping, at-sea incineration, navigation, training, etc. Some of these subjects are discussed elsewhere in the *Handbook*. IMCO has demonstrated its leadership and member governments their concern for controlling hazardous materials. The complexities of international negotiations demand time for acceptance. IMCO can only be as strong as member governments permit by their support of these agreements. The record of the 1970s and early 1980s indicates that a commitment has been made to move hazardous materials safely and with a minimum of environmental effect.

GENERAL REFERENCES

1. *Everyman's United Nations,* ser. E, United Nations, New York, Jan. 4, 1967.
2. *Transport of Dangerous Goods,* recommendations prepared by Committee of Experts on the Transport of Dangerous Goods, vol. 8, no. 2, ST/SG/AC.10/1, United Nations, New York, 1976.
3. *International Maritime Dangerous Goods Code,* vols. 1, 2, and 3, Intergovernmental Maritime Consultative Organization, London, 1976——.
4. *Marine Pollution,* vol. 23, no. 10(b), ser. 1, addendum 1, Maritime Safety Committee, Feb. 3, 1971.
5. *Hearings on 1973 IMCO Conference on Marine Pollution from Ships,* U.S. Senate Commerce Committee, serial 93-52, Nov. 14, 1973.

PART 2
Convention on the Prevention of Marine Pollution by Dumping of Wastes and Other Matter*

John E. Portmann, Ph.D.
Fisheries Laboratory, Burnham on Crouch, Essex, England
U.K. Ministry of Agriculture, Fisheries, and Food

Introduction / 1-13
Main Content / 1-14
Difficulties since Entry into Force / 1-18
Other Developments / 1-19

INTRODUCTION

In the early 1970s, spurred by the recognition in the 1960s that the environment did not have an indefinite capacity to assimilate the products of human industrial and consumption-disposal activities, a series of meetings took place in the course of which a number of international agreements or conventions were negotiated, all of which had the common goal of reducing the input of contaminants into the environment and, wherever practicable, of eliminating pollution. Early in 1972 the countries bordering the northeast Atlantic Ocean negotiated a Convention for the Prevention of Marine Pollution by Dumping from Ships and Aircraft (the so-called Oslo Convention). The negotiation of this regional convention was followed later in the same year by a meeting in London which had as its objective the development of a similar convention to control dumping of wastes in any of the world's seas or oceans.

With this as an objective representatives of 78 nations and observers from several international organizations and agencies met on October 30, 1972, in London. By November 13, agreement had been reached on a Convention on the Prevention of Marine Pollution by Dumping of Wastes and Other Matter, and a significant number of the national representatives duly signed the agreed convention on that day.

By late 1975 the necessary 15 countries had indicated their readiness to adopt the convention, and the first meeting of contracting parties marking the formal entry into force took place in London on December 17–19, 1975. That meeting agreed, among other items, to assign secretariat duties to the Intergovernmental Maritime Consultative Organization (IMCO) as a competent organization as described in Art. XIV of the convention. IMCO in September 1976 duly called the First Consultative Meeting, which 29 contracting parties and observers were invited to attend.

---

*The views expressed herein are those of the author and do not necessarily reflect those of the Ministry of Agriculture, Fisheries, and Food.

MAIN CONTENT

The convention consists of 22 articles, 3 annexes, and a technical memorandum. Articles I and II define the objectives, and Arts. III and IV define what is meant by dumping and various other terms used subsequently in relation to the act of dumping and the licensing of dumping. Articles V through X lay down the principles for control and the objectives of contracting parties. Most of the remaining articles deal with the administrative arrangements for the entry into force of the convention, notification of permits, and submission of returns to a Secretariat, the actual provision of a Secretariat, and the duties attendant upon that organization.

From a technical standpoint the annexes are the important item. Annex I (Fig. 2-1) contains a list of substances and materials which under Art. IV may not be dumped except when present as trace contaminants or when rapidly rendered harmless by physical, chemical, or biological processes in the sea. Annex II (Fig. 2-2) contains a list of substances and materials which require "special care" and which may be dumped only after the appropriate national authority has granted a prior special permit to cover their disposal at sea. Annex III (Fig. 2-3) lays down the factors which must be taken into account

---

ANNEX I.

1. Organohalogen compounds.

2. Mercury and mercury compounds.

3. Cadmium and cadmium compounds.

4. Persistent plastics and other persistent synthetic materials, for example, netting and ropes, which may float or may remain in suspension in the sea in such a manner as to interfere materially with fishing, navigation or other legitimate uses of the sea.

5. Crude oil, fuel oil, heavy diesel oil, and lubricating oils, hydraulic fluids, and any mixtures containing any of these, taken on board for the purpose of dumping.

6. High-level radioactive wastes or other high-level radioactive matter, defined on public health, biological or other grounds, by the competent international body in this field, at present the International Atomic Energy Agency, as unsuitable for dumping at sea.

7. Materials in whatever form (eg solids, liquids, semi-liquids, gases or in a living state) produced for biological and chemical warfare.

8. The preceding paragraphs of this Annex do not apply to substances which are rapidly rendered harmless by physical, chemical or biological processes in the sea provided they do not:

   (i) make edible marine organisms unpalatable, or
   (ii) endanger human health or that of domestic animals.

   The consultative procedure provided for under Article XIV should be followed by a Party if there is doubt about the harmlessness of the substance.

9. This Annex does not apply to wastes or other materials (eg sewage sludges and dredged spoils) containing the matters referred to in paragraphs 1–5 above as trace contaminants. Such wastes shall be subject to the provisions of Annexes II and III as appropriate.

Fig. 2-1   Substances that may not be dumped in substantial amounts.

ANNEX II

The following substances and materials requiring special care are listed for the purpose of Article VI(1)(a).

A. Wastes containing significant amounts of the matters listed below:

arsenic ⎫
lead   ⎪
copper ⎬ and their compounds
zinc   ⎭

organosilicon compounds
cyanides
fluorides
pesticides and their by-products not covered in Annex I

B. In the issue of permits for the dumping of large quantities of acids and alkalis, consideration shall be given to the possible presence in such wastes of the substances listed in paragraph A and to the following additional substances:

beryllium ⎫
chromium ⎪
nickel   ⎬ and their compounds
vanadium ⎭

C. Containers, scrap metal and other bulky wastes liable to sink to the sea bottom which may present a serious obstacle to fishing or navigation.

Fig. 2-2 Materials requiring special care.

ANNEX III

Provisions to be considered in establishing criteria governing the issue of permits for the dumping of matter at sea, taking into account Article IV(2), include:

A. Characteristics and composition of the matter

    1. Total amount and average composition of matter dumped (eg per year).

    2. Form, eg solid, sludge, liquid, or gaseous.

    3. Properties: physical (eg solubility and density), chemical and biochemical (eg oxygen demand, nutrients) and biological (eg presence of viruses, bacteria, yeasts, parasites).

    4. Toxicity.

5. Persistence: physical, chemical and biological.

6. Accumulation and biotransformation in biological materials or sediments.

7. Susceptibility to physical, chemical and biochemical changes and interaction in the aquatic environment with other dissolved organic and inorganic materials.

8. Probability of production of taints or other changes reducing marketability of resources (fish, shellfish, etc).

B. Characteristics of dumping site and method of deposit

1. Location (eg coordinates of the dumping area, depth and distance from the coast), location in relation to other areas (eg amenity area, spawning, nursery and fishing areas and exploitable resources).

2. Rate of disposal per specific period (eg quantity per day, per week, per month).

3. Methods of packaging and containment, if any.

4. Initial dilution achieved by proposed method of release.

5. Dispersal characteristics (eg effects of currents, tides and wind on horizontal transport and vertical mixing).

6. Water characteristics (eg temperature, pH, salinity, stratification, oxygen indices of pollution—dissolved oxygen (DO), chemical oxygen demand (COD), biochemical oxygen demand (BOD)—nitrogen present in organic and mineral form including ammonia, suspended matter, other nutrients and productivity).

7. Bottom characteristics (eg topography, geochemical and geological characteristics and biological productivity).

8. Existence and effects of other dumpings which have been made in the dumping area (eg heavy metal background reading and organic carbon content).

9. In issuing a permit for dumping, Contracting Parties should consider whether an adequate scientific basis exists for assessing the consequence of such dumping, as outlined in this Annex, taking into account seasonal variations.

C. General considerations and conditions

1. Possible effects on amenities (eg presence of floating or stranded material, turbidity, objectionable odour, discolouration and foaming).

2. Possible effects on marine life, fish and shellfish culture, fish stocks and fisheries, seaweed harvesting and culture.

3. Possible effects on other uses of the sea (eg impairment of water quality for industrial use, underwater corrosion of structures, interference with ship operations from floating materials, interference with fishing or navigation through deposit of waste of solid objects on the sea floor and protection of areas of special importance for scientific or conservation purposes).

4. The practical availability of alternative land-based methods of treatment, disposal or elimination, or of treatment to render the matter less harmful for dumping at sea.

Fig. 2-3   Permit criteria for dumping of matter at sea.

by the national authority in granting a permit for any material. Annex III is quite extensive and covers matters related to the characteristics and composition of the material to be dumped, the characteristics of the dumping site, the method of disposal, and a number of general considerations and conditions, including the consideration of alternatives to sea disposal.

The countries involved in negotiating the convention clearly expected that scientific knowledge, and probably public opinion, would change with time. Thus Art. XV laid down the arrangements which apply for amending either the convention articles or the annexes. Should any party wish to propose an amendment to the convention, the proposal is discussed at a meeting of contracting parties and, if agreed to by a two-thirds majority of those present, will be regarded as adopted. The change, however, does not enter into force until 60 days after such time as two-thirds of the accepting parties have formally indicated that they accept the amendment. Subsequently for any other party the change enters into force 30 days after it deposits its instrument of acceptance.

Amendments to the annexes of the convention which have been approved by a two-thirds majority of those present at a consultative or special meeting of contracting parties shall enter into force for each contracting party immediately after its notification of acceptance to the Secretariat and 100 days after approval by the above meeting for all other contracting parties, except for those which before 100 days have elapsed formally declare their nonacceptance to the Secretariat.

At the time of negotiation of the convention two major problems were encountered. In contrast to the regional Oslo Convention, there was a strong body of opinion which wanted radioactive substances to be included in the convention. Within the Oslo Convention, it had been argued that the control of radioactive waste disposal at sea was already well provided for by arrangements under the European Nuclear Energy Agency (now the Nuclear Energy Agency) of the Organization for Economic Cooperation and Development and under the International Atomic Energy Agency. Eventually the opinion of those who wished to have radioactivity included prevailed, and as can be seen from Fig. 2-1, dumping of high-level radioactive wastes or other high-level radioactive matter is prohibited, whereas other radioactive materials may be dumped subject to the advice and recommendations of the competent international body.

The second major problem encountered related to the prohibition of wastes containing more than trace quantities of cadmium or mercury. Most countries were able to agree to this particular provision, largely because in practice they never used that method of disposal. However, at least one of the major industrial nations could not accept the prohibition quite so readily. Eventually a way around the apparent impasse was found by the

---

TECHNICAL MEMORANDUM OF AGREEMENT OF THE CONFERENCE

The Conference agreed, on the advice of the Technical Working Party, that for a period of five years from the date when the present Convention comes into effect, wastes containing small quantities of inorganic compounds of mercury and cadmium, solidified by integration into concrete, may be approximately classified as wastes containing these substances as trace contaminants as mentioned in paragraph 9 of Annex I to the Convention but in these circumstances such wastes may be dumped only in depths of not less than 3500 metres in conditions which would cause no harm to the marine environment and its living resources. When the Convention comes into effect, this method of disposal, which will be used for not longer than five years, will be subject to the relevant provisions of Article XIV(4).

Fig. 2-4 Technical Memorandum of Agreement of the Conference.

inclusion of a technical memorandum (Fig. 2-4). This made provision for a limited period of time for any country to dispose of wastes containing small quantities of inorganic compounds of mercury or cadmium subject to very stringent conditions of disposal.

With these matters safely out of the way or at least provided for in a mutually acceptable way, the convention seemed set for a reasonably smooth entry into force.

DIFFICULTIES SINCE ENTRY INTO FORCE

At the time of negotiation, dumping represented a significant means of disposal of wastes for a number of countries. However, there was an almost 3-year gap between signing of the final act and entry into force. Since then most of those countries have substantially tightened their own mechanisms for control, and a number have declared their intention either to reduce substantially the role that dumping at sea plays in their waste disposal options or even to phase it out completely. Consequently, the resolution of certain terms which are included in the convention has proved perhaps even more difficult than might originally have been anticipated, since it has been necessary to try to reach a compromise between nationally adopted regulations.

Annex I (Fig. 2-1) includes seven groups of substances. It also contains two paragraphs that effectively exempt certain types of substances which would otherwise be included in the generic titles and certain types of wastes which contain small quantities of prohibited substances. Subparagraph 8 thus provides for the exemption of organohalogen compounds which are rapidly rendered harmless by physical, chemical, or biological processes, provided the resultant products do not render edible organisms unsuitable for consumption by humans or domestic animals. Subparagraph 9 exempts from prohibition wastes such as dredge spoil or sewage sludge, which almost inevitably contain trace quantities of, for example, organohalogen compounds and mercury or cadmium compounds.

What is meant by harmless or trace quantities may not seem very difficult to define, but in practice it has not proved very easy to gain a universally acceptable definition. After three consultation meetings and an equal number of meetings of an ad hoc scientific group, however, broad agreement was reached and interim guidelines were adopted.

In effect, the contracting parties to the London Convention adopted, as an interim measure, a definition of trace quantities similar to that which has also been applied to the term "harmless" used by the Oslo Convention countries:

> Annex I substances listed in paragraphs 1, 2, 3 and 5 of Annex I shall not be regarded as "trace contaminants" under the following three conditions:
>
> **(a)** if they are present in otherwise acceptable wastes or other materials to which they have been added for the purpose of being dumped;
>
> **(b)** if they occur in such amounts that the dumping of the wastes or other materials could cause undesirable effects, especially the possibility of chronic or acute toxic effects on marine organisms or human health whether or not arising from their bioaccumulation in marine organisms and especially in food species; and
>
> **(c)** if they are present in such amounts that it is practical to reduce their concentrations further by technical means

This definition will be applied by the contracting party having conducted a number of tests on the toxicity, persistence, and bioaccumulation potential of the waste. When as a result of the tests there is any doubt as to whether the waste complies fully with the definition, provision is made for consultations with other contracting parties (when appropriate, through a regional convention). Also included are provisions that exempt sewage sludges and dredged spoils from the biological tests, provided chemical characterization of the material and knowledge of the receiving area allow an assessment of the environmental impact.

Under the convention, wastes containing Annex II substances in "significant" quantities must be given a special permit, which is notified immediately to the convention's Secretariat and circulated to all contracting parties. The definition of the term "significant" has been agreed on an interim basis to be 0.1 percent by weight or greater, but discussions are proceeding on a definition more directly related to the hazard posed by the waste.

OTHER DEVELOPMENTS

Since 1975, when the convention came into force, there have been several consultation meetings of the contracting parties and their ad hoc scientific groups to deal with particular technical issues. Apart from discussion on and adoption of the various resolutions emanating from the technical group meetings, the consultation meetings have been concerned primarily with the more administrative aspects of the operation of the convention, e.g., with how returns should be made. Because these do not significantly affect coverage of the convention, this section will concentrate on the scientific and technical developments.

Since the convention was negotiated, incineration at sea of certain types of waste by specially developed vessels has become more widespread. This method of disposal is particularly attractive for wastes containing halogenated organic compounds, e.g., residues from the production of vinyl chloride, some of which had been dumped at sea before the convention came into being. It has also proved attractive to countries faced with the difficulty of providing adequate land-based incineration with the associated problem of how to construct and operate satisfactory scrubbing systems to remove the hydrogen chloride gas which forms on burning. Incineration at sea relies on the neutralization of the dense vapor cloud of hydrochloric acid by seawater. Studies have shown that this can be achieved without adversely affecting the marine environment, provided the incineration procedure is strictly controlled. (See Chap. 14, Part 4.)

Most countries were of the opinion that this method of disposal should be properly controlled and that it should be brought under the London Convention either by development of a special annex or by a protocol. Following a series of meetings of the Ad Hoc Group on Incineration at Sea and one of legal experts, a set of regulations was developed and agreed to by the Third Consultative Meeting in October 1978. These regulations are included as an addendum to Annex I of the convention. The addendum is mandatory to all countries party to the convention that intend to utilize incineration at sea as a means of disposing of waste. The regulations have been supplemented by technical guidelines, which provide more detailed guidance on several aspects of incineration at sea, in principle at the Fourth Consultative Meeting in October 1979, but they are of a recommendatory rather than a mandatory character and can be readily updated in the light of scientific and technical developments.

At the Third Consultative Meeting a number of proposals were made for amendments to the originally agreed Annexes I, II, and III. These various proposals were only briefly discussed at that meeting and were referred to an ad hoc scientific group, which met in March 1979. The main proposals were for inclusion in Annex I of the following substances: organic pesticides and their by-products, crude oil and fractions thereof, and lead, plus "materials insufficiently described." There were also proposals to include oxygen-consuming and/or biodegradable organic matter in Annex II along with synthetic organic chemicals, and a series of amendments and clarifications to Annex III was proposed.

Following discussion at the Ad Hoc Scientific Group, the Fourth Consultative Meeting agreed in principle to the amendment of Annex I, Par. 5 (referring to oils), and to a new paragraph to Annex II referring to nontoxic wastes dumped in large quantities. Owing to an agreement to continue discussion in the 1980 session on the more substantial amendments on which agreement could not be reached, it was agreed that the minor changes mentioned above would not be formally incorporated into the convention until a future meeting.

PART 3
Transport of Dangerous Goods by Air

D. W. Dines
IATA Restrictive Lists Manager
British Airways

Action of the International Air Transport Association / 1-21
Status of the IATA Regulations / 1-22
Applications / 1-23
Future Developments / 1-24

No portion of the air transport business is more important than safety. Consequently, it requires greater attention to detail on the part of the shipping public.

Today, approximately 300,000 different commodities classified as hazardous materials are produced. Of these, about 2500 are expressly listed in the international air regulations as safe enough to transport by air. The remainder must be evaluated against a list of hazard definitions and classified accordingly. A large majority of this remainder ultimately is found to be prohibited for carriage by air.

Despite the relatively small number of types of hazardous materials considered safe enough to transport by air, economic imperatives result in thousands of air shipments of such materials each day. Consequently, the shipper's role in this important market is paramount.

Prior to 1951 the carriage of dangerous goods by air was not permitted in practically all countries with the exception of the United States and Canada. Many countries had national regulations on dangerous goods for surface transport such as transport by sea, road, and rail. However, these differed greatly in the classes of goods, manufacture, types of packing, quantities, labeling, etc. Since 1956 a UN Committee of Experts on the Transport of Dangerous Goods has been studying the problem and has published recommendations for the guidance of governments and international organizations so that they can secure common classification, labeling, and packaging. Unfortunately, because environmental conditions vary greatly between different modes of transport, complete standardization has not yet been practicable.

ACTION OF THE INTERNATIONAL AIR TRANSPORT ASSOCIATION

In 1951 the Traffic Conference of the International Air Transport Association (IATA) considered the need for some form of agreed international regulations. It studied the cost of having an expert staff to do the job but found this to be prohibitive, and so it finally set up a Restricted Articles Board of airline specialists in cargo, safety, fire protection, and packaging. The terms of reference of the board were to produce a set of international regulations which would be acceptable to all IATA members, would be approved by their respective governments, and would permit the carriage of certain dangerous goods in safety. The board worked for 3 years to produce the first edition of the *IATA Restricted Articles Regulations,* which was issued in 1956.

The basic philosophy and the primary objective behind the whole project were the safety of passengers, crew, and aircraft. Therefore, items which could not be packed safely with certainty would not be carried (hence the many "not acceptable" items). Items which could be packed safely would not be regarded as dangerous but as "restricted articles," safe to carry under conditions normally incident to transportation, provided they complied with the applicable packaging and net-quantity regulations.

A cardinal principle of the regulations was, and still is, that a carrier or an approving government may be more restrictive than the IATA regulations if it can be shown that such restriction is essential because of local statutory conditions or other reasons. No one may be less restrictive than the regulations except in very special cases, and then only with the express permission of the government agency concerned in the country of registration of the aircraft and of those countries which will be overflown in transit. In practice, mere commercial expediency is not a justifiable reason for such permission. Generally, it is confined to cases of famine, pestilence, riots, etc., or to cases in which no other form of transport is possible.

The basic principles of the IATA regulations are to provide:

1. A method of classifying any hazardous article by reference to a series of precise definitions
2. A method of packaging articles which are acceptable so that they will present no hazard under normal conditions of transportation
3. A method of labeling such articles so that their hazard is clearly shown
4. A method of shipper's certification to ensure that the shipper has complied with the regulations

As a consequence, dangerous goods being transported by air have a simple but highly significant lexicon. The terms are:

Classification	Labeling
Packaging	Documentation
Marking	Aircraft stowage

The hazard classes that warrant the attention of this lexicon are:

Explosives	Irritating materials
Compressed gases	Corrosive materials
Flammable liquids	Magnetized materials
Flammable solids	Radioactive materials
Oxidizing materials	Etiologic agents
Poisonous articles	Other restricted articles; Groups A, B, and C

STATUS OF THE IATA REGULATIONS

Pursuant to the provisions of IATA Cargo Traffic Conference Resolution 618, no IATA member carrier involved in scheduled and/or unscheduled operations shall accept and carry dangerous goods unless the goods comply fully with the *IATA Restricted Articles Regulations,* which form Attachment A to the resolution. All countries of the IATA member carriers have approved the resolution without reservation. However, approval of the resolution by any one country does not necessarily attach legal status to the regulations in that country. Basically, the country's approval of the resolution merely allows the carrier to operate according to the IATA regulations without fear of being in breach of any applicable national law.

One exception is the United States. Although the United States government has approved the IATA resolution without reservation, the regulations have no formal recognition or status within the United States. Carriage of dangerous goods to, from, or via United States territory is governed by 49 CFR 170.

Countries that have adopted the IATA regulations and have incorporated them into their national law impose severe penalties on transgressors of the regulations.

APPLICATIONS

Frequently, there is a chain of people and events between the origin of a consignment and its loading on board an aircraft, and the integrity of air safety depends on the accuracy with which this whole sequence of operations is carried out. Obviously, this process can be prejudiced from the beginning if the goods are wrongly or inadequately described at the point of true origin. Nevertheless, the IATA regulations hold the consignor (shipper) responsible for correctly classifying the goods requiring shipment and for meeting the packaging, labeling, and documentation provisions.

Export transactions are normally founded on a bewildering variety of contracts, commonly known as contracts of international sale of goods. These contracts are usually entwined with other contracts, in particular with the contract of carriage by which the goods are exported. Unfortunately, a clerk tends to regard the export transaction as an indivisible whole and is apt to pay little attention to its constituent parts.

To understand the air carriage regulations, one must analyze the individual contracts which constitute the export transaction overall. To this end, it must be understood that the contract for the sale of goods abroad is the principal and central legal arrangement involved in the export transaction. All other contracts, such as contracts of carriage, have a supporting and *incidental* character.

Generally, a contract for international carriage by air is founded on the Warsaw Convention or on the Warsaw Convention as amended by the Hague Protocol. Under the convention, rights and responsibilities are established for the consignor (shipper), the consignee, and the carrier. Furthermore, the only persons who have rights of action under the convention are the consignor, the consignee, and the carrier. *The owner of the goods has no status and can exercise rights only in his or her capacity of consignor or consignee.*

Article 10 of the convention expressly states:

1) The consignor is responsible for the correctness of the particulars and statements relating to the cargo which he inserts in the air consignment note.

2) The consignor shall indemnify the carrier against all damage suffered by him or any other person to whom the carrier is liable by reason of the irregularity, incorrectness or incompleteness of the said particulars and statements furnished by the consignor.

It necessarily follows that subordinate conditions relating to the goods offered for carriage must also be directed at the consignor.

Part 1, Sec. 1, of the IATA regulations requires a consignor to insert on the air waybill the proper shipping name of the dangerous article, the IATA article number, its class, and the type of label required.

By definition within the airline industry, the consignor is "the person whose name appears on the Air Waybill as the person contracting with the carrier for the carriage of the goods." Consequently, in relation to the hazardous materials regulations, the carrier must rely solely upon the consignor to ensure that hazardous materials are correctly classified, documented, packed, marked, and labeled according to the applicable regulations. It is incumbent upon the consignor to obtain the necessary assurances from third parties so that the consignor can meet his or her responsibilities under the contract of carriage by air and any statutory obligations attached to such contract.

In short, the carrier's contract is with the consignor. It is the consignor who must accept the responsibilities arising from the contract of carriage by air. The consignor must take whatever steps are necessary to ensure that his or her responsibilities attached to the air carriage contract are met. The air carrier is under no obligation whatsoever to give recognition to a party other than the consignor and the consignee.

Having established the proper shipping name of the goods being shipped, the consignor has as his or her next, and perhaps most important, responsibility to ensure that the goods are packed strictly in accordance with the packing provisions of Part 1, Secs. 5, 6, and 7, of the IATA regulations. Section 5 contains general packing requirements, many of which apply equally to nonhazardous cargo; Sec. 6 contains packing notes for passenger and cargo aircraft, and Sec. 7 contains specification packaging standards.

Finally, the consignor must complete accurately the shipper's certification. If the consignor does not pack the goods personally, to meet the obligation in respect of the shipper's certification he or she should obtain an appropriate written assurance from the packer.

Questions are frequently asked concerning the liability of persons signing the *Restricted Articles* shipper's certification and, in particular, of employees of shippers who sign whenever the shipper is a body corporate and not a physical person. The certification contains the following statement concerning liability:

> I acknowledge that I may be liable for damages resulting from any mis-statement or omission [in the certification] and I further agree that any carrier involved in the shipment of this consignment may rely upon this Certification.

It would appear that liability for a carrier's damages resulting from inaccuracies in the certification lies primarily with the shipper. The document is entitled "shipper's certification," and it is the shipper's signature which is required in the signature box. Other persons signing do so only on the shipper's behalf.

A direct action in tort brought by a carrier against an employee of the shipper might be possible in certain jurisdictions, but there would appear to be little advantage in such an action, which would probably lie even without the acknowledgment of liability set forth in the certificate. The employee's liability toward his or her employer is untouched.

FUTURE DEVELOPMENTS

For 25 years IATA has exerted considerable effort to obtain full-scale international acceptance of its *Restricted Articles Regulations* and has failed. One reason is that the regulations lack a broad base of national approval, control, and input in their development, which is necessary to ensure full international acceptance and compliance.

Additionally, the possibility of nations' adopting commercially developed regulations which impact on other industries and which all nations do not directly control raises certain questions of principle. Furthermore, some nations have laws specifically forbidding the direct adoption of industry-developed regulations no matter how legitimate the aims of such regulations may be.

Consequently, the International Civil Aviation Organization (ICAO) is developing a new annex to the Chicago Convention, which will provide a comprehensive set of specifications governing the carriage of dangerous goods by air. First, it is expected to provide governments collectively with more effective enforcement teeth. Second, the technical instructions will contain most, if not all, of the United Nations' recommendations on the transport of dangerous goods, thus providing a set of air regulations which will be truly intermodal with respect to hazard classification, packaging, and labeling.

The ICAO is an instrument of governments. Its aim and objectives are to develop the principles and techniques of international air navigation and to foster the planning and development of international air transport so as to meet, *inter alia,* the needs of the people of the world for safe air transport.

The ICAO annex will contain a set of standards and recommended practices to which all member states will be asked to give effect by incorporating them in their respective national legislation applicable to the subject. It will be supported by a set of technical instructions, which are intended to become the field reference document by which manufacturers, shippers, and airlines, in particular, will be expected to operate.

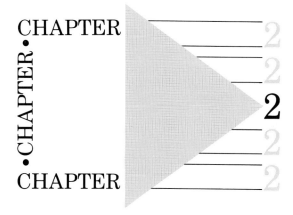

CHAPTER

•CHAPTER•

CHAPTER

2

National Laws and Regulations

PART 1
United States / 2-2

PART 2
Canada / 2-10

PART 3
Europe / 2-14

PART 4
U.S.S.R. / 2-20

PART 1
United States

H. D. Van Cleave
Acting Director, Emergency Response Division
Office of Emergency and Remedial Response
U.S. Environmental Protection Agency

Overview / 2-2
Authority of Federal Agencies to Take Action in Environmental Emergencies /
   2-3

OVERVIEW

This part highlights pertinent United States federal statutes and regulations which can influence spill response operations. Because of the scope and nature of the problem treated in this part of the *Handbook,* a wide variety of emergency situations is covered. The term "environmental emergency" is used broadly to describe the incidents that are covered by federal statutes and regulations.

For federal actions in environmental emergencies, two statutory provisions are dominant: Sec. 311 of the Clean Water Act (CWA) and Sec. 104 of the Comprehensive Environmental Response, Compensation, and Liability Act of 1980 (CERCLA). Section 311 authorizes the federal government to respond to discharges into the environment and to clean them up or to take other mitigating action. This section contains provisions requiring notification of the federal government of such nonpoint discharges and authorizes funding of response activities from a special contingency fund. However, Sec. 311 is limited in two ways: (1) it applies only to discharges of oil and of specifically designated hazardous substances, and (2) it is limited to discharges or threats of discharges to surface water or to adjoining shorelines. Legal authority to overcome the limitations of Sec. 311 was provided in 1980 by the enactment of CERCLA. Section 104 of this law authorizes federal emergency response to *any* release of a material to the environment which endangers public health or welfare.

A number of other legal authorities exist for federal environmental-emergency response. These include provisions of the Safe Drinking Water Act concerning threats to drinking water supplies and a number of statutes authorizing oil spill removal. Important authority exists in the Federal Disaster Relief Act to authorize and fund responses by various federal agencies to major emergencies and disasters. In addition, a number of statutes authorize the federal government to issue emergency orders or regulations and/ or to seek emergency court orders to prevent or abate environmental disasters.

Federal expenses incurred in responding to an environmental emergency are often recoverable from those responsible for causing the emergency. Recovery of the costs of response actions taken under Sec. 311 of the CWA is the subject of elaborate provisions of that section, which states that, subject to certain defenses and limitations, such costs are recoverable from the discharger. Similarly, when emergencies are responded to under Sec. 104 of CERCLA, the expenses incurred are recoverable from the violator. The Trans-Alaska Pipeline Act and the Deepwater Port Act also provide for recovery of oil cleanup costs.

States which respond to spills of oil and designated hazardous substances can be reimbursed from the Sec. 311 contingency fund. In addition, funds to assist states to develop and operate response programs are available under Sec. 106 of the CWA. Funds are also available under Sec. 1442 of the Safe Drinking Water Act to protect drinking-water supplies during emergencies. States may also obtain funds under the Federal Disaster Relief Act and the Deepwater Port and Trans-Alaska Pipeline Acts under certain circumstances.

Individuals have some rights to reimbursement for losses caused by oil spills under the Deepwater Port and Trans-Alaska Pipeline Acts. CERCLA and similar legislation will greatly expand these rights. In addition, dischargers who clean up their own spills can be reimbursed for the cost in limited circumstances under Sec. 311(i) of the CWA.

AUTHORITY OF FEDERAL AGENCIES TO TAKE ACTION IN ENVIRONMENTAL EMERGENCIES

Direct Action

Section 104 of the Comprehensive Environmental Response, Compensation, and Liability Act of 1980 (CERCLA), commonly known as Superfund[1]

Section 104 of CERCLA provides authority for the President to remove or arrange for the removal of hazardous substances, pollutants, or contaminants at any time or to take any other response measures which are considered necessary to protect the public health or welfare or the environment. Under Executive Order 12316 of August 14, 1981, the President delegated authority for federal removal actions primarily to the U.S. Environmental Protection Agency (U.S. EPA) and the U.S. Coast Guard.

CERCLA now provides authority to the federal government to clean up chemical spills as well as hazardous waste sites that threaten the public health and the environment. It also authorizes the federal government to identify responsible parties and to undertake enforcement actions against them. It is the first comprehensive legislation that permits government response to multimedia environmental emergencies; i.e., it applies to releases or threats of release to the environment in general and is not limited to surface waters, as is Sec. 311 of the CWA. The law provides a $1.6 billion hazardous substance response fund to be financed jointly by industry and the federal government.

Section 311 of the Clean Water Act (CWA)

Section 311 is a comprehensive series of provisions dealing with discharges (principally spills) of oil and designated hazardous substances into surface waters. While the provisions concerning oil have been operative since 1970,* those concerning hazardous substances came into effect only in 1979. On March 3, 1978, the administrator of the EPA signed regulations designating an initial list of 271 hazardous substances.[2] The designations became effective on June 12, 1978, with respect to discharges from facilities and on

---

*The oil spill provisions of Sec. 311 were originally enacted as part of the Water Quality Improvement Act of 1970 and were contained in Sec. 11 of the pre-1972 Federal Water Pollution Control Act.

September 11, 1978, with respect to discharges from vessels. The March 3, 1978, regulations also established other regulatory elements needed for implementation of Sec. 311 for hazardous substances.* Court action on Parts 117 and 118 of the March 3, 1978, regulations and subsequent legislative amendments to Sec. 311 of the CWA made it necessary to take certain actions which were published in the *Federal Register* on February 16, 1979. Briefly, these five actions are as follows:

1. Revocation of the existing regulations (Parts 117, 118, and 119). Part 116 remained in effect.

2. Promulgation of changes in definitions for Part 116.

3. Addition of 28 substances to the hazardous list (Part 116).

4. Advance notice of proposed rulemaking to expand criteria for designating a hazardous substance (Part 116).

5. Determination of reportable quantities for hazardous substances (new Part 117).[3]

When the February 16, 1979, regulations became finalized on August 29, 1979,[4] Sec. 311 became operative for hazardous substances. This section is activated only when there is a discharge or a threat of discharge of oil or hazardous substances to the surface water or adjoining shorelines. The waters involved are "navigable waters," meaning all interstate waters with any connection to interstate commerce, and the territorial sea out to 3 nautical mi (5.6 km). Section 311 also applies to the contiguous zone, out to 12 nautical mi (22.2 km), and, pursuant to the Clean Water Act of 1977, for most purposes to water beyond the contiguous zone to approximately 200 nautical mi (370 km).†

Section 311(c) contains the principal provisions relating to federal emergency response activity. It provides that the President may act to remove or arrange for the removal of oil or designated hazardous substances discharged into the "navigable waters" or adjoining shorelines unless he finds that removal will be properly carried out by the discharger. The term "removal" as used in Sec. 311 is broadly defined to include action "necessary to minimize or mitigate damage to the public health or welfare." The relevant legislative history clearly indicates that the following actions are included in the term:

> containment measures, measures required to warn and protect the public of acute danger, activities necessary to provide and monitor the quality of temporary drinking water sources, monitoring for spread of the pollutant, biomonitoring to determine extent of the contamination, physical measures to identify and contain substances contaminated by the discharge, providing navigational cautions while response to the problem is underway, efforts to raise sunken vessels which are the source of the discharge, and implementation of emergency treatment facilities. . . . Mitigation also includes efforts necessary to locate the source of the discharge and identify properties of the pollutants released.

Activities under Sec. 311(c) are carried out in accordance with the National Oil and Hazardous Substances Pollution Contingency Plan developed by the Council on Environmental Quality pursuant to Sec. 311(c)(2).‡ The plan, published in 40 CFR Part 1510, 40 FR 6282–6302 (February 10, 1975), is designed to provide a coordinated federal

---

*These were contained in 40 CFR Parts 117, 118, and 119.

†More specifically, the new authority involves activities under the Outer Continental Shelf Lands Act and the Deepwater Port Act and discharges which affect natural resources under United States ownership or control, including fisheries covered by the Fishery Conservation and Management Act of 1976. Certain penalties do not apply to vessels for violations of Sec. 311 occurring beyond the contiguous zone if the vessels are not "otherwise subject to the jurisdiction of the United States."

‡Authorities under Sec. 311 not statutorily given to the EPA were delegated by the President to various agencies, principally the EPA and the Coast Guard, by Executive Order 11735 (1973).

response to spills of oil and hazardous substances. Among other things, it establishes a national response team, consisting of representatives of various agencies, which is responsible for planning and preparedness for pollution emergencies and for coordination of responses to emergencies. Regional response teams are also established, together with a National Strike Force, which provides on-scene assistance when necessary. Pollution removal actions are supervised by on-scene coordinators (OSCs). An OSC is a federal official responsible for ensuring proper removal and for activating and supervising federal removal efforts when necessary. The U.S. Coast Guard appoints OSCs in coastal and Great Lakes waters, while the EPA appoints OSCs for inland waters. The plan sets out the responsibilities of OSCs and the general pattern of response actions to be taken. It also contains numerous other provisions, including Annex 10, which regulates the use of chemicals in spill cleanup operations.*

To ensure prompt federal awareness of discharges, Sec. 311(b)(5) requires persons who discharge to provide immediate notification to the federal government.† Failure to notify is a criminal offense. Section 311 also provides authority to the Coast Guard to remove or, if necessary, to destroy vessels which present a threat of pollution hazard from an imminent discharge of oil or hazardous substances subsequent to a marine accident. Section 311(k) provides that action taken under Sec. 311 may be financed from a revolving fund maintained at $35 million and administered by the Coast Guard.

Section 115 of the Clean Water Act

Section 115 provides the administrator of the EPA with authority to arrange for removal and disposal of in-place pollutants, with emphasis on toxic pollutants in harbors and waterways; $15 million is authorized to carry out the section. This authority can be used to undertake this form of cleanup action after an environmental emergency.

Safe Drinking Water Act (42 U.S.C. Secs. 1401 et seq.)

Section 1431 gives the administrator of the EPA broad powers to take action when a contaminant present or likely to be present in a public water system may pose an imminent and substantial endangerment to human health and welfare and when state and local authorities have not acted. In such circumstances, the administrator can "take such actions as he may deem necessary in order to protect" human health.‡

Outer Continental Shelf Lands Act (43 U.S.C. Secs. 1331 et seq.)

Regulations issued by the Department of the Interior (20 CFR Sec. 250.43) provide that spills and leaks of oil or waste material from operations conducted by leaseholders on the Outer Continental Shelf must be reported to the federal government and that the leaseholders have an obligation to control or remove any such spills or leaks which damage or threaten to damage aquatic life, wildlife, or public or private property.§ If the lessee fails to conduct cleanup operations, the U.S. Geological Survey is authorized to control and remove the pollution in cooperation with other federal, state, and local agencies and in accordance with the National Oil and Hazardous Substances Pollution Contingency Plan. Costs of governmental cleanup are recoverable from the lessee.

*The National Oil and Hazardous Substances Pollution Contingency Plan also provides for coordination of federal agencies with state and local authorities and invites these authorities to participate in regional response team activities. Annex XI to the plan authorizes such authorities to participate in regional response planning to develop emergency response contingency plans. Section 311(c) of the act authorizes states to conduct cleanup operations and to be reimbursed from the Sec. 311(k) revolving fund for expenses so incurred.

†Procedures for notification have been established by the Coast Guard in 33 CFR Part 153, Subpart B. These procedures include a toll-free 800 telephone number for receipt of notifications.

‡No regulations have yet been issued to implement this authority.

§These regulations are issued under 43 U.S.C. Sec. 1334, which provides broad authority to regulate leasing of the Outer Continental Shelf.

Federal Disaster Relief Act (42 U.S.C. Secs. 5121 et seq.)

Under Secs. 5145 and 5146 of this act the President is authorized to provide assistance to save lives and protect property and public health and safety in emergencies and major disasters.* Assistance is coordinated by the Federal Emergency Management Agency (FEMA) and is provided with the cooperation of various federal agencies. The forms of assistance that may be rendered by the federal government include practically any action deemed necessary to protect life and property. Although the act does not directly concern protection of the environment, actions taken under it will often have that effect. Expenditures by federal agencies which are asked to participate in disaster assistance may be reimbursed by the FEMA. Under agency regulations (24 CFR Part 2205), however, reimbursement is not available for costs incurred by an agency operating under its own authority.

U.S. Army Corps of Engineers Emergency Authorities

The U.S. Army Corps of Engineers (like other federal agencies) is authorized to provide emergency assistance in disasters pursuant to the Federal Disaster Relief Act. In addition, 33 U.S.C. Sec. 701n provides special authorities to the Corps in the event of emergencies caused by floods. These authorities include actions to prevent and combat floods and to restore flood damage. Environmental protection is not an explicit goal but may result indirectly from such actions. The Corps is also, under the umbrella of the Department of Defense, a primary agency under the National Oil and Hazardous Substances Pollution Contingency Plan and may be called in to assist in Sec. 311 cleanup operations. Detailed regulations implementing these authorities and responsibilities are in 33 CFR part 203.† A 1974 amendment to 33 U.S.C. Sec. 701n added authority for the Corps to "provide emergency supplies of clean drinking water . . . to any locality . . . confronted with a source of contaminated drinking water causing or likely to cause a substantial threat to the public health or welfare. . . . " Procedures and policies for implementation of this authority are in 33 CFR Part 214.

Emergency Authorities Limited to Oil Spills

Intervention on the High Seas Act (33 U.S.C. Secs. 1471 et seq.)   This act gives the Coast Guard authority to respond to oil spills or threats of oil spills resulting from ship casualties on the high seas if the spills could cause "a grave and imminent danger" of "major harmful consequences" to the coastline or related interests of the United States. The Coast Guard may supervise or itself take action to remove or eliminate the pollution damage, or it can remove or destroy the ship and cargo which are the source of the danger. Measures taken must be reasonably proportionate to the actual or threatened damage that they are designed to prevent or mitigate. The revolving fund under Sec. 311(k) of the CWA may be used to finance actions taken under the act.

Deepwater Port Act (33 U.S.C. Secs. 1501 et seq.)   Section 1517(c) of the Deepwater Port Act authorizes the Coast Guard to act to remove oil discharged from a deepwater port, near a deepwater port (within the designated "safety zone"), or from a vessel loaded at a deepwater port unless removal would be promptly accomplished by the discharger. Removal actions are to be consistent with the National Oil and Hazardous Substances Pollution Contingency Plan. Funding for removal actions is obtainable from the Deepwater Port Liability Fund established under Sec. 1517(f) of the act. The fund is created from a fee collected on oil loaded or unloaded at deepwater ports and is to be maintained

---

*The terms "emergency" and "major disaster" are defined in Sec. 5122 as catastrophes in which local authorities require federal assistance. The act provides specific procedures whereby catastrophes are designated as emergencies or major disasters.

†See also 32 CFR Part 502 (U.S. Army regulations governing the provision of assistance for disaster relief).

at $100 million. As under Sec. 311(b)(5) of the CWA, oil discharges must be reported to the federal government.

Trans-Alaska Pipeline Act (43 U.S.C. Secs. 1651 et seq.)   Section 1653(b) of the act provides that if an oil spill resulting from pipeline activities which damages or threatens aquatic life, wildlife, or property is not cleaned up by the pipeline company, the secretary of the interior (in cooperation with other federal agencies) may mitigate and remove the oil spill. Removal costs are to be borne by the pipeline company.

Authority to Take Emergency Action in the Form of Administrative or Judicial Orders

Section 504(a) of the Clean Water Act

This section authorizes the administrator of the EPA to seek a remedial federal district court order when "a pollution source or combination of sources is presenting an imminent and substantial endangerment to" health and welfare. The provision probably is limited to water pollution.

Section 311(e) of the Clean Water Act

This provision is similar to Sec. 504(a), described above, except that it is limited to emergencies caused by discharges of oil or hazardous substances designated under Sec. 311(b)(2)(A) into navigable waters.

Section 1431 of the Safe Drinking Water Act

This section, described above, authorizes, among other actions, issuance by the EPA of administrative orders designed to protect the health of users of public water systems and authorizes suits in federal district court to obtain court orders for the same purpose.

Solid Waste Disposal Act (42 U.S.C. Secs. 6901 et seq.)*

Section 7003 of the act authorizes the administrator of the EPA, upon learning that solid waste or hazardous waste handling, storage, treatment, transportation, or disposal is presenting an imminent and substantial endangerment to health or the environment, to bring suit in federal district court to obtain an order immediately restraining the activity causing the danger or to obtain other such relief.

Clean Air Act (42 U.S.C. Secs. 1857 et seq.)

Section 303 provides that when one or more sources of air pollution create an imminent and substantial danger to human health the administrator may issue short-term administrative orders (for no longer than 48 h unless extended by court permission) to protect human health or may seek a federal district court order restraining the emissions or otherwise providing relief.†

Marine Protection, Research, and Sanctuaries Act (Ocean Dumping Act) (33 U.S.C. Secs. 1401 et seq.)

Section 1415(d) of the act authorizes suit for injunctive relief to prevent imminent or continuing violations of the statute, regulations, or permits.

Toxic Substances Control Act (15 U.S.C. Secs. 2601 et seq.)

Section 7 authorizes suit in federal district court to order seizure of imminently hazardous substances or mixtures or for relief against persons manufacturing, processing, distrib-

---

*As amended by the Resource Conservation and Recovery Act of 1976.
†Section 110(a)(2)(F)(i) requires states to have authority comparable to that of Sec. 303.

uting, using, or disposing of such substances or mixtures and articles containing such substances or mixtures. The act also authorizes expedited regulatory or judicial actions against new or existing substances presenting a risk of injury to health or the environment under Secs. 5(e), 5(f), and 6(d)(2).*

Federal Insecticide, Fungicide, and Rodenticide Act (7 U.S.C. Secs. 135 et seq.)

Section 6(c) authorizes suspension of the registration of a pesticide (thus making its sale or distribution illegal) on an emergency basis if it presents an imminent hazard.

Occupational Safety and Health Act (29 U.S.C. Secs. 651 et seq.)

Section 655(c) of the act authorizes promulgation of emergency temporary safety or health standards, without the necessity of prior notice and comment, if necessary to protect workers from grave danger from exposure to toxic or physically harmful substances or agents or from new hazards. Section 662 authorizes the initiation of suits in federal district court for injunctive relief to restrain or remedy conditions causing imminent danger of death or serious physical harm.

Consumer Product Safety Act (15 U.S.C. Secs. 2051 et seq.)

Section 2061 of the act authorizes filing suit in federal district court for seizure of imminently hazardous consumer products or for remedial orders directed to manufacturers, distributors, or retailers of such products to prevent imminent and unreasonable risk of death, serious illness, or severe personal injury.

Hazardous Materials Transportation Act (40 U.S.C. Secs. 1801 et seq.)

Section 1810(b) authorizes the secretary of transportation to seek a federal district court order suspending or restricting the transportation of hazardous material or for such other court order as may be necessary if there is reason to believe an "imminent hazard" exists. An imminent hazard exists if there is substantial likelihood that serious harm will occur prior to completion of administrative proceedings to abate the risk.

Under this act and related authorities the Department of Transportation has established an array of disaster-reporting requirements, which include reports of emergencies involving trucks, aircraft, vessels, pipelines, and railroads. These requirements generally are triggered by injury or threat of injury to persons or property.

Ports and Waterways Safety Act (33 U.S.C. Secs. 1221 et seq.)

Section 1221(6) authorizes the Coast Guard to establish procedures and standards for the emergency removal, control, and disposition of dangerous articles or substances in ports and harbors.

Deepwater Port Act

The secretary of transportation is authorized by Sec. 1511(a) to seek a court order to suspend any deepwater port license if the licensee violates applicable requirements. If the secretary determines that there is danger to public health or safety or imminent and substantial danger to the environment he or she may, under Sec. 1511(b), order a deepwater

---

*Section 8(e) of the Toxic Substances Control Act requires that persons who manufacture, process, or distribute chemical substances or mixtures must notify the administrator of the EPA of any information that the substance or mixture "presents a substantial risk of injury to health or the environment." In some cases such notification could include notice of impending or ongoing environmental emergencies.

port licensee to suspend or alter construction or operation of the port pending completion of judicial proceedings instituted under Sec. 1511(a).

REFERENCES

1. 42 U.S.C. 9601; also known as the Comprehensive Environmental Response, Compensation, and Liability Act of 1980.
2. 43 FR 10474 et seq. (Mar. 12, 1978), 40 CFR Part 116.
3. *Federal Register,* vol. 44, no. 34, Feb. 16, 1979, pp. 10266–10284.
4. *Federal Register,* vol. 44, no. 169, Aug. 29, 1979, pp. 50766–50779.

PART 2
Canada

John F. MacLatchy
Legislative Adviser
Environmental Protection Service
Environment Canada

Transportation of Dangerous Goods Act / 2-10
Fisheries Act / 2-11
Canada Shipping Act / 2-11
Arctic Waters Pollution Prevention Act / 2-12
Environmental Contaminants Act / 2-12
Atomic Energy Control Act / 2-12
Other Federal Statutes / 2-12
Provincial Statutes and Law / 2-12

In Canada the division of legislative powers, as between the federal government and the provincial governments, is similar to the division of powers as between the United States federal government and the states. Thus, the authority to regulate and control various aspects of hazardous materials is divided between federal and provincial governments. The federal government has authority under Sec. 91 of the British North America Act to make law related to the regulation of trade and commerce, navigation and shipping, and seacoast and inland fisheries and generally to make laws for "Peace, Order, and Good Government of Canada" in relation to all matters not exclusively the legislative responsibility of the provinces. The provinces have legislative authority over property and civil rights and generally all matters of a merely local or private nature in the provinces.

Concerns for spills of hazardous materials have resulted in the development of a number of new laws to address this type of problem from various approaches. Those laws related to oil spills and the production and transport of oil will not be considered in this part except to the extent that a general spill requirement may relate to either oil or other hazardous materials.

TRANSPORTATION OF DANGEROUS GOODS ACT[1]

The Transportation of Dangerous Goods Act was passed by the Parliament of Canada in July 1980. It had been in preparation for a number of years, and versions of the bill had been introduced as early as May 1978. The 1979 incident involving the rupture of a railway car carrying chlorine and the evacuation of 240,000 people at Mississauga, Ontario, led to greater awareness of the problems associated with the transportation of hazardous materials. The final version of the act also contained provisions to regulate the international and interprovincial shipments of hazardous wastes through a manifest system with waybills similar to the system in the United States. In the area of prevention of accidents, the act will make it possible to prescribe safety procedures applicable to the handling for transportation and transporting of dangerous goods, safety standards and safety marks for containers and packages, and shipping documents to record compliance with the various safety requirements. The act also contains provisions related to mandatory reporting of spills of hazardous materials, cleanup of spills, and authority for inspectors to direct cleanup when immediate action is necessary. These provisions also cover the emission of ionizing radiation levels or quantities prescribed pursuant to the Atomic Energy Control Act from any container in transit.

The act makes the person in possession of the dangerous goods, along with any persons who caused or contributed to the spill, jointly and severally, liable for the costs and expenses of the federal government if the federal government has to clean up a spill. A person engaged or proposing to engage in the transportation or handling of dangerous goods may be required to show evidence of financial responsibility in the form of insurance or an indemnity bond.

The act is applicable to the handling for transportation and the transportation of dangerous goods in virtually all circumstances. At earlier stages in the drafting the scope of the act was narrower and limited in application to dangerous goods that were being transported by highway across provincial boundaries. However, given the need to have uniform national and international standards in dealing with dangerous goods, the scope was broadened.

FISHERIES ACT[2]

The pollution provisions of the Fisheries Act are concerned with preventing the deposit of deleterious substances in waters frequented by fish. The deleterious substances may originate from any source (i.e., either spills or continuous discharges from industrial operations). While the majority of incidents that have been prosecuted under the Fisheries Act have related to incidents like oil spills, effluent from pulp mills, or effluents from mercury chloralkali plants, the pollution provisions of the Fisheries Act have been applied to cyanide, polychlorinated biphenyl (PCB), and copper sulfate spills. The 1977 amendments to the Fisheries Act contain a number of provisions related to the mandatory reporting of spills, the obligation to clean up spills, directions by inspectors to clean up spills, and civil liability for spills. The 1977 Fisheries Act also allows the minister to issue orders so that the risk of deposits of deleterious substances will be minimized.

The highest fine imposed for an offense under the Fisheries Act to the end of 1980 has been $120,000 for depositing wastes from a pulp mill on 6 different days. While the 1977 amendments increased the maximum possible fine to $100,000 for each day of an offense, the highest fine for a single day has been $20,000. The size of fines and the frequency of prosecutions generally appear to be less in Canada than in the United States.

CANADA SHIPPING ACT[3]

The pollution provisions in Part XX of the Canada Shipping Act are primarily directed at oil spills. However, the *Pollutant Substances Regulations* prescribe a number of substances as pollutants and require reports of any spills of these substances from ships. A ship is civilly liable for spills of pollutants in circumstances and to the limits of liability described in the act.

ARCTIC WATERS POLLUTION PREVENTION ACT[4]

The Arctic Waters Pollution Prevention Act and its regulations contain general provisions concerning the reporting of spills and civil liability for spills. The legislation is generally oriented toward pollution prevention related to shipping, oil spills, and drilling for oil and gas. However, the definition of "waste" in the statute also covers hazardous materials that might be spilled.

ENVIRONMENTAL CONTAMINANTS ACT[5]

The Environmental Contaminants Act provides for assessment of the danger from substances or classes of substances to human health or the environment. If a significant danger exists, the substance can be scheduled under the act as a contaminant. The act further provides for control of these scheduled substances in their use, manufacture, importation, sale, and *willful* release. Spilling is included in the definition of release. The act, however, does not deal with civil liability for spills of scheduled substances. The recent restrictions on such contaminants as PCBs, PCTs (polychlorinated terphenyls), PBBs (polychlorinated biphenyls), and mirex are expected to reduce the problems related to spills of these materials.

ATOMIC ENERGY CONTROL ACT[6]

At the time of writing, the Atomic Energy Control Act provided for controls over atomic energy by the Atomic Energy Control Board. The Nuclear Control and Administration Act was first introduced into the Parliament of Canada on November 24, 1977, to replace the Atomic Energy Control Act.

The new act will provide for broader controls over all aspects of atomic energy including accidents. It will contain provisions giving inspectors authority to deal with emergencies that may endanger the health and safety of persons or endanger the environment.

OTHER FEDERAL STATUTES

The Explosives Act[7] deals with product control of explosives (i.e., manufacturing, importation, storage, and transportation of explosives); it does not deal with spills or civil liability. Various provincial laws deal with the use of explosives. The Pest Control Products Act[8] deals with product control (i.e., manufacturing, importation, and labeling) and not with spills of such products. Likewise, various provincial laws deal with the use of pesticides. The Ocean Dumping Control Act[9] deals with the deliberate dumping of waste material at sea. The act does require the reporting of incidents in which emergency dumping is necessary to avert danger to human life at sea or to any ship or aircraft. Part 8 of the *Air Regulations* under the Aeronautics Act[10] contains general provisions to control the air transport of hazardous materials, while detailed requirements are contained in instruction manuals of the Department of Transport.

PROVINCIAL STATUTES AND LAW

The provinces also have laws dealing with spills of hazardous material within the provinces. The legislation commonly consists of general environmental legislation that deals with any kind of hazardous material which might be involved in a spill and more specific legislation that deals with a particular class of problems. For example, almost every province has specific laws related to the use of pesticides.

Since 1971 almost every province has developed mandatory reporting requirements related to spills. In some provinces the reporting requirements are applicable to any spill that may contaminate the environment, while in other cases provincial spill-reporting regulations relate only to certain spills of petroleum products. However, the combination of federal and provincial laws now requires that virtually every situation involving a spill of hazardous materials be reported under one or another of the mandatory reporting requirements.

Various working arrangements have been established between the federal and provincial governments with respect to the mandatory reporting requirements and other matters related to spills of hazardous material. While federal and provincial legislation on the problems of spills of hazardous material may differ in approach, the various federal and provincial agencies, through practical working arrangements, administer the legislation in a unified manner.

REFERENCES

1. *Transportation of Dangerous Goods Act,* S.C. 1979–1980, chap. 36.
2. *Fisheries Act,* R.S.C. 1970, chap. F-14; amended R.S.C. 1970, 1st Supp., chap. 17, and S.C. 1976–1977, chap. 35.
3. *Canada Shipping Act,* R.S.C. 1970, 2d Supp., chap. 27.
4. *Arctic Waters Pollution Prevention Act,* R.S.C. 1970, 1st Supp., chap. 2.
5. *Environmental Contaminants Act,* S.C. 1974, 1975, and 1976, chap. 72.
6. *Atomic Energy Control Act,* R.S.C. 1970, chap. A-19.
7. *Explosives Act,* R.S.C. 1970, chap. E-15, S.C. 1974, 1975, and 1976, chap. 60.
8. *Pest Control Products Act,* R.S.C. 1970, chap. P-10.
9. *Ocean Dumping Control Act,* S.C. 1974, 1975, and 1976, chap. 55.
10. *Aeronautics Act,* R.S.C. 1970, chap. A-13.

PART 3
Europe

J. Bentley
Land Waste Division
U.K. Department of the Environment

Belgium / 2-15
Denmark / 2-15
Finland / 2-15
France / 2-16
German Democratic Republic / 2-16
German Federal Republic / 2-16
Greece / 2-17
Ireland / 2-17
Luxembourg / 2-17
Netherlands / 2-17
Norway / 2-18
Sweden / 2-18
United Kingdom / 2-18

No European country has one single piece of comprehensive legislation covering all aspects of hazardous material spills, which are considered to include uncontrolled deposits of such materials. There is evidence, however, that certain countries are attempting some rationalization in this field. Currently the powers available to the various authorities involved—the police, the fire services, and the transport, public health, water, and waste disposal authorities—are provided in a range of often unrelated legislation. This legislation includes police, public health, transport, water protection, and waste disposal acts.

The European Economic Community (EEC) has no specific directive on the carriage of dangerous goods or on dangerous and toxic wastes or general wastes. However, the member countries of the EEC—Belgium, Denmark, France, the German Federal Republic, Ireland, Italy, Luxembourg, the Netherlands, and the United Kingdom—are bound by the provisions of certain directives which require the member states to institute legislative measures which could apply to spills.

The directives concerned include the Framework Directive on Wastes of July 15, 1975, which required responding legislation to be implemented by July 15, 1977. Articles 4, 5, 7, 8, 9, and 10 of this directive prohibit uncontrolled deposits and deliberate abandonment of wastes. These general requirements were strengthened in Art. 5 of the subsequent directive of March 20, 1978, the Directive on Toxic and Dangerous Wastes. In addition, Art. 9 of this directive requires that undertakings engaged in the carriage of toxic and dangerous wastes shall be controlled by the designated competent authority of each member state.

Article 14 of the 1978 directive prescribes that when toxic and dangerous wastes are transported, they shall be accompanied by an identification form containing certain specific details. Some member states have responded to this requirement by applying the provisions of international conventions such as the European Agreement Concerning the International Carriage of Dangerous Goods by Road (ADR) or equivalent systems which provide for labeling with respect to hazards and emergency measures to their internal transport.

The situation in various European countries on which information is available is as follows.

BELGIUM

Belgium has no specific law on spills of hazardous materials. General laws on transport and public health provide the transport minister and the interior minister or authorities to whom they have delegated responsibilities with powers to deal with the immediate situation regarding spills.

The special Law on Toxic Wastes (Loi sur les Déchets Toxiques) of July 22, 1974, as modified by a crown order of February 9, 1976, provides measures for the disposal of spilled material. Section 23 of this law contains provisions on transportation and packaging and requires an emergency center (*centre pour secours*) or, alternatively, the police, fire service, etc., to be informed in the event of a spill. Section 34 requires that, in the event of an accident involving toxic wastes, the minister of employment and labor must consult with the minister of public health and other competent ministers and officials on measures required to protect the public and the environment.

The Belgian legislation gives regional and local authorities (governors of provinces and mayors of communes) wide powers to prevent toxic wastes from affecting the health and safety of workers and the general public.

DENMARK

The Law on Environmental Protection, Act No. 372 of June 13, 1973, as amended by Act No. 107 of March 29, 1978, and Statutory Order No. 290 of June 1978, contains general provisions on the storage, treatment, and disposal of waste which in fact also apply to toxic or hazardous chemicals. The more relevant activities, such as the transport, treatment, and disposal of toxic and hazardous wastes, are included in the general definition of the law.

The transport of chemical wastes is specifically controlled by the provisions of the Order on Chemical Wastes No. 121 of March 17, 1978 (Bekendtgorelse om Kemikalieaffald), which require that anyone who stores, transports, or disposes of chemical waste shall be responsible for ensuring that these operations do not result in the pollution of the air, ground, groundwater, or surface water including the sea.

FINLAND

Law No. 510 of June 20, 1974, controls the transport of dangerous substances by road (*Suomen Asetuskokoelma—Finlands Forfattningssamling* of June 25, 1974, Nos. 498–510, pages 906–908). For the purposes of this law, "dangerous substances" are defined to mean substances or objects which by virtue of being explosive, flammable, radioactive, poisonous, or corrosive or by virtue of any other such characteristic may cause damage to persons or property or to the environment. The law provides for the promulgation of subsidiary legislation concerning the transport of dangerous substances by road, designates the competent authorities, and indicates the responsibilities of shippers and carriers.

FRANCE

Emergency situations connected with hazardous materials which cover spills appear to be dealt with under the general provisions of a number of laws, one of which is the Law on Waste Disposal and Waste Recovery (Loi No. 75-633, July 15, 1975, Portant sur l'Élimination des Déchets et la Récupération des Matériaux). In this law, wastes are divided into three categories. The most dangerous wastes are regulated by Art. 3 of a subsidiary document (No. 17-974 of August 19, 1977). Activities relating to materials in the most dangerous category require authorization of the Ministry of the Environment. Article 8 of the law stipulates that anyone who produces, imports, transports, or disposes of toxic waste must notify the origin, chemical and physical characteristics, quantity, and destination to the competent authorities.

GERMAN DEMOCRATIC REPUBLIC

The transport of hazardous materials and provisions for remedial measures with respect to spills involving such materials are covered by the second regulations of May 31, 1977, for the implementation of the Poisons Law, *List of Classified Poisons* (*Gesetzblatt der Deutschen Demokratischen Republik*, Part 1, July 13, 1977, No. 21, pages 279–282), which contains requirements for labeling. The third regulations of May 31, 1977, for the implementation of the Poisons Law, *Transportation of Poisons* (*Gesetzblatt der Deutschen Demokratischen Republik*, Part 1, July 13, 1977, No. 21, pages 282–283), requires that spills into the soil or watercourses during operations covered by the regulations be notified to the environmental protection and water resources department of the relevant district council. The act also contains provisions dealing with decontamination by qualified persons.

Article 20 of the Law on the Protection, Use, and Care of Waters of April 17, 1963, states that waters shall be protected from any harmful effects. In particular, it is forbidden to deposit waste, etc., in waters. Harmful substances including radioactive material must be carried, unloaded, stored, and used in such a way that they cannot pollute water.

The order on substances harmful to water of December 15, 1977, contains provisions dealing with traffic accidents involving vehicles carrying such substances. It lists the authorities with responsibilities in the case of accidents, among them the fire brigade, the Ministry of the Environment, and local and national highway authorities. These authorities are to provide the requisite equipment and personnel to deal with such emergencies. The order does not give detailed information on what to do when spills of hazardous and toxic materials occur.

GERMAN FEDERAL REPUBLIC

In West Germany there are a large number of laws and regulations dealing with transportation, spills, and disposal of hazardous materials and wastes. The most relevant are:

- Law on the Carriage of Dangerous Goods of August 6, 1975, published in the *Federal Gazette (Bundesgesetzblatt)*, Part 1, No. 95, of August 12, 1975.

• Law on Waste Disposal of June 7, 1972, in the text of June 5, 1977. Article 12 of this law relates to collection and transportation licenses and contains provisions for the safeguarding of public safety. Article 13, Sec. 5.2, gives power to make regulations for safeguarding the welfare of the public in transfrontier traffic.

In most incidents, the police acts provide power to deal with the immediate situation, local authorities being responsible for the cleanup of the spills and the disposal of any residual material.

GREECE

At present, measures to deal with spills and uncontrolled deposits of dangerous materials and wastes on or adjacent to national or county roads are provided for only by general police regulations. Any remedial measures in connection with roads in townships and municipalities are the responsibility of their controlling bodies.

A new law will, among other things, make the minister of public works responsible for spills on national roads. It will also give the minister power to make decisions relating to spills or to the deliberate abandonment of hazardous materials.

IRELAND

As in many other countries, the general powers afforded by the Road Traffic Act of 1961, Sec. 91, enable the police of Ireland to prevent hindrance to the emergency services dealing with spills, allowing them to divert traffic and prevent parking. The Water Pollution Act of 1977 gives local authorities power to take such action as is considered necessary when spillage puts water supplies at risk. It also authorizes them to recoup the cost of the necessary remedial works. The Department of the Environment has encouraged local authorities to prepare emergency plans to deal with spillages. Specific enabling legislation on the carriage of dangerous goods is in an advanced state of preparation.

LUXEMBOURG

Luxembourg has no specific law on spills, but a law on the disposal of wastes has been drafted to respond to the EEC Directive on Toxic and Dangerous Wastes. Article 4 of this draft law requires that importers, exporters, and transporters of waste have permits. These permits are granted only if the given operations present no risk to people or the environment. The article further provides for special measures in the interest of public health and safeguarding the environment when these are considered necessary.

NETHERLANDS

Powers afforded by the Nuisance Law (Stb. 1952, 274) and the Law on the Pollution of Surface Waters (Stb. 1969, 536) provide general measures which can be used to deal with spills. The decree of March 20, 1979, implements Art. 9, Pars. 1 and 3, of the Chemical Waste Law (Stb. 1976, 214), which deals with the transportation of wastes.

NORWAY

In 1975 the Ministry of the Environment notified its intention to submit proposals for a new comprehensive pollution control law. This law will contain measures regarding pollution caused by emergency situations. The proposals are based in part on a public health report (*Norges Offentlige Utredninger*) of 1974, No. 25: *Preparedness Systems for Pollution Emergencies.*

SWEDEN

Sections 9 and 13 of Ordinance No. 346 of May 22, 1975, on wastes endangering the environment (*Svensk Forfattningssamling, 1975*, Volume 2, June 17, 1975, page 5) provide for transportation licenses for specified wastes.

UNITED KINGDOM

The law relating to the transport of dangerous goods by road at present has the following requirements:

1. Certain vehicles carrying specified flammable liquids, corrosives, or organic peroxides must display diamond-shaped labels indicating the nature of the substance.
2. Containers carrying prescribed dangerous substances must indicate the name and address of the manufacturer, the importer, and the wholesaler or supplier of the substances in some instances.
3. Tank vehicles and tank containers carrying specified substances must bear hazard-warning panels. The design and coloring of these panels are specified. A variation is allowed for foreign and United Kingdom vehicles going to the European continent.

The relevant legislation regarding these requirements is provided in the following:

- *Petroleum (Inflammable Liquids) Order, 1971* [Statutory Instrument (SI) 1971, No. 1040]
- *Inflammable Liquids (Conveyance by Road) Regulations, 1971* (SI 1971, No. 1061)
- *Inflammable Substances (Conveyance by Road; Labeling) Regulations, 1971* (SI 1971, No. 1062)
- *Petroleum (Corrosive Substances) Order, 1970* (SI 1970, No. 1945)
- *Corrosive Substances (Conveyance by Road) Regulations, 1971* (SI 1971, No. 618)
- *Petroleum (Organic Peroxides) Order, 1973* (SI 1973, No. 1897)
- *Organic Peroxides (Conveyance by Road) Regulations, 1973* (SI 1973, No. 3221)
- *Packaging and Labeling of Dangerous Substances Regulations, 1978* (SI 1978, No. 209).
- *Hazardous Substances (Labeling of Road Tankers) Regulations, 1978* (SI 1978, No. 1702)

The Health and Safety Commission (HSC) is seeking to extend the scope of the last-named piece of legislation and has produced a consultative document, *Proposals for Dangerous Substances (Conveyance by Road) Regulations,* published on March 6, 1979. The HSC proposes new regulations which will be sufficiently comprehensive to cover all hazardous substances (except for explosives and other specified substances) transported by road. They would apply at all times, from loading to unloading, whether or not the vehicle was on a public road.

The draft proposals incorporate and extend the new tank truck regulations which came into force on March 28, 1979. They would adopt similar requirements for the identification and classification of dangerous substances. Among the matters covered by the proposals is a requirement for the notification and reporting of accidents.

In the event of spills, the immediate emergency is normally dealt with by the police, the fire brigade, and other emergency services using powers provided by provisions of general legislation such as the Road Traffic Acts to keep the public away from the danger and to prevent hindrance of the emergency services.

Once the immediate emergency is over, a highway authority may be called upon to remove dangerous substances from the highway. The highway authority has a statutory duty under Sec. 22 of the Control of Pollution Act of 1974 to remove any spillages from the highway when such removal is necessary for the maintenance of the highway or the safety of traffic. Under Secs. 8 and 9 of the Highways (Miscellaneous Provisions) Act of 1964, the highway authority also has power to remove anything deposited on a highway which constitutes a danger to the users of the highway and to recover the cost of removing it from the person responsible.

If the residual material resulting from a spill emergency is not present on a highway, the waste disposal authority or the collection authority could if necessary use the powers given in Sec. 16, Par. 5, of the Control of Pollution Act to remove the material for safe disposal.

When a spill presents a water pollution potential, Sec. 76 of Water Resources Act of 1963 provides the water authority with the power to take any necessary action to prevent, remedy, or mitigate the problem.

PART 4
U.S.S.R.

William J. Lacy
Director of Research and Development
Water and Hazardous Waste Monitoring
U.S. Environmental Protection Agency

Introduction / 2-20
Control of Water Pollution: General Provisions / 2-20
Enforcement and Penalties for Law Violations / 2-23
Institutions and Administration / 2-24
Summary / 2-26

INTRODUCTION

What the Soviet Union, as the world's second leading industrial power, does to the eco-system within its huge territory (which covers one-sixth of the earth's land area) will have an environmental impact in countries around the globe, including the United States. In some areas, the Soviets are less advanced than the United States. Nevertheless, Soviet investment in environmental protection is growing at an impressive rate, and environmental concerns are playing an increasing role in economic policies. Soviet officials have advised American scientists that decisions on their long-talked-about plans for diversion of major rivers to increase agricultural yields are being delayed to permit time for a full study of the environmental impact of such action.

Included in this part are a brief discussion of enforcement and penalties for violations of environmental protection laws and a discussion of the All-Union Soviet agency primarily responsible for water quality protection and research.

The following are highlights of the Soviet Water Law and the basis for the current Soviet environmental program. However, these highlights are not all-inclusive and present just a brief picture of the entire program. The Soviet Water Law provides for the rational utilization and conservation of water in the U.S.S.R. Various statutes, decrees, regulations, instructions, and acts contain rules of the Soviet Water Law, and the U.S.S.R. Constitution proclaims general principles for the Soviet Water Law as a whole.

CONTROL OF WATER POLLUTION: GENERAL PROVISIONS

The law of December 10, 1970, of the U.S.S.R. provides the basis and the Fundamental Principles of the Water Legislation of the U.S.S.R. and the Union Republics:

> This law, which establishes the Fundamental Principles of the Water Legislation of the U.S.S.R. and the Union Republics, contains a number of provisions dealing with the sanitary

protection of waters, including the following: Bodies of water where the water quality meets the prescribed sanitary requirements are to be used for drinking and domestic water supply purposes and to meet the other water needs of the population, while the utilization of groundwater of drinking quality for purposes not connected with the supply of water for drinking and domestic water supply purposes is, in general, not permitted (Sec. 21). Except in special cases, bodies of water classed as medicinal in accordance with the established procedure must be used in the first place for therapeutic and balneological purposes and no wastewaters may be discharged into such bodies of water (Sec. 22). The irrigation of agricultural land with wastewaters requires the permission of the agencies responsible for regulating water utilization and protection, in agreement with the agencies responsible for state sanitary and veterinary surveillance (Sec. 23). Water users utilizing water sources for industrial purposes are required to take the necessary measures to limit wastewater discharges (Sec. 24). Industrial, domestic, drainage, and other effluents may only be discharged into bodies of water if permission has been obtained from the agencies responsible for regulating water utilization and protection, with the agreement of, *inter alia,* the agencies responsible for state sanitary surveillance (first paragraph of Sec. 31). Wastewaters may only be discharged in cases in which this does not cause the content of pollutants in a body of water to exceed established standards and on condition that the water consumer purifies the wastewaters to the levels imposed by the agencies responsible for water utilization and protection (second paragraph of Sec. 31). The procedures governing and the conditions for the utilization of bodies of water for the discharge of wastewaters are to be laid down by legislation enacted by the U.S.S.R. and by the union republics (fourth paragraph of Sec. 31).

All bodies of water must be protected against pollution, obstruction, and depletion liable to be prejudicial to public health, etc., as a result of changes in the physical, chemical, and biological properties of the water, diminution of its capacity for self-purification, and disturbance of its hydrological and hydrogeological characteristics (first paragraph of Sec. 37). In order to protect waters against pollution, obstruction, and depletion and to improve the condition and regime of waters, the undertaking organizations and institutions whose activities affect water conditions are required to take the requisite technical, sanitary, and other measures approved by the competent agencies, including the agencies responsible for state sanitary surveillance (second paragraph of Sec. 37).

The discharge of industrial, domestic, and other forms of waste and residues into bodies of water is prohibited, the discharge of wastewaters being permitted only in compliance with the requirements stipulated in Sec. 31 (first paragraph of Sec. 38). Proprietors of means of transport, pipelines, floating or other installations on bodies of water, timber-rafting organizations, and other undertakings, organizations, and institutions are required to prevent pollution and obstruction of waters caused by oil leakage, loss of timber, or escape of chemical and petroleum products, etc. (second paragraph of Sec. 38). Undertakings, organizations, and institutions are required to prevent pollution and obstruction of the surface of reservoirs, the ice cover of watercourses, or the surface of glaciers, by industrial, domestic, or other wastes, residues and refuse, or petroleum and chemical products, which when released cause deterioration in the quality of surface waters and groundwater (third paragraph of Sec. 38).

The administrative authorities responsible for state water resource systems, collective farms, state farms, and other undertakings, organizations, and institutions must prevent water pollution by fertilizers and poisonous chemicals (fourth paragraph of Sec. 38).

Health protection zones and areas are to be created, in accordance with the legislation of the U.S.S.R. and the union republics, with a view to protecting water utilized for drinking or domestic water supply purposes or for the therapeutic, balneological, and health-promoting needs of the population (fifth paragraph of Sec. 38).[1]

On December 10, 1970, the Supreme Soviet of the U.S.S.R. approved the basic water law, the Fundamental Principles of the Water Legislation of the U.S.S.R. and the Union Republics, which became effective on September 1, 1971. This law involves the most general and basic regulations governing the utilization and protection of rivers.

The decree of the Supreme Soviet of the U.S.S.R. entitled *Measures for Further Improving Nature Protection and Rational Utilization of Natural Resources,* issued on

September 20, 1972, declares:

> An indefatigable concern for protection of nature and the better utilization of natural resources, the strict observance of the legislation of the conservation of land and its riches, forests and waters, plant and animal life, and atmospheric air shall be considered as one of the most vital tasks of the state, bearing in mind that scientific and technical progress must be combined with a regard for nature and its resources and must contribute to the creation of the most favorable conditions for the life, health, work, and leisure of the working people.[2]

The Soviet Water Law also involves and suggests a number of principles, including exclusive state ownership of water resources in the U.S.S.R., rational (multipurpose) utilization of water resources, priority of water supply for drinking and domestic purposes, prohibition of operating new or reconstructed economic projects if they are not provided with adequate facilities to prevent water pollution, improvement of all types of technology for the purpose of water conservation, constant monitoring of water resources, registration and control of all forms of water use, a comprehensive basic approach to water utilization and conservation, active participation of the population in all projects designed to ensure rational utilization and conservation of water resources, and other measures.

The August 12, 1971, *Decree to Implement the Fundamental Principles of the Water Legislation of the U.S.S.R. and the Union Republics*,[3] issued by the Supreme Soviet of the U.S.S.R., establishes procedures for implementing the new water law of December 10, 1970, Fundamental Principles of the Water Legislation of the U.S.S.R. and the Union Republics. It became effective on September 1, 1971. Water laws of the various republics which do not contradict the provisions of the December 10, 1970, Fundamental Principles remain in effect until they are revised in accordance with the All-Union law. Agreements on water use and disposal of sewage which were entered into before the effective date of the new law must be revised in accordance with the December 10, 1970, Fundamental Principles if they violate its provisions.

The July 9, 1976, *Decree on the Development and Ratification of Complex Water Protection and Utilization Scheme*,[4] issued by the R.S.F.S.R. Council of Ministers in accordance with the U.S.S.R. Council of Ministers decree of June 2, 1976, makes the R.S.F.S.R. Ministry of Land Reclamation and Water Resources directly responsible for developing a complex scheme for the protection and utilization of R.S.F.S.R. water resources. Together with the R.S.F.S.R. State Planning Committee (Gosplan), this ministry is also to present proposals to the U.S.S.R. Ministry of Land Reclamation and Water Resources on yearly and long-term development plans. This decree also stipulates that the R.S.F.S.R. Gosplan and the Construction Affairs Committee (Gostroy) work with the ministry in developing this complex scheme.

The December 10, 1976, *Edict on Temporary Procedures to Preserve Living Resources to Regulate Fishing in the U.S.S.R. Coastal Zones*,[5] issued by the U.S.S.R. Supreme Soviet Presidium, establishes temporary measures to preserve living resources and regulate fishing in the U.S.S.R. coastal zones. It also provides procedures to be followed if these temporary measures are not adhered to. This edict is to be in effect until new legislation being formulated by the ongoing International Convention of the Third UN Conference on the Law of the Sea is accepted by all nations concerned.

The Soviet Water Law is based on an established classification of water resources according to physical, economic, and political features. This historical classification, which has assumed a legal meaning, stipulates three main categories of waters: surface waters, underground waters, and glaciers. In turn, these categories have their own subdivisions.

Surface-water resources involve seas, lakes, rivers, reservoirs, other water bodies, and water sources. Inland water bodies are of two general types: those in the exclusive use of definite enterprises and organizations and those in joint (collective) use of an infinite number of enterprises, institutions, organizations, or citizens. Underground-water resources are subdivided into fresh and saline, cold and hot (thermal). All underground waters are considered to be inland.

The given categories of water resources and their subdivisions have different legal implications, which are taken into account when regulating the utilization and conservation of water resources.

ENFORCEMENT AND PENALTIES FOR LAW VIOLATIONS

Enforcement is effected through various criminal, administrative, disciplinary, civil, and special water legislation. For acts that cause pollution, criminal legislation specifies penalties ranging from imprisonment up to 5 years to corrective labor up to 1 year or a fine not exceeding 300 rubles (about $450).

Actions associated with the pollution of the inland seas and territorial waters of the U.S.S.R. as a result of unlawful discharge from vessels of substances harmful to people and marine life or of neglecting to take the necessary measures to prevent such discharge shall be punishable by imprisonment up to 2 years, or by corrective labor up to 1 year, or by a fine of up to 10,000 rubles (about $15,000). The same actions that may cause considerable damage to people or marine living beings shall be punished by imprisonment for a period up to 5 years or by a fine of up to 20,000 rubles.[6]

Anyone causing pollution and littering of surface and underground waters as a result of neglect of rules of conservation and utilization of water resources shall be fined up to 10 rubles (citizens) or up to 50 rubles (government and administrative officials) by administrative order. In addition, the right to use the water may be revoked except in the case of drinking-water uses.

Courts study criminal cases, settle conflicts associated with compensation for damages (if one of the conflicting parties is a citizen or a collective farm), and study complaints of officials and citizens about resolutions of administrative agencies imposing a fine. The arbitration agencies settle property conflicts of state enterprises, institutions, and organizations.

The U.S.S.R. has concluded agreements with all neighboring countries. With different degrees of comprehensiveness, these agreements determine the terms and requirements of joint utilization and conservation of boundary waters. The most comprehensive agreements are between the U.S.S.R. and the Finnish Republic on boundary water systems (April 24, 1964) and between the U.S.S.R. and the Polish People's Republic on water management in boundary waters (July 24, 1964). Some agreements stipulate the procedure of joint utilization and conservation of such rivers as the Tisza, Prut, Araks, and Amu Darya.

The Soviet Water Law is still being studied and improved. In the near future, the Soviets hope to improve the procedures of granting permission for water use, to improve methods of managing multipurpose reservoirs, and to achieve better coordination in solving the problems of conservation of water resources and other elements of the natural environment. They are also working on practical measures to provide better incentives for complying with existing laws.

At present in the U.S.S.R. over 400 harmful substances are defined for drinking water and over 60 harmful substances for fish waters. The maximum allowable concentrations of these substances are based on the most restrictive of three general criteria:

1. Organoleptic (involving senses: taste, smell, color, etc.)
2. Damage to the self-purification capacity of the water body
3. Toxicity to animals or plants

The most restrictive of these criteria is called the "limiting value of harmfulness." Such limits are established by experts after rather exhaustive tests. For potentially toxic substances, three types of animals are used in experiments lasting at least half a year. The animals are given biochemical, clinical, and neurological tests and are examined for changes in internal organs. The choice of animals depends on the type of substance. The tests determine the influence of various substances on the self-purification processes of water bodies including nitrification, ammonificiation, deoxygenation, and reaeration and the influence of pH on the saprophyte population and activity.

Fish-water standards are more severe than drinking-water standards. Two reasons are given: drinking water receives additional treatment prior to consumption, and "fish live in this aquatic environment all of the time, whereas humans drink water only occasionally."

Stream standards specifying a maximum allowable mass of a pollutant (rather than a concentration) are being studied but have not yet been implemented. Nor have biological factors or indicators been used in the Soviet Union. At present, VODGEO is trying to define new and better methods for defining safe water quality by using biological indicators and economic considerations (see Fig. 4-1).

INSTITUTIONS AND ADMINISTRATION

The agency primarily responsible for water quality protection in the Soviet Union is the All-Union Scientific Research Institute of Water Protection (VODGEO or VNIIVO). Its responsibilities include water protection planning, the design of water protection structures, recommending water quality standards, collecting scientific and technical information, and assisting local water protection agencies (inspectorates) in each of the republics.

VNIIVO reports to the Ministry of Land Reclamation and Water Resources and advises and assists republic inspection and protection agencies. Some republics have environmental protection committees which oversee water quality problems. For those that do not, the Ministry of Land Reclamation and Water Resources of the republic or of the U.S.S.R. is responsible.

There are in the U.S.S.R. over 100 river basin organizations, called inspectorates, which are generally based on watersheds. These inspectorates are responsible for monitoring and enforcing the laws concerning the quality and usage of water resources. They also oversee the construction and operation of treatment plants within their jurisdictions. They carry on daily observations of the state of water bodies, water distribution, wastewater discharge, etc. For example, all industrial and other enterprises consuming over 100 m³ of water per day and discharging wastes are under the control of these agencies. They participate in the acceptance tests of newly constructed or reconstructed economic facilities required prior to operation. Each inspectorate has at least one hydrotechnical laboratory; some of these may be mobile.

The final coordination of water quality and quantity management is made by the Ministry of Land Reclamation and Water Resources. VNIIVO's counterpart for advising the ministry on water quantity problems and plans is the Scientific Research Institute for Complex Usage of Water Resources, located in Minsk. Under the All-Union Council of Ministers, Gosplan's Department of Environment, Division of Water Protection, reviews

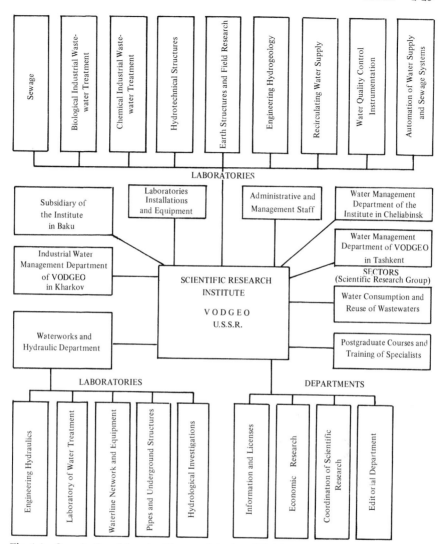

Fig. 4-1   Organization of the scientific research institute VODGEO.

all water protection plans. Gosplan coordinates water protection planning with all other plans and recommends the appropriate level of funding. When water protection plans are approved by the Council of Ministers, they become law. Priorities among sectors of the economy are determined by Gosplan. Gostroy reviews plans for the best available technology.

Other ministries also have responsibilities regarding water management. Hydromet is responsible for monitoring surface-water quantities, the Ministry of Geology is responsible for monitoring groundwater quantities, the Ministry of Health is responsible for monitoring the sanitary condition of drinking water and for specifying appropriate quality standards, the Ministry of Fisheries has similar responsibilities for fish waters, the

Ministry of Agriculture is responsible for soil erosion and agricultural pollution control, and the Ministry of Power monitors radioactive pollutants.

SUMMARY

There is not a specific Soviet water law relating to spills of oil and hazardous materials. Violations of environmental standards seem to be established under a rather broad provision of the fundamental water legislation, which states in part:

> Enterprises, organizations, and establishments are obligated not to allow the pollution and obstruction of surface water bodies, ice covers of water bodies, and the surface of glaciers by industrial, domestic, or other discharges, effluents, or refuse, as well as by oil or chemical products, which will lead to degradation of surface or ground waters.[7]

As far as one can tell, the clause "which will lead to degradation of surface or ground waters" has not been defined as "harmful quantities" under Sec. 311 of the U.S. Federal Water Pollution Control Act. It would probably be defined on a case-by-case basis, severity being dependent on the particular use or status of each water body. If the water body were a specially protected water body such as the Caspian Sea, the Volga River, or Lake Baikal, degradation would be more strictly determined than if the water body were little used. The Soviet equivalent of water quality standards (much more comprehensive than those of the United States) may well be used as the appropriate standard, although that is very difficult to tell.[8]

In addition, there is special legislation dealing with pollution of the sea, aimed specifically at "oil products and other materials hazardous to the health of man or to the living resources of the sea." The law seems to be aimed at foreign vessels in particular, since much higher penalties are to be levied than for inland waters. It requires the listing of:

> materials the discharge of which is forbidden, and also water quality standards for these materials to be established by the Ministry of Reclamation together with the Ministry of Health and the Ministry of Fish Resources of the U.S.S.R.[9]

The Sharikov compilation lays out in a systematic fashion the most important regulatory documents and materials about conservation of nature and water resources and protection of air from industrial emissions. It also presents information about methods of purifying dirty air and water, as well as the organization of control of the quality of the environment.

REFERENCES

1. *Vedomosty,* Supreme Soviet of the U.S.S.R., no. 50, serial no. 566, Dec. 16, 1970, pp. 706–723.
2. *Vedomosty,* Supreme Soviet of the U.S.S.R., no. 39, 1972, p. 346.
3. *Sovetskaya Yustitsiya,* no. 20, 1971.
4. *Sobranie Postanovleniy Pravitel'stva R.S.F.S.R.,* no. 13.
5. *Vedomosty,* Supreme Soviet of the U.S.S.R., no. 50, 1976.
6. *Vedomosty,* Supreme Soviet of the U.S.S.R., no. 10, 1974, p. 161.
7. L. P. Sharikov (comp.), *Environmental Protection: An Information Directory,* Leningrad, 1978, p. 44.
8. Ibid., pp. 203–222.
9. Ibid., pp. 166–167.

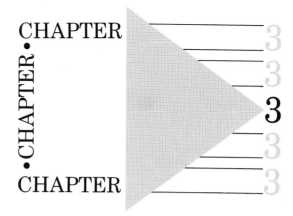

CHAPTER

•CHAPTER•

CHAPTER

3

Information Systems and Reporting

PART 1
U.S. Coast Guard Systems / 3-2

PART 2
Other United States Systems / 3-9

PART 3
European Systems / 3-25

PART 4
Canadian Systems / 3-45

PART 1
U.S. Coast Guard Systems

Mark E. Ives
Richard V. Harding
Michael C. Parnarouskis
Robert J. Embrie
Richard G. Potts
Kirk R. Karwan
Marine Environmental Protection Division
U.S. Coast Guard

Chemical Hazards Response Information System (CHRIS) / 3-2
Spill Cleanup Inventory System (SKIM) / 3-5
Pollution Incident Reporting System (PIRS) / 3-7

The National Oil and Hazardous Substances Pollution Contingency Plan, prepared in accordance with the Federal Water Pollution Control Act, delineates the authorities and responsibilities of various federal agencies in pollution response. The Chemical Hazards Response Information System (CHRIS), a system of quantitative and qualitative chemical-specific and chemical-related response information, has been developed by the U.S. Coast Guard to assist these agencies when they respond to emergencies involving the water transport of hazardous chemicals. Developed primarily for the use of field personnel, the CHRIS system is built around six components: four manuals (COMDTINST M16465.11 through COMDTINST M16465.14), data bases for regional contingency planning, and the Hazard Assessment Computer System (HACS).

CHEMICAL HAZARDS RESPONSE INFORMATION SYSTEM (CHRIS)

A Condensed Guide to Chemical Hazards (CONDTINST M16465.11)

A compact and convenient source of chemical-related information, particularly on bulk-shipped hazardous material, the guide[1] is intended for use by the personnel first reporting to the site of an incident. It assists in determining the proper, responsible actions to be taken to safeguard the immediate population and to mitigate damage to the environment. The guide contains precautionary advice on the chemical in question and any characteristic physical and biological hazards it may present, basic information required in determining subsequent response actions. Since the guide is the only component of CHRIS initially available at the scene, it includes a list of on-scene information needs that the

Hazard Assessment Handbook (COMDTINST M16465.13) and HACS require as inputs.

Hazardous Chemical Data Manual (COMDTINST M16465.12)

This manual,[2] often called the cornerstone of CHRIS, contains detailed, largely quantitative chemical, physical, and biological data necessary for the formulation, evaluation, and implementation of response plans. It is intended for use by the officer of the on-scene coordinator (OSC) and the National Response Center (NRC) and regional response centers. The manual contains the hazard assessment code, essential in selecting the appropriate calculation procedures for the hazard assessment, and lists the physical and chemical property data required to perform the calculations in COMDTINST M16465.13 and HACS. The manual currently contains data for 1017 chemical compounds; about 30 to 35 chemicals are added each year.

Hazard Assessment Handbook (COMDTINST M16465.13)

This handbook[3] contains methods of estimating the rate and quantity of release of a chemical in specific situations. Methods for predicting the potential threat are incorporated in the manual, along with procedures for estimating the concentration of chemicals (in both water and air) for determining potential toxic, fire, and explosive effects after chemical spills.

Response Methods Handbook (CONDTINST M16465.14)

The *Response Methods Handbook*[4] is a compendium of descriptive information and technical data pertaining to response methods for hazardous chemical spills. Written for personnel with prior training or experience in pollution and hazard response, it emphasizes existing or prospectively available methodology with updates being provided as needed.

Data Base for Regional Contingency Plan

Predominantly for use by OSC personnel, the data base contains data pertinent to a region, subregion, or locale. It provides detailed data on resources that might be threatened and the availability of response equipment. Examples of such data are an inventory of physical resources and response personnel, vulnerable or exposed resources (critical water use areas), potential pollution sources, geographical and environmental features, cooperating agencies and organizations, and experts available if required. Most of these data are in the form of regional contingency plans.

Hazard Assessment Computer System (HACS)

HACS[5] is the computerized counterpart of the CHRIS *Hazardous Chemical Manual* and the *Hazard Assessment Handbook*. The objective of the system is to provide rapid and accurate quantitative estimates on the type and extent of hazards associated with large chemical incidents, displaying these estimates in a form useful to response personnel.

A chemical discharged into the environment can create a hazard to life and property because of flammability, toxicity, or both. As dispersion and/or dilution of the spilled chemical occurs, the associated hazards usually decrease and eventually disappear. Knowledge of the physical properties of the chemical and of its characteristics under different environmental conditions provides the basis for estimation of the type and extent of hazards presented by the chemical.

The processes of dispersion, evaporation, combustion, and the like are quite complex and depend upon many variables. HACS offers a systematic and convenient approach to developing a hazard estimate for a particular incident utilizing these variables. The estimate is a function of distances and times over which toxic and/or flammable concentrations may exist in air and water and includes the minimum safe distance from the spill site if the chemical should ignite. HACS currently contains the physical and chemical property data necessary to perform a hazard assessment on 900 commonly shipped chemicals.

To obtain a hazard assessment calculation, HACS users must enter input describing the spill scenario, the type of assessment to be performed, and options to be selected during system operation. The minimum input requirements include a run title, used simply to provide identification for the computer output report, the path codes specifying the assessment models to be executed, and the recognition code identifying the spilled chemical. Execution control, selected by the user, specifies the type of run; for example, it specifies whether a basic assessment run is to be made, or a series of runs is required to estimate the sensitivity of potential hazards to variations in spill conditions.

The recently developed User Interface allows for fully interactive terminal-mode operation, which simplifies the data input process. Typically, user input data include discharge, atmospheric, and marine conditions in the environment of the spill and parameters for HACS output display requests. Additional user input might include observed hazard conditions and chemical property data to be used in place of data from the physical property file.

All input describing the spill conditions is optional in the sense that HACS will automatically select default values for required items not entered by the user. Typically, users will enter data which are immediately available, obtain a hazard assessment, and then review the output report to determine the extent and significance of missing data.

Output is graphic, displaying the relationships among spill concentration, thermal radiation, location, and time. This information may then be transmitted to the user by facsimile. To achieve these capabilities, the following system features are incorporated in HACS:

- The system operates in over-the-counter batch, remote batch, and interactive terminal modes and provides cathode-ray tube (CRT) or line-printer text output and electromechanical or line-printer plot output.

- HACS interfaces with an external data base to obtain automatically chemical-specific data for computational use. The data base currently contains chemical and physical properties of 900 hazardous chemical substances.

- The system incorporates the computerized counterparts of the analytical hazard assessment models developed for CHRIS and extensions of these models.

- The system can plot thermal radiation, concentration, or other variables (e.g., pool radius, temperature of liquid remaining, volume of liquid remaining, etc.) as a function of location and/or time, as appropriate.

- HACS incorporates bounds or limit values which are used to check the reasonableness of input data and estimated hazard levels.

- The system output identifies the spilled chemical, restates the discharge, environmental, and marine conditions entered by the user, and reports the hazard assessment. The output information enables the user to validate his or her input data quickly and accurately.

- These features are provided under the control of an executive program which sequences the input, computation, and output reporting portions of the system.

The installation of a CRT terminal and remote printer within the National Response Center at Coast Guard headquarters in Washington has provided Coast Guard personnel with the capability of conducting the in-depth hazard calculations possible from HACS in both emergency and nonemergency situations. With the addition of the User Interface Module (UIM), turnaround time from receipt of a call for assistance until the information is transmitted to the requesting office will be on the order of 30 to 45 min. For incidents involving simple data input and use of basic program models, the turnaround time will be less; for complex situations, the turnaround time may be greater.

HACS is intended for use by hazard assessment specialists at Coast Guard headquarters who possess a fundamental appreciation for the nature of the physical and chemical processes being modeled. Although NRC personnel are proficient in the operation of HACS, much of the output requires a trained chemist to interpret it.

It is anticipated that HACS will be a totally dynamic system. Presently, plans call for the review, revision, and updating of all the analytical models contained in HACS. As new chemical water reaction models are developed, they will be added to HACS. Introduction of new chemicals to the CHRIS system will require updating the physical property data base for HACS.

SPILL CLEANUP INVENTORY SYSTEM (SKIM)

The National Oil and Hazardous Substances Pollution Contingency Plan (40 CFR 1510) provides for a coordinated and integrated federal response[6] in protecting the environment from the damaging effects of pollution discharges. In achieving this goal, the plan established as one objective the identification, procurement, maintenance, and storage of the response equipment and supplies necessary in mitigating the environmental effects of pollution.

Emphasis was added to the objective by the March 17, 1977, presidential message to Congress, a message that presented in detail an interrelated group of measures designed to reduce the risk associated with the marine transportation of oil. The measures, in part, directed the U.S. Coast Guard, the U.S. Environmental Protection Agency (U.S. EPA), and other responsible federal agencies to begin plans to upgrade their capability to respond to oil spills. The identification and cataloging of existing equipment was an initial step in developing these plans.

Information on spill response equipment is addressed in varying formats and degrees of detail in regional and local contingency plans. Review of these plans indicated several common problem areas: no central repository for equipment data, difficulty in retrieval of specific equipment information, and need for accurate, timely updating of the information.

The Coast Guard's Marine Environmental Protection Division, Pollution Response Branch, undertook the task of developing the Spill Cleanup Inventory System (SKIM), a computer-based equipment inventory available for pollution response in the United States (including Puerto Rico and Guam).[7] This national equipment inventory contains data on public equipment (owned by federal agencies and state and local governments) and cleanup-containment equipment maintained by contractors, cooperatives, and private companies.

Twenty-six individual types of equipment and supplies were identified in designing the structure of the data base for SKIM. The equipment and supply categories are:

Heavy-duty offshore boom	Tank ships
River-harbor boom	Rubber bladder
Skimmer	Boat

Sorbent	Aircraft
Dispersant	Communications equipment
Surface collecting agency	Safety-special equipment
Biological agent	Generators
Transfer-lightering system	Disposal facilities
Pumps	Chemical agents
Hose	Burning agents
Vacuum-pumper truck	Cold-weather equipment
Beach cleanup equipment	Scientific-analysis equipment
Barges	Other equipment

Data Base

The data-gathering effort commenced in January 1978. Coast Guard predesignated OSCs collected data on equipment from the private sector and on Coast Guard–owned equipment for their areas of responsibility, primarily through telephone contacts and facility visits after utilizing information in existing regional and local contingency plans. In a like manner, EPA predesignated OSCs collected information within their jurisdiction for inclusion in SKIM. Other members of the national response team advised their representatives on the regional response teams to ensure that data on equipment maintained by each federal agency was entered in the inventory system. To assist with the data collection from federal resources, Coast Guard district marine environmental protection branches made an initial distribution of inventory forms to the regional response teams and coordinated collection at the regional level.

Participation in SKIM by contractors, private companies, and cooperatives is voluntary. However, strong support for the inventory system has been received from the American Petroleum Institute and the Spill Control Association of America. Information received from members of these organizations and others has enabled the establishment of a data base with more than 10,000 records. The value of the system will continue to increase as new information is added to the data base by field units using dataphone terminals.

Computer Hardware and Software

The Spill Cleanup Inventory System has been implemented on the CYBERNET computer owned by Control Data Corporation and located in Rockville, Maryland. Coast Guard units can access the SKIM data base through portable terminals via telephone connection, enabling effective and timely updating and retrieval of information. Local telephone access numbers are available in a number of major and intermediate cities across the country. Other users can tie into the system on a long-distance basis through one of these numbers.

System 2000, developed by MRI, Incorporated, of Austin, Texas, was chosen as the data-base management system to be used with SKIM. This software package was designed to operate on a variety of hardware systems and had already been tested in conjunction with a number of other Coast Guard data bases. With System 2000, SKIM data are stored in a hierarchical fashion. As a result, very specific retrievals and changes in the data base can be made efficiently. Complex management reports can be generated through a special System 2000 module called Report Writer or through Fortran or Cobol programs and knowledge of the host-language interface.

Computer Searching

Using System 2000, it has been possible to write special-purpose computerized programs which can be accessed by users of SKIM. At this time, users of the system can obtain listings of equipment from the data base by type and area, e.g., listings of all booms located in Coast Guard District 1 or all skimmers in EPA Region IV. Information can also be obtained by distance from a specific location, the location inputted at the terminal by the user. This location, specified in terms of latitude and longitude, will generally represent a spill site or staging site for equipment to be used for a specific spill.

System Objectives

The system is designed to fulfill various functions, among them:

1. Providing current and accurate data to OSCs. When responding to an incident, the predesignated OSC in the area must be able to obtain an accurate picture of available resources so as to provide the best response.
2. Providing rapid and accurate listings of available equipment for local contingency plans.
3. Serving as an additional resource for use by Coast Guard marine safety offices, captains of the ports, and district headquarters managers in budgeting and planning for future requirements and contingencies.

Operating instructions and information about the SKIM system can be obtained from the Coast Guard Office of Marine Environment and Systems, Pollution Response Branch (G-WER-2), Washington, D.C. 20590.

POLLUTION INCIDENT REPORTING SYSTEM (PIRS)

The Federal Water Pollution Control Act (Sec. 311) requires that any discharge of an oil or a hazardous substance in a harmful quantity* be reported to the "appropriate agency of the United States Government." Executive Order 11735 of March 5, 1970, designated the Coast Guard as one of the agencies.

PIRS is a computerized data base developed in December 1971 as an adjunct to the Coast Guard's Marine Environmental Protection Program.[8] Initially designed for the collection and maintenance of discharge data, it was modified in 1972 to permit the inclusion of additional data on cleanup (response) activities and penalty actions. In the accompanying table are the data that PIRS collects.

Discharge	Response	Penalty
Time of occurrence	Cleanup party	Initiating agency
Location	Equipment used	Authority
State	Personnel	Action against
Water body	Cleanup duration	Action date
Source	Amount recovered	Penalty proposed
Cause	Cleanup cost	Penalty collected
Operation		Hearing results
Material		Appeal results
Quantity		Case status
Notifier		

*A harmful quantity is now defined as a reportable quantity in the latest U.S. EPA regulations.

Concurrent with the 1972 modifications, a sophisticated report-generating language (SPECTRAN) was incorporated into PIRS, providing greater flexibility in structuring both report content and format.

The system can be programmed to produce repetitive listings of spills meeting predetermined criteria. Any or all data fields can be retrieved for each spill listed, depending upon the requirements of the eventual user. For applications not requiring such detail, as when statistical analysis or summations involving specific data items are required, SPECTRAN provides the necessary flexibility to generate reports in this type of format.

Requests for data from PIRS are processed at Coast Guard headquarters only, and because the types of data involved are not critical in situations requiring rapid response, the requests are processed in a batch-mode environment. The average length of time required for processing, from receipt of a request until final product, is 3 days.

PIRS functions in a dual capacity: (1) as a resource utilized by the Coast Guard in the management of its Marine Environmental Protection Program and (2) as a resource available to Congress, academia, private firms, and individuals. PIRS availability to academic and private individuals or institutions is on a first-come–first-served basis. Inquiries may be sent to:

Commandant (G-WER-2)
2100 Second Street, S.W.
Washington, D.C. 20593
Attention: PIRS Request

REFERENCES

1. *CHRIS: A Condensed Guide to Chemical Hazards,* U.S. Coast Guard Report COMDTINST M16465.11, Washington, 1978.

2. *CHRIS: Hazardous Chemical Data Manual,* U.S. Coast Guard Report COMDTINST M16465.12, Washington, 1978.

3. *CHRIS: Hazard Assessment Handbook,* U.S. Coast Guard Report COMDTINST M16465.13, Washington, 1978.

4. *CHRIS: Response Methods Handbook,* U.S. Coast Guard Report COMDTINST M16465.14, Washington, 1978.

5. R. V. Harding, M. C. Parnarouskis, and R. G. Potts, "The Development and Implementation of the Hazard Assessment Computer System (HACS)," in G. F. Bennett (ed.), *Proceedings of the 1978 National Conference on Control of Hazardous Material Spills,* Miami Beach, Fla., April 1978, pp. 51–59.

6. C. R. Corbett, "A Dynamic Regional Response Team," in G. F. Bennett (ed.), *Proceedings of the 1978 National Conference on Control of Hazardous Material Spills,* Miami Beach, Fla., April 1978, pp. 4–8.

7. R. J. Imbrie, K. R. Karwan, and C. M. Stone, "The Spill Cleanup Equipment Inventory System (SKIM)," in G. F. Bennett (ed.), *Proceedings of the 1978 National Conference on Control of Hazardous Material Spills,* Miami Beach, Fla., April 1978, pp. 65–68.

8. B. D. Boyd, "The Augmentation of the Pollution Incident Reporting System Designed for Spillage Prevention," in G. F. Bennett (ed.), *Proceedings of the 1978 National Conference on Control of Hazardous Material Spills,* Miami Beach, Fla., April 1978, pp. 46–50.

PART 2
Other United States Systems

George J. Moein
Chief, Emergency Response and Control Section
Region IV, U.S. Environmental Protection Agency

Oil and Hazardous Materials Technical Assistance Data System (OHM-
    TADS) / 3-9
Other Information Systems / 3-11

The federal instrument for response to spills of oil or hazardous materials is the National Oil and Hazardous Substances Pollution Contingency Plan, which fulfills its mission by coordinating the response to these environmental crises. The plan requires that accurate assessments be made of the potential or actual danger that a discharge of oil or hazardous substances may present. It calls for the appointment of predesignated federal on-scene coordinators (OSCs) by the U.S. Environmental Protection Agency (U.S. EPA) and the U.S. Coast Guard to be available for spill response activities. To aid the OSCs and other water quality managers in time of emergency, data banks are valuable tools.

OIL AND HAZARDOUS MATERIALS TECHNICAL ASSISTANCE DATA SYSTEM (OHM-TADS)

As one means of developing an active response posture, the U.S. EPA launched a program to provide rapid, up-to-date technical information for field personnel assigned to counteract spills. In June 1971, it provided the basic structure of information required and contracted with Battelle Memorial Institute's Pacific Northwest Laboratories to develop a pilot data file for use in an automated retrieval system. This system, the Oil and Hazardous Materials Technical Assistance Data System (OHM-TADS), was first used in January 1972; it has since been utilized by EPA response personnel and other interested groups.

OHM-TADS is an automated information retrieval file designated to facilitate the rapid retrieval of information on approximately 1000 chemicals. Data files are constructed so that a systematic query program can prove of great value both for on-line response to spill incidents and for summary evaluations relating to direct enforcement and research activities.

The primary function of the files is to provide immediate information feedback on hazardous materials to spill response personnel. Both numerical data and interpretive comments are contained in the data files. These can serve as background information for decision-making processes and guidelines for the initiation of corrective action.

OHM-TADS is designed to include all information pertinent to spill response efforts related to any material designated as an oil or a hazardous substance. As such, it includes a wide variety of physical, chemical, biological, toxicological, air, land, and water effects and commercial data. However, the greatest emphasis is placed on the deleterious effects that these materials may have on water quality.

Sources of information include articles in journals, books, papers presented at various symposia, compendiums, governmental reports, and basic reference texts. A complete list of more than 1000 references is attached to the file and can be retrieved through the automated system or be referred to manually.

Data are entered into the OHM-TADS files in a form that requires some technical background for maximum benefit. It is assumed that the user is familiar with chemical symbols and common chemical biological terms.

OHM-TADS is an on-line interactive information retrieval system. It is capable of processing structural and unstructural data in an on-line conversational mode, whereby the user can interact with the system in natural language or in abbreviated expressions. Data in the system are condensed to alleviate the need for extensive study by the user. The random-access provision permits the user to solve problems involving unidentified pollutants by searching for color, odor, or other physical or chemical characteristics as observed on scene.

The main characteristic of this system is that it automatically takes each word and processes it into an inverted index file, making each word a search component of the data base. The data themselves are in two files: (1) a serial file consisting of variable block-length character strings plus additional information and (2) an inverted file consisting of the index expression followed by the associated information strings.

Searches are formulated in an English-like language using boolean logic. The system responds with the number of documents meeting the request, and the researcher is then able to refine or restructure the query if necessary. The resulting pertinent information can then be displayed at the user's terminal, listed at a remote medium-speed terminal, or at the central site.

Another way of describing the search technique is to say the data base can be searched in a "free text" mode. That is, the computer will take a word submitted to it and then go through all the material in its data bank to match the given term to relevant stored articles. Users, therefore, do not have to view the data base's structured thesaurus to get material out of the system.

The OHM-TADS system is oriented toward an informational retrieval problem that is characterized by difficult and vague subject definition, extensive variance in term selection, changing scientific and technical terminology, and imprecise search definitions. The system greatly facilitates file browsing.

Search Strategy

OHM-TADS searches are based on the 126 subjects into which all information in the data base has been divided. These subjects are referred to as fields and are defined in Table 2-1. For example, the MAT field contains the material name, and the DRK field contains the recommended drinking-water limits. Every substance in the data base has information in the MAT and ACC (OHM-TADS accession number) fields. However, because of the unavailability of published data, information may not be available on every subject for the material of interest. All the 126 fields for each chemical can be listed on the printout through a command mode, as shown in Fig. 2-1 for ammonium hydroxide, or only a few subject fields can be requested for specific needs. The process also allows the searcher to manipulate the data and to browse in the file without significant cost or unnecessary printing.

OTHER INFORMATION SOURCES

Apart from the EPA's OHM-TADS and the Coast Guard's CHRIS (see Chap. 3, Part 1), there are a number of other information systems whose main function is to provide assistance during hazardous substance spills. There are also information retrieval sources, both computerized and manual, which provide information or a list of titles or abstracts of articles dealing with a specific subject. The availability of an on-line computer usually indicates a short turnaround time for responses. This is often crucial in an emergency situation. There are also available numerous reference texts and handbooks which contain information on the properties of hazardous chemicals. Those likely to be faced with a hazardous spill may find it helpful to obtain one or more of these books for future reference. An attempt is made in this section to present an inventory of these sources and to give a brief overview of each. Any discussion related to evaluation of these systems is the personal opinion of the author and does not reflect his affiliation with the EPA.

Chemical Transportation Emergency Center (CHEMTREC)

CHEMTREC serves as a clearinghouse by providing a single emergency 24-h telephone number for chemical emergencies. Upon receiving notification of a spill, CHEMTREC immediately gets in touch with the shipper of the chemicals involved for assistance and follow-up. CHEMTREC also provides warning and limited guidance to those at the scene of the emergency if the product can be identified by either the chemical or the trade name. There are more than 3600 listings of chemicals available for CHEMTREC users.

TABLE 2-1   Rationale upon Which OHM-TADS Is Based

Segment no.	Mnemonic	Segment title
1	ACC	OHM-TADS accession number: a unique computer-assigned identifier for the data file.
2	CAS	Chemical Abstracts Service registry number: a unique international identifier for material of interest.
3	SIC	Standard industrial code: industry-employed code which can be used to identify manufacturers of material.
4	MAT	Material name: generally, the common name for the material.
5	SYN	Synonyms: alternative identifiers of similar isomers for which the data are valid.
6	TRN	Company trade names: listing of commercial trade names and the associated manufacturer whenever possible.
7	FML	Chemical formula: most common formula or description of nature of materials included in the general heading, such as components of an industrial blend or mixture.
8	SPC	Species in mixture: identification of typical product purity in cases of single-constituent materials or of specific major components of heterogeneous mixtures.
9	USS	Common uses: common uses of materials enumerated.
10	RAL	Rail (percent): percentage shipped by rail (estimate).
11	BRG	Barge (percent): percentage shipped by barge (estimate).
12	TRK	Truck (percent): percentage shipped by truck (estimate).
13	PIP	Pipeline (percent): percentage shipped by pipeline (estimate).
14	CON	Containers: list of types of shipping containers normally used or required by law. Typical shipment size is given when available.
15	STO	General storage procedures: precautions to be taken when storing the material. The rationale for these measures varies from

TABLE 2-1   Rationale upon Which OHM-TADS Is Based   *(Continued)*

Segment no.	Mnemonic	Segment title
		safety considerations to precautions designed to prevent degradation of the material.
16	HND	General handling procedures: precautions to be taken when handling the material. Information relates to both safety considerations and practices designed to prevent degradation of the material.
17	PRD	Production sites: list of major producers and their plant locations.
18	HYD	Hydrolysis, product of: list of hazardous materials which decompose to the material of reference when in contact with water.
19	ADD	Additive (percent): list of typical stabilizers and inhibitors added to the base material.
20	BIN	Binary reactants: list of materials known to react when put in contact with the material of reference.
21	COR	Corrosiveness: general observations on corrosive action to materials commonly used for packaging or equipment that might be required at a spill site.
22	SGM	Synergistic materials: list of other materials and water quality parameters whose presence can increase the toxicity of the material of interest.
23	ANT	Antagonistic materials: list of other materials and water quality parameters whose presence can reduce the toxicity of the material of interest.
24	FDL	Field detection techniques, limit (ppm), reference: three-part segment listing potential field detection techniques, the lower sensitivity limit, and the literature reference from which more data can be obtained. "Field test" generally refers to any gross identification method that can be used at the spill site without elaborate or nonportable equipment. It normally assumes that the material or the chemical class has been identified so that general tests for aldehydes or phenols, etc., are applicable. The two major types of tests listed are inorganic colorimetric reactions and organic spot tests.
25	LDL	Laboratory detection techniques, limit (ppm), reference: format of preceding segment followed for specific tests that can be used for positive identification of material. These tests generally rely on sophisticated laboratory analysis equipment, such as atomic absorption units and gas chromatographs.
26	STD	Standard codes: National Fire Protection Association (NFPA) codes for materials as well as pertinent transportation codes.
27	FLM	Flammability: potential for fire at a spill site summarized. Segment uses the NFPA ranking system described by one of the following modifiers: very, quite, moderate, slight, nonflammable.
28	LFL	Lower flammability limit (percent): listed value, percentage of material in air which is the lower limit of flammability.
29	UFL	Upper flammability limit (percent): listed value, percentage of material in air which is the upper limit of flammability.
30	TCP	Toxic combustion products: occasional listing of specific materials or classes of materials released when the compound of concern is burned or heated to decomposition.
31	EXT	Extinguishing methods: Fire-fighting techniques and unique precautions to be taken, if any.

Segment no.	Mnemonic	Segment title
32	FLP	Flash point (°C): open-cup value when available; otherwise, closed-cup value.
33	AIP	Autoignition point (°C): listed value at which autoignition occurs in the presence of adequate air.
34	EXP	Explosiveness: potential for violent rupture or vigorous reaction at a spill site summarized.
35	LEL	Lower explosive limit (percent): listed value, percentage of material in air which is the lower explosive limit.
36	UEL	Upper explosive limit (percent): listed value, percentage of material in air which is the upper explosive limit.
37	MLT	Melting point (°C): accepted value under standard conditions unless otherwise noted in Segment 38.
38	MTC	Melting characteristics: decomposition, ignition, etc.
39	BLP	Boiling point (°C): accepted value under standard conditions unless otherwise noted in Segment 40.
40	BOC	Boiling characteristics: reduced pressure, etc.
41	SOL	Solubility (ppm, 25°C): typically, the listed value for standard reference conditions.
42	SLC	Solubility characteristics: "slightly" and "moderately" used when a specific value is not given.
43	SPG	Specific gravity: listed value for material in the state in which it is most often shipped. For materials whose boiling point is near ambient temperatures, the liquid state usually was referenced.
44	VPN	Vapor pressure (mm, mercury): pressure characteristic (at any given temperature) of a vapor in equilibrium with its liquid or solid form.
45	VPT	Vapor pressure text: conditions under which measurement is made.
46	VDN	Vapor density: value derived by dividing the mass of the vapor by its volume and measuring at a specific temperature. A value $<$ 1 indicates that the vapor is lighter than air; a value $>$ 1, that it is heavier than air and will give the appearance of a fog, hugging the ground.
47	VDT	Vapor density text: temperature and any other conditions under which measurement is made.
48	BOX	Biochemical oxygen demand (BOD, lb/lb): relative oxygen requirements of wastewaters, effluents, and polluted waters. Biochemical oxygen demand of a pure substance is listed on a lb/lb, or percentage of theoretical demand, basis.
49	BOD	Biochemical oxygen demand text: information listed in Segment 48 displayed. Segment includes duration of test and source of information.
50	PER	Persistence BOD and chemical data interpreted to estimate material life span in a free aquatic system. When possible, degradation products are specified.
51	PFA	Potential for accumulation: data on ability of various organisms to accumulate a material and the specific organs in which concentration is most pronounced.
52	FOO	Food-chain concentration potential: potential for material to be concentrated to toxic levels while it is passed up the food chain. When possible, data given on findings in predator species.
53	EDF	Etiological potential: diseases and ailments initiated or accelerated by exposure to the material of interest.

TABLE 2-1   Rationale upon Which OHM-TADS Is Based   (*Continued*)

Segment no.	Mnemonic	Segment title
54	CAG	Carcinogenicity: results of work directed to isolating carcinoma in test animals. Human data are used when available.
55	MUT	Mutagenicity: finding of tests for mutagenicity.
56	TER	Teratogenicity: finding of tests for teratogenicity.
57	FTX	Freshwater toxicity number (ppm): concentration in parts per million at which test results were reported.
58	FTB	Freshwater toxicity text: Column 1—concentration in parts per million at which test results were reported. Column 2—time of exposure expressed in hours. Column 3—species tested, usually a common name. Column 4—effect on organism tested, often given as TL_M or LD_{50}. Column 5—test environment, including data on water quality and other controlled conditions. Column 6—source of information.
59	CAT	Chronic aquatic toxicity limits (ppm): maximum level in parts per million found to be safe for extended exposure of fish to the material of interest.
60	CAR	Reference for chronic aquatic toxicity: source of information.
61	STX	Saltwater toxicity: toxicity to estuarine or marine animals in parts per million.
62	STB	Saltwater toxicity text: same general format as in Segment 58 followed.
63	ATX	Animal toxicity: doses reported in milligrams of material per milligram of body weight of test animal (unless otherwise noted).
64	ATB	Animal toxicity text: Column 1—doses in milligrams of material per milligram of body weight of test animal. Column 2—time of exposure. Column 3—species. Animals of reference (typically laboratory animals such as rats, guinea pigs, mice, pigs, dogs, and monkeys) are listed. Column 4—parameter, description of exposure. Terms indicate whether the dose caused death or other toxic effects and whether it was administered as a lethal concentration or as a toxic concentration in the inhaled air. Abbreviations are listed in an appendix. Column 5—route. Mode of application is listed. Abbreviations are listed in an appendix. Column 6—reference, source of data.
65	ATL	Chronic animal toxicity limits (ppm): maximum level reported in parts per million thought to be the threshold for extended use on livestock.
66	ATR	Reference for chronic animal toxicity limits: source of information.
67	LVN	Livestock toxicity (ppm): recommended or safe levels of concentration in parts per million for use on livestock.
68	LVR	Reference for livestock: source of information.
69	WAN	Acute waterfowl toxicity (ppm): concentration in parts per million considered to be hazardous to waterfowl upon acute exposure.
70	WAR	Reference for acute waterfowl toxicity: source of information.
71	CWF	Chronic waterfowl toxicity limits (ppm): concentration in parts

Segment no.	Mnemonic	Segment title
		per million considered to be the maximum permissible in water inhabited by waterfowl.
72	CWR	Reference for chronic waterfowl toxicity: source of information.
73	AQN	Aquatic plants (ppm): concentration in parts per million found to be injurious to aquatic flora listed.
74	AQR	Reference for aquatic plants: source of information.
75	IRN	Irrigable plants (ppm): concentration expressed in parts per million found to be injurious to the crop listed.
76	IRR	Reference for irrigable plants: source of information.
77	CPT	Chronic plant toxicity limits (ppm): threshold level expressed in parts per million for extended use as irrigation water.
78	CPN	Reference for chronic plant toxicity limits: source of information.
79	TRT	Major species threatened: segment originally designed to spotlight individual species especially susceptible to the material of interest. Such data are very rare. Consequently, the segment includes specific data on tests run with different species.
80	TIC	Taste-imparting characteristics (ppm): level in parts per million at which the material will impart a taste to the flesh of fish living in the affected waters.
81	TIR	Reference for taste-imparting characteristics: source of information.
82	INH	Inhalation limit (value): generally, the accepted threshold-limit value (TLV), which is that level acceptable for industrial exposure over an 8-h period. It may sometimes be the LC_{50} [lethal concentration which results in the death of 50 percent of the population in the time period specified] for inhalation.
83	INT	Inhalation limit (text): units and source of information for Segment 82.
84	IRL	Irritation levels (value): level at which irritation of the skin and mucous membrane occurs.
85	IRT	Irritation levels (text): reference and explanatory comments for Segment 85.
86	DRC	Direct contact: summary statement indicating corrosiveness or irritation value of material in direct contact with skin, mucous membranes, or eyes.
87	JNS	General sensation: segment designed to identify some of the reactions that people might have (symptoms and effect on the body) when exposed to the designated material, sensation upon breathing the vapors, vapor concentration levels at which noticeable reactions occur, warning properties, and miscellaneous toxicological observations.
88	LOT	Lower odor threshold (ppm): listed value in parts per million.
89	LOR	Lower odor threshold reference: source of infromation.
90	MOT	Medium odor threshold (ppm): listed value in parts per million.
91	MOR	Medium odor threshold reference: source of information.
92	UOT	Upper odor threshold (ppm): listed value in parts per million.
93	UOR	Upper odor threshold reference: source of information.
94	LTT	Lower taste threshold (ppm): listed value in parts per million.
95	LTR	Lower taste threshold reference: source of information.
96	MTT	Medium taste threshold (ppm): listed value in parts per million.
97	MTR	Medium taste threshold reference: source of information.
98	UTT	Upper taste threshold (ppm): listed value in parts per million.
99	UTR	Upper taste threshold reference: source of information.
100	DHI	Direct human ingestion (mg/kg weight): toxic dose levels via

TABLE 2-1   Rationale upon Which OHM-TADS Is Based   *(Continued)*

Segment no.	Mnemonic	Segment title
		human consumption in milligrams of toxicant per kilogram of body weight.
101	DHR-	Reference for direct human ingestion: source of information.
102	DRK	Recommended drinking-water limits (ppm): Public Health Service drinking-water standards cited whenever available.
103	DRR	Reference for recommended drinking-water limits: source of information.
104	BCE	Body contact exposure (ppm): acute contact threshold limits in water when available.
105	BCR	Reference for body contact exposure: source of information.
106	PHC	Prolonged human contact (ppm): safe level for bathing and swimming (prolonged) in parts per million.
107	PHR	Reference for prolonged human contact: source of information.
108	SAF	Personal safety precautions: list of equipment to be employed when working in a spill area. This segment refers to disaster conditions and often presupposes fire or intense heat. Response teams should use their own judgment in deciding when stated precautions are no longer necessary. For most circumstances, eye protection, hard hats, and gloves are recommended.
109	AHL	Acute hazard level: attempts to indicate the level of hazard resulting from a spill. This segment relates to inhalation, ingestion, and contact with the material; it also lists specific water use hazard levels such as fish toxicity and irrigation-water toxicity.
110	CHL	Chronic hazard level: interpretation of the chronic toxicological-biological hazard to life forms subjected to material of interest for extended periods of time.
111	HEL	Degree of hazard to public health: interpretive summary of data from previous segments. This segment focuses on toxicological chemical hazards directly affecting public health.
112	AIR	Air pollution: degree of hazard to people in the vicinity of a spill. This segment may refer to fumes, vapors, mists, or dusts of the material spilled or its combustion and/or decomposition products.
113	ACT	Action levels: interpretive segment designed to aid in initiating response activities. It suggests notification of fire and air authorities if the material poses flammability or air hazards. It recommends alerting civil defense authorities if an explosion hazard exists. When an explosion or severe air pollution exists, evacuation is indicated. If the material in question is highly corrosive or can be absorbed through the skin at toxic levels, affected waterways should be restricted from public access. When flammable materials are involved, ignition sources should be removed. Air contaminants require entry from upwind. If the spill involves solids, attempts should be made to prevent the suspension of dusts in the air. If the material will form a slick on water before dissolving, early attempts at containment will be quite beneficial.
		It is assumed that these actions will be complemented by general defensive responses. These include notifying downstream water users of the spill, stopping all leaks or diverting their flow from reaching surface waters, and removing all bags, barrels, and/or other containers that may still be leaking to the water body.

Segment no.	Mnemonic	Segment title
114	AML	*In situ* amelioration: list of potentially effective treatment methods which could be applied to the body of water for removal of the spilled material. Methods deemed to include hazards equal to or greater than that of the contaminant are systematically excluded. The term "carbon" refers to activated carbon in granular or powdered form.
115	SHR	Beach and shore restoration: segment used mainly to indicate whether the material can be safely burned off beaches. Occasionally a recommendation is made to wash affected area with a neutralizing solution.
116	AVL	Availability of countermeasures material: list of major materials required for countermeasures recommended in Segment 114 and possible local sources for those materials.
117	DIS	Disposal methods: recommended techniques for disposing of spilled materials.
118	DSN	Disposal notification: list of local authorities who should be notified before disposal methods in Segment 117 are initiated.
119	IFP	Industrial fouling potential: potential problems from the use of water contaminated by the material of interest. This segment generally refers to use in boiler feedwater and cooling water.
		Materials with flash points below 50°C are listed as potential rupture hazards when included in boiler feedwater or cooling water.
120	WTP	Effect on water treatment process: potential interaction with typical water and wastewater treatment facilities. Most frequent entries concern the effect of chlorination on the aesthetic properties of contaminated water and the effect of high concentration on sewage organisms.
121	WAT	Major water uses threatened: list of water uses imperiled by a spill. This segment indicates what type of downstream water users should be notified of the spill.
122	LOC	Probable location and state of the material: interpretive segment of physical data designed to assist personnel in identifying the material spilled and its whereabouts. The data attempt to describe the physical appearance of the material as shipped (e.g., a dark red powder, etc.) and its probable location if the spill occurs in or near surface water.
123	DRT	Soil chemistry: general description of the behavior and exchange capacity of various cations and ions in the soil.
124	HOH	Water chemistry: general description of the behavior of the material of interest in an aqueous solution.
125	COL	Color in water: color or appearance of concentrated solutions of the material of interest. In many cases, dilution and material coloring will minimize the visibility of the color listed here.
126	DAT	Adequacy of data: simple classification used to indicate the availability of data.
		Poor—toxicological data sparse if they exist at all.
		Fair—toxicological data found but no aquatic toxicities listed.
		Moderate—toxicological data found along with information on toxicity toward fish.
		Good—both toxicological and aquatic toxicity data found.
		Limited references—materials for which a complete literature survey has not been run.

(1) TECHNICAL ASSISTANCE DATA SYSTEM: 72T16587
(2) CAS REGISTRY NO: 1336216
(3) SIC CODE: 2841; 2819; 2819
(4) MATERIAL: $$$ AMMONIUM HYDROXIDE $$$
(5) SYNONYMS: AMMONIUM-HYDRATE, AQUA-AMMONIA, SPIRIT-OF-HARTSHORN, WATER-OF-AMMONIA
(7) CHEMICAL FORMULA: NH_4OH
(8) SPECIES IN MIXTURE: 28% NH_3
(9)   COMMON   USES: DETERGENT   BLEACHING   AMMONIUM SALTS   ANILINE DYES
(10) RAIL TRANSPORT (%):   58.7
(11) BARGE TRANSPORT (%):   13.2
(12) TRUCK TRANSPORT (%):   24.7
(13) PIPE TRANSPORT (%):   03.4
(14) CONTAINERS: TANK CARS, STEEL DRUMS (110 GAL), CARBOYS (13 GAL), GLASS BOTTLES (5PT)
(15) GENERAL STORAGE PROCEDURES: KEEP COOL IN STRONG GLASS, PLASTIC OR RUBBER-STOPPERED BOTTLES NOT COMPLETELY FILLED. KEEP IN VENTILATED AREA.
(16) GENERAL HANDLING PROCEDURES: AVOID DIRECT SKIN CONTACT.
(17) PRODUCTION SITES:
AIR PRODUCTS AND CHEMS. INC., ESCAMBIA CHEM. CORP. SUBSID., NEW ORLEANS, LA.; PENSACOLA (PACE), FLA.
ALLIED CHEM. CORP., AGRICULTURAL DIV., GEISMAR, LA.; HOPE-WELL, VA.; IRONTON (SOUTH POINT), OHIO; OMAHA (LA PLATTE), NEB.
AMERICAN CYANAMID CO., AGRICULTURAL DIV., FORTIER (NEW ORLEANS), LA.
APACHE POWDER CO., BENSON, ARIZ.
ARKANSAS LOUISIANA GAS CO., ARKLA CHEM. CORP. SUBSID., HELENA, ARK.
ATLANTIC RICHFIELD IO., ARCO CHEM. CO. DIV., FORT MADISON, IOWA
ATLAS CHEM. INDUST. INC., CHEMS. DIV., JOPLIN, MO.
BORDEN INC., BORDEN CHEM. DIV., PETROCHEMICALS, GEISMAR, LA.
CF INDUST. INC. FEL-TEX DIV., FREMONT, NEB.
FIRST NITROGEN CORP. SUBSID., DONALDSONVILLE, LA.
CENTRAL NITROGEN, INC., TERRE HAUTE, IND.
CHEROKEE NITROGEN CO., PRYOR, OKLA.
CITIES SERVICE CO. INC., NORTH AMERICAN CHEMS. AND METALS GROUP, AGRICULTURAL CHEMS. DIV., TAMPA, FLA.
COLORADO INTERSTATE CORP., WYCON CHEM. CO. SUBSID., CHEYENNE, WYO.
COLUMBIA NITROGEN CORP., AUGUSTA, GA.
COMMERCIAL SOLVENTS CORP., STERLINGTON, LA.
CONTINENTAL OIL CO, AGRICO CHEM. CO. DIV., SLYTHEVILLE, ARK.
COOPERATIVE FARM CHEMS. ASSOCIATION, LAWRENCE, FLA.
DIAMOND SHAMROCK CORP., DIAMOND SHAMROCK OIL AND GAS CO., DUMAS, TEX.
THE DOW CHEM. CO., FREEPORT, TEX.

Fig. 2-1   OHM-TADS data printout for ammonium hydroxide.

E.I. DU PONT DE NÉMOURS & CO. INC. EXPLOSIVES DEPT., BEAU-
MONT, TEX. INDUST. AND BIOCHEMS. DEPT., BELLE, S.C. PLASTICS
DEPT., VICTORIA, TEX.

EARLY CALIFORNIA INDUST. INC., SOUTHWESTERN NITROCHEMI-
CAL CORP., SUBSID. OF ARIZONA AGROCHEMICAL CORP., CHANDLER,
ARIZ.

EL PASO NATURAL GAS CO., EL PASO PRODUCTS CO. SUBSID.,
ODESSA, TEX.

FMC CORP., INORGANIC EHCMS. DIV., SOUTH CHARLESTON, S.C.

FARMERS CHEM. ASSOCIATION, INC., TUNIS, N.C.; TYNER, TENN.

FARMLAND INDUST. INC., DODGE CITY, FLA.; FORT DODGE, IOWA;
HASTINGS, NEB.

FELMONT OIL CORP., OLEAN, N.Y.

GOODPASTURE, INC., DIMMITT, TEX.

W.R. GRACE & CO., AGRICULTURAL CHEMS. GROUP, BIG SPRING,
TEX.; MEMPHIS, TENN.

GREEN VALLEY CHEM. CORP., CRESTON, IOWA

GULF OIL CORP., GULF OIL CHEMS. CO. DIV., AGRICULTURAL
CHEMS. DIV., DONALDSONVILLE FAUSTINA, LA.

GULF & WESTERN INDUST. INC., THE NEW JERSEY ZINC CO. SUBSID,
PALMERTON, PA.

HERCULES INC., SYNTHETICS DEPT., HERCULES, CALIF.; LOUISIANA,
MO.

HILL CHEMS. INC., BORGER, TEX.

KAISER ALUMINUM & CHEM. CORP., KAISER AGRICULTURAL
CHEMS. DIV., SAVANNAH, GA.

LONE STAR GAS CO., NIPAK, INC. SUBSID., KERENS, TEX.; PRYOR,
OKLA.

MISCOA, PASCAGOULA, MISS.; YAZOO CITY, MISS.

MOBIL OIL CORP., MOBIL CHEM. CO. DIV. OF MOBIL OIL CORP.,
PETROCHEMICALS DIV., BEAUMONT, TEX.

MONSANTO CO., AGRICULTURAL DIV., LULING, IA.; MUSCATINE,
IOWA

OCCIDENTAL PETROLEUM CORP. CALIFORNIA AMMONIA CO. SUB-
SID., LATHROP, CALIF.

HOOKER CHEM. CORP. SUBSID., INDUST. CHEMS. DIV., TACOMA,
WASH.

OCCIDENTAL CHEM. CO. SUBSID., WESTERN DIV., LATHROP, CALIF;
PLAINVIEW, TEX.

OLIN CORP., AGRICULTURAL CHEMS. DIV., LAKE CHARLES, LA.

PPG INDUST. INC., INDUST. CHEM. DIV. (FORMERLY THE COLUM-
BIA- SOUTHERN CHEM CORP.), NEW MARTINSVILLE (NATRIUM), S.C.

PENNWALT CORP., CHEM. DIV., PORTLAND, ORE.; WYANDOTTE,
MICH.

PHILLIPS PACIFIC CHEM. CO., KENNEWICK, WASH.

PHILLIPS PETROLEUM CO., BEATRICE, NEB.; ETTER, TEX.; PASA-
DENA, TEX.

RESERVE OIL & GAS CO., HANFORD, CALIF.

ROHM AND HAAS CO., DEER PARK, TEX.

ST. PAUL AMMONIA PRODUCTS, INC., EAST DUBUQUE, ILL.

SHELL CHEM. CO., AGRICULTURAL DIV., ST. HELENS, ORE.; VEN-
TURA, CALIF.

Fig. 2-1 *(Con't.)*

J.R. SIMPLOT CO., MINERALS AND CHEM. DIV., POCATELLO, IDAHO

SKELLY OIL CO., HAWKEYE CHEM. CO. SUBSID., CLINTON, IOWA

STANDARD OIL CO. OF CALIFORNIA, EL SEGUNDO, CALIF.; RICH-MOND CALIF,

CHEVRON CHEM. CO. SUBSID., ORTHO DIV., FORT MADISON, IOWA

STANDARD OIL CO. (KENTUCKY) SUBSID., PASCAGOULA, MISS.

STANDARD OIL CO. (INDIANA), AMERICAN OIL CO. SUBSID, TEXAS CITY, TEX.

THE STANDARD OIL CO. (OHIO), VISTRON CORP. SUBSID., AGRICUL-TURAL CHEMS. DEPT., LIMA, OHIO

SUN OIL CO, SUNOCO DIV., MARCUS HOOK, PA.

TENNECO INC., TENNECO CHEMS. INC. (A MAJOR COMPONENT OF TENNECO INC.), TENNECO HYDROCARBON CHEMS. DIV., HOUSTON, TEX.

TENNESSEE VALLEY AUTHORITY, MUSCLE SHOALS, ALA.

TERRA CHEMS. INTERNATIONAL, INC., PORT NEAL SIOUX CITY, IOWA

TEXACO INC., LOCKPORT, ILL.

TRIAD CHEM., DONALDSONVILLE, LA.

UNION OIL CO. OF CALIFORNIA, COLLIER CARBON AND CHEM. CORP., SUBSID., BREA, CALIF.; KENAI, ALAS.

UNITED STATES STEEL CORP., USS AGRI-CHEMICALS DIV., CHERO-KEE, ALA.; CLAIRTON, PA.

VALLEY NITROGEN PRODUCERS, INC., EL CENTRO, CALIF.; HELM NEAR FRESNO, CALIF.

VULCAN MATERIALS CO., CHEMS. DIV., WICHITA, FLA.

(20) BINARY REACTANTS: H2SO4, STRONG MINERAL ACIDS, SILVER NITRATE, SILVER OXIDE, SILVEROXIDE AND ETHYL ALCOHOL

(21) CORROSIVENESS: COPPER, ALUMINUM ALLOYS AND GALVAN-IZED SURFACES.

(23) ANTAGONISTIC MATERIALS: DISSOCIATION OF NH4OH DEPENDS ON PH, NH4OH BEING MORE TOXIC

(24) FIELD DETECTION LIMIT (PPM), TECHNIQUES, REF: .2, SP EL, BNW, 100193

(25) LAB DETECTION LIMIT (PPM), TECHNIQUES, REF: .01, SP, EL, BNW, 100193

(26) STANDARD CODES: EPA 311; NO ICC; NFPA—2,1,0

(27) FLAMMABILITY: NH3 FUMES ARE FLAMMABLE QUITE FLAMMABLE;

(28) LOWER FLAMMABILITY LIMIT (%): 016.

(29) UPPER FLAMMABILITY LIMIT (%): 025.

(30) TOXIC COMBUSTION PRODUCTS: WEAR SELF-CONTAINED BREATHING APPARATUS

(33) AUTO IGNITION POINT (DEG C): 00650.

(34) EXPLOSIVENESS: STABLE;

(37) MELTING POINT (DEG C):   −00077

(40) BOILING CHARACTERISTICS: SOL. ONLY

(41) SOLUBILITY (PPM), 25 DEG C: 1

(43) SPECIFIC GRAVITY: 00.9

(50) PERSISTENCY: NATURAL CO2 WILL SLOWLY NEUTRALIZE AND AMMONIA GAS WILL VOLATILIZE TO SOME EXTENT.

(51) POTENTIAL FOR ACCUMULATION: NEGATIVE;

Fig. 2-1 *(Con't.)*

(52) FOOD CHAIN CONCENTRATION POTENTIAL: NEGATIVE;
(57) FRESHWATER TOXICITY NUMBER (PPM):   37; 2.8; 4.5; 1.9; 6.25; 2.1;
10; 1.8; 13; 13.3; 15; 13.2; 17.5; 16.2; 18.5; 15.9; 20; 30; 30; 30; 60; 32; 20; 12
(58) FRESH WATER TOXICITY TEXT
CONC. (PPM)/EXPOS. (HR)/SPECIF/EFFECT/TEST ENV/REF

37/96/MOSQUITO FISH/TLM/TURBID WATER/C-1
2.8/96/STRIPED BASS/TLM/STATIC 15 DEG C/R-120;
4.5/ /GOLDFISH/LETHAL/ /C-1
1.9/96/STRIPED BASS/TLM/STATIC 23 DEG C/R-120;
6.25/24/TROUT LETHAL/ /C-1
2.1/96/STICKLEBACK/TLM/STATIC 15 DEG C/R-120;
10/ /SUCKERS, SHINERS, CARP, TROUT/LETHAL//C-1
1.8/96/STICKLEBACK/TLM/STATIC 23 DEG C/R-120;
13/24/SUCKERS, SHINERS, CARP/LETHAL/ /C-1
13.3 AS NH4/ /EURASION WATERMILFOIL/IL50 ROOT WT/R-178;
15/48/SUNFISH/TLM/PHIL. TAP WATER 20 DEG./C-1
16.2/ /EURASION WATERMILFOIL/IL50 ROOT LENGTH/ /R-178;
17.5/48/MINNOW/TLM/ /C-1
13.2 AS NH4/ /EURASION WATERMILFOIL/IL50 STEM WT/R-178;
18.5/48/SUNFISH/TLM/PHIL. TAP WATER OXYG. 20 DEG./C-1
15.9/ /EURASION WATERMILFOIL/IL50 STEM LENGTH/ /R-178;
20/.2/SUCKERS, SHINERS, CARP/LETHAL/ /C-1
30/24/SMALL FISH/LETHAL/ /C-1
30/24/CHUB/LETHAL/15-21DEG./C-1
30/ /PERCH/LETHAL/ /C-1
60/24/DAPHNIA/TLM/TEMP CON./E-25
32/48/DAPHNIA/TLM/TEMP CON./E-25
20/96/DAPHNIA/TLM/TEMP CON./E-25
12/48/RANA PIPIENS & BULLFROGS/TOXIC/FIELD/E-187
(61) SALT WATER TOXICITY NUMBER (PPM):   2.8; 2.0; 2.1; 1.5; 5.2; 10.4;
2.4; 2.3
(62) SALT WATER TOXICITY TEXT
CONC. (PPM)/EXPOS. (HR)/SPECIE/EFFECT/TEST ENV/REF

2.8/96/STRIPED BASS/TLM/STATIC BRACKISH 15 DEG C/R-120;
2.0/96/STRIPED BASS/TLM/STATIC SEAWATER 15 DEG C/R-120;
2.1/96/STRIPED BASS/TLM/STATIC BRACKISH 23 DEG C/R-120;
1.5/96/STRIPED BASS/TLM/STATIC SEAWATER 23 DEG C/R-120;
5.2/96/STICKLEBACK/TLM/STATIC BRACKISH 15 DEG C/R-120;
10.4/96/STICKLEBACK/TLM/STATIC SEAWATER 15 DEG C/R-120;
2.4/96/STICKLEBACK/TLM/STATIC BRACKISH 23 DEG C/R-120;
2.3/96/STICKLEBACK/TLM/STATIC SEAWATER 23 DEG C/R-120;
(63) ANIMAL TOXICITY VALUE: 250; 350
(64) ANIMAL TOXICITY TEXT
VALUE/TIME/SPECIES/PARAM./ROUTE/REF

250/ /CAT/LD50/ORL/D-2;
350/14 DAY/RAT/LD50/ORL/R-119;
(75) IRRIGABLE PLANTS (PPM):   000009.8

Fig. 2-1 *(Con't.)*

(76) REF FOR IRRIGABLE PLANTS: (PH) INHIBITS GERMINATION, E-181

(79) MAJOR SPECIES THREATENED: ALL SPECIES

(82) INHALATION LIMIT (VALUE):   0000050.

(87) GENERAL SENSATION: INTENSE, PUNGENT, SUFFOCATING ODOR, ACRID TASTE, STRONG ALKALINE REACTION. ODOR WILL BE DETECTABLE AT LEVELS BELOW HARMFUL. HIGHLY IRRITATING TO EYES AND MUCOUS MEMBRANE.

(108) PERSONAL SAFETY PRECAUTIONS: RUBBER OUTERWEAR, COTTON INNERWEAR. PVA NOT RECOMMENDED FOR GLOVES(R-121); DO NOT USE CANNISTER GAS MASKS. EYE PROTECTION AND CONTAINED BREATHING EQUIPMENT. DO NOT USE COPPER, ALUMINUM OR GALVANIZED EQUIPMENT.

(109) ACUTE HAZARD LEVEL: TOXICITY AND IRRITANT VALUE THREATEN ALL FORMS OF LIFE. TOXICITY IS DIRECTLY RELATED TO THE LEVEL OF UNIONIZED AMMONIA; THRESHOLD CON. FOR FISH .5 PPM FREE NH_3 (E-188).

(110) CHRONIC HAZARD LEVEL: CHRONIC IRRITATION HAZARD EXISTS.

(111) DEGREE OF HAZARD TO PUBLIC HEALTH: HIGHLY IRRITATING AND CAN CAUSE INHALATION PROBLEMS.

(112) AIR POLLUTION: HIGH;

(113)   ACTION   LEVELS: NOTIFY   AIR   AND   FIRE AUTHORITY.   ENTER FROM UPWIND SIDE.   WARNING PROPERTIES ARE GOOD.   RESTRICT DOWNSTREAM CONTACT WITH AFFECTED WATERS AS SOLUTION CAN CAUSE SEVERE BURNS.

(114) IN SITU AMELIORATION: SPURGING MAY RELEASE NH_3 GAS. SODIUM BICARBONATE SHOULD BE USED TO NEUTRALIZE SPILLS. NATURAL ZEOLITES WILL THEN PICK UP AMMONIUM IONS.

(115) BEACH AND SHORE RESTORATION: NEUTRALIZE.

(116) AVAILABILITY OF COUNTERMEASURE MATERIALS: SODIUM BICARBONATE—GROCERY DISTRIBUTORS, LARGE BAKERIES ZEOLITES—LOCAL QUARRIES, SUPPLIERS

(117) DISPOSAL METHODS: POUR INTO LARGE TANK OF WATER, NEUTRALIZE AND ROUTE TO SEWAGE PLANT.

(118) DISPOSAL NOTIFICATION: LOCAL SEWAGE AUTHORITY.

(119) INDUSTRIAL FOULING POTENTIAL: MAY BE CORROSIVE TO PIPING AND PROCESSING EQUIPMENT.

(120) EFFECTS ON WATER TREATMENT PROCESS: CHLORINATION CAUSES FORMATION OF ODOROUS CHLORAMINES.

(121) MAJOR WATER USES THREATENED: ALL USES

(122) PROBABLE LOCATION AND STATE OF MATERIAL: IN SOLUTION ONLY. WILL DISSOLVE; WILL BE DISSOLVED IN WATER. COLORLESS LIQUID.

(124) WATER CHEMISTRY: SEE FILE ON AMMONIA FOR DATA ON SOLUTION CHARACTERISTICS. ALKALINE.

(125) COLOR IN WATER: COLORLESS

(126) ADEQUACY OF DATA: GOOD

Fig. 2-1   *(Con't.)*

CHEMTREC is sponsored by the Chemical Manufacturers Association and is available to members and nonmembers alike.

CHEMTREC's 24-h toll-free telephone number, 800-424-9300, is widely distributed to emergency service personnel, carriers, and the chemical industry. When an emergency call is received by CHEMTREC, the person on duty records the essential information in writing. He or she tries to obtain as much information as possible from the caller. The person on duty will give out information as furnished by the chemical producers on the chemical or chemicals reported to be involved. This includes basic data such as information on hazards of spills, fire, or exposure. After advising the caller, the person on duty immediately notifies the shipper of the chemical by phone, giving the details on the emergency. At this point, responsibility for further guidance passes to the shipper. CHEMTREC also serves as a contact point for the Chlorine Institute, the National Agricultural Chemicals Association (pesticides), and the Department of Energy (radioactive materials).

Chlorine Emergency Plan (CHLOREP)

Chlorine manufacturers in the United States and Canada through the Chlorine Institute have established the Chlorine Emergency Plan (CHLOREP) to handle chlorine emergencies. This is essentially a mutual-aid program whereby the manufacturer closest to the emergency will provide technical assistance even if the emergency involves another manufacturer's product.

The CHLOREP system operates through CHEMTREC. Upon receiving an emergency call, CHEMTREC notifies the appropriate person in accordance with the mutual-aid plan. This person then gets in touch with persons on the emergency scene to determine whether or not a technical team must be sent to provide assistance. Each participating manufacturer has trained personnel and equipment available for emergencies. CHLOREP may be reached on a 24-h basis through CHEMTREC's telephone number.

NACA Pesticides Safety Team Network

The National Agricultural Chemicals Association (NACA) through its members operates a national pesticide information and response network. The network's function is to provide advice and on-site assistance when the spill situation warrants it. It operates through the CHEMTREC office. When CHEMTREC is notified of an emergency involving a pesticide, it gets in touch with the manufacturer. The manufacturer will then provide specific advice regarding the handling of the spill. If necessary, spill response teams are available on a geographical basis to assist at the emergency scene.

Interagency Radiological Assistance Plan (IRAP)

The Interagency Radiological Assistance Plan (IRAP) is designed to assist any person in obtaining technical guidance in coping with radiation emergencies. It operates through the Department of Energy but works closely with other federal, state, military, and regional groups. Under IRAP the United States is divided into eight geographical areas of responsibility, each with a regional coordinating office.

Upon receiving an emergency call, the regional coordinator investigates the situation to assess the potential radioactive hazard. The coordinator tries to get as much information as possible on the telephone on the specifics of the situation and the type of material (e.g., as listed on the shipping papers). Advice will be given over the phone if the potential hazard appears to be minimal. If the spill or leak appears to be serious, a technical response team will be dispatched. This team will work jointly with state civil defense or public health personnel whenever possible. In any case the coordinating office will notify the appropriate state office of the radioactive spill. When the response team is dispatched, the Nuclear Regulatory Commission is notified, especially if the spilled material is licensed. The main functions of the response team are to assess the hazard, to inform people of the hazard, and to recommend emergency actions to minimize the hazard. Responsibility for cleanup rests with the shipper or carrier (the party who possesses the material at the time of the spill).

Access to IRAP is through the regional coordinating offices or through CHEMTREC.

Other Emergency Systems

In addition to the data systems described above, there are individual programs which allow the specific companies involved to assist in the solution of emergencies.

In 1966, E. I. du Pont de Nemours & Co. developed the Transportation Emergency Reporting Procedure (TERP), which provides immediate information on any of Du Pont's 1500 products which might be involved in a spill situation. TERP consists of one hot-line telephone number through which expert advice and action can be channeled when needed.

Other chemical companies have developed similar emergency systems for their products. Among these are the American Cyanamid Co.'s Transportation and Warehouse Emergency Reporting Procedures, the Union Carbide Corp.'s Hazardous Emergency Leak Procedure (HELP) system, the Dow Chemical Company's Distribution Emergency Response System, the Allied Chemical Corporation's Transportation Emergency System, and many others that have been or are being formed as companies realize the hazardous nature and possible repercussions of spills of their material.

Thus, with the proliferation of systems attempting to solve the spill problem, one would imagine that the situation with hazardous materials is practically moot. However, despite the progress that has been made, the continuing lack of coordination and the gaps between programs have left much room for improvement. For example, in some cases there is still likely to be either duplication of effort or the reverse because of lack of knowledge of the materials spilled. There may also be a conflict of authority between emergency units at the scene. In addition, the purpose of some of the programs mentioned above has unfortunately been singular. The programs are devised primarily to protect the general population in the vicinity of a spill and personnel working to contain the material from hazardous properties and consequences. They have devoted too little effort to the total environmental problem of the hazardous spill. What are the long-term effects on the environment and its constitutents? Has the material been allowed to percolate into the ground? Will drainage ultimately into nearby water systems pose a future problem? Has the substance been neutralized so that it will not reappear at some future date? Have the effects on resident flora and fauna been documented to allow assessment of the acceptable level of risk involved? What are the immediate and long-term consequences to the food chain? What would the cost be to restore the insulted area completely to its former state? What are the aesthetic consequences of the spill and the procedures used to limit visual damage? All these questions and many more recur during each spill event involving hazardous substances and should be properly addressed in the context of emergency data.

PART 3
European Systems

R. F. Cumberland
Manager, National Chemical Emergency Centre
Harwell Laboratory

Introduction / 3-25
Chemical Hazards / 3-26
Hazard Information (Labeling) Systems / 3-27
Sources of Specialist Chemical Information / 3-36
Conclusions / 3-43

INTRODUCTION

The uncontrolled release of chemicals into the environment, as in a transportation accident, invariably presents an immediate need for essential data. The nature and depth of the information required will clearly depend on particular circumstances. In general, information systems are designed to provide data for a particular stage in a spill or potential spill situation, for example:

Stage 1 Primary data for first aid (e.g., on the container).

Stage 2 Basic but more detailed information (e.g., written instructions or data bank).

Stage 3 Specialist advice (e.g., data bank or chemical manufacturer).

Many schemes have been introduced both nationally and internationally in an effort to identify rapidly chemical substances, their intrinsic hazards, and/or appropriate emergency countermeasures. Some of the fundamental difficulties in such schemes are those of coping with a multiplicity of synonyms and languages applied to product labeling, particularly in the international movement of chemical products. It is the purpose of this part to outline some of the principal information systems used in Europe for emergency spill response.

The movement of hazardous substances by any mode of transport presents, in general, a greater risk of accidental release than in a static installation. It is more probable that appropriate information will not be so readily available in a transport accident, and in consequence it is this aspect on which attention is focused.

The chemical industry produces materials which vary enormously in both type and degree of hazard, ranging from explosives to inert building materials. Products are transported as solids, liquids, or gases under a wide range of temperature and pressure. The packaging methods used are directly related to the nature and degree of hazard of the product and to the regulations imposed, appropriate to the transport mode. Such regulations, although closely related to packages and vehicle-marking schemes, are discussed more fully in Chap. 6.

CHEMICAL HAZARDS

Hazard Classification

The classification of chemical hazards as recommended by the UN Committee of Experts on the Transport of Dangerous Goods[1] has been widely adopted for the conveyance of hazardous chemicals for all modes of transport. Hazard types are segregated into nine basic classes represented numerically from 1 to 9. Many of these classes are further separated into divisions and subdivisions according to appropriate criteria (see Table 3-1). The criteria adopted for the classification of hazards into divisions may vary from United Nations recommendations according to the mode of transport and the regulatory body concerned, such as the European Agreement Concerning the International Carriage of Dangerous Goods by Road (ADR),[2] the *International Regulations Concerning the Carriage of Dangerous Goods by Rail* (RID),[3] the Intergovernmental Maritime Consultative Organization (IMCO) Dangerous Goods Code,[4] and the International Air Transport Association (IATA) *Restricted Articles Regulations.*[5]

Hazard Identification

Each United Nations hazard class (with the exception of Class 9) has a distinctive diamond-shaped label bearing a pictogram for quick hazard recognition (Fig. 3-1). Each

TABLE 3-1   United Nations Hazard Classification

United Nations	Division	Hazard type
1		Explosives
	1.1	Mass explosion hazard
	1.2	Projection hazard but not mass explosion hazard
	1.3	Fire hazard and minor blast or projection hazard
	1.4	No significant hazard
	1.5	Very insensitive substances
2		Gases
3		Flammable liquids
	3.1	Flash point below 23°C (closed cup) [IMDG, Code*, flash point below −18°C (closed cup)]
	3.2	Flash point between 23 and 60.5°C (closed cup) [IMDG Code, flash point between 0 and 23°C (closed cup)]
	(3.3)	[IMDG Code, flash point between 23 and 60.5°C (closed cup)]
4		Flammable solids
	4.1	Flammable solids
	4.2	Substances liable to spontaneous combustion
	4.3	Substances emitting flammable gases in contact with water
5		Oxidizing substances
	5.1	Oxidizing substances other than organic peroxides
	5.2	Organic peroxides
6		Poisonous (toxic) and infectious substances
	6.1	Poisonous (toxic) substances
	6.1.1	Substances emitting toxic vapors
	6.1.2	Poisonous substances other than those emitting toxic vapors
	6.2	Infectious substances
7		Radioactive substances
8		Corrosive substances
9		Miscellaneous dangerous substances

*International Maritime Dangerous Goods Code.

label also has a characteristic background color:

Explosive	Orange
Flammable	Red
Water-reactive	Blue
Oxidizer	Yellow
Toxic or infectious	White
Radioactive	White or yellow-and-white
Corrosive	Black-and-white

An additional diamond introduced in the United Kingdom to identify "other hazardous substances" (but excluding explosive or radioactive materials) consists of a black exclamation point on a white background. The use of this label for multiloads and hazardous waste is now a statutory requirement under the 1978 *Hazardous Substances (Labelling of Road Tankers) Regulations.*[6]

In addition to the pictogram, hazard-warning diamonds may also bear an approved inscription quoting the hazard and/or the United Nations hazard class number. The basic principle however, is that the shape, color, and pictogram convey a clear message of danger, thus overcoming language difficulties. With international acceptance, the value of such a labeling system when displayed on vehicles and packages is clear:

1. It provides a warning to the general public to keep away.
2. In an accident situation the emergency services are provided with an indication of the primary hazard likely to be encountered.

It is important to recognize that such warning diamonds are intended for use in transport situations. They should not be confused with rectangular labels some of which bear similar pictograms on an orange background. Such labels are intended for storage and use applications for specified hazardous chemicals as required by a directive of the Council of the European Economic Community (EEC).

Detailed requirements and recommendations for labeling chemical products are outside the scope of this part but can be found in the respective regulations already mentioned or in a general guide elsewhere.[7]

Substance Identification

The classification of hazardous chemical products is considered by the group of rapporteurs of the UN Committee of Experts on the basis of information submitted in a prescribed form by manufacturers. If such data meet appropriate criteria, the product is assigned a four-digit identification (or United Nations) number and added to the United Nations list. At present approximately 2000 products (excluding explosives) have been included.

HAZARD INFORMATION (LABELING) SYSTEMS

A need for essential information to be clearly displayed in a transport emergency has always been accepted by both industry and the emergency services. Whether such information should provide, in addition to product identification, hazards and/or emergency action has long been a topic of considerable international discussion. The basis of many

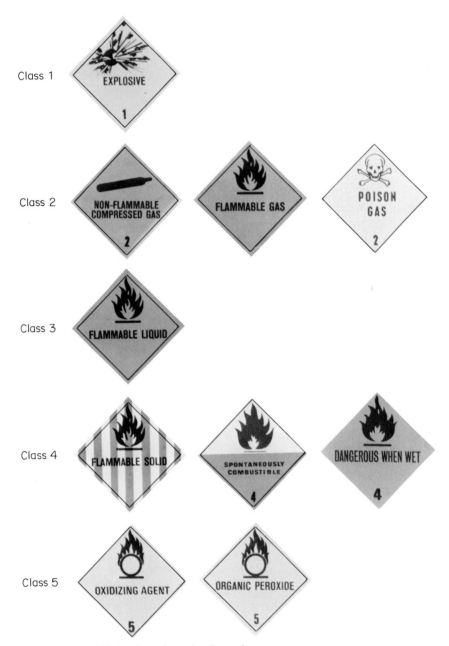

Fig. 3-1 United Nations hazard-warning diamonds.

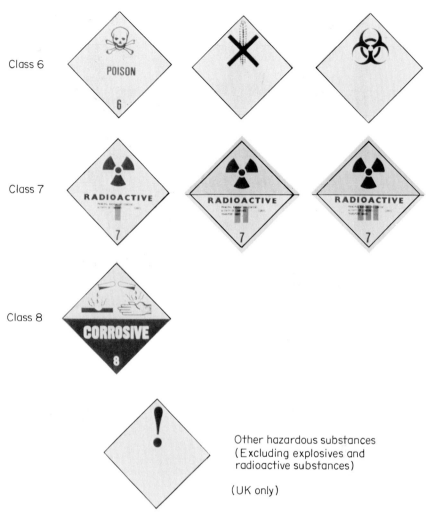

Class 6

Class 7

Class 8

Other hazardous substances
(Excluding explosives and
radioactive substances)

(UK only)

Fig. 3-1   (*Con't.*)

emergency information systems adopted in various parts of the world has been a combination of hazard classification and United Nations substance identification.

ADR/RID

The international movement of hazardous chemicals by road or rail across most frontiers in Europe is controlled by the ADR or RID regulations already mentioned. These regulations require the display of hazard information panels (Fig. 3-2). The panels have an orange background and display two separate numerical codes. The lower code is the United Nations number which identifies the substance. In the upper section a two- or

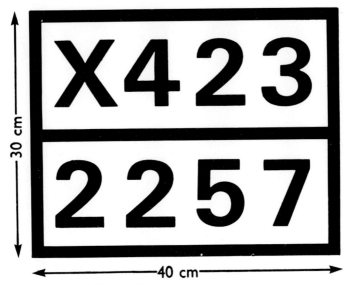

Fig. 3-2 ADR/RID hazard information panel.

three-digit hazard identification number (properties code) is displayed. The first figure of this hazard identification number indicates the primary hazard:

2 Gas

3 Flammable liquid

4 Flammable solid

5 Oxidizing substance or organic peroxide

6 Toxic substance

8 Corrosive

Secondary hazards are indicated by the second and third figures:

0 No meaning

1 Explosion risk

2 Possibility of gas being given off

3 Flammable risk

5 Oxidizing risk

6 Toxic risk

8 Corrosive risk

9 Risk of violent reaction from spontaneous decomposition or self-polymerization

An intensification of the primary hazard is indicated when the first and second figures are the same. For example, 33 means a highly flammable liquid (flash point $< 21°C$); 66, a very dangerous toxic substance; and 88, a very dangerous corrosive substance. When the first two figures are 22, a refrigerated gas is indicated. The combination 42 indicates a solid which may give off a gas on contact with water. When the hazard identification number is preceded by the letter X, the application of water to the product is absolutely prohibited. The example given in Fig. 3-2, therefore, provides the following information in numerical form:

1. The substance is potassium metal.
2. It is a flammable solid which may give off a gas and possess a flammable risk.
3. It must not be allowed to come into contact with water.

This hazard information system must be adopted for international journeys of scheduled chemicals by all signatories to the agreement. At present these signatories include Austria, Belgium, the Federal Republic of Germany, France, the German Democratic Republic, Italy, Luxembourg, the Netherlands, Norway, Poland, Portugal, Spain, Sweden, Switzerland, the United Kingdom, and Yugoslavia. Hazard information systems for domestic traffic, however, may be different.

United Kingdom Hazard Information System (UKHIS)

This system for the movement of certain loads of bulk chemicals in tank trucks became a statutory requirement in the United Kingdom in 1978. UKHIS requires the display of orange-and-black composite labels, an example of which is shown in Fig. 3-3. The UKHIS label consists of five sections:

1. An emergency action code (Hazchem) in the upper orange section
2. An appropriate substance identification number in the lower orange section
3. A hazard-warning diamond
4. A source of specialist advice
5. The supplier's name or symbol (optional)

The source of specialist advice could be the manufacturer, the supplier, or an organization acting on the supplier's behalf such as the National Chemical Emergency Centre

Fig. 3-3 UKHIS hazard information panel.

(NCEC) at Harwell. The Harwell number would, however, be displayed only when contractual arrangements existed between the company and Harwell under the Chemical Emergency Agency Service Scheme.

A significant difference between the United Kingdom and the ADR/RID systems is the action code which appears on the UKHIS label. This simple two- or three-character code (Hazchem) was devised by London Fire Brigade and adopted by the United Kingdom government and the chemical industry. It enables the emergency services reaching the scene of an accident to take immediate action while awaiting specialist advice.

The Hazchem code, by reference to the scale card (Fig. 3-4) carried by emergency service personnel, gives the following information:

1. Whether the spill should be contained or washed away

2. The appropriate extinguishing media to be used in a fire situation

3. Whether a risk of violent reaction exists

4. A need to consider evacuation

5. Appropriate protective clothing

Hazchem codes are allocated by a technical subcommittee of the Joint Committee on Fire Brigade Operations within the Home Office and subsequently ratified by the Health and Safety Executive. This subcommittee includes representatives from the chemical industry, government departments, water authorities, and the Fire Service. Codes are allocated on the basis of a product's physical properties, toxicology, and chemical reactivity. The Home Office issues lists of approved Hazchem codes and United Nations numbers or chemical names to the emergency services. The Chemical Industries Association (CIA) publishes similar information[8] for other organizations.

The use of UKHIS panels, and therefore of Hazchem, is a statutory requirement in the United Kingdom for the conveyance of specified chemicals in tank trucks only. Labeling of vehicles conveying other chemical loads is currently the subject of a government consultative document[9] proposing more general regulations for the transport of hazardous chemicals.

The UKHIS system has been extended on a voluntary basis in the United Kingdom for marking crop-spraying aircraft. This practice was introduced because emergency services attending crashed light aircraft were unaware that potentially hazardous chemicals were being conveyed. The signs used are approximately one-quarter of the full size and bear the Hazchem code 3WE with the United Nations Class 6 diamond.

The United Kingdom chemical industry is also encouraging the use of UKHIS-type panels on tank trucks conveying chemical products of low hazard which do not meet the criteria for statutory labeling. The chemical producers provide their own codes, following a procedure recommended by the CIA.[10] The labels used in this voluntary scheme are black-and-white and hence are clearly distinguishable from the orange labels for hazardous products and provide equally valuable information to the emergency services.

Hazchem is used as a source of rapid information in transport incidents by emergency services in other countries including Sweden, the Netherlands, and Australia.

Hazard information systems applicable to storage situations have been of considerable interest to those concerned with fire fighting for many years. A number of different systems have been designed or adapted for this purpose. A feasibility study is being carried out to assess the suitability of Hazchem for this purpose.

NFPA Hazard Information System

An American system conceived by the National Fire Protection Association (NFPA)[11] places emphasis on hazards (fire, health, explosive, and chemical reactivity) rather than

Notes for Guidance

FOG

In the absence of fog equipment a fine spray may be used.

DRY AGENT

Water **must not** be allowed to come into contact with the substance at risk.

V

Can be violently or even explosively reactive.

FULL

Full body protective clothing with BA.

BA

Breathing apparatus plus protective gloves.

DILUTE

May be washed to drain with large quantities of water.

CONTAIN

Prevent, by any means available, spillage from entering drains or water course.

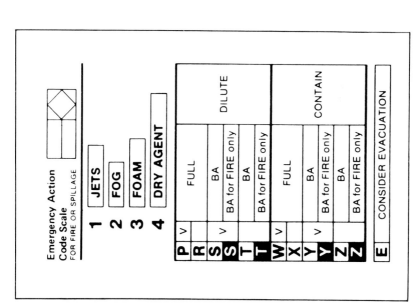

Emergency Action Code Scale
FOR FIRE OR SPILLAGE

1 JETS
2 FOG
3 FOAM
4 DRY AGENT

P	v	FULL	DILUTE
R		FULL	DILUTE
S	v	BA	DILUTE
S		BA for FIRE only	DILUTE
T		BA	DILUTE
T		BA for FIRE only	DILUTE
W	v	FULL	CONTAIN
X		FULL	CONTAIN
Y	v	BA	CONTAIN
Y		BA for FIRE only	CONTAIN
Z		BA	CONTAIN
Z		BA for FIRE only	CONTAIN
E		CONSIDER EVACUATION	

Fig. 3-4 Hazchem code scale.

emergency action. This standard system, frequently referred to as NFPA 704M, was originally intended to safeguard personnel concerned with fires in storage situations where the hazards might not be apparent. Hazard identification is provided by a diamond-shaped label (Fig. 3-5). Three of the label's sections have characteristic colored backgrounds denoting the type of hazard (blue for health, red for flammability, and yellow for reactivity), on which is superimposed a numerical grading of 0 to 4, indicating the order of severity of hazard. The fourth section is used to provide additional information such as water reactivity, oxidizing, or radiation hazard. This standard system is also used in some reference books, notably Hommel,[12] and forms an immediate guide to chemical hazards.

Transport Emergency Cards

The hazard information systems described so far are representative of first-aid schemes by providing brief but essential data to enable immediate action to be taken in an emergency. More detailed information, particularly in the event of a transport accident, may be available in the form of written instructions carried by the vehicle driver. Such instructions are required under ADR for scheduled chemicals conveyed between or through countries which are signatories to the agreement. They must also be in the language of the countries of origin, transit, and destination.

In order to assist chemical suppliers to comply with this ADR requirement and with any domestic legislation within a particular country, a scheme was developed by the European Council of Chemical Manufacturers Federations (CEFIC) which provides such instructions in the form of standard transport emergency cards (Tremcards). These cards (Fig. 3-6) have a common format in A4 size and use standard phrases agreed to internationally by the 14 member countries of CEFIC (Austria, Belgium, Denmark, Finland, Federal Republic of Germany, France, Ireland, Italy, the Netherlands, Norway, Spain, Sweden, Switzerland, and the United Kingdom).

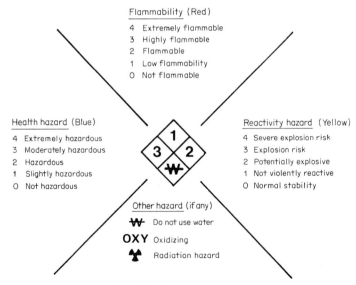

Fig. 3-5   NFPA hazard identification system.

TRANSPORT EMERGENCY CARD (Road)

Cargo

AMMONIA
Liquefied pressure gas with perceptible odour

Nature of Hazard
Heating will cause pressure rise, severe risk of bursting and explosion
The gas is suffocating
Reaction with moist air produces mist which has strongly irritant effect on eyes,
skin and air passages
Contact with liquid causes skinburns
Spilled liquid has very low temperature and, unless contained, evaporates quickly

Protective Devices
Suitable respiratory protective device
Goggles giving complete protection to eyes
Plastic or rubber gloves, boots, suit and hood giving complete protection to head, face and neck
Eyewash bottle with clean water

EMERGENCY ACTION — Notify police and fire brigade immediately

- If possible move vehicle to open ground
- Stop the engine
- Mark roads and warn other road users
- Keep public away from danger area
- Keep upwind
- Put on protective clothing

Spillage
- If vapour cloud drifts towards populated area, warn inhabitants
- Contain leaking liquid with sand or earth ; allow to evaporate
- If this is not practicable use waterspray to "knock down" vapour

Fire
- Keep containers cool by spraying with water if exposed to fire

First aid
- If the substance has got into the eyes, immediately wash out with plenty of water for at least 15 minutes
- Remove contaminated clothing immediately and wash affected skin with plenty of water
- Seek medical treatment when anyone has symptoms apparently due to inhalation or contact with skin or eyes
- Persons who have inhaled the gas must lie down and keep quite still
- Keep patient warm
- Do not apply artificial respiration if patient is breathing

Additional information provided by manufacturer or sender

TELEPHONE

Prepared by CEFIC (CONSEIL EUROPEEN DES FEDERATIONS DE L'INDUSTRIE CHIMIQUE, EUROPEAN COUNCIL OF CHEMICAL MANUFACTURERS' FEDERATIONS) Zürich, from the best knowledge available ; no responsibility is accepted that the information is sufficient or correct in all cases
Obtainable from NORPRINT LIMITED, BOSTON, LINCOLNSHIRE
Acknowledgment is made to V.N.C.I. and E.V.O. of the Netherlands for their help in the preparation of this card

Applies only during road transport　　　　　　　　English

Fig. 3-6 CEFIC transport emergency card.

At present 400 CEFIC Tremcards have been published,[13] covering hazardous substances commonly transported in bulk. These are nominally available in nine languages into which each of the standard phrases has been translated in an agreed form. To limit the number of documents carried on road vehicles, the CEFIC master texts may be printed in four languages, two on each side of an A4 card. In addition to the 400 Tremcards available, a volume of 100 group texts has been published[14] in A5 size to assist companies to prepare their own emergency cards when necessary.

The National Association of Waste Disposal Contractors in the United Kingdom has produced a set of transport emergency cards, an example of which is shown in Fig. 3-7. These cards cover 16 categories of hazardous waste as approved by the Health and Safety Executive. The cards, drawn up in collaboration with the NCEC, adopt CEFIC style with standard phrases, thus permitting simple translation into other languages if required. Such cards, apart from meeting any future regulatory needs, also provide emergency services with essential information in the event of a transport accident.

The information provided on Tremcards is clearly more detailed than that provided by a vehicle-marking scheme. In an accident situation, however, such information must be supplemented by specialist knowledge. An important feature of Tremcards is the facility for incorporating a source of such advice, usually the name and telephone number of the supplier.

SOURCES OF SPECIALIST CHEMICAL INFORMATION

Chemsafe

Prompt, detailed specialist advice, coupled with the availability of expert practical assistance, is essential if emergency situations are to be effectively handled. Experience in the United Kingdom has shown that the best source of such advice is the manufacturer of the product concerned. The CIA adopted this approach as a fundamental principle in setting up the Chemical Industry Scheme for Assistance in Freight Emergencies (Chemsafe),[15] which came into operation in January 1974.

Chemsafe encourages manufacturers and suppliers of all chemical products to include on their package labels the company's name and telephone number to enable emergency services to establish immediate contact if necessary. To cater for situations in which a manufacturer cannot be reached or identified, a national chemical hazard information and advisory center was established as the Chemsafe backup. This service is provided in collaboration with the NCEC[16] at the Harwell Laboratory of the U.K. Atomic Energy Authority in Oxfordshire. The NCEC was established in 1973 by the Home Office and the Department of the Environment. An emergency telephone line at the NCEC is manned continuously, and a team of technically qualified duty officers is always available.

One of the main reasons for the NCEC's being called for advice is the inadequacy of product labels, which frequently display only a trade name or a code mark. The technical knowledge of duty officers in such situations is in itself insufficient to respond adequately without reference to specific data. Very few data, however, are published on such trade products in the open literature. The only satisfactory source of information for recognizing the many thousands of chemical products being transported and their intrinsic hazards is the manufacturer concerned.

Such information is obtained by means of a questionnaire (Fig. 3-8) which has been compiled jointly by the CIA and Harwell. All chemical companies are requested to complete a questionnaire for each chemical, whether manufactured, marketed, or imported. The information sought is based on meeting the needs of the emergency services. The

2WE
7006

TRANSPORT EMERGENCY CARD (Road)

NAWDC TECH.
March 1979

Cargo

HAZARDOUS WASTE

Liquid containing Acid

Nature of Hazard

Corrosive
Contact with liquid or vapour may cause skin burns and severe damage to eyes and air passages
Attacks clothing
Attacks many metals with liberation of hydrogen which is flammable and may form explosive mixture with air
Reaction with combustible substances generates heat and may cause fire or explosion or toxic fumes
May contain dissolved or suspended toxic constituents

Protective Devices

Suitable respiratory protective device
Goggles giving complete protection to eyes
Plastic or synthetic rubber gloves, boots and suit with hood giving complete protection to head, face and neck
Eyewash bottle with clean water

EMERGENCY ACTION — Notify Police and Fire Brigade immediately

- Stop the engine
- No naked lights. No smoking
- Mark road and warn other road users
- Keep public away from danger area
- Keep upwind
- Put on protective clothing

Spillage

- Contain leaking liquid with sand or earth, consult an expert
- Shut off leaks if without risk
- Prevent liquid entering sewers, basements and workpits. Vapour may create toxic atmosphere
- If substance has entered a water course or sewer or contaminated soil or vegetation, advise police
- Warn everybody—evacuate if necessary

Fire

- Keep containers cool by spraying with water if exposed to fire
- Extinguish with water spray, dry chemical or foam

First aid

- If substance has got into eyes, immediately wash out for several minutes
- Remove contaminated clothing immediately and wash affected skin with plenty of water
- Seek medical treatment when anyone has symptoms apparently due to inhalation, swallowing, contact with skin or eyes, or the fumes produced in a fire
- Even if there are no symptoms, send to a doctor and show him this card

Additional information provided by manufacturer or sender

TELEPHONE

WARNING: This information is intended to assist any person involved in recognising the characteristics of this load. It is emphasized that such substances may vary according to conditions; this information, therefore, must be regarded as a general guide.

Applies only during road transport

Fig. 3-7 United Kingdom waste disposal industry transport emergency card.

CHEMICAL INDUSTRIES ASSOCIATION LIMITED
CHEMSAFE — CHEMICAL PRODUCT EMERGENCY INFORMATION

1. NAME OF COMPANY . . . .

   ADDRESS . . . .

   . . . .

2. PRODUCT NAME
   (ie NAME GIVEN
   ON LABEL/PACKAGE)

	Manuf'd.	Marketed	Imported

   ALTERNATIVE NAMES USED. . . . .

   (IF ANY) . . . .

3. CODE MARKS (if any)

   Please ✓ box
   %

4. APPROVED CHEMICAL NAME OF
   CONSTITUENTS.
   (with approx. concn. if mixture)

5. PHYSICAL FORM

SOLID	
LIQUID	
GAS	

   Please ✓ box

   COLOUR:
   OTHER FEATURES:

6. TYPE OF PACKAGING   size and description

SACK	
DRUM	
BULK	
OTHER	

7. HAZARDS (Brief description and handling precautions)

   Flash Point

8. PRODUCT/HAZARD CLASSIFICATIONS (if known)

U. N. SERIAL NO.
U. N. HAZARD CLASSIFICATION (Division & Subdivision)
KEMLER CODE
HAZCHEM CODE

Fig. 3-8 Chemsafe questionnaire.

9. RECOMMENDED EMERGENCY ACTION IN EVENT OF:

a) SPILLAGE

b) FIRE (eg. extinguishing media)

10. FIRST AID TREATMENT

11. NAME OF INDIVIDUAL/ORGANIZATION WITH SPECIALIST KNOWLEDGE

. . . .

EMERGENCY TELEPHONE NUMBER

AVAILABILITY (days and hours)

12. PRINCIPAL TRANSPORT ROUTES:

13. *LITERATURE REFERENCES (eg. Technical data sheets giving additional information)

14. NAME AND TELEPHONE NUMBER OF COMPILER (in event of any queries)

. . . .

N. B. *Wherever possible such publications should be included with this completed form.

Fig. 3-8 (*Con't.*)

completed questionnaires are used as the input format for a computerized data bank (Hazfile), which currently holds more than 10,000 documents.

Information can be retrieved on line from any suitable terminal by using public telephone lines. The data are searchable by using the Status[17] program on the basis of the manufacturer, product name, code mark, physical form, or type of packaging. It is possible also to conduct a search by using part of a product name, provided that at least the first character is known.

A typical computer printout for a fictitious material is shown in Fig. 3-9. It is possible, if desired, to retrieve only those sections of data that may be required, for example, the product name, hazards, and spillage.

Access to the data bank is available 24 h a day by duty officers of the NCEC and also by 15 county fire authorities in England, Scotland, and Wales as part of a pilot study being undertaken by the Home Office. Monitoring and updating the information held on Hazfile (an essential requirement for any operational data bank) are carried out continuously through collaboration with the chemical industry.

Hazfile does not contain medical information beyond basic first aid. Such specialist information is made available to doctors through a Poisons Information Service.[18] Government-sponsored centers for this purpose have been established in London, Cardiff, Edinburgh, Belfast, and Dublin in addition to those provided by hospitals in Leeds and Manchester. These centers hold concise essential information on the nature and toxicity of some 10,000 substances. The emphasis on providing essential but brief information appears to contrast with comparable services in continental Europe, where more detailed information is usually provided to an inquirer who may be a member of the public.

Data banks (complete data) and data bases (bibliographical data) holding environmental information are rapidly becoming widespread in Europe as elsewhere, particularly in the fields of water pollution and toxicology. On-line data retrieval from computerized systems offers immediate and unprecedented access to such information. However, with such rapid growth in their number and range, no doubt influenced by data exchanges and development of transmission processes, it is difficult to keep abreast of the vast amount of information available. Directories published by the Commission of the European Communities[19] and Aslib (formerly the Association of Special Libraries and Information Bureaux)[20] are two notable publications which index sources of such information. It is not possible in this part to discuss many of these data systems, the majority of which are not primarily intended for emergency-type information.

ECDIN

One European on-line data bank containing both environmental and emergency information is the Environmental Chemicals Data Information Network (ECDIN). ECDIN is a research project which began in 1973 and is now nearing the end of its pilot phase. It is being conducted within the EEC's direct-action program at the Joint Research Center at Ispra, Italy, in collaboration with many national research centers.

The basic principle of ECDIN is to store, on computer, information on chemical products having an environmental significance and produced in quantities greater than 500 kg per year.[21] It has been estimated that 20,000 to 30,000 chemicals will eventually be covered, including some mixtures and compounds, although it is not clear how such a comprehensive data bank will be maintained in an up-to-date state. Data are organized in 10 basic categories, subdivided into data fields including identification (various names and synonyms), physical and chemical properties, transport, packaging and handling (spill response), regulatory control, disposal practices, and toxicology.

Potential users of ECDIN, identified during a planning phase, include governmental

```
PERMANATE        #DOC     9999      [Demonstration document only]
SEC2
TRADE–NAME:
          PERMANATE

SEC1
COMPANY–NAME:
          PARTON CHEMICAL CO LTD.
          NORTHERN ROAD, PARTON, WARWICKSHIRE.

SEC3
CODE–MARKS:
          PC-9056/21

SEC5
FORM:
          SOLID: CRYSTALS, DARK PURPLE, METALLIC SHEEN.

SEC6
PACKAGING:
          SACKS: 25 KILOGRAM, PAPER.

SEC7
HAZARDS:
          POWERFUL OXIDISING MATERIAL. SPONTANEOUSLY
          FLAMMABLE IN CONTACT WITH GLYCERINE OR ETHYLENE
          GLYCOL.
          EXPLOSIVE IN CONTACT WITH SULPHURIC ACID OR
          HYDROGEN PEROXIDE.
          REACTS VIOLENTLY WITH FINELY DIVIDED EASILY
          OXIDISABLE SUBSTANCES.
          INCREASES FLAMMABILITY OF COMBUSTIBLE MATERIALS.
          A STRONG IRRITANT DUE TO OXIDISING PROPERTIES.
          USE BREATHING APPARATUS.

SEC8
CLASSES:
          U. N. SERIAL NO. 1490
          U. N. HAZ CLASS 5.1.0.

SEC9A
SPILLAGE:
          WEAR BREATHING APPARATUS WITH FULL PROTECTIVE
          CLOTHING.
          FLOOD WITH WATER.
          BEWARE HAZARD OF CONTAMINATED CLOTHING ON DRYING
          OUT.

SEC9B
FIRE:
          FLOOD WITH WATER.

SEC10
FIRST–AID:

SEC11
KNOWLEDGE:
          MR. H. WESTON.
EMERG–PHONE:
          PARTON 49255, AVAILABLE 0800 – 1730 HRS.
          PARTON 75621, AVAILABLE 1730 – 0800 HRS.

SEC12
ROUTES:

SEC13
REFERENCES:

SEC14
COMPILER:
          MR. J. JONES. PARTON 49312.

DATE:
          JULY 1979.
```

Fig. 3-9 Haxfile document: Permanate.

authorities and international organizations involved in environmental protection and control, police and public services concerned with spill countermeasures, industry, and many others. When fully operational and connected to Euronet, ECDIN will be accessible by all member states of the European Communities.

Federal Republic of Germany

A data bank developed in West Germany by the Institut für Wasserforschung, Dortmund,[22] is designed to provide essential information on chemical substances which could create potential waste pollution problems. This computerized data bank DABAWAS was produced by collaboration with representatives of the federal and state governments, the water engineering profession, and industry. It contains more than 100,000 individual pieces of data on 7000 chemical substances. Management and enlargement of the data bank are handled by the Institute for the Federal Office of the Environment Planning and Information System (UMPLIS).

The data stored include substance identification, chemical composition, hazard class, physical and chemical properties, toxicity, immediate and follow-up measures in the event of accidents, and regulatory and environmental information. The data bank can be interrogated by telephone in emergencies or, alternatively, by using a specially designed inquiry form.

IRPTC

The UN Environment Program is developing the *International Register of Potentially Toxic Chemicals (IRPTC)* in Switzerland. This computerized register is expected to be complementary to ECDIN and to include all chemical substances in use or produced as a result of human activity, with all synonyms but excluding proprietary mixtures. One of its likely attributes will deal with chemical spill control information. Data will be available from a central store and elsewhere via a network of cooperating data sources.

Netherlands

The provision of urgent specialist information on chemical products involved in spills is sometimes organized nationally through a principal source. The Fire Service Inspectorate in the Netherlands provides such a service for its fire officers via a national emergency telephone number in The Hague which is staffed 24 h a day. This number is included on a pocket-sized card, carried by fire officers, which also includes a scale for decoding ADR hazard or information panels and a description of hazard-warning diamonds.

Belgium

Chemical emergency information in Belgium is provided by the senior staff of the Institut voor Chemie-Ingenieurstechniek at the Catholic University of Leuven (Louvain) and is available throughout the 24-h period. This department trains safety staff for industry and holds specialist data to assist in dealing with chemical transport accidents. The development of a national data bank for dangerous substances is under consideration.[23]

France

Chemical hazard data are provided in France through the Centre Opérationnel de la Direction de la Sécurité Civile (CODISC), located in Paris. The center is accessible 24 h a day and covers various emergency situations occurring throughout French metropolitan territory. Detailed information on chemicals supplementary to that conveyed on vehicles is provided by operators who consult specially prepared data sheets on microfiche. If necessary, such data can be transmitted by facsimile to the proximity of an accident by the public telephone network. The data sheets currently cover about 1000 chemicals, and ultimately 13,000 products are expected to be covered.

ERS

A scheme to provide an international emergency response system (ERS) for chemicals in transit has been proposed by the Redwood Petroleum and Petrochemical Division of the Société Générale de Surveillance S.A. (SGS) in Geneva.[24] Initially it proposes to use an existing network of SGS Redwood locations in many European and other countries. Operating on a commercial basis, the scheme would enable appropriate emergency response, by trained staff, to be given on behalf of participating chemical suppliers.

It is anticipated that ERS would provide three basic functions—rapid communications, information gathering, and practical assistance—if required. The information function would embrace chemical product data, appropriate information relating to a variety of national government authorities concerned with chemical transport accidents, and specific requirements of chemical suppliers.

There would appear to be operational similarities between the ERS proposals and the Chemical Emergency Agency Service Scheme operated by the NCEC in the United Kingdom. The NCEC, however, provides an emergency response facility on a commercial basis on behalf of chemical suppliers who cannot provide such a response themselves. The majority of companies participating are therefore chemical traders and importers, as distinct from manufacturers. Emergency response in general to chemical transport accidents occurring in the United Kingdom is provided directly by industry through participation in Chemsafe.

CONCLUSIONS

The availability of information appropriate to chemical spills in Europe is wide but varied. National differences in spill response organization call for schemes designed to meet particular needs. This is notably true of vehicle-marking schemes, but international agreement has done much to improve unification and to overcome language difficulties.

Computerization of data to assist in the identification of chemicals and their hazards in spill situations is increasing. It is inevitable, therefore, that duplication will occur. Perhaps the ideal situation would be for each country to establish a data bank of emergency information for the chemicals it manufactures. In this way, international data exchange schemes would enable appropriate information to be readily obtained on practically all chemical products.

The European information systems described in this part are in no way intended to be exhaustive. Many individual fire brigades maintain brief data files, usually in card index form, covering common chemical products. Other systems directly related to spills no doubt exist but are not so widely published. In addition to European-based systems, data banks and other information sources established elsewhere, notably in the United States and Canada, are often used. Continued international cooperation in this field must enhance the availability of essential information and thus reduce the environmental impact created by hazardous chemical spills.

REFERENCES

1. Committee of Experts on the Transport of Dangerous Goods, *Transport of Dangerous Goods: Recommendations Prepared by the Committee of Experts on the Transport of Dangerous Goods,* rev. ed., ST/SG/AC.10/1/Rev.1, United Nations, New York, 1977.

2. Department of Transport, *European Agreement Concerning the International Carriage of Dangerous Goods by Road (ADR),* H. M. Stationery Office, London, 1978.

3. Department of Transport, *International Regulations Concerning the Carriage of Dangerous Goods by Rail (RID),* H. M. Stationery Office, London, 1978.

4. *International Maritime Dangerous Goods Code,* Intergovernmental Maritime Consultative Organization, London, 1977.

5. *Restricted Articles Regulations,* International Air Transport Association, 22d ed., Geneva, 1979.

6. Health and Safety Executive, *Hazardous Substances (Labelling of Road Tankers) Regulations, 1978,* Statutory Instrument No. 1702, H. M. Stationery Office, London, 1978.

7. *Labelling Chemicals,* Chemical Industries Association, London, 1977.

8. *Hazchem Codings, Allocated by the Joint Committee on Fire Brigade Operations and Confirmed by the Health and Safety Executive,* Chemical Industries Association, London, 1976.

9. *Proposals for Dangerous Substances (Conveyance by Road) Consultative Document,* Health and Safety Commission, London, 1979.

10. *Voluntary Scheme for the Marking of Tanker Vehicles Conveying Substances of Low Hazard,* Chemical Industries Association, London, 1979.

11. *Identification System, Fire Hazards of Materials,* NFPA No. 704, National Fire Protection Association, Boston, 1975.

12. Günter Hommel, *Handbuch der gesährlichen Güter,* Springer-Verlag, Berlin, 1978.

13. *Transport Emergency Cards,* 4 vols., Chemical Industries Association, London, n.d.

14. *CEFIC Group Texts,* Chemical Industries Association, London, 1980.

15. *Chemsafe: A Manual of the Chemical Industry Scheme for Assistance in Freight Emergencies,* 3d ed., Chemical Industries Association, London, 1979.

16. R. F. Cumberland, "Role of the National Chemical Emergency Centre at Harwell, U.K.," in G. F. Bennett (ed.), *Proceedings of the 1978 National Conference on Control of Hazardous Material Spills,* Miami Beach, Fla., April 1978, pp. 60–64.

17. N. H. Price, C. Bye, and B. Niblett, "On Line Searching of European Conventions and Agreements," *Information Storage and Retrieval,* vol. 10, 1974, pp. 145–154.

18. H. Matthew and A. A. H. Lawson, *Treatment of Common Acute Poisonings,* Churchill, Livingstone, Edinburgh, 1975.

19. Environmental Information and Documentation Centres of the European Communities, *Endoc Directory,* pilot ed., Hitchin, England, 1978.

20. J. L. Hall, *On-Line Bibliographic Databases Directory,* Aslib, London, 1979.

21. *ECDIN: A European Communities Data Bank for Environmental Chemicals,* communication of the Commission of the European Communities Joint Research Center, Ispra, Italy, n.d.

22. *DABAWAS Datenbank Wassergefährdende Stoffe,* Institut für Wasserforschung GmbH, Dortmund, communication, n.d.

23. Katholieke Universiteit Leuven, Institut voor Chemie-Ingenieurstechniek, private communication, 1979.

24. *E.R.S. Emergency Response System for Chemicals in Transit,* Société Générale de Surveillance S.A., Redwood Petroleum and Petrochemical Division, Geneva, communications.

PART 4
Canadian Systems

Robert A. Beach
Manager, National Environmental Emergency Centre
Environmental Protection Service
Environment Canada

Response Systems / 3-45
Reporting / 3-48

Three Canadian information storage and retrieval systems are used in dealing with spills of hazardous materials. Two are used in the first stages of the response to a spill and are described below under "Response Systems." The third system is used to record details on hazardous materials spills, largely for analytical purposes, and is described below under "Reporting."

RESPONSE SYSTEMS

When a hazardous material is spilled, the on-scene commander (OSC), emergency response teams, and scientific or technical advisory groups need to know:

1. The chemical and physical properties of the material and its biological effects
2. The nearest or most readily available source of equipment, material, and expertise to combat the spill

In Canada, two multipurpose information systems are used to satisfy these requirements: the Hazardous Materials System (HAZMATS) and the National Emergency Equipment Locator System (NEELS).

Hazardous Materials System (HAZMATS)

HAZMATS was originally designed by the Department of the Environment (DOE) to support implementation of the Canada–United States Great Lakes Water Quality Agreement, then expanded to assist in the formulation of a code for the transportation of dangerous goods. It is now being revised to serve the Canadian Contaminants Program in a broader sense under its enabling legislation, the Environmental Contaminants Act. The system, managed by the Contaminants Control Branch of the DOE, currently contains chemical, physical, and biological effects data on about 3700 hazardous chemicals, 1900 of which are cross-referenced with the United Nations list of dangerous goods.[1] It is envisaged that, with minor modifications, HAZMATS could complement and support

TABLE 4-1   HAZMATS Data Base Description*

1. United Nations *Transport of Dangerous Goods* reference number
2. United Nations *Transport of Dangerous Goods* classification number
3. United Nations *Transport of Dangerous Goods* packaging group
4. United Nations packaging number (reference number)
5. United Nations hazard information code
6. Disposal reference
7. Registered chemical name
8. Synonyms (trade names)
9. Physical phase (at room temperature)
10. Solubility in water (text if exact figures not available)
11. Solubility [decimal numbers, in parts per hundred (i.e., percent), in water]
12. Specific gravity
13. Boiling point
14. Melting point
15. Vapor pressure (mm, mercury)
16. Decomposition point
17. Flash point
18. Sublimation point
19. Refractive index (sodium, 20°C)
20. Cup (in reference to the flash point)
21. Persistence (in water, in weeks)
22. Persistence reference number
23. Chemical class (main use)
24. Chemical origin (United Nations list, Canadian list, United States)
25. Aquatic toxicity dose (ppm)
26. Aquatic toxicity type (K = kill, TIA = LC_{50}, after 1 day, ...)
27. Aquatic toxicity test organism (fish, invertebrate, ...)
28. Aquatic toxicity test organism name abbreviation
29. Aquatic toxicity reference number
30. Is aquatic test organism indigenous to the Great Lakes?
31. Oral toxicity dose (mg/kg)
32. Oral toxicity type (LD_{50}, LD_{Lo} ...)
33. Oral toxicity animal (test organism)
34. Dermal toxicity dose (mg/kg)
35. Dermal toxicity type (LD_{50}, ...)
36. Dermal toxicity animal (test organism)
37. Inhalation toxicity dose (vapor = ppm, dust = mg/kg)
38. Inhalation toxicity type (LC_{50}, ...)
39. Inhalation toxicity animal (test organism)
40. Ecological effect rating
41. Booz-Allen rating
42. Intergovernmental Maritime Consultative Organization pollution category for operational discharge
43. U.S. Coast Guard reference number (CHRIS)
44. OHM-TADS reference number
45. Flammability rating
46. Reactivity rating
47. Health hazard rating
48. Environmental hazard rating
49. Phytotoxicity rating
50. Genetic aberration rating
51. Disposal label (reference)
52. Inhalation toxicity phase (dust or vapor)
53. Chemical Abstracts Service (CAS) number
54. Bioaccumulation

TABLE 4-1   *(Con't.)*

55. Environmental damage
56. United Nations secondary risks No. 1
57. United Nations secondary risks No. 2
58. United Nations special provisions No. 1
59. United Nations special provisions No. 2
60. United Nations secondary risks No. 3

*A comprehensive HAZMATS reference manual describes each field in detail.

the *International Register of Potentially Toxic Chemicals* (*IRPTC*) data base proposed by the UN Environment Program (UNEP).

The existing data base contains 60 fields of information (see Table 4-1), some or all of which are useful to a number of other governmental regulatory departments or agencies, and particularly to spill response organizations. HAZMATS is an interactive system, stored on and accessible through the Department of Energy, Mines, and Resources computer network, using the System 2000 natural language.

National Emergency Equipment Locator System (NEELS)

NEELS was designed to assist government and industry response organizations by standardizing their spill countermeasures inventories and by providing a rapid means for locating equipment in an emergency, using portable terminals that may be brought to the spill scene. Participants in NEELS list their equipment and contact identification information in the data base under specified fields or main information headings (see Table 4-2), and the inventory owner is obliged by agreement to keep that information up to date.

All geographic locations in the data base are identified by latitude and longitude, allowing the system to provide the great-circle distance, in nautical miles, between a search datum and each selected inventory location. In the search mode, the user may ask

TABLE 4-2   NEELS Main Information Headings (Fields)*

1. Province or state
2. Place name
3. Organization
4. Latitude and longitude
5. Facility
6. Phone numbers
7. Booms and fencing
8. Watercraft and aircraft
9. Skimmers, pumps, and fittings
10. Hoses, connections, and portable tanks
11. Vacuum and pumper trucks
12. Special vehicles
13. Communications equipment
14. Sorbents, chemical treating agents, and application equipment
15. Safety equipment and special clothing
16. Generators and lights
17. Earth movers and heavy equipment
18. Other equipment and materials and local resources
19. Disposal facilities
20. Comments and control points

*A NEELS briefing pack provides users with guidance on equipment listings within fields.

for "all" of the equipment, or a specified "type" of equipment (main-heading equipment), or a specified "piece" of (trade name) equipment at the location or locations nearest to the search datum.

The system is interactive (prompts the user with progressive questions) and, in a spill situation, first asks for the latitude and longitude of the spill location. The OSC may provide this location or, since the OSC should be quite familiar with the resource situation in his or her own area of responsibility, may select the location of a major logistics support center that can give support more completely or more rapidly, or both, than locations within the OSC's own area. This latter option is particularly appropriate for spills in the remote northern regions of Canada, but it may also be used by the OSC after resources in the OSC's area have been depleted.

In either case, NEELS searches radially from the given locus and prints out all, type, or piece information on the nearest location. If there is not enough of the desired equipment at that location, the system may be instructed to search for the next nearest location, then the next nearest, and so on, until the OSC's needs are satisfied. At this point, the logistics support group takes over and, through contact information provided by NEELS, makes arrangements for delivery of the equipment to the spill location or assembly point.

Outside the search mode, NEELS may also be interrogated by industry and government planners to list company or agency inventories across Canada or in specified areas and to assist in the preparation of contingency plans or in the assessment of existing plans. Predesignated OSCs and leaders of response teams carry bound photocopies of NEELS listings for field checks in their areas of responsibility and for training exercises.

The system specifications were prepared by the DOE in close cooperation with the Department of Transport and the Petroleum Association for the Conservation of the Canadian Environment (PACE). An international computer services company, I. P. Sharp Associates Ltd., wrote the program, using the APL computer language under the modified Sharp APL system. The data base is stored on one of Sharp's computers, free of charge as a public service, and in an emergency is accessible 24 h per day through Sharp's toll-free telephone lines at several key locations in Canada, the United States, and overseas or through the Trans-Canada Telephone System.

NEELS can be accessed by portable or nonportable terminals through a telephone. Hence portable terminals can be brought as close to the spill scene as a telephone is available.

NEELS has been in operation since 1974 and now contains about 600 inventories at locations across Canada and in some states in the United States. (The Department of Natural Resources of Michigan participates in the program.) The interactive portions of the system are accessible in either English or French, Canada's two official languages, but inventories are entered in the data base in the language chosen by the inventory owner.

Operational experience has shown that the NEELS field structure must be made more precise and that special emphasis must be given to chemical and radioactive spill countermeasures equipment. This information has been "buried" in the existing structure, but a planned revision will highlight these subjects under separate fields or main headings.

REPORTING

As stated by John MacLatchy in Chap. 2, Part 2, "the combination of [Canadian] federal and provincial laws now requires that virtually every situation involving a spill of hazardous materials be reported under one or another of the mandatory reporting requirements." Practical working arrangements among federal and provincial regulatory agencies serve to ensure that most of the significant spills of hazardous materials are reported by the fastest possible means. Passage of the Transportation of Dangerous Goods Act in 1980 was designed to close any existing loopholes in mandatory reporting requirements.

The Department of the Environment, through its Environmental Emergencies Program, has been directed by the federal Cabinet to "implement, making maximum possible use of existing facilities of all agencies, a national system for reporting environmental emergencies and alerting appropriate authorities of the situation." In 1973, a national reporting system was put into operation, in cooperation with several federal and provincial regulatory agencies, for the real-time reporting of environmental emergencies. As a logical extension of the reporting system, the Environmental Emergency Branch (EEB) developed a national inventory of accidents involving spills of hazardous materials. The program is known as the National Analysis of Trends in Emergencies System (NATES).

NATES was designed to assist federal and provincial regulatory agencies in analyzing accident trends and their distribution in order to assess contingency plans, equipment development and requirements, equipment deployment, accident prevention programs, and the effectiveness of laws and regulations. Federal and provincial regulatory agencies contribute their spill information to the DOE in the format used by the reporting agency. This information is then transferred to the NATES coding form (see Figs. 4-1 and 4-2) and forwarded to the EEB in DOE headquarters. Quality control checks are made in the EEB before and after keypunch operations, and then the spill data are batch-entered into the data base. For quality control purposes, only the EEB may input or update data for the system, but any participating government agency may search the data base.

A NATES user's manual provides guidance on the preparation of the coding forms and on procedures to be followed to perform various kinds of searches, depending upon the type of analysis to be conducted. The EEB publishes a comprehensive summary of hazardous materials spills annually in its *Spill Technology Newsletter.*

The system specifications were prepared by the EEB in consultation with potential federal and provincial user agencies. The programming language is APL under the Sharp APL system, and the data base is stored, on line, on one of the I. P. Sharp Associates Ltd. computers in Toronto. The system is interactive and is accessible in both English and French through Sharp's toll-free lines or the Trans-Canada Telephone System.

NATES has been operational since 1974, but all available accident records as far back as 1968 have been entered in the data base. The initial field structure is shown in Figs. 4-1 and 4-2. The current field structure, shown in Figs. 4-3 and 4-4, is the product of two revisions that were made to provide more and better data for the analysts. The revisions improved the "cause" and "reason" fields and reduced the previously high percentage of accident causes and reasons that had been relegated to the "other" category. The new field structure is particularly useful in Canadian accident preventing programs. The data base contains over 14,000 accident records, and, when possible, the older records have been amended to incorporate the more detailed information provided for by the latest NATES coding form.

Data Base Changes

The Canadian information systems are amenable to and, indeed, subject to changes. Any inquiries concerning HAZMATS, NEELS, or NATES should be addressed to the

Director General
Environmental Impact Control Directorate
Environmental Protection Service
Department of the Environment
Ottawa, Ontario, CANADA K1A 1C8

NATES CODING FORM

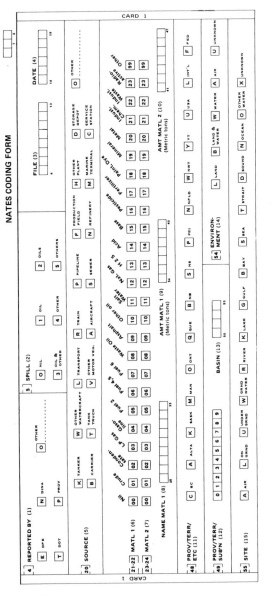

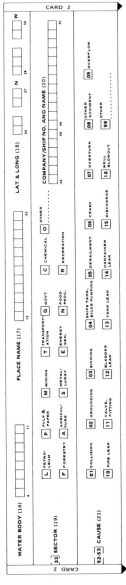

Fig. 4-1 NATES coding form, side 1.

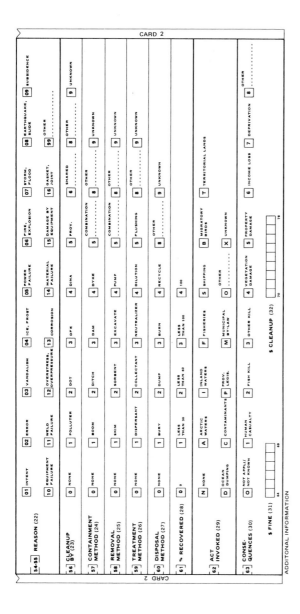

Fig. 4-2 NATES coding form, side 2.

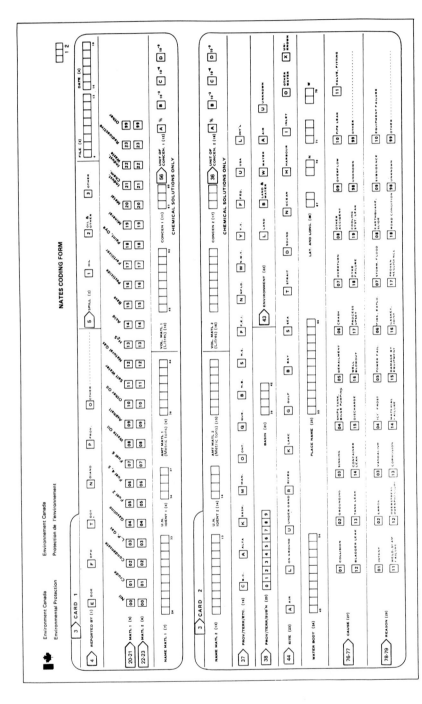

Fig. 4-3 Revised NATES coding form, side 1.

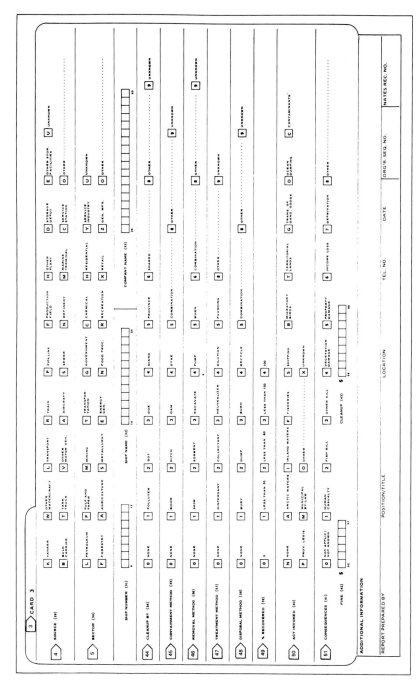

Fig. 4-4 Revised NATES coding form, side 2.

REFERENCE

1. *Transport of Dangerous Goods,* recommendations prepared by Committee of Experts on the Transport of Dangerous Goods, ST/SG/AC.10/1, United Nations, New York, 1976.

CHAPTER

•CHAPTER•

CHAPTER

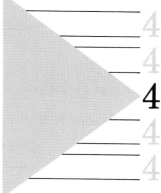

4

Impact

PART 1
Municipal Facilities / 4-2

PART 2
Environmental Effects / 4-15

PART 3
Economic Effects / 4-22

PART 1
Municipal Facilities

C. Joseph Touhill, Ph.D.
President, Baker/TSA, Inc.

Discharges to Municipal Facilities / 4-2
Detection / 4-3
Inventory / 4-5
Effects on Biological Treatment Processes / 4-7
Remedies / 4-8

DISCHARGES TO MUNICIPAL FACILITIES

Planned Disposal

Prior to passage of the Federal Water Pollution Control Act (FWPCA) amendments of 1972, it was common practice for many industries to discharge certain materials to municipal wastewater treatment plants even though these materials were difficult to treat. Benefits expected were (1) dilution within a much larger wastewater stream and (2) possible longer-term acclimation of biological treatment systems so that eventually some removal would occur. In fact, the strategy proved quite successful in numerous instances. For example, phenols and cyanides were degraded readily in biological systems that had treated municipal and steel plant wastewaters jointly for many years.

Thus, planned disposal of potentially hazardous materials to municipal facilities has proven appeal; however, pretreatment requirements emanating from the FWPCA amendments of 1972 have terminated this practice. U.S. Environment Protection Agency (U.S.EPA) pretreatment regulations restrict direct industrial discharges into municipal systems to compatible pollutants, which are defined as those that can be treated routinely without diminishing treatment efficiency. In general, only wastewaters containing biodegradable organics and suspended solids normally are regarded as compatible. Exceptions could include instances in which there were special local circumstances. For example, several physical-chemical treatment plants employ no biological treatment but rather use activated carbon for the removal of organics. Materials which potentially could be toxic in biological systems can be removed effectively by the carbon. Also, many treatment plants have thoroughly acclimated biological systems because of significant continuing industrial flows.

The cases cited above are unusual. Hence, pretreatment guidelines are expected to be applied universally. Even for biodegradable materials, equalization will be required to minimize the adverse potential of shock loads. Nobody should expect to use municipal wastewater treatment systems as a vehicle for the ultimate disposal of hazardous materials. The fact remains, however, that hazardous materials will continue to enter municipal systems through illegal discharges or accidental spills. Because this represents a fait accompli ultimate disposal, the remainder of Part 1 will focus on this mode of entry into municipal wastewater treatment systems.

Emphasis in the sections that follow will be placed upon the practicality of remedial measures. Many methods are available to mitigate the effects of hazardous material spills into municipal facilities, but in actual practice employment of most of them is infeasible. In general, time and simplicity dictate spill response methods.

Spills

There are numerous ways in which spills can occur. However, according to data from several federal studies, most hazardous wastes currently enter municipal wastewater collection and treatment systems in industrial-process-water streams. Federal pretreatment guidelines should help to reduce such entry, but full implementation for all industries will take many years.

Another common way in which hazardous materials enter sewage systems is from the flushing of spilled materials by an emergency response agency such as a fire department. Since explosions and upsets of wastewater treatment plants have often resulted from this practice, alternative practices must be employed to deal with spillage of dangerous or hazardous materials. Educational courses offered by wastewater treatment plant operating personnel have been helpful in informing fire fighters and police officers about the dangers inherent in flushing hazardous materials to sewers.

Adverse effects from the entry of hazardous substances into a municipal system are that (1) flammable or toxic materials pose an explosive as well as a health hazard within the collection and treatment system; (2) the materials can be corrosive or reactive and thus can damage the physical system; (3) biological processes can be upset by the discharged hazardous materials, causing diminished or complete loss of removal efficiency; (4) the hazardous materials can pass through the plant without adverse effect on the processes but can cause violation of National Pollutant Discharge Elimination System (NPDES) permit provisions and water quality standards; and (5) hazardous constituents can accumulate in the biomass and suspended solids, thus creating difficulties in disposing of the resultant contaminated sludge.

Because spillage is not planned and usually there is little forewarning of problems, municipal treatment facilities are at a great disadvantage. To deal effectively with hazardous material spills, there must be some way of detecting them.

DETECTION

Monitoring and Surveillance

Ideally, municipal wastewater collection systems should have continuous-monitoring and surveillance equipment to provide an early warning of hazardous material discharges. In actual fact, few do. There are several reasons why this is so. First, reliable continuous-monitoring equipment is available for only a few parameters. Second, detection equipment for hazardous organic compounds requires great sophistication and is very costly. Third, system maintenance costs are high. Fourth, highly trained personnel are required to operate and maintain the system.

Those pollutant parameters for which continuous-monitoring or control equipment is available are listed in Table 1-1. Even among the parameters listed there, only a few have had reliable continuous monitoring demonstrated over long periods of time; examples are conductivity, pH, and temperature.

TABLE 1-1 Pollutant Parameters for Which Continuous-Monitoring or Control Equipment Is Available

Acidity	Hydrogen sulfide
Alkalinity	Iron (ferrous and ferric)
Amines	Manganese
Ammonia	Nickel
Bromine	Nitrogen (ammonia, nitrate, and nitrite)
Chloride (photometric method)	Oxidation-reduction potential
Chlorine (free, residual, and total)	Ozone
Chromate	pH
Chromium (hexavalent, trivalent, and total)	Phosphates (orthophosphates, polyphosphates,
Color	and total)
Conductivity	Silica
Copper	Silver
Cyanide	Sulfite
Dissolved oxygen	Tannin and lignin
Fluoride	Temperature
Hardness	Turbidity
Hydrazine	

The Ohio River Valley Water Sanitation Commission (ORSANCO) has implemented an early-warning organics detection system for hazardous materials.[1] Eleven stations along the Ohio River use gas chromatography to detect organics which have boiling points of 150°C or less and are relatively insoluble in water. Large amounts of chloroform and carbon tetrachloride have been detected in an Ohio River tributary and the problem sources eliminated as a result of the early-warning system.

However, even large municipalities would have great trouble in reproducing ORSANCO's performance for the following reasons: (1) the river system permits more time for analysis and response, (2) the ORSANCO monitoring system is more compatible with batch or composite sampling, and (3) ORSANCO has a better funding and personnel base from which to draw.

The Allegheny County Sanitary Authority (ALCOSAN) system in Pittsburgh participated in a major program to deal with hazardous material spills into municipal systems.[2] Detection was an important aspect of the program. Five sensors were evaluated for the design of a monitoring and surveillance system for hazardous materials: pH, dissolved oxygen (DO), temperature, conductivity, and oxidation-reduction potential. Initial plans called for monitoring stations at the plant head end and for five field monitoring stations located at strategic points in the interceptor system.[3] If an alarm was triggered at the plant head end, approximately 3 h were available for countermeasures before the biological segment of the plant was reached. To date only the pH, DO, and temperature probes are operational at ALCOSAN.

Lacking instrumentation, most plant operators currently rely upon operational parameters (such as color and smell) and judgment to determine when trouble occurs.

Operational Parameters

For many potentially toxic spills an operator's nose turns out to be the most reliable instrument. In addition to smell, color changes, increased effluent turbidity, lower sludge-dewatering ability, sludge bulking, poor settleability, and foaming can indicate that something is wrong. Subjective measures of performance must subsequently be confirmed by laboratory analyses and complete data evaluation to discover spill arrival time, spill magnitude, material type, and impact upon the treatment plant and receiving waters. Regardless of the sophistication of detection systems and analytical procedures, however, the judgment of plant operators remains the principal line of defense in preventing, discovering, and treating hazardous material spills.

Investigative Procedures

Tenacious investigation of spills is an effective deterrent to further spillage. If a spill is serious enough to be detected, then it is serious enough to pursue. Complete documentation of the spill and its consequences should be kept on file for possible legal action.

If the source of the spill is discovered early, the spiller can be held accountable for cleanup actions and costs. Presumably, laboratory procedures will enable identification of the spilled material. If an inventory of hazardous materials for the affected region exists, access to the inventory could narrow the search for and investigation of the spilling entity.

INVENTORY

Uses

Well-conceived hazardous material inventories can serve several purposes. Two of the more important purposes are to identify dischargers and to estimate risks to municipal wastewater treatment facilities. In the former case, it is possible to isolate spill investigations to a few entities that either store or manufacture certain hazardous materials. Most inventories, however, focus on solid wastes.[4,5,6]

One of the most extensive inventories of discharged and stored hazardous materials was made by ALCOSAN.[2] Over 5000 questionnaires were mailed to industrial and commercial firms to determine the magnitude of hazardous materials, both those discharged routinely to the ALCOSAN system and those stored by industry in the service area. More than 1000 replies were received, the respondents including most of the larger industries in the area. Stored materials were divided into 23 categories. These categories and some of the primary materials found under each of them are shown in Table 1-2.

Information gathered by the ALCOSAN survey was computerized for easy organization, storage, and retrieval. Data sorts were made according to zip code, standard industrial classification (SIC) category, interceptor diversion structure number, alphabetical listing by name, number of employees, number of discharge points, production rate, or any combinations thereof.[3] Inventory data, such as that collected by ALCOSAN, can be very helpful in preparing risk estimations for the municipal wastewater system.

Risk Estimations

To formulate countermeasures for hazardous material spills into wastewater treatment systems, it is helpful to know the magnitude of danger to the facility. Two hypothetical events will be used to indicate degrees of risk. In the first case, assume that two 0.21-m^3 (55-gal) drums of the pesticide 2, 4-dichlorophenoacetic acid (2, 4-D) are spilled accidentally and over a period of about 15 min are flushed to the sewer. Further assume that the treatment plant flow is 1.1 m^3/s. The resulting 2,4-D concentration will be about 210 mg/L if the pesticide is assumed to arrive as a plug of 15-min duration. Pajak et al.[7] report that this concentration could be detrimental to the performance of biological treatment plants.

In the second example, assume that a large tank of acetone fails at night and that 40 m^3 (10,568 gal) escapes in a 4-h period before the leak is stopped. Further assume that

TABLE 1-2   Stored Hazardous Materials Reported through ALCOSAN Survey

Category	Primary materials stored
Elements	Carbon Sulfur
Minerals	Silicon dioxide Magnesium oxide
Salts of low to medium molecular weight	Aluminum sulfate Sodium chloride
Salts of low to medium toxicity	Aluminum hydroxide Sodium nitrate Sodium hexametaphosphate
Salts containing heavy metals	Iron sulfate Nickel sulfate Potassium dichromate
Acids	Hydrochloric acid Sulfuric acid
Short-chain organic acids	Fumaric acid Citric acid
Long-chain and cyclic organic acids	Phthalic acid
Caustics	Sodium carbonate Sodium hydroxide
Oxides	Antimony oxide Lead oxide
Insecticides, fungicides, etc.	Chlorinated and organic phosphorus Insecticides PCP (pentachlorophenol)
Phenols and cresols	Phenol and mixtures
Poisons (metal)	Cyanide products
Poisons (halogenated)	Tetrachloroethylene Methylchloroform
Poisons (organic)	Allyl alcohol
Radioactive material	Cesium
Heavy-metal organics	Miscellaneous compounds
Flammable hydrocarbons	Bunker C fuel oil Diesel oil Gasoline
Nonflammable hydrocarbons	Polyethylene
Flammable hydrocarbon derivatives	Butanol Acetone
Nonflammable hydrocarbon derivatives	Dialkylphthalates Trichloroethylene
Compressed gases	Freon Acetylene Propane
Miscellaneous and special materials	Hydrogen peroxide Miscellaneous

SOURCE: G. A. Brinsko, F. J. Erny, E. J. Martin, A. P. Pajak, and D. M. Jordan, *Hazardous Material Spills and Responses for Municipalities*, EPA-600/2-80-108, U.S. Environmental Protection Agency, July 1980.

all the acetone spills into a combined sewer. Approximately one-half of the acetone vaporizes prior to arrival at the 1.1-m^3/s treatment plant. Therefore, the acetone concentration entering the plant will be 1270 mg/L (plug flow is assumed). Verschueren[8] reports that treatment plant biological nitrification would be decreased substantially at this concentration.

Risk calculations can be useful to plant operating personnel in structuring countermeasures. By using historical information on spill frequency and severity as well as a current inventory of hazardous materials, plant personnel can estimate worst-case examples. Then, these "critical" concentrations of likely materials can help to define and bound solutions. In many instances, remedies will be based upon the degree of effect upon biological treatment processes.

EFFECTS ON BIOLOGICAL TREATMENT PROCESSES

Several references catalog the effects of hazardous material spills on biological treatment processes. Interestingly, most of these were prepared at about the same time. Pajak et al.[7] produced their report in conjunction with the ALCOSAN program cited earlier. The reference was used in preparing for pilot-scale evaluations of biological treatment effects.[2] Pilot-plant studies investigated the effects of spills of the following materials: cadmium chloride, sulfuric acid, sodium hydroxide, methanol, phenol, ammonium chloride, copper sulfate, sulfuric pickle liquor, No. 2 fuel oil, perchloroethylene, and sludge incinerator scrubber water. Influent concentrations ranged from 100 to 1600 mg/L, depending upon the material.

Pilot-plant studies revealed that the activated-sludge process is resilient.[2] Activated sludge appeared to be less sensitive to massive doses of hazardous materials than previous studies had reported. While some loss in efficiency was noted in the first few hours after spill arrival, recovery was fast, the process generally being back to normal in less than 24 h.

The Pajak reference[7] reports on the work of others and does not necessarily indicate inhibitory thresholds. In contrast, the EPA pretreatment guidelines report[9] lends itself well to being a ready source of inhibitory threshold numbers. Even then, the authors caution that the numbers cited may be generalizations and that major decisions require additional inquiry.

Verschueren[8] also has produced a useful handbook which includes biological effects on treatment plants for organic compounds. This handbook is more useful than others in certain respects; i.e., it gives information on physical and chemical properties, air pollution factors, and biological effects in addition to water pollution factors.

Envirex Inc.[10] produced for the EPA a "control manual," one objective of which was to determine the best treatment methods for dealing with hazardous material spills. More than 300 materials were covered in the manual. One aspect discussed for each material was its amenability to biological treatment at municipal sewage treatment plants. As a result, the Envirex work can be compared directly with some of the others.

Data on the effects of certain selected materials on biological treatment processes are given in Table 1-3. This table is designed to illustrate the kinds of information available from the references cited in this section rather than to be exhaustive for any one compound.

The Pajak reference[7] has an extensive summary, and the information on biological effects by Envirex[10] is found within a host of other data. On the other hand, the EPA pretreatment guidelines report[9] has a useful summary table. It is reproduced here as Tables 1-4 and 1-5.

TABLE 1-3   Effects of Selected Materials on Biological Treatment Processes

Material	Effect	Reference no.*
Acetaldehyde	70 to 90 percent removal in activated-sludge process	7
	27.6 percent of theoretical oxygen demand exerted in 24 h	8
	Degradable when diluted	10
Allyl alcohol	100 mg/L inhibitory in anaerobic digestion	9
	75 percent inhibition of nitrification process in activated sludge at 19.5 mg/L	8
Benzidine	Inhibitory to activated sludge at 500 mg/L	7, 9
Cadmium	10 to 100 mg/L inhibitory to activated sludge; 0.02 mg/L inhibitory to anaerobic digestion	9
Chloroform	10 to 16 mg/L inhibitory to anaerobic digestion	9
Cyanide	0.1 to 5 mg/L inhibitory to activated sludge; 4 mg/L inhibitory to anaerobic digestion	9
	2 to 3 mg/L inhibitory to activated sludge	7
	Degradable if diluted and previously acclimated	10
Ethylenediaminetetraacetic acid (EDTA)	25 mg/L inhibitory to activated sludge	9
Phenol	200 mg/L inhibitory to activated sludge	9
	500 mg/L inhibitory to activated sludge	7
	500 to 1000 mg/L inhibitory to activated sludge	8
Zinc	0.08 to 10 mg/L inhibitory to activated sludge; 5 to 20 mg/L inhibitory to anaerobic digestion	9
	2.5 to 15 mg/L inhibitory to activated sludge	7

*Numbers are those of references at the end of Part 1.

REMEDIES

Preventive Measures

Several steps can be taken to help prevent the entry of hazardous materials into municipal collection and treatment systems: (1) flushing spillage to the system can be prohibited, (2) storage of certain designated materials in the sewer service area can be prohibited, and (3) mandatory containment techniques for stored materials can be established. But

TABLE 1-4 Threshold Concentrations of Organic Pollutants That Are Inhibitory to Biological Treatment Processes

Pollutant	Concentration (mg/L)		
	Activated-sludge processes	Anaerobic digestion processes	Nitrification processes
Alcohols			
Allyl		100	19.5
Crotonyl		500	
Heptyl		500	
Hexyl		1000	
Octyl		200	
Propargyl		500	
Phenols			
Phenol	200		410
Creosol			4–16
2, 4-Dinitrophenol			150
Chlorinated hydrocarbons			
Chloroform		10–16	
Carbon tetrachloride		10–20	
Methylene chloride		100–500	
1, 2-Dichloroethane		1	
Dichlorophen*		1	
Hexachlorocyclohexane		48	
Pentachlorophenol*		0.4	
Tetrachloroethylene		20	
1, 1, 1-Trichloroethane		1	
Trichloroethylene		20	
Trichlorofluoromethane*		0.7	
Trichlorotrifluroethane (Freon)		5	
Allyl chloride			180
Dichlorophen			50
Organic nitrogen compounds			
Acrylonitrile		5	
Thiourea			0.075
Thioacetamide			0.14
Aniline			0.65
Trinitrotoluene (TNT)	20–25		
Ethylenediaminetetraacetic acid (EDTA)	25		300
Pyridine			100
Surfactants			
Nacconol	200		
Ceepryn	100		
Miscellaneous organic compounds			
Benzidine	500	5	
Thiosemicarbazide			0.18
Methyl isothiocyanate			0.8
Allyl isothiocyanate			1.9
Dithiooxamide			1.1
Potassium thiocyanate			300
Sodium methyl dithiocarbamate			0.9
Sodium dimethyl dithiocarbamate			13.6
Dimethyl ammonium dimethyl dithiocarbamate			19.3
Sodium cyclopentamethylene dithiocarbamate			23
Piperidinium cyclopentamethylene dithiocarbamate			57
Methylthiuronium sulfate			6.5

TABLE 1-4   Threshold Concentrations of Organic Pollutants That Are Inhibitory to Biological Treatment Processes (*Continued*)

Pollutant	Concentration (mg/L)		
	Activated-sludge processes	Anaerobic digestion processes	Nitrification processes
Benzylthiuronium chloride			49
Tetramethylthiuram monosulfide			50
Tetramethylthiuram disulfide			30
Diallyl ether			100
Dimethylparanitrosoaniline			7.7
Guanidine carbonate			19
Skatole			16.5
			7.0
Strychnine hydrochloride			175
2-Chloro-6-trichloromethylpyridine			100
Ethyl urethane			250
Hydrazine			58
Methylene blue			100
Carbon disulfide			35
Acetone			840
8-Hydroxyquinoline			73
Streptomycin			400

NOTE: Concentrations shown represent the influent to the unit process. When marked with an asterisk, the concentration represents the total plant influent.

SOURCE: *Federal Guidelines, State and Local Pretreatment Programs*, vols. I–III, EPA-430/9-76-017, U.S. Environmental Protection Agency, January 1977.

because spillage is inevitable, the most effective preventive measures are those which permit advanced warning and provide established procedures for reacting to these spills.

Countermeasures, including diversion of flow, become effective options in real time only in conjunction with some sort of early-warning system. The type of system chosen depends upon the mix of domestic and industrial flows, the hazardous material inventory, system objectives, and the instrumentation available.

Contingency plans are important in all phases of hazardous material control, from prevention of spillage to implementation of countermeasures. Many examples of contingency plans are available in the literature, but each must be tailored specifically for the site involved. Experience has shown that the most important aspect of contingency plans is a clear definition of responsibilities for action and specified chains of command.

Operational Procedures

A variety of remedies or countermeasures can be applied to spilled materials in municipal treatment systems. In general, seven types of countermeasures are available for problem mitigation: (1) neutralization, (2) oxidation-reduction, (3) precipitation, (4) adsorption, (5) absorption, (6) physical methods, and (7) biological methods.

Neutralization

This technique involves pH adjustment by using acids or bases. Sulfuric acid is preferred for pH reduction mainly because of cost, while sodium hydroxide (caustic) is the preferred

TABLE 1-5 Threshold Concentrations of Inorganic Pollutants That Are Inhibitory to Biological Treatment Processes

Pollutant	Concentration (mg/L)		
	Activated-sludge processes	Anaerobic digestion processes	Nitrification processes
Ammonia	480	1500	
Arsenic	0.1	1.6	
Borate (boron)	0.05–100	2	
Cadmium	10–100	0.02	
Calcium	2500		
Chromium (hexavalent)	1–10	5–50	0.25
Chromium (trivalent)	50	50–500	
Copper	1.0	1.0–10	0.005–0.5
Cyanide	0.1–5	4	0.34
Iron	1000	5	
Lead	0.1		0.5
Manganese	10		
Magnesium		1000	50
Mercury	0.1–5.0	1365	
Nickel	1.0–2.5		0.25
Silver	5		
Sodium		3500	
Sulfate			500
Sulfide		50	
Zinc	0.08–10	5–20	0.08–0.5

NOTE: Concentrations shown represent the influent to the unit process in dissolved form.
SOURCE: *Federal Guidelines, State and Local Pretreatment Programs,* vols. I-III, EPA-430/9-76-017, U.S. Environmental Protection Agency, January 1977.

strong base. Even though lime is considerably cheaper, it usually is avoided because of the formation of calcium sulfate sludges. Carbonates and bicarbonates also have been used to control acid spills. Whenever possible, bicarbonates should be used for pH adjustment. It is difficult to overdose with bicarbonates, whereas close operator control and more expensive mixing tanks and chemical feed equipment are necessary for caustics, lime, and carbonates.

Oxidation-Reduction

This method can serve several purposes. For example, oxidation with ozone, peroxide, or permanganate can be a final treatment step (oxidation of cyanides), or it can be used to break down longer-chained organics to make them more amenable to further treatment by other methods. However, strong oxidants are difficult to work with and require close operator control. Aeration is the best and safest oxidizer. With aeration there is no danger of overdosing and certain other benefits could be achieved, among them precipitation of ferrous iron, oxidation of sulfites and hydrogen sulfide, and stripping of volatile constituents. Sodium thiosulfate and calcium sulfite both are good reductants. In fact, calcium sulfite often is used to deal with chlorine spills. But the best reductant is domestic sewage itself. The higher the biochemical oxygen demand (BOD) or the chemical oxygen demand (COD), the greater is sewage's reducing power.

Precipitation

This is a common technique. Hydroxides can be added to precipitate hydrous oxides of various metals. If sludge volumes are not a major consideration, lime easily is the pre-

TABLE 1-6 Countermeasures as a Function of Material Spilled

Material		Countermeasures																	
	Dilute	Blend	Add base	Add acid	Add reductant	Add oxidant	Aerate	Add ion exchanger	Add resin	Add carbon	Add precipitant	Add fibers	Add foams	Skim	Settle	Increase aeration	Aerate sludge prior to recycling	Change sludge return rate	Change wasting
Acid substances	X	X	X																
Alkaline substances	X	X		X															
Oxidants	X	X			X														
Reductants	X	X				X	X												
Mineral salts (low toxicity)	X	X						X											
Heavy-metal salts	X	X						X	X	X	X				X				
Insoluble organics	X	X							X	X		X	X	X	X				
Soluble biodegradable organics	X	X								X						X	X	X	
Soluble nonbiodegradable organics	X	X																	X

4-12

ferred choice for hydrous oxide precipitation. Sulfides have been successful in precipitating heavy metals, but close operator control is required for effective use of sulfide precipitation.

Adsorption

This method includes a number of highly effective techniques, among them activated carbon (powdered or granular), ion exchangers, and macroreticular resins. Some intermediate products such as carbonaceous resins have become more important in spanning from inorganic to organic sorption. Adsorption is expensive to apply and usually is reserved for situations in which the need for quick and efficient action justifies the high cost.

Absorption

This technique features porous foams and fibers for the pickup of oil and other hydrocarbon materials. Cellulose, polyurethane, olefin polymer, and polypropylene have been used for such applications. In many cases, the sorbed material can be recovered from the sorbent.

Physical Methods

Among these methods are dilution, blending, diversion, storage, skimming, settling, filtration, and flotation. Often they are used in conjunction with other processes or until a permanent solution can be reached.

Biological Methods

These methods depend upon the way in which the treatment plant is operated. A well-operated activated-sludge plant can accommodate broad ranges of influent concentrations through increased aeration and changes in sludge waste and return rates.

Summary

Countermeasures to be used in municipal systems for different types of spilled materials are summarized in Table 1-6. The best countermeasures are those in which materials on hand can be applied quickly and efficiently.

Management Options

Management options for dealing with hazardous material spills include the adoption of restrictive ordinances, the establishment of a system of fines and penalties, and the technical option of flow routing. The first two options, ordinances and fines, are intended to be deterrents rather than remedies. Flow routing, on the other hand, could provide a viable remedy.

In flow routing, computer-controlled programs distribute sewage flow primarily for purposes of controlling storm-water runoff. Hydrologic and flow data are telemetered from remote field sites to the central computer, which can signal and activate fabric-type in-line dams to route high flows to areas with excess capacity. This technique also can be helpful in managing hazardous material spills.

REFERENCES

1. L. Weaver and M. Chaudhry, "Monitoring: An Early Warning Experience," *Symposium on Organics in Our Water,* Robert Morris College, Pittsburgh, Nov. 13, 1978.
2. G. A. Brinsko, F. J. Erny, E. J. Martin, A. P. Pajak, and D. M. Jordan, *Hazardous Material*

Spills and Responses for Municipalities, EPA-600/2-80-108, U.S. Environmental Protection Agency, July 1980.

3. A. P. Pajak, C. J. Touhill, E. J. Martin, G. A. Brinsko, F. J. Erny, and J. E. Brugger, "Management of Hazardous Material Spills in Municipal Wastewater Systems," in G. F. Bennett (ed.), *Proceedings of the 1974 National Conference on Control of Hazardous Material Spills,* San Francisco, August 1974, pp. 58–64.

4. M. W. Stradley, G. W. Dawson, and B. W. Cone, *An Evaluation of the Status of Hazardous Waste Management in Region X,* EPA-910/9-76-024, U.S. Environmental Protection Agency, Region X, Solid Waste Branch, December 1975.

5. Battelle Memorial Institute, Pacific Northwest Laboratories, "The Impact of Hazardous Waste Generation in Minnesota," prepared for Minnesota Pollution Control Agency, Roseville, Minn., October 1977.

6. *Draft Report on Hazardous Waste Disposal in Erie and Niagara Counties, New York,* Interagency Task Force on Hazardous Wastes, March 1979.

7. A. P. Pajak, E. J. Martin, G. A. Brinsko, and F. J. Erny, *Effect of Hazardous Material Spills on Biological Treatment Processes,* EPA-600/2-77-239, U.S. Environmental Protection Agency, December 1977.

8. K. Verschueren, *Handbook of Environmental Data on Organic Chemicals,* Van Nostrand Reinhold Company, New York, 1977.

9. *Federal Guidelines, State and Local Pretreatment Programs,* vols. I–III, EPA-430/9-76-017, U.S. Environmental Protection Agency, January 1977.

10. K. R. Huibregtse, R. C. Scholz, R. F. Wullschleger, J. H. Moser, E. R. Bollinger, and C. H. Hansen, *Manual for Control of Hazardous Material Spills,* vol. I: *Spill Assessment and Water Treatment Techniques,* EPA-600/2-77-227, U.S. Environmental Protection Agency, November 1977.

11. W. H. Bauer, D. N. Borton, J. J. Bulloff, and J. R. Sinclair, "Agents for Amelioration of Discharges of Hazardous Chemicals on Water," in G. F. Bennett (ed.), *Proceedings of the 1976 National Conference on Control of Hazardous Material Spills,* New Orleans, April 1976, pp. 277–287.

12. R. J. Pilie, R. E. Baier, R. C. Ziegler, R. P. Leonard, J. G. Michalovic, S. L. Pek, and D. H. Bock, *Methods to Treat, Control and Monitor Spilled Hazardous Materials,* EPA-670/2-75-042, U.S. Environmental Protection Agency, June 1975.

PART 2
Environmental Effects

John Cairns, Jr., Ph.D.
University Distinguished Professor and Director,
Center for Environmental Studies
Virginia Polytechnic Institute and State University

Introduction / 4-15
Estimating the Vulnerability of Aquatic Ecosystems to Spills of Hazardous
 Materials / 4-17
Prospects for Recovery Following a Spill of Hazardous Materials / 4-17
Management Practices for Perturbed Ecosystems / 4-19
Spill Technology / 4-20
Summary / 4-20

INTRODUCTION

Spills of hazardous materials have three major effects on aquatic ecosystems: (1) they destroy part or all of the indigenous biota; (2) they impair ecosystem function in a variety of ways such as destruction of habitat, elimination of organisms, and chemical-physical effects; and (3) they place the ecosystem in disequilibrium, an effect that may be extended if persistent residual toxicants remain.

Aquatic ecosystems provide a variety of benefits, which include:

1. Transport and fragmentation of various materials (essential to biochemical cycles)
2. Chemical and physical reactions that bind nutrients such as phosphorus, nitrogen, and frequently heavy metals to these minute particles
3. Biological decomposition of complex organics
4. Assimilation of soluble organics

These benefits are impaired or eliminated by any spill which seriously perturbs the structure and/or function of the receiving ecosystem. Among the structural changes likely to occur as a result of a spill is a decreased diversity or reduction in the number of species present.[1] Reducing the number of species impairs the stability of the system. A healthy system is characterized by a large number of species with relatively few individuals per species, whereas a polluted and stressed system (resulting from a spill) is characterized by a large number of individuals in relatively few species. The loss of a single species through natural or societal change is not likely to affect the system with a high diversity, while the loss of one of the very abundant species from the severely stressed system could mean a loss in total biomass of as much as one-third to one-half of the total.

Since the indigenous organisms, especially the microorganisms, degrade wastes, produce oxygen, and carry out many other beneficial activities, such a loss very probably has a major impact on the nondegrading-waste assimilative capacity of the system.[2] Functional parameters such as nitrogen and sulfur cycling and a variety of other activities might be impaired.[3]

It is a sine qua non that no material should be manufactured, transported, or used without a hazard evaluation, and protocols for this purpose are available.[4,5,6] Few guidelines are available to furnish advice in resolving this problem. The chapter "Episodic Exposures" in *Principles for Evaluating Chemicals in the Environment*[7] contains some useful suggestions.

Since the British have been involved with water problems for more years than Americans, their present attitude is worth examining. Officials of the Thames Water Authority believe that failure to report a spill of any sort (e.g., in transportation or manufacturing) is a worse offense than the spill itself. This is true because officials can take action to protect public water supplies, human health, etc., only after being informed. This seems to be a reasonable approach if all concerned parties use common sense. If not all spills are reported, criteria for deciding when to report and when not to do so are difficult to determine on a countrywide basis.*

Some types of spills should be reported regardless of circumstances:

1. Spills containing compounds which undergo biological accumulation. Partition coefficients and other evidence may be used in making the determination. Materials subject to biological accumulation represent a threat to both human health and the environment and should receive immediate serious attention.

2. Large-volume spills which even temporarily exceed 10 percent of the receiving water volume. For toxic material, the limit should be 1 percent or less.

3. Spills likely to enter public water supplies, affect livestock or edible materials, or endanger human health in bathing areas, etc.

4. Nontoxic or toxic spills likely to cause flavor impairment, objectionable taste and odors, and the like.

5. Spills occurring in areas containing rare and/or endangered species, spawning grounds of commercially valuable species, or an ecological preserve.

A decision not to report a spill should be revised immediately under the following conditions:

1. If a significant kill or effect on any important indigenous species becomes evident. For a manufacturing plant, this decision requires some knowledge of the local biota, which should be obtained before the plant begins operating or as soon as possible thereafter.

2. If the effects of the spill (e.g., the number of dead fish) become more noticeable with time, indicating a "steep" dose-response curve.

3. If information becomes available that another spill has occurred in the vicinity (which is quite possible in bad weather in a heavily industrialized area), raising the possibility of synergistic interactions.

4. If personnel tracking the spill find a significant change in environmental quality (e.g., in dissolved oxygen concentration or pH). This determination will require a priori knowledge of the local ecosystem.

More quantitative and specific guidelines should be developed for each industrial plant on the basis of site-specific information. In all cases, a program of action (e.g., reporting,

---

*Legal requirements for spill reporting are discussed in Chapter 2.

data collecting) should be developed *before* an emergency occurs. A series of observation sites should be selected for the most likely areas to be affected. Baseline sampling should be carried out before the spill by those individuals who would take action if a spill should occur. Vehicles, sampling equipment, and access to sampling locations should be prearranged.

ESTIMATING THE VULNERABILITY OF AQUATIC ECOSYSTEMS TO SPILLS OF HAZARDOUS MATERIALS

It is well known that some ecosystems are more vulnerable to spills of hazardous materials than others.[8] Some ecosystems are exceedingly fragile, and others, by comparison, are much tougher. A prudent management policy should include an analysis of the vulnerability of aquatic and terrestrial ecosystems to damage. Admittedly, such estimates lack precision because of the inadequate data base, but most ecologists can determine which ecosystems are fragile and can even make some estimate of the ecological inertia (resistance to loss of biological integrity), elasticity (ability to recover integrity following displacement), and resilience (ability to snap back after successive displacements). Discussions of biological integrity (maintenance of structure and function of the aquatic community characteristic of a particular locale) may be found in Cairns,[3] and discussions of inertia, elasticity, and resilience for aquatic ecosystems in Cairns's 1976 article.[9] Tables 1 and 2 from the 1976 paper provide a means of estimating these characteristics on the basis of an array of criteria. Cairns et al.[10] have used the Youghiogheny River to illustrate how these estimates might be made. The inertia index can be estimated without any knowledge of the type of spill or pollutional stress likely to occur, but the estimation of elasticity is improved markedly if such information is available.

Industries planning to build new plants on sites with a high degree of ecological vulnerability or those with plants already on such sites would be well advised to take greater precautions to prevent spills than those with plants on less vulnerable sites. Estimations of vulnerability, although crude, are far preferable to an assumption that all ecosystems are identical. In addition, those individuals involved in developments on particularly vulnerable sites might wish to consider the installation of some form of biological monitoring. Such monitoring can involve very sophisticated computer-interfaced systems, such as that described by van der Schalie et al.,[11] or can be based on simple bioassays or simple diversity index analyses of aquatic communities.

PROSPECTS FOR RECOVERY FOLLOWING A SPILL OF HAZARDOUS MATERIALS

A few books and publications analyzing the prospects for ecosystem rehabilitation following severe damage[12,13,14,15,16] have recently appeared. These books contain some information on case histories following catastrophic spills. More important, however, they show, through the use of case histories, the problems associated with the rehabilitation of damaged ecosystems. Any individuals associated with the manufacture, transport, use, or disposal of hazardous materials should read some of this literature to become more knowledgeable about the effort involved in restoring a damaged ecosystem. This knowledge might well influence involved personnel to take greater precautions to avoid spills.

Compelling evidence exists to show that certain types of spills have much longer-lasting effects than others and that the location of a spill may strongly influence the speed with which the potentially toxic material is dispersed regionally. Spills may be categorized as four major types: (1) spills of nonpersistent materials into ecosystems in which rapid dispersal from the site of a spill is highly probable, (2) spills of nonpersistent materials into ecosystems in which rapid dispersal from the site of a spill is unlikely; (3) spills of highly persistent materials when rapid dispersal from the site of a spill is highly probable; and (4) spills of highly persistent materials when rapid dispersal from the site of a spill is improbable.

Spills of nonpersistent materials into ecosystems in which rapid dispersal from the site of a spill is highly probable are not likely to create major ecological damage. This is particularly true if mobile organisms such as fish, birds, or mammals are likely to avoid the material (i.e., have a negative reaction toward it) when they detect it. Rapid dilution may quickly lower the concentration, thereby avoiding adverse biological effects, especially for materials not particularly toxic even in high concentrations. Thus, most of the damage is likely to be done to microbial species and the smaller macroinvertebrates and plants, most likely in a restricted area. These effects may be serious because the affected organisms may be rare and endangered species or species that are crucial to the ecosystem in the area. However, this is probably the least environmentally damaging of the four categories. An example of this category is a spill of a rapidly degrading material (e.g., phenol) in a stream or coastal region of the ocean where rapid mixing occurs and large volumes of water are available for dilution.

The second type of spills (spills of a nonpersistent or highly degradable toxicant into an area where extensive dispersal from the site of a spill is not likely) will probably have more severe effects at the site than would occur in the previously discussed type. The reason is that the material is more likely to remain at concentrations adverse to various organisms than if it were rapidly diluted. Even nonpersistent materials may remain at lethal concentrations for days or even weeks, causing havoc in a restricted region. If the area is a small wetland, marsh, pond, or terrestrial ecosystem (particularly one containing rare and endangered species or organisms relatively uncommon to the region), the loss will be particularly grievous. However, if the spill occurs in an area occupied by organisms that are abundant throughout the region, reinvasion may take place as soon as the concentrations have been reduced below those at which adverse biological effects occur. Even when this has happened, however, ecosystem recovery may be quite slow (examples are given in Cairns et al.[13]).

Spills of persistent chemicals in ecosystems in which rapid dispersal from the site of the spill is likely are probably the most dangerous and unfortunate of all. Persistent materials, particularly those that are fat-soluble (e.g., chlorinated hydrocarbons), are likely to undergo biomagnification (progressive concentration as the material moves upward through the food chain, ending with the ultimate consumers, humans). The biomagnification potential negates the primary biological advantages of rapid dispersal and makes the material more difficult to retrieve and treat because of its wider distribution. For example, Kepone in the James River and adjacent ecosystems and PCBs in the Hudson River may be present in various concentrations in the sediment, in the water column, and in associated organisms of all kinds. Even the areas with the heaviest concentration of these materials still contain rather small (but not necessarily harmless) concentrations (often in micrograms per milliliter or less).

Nevertheless, the possibility of reconcentration over a period of years, ultimately affecting the health of indigenous organisms and consuming humans, should not be ignored. Removal of millions of cubic meters of sediment for treatment would be an enormously costly and difficult process. Estimates of the costs of treating sediments in the James River for Kepone with ozone and ultraviolet radiation range upward to billions of dollars. These costs are careful estimates based on known technology and data indicating dispersal of the material in the James River area.

It is also possible that the removal of this quantity of material for treatment would inadvertently release material into the water column that might otherwise not be released or that the treatment itself might produce nitrogenous compounds that would represent a hazard. The ecological disturbance caused by this massive movement of materials might be far greater than the benefits accrued from it. Consequently it may well be that the best course is to do nothing.

The consequences of exercising this option, however, are tempered by the fact that it may take more than 100 years for the material to decay to a point of ecological harmlessness. During this period, routine dredging to keep navigation channels in the river open will undoubtedly continue, thereby releasing some of the toxicants to the environment. The point of this attenuated discussion of a complex problem is that sometimes when a spill involving relatively persistent and quite hazardous materials occurs, present technology may not permit treatment without causing more problems than already exist even if economic factors are not a consideration. It is obvious that if any retrieval of such hazardous materials is to be undertaken, it should be carried out at the earliest possible moment before extensive dispersal occurs. This is the most economical and ecologically sound management practice.

The last spill type involves the release of persistent materials in ecosystems in which rapid dispersal from the spill site is not probable. In this situation, rapid action is highly desirable, particularly if the material is likely to be accumulated by organisms and dispersed in that fashion. Even a spill on land in a terrestrial ecosystem is likely to be dispersed more rapidly following a heavy rain or a rapid snow melt than would ordinarily occur. There are few ecosystems in which the opportunity for rapid dispersal of hazardous material is not present at some time, although such circumstances occur infrequently. Nevertheless, by using the worst-case approach, prompt action to retrieve the material and to contain it in clay-lined pits, etc., would be prudent, highly desirable, and comparatively economical. The movement of organisms and water in ecosystems is often rapid, and it would be well to err on the side of caution. Sampling procedures to ascertain the efficacy of the removal and the degree of contamination of indigenous biota and the ecosystem in general should be maintained for a reasonable period, perhaps for as long as 5 or 6 years in cases involving persistent materials.

MANAGEMENT PRACTICES FOR PERTURBED ECOSYSTEMS

There is a paucity of literature on the application of procedures coupled with optimization techniques for managing perturbed ecosystems. However, Fiering and Holling[17] provide a conceptual framework for developing management practices and procedures. Their paper emphasizes the element of recovery time and its manipulation in ecosystem management and espouses the concept of an "environmental zoo," which they describe as a physical entity in which living organisms are stocked to replenish depleted numbers in the field, a concept equivalent to the epicenter for dispersal described by Cairns.[9] This possibility for enhancing the recovery prospects of damaged ecosystems should receive greater attention. The basic strategy would be to establish, at frequent intervals throughout the United States, ecological preserves afforded the highest possible protection from which organisms could disperse to reinvade damaged adjacent ecosystems. With proper management, these preserves could be used for other purposes such as recreation. The important point is that the location of such preserves would be primarily determined by the distance from other such preserves rather than by other considerations.

SPILL TECHNOLOGY

The literature frequently announces various types of cleanup devices, dispersant techniques, etc., that are being developed for controlling oil and chemical spills. When developing these technologies, it would be advisable to carry out tests to determine their impact on the organisms and ecosystems that they are designed to protect. In some cases, the cure may be worse than the disease. In addition, if damage has been done to an ecosystem, the most efficient and efficacious cleanup will have severely limited ecological benefits unless organisms are able to reinvade from an environmental zoo or an ecological epicenter.

SUMMARY

It is evident that the environmental effects of spilled hazardous materials deserve greater attention than has been given this important subject in the past. Highest priority should be given to (1) the process of hazard evaluation, (2) mapping particularly vulnerable ecosystems, (3) perfecting methods for estimating ecosystem vulnerability, and (4) providing better management techniques for the rehabilitation of damaged ecosystems. Available information in these fields is adequate to develop management practices far better than those in place. Underutilization of existing information appears to be a serious problem.

REFERENCES

1. J. Cairns, Jr., and K. L. Dickson, "A Simple Method for the Biological Assessment of the Effects of Waste Discharges on Aquatic Bottom-Dwelling Organisms," *The Journal of the Water Pollution Control Federation,* vol. 43, no. 5, 1971, pp. 755–772.

2. J. Cairns, Jr., "Aquatic Ecosystem Assimilative Capacity," *Fisheries,* vol. 2, no. 2, 1977, pp. 5–7, 24.

3. J. Cairns, Jr., "Quantification of Biological Integrity," in R. K. Ballentine and L. J. Guarraia (eds.), *Integrity of Water,* 055-001-01068-1, U.S. Environmental Protection Agency, Office of Water and Hazardous Materials, 1977, pp. 171–187.

4. J. Cairns, Jr., and K. L. Dickson, "Field and Laboratory Protocols for Evaluating the Effects of Chemical Substances on Aquatic Life," *The Journal of Testing and Evaluation,* vol. 6, no. 2, 1978, pp. 81–90.

5. J. Cairns, Jr., K. L. Dickson, and A. Maki (eds.), *Estimating the Hazard of Chemical Substances to Aquatic Life,* American Society for Testing and Materials, Philadelphia, 1978.

6. K. L. Dickson, J. Cairns, Jr., and A. Maki (eds.), *Analyzing the Hazard Evaluation Process,* American Fisheries Society, Wahington, 1979.

7. *Principles for Evaluating Chemicals in the Environment,* National Academy of Sciences, Washington, 1975.

8. J. Cairns, Jr., "Waterway Recovery," *Water Spectrum,* vol. 10, no. 4, 1978, pp. 26–32.

9. J. Cairns, Jr., "Heated Wastewater Effects on Aquatic Ecosystems," in G. W. Esch and R. W. McFarlane (eds.), *Thermal Ecology II,* Technical Information Center, Springfield, Va., 1976, pp. 32–38.

10. J. Cairns, Jr., J. R. Stauffer, and C. H. Hocutt, "Opportunities for Maintenance and Rehabilitation of Riparian Habitats: The Eastern United States," in R. R. Johnson and J. F. McCormick (eds.), *Strategies for Protection and Management of Floodplain Wetlands and Other Riparian Ecosystems,* U.S. Department of Agriculture, Forest Service, Office of Biological Sciences, Washington, 1979.

11. W. H. van der Schalie, K. L. Dickson, and J. Cairns, Jr., "Continuous Automatic Monitoring

of Waste Effluents Using Fish: A Review of Several Methods and a Field Test of One System," *Environment Management*, vol. 3, no. 3, pp. 217–325.

12. J. Cairns, Jr. (ed.), *The Recovery Process in Damaged Ecosystems*, Ann Arbor Science Publishers, Inc., Ann Arbor, Mich., 1980.

13. J. Cairns, Jr., K. L. Dickson, and E. E. Herricks (eds.), *The Recovery and Restoration of Damaged Ecosystems*, University Press of Virginia, Charlottesville, 1977.

14. M. W. Holdgate and M. L. Woodman (eds.), *The Breakdown and Restoration of Ecosystems*, Plenum Press, New York, 1978.

15. D. E. Samuel, J. R. Stauffer, C. H. Hocutt, and W. T. Mason, Jr., *Surface Mining and Fish/Wildlife Needs in the Eastern United States*, FWS/OBS-78/81, U.S. Department of the Interior, 1978.

16. A Thorhaug (chairman), *Proceedings of the Symposium on Restoration of Major Plant Communities in the U.S.A.*, reprinted from *Environmental Conservation* and published for the Foundation for Environmental Conservation, 1978.

17. M. B. Fiering and C. S. Holling, "Management and Standards for Perturbed Ecosystems," *Agro-Ecosystems*, vol. 1, 1974, pp. 301–321.

PART 3
Economic Effects*

Jonathan E. Amson
Biological Science Administrator
Office of Water Regulations and Standards
U.S. Environmental Protection Agency

Legislative and Regulatory Background / 4-22
Previous Economic-Impact Analyses / 4-25
Present Economic-Impact Analysis / 4-26
Summary / 4-35
Conclusion / 4-37

LEGISLATIVE AND REGULATORY BACKGROUND

The first attempts by Congress to control the discharge of hazardous substances to the waters of the United States appeared in Sec. 12 of the Water Quality Improvement Act of 1970 (Public Law 91-224). In this section of the act, the President was directed to "develop, promulgate, and revise as may be appropriate, regulations designating as hazardous substances, ... such elements and compounds which, when discharged in any quantity into or upon the navigable waters of the United States ..., present an imminent and substantial danger to the public health or welfare. ..."[1] The President was further directed to remove or arrange for the removal of any hazardous substance whenever discharged into or upon the navigable waters of the United States, unless the removal was immediately undertaken by the owner or operator of the vessel or onshore or offshore facility from which the discharge occurred. In addition, the President was directed to submit a report to Congress "on the need for, and desirability of, enacting legislation to impose liability for the cost of removal of hazardous substances discharged from vessels and onshore and offshore facilities. ..."[1] In preparing the report, the President was further directed to conduct a study which would include "the most appropriate measures for (1) enforcement (including the imposition of civil and criminal penalties for discharges and for failure to notify), and (2) recovery of costs incurred by the United States if removal is undertaken by the United States."[1] Responsibility for conduct of the study and completion of the report was delegated by the President in Executive Order 11548 to the secretary of transportation in consultation with the secretary of the interior. Responsibility for the preparation of the required materials was further delegated by the respective secretaries to the U.S. Coast Guard and the Federal Water Quality Administration.

The President submitted the required report and study to Congress on March 16, 1971.[2] The report [commonly known as "the 12(g) study," since the charge to the Pres-

---

*The contents of this part represent solely the work of the author and do not necessarily reflect the views and policies of the U.S. Environmental Protection Agency.

ident to prepare the study and report was contained in Sec. 12(g) of the Water Quality Improvement Act] contained proposed regulations designating 327 substances as hazardous, recommendations for legislation to impose unlimited liability for removal on discharges of hazardous polluting substances, and recommendations for the imposition of several penalties for the discharge of hazardous polluting substances, including a $10,000 civil penalty for failure to notify the appropriate agency of the United States government of a discharge and both a $10,000 civil penalty and a $25,000 criminal penalty for knowingly discharging a hazardous polluting substance. The report, however, made the recommendation that no penalty be imposed for the accidental discharge of a hazardous polluting substance.

Before the proposed regulations were issued in final form, Congress passed extensive amendments to the Federal Water Pollution Control Act (Public Law 92-500) on October 18, 1972.[3] In those amendments, Congress considerably expanded the mandate for control of hazardous substances that had appeared in Sec. 12 of the Water Quality Improvement Act. In the revised and expanded legislation, responsibility for control of oil and hazardous substances (now identified as Sec. 311 and entitled "Oil and Hazardous Substance Liability") was given to the administrator of the U.S. Environmental Protection Agency (U.S. EPA).

Under Sec. 311, the administrator was given a number of major responsibilities, four of which were crucial to the control of hazardous substances. The first required the administrator to promulgate regulations designating as hazardous substances those elements and compounds which, when discharged into the navigable waters of the United States, presented an imminent and substantial danger to the public health or welfare. (This mandate to the administrator was essentially identical to the one given the President in Sec. 12 of Public Law 91-224.) The second required the administrator to include in any designation of hazardous substances a determination whether those substances "can actually be removed."[3] The third responsibility required the administrator to establish units of measurement and rates of penalty for the discharge of hazardous substances determined to be nonremovable. Such units of measurement and rates of penalty were to be based on the toxicity, degradability, and dispersal characteristics of the substances designated. Finally, the fourth responsibility required the President (the requirement was delegated to the administrator by Executive Order 11735) to determine those quantities of hazardous substances "the discharge of which, at such times, locations, circumstances, and conditions, will be harmful to the public health or welfare. . . ."[3]

On August 22, 1974, the U.S. EPA issued an advance notice of proposed rulemaking (ANPR) which proposed to designate 371 elements and compounds as hazardous substances and which further proposed to make the determination that not all the substances identified were actually removable.[4] Included in the ANPR were the proposed selection criteria used to describe quantitatively the concept of "imminent and substantial danger to the public health or welfare" and the proposed spill potential criteria used to describe the potential for discharge into a water body, as well as the proposed factors considered in the determination of actual removability. However, the ANPR did not propose to establish units of measurement or rates of penalty, nor did it propose to determine harmful quantities for those substances proposed to be designated.

Following consideration of the public comments received and revision of the spill potential criteria identified in the ANPR, the EPA issued a notice of proposed rulemaking (NPR) for hazardous substances on December 30, 1975.[5] Included in the NPR were the designation of 306 elements and compounds as hazardous substances (65 substances were eliminated from the listing in the ANPR), the determination that substances so designated could actually be removed, and the determination of units of measurement, rates of penalty, and harmful quantities for the 306 substances proposed to be designated.

Before the regulations were issued in final form, however, Congress passed the Clean

Water Act (Public Law 95-217) on December 15, 1977.[6] This legislation expanded the jurisdiction of Sec. 311 in several important areas and, as a consequence, required the revision of the proposed final rulemaking for hazardous substances then under consideration by the EPA.

On March 13, 1978, the EPA issued a notice for final rulemaking (NFR) for hazardous substances.[7] Included in the NFR were the four essential requirements for the control of hazardous substance discharges: (1) the designation of 271 elements and compounds as hazardous substances (35 substances were eliminated from the listing in the NPR); (2) the determination that 261 of the substances so designated could actually be removed (10 substances, because of their limited water solubility, relatively cohesive mass, and density less than unity, resembled petroleum oils in their behavior when discharged into water and were determined to be actually removable under most conditions of discharge); (3) the determination of units of measurement and rates of penalty; and (4) harmful quantities for the 271 substances designated. At the time of the promulgation of the NFR, the EPA proposed the designation of 28 additional substances as hazardous, with concurrent determinations of removability, units of measurement, rates of penalty, and harmful quantities.

However, prior to their effective date, the hazardous substance regulations were challenged in several lawsuits. The plaintiffs questioned, among other things, the determination by the EPA that only 10 substances could actually be removed (believing that under certain conditions of discharge many of the remaining 261 substances could actually be removed) and, further, questioned whether in its determination of harmful quantities the EPA had adequately considered the requirement that "times, locations, circumstances, and conditions" be included in a determination of harm following the discharge of a hazardous substance.

On August 14, 1978, the District Court for the Western District of Louisiana (where the regulations had been enjoined) declared certain portions of the regulations invalid and unenforceable.[8] In its order, the court declared the EPA's determinations of removability and harmful quantities invalid; the order resulted indirectly in the invalidation of the EPA's determinations of units of measurement and rates of penalty. The designation of hazardous substances was neither challenged by the plaintiffs nor affected by the order of the court.

On November 2, 1978, Sec. 311 of the Clean Water Act was amended by Public Law 95-576.[9] This amending was the result of intensive negotiations between senior EPA management, the staffs of several congressional committees, and representatives of the various industrial groups affected by the proposed legislation. The negotiations resulted in a compromise on the applicability of the hazardous substance regulations to permitted discharges under the National Pollutant Discharge Elimination System (NPDES) that met most of the objections of the plaintiffs in the litigation and was acceptable to both Congress and the EPA. The amended legislation no longer required the EPA to make determinations of removability or units of measurement for establishing rates of penalty. In addition, the basis for determining harm, viz., the discharge of a "harmful quantity" which "will be harmful" at the "times, locations, circumstances, and conditions" of discharge, was changed to those quantities of a substance which "may be harmful," and the legislative reference to specific times, locations, circumstances, and conditions of discharge was removed. The legislative history of Public Law 95-576 made it clear that Congress intended that the determination of reportable quantities did not require an assessment of actual harm in the many circumstances in which a hazardous substance might be discharged; rather, it intended that the determination be based on the chemical and toxic properties of the substance itself, not the circumstances surrounding the release.

Moreover, the potential penalties to be assessed by the regulations were considerably reduced from those originally promulgated. The original mandate established a maximum penalty of $500,000 for a discharge of a nonremovable substance from an onshore or an offshore facility and a maximum penalty of $5 million for a discharge of a nonremovable substance from a vessel. The revised penalties established a maximum penalty of $50,000 for any discharge equal to or in excess of a reportable quantity, except when it could be shown that the discharge was the result of willful negligence or willful misconduct. In that case, the maximum penalty was established at $250,000.

On February 16, 1979, the EPA issued an NPR that incorporated the legislative changes mandated by Public Law 95-576.[10] At the same time, the agency finalized the designation of the 28 substances that had been proposed in the NFR of March 13, 1978. On August 29, 1979, the EPA issued an NFR for hazardous substances that included the determination of reportable quantities as required by Public Law 95-576.[11] The regulations became effective on September 28, 1979; on that date, quantities of designated hazardous substances equal to or greater than reportable quantities, when discharged into or upon the navigable waters of the United States, upon adjoining shorelines, into or upon the contiguous zone, or beyond the contiguous zone, as provided in Sec. 311(b)(3) of the Clean Water Act, must be reported to the appropriate agency of the United States government. In this case, the appropriate agency is the National Response Center of the U.S. Coast Guard; the toll-free telephone number of the center is 800-424-8802.

PREVIOUS ECONOMIC-IMPACT ANALYSES

In the development of the December 30, 1975, proposed rulemaking for hazardous substances, an analysis of the expected areas of economic impact was prepared. A modified version of that analysis was published in the *Proceedings of the 1976 National Conference on Control of Hazardous Material Spills.*[12] The analysis examined three areas of expected economic impact: (1) the cost of civil penalties to be assessed, (2) the cost of spill response, cleanup, and damage mitigation, and (3) the cost of spill prevention equipment installed and techniques developed. The expected economic impact was approximately $4,250,000 per year for the costs of applicable civil penalties, approximately $8,300,000 per year for cleanup and mitigation costs, and approximately $78,500,000 for the 20-year period 1976–1996 for spill prevention equipment. The figure for expenditures on spill prevention equipment was derived by using a capital discount approach of the moneys that the chemical industry should be willing to spend to avoid the costs of penalties, cleanup, and mitigation for the 20-year period. Although the figures derived in the analysis are no longer valid (the basis for the assessment of civil penalties was changed from the concept of "gross negligence" to the concept of "gravity of the violation," the cleanup and mitigation costs have increased substantially since the original analysis, and the projected cost of capital to the industry has increased dramatically as a result of the inflationary events of the late 1970s), the analysis nevertheless is one of the few papers to have examined in considerable detail the factors involved in the economic impact of hazardous substance discharges.

The economic-impact analysis for the March 13, 1978, final rulemaking reexamined the figures derived in the December 1975 analysis. As noted above, the later regulations abandoned the concept of gross negligence in the assessment of civil penalties and substituted the concept of gravity of the violation. In addition, a list of factors to be considered in deciding which penalty option to use was specified. These factors were (1) the size of the discharge, (2) the culpability of the owner or operator, (3) the nature, extent, and degree of success of any efforts by the owner or operator to minimize, mitigate, contain, or remove the effects of the discharge, and (4) any other factors deemed appropriate under the circumstances of the discharge. Moreover, the Clean Water Act of 1977 extended the scope of applicability of the regulations to include activities covered by the Deepwater

Port Act (DPA) of 1974,[13] the Fishery Conservation and Management Act (FCMA) of 1976,[14] and the Outer Continental Shelf Lands Act (OCSLA) amendments of 1978.[15]

In the case of the DPA, applicability was extended to cover those deepwater ports (such as the proposed Louisiana Offshore Oil Platform) covered by the DPA; in the case of the FCMA, applicability was extended to cover fishery conservation zones from the baseline of the territorial sea out to 200 nautical mi (370 km); and in the case of the OCSLA amendments, applicability was extended to cover those offshore facilities (such as drilling rigs and platforms) covered by the OCSLA amendments. The reexamination of the December 1975 economic-impact analysis revealed that no substantial changes in the conclusions of that analysis were necessary for the March 1978 final rulemaking.

The economic-impact analysis for the February 16, 1979, proposed rulemaking stated that since the regulations did not require the construction or purchase of spill prevention equipment or prohibit the manufacture, use, or transport of any designated hazardous substance, compliance with the regulations would not result in any direct costs to regulated parties. Two types of expenses may be incurred as a result of violation of the regulations: (1) costs of civil penalties and (2) costs of mitigation and cleanup. However, since economic-impact analysis is based on costs of compliance rather than on costs resulting from failure to comply, these two costs are not considered to be direct costs.

As a result of the amendment of the act by Public Law 95-576, the potential penalties that may be assessed under the statute were considerably reduced from those originally promulgated. Consequently, the February 1979 economic analysis concluded that the potential amount of civil penalties would be substantially less than those projected in the March 1978 final rulemaking; that rulemaking concluded that no substantial changes in penalty costs occurred from those projected in the December 1975 proposed rulemaking. Thus, the February 1979 economic analysis projected substantially lower civil penalty costs than those of the December 1975 economic analysis. The extent of the reduction was not specified. In addition, the February 1979 analysis made no estimate of any projected changes in costs of mitigation and cleanup from the December 1975 analysis.

The economic-impact analysis for the August 29, 1979, final rulemaking again stated that compliance with the regulations would not result in any direct costs to the regulated parties. The analysis examined the incremental treatment costs for those parties subject to NPDES permit limitations and concluded that, in the majority of cases, the treatment used to control toxic and nonconventional pollutants in a permit limitation would also be capable of controlling hazardous substances in a nonroutine discharge situation. Thus, the analysis concluded that incremental treatment costs for hazardous substance discharges would be insignificant, since in most cases the best available technology economically achievable used to treat toxic and nonconventional pollutants would also remove hazardous substances.

The August 1979 economic analysis also projected substantially lower civil penalty costs than those of the December 1975 analysis, as a result of the reduction in the penalty scheme. The extent of the reduction, as in the February 1979 analysis, was not specified, nor was an estimate made of any projected changes in the costs of mitigation and cleanup.

PRESENT ECONOMIC-IMPACT ANALYSIS

The mitigating or primary causes of hazardous substance spills can generally fit into one of five categories:

1. Discharges resulting from rupture of or leakage from a storage or transporting container
2. Discharges resulting from failure of storage, transfer, or transporting equipment

3. Discharges resulting from an in-plant or transportation accident

4. Discharges resulting from personnel error in loading or unloading operations or other improper handling

5. Discharges resulting from intentional events such as vandalism, sabotage, or deliberate release

Some initiating causes may be combinations of two or more of these factors, such as a discharge resulting from the accidental spearing of a storage tank with the tine of a forklift by a careless employee.

It should be noted at the outset of any discussion of the economic impact of hazardous substance spills that there is a paucity of data on both the number of hazardous substance discharges and the costs of such spills. Other than the mandatory reporting since September 28, 1979, of those designated hazardous substance discharges required by Sec. 311 of the Clean Water Act, industrial producers, handlers, shippers, and users of hazardous substances are quite reluctant to provide voluntarily information on either the number and the volume of discharges or the costs of such discharges. In addition, in-plant discharges from production processes or from storage areas are extremely difficult to document. Such discharges are not required by Sec. 311 regulations to be reported to the federal government and may not even be reported to the company on whose property the discharge occurs. Reasons for not reporting such discharges include:

1. The substance or quantity discharged may not be considered hazardous by an employee.

2. The value of the substance discharged and the associated interruption of the production process are below the deductible limit for an insurance claim, and thus the insurer is not notified.

3. The discharge may be limited to corporate boundaries and thus is not required to be reported.[16]

Although a number of case studies of individual spills have been conducted, little work in the technical literature has addressed the identification and quantification of the major parameters which affect the economic impact of hazardous substance spills. Two papers which have addressed the subject in detail are by the author; [12,17] another extensive review has explored the economic costs of the blowout of the Santa Barbara oil platform that began on January 28, 1969.[18] Despite the general lack of published literature on the subject, it is possible to identify and quantify to a certain extent those factors that affect the economic impact of hazardous substance spills. These factors can be identified as follows:

1. Number of discharges occurring annually

2. Volume of the discharge

3. Value of product lost, less the value of product recovered suitable for further commercial use

4. Cost of retrieval and cleanup efforts, including cost of the mitigation chemicals used

5. Loss due to disruption of the production process

6. Cost of equipment and property damaged and destroyed

7. Cost of damage to nonplant property

8. Cost of damage to the environment, including loss of commercial fishing and shellfish-harvesting revenues

9. Cost of compensation and/or injury to employees, including cost of lost time due to injury

10. Cost of lost water usage, including industrial, municipal, agricultural, recreational, household, and drinking-water supply and aesthetic water uses

In the discussion that follows, these factors will be examined individually.

Number of Discharges Occurring Annually

The number of discharges occurring annually is one of the factors essential to any derivation of the economic impact of a hazardous substance discharge. In the 1976 paper on the economic impact of the hazardous substance regulations, an estimated spill rate of 700 spills per year was derived.[12] The data base from which the annual number of discharges was derived represented voluntarily reported or fortuitously discovered spills documented by the EPA. This data base spanned 2½ years of discharge records from June 1972 to December 1974 and included 174 spills in which proposed hazardous substances were determined to have entered a surface water body, probably in excess of the proposed harmful quantity. This was the equivalent of 70 spills per year reported or discovered.

To compensate for the lack of mandatory reporting, however, it was assumed that those spills accounted for only 10 percent of the actual number of discharges; thus, the estimated spill rate was 700 spills per year. This estimate of 10 percent of the actual levels is consistent with the experience of the oil-spill-reporting program established in 1970, in which the mandatory reporting requirements of Sec. 11(b)(4) of the Water Quality Improvement Act produced at least a 10-fold increase in reported spills over the previously voluntary oil-spill-reporting program.

More recent evidence, however, has indicated that this estimate of 700 spills per year may be too high. A study of hazardous material incidents voluntarily reported to EPA regional offices from October 1977 through September 1979 revealed a total of 3076 incidents in that 2-year period.[19] The record contained reported discharges of hazardous substances as well as of oil discharges from active and inactive waste disposal sites, discharges due to the failure of manufacturing processes or waste treatment systems, and a number of discharges for which the cause was unknown.

A review of the 3076 incidents revealed 136 confirmed spills in which designated hazardous substances were discharged into water in quantities equal to or in excess of the reportable quantity for those substances for the period of October 1977 through September 1978, and 159 confirmed spills with the same restrictions for the period of October 1978 through September 1979. Since some of the 3076 reported incidents involved the release of oil, the discharge of which was required by law to be reported, it is assumed that the 136 confirmed spills for the fiscal year 1978 and the 159 confirmed spills for the fiscal year 1979 represented only 50 percent of the actual number of hazardous substance spills for those years. Thus, an estimated number of 272 spills for the fiscal year 1978 and an estimated number of 318 spills for the fiscal year 1979 may be derived. However, it should be noted that reports of a number of the 3076 discharges provided no information on the quantity discharged or the environmental medium affected. Thus, the numbers derived above would probably be substantially higher if such information were known.

The numbers derived for fiscal years 1978 and 1979 are supported by an examination of the discharge reports at the National Response Center, U.S. Coast Guard headquarters, for the period of September 28, 1979, through March 31, 1980.[20] Those 6 months represented the first period in which discharges of designated hazardous substances that entered the navigable waters of the United States in quantities equal to or in excess of the reportable quantity for those substances were required by law to be reported. The examination revealed a total of 147 spills for the 6-month period. Thus, an estimated 294 hazardous substance spills would occur annually. However, because of a required revision by the Department of Transportation (DOT) of applicable labeling and packaging regulations, discharges of designated hazardous substances from common carriers were not required to be reported during this period. The required revision was published by the DOT on May 20, 1980, and the reporting of hazardous substance discharges by common carriers became mandatory on November 20, 1980. Thus, the figures calculated above will be substantially higher when discharges of designated hazardous substances from common carriers are included.

Since such data are not currently available, it may be assumed for the purposes of this analysis that there occur annually approximately 300 hazardous substance spills in which the quantity of the substance discharged reaches water in amounts equal to or in excess of the reportable quantity for that substance.

Volume of the Discharge

Data on the volume of the discharge are available from several sources. In the 1978 paper by the author on the economic impact of spill prevention costs on the chemical industry, an average spill size was derived.[17] A previous study by the DOT[21] found that the relationship between the annual spill frequency SF and an annual volume V in tons of chemical production for a number of facilities could be represented as

$$SF = 0.004 \times V^{1/3}$$

Further, the expected number of discharges EN over a period of y years is

$$EN = (y)(SF)$$

Thus, the expected number of discharges may be defined as

$$EN = y \times 0.004 \times V^{1/3}$$

These relationships represented a sample of reported discharges of some 30 different hazardous chemicals and averaged out any possible differences in frequency related to variations in chemical production processes.

Data on oil discharges reported to the American Petroleum Institute for the calendar years 1972 through 1974 illustrate the relationship between the number of spills and discharge volume.[12] Discharges of less than 1000 gal (3.8 m³) represent 83 percent of the number of discharges but accounted for only 5 percent of the volume of oil discharged. In strong contrast, discharges of greater than 10,000 gal (37.9 m³) represent only 4 percent of the number of discharges but accounted for 78 percent of the volume of oil discharged. Finally, discharges of oil between 1000 and 10,000 gal represent 13 percent of the number of discharges and accounted for 17 percent of the volume discharged. Since this oil spill data base represents 12,725 discharges over a full 3-year period, with a total discharge of 35,838,482 gal (135,663 m³), it provides a solid base from which to extrapolate to discharges of hazardous substances, if it is assumed that the same percentage distribution applies to hazardous substance discharges.

If the figures above are then plotted in lognormal distribution of size of discharge, the graphical representation reveals that the average spill size predicted by the distribution is 4000 gal (15.1 m³).

Other data supporting the figure derived above are found in an annual publication of the U.S. Coast Guard.[22] The Pollution Incident Reporting System (PIRS) serves as the information management system for all spills reported to or detected by the Coast Guard. In the report for the calendar year 1977, PIRS identified 280 chemical discharges with a total discharge volume of 1,427,992 gal (5506 m³). An average discharge size can therefore be calculated as 5100 gal (19.3 m³). Since the reporting period covered was prior to the initiation of mandatory reporting for hazardous substance discharges, the data cited above represent voluntarily reported or fortuitously discovered discharge events documented by the Coast Guard. However, there is no reason to believe that since the institution of mandatory reporting, calculated average discharge volumes will be statistically different from the figure derived above.

Additional data that strongly corroborate the two figures derived above are found in a recent publication of the Industrial Environmental Research Laboratory of the EPA.[16] Data from 272 confirmed spills of seven of the most widely used industrial chemicals (sulfuric, nitric, and hydrochloric acids, sodium hydroxide, toluene, xylene, and phenol) reveal that the calculated average discharge size is 6750 gal (25.6 m^3). Since the seven are among the most amply produced as well as widely used industrial chemicals and are also among the most frequently reported to be discharged, the average discharge size for all hazardous substance discharges is probably somewhat smaller than the figure derived immediately above. However, the specific amount of decrease is unknown.

Therefore, it may be calculated that, on the basis of the three figures separately derived above, the average size of hazardous substance spills is approximately 5300 gal (20.1 m^3). However, the most representative of the three figures derived above is probably the data derived from PIRS of the Coast Guard. Thus, it may be assumed for the purposes of this analysis that the average size of hazardous substance spills is 5100 gal.

Value of Product Lost

The total value of industrial organic and inorganic chemical production in 1972* was approximately $18 billion.[23] The total amount of industrial organic and inorganic chemicals produced in 1972 was approximately 4.3 × 10^{11} lb (1.95 × 10^{11} kg).[24,25] Thus, the average value of chemical product per pound may be calculated to be approximately 4 cents. If it is assumed that there are 300 discharges per year in which the quantity that reaches water is equal to or in excess of the reportable quantity, that the average discharge is 5100 gal, and that the average specific gravity of discharged materials is 1.0, then the annual value of product lost would be approximately $512,000. This figure should be viewed as a lower bound, since the average value of chemical product per pound will have increased substantially since 1972.

Another measure of the value of product lost comes from a study done for the Office of Domestic Shipping of the Maritime Administration in 1974.[26] A portion of the study examined 104 barge accidents between July 1968 and June 1973 in which there was a loss of product carried. The value of product lost from the 104 discharges was approximately $630,000. This calculates to an average loss per discharge of $6050. If this figure is assumed to be typical of all hazardous substance discharges and is multiplied by an assumed 300 discharges per year, the calculated annual amount of product lost from hazardous substances discharges is $1,815,000. This figure should probably be viewed as an upper bound, since it is unlikely that the average value of product lost per discharge will exceed $6000.

Cost of Retrieval and Cleanup Efforts

The discharger of a designated hazardous substance that enters the navigable waters of the United States is required to take appropriate actions to remove the discharged material and to mitigate the effects of any material remaining in the water body. Any removal

---

*These figures come from the 1972 *Census of Manufactures,* issued by the Bureau of the Census. The *Census of Manufactures* is a quinquennial index, and a considerable period of time elapses between the collection of data and the issuance of the volume. The 1972 *Census of Manufactures* was issued in 1976; the 1977 *Census of Manufactures,* in 1981.

or mitigation actions, such as neutralization, containment or surface removal, or dredging or vacuum pumping, will depend on the time, location, and circumstance of the discharge. Additional factors that will affect removal or mitigation actions include the nature of the substance discharged, the physical characteristics of the water body, and environmental conditions, such as cold temperatures, high winds, or heavy current or wave action.

Various estimates have been made of the costs of removal, mitigation, and cleanup. The 1976 paper by this author yielded a figure of $1 per pound for mitigation of soluble substances such as acids or bases.[12] The same paper yielded a figure of 84 cents per pound for cleanup and mitigation of insoluble floating hazardous substances that are oil-like in their physical-chemical characteristics. An analysis by the state of Washington of a hypothetical response to a discharge of oil or oil-like substances into Puget Sound yielded a mitigation and cleanup cost of $1 per pound.[27] A chemical cleanup firm on the west coast estimated that direct removal costs were approximately $1 per pound for hazardous substances, but that ultimate disposal costs could add $3 per pound to this figure.[28]

However, removal, mitigation, and cleanup costs have increased substantially since these estimates were derived, in most cases by at least a factor of 10 and in some instances by two orders of magnitude. This increase in costs is attributable to a considerable improvement in recent years in cleanup and removal techniques, to the point that substantial removal of discharged hazardous substances is now feasible. However, these improved technologies have also resulted in substantially increased equipment, material, and labor costs.

If an average cost of cleanup of $10 per pound is assumed and the assumed figures of 300 discharges per year, 5100 gal per discharge, and an average specific gravity of 1.0 are utilized, the calculated annual removal, mitigation, and cleanup costs are approximately $128 million. This figure may be conservative, since some removal or mitigation actions will be adversely affected by such factors as cold temperatures, high winds, or heavy currents, which will increase the required expenditures. However, some hazardous substance discharges will undoubtedly be removed or mitigated for less than $10 per pound, thus reducing the figure derived above.

Loss Due to Disruption of the Production Process

A discharge of a hazardous substance will not necessarily cause a disruption of the production process; such a disruption will depend on the nature of the chemical discharged, the volume of the discharge, and the location of the discharge within the chemical production or handling facility. However, in spills in which there is a release of toxic fumes, a discharge of substantial size, a hazard of fire, or other life-endangering circumstances, partial or total operational shutdown of the facility may be required during cleanup efforts. It has been estimated that the average loss of production time per discharge is approximately 8 h.[17] If it is assumed that each discharge results in the loss of one 8-h shift of production time, the loss due to disruption of production processes would be

> Annual production of chemicals (in pounds)
> $\times$ (number of chemical production facilities)$^{-1}$
> $\times$ average cost per pound of chemicals produced
> $\times$ annual number of discharges $\times$ (number of hours per year)$^{-1}$
> $\times$ loss of 8 h of production per discharge
> = loss due to disruption of production process

Another method of determining this loss is to estimate the percentage of time in which chemical production facilities are not producing because of unscheduled outages and to estimate the percentage of outage time that is due to hazardous substance discharges. It has been estimated that chemical production facilities are not producing 0.5 percent of the time, owing to unscheduled outages and that hazardous substance discharges are the cause of 5 percent of outage time.[17] Using these estimates, the loss due to disruption of production processes would be

>Annual production of chemicals (in pounds)
> $\times$ average cost per pound of chemicals produced
> $\times$ unscheduled outages as a percentage of production time (0.005)
> $\times$ percentage of unscheduled outages caused by discharges (0.05)
>   = loss due to disruption of the production process

The 1978 paper by the author derived values of $3.9 million and $4.3 million respectively for the two methods described above. These values may no longer be accurate, since the cost of discharged hazardous substances is generally more than the 4 cents per pound used in the analyses. However, since many facilities experience no loss of production time following a discharge, the change may be minimal. Nevertheless, it is believed that the figure of approximately $4.1 million is a reasonable estimate for the value of production lost due to disruption of production processes by hazardous substance discharges.

Cost of Equipment and Property Damaged and Destroyed

It is difficult to identify accurately the value of equipment and property damaged or destroyed following hazardous substance discharges, since specific data on such costs are rarely available. Chemical industry facilities are usually insured for damages due to chemical discharges that occur on the facilities' properties, and most insurance carriers generally consider compensation settlements to be policyholder-confidential information. Thus, these data are usually only minimally available.

Nevertheless, it is possible to make an estimate of the cost of equipment and property damaged or destroyed as a result of hazardous substance discharges. The organic and inorganic chemical industries had approximately $17 billion invested in plant equipment and property in 1972; this is an average of approximately $6 million per facility.[23] If it is assumed that 10 percent of hazardous substance discharges involved a 10 percent loss to an average facility's assets in terms of equipment and property, then an annual loss of $18 million in equipment and property damaged or destroyed would be sustained by all chemical industry facilities. This estimate does not include losses incurred as the result of catastrophic events such as massive explosions or fires.

Another source of data on equipment and property damaged or destroyed in the railroad industry is available from the Association of American Railroads.[29] In a report prepared for the Interindustry Task Force on Rail Transportation of Hazardous Materials, information is presented for the 2-year period 1975–1976, as shown in Table 3-1.

Other data in the report indicate that only 1 percent of all railroad accidents involve a release of hazardous material. However, derailments with a release of hazardous material account for 91 percent of all railroad accidents with a release of hazardous material; thus, another 9 percent of damage costs must be added. The result is a figure of $3,046,495, which represents damage to railroad equipment, track, and structures from train accidents with a release of hazardous material. However, this figure is for the 2-year period 1975–1976; thus, it must be halved, resulting in an approximate annual cost of $1,500,000 for railroad equipment, track, and structures damaged or destroyed as the result of an accident with a concomitant release of hazardous material.

Cost of Damage to Nonplant Property

As with damage and destruction of chemical facility equipment and property, it is difficult to identify accurately costs of damage to nonplant property caused by hazardous substance

Table 3-1 Number of Mainline Derailments by Speed and Damages, 1975–1976

Speed at derailment (kilometers per hour)	Number of derailments	Average damage per derailment*	Total damage*
1.6–16	2670	$ 17,000	$ 45,390,000
17–32	1618	$ 22,000	35,596,000
33–48	1659	$ 42,000	69,678,000
49–64	760	$ 72,000	54,710,000
65–80	471	$ 85,000	40,035,000
81–96	213	$110,000	23,430,000
96+	66	$127,000	8,382,000
			$277,231,000

*Damage to railroad equipment, track, and structures only.
SOURCE: Report No. R-344, Association of American Railroads, Chicago, December 1978.

discharges, since valid or detailed data on such incidents rarely are available. Nonplant property damage from a discharge typically is caused by airborne vapor carried downwind of the discharge, resulting in the peeling or chemical burning of painted surfaces such as those on residences or automobiles or damage to trees and shrubbery. Very rarely does substantial damage occur to residences or property located beyond the facility where the discharge occurred. Explosions and fires at chemical facilities must be massive, as at Flixborough, England, before they cause substantial damage to nonplant property.

Personnel in the insurance industry have estimated that average nonplant damage claims are approximately $5000 per discharge when claims are actually made. If it is assumed that 10 claims are made for each of half of the estimated spills, damage claims would result in $7.5 million per year in compensation by the chemical industry for damages to nonplant property. This figure is less than the real cost of damages, since the analysis does not take into account any environmental damages caused by spills. One chemical facility manager stated that such costs depend strongly on the environmental value of the water body or area in question and, more importantly, on the environmental consciousness of the population surrounding the chemical facility.[17]

Cost of Damage to the Environment

The costs resulting from damage to the environment from hazardous substance discharges are difficult to determine accurately, since such costs are rarely identified or quantified. Even when there are visible and identifiable damages, such as demonstrable fish kills, there often is disagreement among planners of natural resources or environmental managers of the replacement value of the resource. When a fish or wildfowl kill can be attributed to a specific hazardous substance discharge, the discharger may agree to pay either the hatchery and distribution costs or a nominal amount, such as $1 per animal destroyed. In other cases, the discharger may be held liable under Sec. 311(f)(4) of the Clean Water Act, which states, "The costs of removal of oil or a hazardous substance . . . shall include any costs or expenses incurred . . . in the restoration or replacement of natural resources damaged or destroyed as a result of a discharge . . . ."[6]

In a case dealing with damages caused by an oil spill from a vessel into an estuarine mangrove forest, a federal judge in Puerto Rico ruled that the vessel's insurance underwriters had to compensate the Commonwealth of Puerto Rico with $6.1 million for damages to the environment.[30] His determination was based on the catalog value from biological supply laboratories of approximately 92 million animals over an estimated area of 20 acres (8.1 ha) that had been killed by the oil and on the cost of replanting 23 acres (9.3 ha) of destroyed mangrove trees. However, there was no guarantee that the Commonwealth of Puerto Rico would be able to restore the damaged environment with the $6.1 million that it was to receive or even with the $10 million that had been originally requested. Even though a federal appeals court later set aside the financial award to the Commonwealth of Puerto Rico as being too high and remanded the case to the original district court for a more "appropriate" award, it fully sustained Sec. 311(f)(4) of the Clean Water Act and government's right to sue for the "total value" of damages caused to the environment and natural resources by pollutional discharges.[31]

Damages to sessile populations of organisms such as clams or oysters, which have commercial value, may be more difficult to assess. In the absence of an extensive survey of the damaged area, it may be difficult to determine accurately the extent of damages in order to assess costs. For example, large areas of shallow Chesapeake Bay bottoms produce in excess of 30 bushels of oysters per acre per year. The average price for oysters to the harvester during the summer of 1979 at Chesapeake Bay dockside landings was $80 per bushel. Thus, the value of such bottoms in annual yields is $2400 per acre. But such ecosystems are naturally renewing resources; thus, the actual worth of such a resource is its capitalized value, which is the amount of capital, in real-dollar terms, which must be invested at a given market interest rate in order to return a perpetual dividend equal to a fixed annual net profit. For example, if a particular estuary provides $1000 in net social benefits per year from its shellfish resources, then its capitalized value at a market interest rate of 5 percent is $20,000. (The calculation is derived from the formula for capitalized value $C = V/i$, where C is the capitalized value, V is the annual net profit, and i is the interest rate.) If a 5 percent interest rate for the calculation of income capitalization is assumed, an acre of Chesapeake Bay bottom is worth $48,000 for shellfish production alone. Yields approaching a $150,000 capitalized value per acre (in 1973 prices) have been recorded for clam beds in Maine.[32] A hazardous substance discharge that destroys 10 acres (4 ha) of productive shellfish beds may thus cause environmental damages approaching $1.5 million. The cost of planting seed oysters in the harbor at New Haven, Connecticut, in 1967 was $578,000, including the cost of the oyster spat. Destroyed owing to pollutional discharges, the beds would otherwise have yielded a potential income in succeeding years of $6.7 million.[33]

Further, environmental or ecological damages from hazardous substance discharges are often sublethal and long-term, and disruptions to the ecosystem may not be revealed for several months or even longer. A hazardous substance discharge causing sublethal effects to a species essential to a particular food chain may have damaging effects on the other organisms in the chain, and such disruptions might not be revealed for several months. For example, certain migratory waterfowl that winter in the wetlands of Chesapeake Bay consume large quantities of snails, clams, and marine worms as part of their diet. A hazardous substance discharge that resulted in decreased productivity of these bottom-dwelling organisms might have long-term effects on the survival of the wintering waterfowl, should a replacement food source not be available.

Cost of Compensation and/or Injury to Employees

In 1977, there were 187,000 industrial chemical production workers[34] of whom 10 percent experienced a job-related injury or illness. Assume that one-quarter of those injuries or illnesses were the consequence of hazardous substance discharges. Medical costs for treating those disabled as a result of discharges, at an assumed cost of $200 per incident for 90 percent of the cases and $2000 per incident for 10 percent of the cases, would then be approximately $1.8 million per year. This estimate does not include the value of lost salary by workers forced by injury or illness to remain off the job, since such costs are not direct expenditures by the chemical industry from hazardous substance discharges.

However, the associated incidental costs of industrial accidents have been found to be 4 times as large as the immediate compensation and medical payments.[35] Expressed in another way, compensation and medical payments constitute only 20 percent of the total employer accident costs. Thus, the total costs to the chemical industry for job-related injuries or illnesses resulting from hazardous substance discharges may be calculated to be approximately $8.9 million per year. The associated incidental costs of industrial accidents include such factors as:

1. Cost of lost time and productivity of an injured employee
2. Cost to the employer under employee welfare and benefit programs
3. Cost of time spent on the case by industrial medical attendants or hospital personnel when such costs are not covered by insurance
4. Cost of lost time of supervisors assisting an injured employee, investigating the cause of the accident, or preparing accident reports
5. Cost of lost time of other employees who stop work out of curiosity or sympathy to assist the injured employee

These factors as well as others, such as the cost due to disruption of the production process and the cost of plant equipment and property damaged or destroyed, contribute significantly to the increase in total cost to the chemical production industry for job-related injuries or illnesses resulting from hazardous substance discharges.

Cost of Lost Water Usage

Very little information exists in the technical literature concerning the cost of lost or disrupted water usage by industrial, municipal, agricultural, recreational, or other water users as a result of hazardous substance discharges. Often the user may not even be aware that the water supply has been affected by a hazardous substance discharge; the only evidence may be a chemical process disrupted for a period of time by a too alkaline or a too acidic water supply, an agricultural commodity whose growth appears to have been retarded, or a complaint that the water in lakes or rivers or from household faucets "tastes funny." Nevertheless, the costs of such lost or disrupted water usage, even though not presently quantifiable, must be a part of the economic impact of hazardous substance spills. It is hoped that the results of future research will be able to assist in the quantification of the costs of such lost water usage.

SUMMARY

A summary of the costs derived in Part 3 can thus be tabulated on an annual basis for the chemical industry:

Factor	Cost
Value of product lost	$ 1,165,000*
Cost of retrieval and cleanup efforts	$128,000,000†
Loss due to disruption of the production process	$ 4,100,000*
Cost of equipment and property damaged and destroyed	$ 18,000,000†
Cost of damage to nonplant property	$ 7,500,000
Cost of damage to the environment	. . . . .‡
Cost of compensation and/or injury to employees	$ 8,900,000
Cost of lost water usage	. . . . .‡

*Average of two figures cited in the preceding section.
†Estimate that is probably higher than actual expenditures.
‡Not determined on an annual basis for the chemical industry as a whole.

If the above figures in the table are totaled and divided by the estimated annual spills, an average statistical cost per discharge may be determined. This cost exceeds $500,000 per spill. While this figure is undoubtedly too high for a number of hazardous substance discharges that reach water in amounts equal to or in excess of the reportable quantity for the particular substance, it may well underestimate the true economic impact of large, particularly damaging hazardous substance spills.

A contract report for the EPA developed an economic decision model to aid chemical facility managers in arriving at a determination about spill prevention investment costs.[28] The model provided a methodology for a chemical plant manager to determine two decision points:

1. The breakeven level of investment for an individual facility above which the cost of spill prevention exceeds the penalty cost of spillage

2. The point in time at which the investment should optimally be made for the facility

The model developed by the report is quite lengthy and complex, and a full discussion of it is beyond the scope of this part. For a complete description and analysis of the model, the reader is referred to the final contract report.[28]

There are a number of methods which the federal government can use to promote a high standard of care in the manufacture, storage, handling, transportation, and use of hazardous substances. These incentive systems should:

1. Encourage prevention of discharges, as well as better preparation for and implementation of removal and mitigation measures following spills

2. Have no unreasonable economic impacts

3. Be feasible to implement and enforce

4. Encourage responsible activity, as well as have direct and certain penalties for irresponsible actions

5. Be equitable for both large and small firms and for all transportation modes

6. Be fair in both intent and application

Incentive systems can take the form of either direct controls or financial inducements.

Direct Controls

Direct controls can be either specific standards and regulations or expanded damage liabilities. Specific standards and regulations require compliance with stipulated language fixed by law. All potential dischargers must comply with specified conditions that are directly enforceable by legal means. The advantage of specific standards and regulations is that if they clearly define the objective to be obtained, the environmental benefits to be gained are also clearly defined. Further, once potential dischargers have complied with the specified standards and regulations, there usually are no further charges to pay. The disadvantage of standards and regulations is that the cost of compliance can be quite high and, further, that no real incentive is provided for potential dischargers to surpass the specified level of compliance. Moreover, potential dischargers who do not comply can offer products at a lower cost than those making an effort to comply.

Expanded damage liabilities, which make the producers, handlers, or transporters of hazardous substances liable for all damages resulting from spills, regardless of fault, provide a strong incentive for the safe handling of hazardous substances at all levels of usage. The advantage of this approach is that liability for environmental damage or insult is made signally clear; the disadvantage is that substantially larger deductible limits and total liability coverage must be borne by the producer, handler, or transporter of the substances regulated.

Financial Inducements

Financial inducements can be either positive, such as direct payments or subsidies, or negative, such as taxes or pollution charges. Direct payments invert the principle of "the polluter pays" by granting the potential discharger of hazardous substances a payment in exchange for an agreement by the potential discharger to abate discharges. In essence, the potential discharger sells the right to pollute in exchange for the payment. The advantage of direct payments is that there is no reason for the potential discharger not to make use of the payment to implement improved discharge prevention and response measures. The disadvantage is that the public, through its collected taxes, must bear the cost of the potential discharger's antipollution measures; further, to meet the criterion of fair intent and application, payments would have to be made to potential dischargers who implement improved discharge prevention and response measures on their own.

Subsidies relieve potential dischargers of part or all of the costs of discharge prevention and response measures with which potential dischargers are obliged to comply. Typically they take the form of tax credits, accelerated amortization, or low-cost loans. The advantage of subsidies is that they encourage more rapid implementation of discharge prevention and response measures. The disadvantage is that subsidies are rarely large enough to make pollution control measures profitable; further, they can be an incentive for potential dischargers to inflate pollution control costs in order to obtain additional subsidized aid.

Taxes can be imposed by the government on the products or activities of potential dischargers that may result in discharge events. The advantage of taxes is that the revenues collected can be utilized by the government to create economic incentives for discharge prevention and response measures such as direct payments, subsidies, low-cost loans, or tax credits. The disadvantage is that taxes may be regressive or inflationary or may result in increased unemployment or economic dislocation. Further, a tax structure can be distorted into a purely revenue-collecting mechanism if the collected funds are not returned to the system as an economic stimulus.

Pollution charges require the potential discharger to pay a penalty proportional to the amount of pollution discharged. If the penalty is directly related to the damage caused by the spill, the discharger should pay only for the damage resulting from the spill. The advantage of pollution charges is that the potential discharger either pays for discharge prevention and response measures in advance or for damages caused by the discharge subsequent to the spill. The charges also encourage a potential discharger to develop improved solutions for spills, thereby reducing the magnitude and frequency of pollution charges. The disadvantage of this approach is that an equitable relationship between actual damages caused by a discharge and the penalties assessed is difficult to develop. The level of charges assessed is also difficult to establish equitably. Pollution charges that are excessively low allow a potential discharger to pollute at a financially advantageous level; charges that are unduly high are regressive and generally result in economic dislocation.

CONCLUSION

Hazardous substance spills may have a variety of outcomes, ranging from minimal effects to olfactory insults, chemical burns or poisoning, destruction of both human and aquatic life, destruction of tangible property, and modification of ecosystems. This part has utilized available information to derive various aspects of the economic impact of hazardous substance spills.

Although the figures derived in these analyses are not to be regarded as definitive, they are the best estimates obtainable on the basis of available data. However, some of the data utilized are dated, such as the value of industrial organic and inorganic chemical production cited in the subsection "Value of Product Lost." Other portions of the analysis, such as the number of spills occurring annually, are based on professional judgment, which inherently incorporates arbitrary discretion. In addition, some portions of the analysis are flawed by a lack of such data as the cost of equipment and property damaged or destroyed from hazardous substance spills in the chemical trucking industry.

All these factors contribute to a moderation of the impact of this analysis. However, additional data possessed by the chemical production industry, discharge cleanup contractors, or producers of discharge prevention equipment may lead to improvement in these analyses.

REFERENCES

1. *Water Quality Improvement Act of 1970,* Pub. L. 91–224, 91st Cong., 1st Sess., Apr. 3, 1970.

2. *Control of Hazardous Polluting Substances: A Message from the President of the United States Transmitting a Report on Control of Hazardous Polluting Substances, Pursuant to Section 12(g) of the Federal Water Pollution Control Act, as Amended,* 92d Cong., 1st Sess., House Document 92-70, 1971.

3. *Federal Water Pollution Control Act Amendments of 1972,* Pub. L. 92–500, 92d Cong., 2d Sess., Oct. 18, 1972.

4. *Federal Register,* vol. 39, Aug. 22, 1974, pp. 30466–30471.

5. *Federal Register,* vol. 40, Dec. 30, 1975, pp. 59960–60017.

6. *Clean Water Act of 1977,* Pub. L. 95–217, 95th Cong., 1st Sess., Dec. 15, 1977.

7. *Federal Register,* vol. 43, Mar. 13, 1978, pp. 10474–10508.

8. *Manufacturing Chemists Association et al. v. Costle et al.,* 455 F. Supp. 968 (W.D. La. 1978).

9. Pub. L. 95–576, 95th Cong., 2d Sess., Nov. 2, 1978.

10. *Federal Register,* vol. 44, Feb. 16, 1979, pp. 10266–10284.

11. *Federal Register,* vol. 44, Aug. 29, 1979, pp. 50766–50786.

12. J. E. Amson, "An Analysis of the Economic Impact of Hazardous Substances Regulations," in G. F. Bennett (ed.), *Proceedings of the 1976 National Conference on Control of Hazardous Material Spills,* New Orleans, 1976, pp. 49–54.

13. *Deepwater Port Act of 1974,* Pub. L. 93–627, 93d Cong., 2d Sess., Jan. 3, 1975.

14. *Fishery Conservation and Management Act of 1976,* Pub. L. 94–265, 94th Cong., 2d Sess., Apr. 13, 1976.

15. *Outer Continental Shelf Lands Act Amendments of 1978,* Pub. L. 95–372, 95th Cong., 2d Sess., Sept. 18, 1978.

16. J. L. Buckley and S. A. Wiener, *Hazardous Material Spills: A Documentation and Analysis of Historical Data,* EPA-600/2-78-066, U.S. Environmental Protection Agency, Industrial Environmental Research Laboratory, April 1978, p. 21.

17. J. E. Amson and J. L. Goodier, "An Analysis of the Economic Impact of Spill Prevention Costs on the Chemical Industry," in G. F. Bennett (ed.), *Proceedings of the 1978 National Conference on Control of Hazardous Material Spills,* Miami Beach, Fla., 1978, pp. 39–45.

18. W. J. Mead and P. E. Sorensen, "The Economic Cost of the Santa Barbara Oil Spill," *Santa Barbara Oil Symposium: An Environmental Inquiry,* Santa Barbara, Calif., 1970, pp. 183–226.

19. *Hazardous Material Incidents Reported to U.S. Environmental Protection Agency Regional Offices from October 1977 through September 1979,* EPA-430/9-79/019, U.S. Environmental Protection Agency, Oil and Special Materials Control Division, January 1980.

20. *National Response Center Discharge Reports,* U.S. Coast Guard headquarters, Washington, 1980.

21. *Survey Study to Select a Limited Number of Hazardous Materials to Define Amelioration Requirements,* Report No. CG-D-46-75, U.S. Department of Transportation, Office of Research and Development, March 1974.

22. *Polluting Incidents in and around U.S. Waters: Calendar Year 1977,* COMDTINST M16450.2, U.S. Coast Guard, Department of Transportation, July 1978.

23. *Census of Manufactures: 1972,* U.S. Bureau of the Census, 1976.

24. *Assessment of Industrial Hazardous Waste Practices: Organic Chemicals, Pesticides, and Explosives Industries,* Publication PB-251307, U.S. Environmental Protection Agency, Solid Waste Programs Office, April 1975.

25. *Assessment of Industrial Hazardous Waste Practices: Inorganic Chemical Industry,* Publication PB-244832, U.S. Environmental Protection Agency, Solid Waste Programs Office, March 1975.

26. "A Modal Economic and Safety Analysis of the Transportation of Hazardous Substances in Bulk," *Final Contract Report to the Office of Domestic Shipping, U.S. Maritime Administration,* Report C-76446, Arthur D. Little, Inc., Cambridge, Mass., July 1974.

27. *Background to Requested Oil Spill Contingency Fund,* State of Washington, Office of Program Planning and Financial Management, April 1975.

28. "Response by Regulated Industry in Terms of the Degree of Prevention and Associated Cost Relative to Section 311 of P.L. 92–500," *Final Contract Report to the Office of Water Planning and Standards, U.S. Environmental Protection Agency,* Report C-79212, Arthur D. Little, Inc., Cambridge, Mass., December 1976.

29. *Phase I Final Report of the Systems Safety Analysis Subcommittee, Interindustry Task Force on Rail Transportation of Hazardous Materials,* Report No. R-344, Association of American Railroads, Chicago, December 1978.

30. *Commonwealth of Puerto Rico v. the S.S. Zoe Colocotroni et al.,* 456 F. Supp. 1327 (D.P.R. 1978).

31. *Commonwealth of Puerto Rico v. the S.S. Zoe Colocotroni et al.,* 628 F.2d 652 (1st Cir. 1980).

32. R. L. Dow, *Some Economic and Educational Values of Estuaries,* Atlantic States Marine Fisheries Commission, Augusta, Me., 1966.

33. Dennis P. Tihansky and Norman F. Meade, "Establishing the Economic Value of Estuaries to U.S. Commercial Fisheries," *Estuarine Pollution Control and Assessment: Proceedings of a Conference,* Washington, 1976, pp. 671–684.

34. *Statistical Abstract of the United States: 1978,* U.S. Bureau of the Census, 1978.

35. H. W. Heinrich, *Industrial Accident Prevention,* 4th ed., McGraw-Hill Book Company, New York, 1959.

Assessment
of Hazard and Risk

V. C. Marshall

Director of Safety Services
University of Bradford

How Containment May Be Lost / 5-2
Consequences of Loss of Containment / 5-3
Spillages of Gases, Liquids, and Solids / 5-4
Special Problems Relating to Liquefied Gases / 5-4
Formation of Vapor Clouds / 5-7
Differences between Flammable Clouds and Toxic Clouds / 5-8
Toxicity / 5-8
Flammable Limits / 5-10
Fireballs / 5-11
Unconfined Vapor Cloud Explosions / 5-11
Special Problems of Reactors / 5-14
Lethal Effects and Damage Effects of Vapor Clouds / 5-16
Quantifying the Hazard / 5-16
Question of Risk / 5-19
Question of Damage / 5-22

The objectives of this chapter are:

1. To characterize the principal hazards associated with spillages
2. To recognize the agencies by which these hazards may realize their potential
3. To quantify the various hazards with regard to their capacity to injure life or limb or to damage property
4. To provide a select list of major spillages (see Table 4 at the end of the chapter). Such spillages are marked in the text with an asterisk.

Spillages which are the province of the nuclear engineer, such as radioactive fallout, and spillages which are the province of the civil engineer, such as those of water impounded in a dam or the movement of spoil heaps, as, for example, that which occurred in the Aberfan, Wales, disaster of 1966, will be excluded, as will be spillages of crude oil in transit. Nor will this chapter be concerned with the chronic effects of long-continuing releases which fall under the heading of environmental pollution.

For the purposes of this chapter a hazard will be defined as "a physical situation with a potential for injury to life, limb, or property." A risk will be defined as "the probability of the realization of a hazard at any given level of injury." "Spillage," for the purposes of this chapter, will be defined as a loss of containment.

HOW CONTAINMENT MAY BE LOST

Loss of containment may occur at any point on a continuum from a pinhole leak to a catastrophic failure. Losses of the former type are of frequent occurrence; those of the latter type are rare. But catastrophic failures do occur in spite of all the efforts of designers to avoid them, and they are, for obvious reasons, very dangerous as they give little or no time for evacuation. Catastrophic failure, regardless of the nature of the contents, is generally very dangerous as it will almost certainly give rise to a blast wave and may also give rise to the projection of missiles. Examples of catastrophic failures include a railcar of dimethyl ether at Ludwigshafen, Germany, in 1948* (207 fatalities) and a tank of ammonia at Potchefstroom, South Africa, in 1973* (18 fatalities). In such cases the time taken to release the contents is very short and is almost impossible to calculate.

At the other end of the scale, the rate of loss is relatively easy to calculate for apertures which are small in relation to the surface area of the container. The failure at Flixborough, England, in 1974,* at which approximately one-third of the contents of a reaction system containing about 120 metric tons escaped in about 45 s is an intermediate case. Gas flows from apertures are limited by sonic velocities, and beyond a certain critical figure an increase in internal pressure does not increase the rate of egress.[1]

Types of Spillage

It may be useful to distinguish two basic situations: spillage from storage and spillage from reaction. The former type includes all cases of storage including the transport of materials in containers. The latter includes cases in which spillage occurs from reaction systems either through failure at operating pressure as at Flixborough or, more usually, through failure of control of reaction as at Seveso, Italy, in 1976.*

CONSEQUENCES OF LOSS OF CONTAINMENT

Factors Governing Severity

Five principal groups of factors govern the severity of the consequence of spillage:

1. Intrinsic properties: flammability, toxicity, and instability
2. Dispersive energy: pressure, temperature, and state of matter
3. Quantity present
4. Environment factors: topography and weather
5. Population density in the vicinity and proximity of property

Size and Shape of Affected Area

The shape of the area affected by an explosion will approximate to a circle, distorted perhaps by local topography, and the area affected will in general be roughly proportional to $M^{2/3}$, where M is the mass of the exploding material. The size and shape of the area affected by a sudden catastrophic fire will be similar.

The shape and orientation of the area affected by a toxic release will be governed by the wind speed and direction. Only in conditions of calm will the shape approximate to a circle. Generally there will be wind, which will produce a roughly triangular shape. The higher the wind speed, the smaller the apex angle. There does not seem at present to be any simple law which will relate the area affected to an exponential function of the mass of material spilled.

*See Table 4 for details.

SPILLAGES OF GASES, LIQUIDS, AND SOLIDS

Gases

Gases under pressure will, if containment is lost, disperse into the surrounding atmosphere until the pressure in the burst container is equal to the atmospheric pressure. During and after release they will mix with the surrounding atmosphere by turbulence and diffusion.

Liquids

The behavior of liquids on loss of containment depends much upon whether they are being stored at a temperature below or above their boiling point at atmospheric pressure. If they are stored at a temperature below their atmospheric-pressure boiling point, it is ruptures below the free surface of the liquid which are significant. From these ruptures the liquid will escape at a rate governed by the hydrostatic head available, by the size and shape of the aperture, and by the flow properties of the liquid.

Liquids stored above their atmospheric-pressure boiling point will escape from containers ruptured below the free surface at a rate which is governed by the excess pressure plus the hydrostatic head. (The subsequent behavior of such a spilled liquid will be described later.) When the container is ruptured above the free surface, vapor will escape, thus lowering the pressure, which will cause the liquid to boil, perhaps with great violence, and resulting in the expulsion of entrained liquid with the vapor.

Solids

The behavior of solids is more complex than that of liquids. In some cases solids can be scattered by localized explosions or runaway reactions (as at Seveso), or they may be scattered by loss of containment during transit. Much more rarely solids may become fluidized, and a heap of solid matter may shift to engulf people (as at Aberfan).

SPECIAL PROBLEMS RELATING TO LIQUEFIED GASES

The question of liquefied gases is so important that their behavior on spillage deserves special treatment. This is so for two reasons: (1) a number of such gases are now being manufactured or handled on a world scale ranging from millions to hundreds of millions of metric tons annually, and (2) the loss of containment of such gases can lead to serious loss of life or serious damage, or both. The vapor pressure–temperature relationships of the principal liquefied gases are shown in Fig. 1.

Generally, there are two divisions of liquefied gases, toxic and flammable, although some of these gases have both properties. Both classes behave similarly on loss of containment, in that partial vaporization rapidly occurs under adiabatic conditions. This vaporization is accompanied by a decrease in temperature down to the atmospheric-pressure boiling point. Thereafter the rate of vaporization is determined by the rate of heat input from the surroundings and by the rate at which the vapor mixes with the ambient air. The latter rate is greatly influenced by wind velocity.

Flashing Characteristics

As noted above, liquefied gases will undergo flash vaporization on loss of containment. The fraction which may theoretically flash is governed by (1) the difference between the

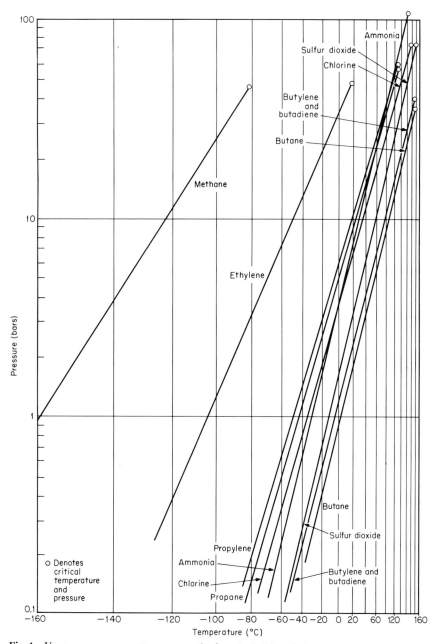

Fig. 1 Vapor pressure versus temperature plot for principal liquefied gases.

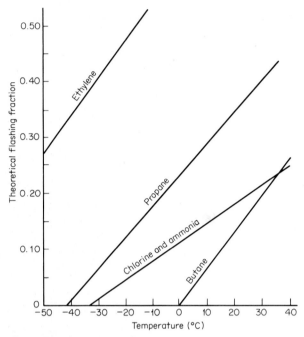

Fig. 2   Theoretical flashing fraction of principal liquefied gases.

temperature at which the liquid is stored and the atmospheric-pressure boiling point and (2) the ratio between the specific heat of the liquid and its latent heat.

The fraction evaporated x may be calculated by using the following formula:

$$x = (A - B)/(C - B) \qquad (1)$$

where A = specific enthalpy at a temperature T° above the atmospheric-pressure boiling point

B = specific enthalpy of the liquid at the atmospheric-pressure boiling point

C = specific enthalpy of the vapor at the atmospheric-pressure boiling point

The fraction theoretically vaporized has been plotted as a fraction of temperature for a number of liquefied gases in Fig. 2. Flashing may be so rapid as to be almost explosive in its violence, and in consequence slugs of liquid may be propelled for considerable distances. Such slugs will themselves tend to disintegrate through vapor formation, and the droplets produced may undergo evaporation by contact with the surrounding air. This evaporation will cool the drops and will also cool the air, contributing to its negative buoyancy.

It is virtually impossible to compute exactly how much liquid will be entrained in any given situation, and only a rule of thumb can be suggested, namely, that entrainment under conditions of rapid loss of containment be taken as equal to the theoretical fraction evaporated. If this rule is accepted, some of the liquefied gases under consideration may be expected to be almost completely dispersed.[2]

Vaporization from Pool

If a pool remains, the evaporation rate will depend upon wind speed, upon certain physical characteristics of the liquid, and/or upon the rate at which heat is transferred into the pool from the surroundings. The calculation of rates of evaporation is relatively well understood compared with some of the other problems under discussion and may best be summed up in Fig. 3, which is taken from the work of Clancey[3] and which assumes a constant surface area. This constant-rate evaporating phase is usually the less dangerous stage in a spillage.

However, in the case of cryogenic liquids, in which there is no flash evaporation stage, there is an initial stage at which the surroundings are much hotter than the liquid and are capable of inducing very rapid boil-off with some entrainment. Later, when the surroundings have become chilled, evaporation becomes constant in rate. When spillage occurs onto a water sufrace such as the ocean, there may be effectively only one stage. From a water surface the boil-off is much more rapid than shown by Clancey's graph.

FORMATION OF VAPOR CLOUDS

The behavior of the released vapor follows a number of stages of which three are the most significant. Stage 1 is the initial phase, in which a cloud substantially unmixed with air forms in the immediate vicinity of the spillage. Most if not all of the vapors under consideration are heavier than air on release. This is so because even vapors with a molecular

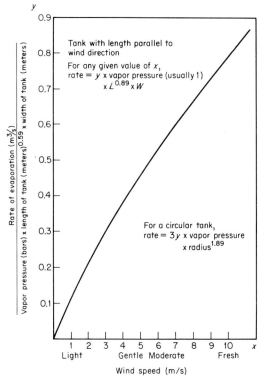

Fig. 3 Rate of evaporation from a pool of liquefied gas on land. [*Based on V. J. Clancey,* The Evaporation and Dispersal of Flammable Liquid Spillages, *Institution of Chemical Engineers Symposium Series, no. 39a, Rugby, England, 1974. Used by permission.*]

weight of less than 28.8 have a mean temperature on release substantially below the ambient temperature.

In Stage 2, the clouds slump under the influence of gravity, mix with the ambient air, and become propelled by the wind. The behavior of dense clouds of this character is not as yet well understood, although it is now the subject of considerable research.[4]

Stage 3 is the stage at which mixing has so far advanced that the cloud no longer differs greatly in its density from air; i.e., it is of neutral buoyancy. The behavior of such clouds forms an integral part of the science of meteorology. The elucidation of the laws governing these clouds owes much to Pasquill.[5]

DIFFERENCES BETWEEN FLAMMABLE CLOUDS AND TOXIC CLOUDS

There is a considerable difference between the properties of clouds containing flammable vapors and those of clouds containing toxic vapors. The lower limit of flammability, which is of the order of 1 to 3.5% by volume for a hydrocarbon at atmospheric pressure and which bears a roughly inverse relationship to the molecular weight of the hydrocarbon, lies in the region at which the density of the cloud differs markedly from that of the surrounding air. (See Fig. 4 for the flammability limits of some common gases.) Thus when diluted with the air, the gas has ceased to be dangerous long before the cloud has become neutrally buoyant.

This situation is by no means true of toxic clouds. For chlorine, exposure of 10 min to 1000 ppm may prove fatal, but air containing 1000 ppm of chlorine may well be neutrally buoyant. Prediction of the dangerous zone for toxic clouds must take account of their behavior in a state of neutral buoyancy until the point at which they have become so dilute as not to be dangerous.

TOXICITY

Toxicities of common gases are shown in Table 1,[7,8,9,10] but the figures given must be treated with caution. There is no unique figure for the concentration which is lethal to

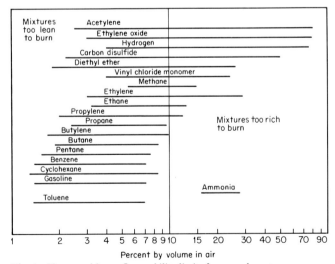

Fig. 4   Upper and lower flammability limits for several gases.

TABLE 1 Comparative Toxicities of Selected Gases

Gas	Odor threshold (ppm)*	Threshold limit values (1975 values) ppm	mg/m³	Lethal concentration	Lethal dose (as ppm × min)†	Lethal dose concentration	Lethal dose (as ppm × min)†	Lethal dose (as ppm × min)†	Toxicity ratings ...†,‡	...§	...¶
Phosgene (COCl₂)	1	0.10	0.40	90 ppm rapidly fatal; 25 ppm for 30 min dangerous	750 dangerous	12.4 ppm for 20 min; 100 ppm for 30 min	240–3000	790	1	Highly toxic	
Hydrogen cyanide (HCN)	2–5	10	11	181 ppm for 10 min; 135 ppm for 30 min	1810–4005	180 ppm for 5 min; 180 ppm for 10 min	900–1800		1	5 z	
Hydrogen sulfide (H₂S)	0.3	10		400–700 ppm for 30–60 min	12,000–42,000	500 ppm for 1 min; 900 ppm for 5 min	500–4500		1–2	5 z	
Chlorine (Cl₂)	5	1	3	610 ppm for 60 min	37,000	280 ppm for 24 min; 630 ppm for 66 min	8700–40,000	4200	2–3	5 z	
Sulfur dioxide (SO₂)	ca. 1	5	13			800 ppm for 20 min; 600 ppm for 5 h	15,000–180,000		2–3	4 z	Group 1
Ammonia (NH₃)	5	25	18	5000–10,000 ppm for 30 min dangerous	150,000–300,000 dangerous	10,000 ppm for 10 min 100 ppm for 60 min	100,000–600,000		3–4	5 z	Group 2

*Odor threshold varies greatly from individual to individual and with training. High concentrations of some gases such as hydrogen sulfide can paralyze the sense of smell. Sulfur dioxide can be detected by taste before it can be smelled.

†Computed by the writer from the data given.

‡Ratings are concentrations producing death from 4-h inhalation of two to four rats out of six. 1 = < 10, 2 = 10–100, 3 = 100–1000, 4 = 1000–10,000 ppm. Thus rating 5 = log of upper figure in band.

§Rating 5 = 4-h inhalation of 250 ppm kills five to six of six standard rats. Rating 4 = 4-h inhalation kills from no to four of six standard rats. z denotes that experience with humans has yielded a judgment of hazard differing from that based on animals.

¶Underwriters Laboratories classification. Group 1: gases of vapors which in a concentration of 5000 to 10,000 ppm for 5 min are lethal or produce serious injury; Group 2: the same as Group 1, but exposures of 30 min or more needed.

human subjects. Humans vary in sex, age, physique, and state of health; also the concentration necessary to produce death in a given time will depend upon the rate of breathing and thus upon the degree of exertion. Moreover, direct experimentation is impossible. Reliance usually has to be placed upon animal experiments, which cannot always be reliably extrapolated to human subjects. Sometimes, however, there is a degree of confirmation from accidents with fatal results. Then, again, the time factor is crucial. Only a short exposure will suffice with a high concentration, whereas a long exposure will be required for a low concentration. The product, concentration-exposure time, is not a constant, generally speaking, for lethal effects. The product is less for the couplet high concentration–low time than it is for low concentration–high time.

There is a threshold value for concentration below which even indefinite exposure will not produce adverse, let alone fatal, results.[6] This threshold value may be higher for systemic poisons than for less toxic irritants; for example, compare the threshold value for hydrogen cyanide with the value for chlorine in Table 1.

Role of the Toxicologist

The laws which govern the subject do not fall within the body of engineering knowledge, and it cannot be too strongly stressed that in predicting the consequences of toxic spillages engineers must seek the guidance of professional toxicologists.

FLAMMABLE LIMITS

As previously discussed, flammable vapors have a lower limit of flammability and thus are diluted below the level of danger long before toxic clouds. Toxic clouds obviously have no upper toxic limit, but in the region close to an escape clouds of potentially flammable materials may well have concentrations beyond the upper limit of flammability, i.e., the concentration at which the oxygen concentration is the determining factor. The upper and lower flammable limits for a number of gases are shown in Fig. 4. Hydrogen, acetylene, and ethylene oxide have very wide limits, the two latter being relatively unstable molecules. A vapor cloud will typically have three zones: a lean zone, a flammable zone, and an overrich zone, as idealized in Fig. 5. In practical circumstances the zones will be less well defined, with a pocket of one condition being surrounded by a zone of a different condition.

Flammable Clouds May Also Be Toxic

The overrich zone and, in some cases, the flammable zone may asphyxiate through oxygen deficiency (a 50% vapor concentration implies 50% oxygen depletion) as well as having narcotic effects. Thus a 25% concentration of hydrocarbon vapor will kill in minutes by combined narcotic and asphyxiatory effects. In addition, the cryogenic effects may also cause injury or death.

Fig. 5   Idealized section of flammable cloud.

FIREBALLS

The ignition of a cloud of flammable vapor may give rise to a fireball. Such fireballs are dangerous in the extreme. When formed of hydrocarbons, they are luminous, and they radiate heat which can cause fatal burns to bystanders. As they rise to form a mushroom cloud, the stalk consists of a violent upward convection current which can suck up debris, ignite it, and scatter burning brands over a wide area.

Size and Duration of Fireballs

Attempts have been made to predict the size and duration of fireballs. Formulas attributed to the National Aeronautics and Space Administration have been modified by the writer as follows:

$$D = 55 \times \sqrt[3]{M} \text{ (approximately)} \tag{2}$$

where D = diameter of fireball (in meters)

M = mass of hydrocarbon of assumed formula C_nH_{2n} (in metric tons)

and

$$T = 3.8 \times \sqrt[3]{M} \text{(approximately)} \tag{3}$$

where T = duration in seconds

M = mass of hydrocarbon of assumed formula C_nH_{2n} (in metric tons)

It is not only the energy release which is significant but the rate at which it is released, i.e., its power. If the energy released is a linear function of M and the time taken to release it is a function of M, then power = $f(M^{2/3})$. Taking a release of 50 metric tons of hydrocarbon,

$$T = \text{ca. 14 s}$$
$$P = \text{ca. 170,000 MW}$$

The size of the cloud and the rate of combustion are difficult to imagine. Perhaps it might be helpful to say that the cloud would have a volume equal to several times the volume of St. Paul's Cathedral in London, and the combustion rate, about 3.5 metric tons/s, compares with 0.1 metric ton/s, which is the rate of combustion in the boiler of a 1000-MW station and is equal to the combustion rate of 1 million automobiles. Such phenomena are completely outside the range of conventional fires.

Fireballs can clearly carry fire far beyond the normal safe distance. An example of such a fireball was the San Carlos, Spain, disaster of 1978,* in which a tank truck containing about 20 metric tons of propylene was split open. The resulting fireball killed more than 200 people.

UNCONFINED VAPOR CLOUD EXPLOSIONS

For reasons as yet imperfectly understood, ignition of vapor clouds may give rise to unconfined vapor cloud explosions (UVCEs). Such explosions may have devastating effects, as at Ludwigshafen in 1948* or at Flixborough in 1974.* They are capable of destroying strong buildings, of killing everyone within 100 meters or so of the explosion center (or epicenter, as it is convenient to call it), and of causing damage in the tens of millions of dollars.

The theoretical difficulties in formulating a mechanism to explain the phenomenon are well set out in Gugan's book *Unconfined Vapour Cloud Explosions.*[11] Briefly, the problem is to explain how it can happen that the flame velocity of a hydrocarbon gas can accelerate from the normal 3 m/s or so up to several hundred meters per second (i.e., to near sonic velocity) in traveling a distance not exceeding the diameter of the cloud.

Detonation

The phenomenon of detonation in which the blast wave travels at supersonic velocities has been produced under experimental conditions, sometimes for military purposes. However, such detonations have always been primed by a small charge of solid explosive, and there is no convincing evidence that detonation has ever been a feature of an industrial vapor cloud explosion.

Scaled Distance

The question as to whether UVCEs can be given a rating in terms of conventional explosives is being vigorously debated. The effects of conventional explosives have been extensively studied, especially for military reasons, and the most important conclusions to emerge are the various scaling laws. The earliest of these and the one most useful for the purpose is that attributable to Hopkinson,[12] which in broad terms declares that for any given explosive in a given medium the relative distances at which various properties are numerically equal are in the ratio of the cube roots of the masses of the explosive charges. Thus, let R_1 be the distance at which the overpressure (measured at right angles to the blast wave) from an explosive charge of mass M_1 in air is equal to Z units of pressure above atmospheric pressure; then the distances R_2 at which the overpressure from an explosive charge of the same explosive and of mass M_2 is also equal to Z units of overpressure is equal to $R_1 \times \sqrt[3]{M_2/M_1}$.

The most important properties so scaled are P_o (overpressure), which may be positive or negative, T_I, the duration of the pressure exceeding atmospheric pressure (positive duration) or falling below atmospheric pressure (negative duration), and I, the impulse, which is the integral of P_o and T_I. Generally, the algebraic sum of the positive and negative impulses is zero for conventional explosives. For many purposes, it is useful to have the concept of a "scaled distance." In SI units a scaled distance R_S may be expressed in meters and is the radius at which a given effect is observed from the detonation of 1 kg of a given explosive (conveniently, TNT). For a different charge of the same explosive M, the same effect will be produced at a distance R, where $R = R_S \times M^{-1/3}$.

It might be hoped, therefore, that if the TNT equivalent of a given mass of vapor is known, damage effects will be predictable. Thus an important scaled distance for TNT is the R_B distance, the radius at which Class B damage is done to houses. [Class B damage is damage such that houses are not totally destroyed but are beyond repair. The overpressure required is about 0.3 bar (3 kPa). R_B has the value of $7 \times M^{-1/3}$, where M is expressed in kilograms.]

TNT Equivalence

Is it possible, therefore, to express the effects of UVCEs in terms of TNT equivalence? The answer is that it is impossible in exact terms. The most immediate objection is that the UVCEs do not possess the "brisance," or shattering effect, of dense explosives which are capable of exerting pressures up to 400,000 bars (4000 MPa).[13] There is no exact knowledge of the maximum pressure exertable by a UVCE. Marshall[14] has suggested 1 bar (10 kPa); Gugan[11] has suggested up to 10 bars (100 kPa). The official report of the Beek disaster[15] has suggested that the general pressure in the cloud was of the order of 1 bar but that higher pressures existed locally in pockets. Certainly unconfined vapor explosions have never given rise to craters. The point of escape at Flixborough is shown

Fig. 6 Point of release at Flixborough, England.

in Fig. 6. Nothing has been shattered, and much has survived with little distortion except that which arose from the subsequent fire. Detailed investigations at Flixborough suggest that the overpressure at the edge of the cloud was approximately 1 bar and that thereafter it decayed approximately as would be expected for a TNT overpressure of this magnitude.

Sadee, Samuels, and O'Brien[16] have sought to approximate the ground-level effects by a model in which the TNT is assumed to be detonated in the air at a level that would not cause a crater. Such an approximation cannot accurately reproduce a UVCE. It is believed, too, that there are other dissimilarities between the blast waves from dense explosives and those from UVCEs, but to discuss these in detail is beyond the scope of this chapter. These concern such questions as the duration of the blast wave and its impulse and the "shape" of the wave. Nevertheless, TNT equivalents, however inaccurate, do provide some measure of prediction.

Quantification of Equivalence

Even if the TNT model is valid, the problem remains as to how to predict the quantity of TNT equivalent to a given loss of containment.

Theoretical Prediction

All that theory in its present state can say is that there is an upper limit imposed by the maximum energy release which would be associated with the flammable substance. This maximum energy is the heat of combustion, and if all this were to be realized as blast-wave energy, it would yield, for most hydrocarbons, a value for TNT equivalence of about 10 metric tons of TNT per metric ton of hydrocarbon. In industrial practice (as distinct from military experimentation), only a small fraction of this theoretical energy has ever been released as blast.

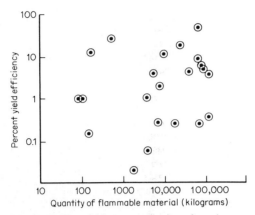

Fig. 7 TNT equivalence as a function of metric tons of hydrocarbon released. [*Taken with permission from K. Gugan,* Unconfined Vapour Cloud Explosions, *Institution of Chemical Engineers and George Godwin, Ltd., Rugby, England, 1979.*]

Analysis of Actual Incidents

Equivalence as deduced from actual incidents by careful observers shows a great deal of scatter, as may be deduced from Fig. 7, in which Gugan attempts to relate TNT equivalence as metric tons of TNT per metric ton of vapor released to the quantity of carbon atoms in the material released. However, only in very few cases has the quantity spilled been accurately estimated, nor has the TNT equivalence.

Warner[17] has claimed that to date methane has never been associated with a UVCE and that there is little evidence concerning ethane. All the other commonly handled hydrocarbons containing two, three, or four carbon atoms have been associated with a UVCE. So has the C_6 hydrocarbon cyclohexane at Flixborough. There have also been serious explosions involving dimethyl ether* and ethylene oxide.*

It is probably safe, therefore, to say that for many low-molecular-weight hydrocarbons, except methane, and their simple derivatives, there is definite evidence of blast effects associated with UVCEs, but the evidence does not support any definite relationship between chemical composition and TNT equivalence. It is suggested in Fig. 7 that there is no correlation between TNT equivalence and the size of release, although on a priori grounds it might be deduced that very small releases will not give rise to UVCEs and that with large releases an ever-increasing proportion of vapor will be present as an overrich central core.

SPECIAL PROBLEMS OF REACTORS

Batch-operated chemical reactors present special problems if the reaction is exothermic in nature and if one or more of the reactants are volatile. The general problem is that increase in temperature has only a linear effect upon the rate of heat transfer but exercises an exponential increase on the rate of reaction and hence upon the rate with which heat is generated by the reaction.

It is seldom considered to be economic to provide a reactor which is sufficiently strong to resist any calculable pressure rise resulting from a breakaway reaction (though this might be necessary for toxic reactants), and recourse is, instead, to a variety of techniques for giving pressure relief, usually by means of relief valves or bursting disks. The mode of action of these devices and their various advantages and disadvantages, together with the problems associated with the design of overall relief systems, lie outside the scope of this chapter, and the reader is referred to a recent review of the problems.[18]

Disposing of Ejected Contents

When reactors were small, a solution was to "put it on the roof," i.e., simply to vent a discharge to the atmosphere without special precautions. Today, with much larger reactors, there is a growing realization that this solution is unacceptable, and containment vessels are usually provided, or release is through a tall stack or a flame stack. However, the ejected contents are seldom merely gases, and sometimes they are multiphase systems which may be difficult to handle. Therefore, failing the provision of a guaranteed trapping system, a runaway reaction can give rise to the same sort of calamity which has already been discussed, for such ejected material can possess in large measure three of the factors previously listed: toxicity and/or flammability, high velocity of ejection, and multiton inventories.

Case of Seveso

This incident has become a classic example of the serious consequences of a runaway reaction. Unfortunately there is not as yet any official report on this accident, although much can be learned from Part 3 of Chap. 11 of the *Handbook*, as well as from a recent book, *Dioxin: Toxicological and Chemical Aspects.*[19] The latter book very usefully supplements information previously available on the incident, which occurred on July 10, 1976, in the vicinity of Seveso,* a small town in the municipality of Milano, Italy.

The release took place in the course of reacting 1,2,4,5-tetrachlorobenzene with sodium hydroxide in the presence of ethylene glycol as a solvent at a temperature of about 170 to 180°C. Unfortunately it is possible for exothermic polymerization of the glycol to begin at this temperature, with a consequent rise in temperature, permitting the formation of 2,3,7,8-tetrachlorodibenzo-p-dioxin (TCDD, or dioxin), a substance of remarkable hostility to animal life. Four similar occurrences involving in each case from 50 to 229 workers have been documented in the past.

The uncontrolled reaction resulted in an explosive release of reaction products as a toxic cloud which contaminated, to a greater or lesser degree, approximately 400 ha. From the most heavily contaminated zone of about 100 ha, approximately 700 people had to be evacuated. The total amount of TCDD released has been estimated as between 0.5 and 3.0 kg. If it is accepted that the area requiring evacuation was about 1 km², it would follow that if the area were triangular in shape, the effect of the incident must have been felt in its most severe form over several kilometers. The B zone, not evacuated, covered a further 2.5 km².

Toxicology of TCDD

There is nothing to suggest, on a priori grounds, that TCDD should be toxic. It is, in fact, chemically very inert, reading neither with acids nor with bases, being very resistant to bacterial degradation and stable up to 800°C, which is most unusual for an organic compound. All these factors make decontamination of affected areas very difficult.

The median lethal oral dose for small animals varies from 0.6 μg/kg for male guinea pigs to 120 μg/kg for mice and rats, compared with about 400 to 1000 μgm/kg for strychnine in humans. No fatal cases have been reported in humans, but minute quantities are capable, when absorbed through the skin, of causing chloracne, inflammation of the eye-lids, and loss of hair. TCDD is believed also, on the basis of animal experiments, to be teratogenic, i.e., to damage an unborn fetus. It is these latter features which have been chiefly responsible for the medical problems at Seveso. Many animals, however, died as the result of contact with ground and vegetation which were also heavily contaminated by caustic soda. There were also a number of nonfatal human casualties arising from contact with caustic substances.

LETHAL EFFECTS AND DAMAGE EFFECTS OF VAPOR CLOUDS

Toxic clouds would be expected to kill in one specific way which is their characteristic lethal effect. Chlorine, for example, kills by edema of the lungs. Only rarely will sudden releases of toxics inflict appreciable damage to property, but this has occurred in rare instances such as that discussed above, producing an irremovable contamination of the surrounding area, which then becomes sterilized.

   With clouds of flammable vapor the case is different. The gas may kill in high con-centrations by asphyxia and anesthesia and in a few cases (e.g., ethylene oxide) is directly toxic. If ignited, the gas may kill by direct flame or by radiation or, should explosion occur, by collapse of buildings or overhead pipelines or possibly by missiles. It is possible, too, that fire or explosion may release toxic gases or generate toxic combustion products. The damage inflicted may be direct or sequential (by producing further spillages).

QUANTIFYING THE HAZARD

A valuable attempt to quantify hazards of fire and explosion is contained in the Dow Chemical Company's *Safety and Loss Prevention Guide,*[20] which is based upon its *Process Safety Guide,* first published in 1966. The guide, which does not cover the handling of conventional explosives, is intended, to quote its introduction, "to be used to numerically rate a chemical process unit for hazards, discover the particular hazards and indicate measures that may be taken to neutralize or minimize them." It proposes that a plant should be divided into units, which are parts of a plant that can be readily and logically characterized as a separate entities. In this way the entire plant is not rated by its most dangerous unit.

   Unfortunately, when the index is applied to three disasters (see Table 2), weaknesses appear. Case 1 in the table is the storage of 30 metric tons of dimethyl ether in a railcar at Ludwigshafen.* Case 2 is the storage at 8 atm of 120 metric tons of cyclohexane in the reactor system at Flixborough,* and Case 3 is the storage of 20 metric tons of propylene in a tank truck at San Carlos, Spain.* In none of these cases would the application of the Dow index have led to the anticipation of more than moderate hazard. Nonetheless, the index is valuable, as it encourages a systematic, quantitative approach to hazard, although it needs to be modified to take much more account of the special process hazards involving liquefied gases.†

Mortality Index Approach

A different approach has been taken by Marshall.[21] His approach covers the hazards of toxic release, catastrophic fire, and explosions, whether conventional explosions or

---

†The most recent edition of the Dow index now does take account of the special dangers associated with liquefied gases.

TABLE 2 Dow Index Applied to Known Disasters

| Case | Location | Material and quantity | Material factor | Special material hazards | General process hazards | | Special process hazards | | | Special process hazards multiplier i.e., $1 + d + e + f$ | Overall multiplier $b \times c \times g$ | Dow rating ($a \times h$) |
					Loading or unloading operations	Exothermic reactions	High temperature	Greater than average explosion hazard	Large quanitites of flammable fluids			
1	Ludwigshafen, 1948	Dimethyl ether (30 metric tons)	12.4	None	× 1.25		0.20	0.40	0.65	× 2.25	× 2.81	35 (light)
2	Flixborough, 1974	Cyclohexane (120 metric tons)	18.7	None		× 1.25	0.20	0.40	0.90	× 2.50	× 3.125	58 (moderate)
3	San Carlos, 1978	Propylene (20 metric tons)	19.7	None	× 1.25		0.20	0.40	0.55	× 2.15	× 2.68	58 (moderate)
Factor			a	...	b	c	d	e	f	g	h	

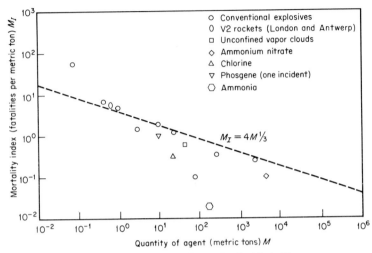

Fig. 8   Mortality index versus quantity of agent involved in incident.

UVCEs. He attempts to derive a "mortality index," or specific mortality, by analyzing past events, deriving the index by dividing the number of fatalities by the appropriate tonnage released. When there are sufficient data, they are divided into classes, the maximum quantity in each class being the maximum quantity in the previous class multiplied by $\sqrt[2]{10}$. The principal findings are displayed in Fig. 8, which is an updated version of the chart in Marshall's article.

Ratio of Injuries to Fatalities

The ratio of injuries to fatalities in factory accidents is roughly 500 minor and 100 major injuries to 1 fatality. It would be grossly misleading to suppose that this ratio applies to injury and death from major spillages, as the mechanisms involved are quite different. The ratios for industrial explosions are probably similar to wartime figures for aerial bombardment, in which there were 5 minor and 3 major injuries per fatality. The ratios at Ludwigshafen in 1948 were $17:2.5:1$, and at Flixborough they were $2.7:0.5:1$. Minor injury is hard to define, and sometimes not all such injuries treated by first-aid workers are recorded, which seems to have been the case at Flixborough. Figures for chlorine in World War I suggest that the ratio of nonfatally gassed to fatally gassed persons ranged from $4:1$ to $10:1$. Though there are considerable doubts about the accuracy of these figures, they are clearly of a lower order of magnitude than the ratio from factory accidents. The ratios discussed above may be used in conjunction with the mortality index to predict nonfatal injuries.

Mortality Index Applied to United Kingdom Notifiable Levels

The U.K. Advisory Committee on Major Hazards made certain recommendations on storage amounts which would merit notification to the authorities concerned. These amounts were then published in a consultative document.[22] The storage amounts are given

TABLE 3   Mortality Indices Compared with Storage Amounts

Substance	Consultative document inventory (metric tons)	M_I from Fig. 8	Likely fatalities ($M_I \times$ size of release)
Chlorine	10	0.30	3
Ammonia	100	0.02	2
Liquefied flammable gases	20	0.60	12.0
Unstable substances (equivalent to 3 metric tons of TNT)	5	1.50	4.5

in Table 3 together with mortality indices from Fig. 8. (It is assumed that 5 metric tons of intrinsically unstable substances in the document is equivalent to 3 metric tons of TNT and, in the absence of scaling laws, that 10 metric tons of chlorine has the same mortality index as 20 metric tons in Fig. 8.)

The figures for amounts stored were established on the basis of practical experience and not on theoretical grounds. Nevertheless, they seem to be buttressed by the application of the mortality index methodology and shown to be of a reasonable order of size. However, it would appear that the consultative document is probably too severe on toxics and too lenient on flammable vapors.

It would appear from Fig. 8 that vapor clouds are about as lethal, ton for ton, as high explosives, although their TNT equivalent is almost always less than the mass of vapor in the cloud and sometimes they burn without blast effects. This high mortality is probably accounted for by the many different lethal mechanisms displayed by vapor clouds.

Influence of Population Density

The mortality index has the advantage of being based on real rather than hypothetical incidents, but as it stands, it does not fully disclose the variance resulting from differing population densities. A later study[23] which included consideration of population densities led to the following correlation for conventional explosives:

$$M_I = P_D \times M^{0.333} \qquad (4)$$

where M_I = mortality index (fatalities per metric ton of lethal agent)
P_D = population density, in thousands per square kilometer
M = mass of lethal agent in metric tons

QUESTION OF RISK

The quantitative assessment of risk is at least as difficult as the assessment of hazard. It is true that by now there is a considerable body of experience which enables the reliability (i.e., the reciprocal of the risk of failure) of individual components to be assessed. In the United Kingdom much work has been done in this field, particularly by the Safety and Reliability Directorate of the U.K. Atomic Energy Authority. The data which exist on individual components may then be treated by established techniques of risk analysis. Such an established method is that of fault-tree analysis. A systematic approach to the problem would involve first the analysis of hazard by such a technique, which entails a systematic examination of departure from design conditions for every pipe, valve, and vessel to identify those areas which are most hazardous.[24] It is at this point that it becomes worthwhile to estimate risk by means of a fault-tree analysis.

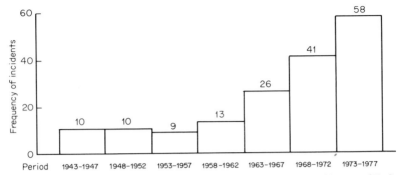

Fig. 9   Worldwide trends in number of major incidents involving flammable gases. [*V. C. Marshall, "The Hazards: Their Probable Outcome and Frequency,"* Oyez Conference: Bulk Storage and Handling of Flammable Gases and Liquids, *London, Apr. 18, 1979.*]

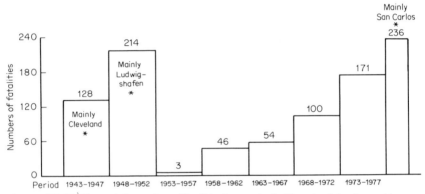

Fig. 10   Worldwide trends in numbers of fatalities from major incidents involving flammable vapor clouds. [*V. C. Marshall, "The Hazards: Their Probable Outcome and Frequency,"* Oyez Conference: Bulk Storage and Handling of Flammable Gases and Liquids, *London, Apr. 18, 1979.*]

Worldwide Trends

An alternative approach is to examine worldwide trends; these are shown in two histograms (Figs. 9 and 10). The first plot shows a trend in the number of incidents which has roughly kept pace with the scale of handling hydrocarbons, while the second plot shows the number of fatalities.

There is an inherent bias in the reporting of incidents toward the reporting of the most spectacular events, and diagrams of this sort tend to underestimate the number of minor incidents and certainly to underestimate the number of incidents which do not involve fatal results. A study by Marshall[25] of 185 incidents in which there were leakages of a metric ton or more of flammable agents on a worldwide scale suggested the following conclusions:

1. Only 4 percent of the incidents studied did not result in ignition.

2. Of the cases of ignition approximately one-third resulted in fires.

3. The remaining two-thirds were accompanied by blast effects either before or during the fire.

TABLE 4 Select List of Major Chemical Spillages

Place	Date	Agent responsible	Origin of spillage	Quantity spilled	Nature of subsequent event	Number of fatalities	Remarks
Zărneşti, Romania	Dec. 24, 1939	Chlorine	Storage tank ruptured	24 metric tons	Toxic vapor cloud formed	60	LNG ran into sewage system. The event set back LNG technology for a decade.
Cleveland, Ohio	Oct. 20, 1944	Liquefied natural gas (LNG)	Bulk storage collapsed	> 2000 metric tons	Fires and confined explosions	ca. 130	
Ludwigshafen, West Germany	July 28, 1948	Dimethyl ether	Railcar ruptured	30 metric tons	Unconfined vapor cloud explosion	207	It is believed that the railcar overfilled and burst owing to hydraulic pressure when it warmed up. Damage to factory was very severe.
Antwerp, Belgium	June 4, 1964	Ethylene oxide	Reflux vessel blown up	1 metric ton	Unconfined vapor cloud explosion	4	
Potchefstroom, South Africa	July 13, 1973	Ammonia	Storage tank ruptured	30 metric tons	Toxic vapor cloud formed	18	
Flixborough, England	June 1, 1974	Cyclohexane	Bypass in series of reactors collapsed	40 metric tons	Unconfined vapor cloud explosion	28	Factory was severely damaged. Control room collapsed, killing 19.
Beek, Netherlands	Nov. 7, 1975	Propylene	Pipe system ruptured	5 metric tons	Unconfined vapor cloud explosion	14	Control room was severely damaged, killing 6.
Seveso, Italy	July 10, 1976	TCDD (dioxin)	Runaway reaction	0.5–3.0 kg	Toxic material scattered over several square kilometers by explosion	None	Many people were affected by severe chloracne; 1 km² was evacuated semipermanently.
San Carlos, Spain	July 11, 1978	Propylene	Tank truck ruptured	20 metric tons	Fireball	>200	Rupture occurred on a road near a holiday camp. The cause may have been similar to that at Ludwigshafen.

4. Approximately 40 percent of the incidents involved propane, butane, and their derivatives, i.e., liquefied gases, to which frequent reference has been made in this chapter.

QUESTION OF DAMAGE

When fatalities have occurred in incidents involving the release of flammable vapors, there has been almost invariably severe damage, but even in cases in which there were no fatalities or even injuries damage losses have been very serious. The Ludwigshafen incident of 1948, which resulted in more than 200 deaths, produced damage comparable to that of a well-directed air raid. The Flixborough disaster in 1974 caused damage of about $100 million.

The most extensive studies in this area have been those of Davenport,[26] who has listed estimated losses and attempted to relate them to the product of the quantity spilled and its heat of combustion. In this way a form of damage index has been derived typically for an energy content of 10^9 Btu (10^{12} J). The estimated damage in 1976 dollars is $20 million. Davenport's damage index, like Marshall's mortality index, displays considerable variance, which might be reduced if it could be corrected for the density of investment in the area in question. Marshall[23] has suggested that damage may be related to fatalities as 125×10^6 (1976) per fatality.

REFERENCES

1. R. H. Perry and C. H. Chilton (eds.), *Chemical Engineers' Handbook,* 5th ed., McGraw-Hill Book Company, New York, 1973, pp. 5-11–5-13.

2. T. A. Kletz, "Some Myths on Hazardous Materials," *Journal of Hazardous Materials,* vol. 2, no. 1, 1977.

3. V. J. Clancey, *The Evaporation and Dispersal of Flammable Liquid Spillages,* Institution of Chemical Engineers Symposium Series, no. 39a, Rugby, England, 1974, p. 80.

4. A. P. van Ulden, "On the Spreading of a Heavy Gas Released near the Ground," *First International Symposium: Loss Prevention and Safety Promotion in the Process Industries,* The Hague and Delft, May 1974.

5. F. Pasquill, "The Estimation of the Dispersion of Windborne Material," *Meteorological Magazine,* vol. 90, 1961, pp. 33–49.

6. *Threshold Limit Values, 1976,* American Conference of Governmental Industrial Hygienists, Cincinnati, published with introduction for British readers as Guidance Note EH 15/76 by Health and Safety Executive, H. M. Stationery Office, London, 1976.

7. F. A. Patty (ed.), *Industrial Hygiene and Toxicology,* vol. 2, 2d rev. ed., Interscience Publishers, Inc., New York, 1963.

8. W. S. Spector (ed.), *Handbook of Toxicology,* vol. 1, W. B. Saunders Company, Philadelphia, 1956.

9. N. V. Steere (ed.), *CRC Handbook of Laboratory Safety,* 2d ed., Chemical Rubber Co., Cleveland, 1971.

10. R. C. Weaste, *Handbook of Chemistry and Physics,* 53d ed., Chemical Rubber Co., Cleveland, 1972, p. 30.

11. K. Gugan, *Unconfined Vapour Cloud Explosions,* Institution of Chemical Engineers and George Godwin, Ltd., Rugby, England, 1979.

12. B. Hopkinson, *British Ordnance Board Minutes,* 13656, 1915.

13. R. Houwink, *Sizing Up Science,* Scientific Book Club, London, 1976, p. 115.

14. V. C. Marshall, *The Siting and Construction of Control Buildings: A Strategic Approach,* Institution of Chemical Engineers Symposium Series, no. 47, Rugby, England, 1976.

15. *Rapport over de explosie bij DSM te Beek (L),* Directoraat General van de Arbeid van het Ministerie van Social Zaken, Postbus 69, Voorburg, Netherlands, 1976. (In Dutch.)

16. E. Sadee, D. E. Samuels, and T. P. O'Brien, "The Characteristics of the Explosion of Cyclohexane at the Nypro (UK) Plant," *Journal of Occupational Accidents,* vol. 1, no. 3, 1977.

17. Sir Frederick Warner, *Safety and Natural Gas,* Institution of Chemical Engineers Symposium Series, no. 44, Rugby, England, 1976.

18. *The Safe Venting of Chemical Reactors,* symposium by Chester Centre of North West Branch of the Institution of Chemical Engineers, at Chester, England, March 1979, Institution of Chemical Engineers, Rugby, England.

19. F. Cattabeni, A. Cavallora, and G. Galli (eds.), *Dioxin: Toxicological and Chemical Aspects,* Spectrum Publications, Halsted Press, New York and London, 1978.

20. *Dow's Safety and Loss Prevention Guide,* 5th ed., American Institute of Chemical Engineers, New York, 1981.

21. V. C. Marshall, "How Lethal Are Explosions and Toxic Releases?" *The Chemical Engineer,* August 1977, pp. 573–577.

22. *Hazardous Installations (Notification and Survey) Regulations 1978,* consultative document, H. M. Stationery Office, London, 1978.

23. V. C. Marshall, "The Physical Implications of Major Chemical Hazards," *Proceedings of Major Chemical Hazards Seminar,* Harwell, England, April 1978, pp. 197–210.

24. *A Guide to Hazard and Operability Studies,* Chemical Industries Association, London, 1977.

25. V. C. Marshall, "The Hazards: Their Probable Outcome and Frequency," *Oyez Conference: Bulk Storage and Handling of Flammable Gases and Liquids,* London, Apr. 18, 1979.

26. J. A. Davenport, *The Study of Vapor Cloud Incidents,* 83d National Meeting, American Institute of Chemical Engineers, Houston, March 1977.

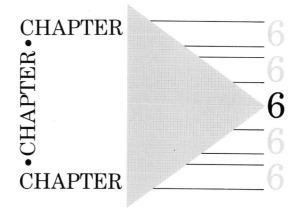

CHAPTER

•CHAPTER•

CHAPTER

6
6
6
6
6

Prevention

PART 1
Plant Operations / 6-2

PART 2
Maritime Transportation / 6-25

PART 3
Rail Transportation / 6-43

PART 4
Tank Truck Transportation / 6-51

PART 1
Plant Operations

William B. Katz
President, Illinois Chemical Corporation

Introduction / 6-2
Benefits of a Prevention Program / 6-3
Requirements for a Prevention Program / 6-3
Factors to Be Considered in Planning a Prevention Program / 6-4
Facility Inspection / 6-7
Decision Making / 6-22

INTRODUCTION

Spills are accidents. Spill prevention should be approached in the same fashion as accident prevention: identify as many probable causes of spills as possible, and eliminate them. In most spill situations, a chain of events leads to the spill. When similar spills have occurred over a period of time, common factors can often be identified in all the occurrences. Changing or correcting those factors can reduce the chance of future spills. Since one can never be certain that all possible causes of a spill have been identified, zero spill potential probably will never be attained.

Definition

Spill prevention, therefore, can be defined as the determination of the probable factors and events that can lead to an actual spill and, by changing those factors and events, the reduction in the odds that such a spill will occur.

Assumptions

One can make some assumptions about the conditions leading to spills:

1. Large manufacturing plants generally have a good idea of the problems involved in handling hazardous materials and do a fairly good job of seeing that *obvious* preventative measures are taken. Small plants, using small quantities of hazardous materials, may not have a good idea of the problems involved; generally, personnel in such plants need to be educated in spill prevention.

2. When there has been some recognition of spill problems, the obvious causes have usually been recognized and some action taken. Hence one usually is concerned with finding and dealing with low-probability, nonobvious causes of spills.

3. Spill probabilities can only be estimated ("guesstimated" might be a better word).

BENEFITS OF A PREVENTION PROGRAM

There are a number of benefits to be derived from an active spill prevention program; some of these are rather obvious, and others are not. One immediate benefit is reduced loss of product, which is a direct economic gain. If a spill occurs, there are immediate costs for direct cleanup, and there may be a loss from downtime if the spill site is close enough to an active production area to require evacuation of personnel or curtailment of production activities.

More indirect benefits include a reduction in capital investment for cleanup equipment and materials, lowered out-of-pocket charges for cleanup contractors' services, and a reduction in fines and penalties associated with a spill. Frequently, increased consciousness of spill potential, as with accidents of any kind, makes workers more careful in their operations, with a reduction in injuries and lost-time accidents. The fact that none of these benefits can be calculated accurately is one of the difficulties in trying to justify a particular level of expense for prevention.

REQUIREMENTS FOR A PREVENTION PROGRAM

When people operate in a daily set routine, they develop an attitude of acceptance rather than an attitude of questioning. What is needed to identify the potential causes of spill accidents and to suggest solutions that will reduce the chance of those spills' occurring is a "why" attitude, or a "what if" kind of approach to surroundings, jobs, procedures, and policies.

Records of past spills should always be made available to those responsible for spill prevention. An accurate record-keeping and reporting procedure should be established as one of the first steps in spill prevention. There are two reasons for this. The most obvious is the need to have available a record for the study of causes, with a view to making corrections that will eliminate those causes. A less obvious reason is protection in the event of lawsuits, hearings, or fine appeals. In the excitement and turmoil of a spill problem, chronology is forgotten or warped. Memories become dim after a passage of time, and when a lawsuit finally is brought to court, defense becomes difficult without some written, legally acceptable reminders of what occurred.

It is not always possible, but is certainly highly desirable, to have some one person involved with the spill as a technically qualified observer whose function is to record on film or tape *and* in writing what is occurring. Such records should be made in the form determined by legal counsel to be valid for legal purposes.

Personnel

Personnel for a spill prevention program may come from the plant staff or from another plant whose personnel are not familiar with the plant being studied. In either event, the person or persons selected should have the kind of mind that might set them apart as "different" in a routine situation. Such people are not easy to find, since in routine production jobs they tend to upset applecarts, and often they are not around very long. Frequently such people are to be found in R&D activities, in which they dream of things as they are not. Once a prevention program has been established, these people usually are not the best people to run it, but a regular reapplication of their special approach will prove useful and profitable.

Outside Help

Outside consultants are available from a number of sources. Their aid may cost a lot or very little.

Consulting firms that specialize in this type of work are not yet common, but there are some with a background in spill prevention work, and the number will undoubtedly grow.

Cleanup contractors often can be hired to do this kind of survey work. If they are employed on a retainer to provide the cleanup service needed in the event of a spill, they may do survey work and help to set up a prevention program as part of their retainer fee. Most firms specializing in this field are members of the Spill Control Association of America, currently headquartered at 1040 North Park Plaza, 17117 West Nine Mile Road, Southfield, Michigan 48075.

Insurance agencies and suppliers may provide engineering service, free or for a fee, as a means of reducing their risk exposure.

Federal and state environmental agency people may be helpful in making suggestions, although normally they are not available to do this kind of work while in government employ.

Avoidance of Upsets

One of the normal results of an active prevention program is change—in procedures, in equipment, in policies, and occasionally in jobs. Such change is usually resisted. Management, especially middle management, tends to resist because a prevention program requires spending money and interrupting procedures that they hope are efficient and because the net result is a reduction in profit with no visible signs of gain. Plant workers whose jobs or accustomed ways of doing things are threatened may find ways to be uncooperative or even try to stop the installation of a spill prevention program. Hence it is necessary first to have the active support of the members of top management, who must understand the aims and goals of such a program and agree to the cost and the timetable. Unions, if involved, must be informed and sold on the positive gains for their members: safer jobs, better working conditions, or anything else that is applicable. Any changes resulting from a prevention program must be planned carefully and, if possible, implemented with the help of involved personnel in the planning stages.

FACTORS TO BE CONSIDERED IN PLANNING A PREVENTION PROGRAM

Personnel

The human element is almost always a factor in spills. This involvement can be direct, as the result of carelessness, indifference, or sabotage, or indirect, as the result of inadequate knowledge of job or procedures, poor training, physical inability, or another cause. Hence any good spill prevention program must start with a careful evaluation of personnel and correction of deficiencies in attitude, motivation, training, and knowledge.

Fig. 1-1   Unprotected tank car unloading area with a storm drain alongside.

Equipment

Accidental spills are frequently the result of inherent flaws in equipment design. Much equipment now in use was designed, built, and installed with ends in mind very different from minimizing spills. The changed climate of regard for spill control may well necessitate modifications in equipment.

As one example from the past, the filling of tank trucks with flammable materials was often done with little regard for vapor emission. Tight vapor emission control has dictated the design of new loading equipment, and procedures for the use of that equipment, to eliminate or at least to decrease drastically the emission of vapor during tank loading.

As another example, hose connections to the bottom of tank cars containing toxic or dangerous products often were made with loose control of spillage from hose drainage. Frequently the hoses were just drained onto the ground. Modern improvements call for installation of drain pans beneath rail unloading facilities, plus the use of quick shutoff fittings on hoses and piping, to prevent such spills and to collect those that do occur. (See Figs. 1-1, 1-2a, and 1-2b.)

Environment

Proper security for both equipment and personnel is a major need in spillage control. This includes prevention of nontrained personnel (such as outside delivery drivers) from operating equipment, control of personnel movements into unauthorized areas, and protection from vehicular accidents that could damage permanent equipment (because of driveways located too close to tanks, for example). In this particular period, protection against deliberate acts of vandalism, sabotage, and terrorism must be considered in planning for security protection.

Prevailing weather must be considered as a contributing cause of accidental spills and must be studied to determine possible effect on spill occurrence. Examples are plant location in regular tornado or hurricane areas or in areas with heavy winter snowfalls or regularly occurring ice and sleet storms. Some attempt must be made to reduce the pos-

Fig. 1-2*a*.

Fig. 1-2*b*.   Rail track spill collection pans installed.

sible impact of weather on facilities that could result in spills, usually under conditions that make effective control and cleanup of a spill difficult or impossible. National Oceanic and Atmospheric Administration records on rainfall, snowfall, winds, and floods all can be useful in this regard. (See Fig. 1-3.)

Plant geography must be studied (or at least considered) in planning to prevent or control spills. Ground elevations will affect water drainage patterns. Water-table depth and soil porosity may be such that a spill will be essentially nonrecoverable (even at tre-

Fig. 1-3 A potentially dangerous spill situation. A flood over the high ground created by the railroad embankment could wash away the drums stored at center right and conceivably could damage the small diked tank.

mendous expense) unless seepage into the water table can be prevented. Proximity of storm sewers and watercourses, either flowing rivers or creeks or ditches that may contain water during thaw or rainfall conditions, may determine in part the preventative measures required. Close proximity of watercourses obviously increases exposure to rapid spreading of a spill, with a vastly increased potential recovery cost. Hence more effort and money may be justified to anticipate spill sites and to install control measures of a possible spill flow path. The location of all storm drains should be entered on the facility plot plan.

FACILITY INSPECTION

The first step in starting a program of spill control is a detailed facility inspection to survey potential problems. The questions below should be answered as the facility is systematically covered completely. These questions may not be complete (such a list never can be complete), but they should give the user a point of view that will stimulate additional questions as the inspection proceeds.

Records

Written records should be kept of such inspections, especially since reinspections (which should be made at perhaps 6-month intervals) will show by comparison with past records what progress is being made in reducing spill exposure. Such inspections can profitably be made by a small team (consisting of no more than three persons) which includes both operating personnel (with some authority) and the specialist (in-plant or outside consultant) working together. Experience has shown that despite the demands on his or her time and attention a plant manager or an assistant plant manager makes a better team member than an operating supervisor, since on-scene inspection impresses the need for action better than secondhand reports. In practice, however, this function is usually delegated (unless a spill incident has resulted in a major crisis) to someone of lesser authority. The more authority that person has, the better.

It is difficult to design an inspection report form that will suffice for all plants. However, a specific form prepared for a specific facility, following the guidelines of the questions posed in the following subsections of this part, will assure that a series of inspections over a period of time will cover the same ground. Any such form should itself be reviewed for completeness once a year or so as plant conditions change. Illustrations attached to the report will help those who have not participated in the actual survey to understand the records.

Product Knowledge of Personnel

It is fundamental, of course, that every person having anything to do with handling products have full and complete product knowledge. This must be true for every product handled in the facility.

What are the products' physical properties? Are they heavier than water? Lighter? Soluble or nonsoluble? Are the products toxic? Flammable? Explosive? Are the products reactive, and if so, with what will they react? If they come into contact with commonly used substances (such as oil, gasoline, or water), will there be an adverse effect? Will special safety precautions be required in the event of a spill? What about on-scene personnel not involved in the spill? Is evacuation necessary? Special protective clothing? Is it available?

Outside Survey

Roadways inside Facility, Access Roads, Parking Areas, and Storm Drains

Drainage   Are all roads paved and/or guttered? Would a spill be retained or drained off onto the ground? Are there storm drains; if so, where do they lead? Would a spill on the roadway reach a watercourse via these storm drains? Can provision be made to shut the drains in an emergency or to bypass the flow through a retention basin that might hold a spill? Are all the drains really necessary? Are they in the right locations to prevent spread of a spill because rainfall does not properly flow away from a high-potential spill area? Are access ways to public roads near public storm drains?

Traffic Control   Are roads protected against traffic accidents, or should controls be installed to protect the movement of vehicles carrying hazardous materials both within the plant and at the point where they enter and leave public roads? Is there sufficient clearance so that all vehicles using the roads cannot possibly damage buildings or equipment? (See Fig. 1-4.)

Vehicle Inspection   What kind of inspection is made of employees', suppliers', or visitors' vehicles (usually none at all) to ensure that there is no leakage from transmissions or fuel tanks? If the facility has a visitors' center, are visitors' actions controlled while their cars are parked there? (Visitors have been known to drain motor oil onto a parking area while visiting a public information center, then refill, and drive away from the mess created.)

Are tank trucks and other supply or delivery vehicles inspected as they enter or leave the facility to be sure that there is no leakage from the tanks or from containers on the vehicles? (See Fig. 1-5.)

Fig. 1-4 Storage tanks with a low dike, piping, pumps, and electrical equipment unprotected from truck travel on the roadway alongside.

Fig. 1-5 Tank car with caps for discharge lines dangling loose. An improperly closed or leaky valve could cause a leak along miles of right-of-way.

Loading and Unloading Facilities

Included in this category are rail switch tracks and rail tank car loading racks; tank truck loading racks, both overhead and bottom-loading; building loading platforms (docks); river docks; pipeline connections; valve manifolds; and perhaps other areas where product is loaded or unloaded in either bulk quantities or drum containers. These areas are high-probability sources of spills, mostly small but occasionally large.

An immediate consideration at any location is whether or not a spill will require cessation of operations, evacuation of the area, and emergency procedures to control or contain the spill. Each location should be inspected with that basic thought in mind. Loading areas especially should have some means of such containment and control, since outside personnel without the high degree of training of in-plant personnel may be involved.

Fig. 1-6   Hose line running alongside a storm drain.

Fig. 1-7   Loading dock with hose connections beneath and no protection for the hoses.

Are there materials at hand for spill control and containment? Is there safety equipment for personnel who may be involved? Where will a spill go if it occurs? Is drainage provided, leading to a controlled collection area? What kind of communication is provided to stop transfer action in the event of a spill or to summon help? Are there automatic shutoff devices? Can a driver pull away from a loading area while the vehicle is still connected to the loading equipment? (See Figs. 1-6 and 1-7.)

How much material could be spilled if the worst-case accident occurred, and is the containment provision adequate to handle such a spill? [A mile of 254-mm- (10-in-) diameter pipeline, for example, holds approximately 83.3 m^3 (22,000 gal) of product, which could all be lost in the event of a valve failure and gravity drainage of the line.]

Are the materials used for construction of the facilities proper for the materials being handled? (This situation may have changed between the time when the facility was built and the time of survey.) Are products handled or stored in the area that could react in a dangerous fashion in the event of a spill? If so, where else can they be stored, or what procedures can be instituted to minimize such contact? What maintenance procedures are used for loading and unloading equipment? How often are safety inspections made? What criteria are used for testing equipment? Are proper test records being kept? What rules and regulations must be complied with?

Tank Storage

Are tank areas properly diked in accordance with existing codes covering size, height, and distance? Are tank dikes and dike areas impervious to a spill? If not, could or must they be made so by the use of an impervious plastic lining, clay, or another means? Is the dike itself proof against penetration by burrowing animals that might impair its integrity (see Fig. 1-8)? Are small horizontal tanks far enough from dike walls so that a leak high on the tank will not spray over the dike?

How often are tanks inspected for integrity? How often are they cleaned and inspected internally?

How are below-ground tanks inspected, if at all? Are underground tanks and piping pressure-tested (local codes vary widely in this regard)? Do any tanks have cathodic protection? Is it necessary? Do dike areas have provisions for water drainage? Is there a separator or other protective device to ensure that there is no discharge of product from the dike area while water is being discharged? (See Fig. 1-9.)

What provision is made for overfill control such as high-level alarms or cutoffs? Are valve connections such that operator error is minimized or eliminated, reducing the chance that product will be put into a full tank, or that products which can react dangerously cannot accidentally be mixed? Are there valve interlocks, padlocked valves (with proper control of keys), color-coded valving and piping, noninterconnectable fittings, and proper supervision and connection inspections before transfers are started?

Fig. 1-8   An animal burrow has destroyed the integrity of this dike.

Fig. 1-9 Unlocked dike drain valves, unprotected from plugging by solids, are a frequent source of leaks from diked areas.

Pipe Alleys

Pipelines into a facility and pipe alleys within are low-potential sources of spills. However, they should be protected if possible by diking when above ground (see Fig. 1-10). This diking serves both as a containment for a spill if it occurs and as protection against damage fron contact. Are pipelines protected cathodically against corrosion? Are the lines pressure-tested? If so, how often? If the pipelines are no longer in use, are they empty of product (blown with an inert gas and blind-flanged)? (See Fig. 1-11.)

Fig. 1-10 Undiked pipe alley alongside a creek. This is a low-probability source of a leak, but a potentially costly one if a leak occurs.

Fig. 1-11 Unused piping properly plugged or capped.

Are there shutoff valves at both ends of the lines, for use if either end is inaccessible because of a spill or a fire?

Drum Storage

If the plant product is packaged in drums, is any outdoor storage area for filled drums roofed or open? Is there protection against leakage and/or damage (as from a fork truck) in the actual storage area (such as a concrete pad drained through a containment device of some sort) and also along the path by which drums must be transported from filling to storage to shipping? (See Fig. 1-12.)

Fig. 1-12 Filled drums stored outdoors on a roofed and drained concrete pad.

Are drums for filling, if new, properly empty and clean? If reconditioned, are they inspected before filling to ensure that they are clean and empty?

Are drums accepted as returns from any source? If so, are they empty, properly closed, and nonleaking? Is there a policy in this regard that has been properly communicated to customers, to those making pickups, and to those accepting the drums at the facility? What happens to rejected returns in handling, storage, and disposal? Is there inspection to ensure that the returned drums are not mislabeled and do not contain hazardous materials, either in themselves or from the standpoint of being mixed accidentally with products handled by the facility?

Where are returns stored? If on a drained pad, is it roofed to minimize rainwater contamination of a product containment device and overload of its capacity? (See Figs. 1-13, 1-14, 1-15.)

Fig. 1-13.

Fig. 1-14   Improper storage of returned "empty" drums.

Fig. 1-15 Returned drums stored on a drained pad. It would be better if this area were roofed, to prevent rain from reaching the pad drain, and if a storm drain were not so close.

For both inbound and outbound drums, are the storage pads properly curbed and drained? Are water-soluble products stored with water-insoluble ones so that an accident might upset separator operation?

Retention Basins, Containment Ponds, and Product Separators

Are there ponds, basins, and/or separators that might accumulate products by design or accident? What provision has been made for removal of accumulated products manually, automatically, continually, or at irregular intervals? If outside contractors are used for this purpose, are they fully aware of product properties? If there is a separator, is it of adequate capacity? Is it functioning properly, and what provision has been made for performance testing? Is there any problem with suspended solids that might affect separator performance? Do retention ponds have baffles or weirs designed to prevent accidental escape of product in the event of flooding or massive product discharge?

Temporary Construction and Other Facilities

Is there contractor activity that might affect or be affected by routine plant operations? Are the contractor's personnel properly trained and informed about spill potential affected by their operations? Are the contractor's storage facilities (usually fuel, stored high for gravity feed close to roadways) properly protected against accident?

Inside Survey

Small Leaks from Fittings, Valve Packing Glands, Bearings, and Seals

Where do such leaks go? If they go into a floor drain, does the drain lead to a sump from which product can be recovered or to a storm sewer or drain? Can such leaks be controlled by better maintenance or by the use of drip pans, sorbents, or other means? If the product flows to a sump, where does the sump effluent go? If to an in-plant separator, is that working properly, and how is proper working determined? Can a surge of water or product overload the separator or sump? Are pumps used to transfer sump contents of a kind that will form emulsions or dispersions of product?

Vapor Exhausts; Safety Blowoff Areas

Where do vents from compressors, exhaust blowers, and emergency vents (rupture disks) lead? If they lead to the outside of the building, do the discharges condense onto the ground or onto roofs, where rainfall might wash them into sewers? If they are discharged into the in-plant atmosphere, will they condense into accumulations that can come in contact with water or steam and be washed into a plant sewer or drain?

Floor Drains

Most plants have floor drains. In the event of an accident, where will these drains take spilled product? Are all the drains really necessary? Can they be closed or relocated so that only the really essential drains are open for use? Can areas where floor drains are essential be subdivided by the use of curbing to segregate water drains from possible contact with spilled product?

Tanks

Are inside storage tanks, including reservoirs in equipment, protected against accidental leaks of product in large amounts? If a rupture occurs, where will the product flow? Can such tanks be curbed, diked, or segregated behind walls to contain spills? If the diked or curbed areas have drains, are they necessary? Could they be closed and spills cleaned up by hand? Are the tanks vented, and if so, to where? If they are vented to the outside of the buildings, do the tanks have high-level alarms or shutoffs to prevent spillage out of the vent caused by an overfill? Are the vents high enough to provide a head greater than the head capacity of the pump feeding the tank? If a discharge through a vent to a roof cannot assuredly be prevented, can the roof itself be safely used as a containment basin? Can it be equipped for such a purpose?

Are process and storage tanks inspected on a regular basis and repairs made promptly when required?

Piping

Are walls and floors free from holes caused by removal of old equipment and piping? Are there cracks alongside pipes extending through walls or floors? Are such openings for pipes properly caulked to prevent leaks if they represent potential exit points of a spill? Is plant piping well marked, and are operators trained so that an operating error could not result in an accidental product discharge? Is piping regularly tested for leaks or inspected for corrosion? (See Figs. 1-16 and 1-17.)

Drum Product and Raw-Material Storage

Is internal drum storage of finished product or raw materials so segregated that an accident in the storage area will not spread through the operating area? Is it possible to install curbing around storage areas to prevent this? (See Fig. 1-18.)

Valves and Fittings

Are valves properly identified, by color coding or naming, so that operating errors are minimized? Are valves not in daily use locked against accidental change in setting? In the event of an accident, are there alternative locations where valving or controls can be operated to minimize the spill by shutting off the flow?

Along the normal work path within the facility (where drums are moved on hand trucks, fork trucks travel, or equipment and machinery may be moved) is all piping and valving safe from accidental contact that could cause a spill by damage to that piping and valving? Are there drains, floor drainage slopes, or other conditions that could cause an accident along the normal work paths to be uncontrollable?

Fig. 1-16 Openings through the floor near a filling area are a potential source of leaks into the area below. Valves are uncoded, unmarked, and a potential source of operator error.

Fig. 1-17 Indoors or outdoors, such a "bird's nest" is not unusual. Unlocked valves, lack of coding, and poor access to piping for maintenance and repair are usually the result of unplanned growth and constitute a high-potential source of small leaks and operator error.

Fig. 1-18   An unprotected storage area inside a building can cause an outdoor leak. Obvious seepage from inside a building can run directly into an external storm drain.

Maintenance Operations

Is the handling of product collected in the maintenance of building and machinery (flushing of equipment, washing of walls or machines, or cleaning of storage areas) properly controlled to prevent discharge of product? In other words, does the cleanup porter dump a bucket of dirty water containing toxic chemicals down the most convenient plant floor drain? How is waste handled and disposed of to prevent discharges? Are the solutions normally used in cleaning and routine building maintenance such that, if improperly disposed of, they might upset the operation of a separating system? Can specific disposal areas be provided for such solutions and segregation of such solutions be assured?

Available Control Materials

It is assumed that, for specific chemicals requiring neutralization or counteraction of some sort if spilled, a plant will have available proper materials and instructions to carry out such counteraction. The plant survey should assure that this is so.

There are some general types of materials for use in many product spill situations which should be considered during a survey. If these are applicable, the survey should indicate which kinds are usable and where they should be located.

Sorbents

Despite some problems with definitions (whether products are absorbents, adsorbents, or just plain sorbents), when there is a spill, sorbents are useful tools to mop up and help dispose of spilled material. There are several classes of sorbents, as described below.

Natural Products   This class includes such materials as hay, straw, rice bran, peanut hulls, cotton linters, pine boughs, and other naturally occurring products (in the sense that they are not processed products designed for sorbent use). Some of these are by-products of other industries and often are wasted. These materials can be used in spill cleanup, but generally with rather poor efficiency and with handling and storage problems; in an emergency, when one is (literally) grasping at straws, one uses what is available. Products made for specific sorbent use, however, are generally more efficient, are less costly per unit of recovered product, are easier to store, and are easier to use and dispose of.

Selective Sorbents: Removing Product from Water   These sorbents are designed to remove product from on top of water, leaving the water behind. Several different types are available.

Cellulosic Sorbents   These are made from cellulose fiber treated to render it water-repellent. They come in pads, sheets, particulates, and rolls. They generally possess good sorption for oil, but they present difficulties in long-term storage (generally covered storage is required). While they will sorb hazardous materials, care in handling and disposal is required.

Synthetic Fibers   These sorbents generally are made from polypropylene fiber that is felted into rolls and sheets or shredded into particles used to form small bags or longer boom sections. Sorption is good for oil and generally is good for a range of hazardous materials. Storage characteristics are good, no special storage protection being required.

Synthetic Foams   These sorbents generally are polyurethane foams but occasionally are polyethylene foams. They offer good pickup of oil but generally pick up more water than the other sorbents described above. Storage is fairly good.

Imbibitive Polymers   These are copolymers that will react with certain classes of organic product, "drinking" them into the molecular structure, forming first sticky masses and then, as saturation nears, a jellylike material which can be cut without losing product. These materials can be used to reduce the vaporization rate and to "solidify," and hence make easier to handle, a wide variety of hazardous and toxic chemicals. In proper form, valves can be constructed that will pass water but will swell and stop the flow when product is encountered. They are very useful for draining water from tanks and out of dike areas. These products cannot be used as in-line filters, since swelling will stop the flow.

Selective Sorbents: Product Dissolved in Water   Many hazardous materials are water-soluble. Activated-charcoal filters have been used successfully in removing such dissolved materials, usually at high cost.

Nonselective Sorbents   A highly porous nonreactive silicate has come onto the market. This product, with a density of about 32 kg/m^3 (2 lb/ft^3), will pick up whatever it touches. Therefore, it is not usable to remove product floating on water, but it can be used on land to sorb water, aqueous solutions of any pH, and pure product, including oil and hazardous materials.

Sorbent Availability   Sorbents are produced by several manufacturers and are generally stocked and sold through local distributors both in the United States and in a number of other countries. Most commercial sorbents are fairly light in weight, are easy to handle, and require only moderately careful storage. Some, as noted, require enclosed space for long-term storage.

Sorbent Disposal   It is not always easy to dispose of sorbent saturated with product. Although most manufacturers make a point of reusability, the cost of squeezing out recovered product, combined with the inherent messiness of such an operation and, depending on the product, the possible handling hazard, makes this characteristic more of a paper benefit than a real one.

Generally, saturated sorbent is put into drums, plastic bags, or, occasionally, open trucks for disposal either by burning (in a properly designed incinerator) or by burial in an approved landfill. Occasionally, especially with oil, the product may be burned for fuel if it is not too heavily contaminated with nonburnable solids. Even more infrequently, synthetics can be dissolved in hot oil and used for fuel as part of the oil.

Special Sorbents Depending on circumstances, sand, snow (in cold climates), dirt, fuller's earth, and a variety of other materials have occasionally been used as sorbents, mostly because of quick availability in a situation demanding immediate action. Proper spill control planning would seem to dictate storage of some sort or sorts of sorbents and prior arrangement for quick supplies of additional sorbents if needed.

Chemicals

"Chemicals" can be a general term covering dispersants, neutralizers, gelling agents, fire-fighting foams, sinking agents, and a host of other products. At this time, Committee F-20 of the American Society for Testing and Materials (ASTM)[1] has a division grappling with the problems of chemical control of chemical spills. The conditions under which chemicals can be used, the proper methods of such use, and the proper determination of relative risk (Is it better to try to neutralize a spill or simply to rely on dilution with water? There may be different decisions in a plant or a fishpond) are simply not as yet well defined. Beyond the obvious need to keep soda ash on hand for neutralizing small spills of acid, for example, little general advice can be presented now for handling large spills.

Equipment

In many instances specialized equipment is available to contain and recover spills of hazardous materials. Most of this equipment was developed for use with spills of petroleum products; some is usable on other kinds of product spills. New equipment is reaching the market all the time. Many members of ASTM F-20 and of the Spill Control Association of America[2] represent companies engaged in the development and manufacture of such equipment in the United States. Contact with these two organizations should provide specific names of such companies.

Much of this equipment relates to recovery and will be discussed in detail in Chap. 9. Such equipment includes various kinds of skimmers for removing spilled product. However, it is proper to discuss the use of some kinds of equipment under the heading "Prevention," and that will be done here.

Containment Materials Preventing the spread of spilled product greatly simplifies cleanup, provided it can be done quickly and safely. Especially, prevention of the spread of spills on water greatly reduces the contaminated area and also reduces the time, cost, and hazard of recovery.

Materials at Hand If a spill occurs on nonporous ground (paved areas, frozen ground, clay soil), it can frequently be surrounded by a dike of earth, dug from the nearest available area with shovels, a backhoe, or other mechanical device, depending on need. Often advance provision can be made in an area of "high" risk by prefilling sandbags and storing them outside, properly protected by tarpaulins, near the potential use site. The same sandbags can be used to shut off storm drains ahead of a spill, provided there is sufficient time after the spill occurs.

Existing ditches, normally used for water draining, can be used to confine a spill by piling earth across the ditch and compacting it to form a dam. Floating product can be retained, if there is water in the ditch and the product is insoluble, by supporting a pipe on stones with the upstream end lower than the downstream end and covering this with dirt to form an underflow dam or weir.

Manufactured Materials: Quick-Setting Foams Several manufacturers of insulation sell aerosol containers of polyurethane foam. While rather expensive and of relatively short shelf life (perhaps a year), these foams rapidly swell and solidify (usually within 1 min of application) to form an impervious barrier which will resist many chemical products and which is nominally fireproof (though most foams will burn if they are ignited at a thin cut edge or come in contact with a hot fire).

Manufactured Materials: Containment Booms   There are a number of manufacturers of containment booms, which generally consist of a flotation section supporting a skirt, at the bottom of which is ballast, usually link chain. The flotation section is often made of polyethylene and hence is relatively impervious to a wide range of hazardous materials. Skirt material is usually a webbing of woven synthetic fiber such as nylon or polyester, coated with a material such as polyvinyl chloride. These booms are quite impervious to petroleum products and to a large range of hazardous chemicals. Other materials are available for special uses, as are special metals for boom fittings, ballast, connectors, etc.

Water-immiscible product can be confined very well on still water such as a pond, a lake, or a quiet bay or, with a difference in technique and boom configuration, on flowing water such as a river (within limits). Booms can thus be used to reduce the spread of a spill reaching water and to aid in cleanup.

Safety Gear

Handling a hazardous material spill requires proper safety equipment. Anyone engaging in containment, control, or recovery activities should be provided with the proper equipment and be well trained in its use.

Storage of Control Material

Material to be used for spill control should be stored as near as practicable to the site where it might be used. Proper protection from deterioration during storage, theft and vandalism, pilferage, and use for other purposes must be provided. Some estimate of possible quantities must be made so that emergency materials are available in sufficient quantity to function efficiently if need arises. It is generally wise to provide separate storage for such materials either in a permanent building or in a small emergency trailer if several areas need protection and this method proves more economical than duplication of materials and storage facilities. Care must be taken that security in storage is not so tight that entry to the stored materials is difficult when need arises. Several persons should be authorized to have keys and access to the stored material.

Emergency Procedures

While Chap. 7 of the *Handbook* covers response plans and procedures, some mention of emergency procedures should be made here.

Contingency Planning

Contingency planning is what spill prevention is all about. The "what if" attitude should disclose most foreseeable (though never all) problems and set forth in detail some method of dealing with them. Things never occur as planned, but thinking about them helps one to prepare for the unexpected.

A contingency plan should be in written form, patterned much after the requirements of the U.S. Environmental Protection Agency's Spill Prevention Control and Countermeasure Plans.[3,4] Copies should be available at potential spill sites and be used as training aids for all personnel.

Of particular importance in responding to any spill emergency is response time. If a spill can be found while it is small and response is rapid, its effects can usually be minimized. Thus an essential part of any contingency plan is provision for regular security checks of both high- and low-risk areas so that if spills do occur, they are found promptly.

Training for Action

Regular training drills are necessary for all personnel who are to be involved in cleanup and control. A training manual should be developed, updated as often as necessary, and combined with regular classroom and field hands-on training, including carefully controlled actual spills of hazardous materials. One must learn to handle problems with advance practice, not by on-the-job training. Response time to a spill will be greatly reduced if such regular training is employed.

Notification

A schedule of notification procedures to be followed in the event of a spill should be established. It should include all legally required notifications (federal, state, and local authorities) with an indication of the person who is to make the call, plant personnel, company personnel not at the plant, and outside cleanup contractors. Notification lists should be updated and reviewed on a regular schedule and reissued whenever any changes are required. This is normally a monthly task.

Outside Help

The equipment and expertise to handle large hazardous material spills is rarely available when needed within a plant or company. Unless there is a continual number of such large spills, it simply is not economically feasible to be prepared for eventualities of low potential. There are a number of highly competent cleanup contractors who specialize in this work, and their ranks are growing with increasing awareness of the need for such services. Competent contractors are expensive; the investment in equipment and training maintained by good contractors is very high, and hence their charges also are high. It is a prudent decision to make advance contact with such contractors and to have advance understanding of their charges, capabilities, and availability. In many instances, a yearly retainer will assure preferred service, plus the help of the contractor in assessing in-plant exposures, corrective measures, and possibly assistance in training personnel.

DECISION MAKING

Spill prevention, like accident prevention, is a matter of identifying potential causes and taking corrective action. Rarely is it possible to base a decision on a mathematical procedure, balancing the cost of taking some action against the savings that accrue because of reduced cleanup costs, product savings, reduced downtime, or perhaps no spill at all.

What kinds of decisions are needed? They are those that will reduce the probability (not determinable) of having a spill by an amount in keeping with the money expended to accomplish that result.

Legal Decisions

There is not much choice in legal matters. Despite the best of goodwill on the part of someone who might have a spill of hazardous material and the deepest commitment (moral and financial) to compliance with the law, there will still be legitimate areas of disagreement about what must be done. If there is a court decision, there is no legal problem, though perhaps a financial one. What the court determines you must do, you must do (after exhausting all legal appeals). Obviously, a public relations problem is also involved. Even though you may be legally right in a position you take, a battle through the court system, with attendant publicity, may cost more in the long run in public ill will and reduced sales than spending the money to do what is not legally required. The type of decision then moves from a legal one to a logical one.

Logical Decisions

"Logical" decisions are based on experience, intuition, and perhaps even engineering appraisal. They tend to be the kinds of decisions that all persons make, one way or another, during most of their lives.

For example, suppose that a plant has an undiked tank situated in a large open area near a road, with a storm drain (leading directly to a river) located a short distance downhill on the road. The tank contains a highly toxic chemical. Logic dictates that the gain from installing a relatively simple earthen dike to protect the river from the result of a low-probability tank leak far outweighs the relatively small cost of installing the dike.

Move that tank inside or to a location where space is at a premium and there is no room for a simple dike properly distanced from the tank, and the cost of protection rises. The decision requires other input. How often does a tank of this sort leak? When was it last inspected? Can it be safely inspected and some sort of regular program established to reduce the chance of an accidental leak? What will a high concrete dike close to the tank cost, and how will it affect ability to operate? If there is a leak in these close quarters, will the only loss be product and the only damage environmental, or will personnel be injured or perhaps killed?

Under these conditions the decision, while still based on "logic," may be different. The cost may be so high and the risk (based on past experience) low enough so that no dike is installed. The tank may be relocated. Or nothing may be done at all.

What is absolutely necessary in such instances is that whatever the decision, it be established as a part of the company records. In today's climate of environmental responsibility, it is essential that decisions be deliberate and recallable. One simply cannot afford to be in the position of having to say "We didn't realize this was a problem." In balancing the many needs for money, one must be able (or at least try) to justify a decision not to do something because of inability to do everything and because of greater need or greater danger to be allayed elsewhere.

Economic Decisions

Economic considerations are a part of most business decisions. In any endeavor with some potential for spills of hazardous materials, a budget must contain some amount for spill control training, planning, and prevention. It may not be possible to reach numerical answers on specific questions. It is possible, as a matter of business judgment, to allocate a portion of an operating budget to spill control matters, just as is done for research and development, sales effort, public relations, or advertising. Spill control is a necessary part of today's world. A good way to approach a decision on how much money to allocate to spill prevention is to make the surveys previously suggested.

It is rare that such a program will not disclose many ways to spend money on fixing up, repairing, or replacing equipment. A program can be established within the normal maintenance schedule of any facility, over a reasonable period of time (up to 5 years, with modification as experience and resurvey dictate), and an annual amount allocated in the budget for spill prevention maintenance. The personnel to be involved and their time for training on a regular basis are relatively easy to determine, and the cost can be added to the maintenance amount. If an outside contractor is to be retained, that fee can also be determined and added to the total.

The mere inclusion of these amounts in a budget may be of great help, if a spill occurs, in reducing fines and obtaining the cooperative assistance of governmental authorities, since one will have demonstrated a commitment to spill prevention and a commitment that is actively in progress.

Technical Decisions

Technical decisions are often involved in choosing between alternative methods of reducing spill probabilities. As an example, should one install a high-level alarm to prevent a tank overfill or raise the overflow pipe height and reduce the capacity of the feed pump head? Both cost and effectiveness of method enter into such decisions, and provided the proper information is available on which to base such decisions, technical decisions are reached more easily than legal, economic, and logical ones.

Much of the equipment now available for cleanup, control, and prevention of oil spills and the techniques of properly using that equipment are under study by ASTM Committee F-20 on Hazardous Substances and Oil Spill Response. The standards for testing such equipment and standard practices for using it will be issued from that committee, and it is hoped that evaluation of equipment will be far easier than it has been in the past. The work of the committee has now been extended to hazardous substances. As standards are issued and papers published from symposia, they will be available through the ASTM.

REFERENCES

1. American Society for Testing and Materials, Committee F-20: Hazardous Substances and Oil Spill Response, 1916 Race Street, Philadelphia, Pa. 19103.
2. Spill Control Association of America, 1040 North Park Plaza, 17117 West Nine Mile Road, Southfield, Mich. 48075.
3. "Oil Pollution Prevention—Non-transportation Related Onshore and Offshore Facilities," 40 CFR Part 112, *Federal Register,* Dec. 11, 1973, pp. 34165–34170.
4. *Suggested Procedure for Development of Spill Prevention Control and Countermeasure Plans,* Bulletin D16, 1st ed., American Petroleum Institute, Washington, 1974 (reissued 1976).

PART 2
Maritime Transportation

William A. Creelman
President, Transport Division
National Marine Service, Inc.

Vessel Design / 6-25
Operating Personnel / 6-28
Safe Navigation / 6-33
Response to Emergencies / 6-34
The Record / 6-34

There are four principal elements in the prevention of hazardous materials spills in marine transportation. The first and most important is vessel design. In the United States, tank vessel design for hazardous materials is regulated under U.S. Coast Guard Subchap. O (46 CFR 151). Subchapter O is a comprehensive set of regulations dealing with hull design criteria, piping, venting, gauging, and compartmentation as well as with detailed special requirements for particularly toxic or corrosive products.

The second element of pollution prevention consists of operating personnel, their training, certification, and operating procedures. In the United States, Coast Guard tankerman regulations apply, and there are ongoing efforts to amend or replace these regulations with a more comprehensive set to cover specifically the special requirements and problems of hazardous materials transfer.

The third element is navigational safety. Here again the qualifications of personnel are vital, and the regulations applying to pilothouse personnel are those of the Coast Guard. In addition to the training and licensing of personnel, navigational safety involves all aids to navigation including the system of buoys and lights, bridge-to-bridge communication by VHF radiotelephone, and Coast Guard–operated vessel traffic services such as those that exist in San Francisco, Seattle, Houston, and New Orleans.

The fourth element of spill prevention concerns response to emergencies such as groundings, collisions, or other incidents which threaten the integrity of the hull or the cargo containment. At the early stages of such an emergency there usually has been no spill, but the potential for a spill is normally present if the response to the emergency is not successful.

VESSEL DESIGN

Vessel design is the first element of spill prevention. The Coast Guard's Subchap. O divides the various hazardous cargoes into three basic categories. (Table 151.05, "Summary of Minimum Requirements," of Subchap. O is included as an appendix to Part 2 of this chapter.) The least hazardous materials are permitted to be moved in Hull Type III. Cargoes with significantly greater hazards require Hull Type II, and the most hazardous cargoes of all, including chlorine and poison cargoes, require Hull Type I.

Hull Type I provides the greatest protection to the cargo containment area of the barge by requiring a 1.2-m- (4-ft-) wide wing wall of void space on the sides, a 7.6-m (25-ft) void tank between the bow of the vessel and the first cargo tank bulkhead, and a full double bottom. In addition, the hull must be able to survive damage at the intersection of a transverse and a longitudinal bulkhead during the flooding of four compartments without causing the barge to sink when fully loaded with cargo.

The Type I hull must also be capable of sustaining a pinnacle grounding without structural failure of the hull girder. The pinnacle-grounding criterion, a very severe test, consists of calculating the deck and bottom stresses experienced by the structure when fully loaded with the bow aground at the surface-water level at the point where the bow rake and bottom intersect. This test assures a hull structure that can withstand the most severe practical grounding stresses which are likely to occur during the life of the vessel.

The 1.2-m wing wall and the 7.6-m bow-to-collision-bulkhead requirement are both intended to provide a high degree of collision protection for the cargo containment area of the barge. They also provide a void-space area to contain any cargo which might leak from the cargo containment area and thereby to prevent it from contaminating the waterway. Similarly, any hull leakage which might develop through minor casualties or through wear and tear will result in the leakage of river or seawater into the void-space area without resulting in any contamination of the cargo. The cargoes shown in Table 2-1 require Type I hulls.

Requirements for Type II hulls are very similar to those for Type I, except that the wing-wall width is 0.9 m (3 ft) instead of 1.2 m and the flooding requirement calls for the barge to withstand damage on any longitudinal *or* transverse bulkhead. Accordingly, the Type II barge must survive two-compartment flooding, whereas the Type I barge must survive four-compartment flooding. The flooding criteria for both Type I and Type II hulls assume full-cargo loading in both cases. Cargoes which require a Type II hull for carriage are listed in Table 2-2.

The Type III hull is specified for those cargoes under Subchap. O which are described as having the least hazard. A Type III barge is essentially any barge which does not meet the requirements of Type I or Type II. Normally, it is a barge with a single skin rather than a double skin, and there are no wing walls or double-bottom spaces. The cargo containment is the outer shell of the hull.

Most oil barges are built in this way, as are most ocean ships, including the largest supertankers afloat today. However, such barges do have large bow and stern void spaces with which to absorb collision impact, and they are built with structural and plating specifications to meet Coast Guard or classification society specifications. The United States classification society is the American Bureau of Shipping.

TABLE 2-1   Chemical Cargoes Requiring a Type I Hull

Acetone cyanohydrin	Hydrofluoric acid
Allyl alcohol	Hydrogen chloride
Allyl chloride	Hydrogen fluoride
Aniline	Methyl bromide
Carbolic oil	Motor-fuel antiknock compounds
Chlorine	Phenol
Chlorohydrin (crude)	Phosphorus (elemental)
Epichlorohydrin	Sulfur dioxide
Ethylene oxide	

SOURCE: 46 CFR 151 (1977), pp. 152–159.

TABLE 2-2   Chemical Cargoes Requiring a Type II Hull

Acetaldehyde	Dichloropropane
Acrylonitrile	Dimethylamine
Adiponitrile	Ethyl chloride
Ammonia (anhydrous)	Ethyl ether
Ammonium hydroxide (not to exceed 28% NH$_3$)	Methyl chloride
Butadiene (inhibited)	Propylene oxide
Camphor	Vinyl chloride
Carbon bisulfide	Vinylidene chloride (inhibited)
Crotonaldehyde	

SOURCE: 46 CFR 151 (1977), pp. 152–159.

In short, Type I and Type II hulls are full double-skin hulls (see Fig. 2-1), and the Type III hull is a single-skin hull (see Fig. 2-2) or one with only a partial double skin, such as a double-side–single-bottom hull or a single-side–double-bottom hull. But if any part of the cargo containment area other than the deck is also part of the outer shell of the hull itself, the barge is described as having a Type III hull.

Typically, the double-skin configuration has all the side and bottom structural members in the void space, thus leaving the interior of the cargo space with a smooth surface. On the other hand, in the single-skin configuration all the structural members of the hull are located within the cargo tank, and the cargo tank interior is cluttered and anything but smooth.

While the smooth-interior tank of the double-skin barge is easier to clean than the single-skin barge, the double-skin void space is a particularly difficult area to clean should cargo enter it through leakage of the cargo tank. Further, if the outer hull is damaged, the double-skin barge tends to take on more draft and to lose buoyancy in the area of the damage. Although the single-skin barge will not lose buoyancy when damaged in the cargo area, it may lose cargo to the exterior of the barge, thereby causing a spill. During grounding or a heavy landing against a dock, bridge, or lock, the structure of the single-skin barge will generally be more flexible, while the double-skin barge will be more rigid,

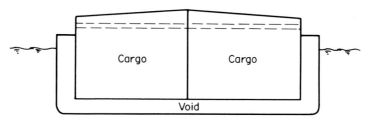

Fig. 2-1   Double-skin tank barge.

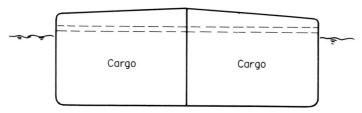

Fig. 2-2   Single-skin tank barge.

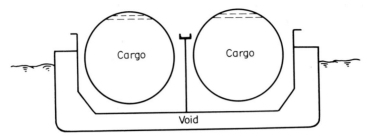

Fig. 2-3   Independent tank barge.

and there may be a tendency for the structural members in the void space to puncture the cargo containment area, causing cargo leakage to the outer hull. Such leakage will be difficult to clean and remove but will usually be contained within the outer hull unless the outer hull itself also is damaged.

In addition to the three basic hull categories described above, there is the independent tank barge (see Fig. 2-3), usually consisting of one or more cylindrical tanks mounted on two or more saddles within the outer hull. While most of these barges do not have full double-skin hulls, they are nonetheless classified as either Type I, Type II, or Type III, depending upon their ability to survive certain damage criteria. These criteria are as described above with respect to the distance between the bow and the collision bulkhead and to the wing-wall widths and only slightly different with respect to floodability.

Cargoes which are frequently transported in cylindrical tank barges are refrigerated anhydrous ammonia, chlorine, propane, propylene, asphalt, liquid sulfur, phosphoric acid, and naphthalene. The cylindrical tanks make it reasonably simple to insulate for either very cold cargoes (e.g., refrigerated anhydrous ammonia) or very hot cargoes (e.g., asphalt). Similarly, the cylindrical tank configuration facilitates the installation of cargo tank lining, such as the rubber lining used for the carriage of various acids. The cylindrical configuration also permits the tank to have a significantly greater working pressure than would be possible in the typical gravity tank. Products which must be carried under pressure such as chlorine, nonrefrigerated ammonia and propane, butadiene, and butane are normally carried in cylindrical tank barges.

OPERATING PERSONNEL

After a well-designed and -constructed vessel has been obtained for the secure carriage of a particular hazardous material, the next factor to be considered in spill prevention is providing a well-trained operator to handle the loading and unloading of the vessel. Coast Guard statistics indicate that approximately 85 percent of the individual spills which occur in United States waterways are the result of tankerman failure or error in the loading and unloading of tank vessels. The Coast Guard certifies tankermen after they have acquired the necessary on-the-job experience aboard tank vessels and have passed a written examination confirming their knowledge of the applicable rules and operating procedures to be followed in the course of their duties.

The Coast Guard publishes manuals which are useful in the training of tankermen, and they also publish the tankermen regulations which apply to the transfer of Subchap. D cargoes. These are petroleum cargoes rather than chemical cargoes; under present United States regulations the transfer of chemical cargoes can be handled by "a competent person" as determined by the employer without any tankerman certificate's being required. Instead of a tankerman certificate a letter of competency, attesting to the qualifications and experience of the individual, is filed with the appropriate Coast Guard offices by the employer. This is a rather loose procedure and one which has been the subject of a great deal of study and draft regulation writing on the part of the Coast Guard and interested industry groups.

The Coast Guard was on the brink of publishing new tankerman regulations for Subchap. O cargoes in 1976, when an apparent conflict with draft Intergovernmental Maritime Consultative Organization regulations caused the new Subchap. O tankerman regulations to be shelved. Accordingly, the letter-of-certification procedures are still being used, and there are no formal regulations or certification procedures for specific Subchap. O cargoes. Because of the absence of formal regulations for these cargoes, operators of chemical tank vessels normally employ as tankermen individuals who have already been certificated as such under Subchap. D by the Coast Guard.

Employers provide in-house training as required to familiarize Subchap. D tankermen with facts about the specific chemical cargoes that they will be handling and the unique hazards of such cargoes. Many companies have their own detailed cargo transfer procedure manuals as well as their own detailed safety procedures for their particular vessel equipment. Coast Guard regulations do require that a cargo transfer procedure (see Figs. 2-4 and 2-5) showing basic data about the cargo piping and valving as well as the arrangement for cargo headers and pumps be aboard each tank vessel. In addition, operating personnel, both ashore and afloat, must file a form entitled "Declaration of Inspection" (see Fig. 2-6).

The declaration of inspection is particularly useful since it requires the individual in charge of cargo transfer on the vessel and the counterpart on the dock to communicate and together attest that all critical items involved in the cargo transfer have been checked and are in good order before actual transfer begins. Each of them must sign the document, and when crews change, each relief person must similarly endorse the document. This contact between the two key individuals in the cargo transfer operation assures good communication between the shore facility and the vessel. It also provides an opportunity to determine the rate of loading that is to apply and to identify any special problems, either ashore or afloat, which may need to be recognized before beginning the operation.

Normal procedures for cargo loading begin with the hookup, or connection of the hose or loading arm between the piping system of the shore facility and that of the vessel. Therefore, attention must be given to adequate support of the hose, and the connections must be carefully checked to ensure that gaskets are properly in place and adequate bolts or connecting cams have been properly adjusted to assure a leakproof system. However, if any leaks do occur at the point of connection, both the vessel and the dock are equipped by regulation with large-capacity drip containers. These prevent the introduction of any cargo into the water and permit cargo transfer from the drip container to storage tanks either aboard the vessel or at the shore facility.

Once the appropriate valves have been opened and the signal to begin operations has been given, cargo transfer usually takes place at a slow rate until the tankerman has had an opportunity to check the entire piping system and its valves to be sure that no leaks exist and that the cargo is routed to the appropriate tanks for loading. After this has been determined, the tankerman will signal the shore facility to speed up the loading to the previously agreed rate. The tankerman will normally bring up the cargo level in the several tanks aboard the barge so that the tanks can be topped off one by one to the desired fill level. At the beginning of the topping-off operation the tankerman will have communicated carefully with the shore supervisor to slow down the rate of loading to a rate which can be safely handled without risk of spillage. Topping off is a time for exercising the greatest care to be sure that the process does not get ahead of the tankerman's ability

Before beginning any loading or discharging of oil, or the stripping of any oily slops or cleaning wastes, consult the adjacent diagram of this barge. Each operator and/or tankerman must be thoroughly familiar with the location of each valve, pump, control device, vent and overflow.

At no time during the transfer operations will there be less than one responsible person on duty. The certificated tankerman assigned shall be in charge and responsible for the safe transfer of cargo. His duties include, but are not limited to, the following:

1. Proper connection of the grounding cable.
2. Proper and tight connection of the header line hoses.
3. Checking the bow and stern rakes and all void spaces, making sure that they are dry and free of product and water.
4. Maintaining proper security of the barge while it is under his cognizance.
5. Taking all precautions to guard against the accidental discharge of oil.
6. Should an accidental discharge of oil occur, stop transfer operations, and report the incident to the proper authorities immediately.
7. Check and sign the declaration of inspection prior to transfer.

During transfer operations, the mooring lines will be checked and adjusted as necessary at half hour intervals. In situations where moored barges are subject to surging due to passing vessels, or where high wind conditions exist, additional mooring lines will be used to insure a safe mooring.

In the event of an emergency, transfer operations can be stopped by pulling the remote shutdown cable. The remote emergency engine shutdown station is located on the forward end of the barge on the port side. Familiarize yourself with the location and the operation of this control prior to transfer.

In the case of a single cargo loading, tanks will be topped off according to their distance from the loading header. That is, the tanks the farthest from the loading header will be topped off first and the tanks the nearest the loading header will be topped off last. The flow of oil while topping off will be controlled by the compartment valve of the compartment being finished. Definite agreement must be reached with the shore personnel concerning the rate of flow during the loading, while topping off, and for final shutdown. Be sure that the tanks are not topped off so full that the maximum or desired draft is exceeded, and so that there is enough space in the tank to allow for the expansion of the product.

As each tank is topped off, the compartment valve should be closed and remain closed until the system is lined up for discharge at the off-load port. After a tank has been finished, it should be checked at frequent intervals to insure that the valve is not leaking, which could result in an overflow. After the loading is complete, close the loading valve and the header line valve.

The deck discharge containment system consists of two large containment tanks, one located at each end of the header line. Each discharge containment tank is designed for a cargo transfer hose of 8″ in diameter. The tanks are equipped with a drain leading directly into the cargo tank and a drain for the removal of slops to the waste tank aboard the towboat or to shoreside facilities. Never drain the containment tanks onto the deck.

Should an accidental discharge of oil on the water occur, notify the nearest Coast Guard office by the fastest means available. The National Marine Service representative in your area must also be notified. The numbers of the National Marine Service offices are:

St. Louis	314-968-2700	This cargo transfer procedure
Houston	713-688-1481	has been accepted by the OCMI
Chicago	312-598-2202	USCG, St. Louis, Missouri, on
NORCO	504-764-6140	19 April, 1974.

Fig. 2-4 Oil transfer procedures. [*NMS 1317.*]

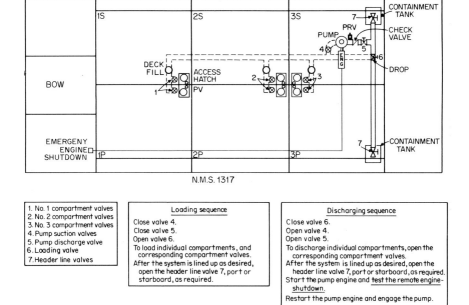

N.M.S. 1317

| 1. No. 1 compartment valves
2. No. 2 compartment valves
3. No. 3 compartment valves
4. Pump suction valves
5. Pump discharge valve
6. Loading valve
7. Header line valves | Loading sequence

Close valve 4.
Close valve 5.
Open valve 6.
To load individual compartments, and
  corresponding compartment valves.
After the system is lined up as desired,
  open the header line valve 7, port or
  starboard, as required. | Discharging sequence

Close valve 6.
Open valve 4.
Open valve 5.
To discharge individual compartments, open the
  corresponding compartment valves.
After the system is lined up as desired, open the
  header line valve 7, port or starboard, as required.
Start the pump engine and test the remote engine-
  shutdown.
Restart the pump engine and engage the pump. |

Fig. 2-5 Cargo-loading and -discharging sequences. [*NMS 1317.*]

to react. If that should occur, the result could be an overflow spill of hazardous material to the deck and ultimately to the water.

The loading operation which has been described is clearly critical to spill prevention. While the rate of loading and the sequence of tank filling are important to the operation, however, determination of the cargo level is also critical during loading. Most cargoes, even hazardous ones, are loaded by using open gauging techniques in which the actual cargo level is observed by the tankerman during loading. Cargoes with more hazardous vapors require restricted or closed gauges, often float-operated devices which can be read either at the cargo tank or remotely by the operator without exposure to the vapors. While closed gauging devices are extremely useful tools, they are mechanical devices which can fail and, when used, should be backed up by a redundant system.

A careful tankerman will observe the draft of the vessel during the loading operation and ensure that the vessel is being kept in reasonable trim and that the draft is consistent with the quantity loaded. If an inconsistency develops, the tankerman can be sure that some unplanned event has occurred (e.g., a leak in a void tank, causing the vessel to ride deeper in the water than planned for the amount of cargo actually loaded). Accordingly, the careful tankerman will also regularly check all void spaces, both before loading begins and throughout the loading operation, to ensure that they remain empty and dry and that there are no leaks.

Some closed gauging devices are equipped with automatic level alarms which sound when the cargo reaches a predetermined level. This level can be selected so that the alarm sounds at the beginning of a topping-off operation or at its end as a signal for normal valve closure. In some cases, the alarm circuit is used to trigger an automatic valve actuator. Clearly, these devices can be quite useful to the tankerman, but because they are mechanical, they can fail. Their failure can result in very serious spills unless the tankerman is monitoring the devices continuously through redundant procedures.

VESSELS _____ _____
_____ _____
_____ _____

TRANSFER FACILITY _____
LOCATION _____

The following list refers to requirements set forth in detail in 33 CFR 156.150 and 46 CFR 35.35-30 (printed on reverse). The spaces adjacent to items on the list are provided to indicate that the detailed requirement has been met.

	DELIVERER	RECEIVER
1. Communication System/Language Fluency. (156.120 (m)(p))		
2. Warning Signs and red Warning Signals. (35.35-30)		
3. Vessels Moorings (35.35-30)		
3. Vessels Moorings. (156.120 (a))		
4. Transfer System Alignment. (156.120 (d))		
5. Transfer System; unused components. (156.120 (e))		
6. Transfer System; fixed piping. (156.120 (f))		
7. Overboard Discharges/See Suction Valves. (156.120 (g))		
8. Hoses or Loading Arms condition. (156.120 (h) (156.170)		
9. Hoses; length and support. (156.120 (b) (c))		
10. Connections. (156.130)		
11. Discharge Containment System. (156.120 (j)(1)		
12. Scuppers or Drains. (156.120 (k)		
13. Emergency Shutdown. (156.120 (n)		
14. Repair Work Authorization. (35.35-30)		
15. Boiler and Galley Fires Safety. (35.35-30)		
16. Fires or Open Flames. (35.35-30)		
17. Lighting (sunset to sunrise). (156.120 (t))		
18. Safe Smoking Spaces. (35.35-30)		
19. Spill and Emergency shutdown procedures. (156.120 (q))		
20 Sufficient Personnel. (156.120 (o)(s))		
21. Transfer Conference. (156.120 (q))		
22. Agreement to begin transfer. (156.120 (r))		

I do certify that I have personally inspected this facility or vessel with reference to the requirements set forth in Section 35.35-30 and that opposite each of them I have indicated that the regulations have been complied with.

PERSON IN CHARGE RECEIVING UNIT   TITLE   TIME & DATE
_____ _____ _____
_____ _____ _____
_____ _____ _____
_____ _____ _____

PERSON IN CHARGE DELIVERING UNIT
_____ _____ _____
_____ _____ _____
_____ _____ _____
_____ _____ _____

TIME & DATE COMMENCED _____ , _____
TIME & DATE COMPLETED _____ , _____

Fig. 2-6 Declaration of inspection prior to bulk-cargo transfer. [*NMS form 305.*]

Because of the hazard presented by many cargoes, the Coast Guard has required remote shutdowns for certain ones. Such a device consists of a mechanically actuated valve with a remote trigger so that personnel at some distance from the critical operating area can stop cargo transfer by closing valves and shutting down pumps. Such systems should be tested as a part of pretransfer procedures at each port.

While regulations, procedures, and hardware can and do make a significant contribution to spill prevention in cargo transfer procedures involving tank vessels, there is no substitute for an alert, intelligent, well-trained tankerman who systematically follows comprehensive procedures. Enhancement of tankerman performance through better training, improved procedures, and experience could reduce the number of spills by up to 85 percent without major capital investment for equipment changes.

SAFE NAVIGATION

After obtaining a well-designed vessel and competent tankerman personnel to load and unload it, the next vital item in hazardous spill prevention is the safe navigation of the vessel from its origin to destination. When groundings, collisions, or hull damage from contact with locks, docks, piers, and other waterway structures threaten the integrity of the craft, spills can occur, and they can be serious. Accordingly, vital to the prevention of hazardous materials spills from tank vessels is the competence of the individual in the pilothouse responsible for the safe movement of the vessel. That person has been licensed by the Coast Guard and operates in waters which have been equipped with aids to navigation provided and maintained by the Coast Guard and charted by either the National Oceanic and Atmospheric Administration, the U.S. Corps of Engineers, or the Great Lakes Survey. The pilothouse is equipped with at least two bridge-to-bridge VHF radiotelephones, which make possible instantaneous communication with other vessels as well as with movable bridges, locks, and vessel traffic services. Comprehensive vessel traffic systems operating somewhat like airport control towers exist in a number of United States ports, including San Francisco, Puget Sound, Houston, and the 370-km (230-mi) stretch of the lower Mississippi River between Baton Rouge and the passes at the mouth of the river.

The loading draft for a vessel under the control of the navigator is limited to permit safe passage through the shallowest portion of the route. If the marine unit is a tow consisting of a number of barges and a tug or towboat, the size of the towing vessel is carefully matched to that of the tow in consideration of the route to be traversed. For instance, if the tow is bound for a destination up a swiftly moving river like the Missouri, more horsepower will be required to handle the assigned barges than if the destination is in a relatively slow-flowing river like the upper Mississippi above the locks and dams which begin just north of St. Louis. Similarly, a tow operating on the lower Mississippi with its relatively swift current will need more horsepower than one which is operating on the Intracoastal Waterway, where there is very little current.

The procedure for training navigators is too complex a subject to deal with at length in this part. However, skillful vessel handling is essential for the prevention of hazardous material spills. Similarly, vessels which may be carrying "safe" cargoes (e.g., sand, cement, or coal) must also be navigated prudently, particularly when they operate in the same area with vessels carrying hazardous materials. One potential advantage of vessel traffic systems in congested areas is the opportunity to identify craft handling hazardous materials and so facilitate their safe passage. Similarly, the communication capability of a vessel traffic system permits notifying all craft in the area of the location of hazardous material transfer operations.

Navigational accidents tend to involve groundings or collisions with other vessels or with fixed structures such as bridge piers, locks, and docks. These accidents may involve hull damage with the threat that the cargo containment area of the vessel will be damaged, thus causing a spill of hazardous cargo into the water. When a spill occurs or is possible (as in a grounding with no apparent hull damage), the first remedy is to obtain another vessel which can be used to remove cargo from the grounded or damaged vessel, permitting that vessel to be refloated. Removing cargo from damaged compartments will also end the risk of a spill. In other cases, the grounded vessel may be entirely undamaged and can be refloated by using additional tugboats to free it or by employing the wash from other tugboats to wash the sandy bottom away from the hull. In extreme cases, it may be necessary to bring in dredging equipment to cut a channel to deeper water before refloating can be accomplished. Such cases are frequent when vessels are grounded during a period of rapidly falling river levels.

RESPONSE TO EMERGENCIES

When groundings or spills related to navigation accidents occur, an on-scene coordinator (OSC) is assigned in accordance with the clean-water provisions of the Clean Water Act of 1977, as amended. When an incident takes place in United States coastal waters, the OSC is assigned by the Coast Guard, but when it occurs away from the coasts on inland waters, the OSC is usually assigned by the U.S. Environmental Protection Agency. When a response is required to the threat of a spill during a lightering operation from one vessel to another or the removal of a grounded vessel to deeper water, a marine expert is usually required. Because of its responsiblity for marine safety, the Coast Guard has extensive facilities and vessels as well as expert personnel deployed throughout the system of inland and coastal navigable waterways.

As long as hazardous materials are carried by any mode of transportation, there will be a threat of spills, and occasionally there will be serious spill incidents. To respond to the threat of spills as well as to actual ones, tankermen and navigators should have available contingency plans describing step-by-step response procedures. Those plans have been slow to develop because they require the cooperative efforts of several disciplines working closely together. Both environmental knowledge and chemical and marine expertise are essential.

In conclusion, hazardous material spills in marine transportation can be prevented by (1) excellent container and vessel design, (2) well-qualified tankermen to handle cargo transfers, and (3) well-qualified navigators to handle vessel movement. Designers, tankermen, navigators, and salvage contractors all must be provided with careful plans and procedures to follow. The regulations which establish standards of vessel design, operation, maintenance, and personnel must be comprehensive and practical.

THE RECORD

While the record of hazardous materials spills in water transportation on an incident-by-incident basis is sketchy because of the absence until recently of reporting requirements, there does exist a record of injuries and deaths maintained over the years. Statistics for injuries and deaths from hazardous materials transportation in 1977 are shown in Table 2-3.

An examination of Table 2-3 shows that for the year 1977 marine transportation of hazardous materials resulted in no injuries and no deaths. This, however, does not imply

TABLE 2-3 Deaths and Injuries from Hazardous Materials Transportation in 1977

Hazard class	Commodity	All modes Death	All modes Injuries	Highway Death	Highway Injuries	Rail Death	Rail Injuries	Air Death	Air Injuries	Water Death	Water Injuries
Flammable liquid	Acetaldehyde	0	1	0	0	0	0	0	1	0	0
Corrosive	Acetic anhydride	0	5	0	1	0	4	0	0	0	0
Flammable liquid	Alcohol: (NOS)	0	4	0	1	0	2	0	1	0	0
Nonflammable gas	Anhydrous ammonia	5	81	3	24	2	57	0	0	0	0
Flammable liquid	Asphalt	1	3	1	3	0	0	0	0	0	0
Corrosive	Bromine	0	101	0	101	0	0	0	0	0	0
Corrosive	Calcium hypochlorite (dry)	0	3	0	1	0	0	0	2	0	0
Nonflammable gas	Chlorine	0	44	0	14	0	30	0	0	0	0
Flammable gas	Compressed gas (NOS)	2	3	2	3	0	0	0	0	0	0
Corrosive	Corrosive cleaning liquid	0	16	0	16	0	0	0	0	0	0
Corrosive	Corrosive liquid (NOS)	0	16	0	9	0	6	0	1	0	0
Flammable liquid	Crude oil	3	0	3	0	0	0	0	0	0	0
Other regulated material	Dichloromethane	0	1	0	0	0	0	0	1	0	0
Flammable liquid	Flammable liquid (NOS)	1	25	1	22	0	3	0	0	0	0
Combustible	Fuel oil	1	5	1	4	0	1	0	0	0	0
Flammable liquid	Gasoline	17	31	17	31	0	0	0	0	0	0
Corrosive	Hydrochloric acid	0	17	0	9	0	6	0	2	0	0
Oxidizer	Hydrogen peroxide	0	8	0	4	0	4	0	0	0	0
Flammable gas	Liquefied petroleum gas	1	22	1	10	0	12	0	0	0	0
Oxidizer	Mercurous nitrate	0	2	0	0	0	0	0	2	0	0
Flammable gas	Methyl mercaptan	0	5	0	0	0	5	0	0	0	0
Flammable liquid	Methyl methacrylate	1	7	1	7	0	0	0	0	0	0
Corrosive	Nitric acid	0	43	0	43	0	0	0	0	0	0
Corrosive	Sodium hydroxide (liquid)	0	17	0	6	0	11	0	0	0	0
Corrosive	Sulfuric acid	0	83	0	33	0	50	0	0	0	0
	Subtotal	32	543	30	342	2	191	0	10	0	0
	All other commodities	0	201	0	161	0	40	0	0	0	0
	Total deaths and injuries	32	744	30	503	2	213	0	10	0	0

NOTE: NOS = not otherwise specified.
SOURCE: U.S. Department of Transportation.

6-35

that there were no spills during 1977 by this mode of transportation. Indeed, there were, and there is definitely room for improvement. The greatest potential for improvement appears to be in the area of tankerman performance in cargo transfer operations. A study of the record of those accidents shows that many are clearly preventable through better training, better procedures, and the use of an alert, well-motivated tankerman working in close cooperation with a similar individual on the shore side of the operation.

APPENDIX

Appendix runs on pages 6-37 through 6-42.

TABLE 151.05 Summary of Minimum Requirements

| Cargo identification | | | Hull type | Cargo segregation tank | Tanks | | | Cargo transfer | | Environmental control | | Fire protection required | Special requirements (section) | Electrical hazard group-class | Temperature control installed | Tank internal inspection period (years) |
Name	Pressure	Temperature			Type	Vent	Gauging device	Piping class	Control	Cargo tanks	Cargo handling					
Acetaldehyde	Pressure	Ambient	II	1NA 2ii	Independent pressure	SR	Restr.	II	P-1	Inerted	Vent F	Yes	151.55-1(h)	I-C	NA	G
Acetic acid	Atmospheric	Ambient	III	11 2ii	Integral gravity	Open	Open	II	G-1	NR	Vent N	Yes	151.55-1(g)	I	NA	G
Acetic anhydride	Atmospheric	Ambient	III	1i 2ii	Integral gravity	PV	Restr.	II	G-1	NR	Vent F	Yes	151.55-1(g)	I	NA	G
Acetone cyanohydrin	Atmospheric	Ambient	I	1ii 2i	Integral gravity	PV	Closed	I	G-1	NR	Vent F	Yes	151.50-5	I	NA	G
Acetonitrile	Atmospheric	Ambient	III	1i 2ii	Integral gravity	PV	Restr.	II	G-1	NR	Vent F	Yes	No	I—D	NA	G
Acrylonitrile	Atmospheric	Ambient	II	1ii 2ii	Integral gravity	PV	Closed	II	G-1	NR	Vent F	Yes	151.55-1(e)	I-D	NA	G
Adiponitrile	Atmospheric	Ambient	II	1ii 2ii	Integral gravity	PV	Open	II	G-1	NR	Vent F	Yes	No	I	NA	G
Allyl alcohol	Atmospheric	Ambient	I	1ii 2ii	Integral gravity	PV	Closed	I	G-1	NR	Vent F	Yes	151.50-5	I	NA	G
Allyl chloride	Atmospheric	Ambient	I	1ii 2ii	Integral gravity	PV	Closed	I	G-1	NR	Vent F	Yes	151.50-5	I-D	NA	G
Aminoethyl-ethanolamine	Atmospheric	Ambient	III	1i 2i	Integral gravity	Open	Open	II	G-1	NR	Vent N	Yes	151.55-1(b)		NA	G
Ammonia (anhydrous)	Pressure	Ambient	II	1NA 2ii	Independent pressure	SR 250 psi	Restr	II	P-2	NR	Vent F	No	151.50-30 151.50-32	I-D	NA	8
Ammonia (anhydrous)	Atmospheric	Low	II	1NA 2ii	Independent gravity	PV	Restr.	II-L	G-2	NR	Vent F	No	151.50-30 151.50-32	I-D	151.40-1(b)(1)	8
Ammonium hydroxide (not to exceed 28% NH)	Atmospheric	Ambient	III	1i 2i	Integral gravity	PV	Restr.	II	G-1	NR	Vent F	No	No	I-1	NA	G
Aniline	Atmospheric	Ambient	I	1ii 2ii	Integral gravity	PV	Closed	I	G-1	NR	Vent F	Yes	151.50-5	I-D	NA	G
Benzene	Atmospheric	Ambient	III	1i 2ii	Integral gravity	PV	Open	II	G-1	NR	Vent F	Yes	No	I-D	NA	G

TABLE 151.05 Summary of Minimum Requirements (*Continued*)

Name	Pressure	Temperature	Hull type	Cargo segregation tank	Type	Vent	Gauging device	Piping class	Control	Cargo tanks	Cargo handling	Fire protection required	Special requirements (section)	Electrical hazard group-class	Temperature control installed	Tank internal inspection period (years)
Butadiene (inhibited)	Pressure	Ambient	II	1NA 2ii	Independent pressure	SR	Restr.	II	P-2	NR	Vent F	Yes	No	I-B	NA	8
n-Butyl acrylate	Atmospheric	Ambient	III	1i 2ii	Integral gravity	PV	Open	II	G-1	NR	Vent N	Yes	No	I	NA	G
Isobutyl acrylate	Atmospheric	Ambient	III	1i 2ii	Integral gravity	PV	Open	II	G-1	NR	Vent N	Yes	No	I	NA	G
Butyraldehyde ( )	Atmospheric	Ambient	III	1i 2ii	Integral gravity	PV	Open	II	G-1	NR	Vent F	Yes	151.55-1 (h)	I	NA	G
Isobutyraldehyde	Atmospheric	Ambient	III	1i 2ii	Integral gravity	PV	Open	II	G-1	NR	Vent F	Yes	151.55-1 (h)	I	NA	G
Camphor oil (light)	Atmospheric	Ambient	II	1ii 2ii	Integral gravity	Open	Open	II	G-1	NR	Vent N	Yes	No	I	NA	G
Carbolic oil	Atmospheric	Ambient	I	1ii 2ii	Integral gravity	PV	Closed	I	G-1	NR	Vent F	Yes	151.50-5	I	NA	G
Carbon bisulfide	Atmospheric	Ambient	II	1NA 2ii	Independent gravity	PV	Restr.	II	G-1	Inert	Vent F	Yes	151.50-40 151.50-41	151.50-41 (g)	NA	G
Carbon tetrachloride	Atmospheric	Ambient	III	1i 2i	Integral gravity	PV	Open	II	G-1	NR	Vent N	No	No	I	NA	G
Caustic potash solution	Atmospheric	Ambient elev.	III	1i 2i	Integral gravity	Open	Open	II	G-1	NR	NR	No	151.55-1(j)		NA	G
Caustic soda solution	Atmospheric	Ambient elev.	III	1i 2i	Integral gravity	Open	Open	II	G-1	NR	NR	No	151.55-1(j)		NA	G
Chlorine	Pressure	Ambient	I	1NA 2ii	Independent pressure	SR 300 psi	Indirect	I	P-2	NR	Vent F	Yes	151.50-30 151.50-31	I	NA	2
Chlorobenzene	Atmospheric	Ambient	III	1i 2ii	Integral gravity	PV	Open	II	G-1	NR	Vent N	Yes	No	I	NA	G
Chloroform	Atmospheric	Ambient	III	1i 2i	Integral gravity	Open	Open	II	G-1	NR	Vent F	No	No		NA	G
Chlorohydrins (crude)	Atmospheric	Ambient	I	1ii 2ii	Integral gravity	PV	Closed	I	G-1	NR	Vent F	Yes	151.50-5	I	NA	G

Chemical	Pressure	Temp.														
Chlorosulfonic acid	Atmospheric	Ambient	III	1ii 2ii	Integral gravity	PV	Open	II	G-1	NR	Vent N	No	151.50-20 151.50-21	I	NA	G
Cresois	Atmospheric	Ambient	III	1i 2i	Integral gravity	Open	Open	II	G-1	NR	Vent N	Yes	No	I-D	NA	G
Crotonaldehyde	Atmospheric	Ambient	II	1ii 2ii	Integral gravity	PV	Restr.	II	G-1	NR	Vent F	Yes	151.55-1(h)	I	NA	G
Dichlorodifluoromethane	Pressure	Ambient	III	1NA 21	Independent pressure	SR	Restr.	8	P-1	NR	NR	No	No		NA	8
Dichloropropane	Atmospheric	Ambient	III	1i 2ii	Integral gravity	PV	Restr.	II	G-1	NR	Vent F	Yes	No	I	NA	G
Dichloropropene	Atmospheric	Ambient	II	1ii 2ii	Integral gravity	PV	Restr.	II	G-1	NR	Vent F	Yes	No	I-D	NA	G
Diethanolamine	Atmospheric	Ambient	III	1i 2i	Integral gravity	Open	Open	II	G-1	NR	Vent N	Yes	151.55-1(c)	I	NA	G
Diethylene triamine	Atmospheric	Ambient	III	1i 2i	Integral gravity	Open	Open	II	G-1	NR	Vent N	Yes	151.55-1(c)	I	NA	G
Diisopropanolamine	Atmospheric	Ambient	III	1i 2i	Integral gravity	Open	Open	II	G-1	NR	Vent N	Yes	151.55-1(c)	I	NA	G
Dimethylamine	Pressure	Ambient	II	1NA 2ii	Independent pressure	SR	Restr.	II	P-2	NR	Vent F	Yes	151.55-1(c)	I-D	NA	8
Epichlorohydrin	Atmospheric	Ambient	I	1ii 2ii	Integral gravity	PV	Closed	I	G-1	NR	Vent F	Yes	151.50-5	I	NA	G
Ethyl acrylate	Atmospheric	Ambient	III	1i 2ii	Integral gravity	PV	Restr.	II	G-1	NR	Vent F	Yes	No	I	NA	G
Ethyl chloride	Pressure	Ambient	II	1NA 2ii	Independent pressure	SR	Restr.	II	P-2	NR	Vent F	Yes	No	I	NA	8
Ethyl ether	Atmospheric	Ambient	II	1NA 2ii	Independent pressure	PV	Closed	II	G-1	Inert	Vent F	Yes	151.50-40 151.50-42	I-C	NA	G
2-Ethyl, 3-propyl acrolein	Atmospheric	Ambient	III	1i 2i	Integral gravity	PV	Restr.	II	G-1	NR	Vent N	Yes	No	I	NA	G
Ethylene cyanohydrin	Atmospheric	Ambient	III	1i 2ii	Integral gravity	Open	Open	II	G-1	NR	Vent F	Yes	No	I	NA	G
Ethylene diamine	Atmospheric	Ambient	III	1i 2ii	Integral gravity	PV	Restr.	I	G-1	NR	Vent F	Yes	151.55-1(c)	I	NA	G
Ethylene dichloride	Atmospheric	Ambient	III	1ii 2ii	Integral gravity	PV	Restr.	II	G-1	NR	Vent F	Yes	No	I	NA	G

TABLE 151.05 Summary of Minimum Requirements (*Continued*)

| Cargo identification | | | Hull type | Cargo segregation tank | Tanks | | | Cargo transfer | | Environmental control | | Fire protection required | Special requirements (section) | Electrical hazard group-class | Temperature control installed | Tank internal inspection period (years) |
Name	Pressure	Temperature			Type	Vent	Gauging device	Piping class	Control	Cargo tanks	Cargo handling					
Ethylene oxide	Pressure	Ambient	I	1NA 2ii	Independent pressure	SR	Restr.	II	P-2	Inert	Vent F	Yes	151.50-10 151.50-12	I-B	151.40-1(c)R 2	4
Formaldehyde solution (37–50%)	Atmospheric	Ambient	III	1ii 2ii	Integral gravity	PV	Restr.	II	G-1	NR	Vent F	No	151.55-1(h)	I	NA	G
Formic acid	Atmospheric	Ambient	III	1ii 2i	Integral gravity	PV	Restr.	II	G-1	NR	Vent F	Yes	151.55-1(i)	I	NA	G
Furfural	Atmospheric	Ambient	III	1ii 2i	Integral gravity	PV	Restr.	II	G-1	NR	Vent F	Yes	151.55-1(h)	I	NA	G
Hydrochloric acid	Atmospheric	Ambient	III	1NA 2ii	Independent gravity	Open	Open	II	G-1	NR	Vent F	No	151.50-20 151.50-22		NA	4
Hydrofluoric acid (3)	Atmospheric	Ambient	I	1NA 2ii	Independent gravity	PV	Closed	II	G-1	NR	Vent F	No	151.50-24		NA	G
Hydrogen chloride (3)	Pressure	Ambient or low	I	1NA 2ii	Independent pressure	SR	Closed	I or 1-L	P-2	NR	Vent F	No	151.50-33 151.55-1(c)		151.40-1(b)(2)	2
Hydrogen fluoride(3)	Pressure	Ambient	I	1NA 2ii	Independent pressure	SR	Closed	I	P-2	NR	Vent F	No	151.50-65		151.40-1(b)(2)	G
Isoprene	Atmospheric	Ambient	III	1i 2ii	Integral gravity	PV	Open	II	G-1	NR	Vent F	Yes	No	I-C	NA	G
Methyl acrylate	Atmospheric	Ambient	III	1i 2ii	Integral gravity	PV	Restr.	II	G-1	NR	Vent F	Yes	No	I	NA	G
Methyl bromide	Pressure	Ambient	I	1NA 2ii	Independent pressure	SR	Closed	I	P-2	NR	Vent F	Yes	151.50-5	I-D	NA	2
Methyl chloride	Pressure	Ambient	II	1NA 2ii	Independent pressure	SR	Restr.	II	P-2	NR	Vent F	Yes	151.55-1(c)	I-D	NA	8
Methyl methacrylate	Atmospheric	Ambient	III	1i 2ii	Integral gravity	PV	Restr.	II	G-1	NR	Vent F	Yes	No	I-D	NA	G
Monochlorodifluoromethane	Pressure	Ambient	III	1NA 2i	Independent pressure	SR	Restr.	I	P-1	NR	NR	No	No	I	NA	8
Monoethanolamine	Atmospheric	Ambient	III	1i 2i	Integral gravity	Open	Open	II	G-1	NR	Vent N	Yes	151.55-1(c)		NA	G

Monoisopropanolamine	Atmospheric	Ambient	III	1i 2i	Integral gravity	Open	Open	II	G-1	NR	Vent N	Yes	151.55-1(c)	I	NA	G
Morpholine	Atmospheric	Ambient	III	1i 2ii	Integral gravity	Open	Open	II	G-1	NR	Vent N	Yes	155.55-1(c)	I	NA	G
Motor-fuel antiknock compounds	Atmospheric	Ambient	I	1ii 2ii	Independent gravity	PV	Closed	I	G-1	NR	Vent F	Yes	151.50-6	I	NA	151.50-6
Oleum	Atmospheric	Ambient	III	1ii 2ii	Integral gravity	Open	Open	II	G-1	NR	Vent N	No	151.50-20 151.50-21	I	NA	4
Phenol	Atmospheric	Ambient	I	1ii 2i	Integral gravity	PV	Closed	I	G-1	NR	Vent F	Yes	151.50-5	I-D	NA	2
Phosphorus (elemental)	Atmospheric	Elev.	I	1ii 2ii	Integral gravity	PV	Closed	I	G-1	Water pad	Vent F	Yes	151.50-50		NA	8-4
Phosphoric acid	Atmospheric	Ambient	III	1ii 2i	Integral gravity	Open	Open	II	G-1	NR	Vent N	No	151.50-20 151.50-23	I	NA	4
Propionic acid	Atmospheric	Ambient	III	1i 2ii	Integral gravity	Open	Open	II	G-1	NR	Vent N	Yes	151.55-1(g)	I	NA	G
Propylene oxide	Pressure	Ambient	II	1NA 2ii	Independent pressure	SR	Restr.	II	P-1	Inerted	Vent F	Yes	151.50-10 151.50-13	I-B	NA	4
Styrene monomer	Atmospheric	Ambient	III	1i 2ii	Integral gravity	Open	Open	II	G-1	NR	Vent N	Yes	No	I-D	NA	G
Sulfur (liquid)	Atmospheric	Elev.	III	1i 2ii	Integral gravity	Open	Open	II	G-1	Vent N	Vent N	Yes	151.50-55	I-C	151.40-1(f)(1)	G
Sulfur dioxide	Pressure	Ambient	I	1NA 2ii	Independent pressure	SR	Closed		P-2	NR	Vent F	No	151.50-30 151.50-35 151.55-1(j)	NA	NA	2
Sulfuric acid	Atmospheric	Ambient	III	1ii 2ii	Integral gravity	Open	Open	II	G-1	NR	Vent N	No	151.50-20 151.50-21	I	NA	4
Sulfuric acid (spent)	Atmospheric	Ambient	III	1ii 2ii	Integral gravity	Open	Open	II	G-1	NR	Vent N	No	151.50-20 151.50-21	I	NA	4
Triethanolamine	Atmospheric	Ambient	III	1i 2i	Integral gravity	Open	Open	II	G-1	NR	Vent N	Yes	151.55-1(b)	I	NA	G
Triethylene tetramine	Atmospheric	Ambient	III	1i 2i	Integral gravity	Open	Open	II	G-1	NR	Vent N	Yes	151.55-1(b)	I	NA	G
Vinyl acetate	Atmospheric	Ambient	III	1i 2ii	Integral gravity	PV	Open	II	G-1	NR	Vent	Yes	No	I-D	NA	G
Vinyl chloride	Pressure	Ambient	II	1NA 2ii	Independent pressure	SR	Closed	II	P-2	NR	Vent F	Yes	151.50-30 151.50-34	I-D	NA	8

TABLE 151.05 Summary of Minimum Requirements (*Continued*)

| Cargo identification | | | Hull type | Cargo segregation tank | Tanks | | | Cargo transfer | | Environmental control | | Fire protection required | Special requirements (section) | Electrical hazard group-class | Temperature control installed | Tank internal inspection period (years) |
Name	Pressure	Temperature			Type	Vent	Gauging device	Piping class	Control	Cargo tanks	Cargo handling					
Vinyl chloride	Atmospheric	Low	II	1NA 2ii	Independent gravity	PV	Closed	II-L	G-2	NR	Vent F	Yes	151.50-30 151.50-34	I-D	151.40-1(b)(1)	8
Vinylidene chloride inhibited	Atmospheric	Ambient	II	1NA 2ii	Independent gravity	PV	Closed	II	P-2	Padded	Vent F	Yes	151.55-1(f)	I	NA	G
For requirements see these sections			151.10	151.13-5	151.15-1	151.15-5	151.15-10	151.20-1	151.20-5	151.25-1	151.25-2	151.30		111.80-5 (Subchap. J)	151.40	151.04-5

FOOTNOTES:
1 Group D required as indicated in Subpart 111.60, Subchap. J of this chapter.
2 May be necessary under certain climatic conditions in order to maintain temperature.
3 Special requirements not completed as of this date.

TERMS AND SYMBOLS:
Segregation—Tank—
  Line 1—Segregation of cargo from surrounding waters:
    i = skin of vessel (single skin) only required. Cargo tank wall can be vessel's hull.
    ii = double skin required. Cargo tank wall cannot be vessel's hull.
  Line 2—Segregation of cargo space from machinery spaces and other spaces which have or could have a source of ignition:
    i = single bulkhead only required. Tank wall can be sole separating medium.
    ii = double bulkhead required. Cofferdam, empty tank, pump room, tank with Grade E liquid (if compatible with cargo) is satisfactory.
Internal tank inspection—
  G—Indicates cargo is subject to general provisions of Sec. 151.04-5(b).
    Specific numbers in this column are changes from the general provisions.

ABBREVIATIONS USED:
Vent:
  PV = pressure vacuum valve.
  SR = safety relief.
Gauging device: Restr. = restricted.
General usage:
  NR = no requirements.
  NA = not applicable.

SOURCE: (49 CFR 1.46(b) and (A)(4)). [CGFR 70-10, FR 3714, Feb. 25, 1970, as amended by CGD 73-275R, 41 FR 3086, Jan. 21, 1976; CGD 75-226, 42 FR 8378, Feb. 10, 1977.]

PART 3
Rail Transportation

Deborah K. Shaver
Manager of Hazardous Materials Programs
Systems Technology Laboratory, Inc.
Robert M. Graziano
Vice President, Marketing
O. H. Materials, Inc.

Prevention Overview / 6-44
General Prevention Program / 6-44
Hazardous Materials / 6-47
Conclusion / 6-49

The railroad industry has successfully coped for many years with the problem of the safe handling and transportation of all materials including explosives, radioactive materials, and other materials designated as hazardous by the Department of Transportation (DOT). Public sensitivity to this problem has been heightened by serious accidents involving hazardous materials, and "emergency response" has become a catch phrase within the transportation community. But what is known or publicized about the equally important side of the issue, prevention? It is true that any transportation entails risk and that the movement of hazardous materials poses even greater risk because of their intrinsic properties. However, these materials are part of modern commerce, and they must continue to be moved efficiently and safely by rail. Thus, the risk involved must be understood on two levels:

1. Hazardous materials pose an exposure risk to both the general public and the carrier industry.

2. Hazardous materials transport is necessary to society and to the railroad industry. Society needs liquefied petroleum gas (LPG) to heat homes, schools, and industrial facilities and chlorine to purify municipal water supplies; railroads need the revenue derived from transporting these materials.

The reduction of risk is therefore a necessity in the rail transportation of hazardous materials, and prevention is the keystone to minimizing risk and achieving safe transportation.

6-43

PREVENTION OVERVIEW

The railroad industry has exacting prevention and safety standards dating to the early 1900s. The nature and problems of rail transportation have evolved from the time when railroads were the modern transportation mode for both freight and passengers. By 1922 more than half a billion passengers were carried annually by rail.[1] Railroad technology was improving with such advances as interlocking and electric signal systems, the Westinghouse brake, and new car design. However, these technologies were not fully and uniformly utilized until they were required by federal regulation.

From 1893 to 1921 several rail safety laws were adopted, implemented, and enforced by the Interstate Commerce Commission (ICC). The early laws usually addressed specific, well-defined problems. Explosives and hazardous materials were regulated by the 1908 Transportation of Explosives Act, which was revised in 1960 to designate "explosives and other dangerous articles." This revision broadened the act to cover radioactive materials and etiological agents and centralized authority for the regulation of transportation of such materials in the ICC. In 1965–1966, these various safety functions were transferred to the Department of Transportation.

The increase in both amount and types of hazardous materials shipped by rail in recent years has impacted both carrier and public exposure risk. At the turn of the century, the main hazardous materials of concern were explosives and weapons. In the first quarter of 1978, the following were among the top 25 hazardous materials in number of carloads shipped by rail: ammonia, caustic soda, sulfuric acid, chlorine, liquefied petroleum gases, propane, ammonium nitrate, gasoline, and phosphoric acid.

Concern with safety has broadened from the casualty problems associated with passenger travel to include property and lading loss and damage due to the involvement of hazardous materials in accidents. Although freight traffic by rail has been on the rise, the greater increase in traffic by the trucking and barge industries has resulted in reduced railroad transportation dominance. Whereas railroads carried 75 percent of the freight in 1929, they handled 37 percent in 1976 and approximately 33 percent in 1980.

The Federal Railroad Safety Act of 1970 addresses all aspects of rail safety, and the Hazardous Materials Transportation Act of 1974 addresses the special problems of shipping hazardous materials. These laws were enacted in response to contemporary rail safety problems.

GENERAL PREVENTION PROGRAM

The railroad industry's prevention program has three major areas of emphasis:

1. Monitoring activities which consist of track and equipment inspections
2. Training programs with continuing education of operations personnel
3. Track and equipment maintenance and upgrading programs

Railroad monitoring and inspection activities cover all aspects of federal and state inspection programs. Federal standards for track, cars, locomotives, signal and train control systems, and train operation have resulted in uniformity among railroad inspection programs. This uniformity stems from specifications in the regulations that determine such items as frequency of inspection and the length of time during which employees can be on duty.

Railroads use the inspection process to serve as a prevention program in addition to ensuring compliance with federal regulations. Inspection provides railroads with data on where preventive maintenance or modification and redirection of other operating practices may be required. Track geometry, signal systems, and car inspections, for instance, all involve both preventive maintenance and safety promotion goals. The inspection system is monitored internally by management, its effectiveness being gauged by both accident prevention and prevention of Federal Railroad Administration (FRA) enforcement and fines.

Railroad inspections are far more frequent than those of the FRA or the states because they are designed to detect defects before they become serious enough to cause damage or violate standards. The two main areas of concern are track and equipment. Equipment inspected includes freight cars, locomotives, and signals.

Track

Track is inspected on the basis of both track class and type, shown in Table 3-1. In addition, inspections are made in accordance with the following FRA regulations:

1. Each switch and track crossing must be inspected on foot at least monthly, except for track used less than once a month, in which case inspection must be made before it is used (49 CFR 213.235).
2. Once a year a continuous search for internal defects must be made of all jointed and welded rails in Classes 4 through 6 track over which passenger trains operate. If new rail is inductively or ultrasonically inspected over its entire length and all defects are removed before or within 6 months after installation, the next continuous search for internal defects need not be made until 3 years later (49 CFR 213.237).
3. Special inspection must be made of track involved in a fire, flood, severe storm, or other occurrence which might have damaged track structures as soon as possible after the occurrence (49 CFR 213.239).

Freight Cars

The general practice in the industry is to inspect freight cars at interchange points, in major yards or terminals, and as required by the 800-km (500-mi) inspection rule.[2] Cars are inspected visually at these points.

As part of the inspection performed on cars at points where cars are placed in trains to detect such defects as those listed below, dates stenciled on the sides of cars are noted

TABLE 3-1   Track Inspections

Class of track	Type of track	Frequency of inspection
1, 2, 3	Main track and sidings	Weekly with at least 3 calendar days between inspection or before use of track if used less than once a week, or twice weekly with at least 1 calendar day between inspections if the track carried passenger trains or more than 10,160,000 gross metric tons of traffic during the preceding calendar year
1, 2, 3	Other than main track and sidings	Monthly with at least 20 calendar days between inspections
4, 5, 6		Twice weekly with at least 1 calendar day between inspections

SOURCE: *Track Safety Standards,* Federal Railroad Administration, Office of Safety, 49 CFR 213, October 1980.

to determine if any time limits, as prescribed by FRA and/or the Association of American Railroads (AAR),[3] have expired with respect to car age as well as to such periodic attention as:

1. Detailed inspection of truck components (wheels, axles, bearings, etc.), couplers, cushioning units, center sills, body bolsters, and center plate
2. Single-car testing of air brakes (in-date test, or IDT)
3. Cleaning, oiling, and single-car testing of air brakes (clean, oil, test, and stencil, or COT&S)
4. Replacement of plain bearing lubricators
5. Lubrication of roller bearings

Further, lading on open cars, such as flats and bulkhead flats, is inspected to see that it has not shifted and that it is properly secured, and closed cars are opened for such inspection when there is evidence, such as leaning of the car, that the lading may have shifted. Cars are usually inspected in train yards by regularly assigned car inspectors who either ride slowly on a special cart or walk along each side of a group of cars.

Detailed inspection, as well as any necessary repair or replacement of the components, is made on a repair track or at a car shop. This inspection is made on high-utilization cars within 24 months after construction or reconditioning and within each succeeding 12-month interval and on other cars within 96 months after construction or reconditioning and within each succeeding 48-month interval.

After cars have been assembled for movement in an outbound train, the air-brake system is tested for leaks by charging the system and observing a gauge to ensure that air-pressure losses remain within limits specified by the FRA. Such a leakage test, as well as inspection of the air-brake cylinder on each car for piston travel which exceeds FRA-specified limits, is also made at 800-km intervals or at railroad property interchanges.

Locomotives

The locomotive inspection regulations in 49 CFR 229 Subpart B consist of the following five sections which govern tests and inspections for locomotives: (1) daily inspections, (2) periodic inspections, (3) annual tests of air-brake systems, (4) biennial tests of air-brake systems, and (5) hydrostatic main reservoir tests. Each of these sections specifies various tests and inspection intervals for certain of the components that it covers.

Signals

Signal mechanisms, switch circuit controllers, shunt fouling circuits, electric locks, relays, and lightning arresters are inspected and tested in accordance with 49 CFR 236. Track switches equipped with a circuit controller connected at the switch point are maintained to ensure that the control circuits will be open or shunted, or both, when the switch point is not in the proper position. Test specifications for dry wire and cable insulation, to determine if the resistance is within the minimum limits allowable by FRA regulations, are found in 49 CFR 236.108.

Training

Training and education programs are another important aspect in the railroad industry's prevention program. Railroads use training methods and techniques varying from on-the-job training to classroom training to combinations of both. Awareness programs for both carrier employees and the general public are another key element. For example, the AAR publishes posters, booklets, and other materials related to safety and has cosponsored seminars to bring local railroad officials an understanding of federal standards and regulations. Railroad representatives give lectures at schools, to local fire and emergency personnel, and to customer personnel on various safety issues. This aspect of prevention takes on additional significance in the area of hazardous materials because of the enormous consequences which are associated with their mismanagement or poor handling.

Track and Equipment Maintenance and Upgrading Programs

Railroads have invested and continue to invest large amounts of capital in maintaining and upgrading track and equipment. In 1979 the rail industry invested an all-time high of $14 billion in maintenance and capital improvement. Of this amount, $4.6 billion was spent on track, roadbed, and facilities and $6 billion for equipment maintenance.

HAZARDOUS MATERIALS

Hazardous materials add another dimension to the issue of safe railroad transportation. Hazardous materials are those which in quantity or form or because of intrinsic characteristics such as flammability, corrosivity, explosivity, or toxicity pose an unreasonable risk to health and safety or property when transported in commerce. According to the AAR, approximately 1 million carloads of hazardous materials were shipped by rail in 1978. It is reasonable to assume that there will be an increase in the volume of hazardous materials shipped in the future, with an associated rise in the risk. Concern for safe transport of hazardous materials and risk reduction via all modes is the prime purpose of the hazardous materials regulations of the Department of Transportation.[4]

All persons involved in the shipment of hazardous materials (i.e., shipper, carrier, and consignee) must act to ensure safe rail shipment. Packaging, classifying, and loading requirements are performed by the shipper in accordance with Part 173, Subparts A and B, of 49 CFR. These subparts also set forth in detail the proper maintenance and loading of tank cars and the proper packaging and loading of intermodal shipments. The shipper must also prepare correct shipping papers (Sec. 172.200) and properly placard railcars (see Fig. 3-1). Further, shipper employees must be instructed in these requirements.

The railroad's responsibility for safe transport of hazardous materials lies in ensuring compliance with railroad equipment and safety standards and with DOT hazardous materials regulations, ensuring receipt of properly executed shipping papers (Sec. 174.24, 49 CFR), and training personnel in FRA and Materials Transportation Bureau (MTB) regulations. Rail carriers must be careful in the handling of placarded cars (those containing hazardous materials) and must maintain vigilance concerning proper loading, blocking and bracing of trailer-on-flatcar (TOFC) and container-on-flatcar (COFC) movements. Railroads must also notify the DOT in the event of certain hazardous materials incidents (Secs. 171.15 and 171.16).

The railroad industry today is involved in inspection, data collection, computerized tracking systems, training, and emergency response as part of an extensive hazardous materials program. Some of the inspection activities are conducted through the AAR's Bureau of Explosives, which has inspectors on railroad property and at shipper facilities who perform safety audits for compliance with the regulations. These inspectors also respond to rail emergencies involving hazardous materials as needed. In 1975, the AAR developed an identification system for hazardous materials shipments by rail—the 49 series of the Standard Transportation Commodity Code. In addition, data bases have been

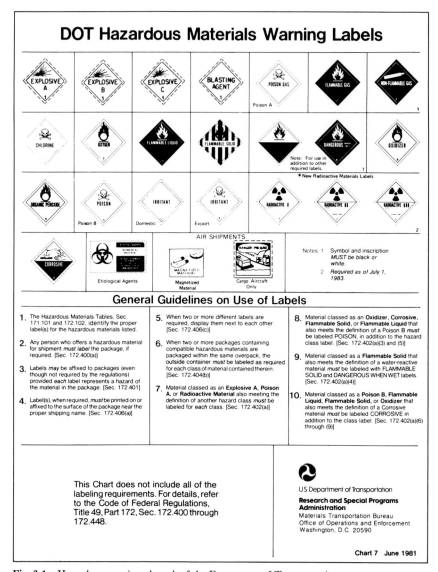

Fig. 3-1   Hazardous warning placards of the Department of Transportation.

developed and are being maintained by the AAR to try to spot problem areas in hazardous material transportation for countermeasure application to prevent catastrophic accidents.

The AAR and individual railroads are very much involved in training programs for both shipper and carrier personnel on the rules, regulations and standards, safe handling, and initial response to incidents dealing with hazardous materials. Railroads also have programs to instruct local fire and emergency personnel in what to do and what to avoid in dealing with hazardous materials emergencies.

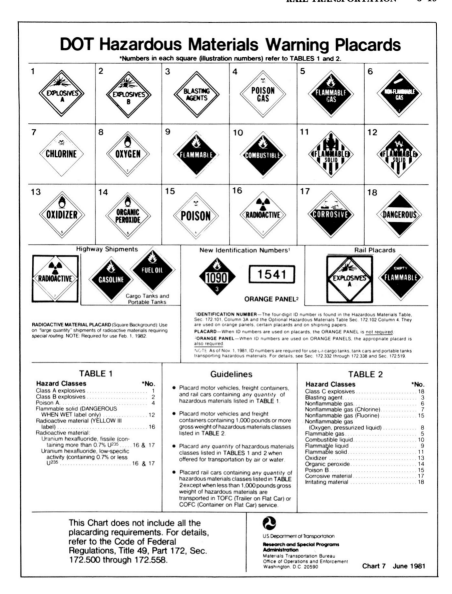

DOT Hazardous Materials Warning Placards

*Numbers in each square (illustration numbers) refer to TABLES 1 and 2.

1 EXPLOSIVES A	2 EXPLOSIVES B	3 BLASTING AGENTS
4 POISON GAS	5 FLAMMABLE GAS	6 NON-FLAMMABLE GAS
7 CHLORINE	8 OXYGEN	9 FLAMMABLE
10 COMBUSTIBLE	11 FLAMMABLE SOLID	12 FLAMMABLE SOLID
13 OXIDIZER	14 ORGANIC PEROXIDE	15 POISON
16 RADIOACTIVE	17 CORROSIVE	18 DANGEROUS

Highway Shipments

RADIOACTIVE   GASOLINE   FUEL OIL
Cargo Tanks and Portable Tanks

New Identification Numbers[1]

1090   1541

ORANGE PANEL[2]

Rail Placards

EXPLOSIVES A   EMPTY FLAMMABLE

RADIOACTIVE MATERIAL PLACARD (Square Background) Use on "large quantity" shipments of radioactive materials requiring *special routing.* NOTE: Required for use Feb. 1, 1982.

[1]IDENTIFICATION NUMBER—The four-digit ID number is found in the Hazardous Materials Table, Sec. 172.101, Column 3A and the Optional Hazardous Materials Table Sec. 172.102 Column 4. They are used on orange panels, certain placards and on shipping papers.
PLACARD—When ID numbers are used on placards, the ORANGE PANEL is not required.
[2]**ORANGE PANEL**—When ID numbers are used on ORANGE PANELS, the appropriate placard is also required.
NOTE: As of Nov. 1, 1981, ID numbers are required for use on cargo tanks, tank cars and portable tanks transporting hazardous materials. For details, see Sec. 172.332 through 172.338 and Sec. 172.519.

TABLE 1

Hazard Classes	*No.
Class A explosives	1
Class B explosives	2
Poison A	4
Flammable solid (DANGEROUS WHEN WET label only)	12
Radioactive material (YELLOW III label)	16
Radioactive material:	
Uranium hexafluoride, fissile (containing more than 0.7% U235	16 & 17
Uranium hexafluoride, low-specific activity (containing 0.7% less U235	16 & 17

Guidelines

- Placard motor vehicles, freight containers, and rail cars containing *any quantity* of hazardous materials listed in TABLE 1.

- Placard motor vehicles and freight containers containing 1,000 pounds or more gross weight of hazardous materials classes listed in TABLE 2.

- Placard *any quantity* of hazardous materials classes listed in TABLES 1 and 2 when offered for transportation by air or water.

- Placard rail cars containing *any quantity* of hazardous materials classes listed in TABLE 2 except when less than 1,000 pounds gross weight of hazardous materials are transported in TOFC (Trailer on Flat Car) or COFC (Container on Flat Car) service.

TABLE 2

Hazard Classes	*No.
Class C explosives	18
Blasting agent	3
Nonflammable gas	6
Nonflammable gas (Chlorine)	7
Nonflammable gas (Fluorine)	15
Nonflammable gas (Oxygen, pressurized liquid)	8
Flammable gas	5
Combustible liquid	10
Flammable liquid	9
Flammable solid	11
Oxidizer	13
Organic peroxide	14
Poison B	15
Corrosive material	17
Irritating material	18

This Chart does not include all the placarding requirements. For details, refer to the Code of Federal Regulations, Title 49, Part 172, Sec. 172.500 through 172.558.

US Department of Transportation
Research and Special Programs Administration
Materials Transportation Bureau
Office of Operations and Enforcement
Washington, D.C. 20590

Chart 7   June 1981

CONCLUSION

The railroad industry has and will continue to move ahead with a vigorous program for prevention of problems in the transport of all commodities by rail with special emphasis on hazardous materials. Railroads are concerned with this problem because accidents mean not only potential injuries and fatalities, loss of revenue, loss of operating time, and loss of equipment but also loss of their public image as an efficient and modern transportation mode. Prevention has been a key element in railroad operations from the beginning, and the more recent efforts to identify and minimize risk in transporting hazardous materials are a complex task to which the railroad industry has dedicated itself for the future.

REFERENCES

1. *An Evaluation of Railroad Safety,* Office of Technology Assessment, Washington, May 1978.
2. *Railroad Freight Car Safety Standards,* 49 CFR 215, Federal Railroad Administration, Office of Safety, October 1980.
3. *Field Manual of the Interchange Rules,* Association of American Railroads, Washington, Jan. 1, 1977.
4. R. M. Graziano, Tariff No. 32, *Hazardous Materials Regulations of the Department of Transportation by Air, Rail, Highway, and Water,* Dec. 15, 1978.

PART 4
Tank Truck Transportation

Clifford J. Harvison
Managing Director, National Tank Truck Carriers, Inc.

Carrier Spill Prevention Program / 6-52
Loading and Unloading / 6-52
Conclusions / 6-52

Tank trucks are constantly transporting hazardous materials, ranging from flammable liquids and gases to complex chemicals. A significant problem area in the safe shipment of these commodities involves communication, since the tank truck carrier is totally dependent upon the shipper for information about the product. If the shipper's instructions and information on the product's handling characteristics are not accurate, an improper tank or hose may be used, or off-loading into the wrong pipe can occur, resulting in a dangerous mixture of reacting chemicals.

The shipper should provide technical information for emergency response and spill cleanup. Accurate information and cooperation are vital to environmentally safe transportation. The development of CHEMTREC by the Chemical Manufacturers Association and the establishment of emergency response programs by chemical companies and their carriers have contributed to solving the communications problem.

The transporter must convey hazardous materials safely from one point to another without damage to the cargo, the public, or the environment. In view of the tonnage and distance traveled, tank truck carriers have maintained a good safety record. In 1977, industrial tank truck carriers averaged 1.1 accidents per 1 million mi, or 1.6 million km [defined as U.S. Department of Transportation (DOT) accidents/million miles: a DOT accident is one that results in a death, an injury, or over $2000 in property damage].

To assure safe transportation of hazardous substances, a cooperative program has been maintained between industry and the DOT to create a strong regulatory structure governing the transportation of these commodities. The center of this program is the DOT vehicle specification and product classification system. This system requires that each commodity to be transported in interstate commerce be tested to determine any dangerous properties. If the commodity is found to be hazardous, transportation must be performed in packages or vessels constructed to specified engineering criteria developed and approved by the DOT. The present cargo tank transporting hazardous materials is a reliable vehicle. Its design is specified by the DOT, from shell thickness to valving. The updating of this specification is a continuous process as technological advances occur.

CARRIER SPILL PREVENTION PROGRAM

An essential item for safe carrier operation is an active company spill prevention and management program. Hooper et al. (1976)[1] have described the historical development and active maintenance of one such program, which encompasses personnel safety programs, equipment maintenance programs, and tractor-trailer design improvement. The safety program utilizes several components: a driver operating manual that stresses accident and cargo loss prevention, safety design meetings with drivers and terminal personnel, driver safety award programs, and a training school. The maintenance program includes periodic tractor inspections, a 30-day corrosive-liquid trailer-hose-fitting special inspection, a 60-day general trailer inspection, a 2-year trailer retest and inspection, and a pretrip vehicle condition report.

Additionally, a company environmental program was created to address the areas of spill control, treatment, etc. An analysis of the company's spill history was performed to determine frequency, locations of spills (at terminals or over the road), disposition of product spilled, and cause of spills (driver error, equipment failure, etc.). This analysis was used to modify and enhance the company's spill prevention program. Spill analysis indicated that most driver-error-caused spills were a result of storage tank overfills, with open valves and poor hose connections running in second and third place respectively.

Subsequent to the spill analysis, personnel awareness and training were made the key to the success of the prevention program. Company drivers participating in this program receive a driver operating manual with instructions on accident and cargo loss prevention, spill reporting, prevention, and response.

LOADING AND UNLOADING

There are three areas in which discharges of hazardous materials occur in tank truck operations: loading, in-transit traffic accidents, and unloading. For this discussion, loading and unloading are much the same function, since the operations and equipment used are similar in both situations. In loading and unloading, proper maintenance and use of product transfer equipment and effective training of employees will preclude the large majority of accidental discharges. A recent survey of "product loss' accidents in the tank truck industry demonstrated that over 70 percent of such incidents occurred at loading and unloading sites.

Employees, whether full-time loading-rack personnel or tank truck drivers, should be fully trained in handling the equipment they are expected to operate. Special emphasis should be placed upon actions to be taken in case of fires, spills, or leaks in the loading and unloading process. The need for routine, continuous use of all protective equipment such as goggles, rubber suits, and gear cannot be overstated. A checklist of fundamental preventive steps to ensure safe product transfer operations is given in Table 4-1.

CONCLUSIONS

Traffic accidents are a major cause of spills from tank trucks. Safe driving will avoid the great majority of spills. Prior to dispatch, a visual inspection of the vehicle should be made. The carrier's pretrip procedures should be followed exactly. Valves and dome covers must be securely shut before the vehicle is driven. Another item to check is gasket condition; in an overturn, the product can leak out of a dome cover because of a faulty gasket.

Table 4-1   Checklist for Personnel Involved in Product Transfer Operations

1. Place chock blocks under the wheels, or use parking brakes to ensure that the vehicle does not move while product transfer is being accomplished.
2. Check hoses, valves, couplers, and other equipment.
3. Be sure that all external and internal valves are operational before loading begins.
4. Make sure that the cargo tank vehicle and, likewise, the receiver's tank can hold the amount of product you intend to transfer. A 28-m³-capacity tank with 4 m³ already on board will not hold an additional 28 m³. Avoid overfills by checking before loading and unloading to make sure that the receiving vessel can hold the load.
5. When loading cargo tanks for top-loading, close the internal and external valves. For bottom-pressure loading, open all internal and external valves and open pressure relief valves.
6. Make sure that loading lines are securely attached and couplers are connected properly.
7. Make certain that all operating instructions and company safety procedures for the product involved are followed completely; shortcuts lead to accidents.
8. Never leave the vehicle unattended during loading or unloading unless you are specifically relieved by an agent of the shipper or receiver.
9. After transfer has been completed, remove all loading arms, ground wires, or loading-unloading lines.
10. Make sure that all internal and external and dome-cover valves are securely closed before moving the vehicle.
11. Remove any wheel chocks that have been used.
12. Check to make sure that placards are correct (according to DOT rules) and as indicated on the bill of lading.
13. Drive to the destination safely.
14. Whenever a vehicle is loading or unloading at a new or complicated site, make sure that, upon arrival, you get in touch with a person authorized to receive the load. If at all possible, that person should direct the unloading. Checks should always be made for leaks in valves and fittings before and during product transfer.
15. If no one is on the site, get in touch with the carrier's terminal. If malfunctions or leaks are observed, immediately get in touch with the shipper or receiver's personnel or the carrier's terminal.

A good maintenance program for power units and cargo tanks will help to avoid accidents caused by mechanical failures. Hoses should be tested and inspected for worn areas, kinks, and other evidence of damage. Fittings, couplers, and other equipment must be kept in good condition. The threads on couplers should be periodically inspected for wear. Proper preventive maintenance is vital to the avoidance of accidents, spills, and leaks.

Hazardous materials are too dangerous and the costs involved in an accident are too expensive for chances to be taken. Carriers must establish, train, and enforce quality safety programs to prevent accidents of all types.

The real opportunity to eliminate accidental discharges of hazardous materials lies in the awareness and safety commitment of the shipper, carrier, and receiver. All parties must work together, communicate, and cooperate with each other to avoid the dangers and problems associated with the mishandling of hazardous materials.

REFERENCE

1. M. W. Hooper, J. E. Warner, and A. M. Galen, "Over the Road/In-Plant Tank Truck Hazardous Materials Spills—One Company's Management Program," in G. F. Bennett (ed.), *Proceedings of the 1976 National Conference on Control of Hazardous Material Spills,* New Orleans, April 1976, pp. 135–145.

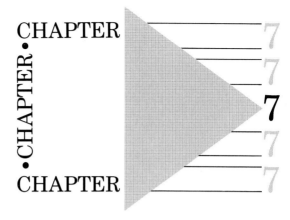

CHAPTER • CHAPTER • CHAPTER • CHAPTER

7 7 7 7 7 7

Response Plans

PART 1
United States Federal Governmental Plans / 7-3

PART 2
United States Local Governmental Plans / 7-15

PART 3
European Governmental Plans / 7-29

PART 4
United States Industrial Plans / 7-43

PART 5
European Industrial Plans / 7-56

PART 6
Land Transportation Plans / 7-64

PART 7
British Maritime Transportation Plans / 7-72

PART 8
Coast Guard Organization for United States
Coastal Areas / 7-77

PART 9
Fire Service Role / 7-86

PART 1
United States Federal Governmental Plans

Kenneth E. Biglane
Director, Hazardous Response Support Division
U.S. Environmental Protection Agency

James J. Yezzi, Jr.
Physical Scientist, Oil and Hazardous Materials Spills Branch
U.S. Environmental Protection Agency

Background / 7-3
Responsibility / 7-4
Organization / 7-5
Plans / 7-9
Operation / 7-11
Reports / 7-14

Discharges of oil and hazardous substances frequently create environmental emergencies that require the immediate implementation of pollution abatement measures. Natural disasters (caused by hurricanes, tornadoes, and floods) and emergencies such as radiological or chemical discharges often present a substantial threat to public health and safety and to environmental quality. These unexpected occurrences are generally accompanied by initial confusion as to location, extent, and severity.

When these emergencies occur, reliable information is needed to evaluate the magnitude and severity of the situation so that immediate and appropriate action can be taken to protect both the public and the environment effectively. Since numerous federal, state, and local agencies may become involved in emergency response activities, efficient response is contingent upon preplanning and the identification and delegation of responsibilities at all levels of government. The National Oil and Hazardous Substances Pollution Contingency Plan, 40 CFR 1510,* provides a mechanism for coordinating federal-state actions to prevent discharges of oil and hazardous substances and to protect the environment when these discharges occur.[1]

BACKGROUND

Recognizing the need to prevent discharges of oil and hazardous substances, the President, in June 1968, directed the secretary of the interior to develop appropriate contingency plans. In November 1968, the United States published its first National Oil and Hazard-

*As amended, March 1980.

ous Substances Pollution Contingency Plan.[2] The plan became part of the *Code of Federal Regulations* in June 1970,[3] in accordance with the Water Quality Improvement Act of 1970.

Although the plan has undergone five revisions since 1968, its basic concepts have remained the same. The mandates of the Comprehensive Environmental Response, Compensation, and Liability Act of 1980,[4] referred to as Superfund, will necessitate another revision of the plan.

The purpose of the plan is to coordinate federal actions to prevent discharges of oil and hazardous substances and to mitigate damages to the environment when discharges occur. The plan provides a framework for the development and coordination of regional and local contingency plans. It is effective for the navigable waters of the United States and adjoining shorelines, including coastal territorial waters, the contiguous zone, and the high seas beyond this zone where there exists a threat to United States waters or shelf bottom.

RESPONSIBILITY

Federal Responsibility

A primary objective of the plan is to "encourage" the party responsible for a spill to clean up the discharge or remove its threat. When appropriate removal actions are taken, federal involvement is generally limited to monitoring the situation. However, when removal actions are not initiated promptly by the responsible party or are inadequate, further federal response actions are taken.

Each of the participating federal agencies which have facilities or other resources that may be useful in federal response situations is required to make those facilities or resources available for use.

The plan identifies those federal agencies having specific duties and responsibilities for responding to discharges of oil or hazardous substances; briefly stated, they are:

1. The Council on Environmental Quality (CEQ) prepares, publishes, revises, and amends the plan.

2. The Department of Agriculture (USDA) provides expertise regarding the effects of pollutants on, through, and over soil and assists in the selection of landfill disposal sites.

3. The Department of Commerce (DOC) provides scientific expertise on living marine resources [through the National Oceanic and Atmospheric Administration (NOAA)].

4. The Department of Defense (DOD) assists in the maintenance of navigation channels and in salvage operations.

5. The Department of Energy (DOE) ensures compliance with the Interagency Radiological Assistance Plan (IRAP) during discharge episodes involving radioactive materials.

6. The Department of Health and Human Services (HHS) provides expert advice and assistance for those discharges that threaten public health and safety.

7. The Federal Emergency Management Agency (FEMA) participates in the development and evaluation of national, regional, and local contingency plans, monitors response actions to such plans, and evaluates requests for presidential declarations of major disasters or emergencies.

8. The Department of the Interior (DOI), through the U.S. Geological Survey (USGS), provides expertise in the areas of oil drilling, producing, handling, and transporting by pipeline. The Fish and Wildlife Service (FWS) provides technical expertise on fish and wildlife.

9. The Department of Justice (DOJ) provides advice on legal questions arising from discharges and federal agency response.

10. The Department of Labor (DOL), through the Occupational Safety and Health Administration (OSHA), provides advice regarding the precautions that are necessary to protect the health and safety of spill response personnel.

11. The Department of Transportation (DOT) provides advice on the transport of oil and hazardous substances. Through the U.S. Coast Guard (USCG), the DOT offers expertise in port safety. For those areas in which it provides the on-scene coordinator (OSC), the USCG chairs the regional response team (RRT).

12. The Department of State (DOS) develops joint international contingency plans and coordinates international pollution response activities.

13. The Environmental Protection Agency (EPA) determines the environmental effects of pollution, advises on what degree of hazard a discharge poses to the public health and safety, and coordinates scientific support, including assessment of damages, in inland regions. For areas in which it provides the OSC, the EPA chairs the RRT. The EPA, in conjunction with the USCG, prepares regional and local contingency plans.

Nonfederal Participation

Designated state representatives participate in all facets of RRT activities. This practice ensures that federal contingency plans are coordinated with plans developed by state and local governments. Federal regional and local contingency plans identify those specific resources that have been committed by state industrial groups, academic organizations, and others for use during pollution removal operations.

ORGANIZATION

Emergency Response Activities and Coordination

Emergency response activities are coordinated by the OSC through the facilities of the National Response Center (NRC) and the regional response centers (RRCs). The OSC is predesignated by the regional plan and is responsible for federal on-scene coordination.

The OSC reports to the RRT, which is composed of representatives from participating agencies and from state and local governments. National coordination is accomplished through the national response team (NRT), which provides advice and guidance to the RRT. The organization of the National Oil and Hazardous Substances Pollution Contingency Plan is shown in Fig. 1-1.

National Response Center (NRC)

The NRC is the national communication center for receiving and distributing reports regarding discharges of oil and other hazardous substances. This continuously staffed center is located at the headquarters of the USCG in Washington. Reports of pollution

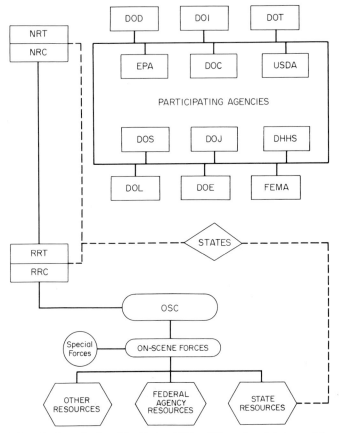

Fig. 1-1 Organization of the National Oil and Hazardous Substances Pollution Contingency Plan. [Federal Register, *vol. 45, no. 55, 1980.*]

incidents are made immediately to the NRC through the NRC duty officer. Notification is accomplished by using a toll-free telephone exchange, 800-424-8802 or 800-426-2675, within the Washington local calling area. Once alerted, the duty officer immediately notifies the appropriate OSC.

National Response Team (NRT)

The NRT, which is located in Washington, consists of representatives from the participating agencies. It functions as the *national* body for planning and preparedness actions prior to a pollution discharge and for coordination and advice during a pollution emergency. Except when activated because of a pollution incident, the representative of the EPA serves as chairperson of the NRT, and the representative of the DOT (represented by the USCG) serves as vice-chairperson. The vice-chairperson maintains records of NRT activities along with national, regional, and local contingency plans for pollution response. When the NRT is activated during a pollution incident, the representative of either the EPA or the DOT serves as chairperson, depending upon the area in which the response is taking place.

Regional Response Center (RRC)

The RRC performs functions parallel to those of the NRC. It receives reports of pollution episodes, notifies the OSC, and provides facilities and personnel for coordinating the response to the pollution incident.

Regional Response Team (RRT)

The RRT serves as the *regional* body for planning and preparedness actions before a pollution discharge and for coordination and advice during a discharge. It consists of regional representatives of the participating agencies and state and local government representatives as appropriate. The full participation of senior-level representatives from state and local governments with major ports and waterways is desired. Regional contingency plans identify the representatives of participating federal and state agencies.

RRT members work with the regional OSCs in developing local contingency plans, providing for the use of agency resources, and responding to pollution incidents.

Representatives of each state lying within a federal region are invited to work with the RRT and OSCs in developing regional and local plans, plan for and make available state resources, and serve as the contact point for coordination with local government in responding to pollution incidents. When the RRT is activated for a pollution emergency, affected states are invited to participate in all RRT deliberations. Any state or local government representative who participates in the RRT has the same status as any federal member of the RRT.

Except when the RRT is activated for a pollution incident, the representatives of the EPA and the DOT serve as cochairpersons. When the RRT is activated for a pollution incident, the representative of either the EPA or the DOT serves as chairperson, depending upon the area of the spill and the response.

The RRT is activated as an emergency response team when a discharge (1) exceeds the response capability available to the OSC in the place where it occurs, (2) transects regional boundaries, or (3) poses a substantial threat to public health and welfare or to regionally significant amounts of property. Regional contingency plans specify detailed criteria for the activation of RRTs.

The RRT may be activated during a pollution emergency by request from any RRT representative to the chairperson of the team. Each representative is notified immediately by telephone when the RRT is activated. The plan identifies three classes of discharges:

1. *Minor discharges.* A discharge of less than 1000 gal (3.8 m³) of oil to inland waters, or a discharge to coastal waters of less than 10,000 gal (37.9 m³) of oil, or a discharge of a hazardous substance in a quantity less than that defined as reportable by regulation.[5]

2. *Medium discharge.* A discharge of 1000 to 10,000 gal of oil to inland waters, or a discharge of 10,000 to 100,000 gal (378.5 m³) of oil to coastal waters, or a discharge of a hazardous substance equal to or greater than a reportable quantity as defined by regulation.

3. *Major discharge.* A discharge of more than 10,000 gal of oil to inland waters, or more than 100,000 gal of oil to coastal waters, or a discharge of a hazardous substance that poses a substantial threat to public health or welfare or results in critical public concern.

Discharge classifications are used as guidelines by the OSC and serve as criteria for initiating the operational response actions described in the section "Operation." They do not imply associated degrees of hazard to public health or welfare, nor are they a measure of environmental damage. The RRT is activated automatically in the event of a major or a potentially major discharge.

When activated for a pollution incident, agency representatives on the RRT meet at the call of the chairperson and (1) request other federal, state, or local governments or private agencies to consider providing resources under their existing authorities to mitigate a discharge or monitor response operations and (2) assist the OSC in preparing information releases to the public and for communication with the NRT. During a pollution emergency, the members of the RRT ensure that the resources of their agencies are made available to the OSC as specified in the regional and local contingency plans.

When not activated for a pollution incident, the RRT serves as a standing committee to (1) recommend needed policy changes in the regional response organization, (2) revise the regional plan, and (3) evaluate the preparedness of the agencies and the effectiveness of local plans for the federal response to pollution incidents.

Each coastal RRT is required to conduct an annual training exercise in which response equipment is actually deployed. These exercises utilize all existing capabilities in the local port area. RRTs for inland regions are strongly encouraged to conduct an annual training exercise in which response equipment is actually deployed.

On-Scene Coordinator (OSC)

Coordination and direction of federal pollution control efforts at the scene of a discharge or potential discharge are accomplished through the OSC. The OSC is predesignated by the regional plan to coordinate and direct pollution control activities in each area of the region.

In the event of a spill of oil or hazardous substances the OSC determines pertinent facts about (1) the discharge, such as its potential impact on human health and welfare, (2) the nature, amount, and location of material discharged, (3) the probable direction and time of travel of the material, and (4) the resources and installations which may be affected and the priorities for protecting them. The OSC initiates and directs response operations and subsequently provides support and documentation for the recovery of damages and other possible litigation.

The USCG has the responsibility to furnish or provide OSCs for coastal waters and for Great Lakes waters, ports, and harbors. The EPA provides or furnishes OSCs for inland waters.

Special Forces Available to the OSC

On-scene coordinators may request the activation or involvement of various special forces. The following units can provide specialized equipment and scientific support:

1. The National Strike Force of the USCG comprises (1) the strike teams established on the east, west, and gulf coasts of the country and (2) emergency task forces located at major ports. The strike teams have expertise in ship salvage, damage control, and diving. When activated by the OSC, they provide communications support, advice, and specialized equipment to assist response activities. The emergency task forces consist of trained regional personnel with supplies of pollution control equipment and materials and detailed contingency plans for their areas of responsibility.

2. The environmental response team (ERT), established by the EPA, provides technical support to the OSC and access to the EPA's Environmental Emergency Response Unit (EERU). EERU's prototype equipment and methods, developed by the EPA's Office of Research and Development, represent the state of the art in hazardous substances spill prevention and control technology.

3. The scientific support coordinators (SSCs), provided by the EPA and the DOC, furnish scientific support, perform environmental damage assessments, and coordinate on-scene scientific activity.

Public Information Network

When the NRT is activated during a major pollution incident, the team chairperson requests the establishment of a national news office. The director of the national news office is provided by the agency providing the OSC. The director of this office is responsible for the overall supervision of public information activities of the NRT.

An on-scene news office is established, at the request of any agency participating on the RRT or by the OSC, to coordinate media relations and issue official federal information during pollution incidents. A public information assistance team (PIAT) is available to help OSCs and regional offices disseminate public information during a major pollution incident or threatened incident.

PLANS

The National Oil and Hazardous Substances Pollution Contingency Plan is the basis for federal action to minimize pollution damage from discharges of oil and hazardous substances. It provides the overall policy and guidelines for the development of *regional* and *local* contingency plans.

Regional Contingency Plans

The purpose of these plans is to provide for the coordinated, timely, and effective response to pollution incidents by various federal agencies and other organizations. Regional plans are broad in scope and contain information on useful facilities and resources from government, commercial, academic, and other institutions. Each RRT is required to develop a *regional* contingency plan for its respective federal region. Figure 1-2 illustrates the 10 standard federal regions and the location of the regional headquarters.

Local Contingency Plans

The National Oil and Hazardous Substances Pollution Contingency Plan requires the OSC to develop local contingency plans. Local plans identify (1) environmentally sensitive areas, (2) the most probable locations for pollution incidents, and (3) the kinds and locations of resources that would be needed to respond to spill incidents, plans of actions for protecting vulnerable resources, sites for disposing of recovered oil and hazardous substances, and local organizational structures for spill response.

Nongovernmental Plans

In September 1978, the EPA proposed (under 40 CFR Part 151)[6] to require spill prevention control and countermeasure (SPCC) plans to prevent discharges of hazardous substances from those onshore and offshore facilities that are subject to permitting requirements under the National Pollutant Discharge Elimination System of the Clean Water Act. These regulations were proposed concurrently with regulations under Sec. 402 of the Clean Water Act (40 CFR Part 125, Subpart L).[7]

Under the proposed Sec. 402 regulations, permittees would be required to develop plans for best management practices (BMP) to prevent the release of toxic and hazardous

Fig. 1-2 Standard federal regions. [Federal Register, *vol. 45, no. 55, 1980.*]

• Regional Headquarters

pollutants to surface waters. Compliance with SPCC plan requirements (40 CFR Part 151) would be established as a minimum level of control for BMP plans (40 CFR Part 125).

The proposed approach for the prevention of hazardous substances pollution is similar to the one developed and used in EPA's "Oil Pollution Prevention" regulation, 40 CFR Part 112.[8]

Spill Cooperative Plans

Some facilities have elected to form spill cooperatives. Generally, cooperatives purchase and maintain the group's spill containment, cleanup, and disposal equipment. When a discharge occurs, the cooperative's equipment is deployed by facility employees who are trained to use the equipment. The cooperative develops and maintains facility-specific response plans to ensure a rapid and cost-effective spill response operation.

International Contingency Plans

Canada–United States

The joint Canada–United States Marine Pollution Contingency Plan provides for coordinated and integrated responses to pollution incidents on the Great Lakes, their interconnecting waterways and major tributaries, and the international section of the St. Lawrence River. The joint plan provides for predesignated OSCs and a joint response team. When an international spill episode occurs, an OSC from one country and a deputy OSC from the other country coordinate response activities. The joint response team, comparable to the RRT in the United States, provides advice and assistance to the OSC. The international plan establishes alerting, notification, surveillance, and funding procedures.

Mexico–United States

In 1977, the United States and Mexico began discussions concerning the development of a joint Mexico–United States contingency plan. Several objectives of the draft plan, prepared in 1978, are the development of a discharge-reporting system, a joint program for containing spills, and the expertise and equipment need for spill cleanup and mitigation.

In 1980, both nations agreed to the establishment of a response team in each country to identify potential pollution situations. Specific contingency plans are to be developed in the future.

National Inventory of Commercial and Governmental Pollution Response and Support Equipment

A computerized national inventory of commercial and governmental pollution response and support equipment is available to OSCs and members of the RRTs. The automated Spill Cleanup Inventory System (SKIM) allows response personnel to gain rapid access to resources during emergencies. The SKIM system can be accessed through the NRC and the regional offices of the USCG.

OPERATION

Coordination

Delegation of Authority

As mentioned in the section "Organization," the EPA and the USCG are responsible for designating OSCs for all areas in each federal region. However, the first federal official from a participating agency to arrive at the pollution site is responsible for coordinating activities under the plan until the OSC arrives. All federal agencies are required to develop emergency plans and procedures to prevent and mitigate discharges of oil and hazardous substances from facilities or vessels under their jurisdiction. These agencies are required to designate the office that will coordinate response actions in accordance with the plan. When the responsible federal agency fails to act promptly or to take appropriate action during a pollution episode, the EPA or the USCG will assume the OSC functions.

Multiregional Actions

There is only one OSC at any time during a response operation. When a discharge affects an area that is covered by two or more federal, regional, and local contingency plans, the EPA and the USCG will agree upon and designate an OSC.

General Pattern of Response

When reports of discharges of a hazardous substance are received by the NRC, the duty officer immediately notifies the appropriate regional OSC. The OSC then proceeds to investigate the report to establish its validity and to determine pertinent information such as the threat to public health and welfare and the source, quantity, and composition of the material discharged. The OSC evaluates the information and determines what response actions are required.

Actions taken to respond to a pollution discharge can be separated into five relatively distinct classes or phases, namely, Phase I, discovery and notification; Phase II, evaluation and initiation of action; Phase III, containment and countermeasures; Phase IV, removal, mitigation, and disposal; and Phase V, documentation and cost recovery. Elements of any one phase may take place concurrently with those of one or more other phases.

Response Phases

Phase I: Discovery and Notification

A discharge may be discovered through (1) a report submitted by a discharger in accordance with statutory requirements, (2) a deliberate search by vessel patrols and aircraft, and (3) random or incidental observations by government agencies or the general public. In the event of discovery through deliberate search, the discharge is reported directly to the NRC. Random observations are reported to the NRC or to the nearest USCG or EPA office.

Regional and local contingency plans require that the aforementioned reports be forwarded to the NRC, RRC, and designated state agency. Reports of major and medium discharges, received by either the EPA or the USCG, are expeditiously relayed by telephone to appropriate members of the RRT. Reports of minor discharges are exchanged between the EPA and the USCG as agreed to by the two agencies.

The agency furnishing the OSC for a particular area is responsible for implementing the appropriate phase activities in that area.

Phase II: Evaluation and Initiation of Action

The report of a discharge is immediately investigated by the OSC. After evaluating all available information, the OSC (1) determines the magnitude and severity of the discharge, (2) evaluates the feasibility of removal, and (3) assesses the effectiveness of removal actions.

The OSC expeditiously notifies the RRC of the need to initiate additional federal response activities. Subsequent actions may be limited to notifying the RRT, or they may range from additional surveillance to the initiation of Phases III and IV.

The OSC ensures that adequate surveillance is maintained to determine that removal actions are being performed properly. When removal efforts are inadequate, the OSC advises the responsible party. If the responsible party has been advised and does not initiate proper removal action, the OSC takes the necessary actions to remove the pollutant. Resources for these actions are expended by utilizing appropriate federal funding authorities.

When the discharger is unknown or otherwise unavailable, the OSC proceeds with removal actions.

Phase III: Containment and Countermeasures

Containment and countermeasure actions are initiated as soon as possible after the discovery and notification of a discharge. These defensive actions are taken to protect public health and welfare, to control the source and spread of pollution, and to protect specific installations or areas.

Phase IV: Cleanup, Mitigation, and Disposal

Cleanup, mitigation, and disposal are those actions taken to recover the pollutant from the water and affected shorelines. Recovered pollutants and contaminated materials are disposed of in accordance with regional and local contingency plans.

Phase V: Documentation and Cost Recovery

The OSC furnishes the documentation required to support the recovery of federal removal costs and to recover damages done to federal, state, or local government property.

Special Considerations

Safety of Personnel

Until the nature of the substance discharged is known, the OSC exercises great caution in allowing civilian or government personnel into the affected area. Local contingency plans are especially valuable since they identify anticipated hazards, precautions, and the requirements to protect personnel during response operations.

Volunteers

Volunteers are not permitted at on-scene operations when the discharged material is toxic to humans. Local plans provide for the direction of volunteers and identify specific areas in which volunteers can be utilized.

Waterflow Conservation

Regional and local contingency plans identify organizations or institutions that are willing to participate in waterflow dispersal, collection, cleaning, rehabilitation, and recovery activities.

Funding

Federal discharge removal actions are initiated when (1) the person responsible for the discharge does not act promptly or fails to take proper removal actions or (2) the person responsible for the discharge is unknown. Federal discharge removal actions are performed under Sec. 311 of the Clean Water Act, as amended.

REPORTS

Pollution Reports (POLREPs)

Whenever an RRT is activated, a pollution report (POLREP) is forwarded to the NRC. The initial POLREP indicates the time when the RRT was activated, the method of activation (e.g., telephone), and the place of assembly. During the activation, POLREPs are submitted to the NRC on a daily basis.

A POLREP consists of four basic sections: situation, action taken, future plans and recommendations, and status. The "situation" section describes the spill details in full; it includes spill location, events, nature and quantity of materials, personnel involved, area impacted, areas threatened, predicted movement, effectiveness of control efforts, and prognosis. The "action taken" section summarizes all actions taken by the responsible party and by federal, state, and local forces. The "future plans and recommendations" section identifies all planned response actions of the federal, state, and local forces and any recommendations that the OSC has pertaining to the response. The "status" section indicates that the case is closed, open, or pending or that federal participation has or has not been terminated. When the RRT is deactivated by agreement between the EPA and USCG team members, the time of deactivation is included in the POLREP.

OSC Report

At the conclusion of a major pollution discharge, the OSC submits to the RRT a complete report on the response operation and the actions taken. The OSC's report is prepared according to the format established by the contingency plan. In general, the OSC report is a detailed accounting of the situation as it developed, actions taken, resources committed, problems encountered, and recommendations. The OSC's recommendations, based on response experiences, are a source for new procedures and policy.

REFERENCES

1. Council on Environmental Quality, "National Oil and Hazardous Substances Pollution Contingency Plan; Final Revision," *Federal Register,* vol. 45, Mar. 19, 1980, pp. 17832–17860.

2. *National Multi-Agency Oil and Hazardous Materials Pollution Contingency Plan,* November 1968.

3. Council on Environmental Quality, "National Oil and Hazardous Materials Pollution Contingency Plan," *Federal Register,* vol. 35, June 2, 1970, pp. 8508–8514.

4. *Comprehensive Environmental Response, Compensation, and Liability Act of 1980,* 96th Cong., Pub. L. 96-510, Dec. 11, 1980.

5. U.S. Environmental Protection Agency, "Water Programs; Determination of Reportable Quantities for Hazardous Substances," *Federal Register,* vol. 44, Aug. 29, 1979, pp. 50766–50786.

6. U.S. Environmental Protection Agency, "National Pollutant Discharge Elimination System; Proposed Requirements for Spill Prevention Control and Countermeasure Plans to Prevent Discharges of Hazardous Substances from Certain Facilities," *Federal Register,* vol. 43, Sept. 1, 1980, pp. 39276–39280.

7. U.S. Environmental Protection Agency, "Criteria and Standards for the National Pollutant Discharge Elimination System; Subpart K—Criteria and Standards for Best Management Practices Authorized under Section 304(e) of the Act," *Federal Register,* vol. 44, May 19, 1980, p. 33290.

8. U.S. Environmental Protection Agency, "Oil Pollution Prevention," *Federal Register,* vol. 41, Mar. 26, 1976, p. 12657.

PART 2
United States Local Governmental Plans

Larry R. Froebe, Ph.D.*
Regional Response Team Manager
Ecology and Environment, Inc.

Distribution of United States Spill Response Capabilities / 7-15
Local Response Plans / 7-19
Some Approaches to Local Response Programs and Plans / 7-22
Additional Local Plan Functions / 7-23
Appendix / 7-24

The basic feature of a hazardous chemical emergency is that no matter where it occurs, who responds, or how fast the response is, there is the simple fact that the incident always occurs at a locality. This part of Chap. 7 describes how individual localities can provide for their own immediate protection through local response plans which are compatible with state and national plans and which utilize local response units. Further, it describes the important role which local response plays in the United States spill response capability.

In addition, this chapter provides guidance and perspective to:

- Those interested in formulating local response plans

- Those seeking to learn the capabilities of the fire service, the traditional first hands-on response, for spill control

- Those seeking to learn the capabilities of local environmental response units working concurrently with fire service and law enforcement units in spill response

DISTRIBUTION OF UNITED STATES SPILL RESPONSE CAPABILITIES

United States National Spill-Fighting Network

There are in the United States the elements of a dynamic, cost-effective national spill-fighting network. Depending on the severity and the impact of a hazardous chemical

*Dr. Froebe was the original coordinator-developer and supervisor of a local environmental emergency response chemical spill team at the Montgomery County Combined General Health District and Regional Air Pollution Control Agency, Dayton, Ohio, from 1975 to 1977.

The author acknowledges with gratitude the aid of Mr. James J. Buchanan, formerly of the Dayton local response team and presently of the Kansas City office of Ecology and Environment, Inc., for helpful discussions on the initial drafts of the manuscript and for providing the equipment listing in the Appendix to this part.

7-15

incident, any or all of these elements can participate in mitigation of the threat. The majority of the response organizations forming the network are government-based, although the Chemical Transportation Emergency Center (CHEMTREC; 1-800-424-9300) performs an important national resource function under the sponsorship of the Chemical Manufacturers Association.[1] Also, the number of regionally based commercial response teams has increased severalfold in recent years; these teams play an important role in actual cleanup operations. Because legislated responsibility for hazardous materials control resides with government agencies, it is sufficient for a description of the national response network to compare the types of response in terms of the level of government at which teams are based, the level of technical capability, and the response time to deliver that capability.

The relative "response-ability" of the various units of the national spill-fighting network compared with the ideal response of infinite technical capability delivered in zero response time is shown in Fig. 2-1. The diagram is qualitative but provides important information in setting up cost-effective local response programs. For instance, although national response and regional response are different arms of the same organization, a regional response with the technology and authority of the national agency can come from a regional office just a few states away from the spill locality. National response involves the ultimate technology available and one-of-a-kind equipment, such as a mobile water treatment unit on a semitrailer truck bed, which may require many days for delivery.[2] A similar relation exists on a reduced scale between state response and district response. In this case, a district response may be quicker, but special generators, booms, skimmers, or similar containment and mitigation equipment may be delivered only from state headquarters because of cost-effective budget constraints.

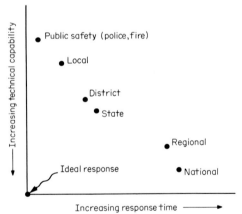

Fig. 2-1 "Response-ability" of United States spill response units. This representation does not reflect the special case in which a district, state, regional, or national spill response team is centered in the locality where a spill occurs. [*Reprinted from* AWWA Seminar Proceedings—Hazardous Material Spills *by permission of the Association. Copyrighted 1977 by the American Water Works Association, Inc., 6666 West Quincy Avenue, Denver, Colorado 80235. L. R. Froebe, "State and Local Response Capabilities for Material Spills Hazardous to a Water Supply."*]

Local Response and the Fire Service

The relationship of local response representing environmental and health protection, among other responsibilities, and local public safety response (police, fire, emergency medical service) is critical to the success of a local response plan. It can be closer and potentially more individualized than any other in the spectrum of relations in the national spill-fighting network. The fire service can make a substantial hands-on response, including the following techniques:

1. Rapid dispatch and containment of spills with dikes and dams before environmental response teams arrive
2. First hands-on response for the application of sand or sawdust to small oil and/or gasoline spills from accidents for cleanup rather than flushing to a sewer
3. Rapid problem evaluation with reference to other response units including environmental teams, manufacturer teams (through CHEMTREC), road crews, emergency medical service, transportation inspectors, railroad scheduling, etc.
4. Use of 63.5-mm (2½-in) hose sections filled from air bottles, capped, and lashed with rope as a floating containment boom for calm waters
5. Use of master streams, water walls, or snorkel streams for vapor entrainment and suppression (a technique that can virtually suppress hydrogen chloride or ammonia vapor and can mitigate chlorine vapor and raise the lower explosion limit of hydrocarbon vapors to gain time for evacuation and protection of life)
6. Use of a foam truck with various foams for the suppression of vapors from spills
7. Use of a foam truck *without foam* for the ventilation of buildings and sewers
8. Use of fog streams pointed from building windows to ventilate odors
9. Use of acid suits for full-body protection in dealing with corrosive vapor and liquid leaks (including the use of hose streams for the decontamination of acid suits or the washing of persons exposed to irritating or corrosive vapors)
10. Use of chlorine kits to plug leaks
11. Use of 2¼-kg (5-lb) packs of sealant compound (of the Dux-seal type) or other wood plugs for the temporary sealing of small leaks and cracks
12. Pumping of water into a storage tank of a lower-density immiscible product that has a leaking bottom seal, seam, or bottom valve (the water will float the product until it can be off-loaded[3])
13. Establishment of preplans for hazardous chemical emergency response

While a local public safety response group is almost invariably the first response group to arrive at a hazardous chemical incident, a local environmental response team can arrive shortly thereafter with important and crucial support functions for the protection of health and the environment. A local environmental response team can perform the following functions in a spill:

1. Consultation with fire departments on how the techniques listed above might be applied to spills of various chemicals
2. Provision of technical information which may make containment and cleanup easier (for instance, it may be easier to freeze a benzene spill with dry ice and then to shovel up the solid)
3. Construction of dikes (including weir skimmers), dams (including underflow and straw dams), and reservoirs to contain spills until mitigation can be completed

4. Meteorological monitoring and determination of hazard areas from vapor dispersion calculations
5. Environmental monitoring to confirm calculated hazard areas from vapor dispersion and to map the impact on water quality of impacted streams and lakes
6. Liaison and coordination between the fire chief (the local officer in charge) and other, later-arriving environmental response units

The fire service and other local public safety organizations play the central role in the mitigation of hazardous chemical emergencies with threats to public health and safety. But there are concurrent functions also critical to the protection of public health and safety that public safety units will *never* be able to provide cost-effectively (unless the mission of the traditional public safety service is changed). The fire service, which above all else must provide rapid response, cannot also provide lengthy involvement in spills requiring several days for mitigation. This is the basis of the relatively new symbiotic relationship between local public safety response units and environmental response teams. In addition, directives from higher levels of government (i.e., state environmental protection agencies such as that of Ohio[4] and the National Oil and Hazardous Substances Pollution Contingency Plan[5]) seek to promote a response to spills by the local environmental agency.

Cost-Effective Network Response

The spiller is responsible by law[6, 7, 8] for cleaning up the spill and removing any associated threat to the environment.* The spiller is also responsible by law for notifying the National Response Center (NRC; 1-800-424-8802), which will notify the appropriate regional Environmental Protection Agency (EPA) or Coast Guard unit for response.† In this case notification might go from the spiller (or some reporting agent) to the national level (NRC), the regional level, the state level, and finally the local level (all within minutes) for a response. A local environmental response almost certainly will involve the fire service and law enforcement even though public safety is not directly threatened.

Notification might also go from the spiller to the local fire service (or law enforcement body) and other local response units, especially in the case of a threat to public safety or health. Notification of a state or local environmental agency does not meet the obligation to notify the NRC, but it will probably shorten the time required to mitigate the spill.

If all elements of the spill-fighting network are notified at the same time, the local police and fire service will be the first to arrive.[9] These units will begin an attack on the threat: fire suppression, evacuations, containment of flow, control of the leak source, and identification of the material. At a point early in the incident history, the previous meetings and training sessions between local and state environmental response units and the fire service will begin to pay off. The fire fighters realize that additional information and techniques are necessary to resolve the problem or that there is environmental impact, and the local response team is notified. The local environmental team arrives within minutes to an hour of notification (depending on response distance) and reports to the fire chief. If the incident is large enough to last several hours before resolution, the local team notifies the state (and its district) response team. The state and local teams confer on the capabilities needed for mitigation, and the state decides whether or not they will respond. The state then advises the regional team, and together they decide whether or not a

---

*Compare individual state environmental regulations.
†The EPA response effort is augmented by technical assistance teams (TAT) in each region, the emergency response team (ERT) headquarters in Cincinnati, and the EPA headquarters in Washington. First response by the EPA may be carried out by a TAT.

response, a standby, or a notification followed by a report is sufficient for the regional team. In very large incidents, specialized expertise or equipment may be called in under the national response.*

An important concept for the cost-effective function of the network is containment first, with mitigation (resolution) once the unpredictable nature of the spill has been resolved. In small "nuisance" spills (such as a leaking gasoline tank or oil pan from an automobile accident), the fire department usually resolves the spill by spreading sand to improve roadway traction until the material has been collected or evaporates. When the fire department cannot resolve the entire threat, it contains the spill if possible and calls the local spill team. The local team aids with containment if necessary and mitigation if possible. This is the philosophy that continues up the response chain.

LOCAL RESPONSE PLANS

Cost-Effective Local Response

The local response plan coordinates the activities of and provides the lines of authority and responsibility for local response units in hazardous chemical incidents. It also describes contact and cooperation with state and national response teams. A recommended format for a local contingency plan is reproduced in Table 2-1.

Figure 2-2 is a simple schematic of a local plan for spill response. Distributed about the spill (at the center) are the various response units, occupying the area of the circle roughly in proportion to the capabilities they can deliver. The label "Local EPA" represents a local environmental agency (air, water, general pollution control) and/or the health department. The "Other" designation represents consultants, cleanup contractors, university experts, industry representatives, and other local public service agencies. The local plan incorporates state and national plans as natural extensions of local capabilities. Depending on the requirements to resolve any given spill, any of the response units and their capabilities can become all-encompassing, but Fig. 2-2 represents the general relation of the national spill-fighting network elements in the context of the local plan.

Not every locality or population center has access to all the elements of a local response plan; one local response plan will not fit the different cost-effective constraints of urban, suburban, and rural localities. For instance, Ohio is fifth in the United States in annual chemical usage, and local response plans utilizing the response units in Cincinnati, Cleveland, Columbus, Dayton, Toledo, and Akron-Canton are justified. However, local response plans for the rural areas of Ohio might specify only mutual aid from nearby communities for the fire department and direct notification to the state EPA for a direct response.† Of course, rural areas near urban areas (for instance, Miami County north of Montgomery County and Dayton, Ohio) may specify local response teams from those adjacent areas. The situation is analogous in all the other states of the United States.

Basis of the Local Plan

Formulating a local spill plan can begin with any group that sees the need. Eventually, the implementation efforts must be presented to some authorizing body able to span many

---

*In spills threatening navigable waters in which cleanup efforts are inadequate or responsibility is not assigned, the federal on-scene coordinator can initiate a federal takeover with the use of the federal revolving fund (the "311 fund") under the Clean Water Act for the cleanup. A federal takeover does not remove responsibility from the spiller, and 311 costs are further assigned to the spiller through administrative and legal procedures.

†Ohio environmental law requires notification of the Ohio EPA for spills.

TABLE 2-1   Recommended Format for a Local Contingency Plan*

Letter of promulgation
Record of amendments
Table of contents
List of effect pages
100   Introduction
    101   Authority
    102   Purpose and objective
    103   Scope
    104   Abbreviations
    105   Definitions
200   Policy and responsibility
    201   Federal policy
    202   Related state policy
    203   Multinational policy
    204   On-scene coordinator responsibility
    205   Nonfederal responsibility
300   Planning and response considerations
    301   Oil and hazardous substances transportation pattern
    302   Transfer storage and processing facilities
    303   Historical spill considerations
    304   Hydrological and climatological considerations
    305   Local geography
    306   Highly vulnerable areas
    307   Local response resources
    308   Waterfowl conservation
    309   Endangered species
400   Response organization
500   Operational response actions
600   Coordination instructions
    601   Delegation of authority
    602   Notification
    603   Coordination with special forces
    604   Termination of response activities
    605   Resolution of disputes
700   Procedures for reviewing and updating the local contingency plan

Annex I	1100	Distribution
Annex II	1200	Pollution response personnel assignments
Annex III	1300	Geographical boundaries
Annex IV	1400	Notifications, communications, and reports
Annex V	1500	Public information
Annex VI	1600	Documentation for enforcement and cost recovery
Annex VII	1700	Funding
Annex VIII	1800	Cleanup techniques and policies
Annex IX	1900	Arrangements for nonfederal groups
Annex X	2000	Interagency support
Annex XI	2500	Geographical-action directory
Annex XVI	2600	Response-assistance directory

*Abstracted from "National Oil and Hazardous Substances Pollution Contingency Plan," *Federal Register*, vol. 45, no. 55, part III, Mar. 19, 1980, p. 17850.

local emergency response units in order to gain community and political support. A study[10] at the University of Dayton on local spill response recommended a hazardous material control council made up as shown in Table 2-2. In fact, in addition to some specific recommendations on equipment and operational procedures the university report made a number of recommendations indicative of concepts for local response plans:

1. A hazardous material spills response system should be established.

2. The response system should be housed in the health district.

3. A person knowledgeable in chemical analysis should be assigned as coordinator with this as his or her primary function.

4. Six additional persons (employee volunteers), also knowledgeable in chemical analyses, should be trained as response specialists.

5. A hazardous material control council should be formed; its function would be to establish policies, set up training programs, provide for program funding, publicize team capabilities to other agencies and the public, and promote working relationships with other agencies and response units.

6. The fire chief of the district involved should remain the commander at the scene. (In threats to public health, the broad powers of the health commissioner may be invoked concurrently with the authority of the fire chief.)

The majority of these recommendations were the basis of a local spill response team in Dayton[11,12] from 1975 to 1979 and of a local response plan in Jefferson County, Kentucky.[13]

Local response planning has an analogy in disaster response planning.* The association is natural, as can be attested by the Jefferson County program which has grown out

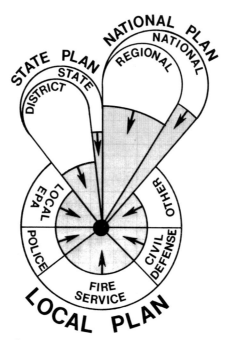

Fig. 2-2 Distribution of response capabilities in the local spill response plan. The dot at the center represents the spill; it is surrounded by the various response units.

*Extensive research in chemical disaster planning has been conducted at the Disaster Research Center, Ohio State University, Columbus, Ohio 43210.

TABLE 2-2 Suggested Hazardous Material Control Council Membership*

Health and service	Industry and commerce
Health Department (includes local environmental response)	Transportation
Civil defense	Manufacturing
American Red Cross	Commercial
Hospitals	Labor unions
Public safety	Individual professionals
Police departments	Chemists
Fire departments	Engineers
Rescue squads (emergency medical services)	Industrial hygienists
Elected officials	Physicians
	Meteorologists

*A Hazardous Material Spill Response System, University of Dayton Design of Systems Report, April 1975. Copies of this report may be obtained from Prof. Robert L. Mott, chairman of the Department of Mechanical Engineering Technology, University of Dayton, Dayton, Ohio 45469.

of the original administration by the Louisville and Jefferson County civil defense. Civil defense has access to considerable equipment, vehicles, and communications gear that can serve double duty in a local response program. A local spill response program also provides an opportunity to maintain a higher level of training with this gear for the isolated time during which it will be needed in a civil defense emergency or a disaster.

Information on equipment and response capabilities for local plans can be obtained from regional and state EPA offices. A typical listing of equipment for a local response team is included as an appendix to this chapter.

Legal Questions

A spill response plan implies new duties and therefore new liabilities. Every response person is responsible for the consequences of his or her decisions and actions, and the potential for litigation is greater in these new and different situations. The guiding phrase in spill response is "reasonable and prudent"; decisions and actions by an individual of reasonable and prudent judgment are, in general, defensible.

However, a local response plan utilizing units of various agencies and organizations must provide additional protection. Employed response specialists must have health insurance, death and dismemberment insurance, and workman's compensation. Volunteer spill responders, typified by many volunteer fire, medical, and rescue units, could qualify for workman's compensation if they received token compensation for their participation.[14] Workers who respond to chemical spills on a consistent basis should have a health screening program for both worker health and organizational liability protection.*

SOME APPROACHES TO LOCAL RESPONSE PROGRAMS AND PLANS

Some local hazardous chemical incident response programs undertaken in the United States illustrate problems associated with local programs and successful concepts.

*Ecology and Environment, Inc., has developed a copyrighted health screening program which is being used for employees working on EPA contracts for technical assistance teams (emergency response) and field investigation teams (hazardous waste site inspections). For more information, get in touch with Mr. Gerard A. Gallagher, Ecology and Environment, Inc., 195 Sugg Road, P.O. Box D, Buffalo, N.Y. 14225 (716-632-4991).

Dayton, Ohio

The city of Dayton had an active response team between 1975 and 1979. The team was organized to aid Dayton-area fire departments with problems involving chemicals. The team was made up of employee volunteers of the Montgomery County Combined General Health District and its regional air pollution control agency. Crucial support of the response program was provided by the Box 21 Rescue Squad, itself a volunteer organization of emergency medical technicians and paramedics, that delivered spill response gear without jurisdictional boundaries on one of its heavy-rescue vehicles 24 h a day. While the team served primarily Dayton-area fire departments and emergency response units, it also provided assistance in the surrounding less urban counties.

The only University of Dayton recommendations not implemented in the program were the formation of a hazardous material control council and the appointment of a coordinator serving with the program as that individual's primary function. Funding for this team was in the form of a $10,000 implementation grant to the health district from the Board of Montgomery County Commissioners. The program became inactive because of lack of additional funding coupled with difficulty in developing an equitable organizational structure that included the Box 21 organization.

Jefferson County, Kentucky

This county has a program similar in concept to the Dayton program. It formed a hazardous materials response group (the forerunner of a hazardous materials control council) made up of representatives of city, county, and locally based federal response agencies, plus industry and the University of Louisville. Representatives of the response group visited Dayton to confer on response capabilities during the initial implementation. The Jefferson County hazardous materials response operation is based on first response by law enforcement officers, who trigger the spill response. This seems to be a cost-effective trend in local spill response. (The Dayton program also relied heavily on first response by the health district's environmental patrol officers.) A drawback which must be recognized is the potential inability of nontechnical response specialists to see a subtle danger[15] which may arise after the initial incident. For instance, nitrogen oxides from nitric acid in fires or chemical reactions are merely obnoxious fumes until some time after an exposure, when the health effects may be fatal.

Jacksonville, Florida

The response system of Jacksonville's fire department[16] illustrates still another worthwhile approach. The department has improved local hazardous material spill response by expanding capabilities within the fire service. Training programs for fire departments such as those in Jacksonville, those provided by state fire academies, and those provided by the National Fire Protection Association[17] and the Environmental Protection Agency[18] are a significant aid to the larger metropolitan departments, where technical knowledge is more readily usable. However, regardless of department size, environmental considerations in hazardous spills imply a concurrent environmental response, and fire departments should be prepared to work side by side with other environmental response people from any of the levels of the national spill-fighting network.

ADDITIONAL LOCAL PLAN FUNCTIONS

Along with the evolution of cost-effective local spill response plans as part of the national spill-fighting network, some specialized associated needs can be met. As specified by the

Dayton report, a local response plan needs a full-time person to oversee the short-term day-to-day needs of the plan, but in the long term the project coordinator is not needed full time every day. There are several systematic needs involving chemicals and public safety protection which the project coordinator can fulfill. For instance, fire departments would benefit from additional chemically oriented expertise in fire prevention inspections, formulation of emergency preplans for chemical-containing facilities, formulation of chemical accident prevention procedures in local industries, and risk analysis for certain facilities involving chemicals. This requires expertise not normally found in the fire service but which could be supported under the more generally based local response plan. The local spill response team can also serve a centralizing function for the fire departments of a locality through its position as essentially a mutual-aid unit with additional expertise and specialized spill-fighting equipment (portable corrosion-resistant pumps, special leak sealants, acid suits, etc.). Finally, the local plan requires continuing response training. Instead of incurring the cost of sending personnel to schools and courses, the local project coordinator can survey, organize, and administer training for all members under the local response plan, including fire departments in the area (especially in view of the turnover experienced by volunteer departments).

The public wants and needs to know more about the chemicals that are used each day as wonder products without any realization of the liability incurred in their use. A community benefit of a local response plan could be practical information on everyday chemicals to promote better living. The coordination and implementation of local talents and capabilities in the form of a local response plan not only improves the efficiency of the national spill-fighting network and saves money through a cost-effective response function but also provides a positive spirit for local self-help, self-assured cooperation with state and regional plans (teams), and new community enlightenment about the chemicals that support the modern way of life.

APPENDIX

Local Spill Team Equipment Reference Material Listing*

Description	Unit cost	Purchase date
Personal Protection Gear		
Jump suits, fire-resistant	$ 19.85	January 1976
Lettering	4.50	January 1976
Hard hat	3.50	January 1976
Safety shield visor	6.40	January 1976
Safety boots, OSHA-approved	26.95	November 1976
Gloves, leather	5.99	November 1976
Gloves, monkey-grip, chemically resistant	3.00	November 1976
Gloves, neoprene, 16-in, cuffed gauntlet	7.50	January 1976
Overboots, neoprene, 18-in	11.00	January 1976
Rain suit, heavy-duty; jacket with hood and pants	18.00	January 1976
Goggles, antifog (fit over prescription glasses)	2.80	January 1976

Description	Unit cost	Purchase date
Personal Protection Gear		
Respirator, with filters for dusts and mists	8.20	January 1976
Fire coat, Nomex, three-quarter length	75.00	January 1976
Fire boots, three-quarter length, insulated	45.00	November 1976
Tote bag	7.95	November 1976
Self-contained breathing apparatus (SCBA), Scott Air Pack II	410.00	January 1976
Acid suit, full-body protection, to accommodate SCBA	320.00	January 1976
Communications Gear		
Portable radio, Motorola, 4-channel	$ 935.00	October 1974
4 crystals	13.00	October 1974
Scanner, 10-channel, UHF-VHF	121.40	March 1976
10 crystals	33.75	March 1976
Pager service, 1 calendar year, voice page†	330.00	January 1976
Insurance, 1 year	12.00	January 1976
Mobile radio, 4-channel with 4 crystals, Motorola type	About 1100.00	January 1976
Response Gear		
Camera, 35-mm	$ 79.97	March 1976
Strobe flash attachment	29.87	March 1976
Cassette recorder, portable	29.97	March 1976
Tapes, cassette, 45-min	2.00	March 1976
Calculator, battery-powered, scientific	75.41	January 1976
Binoculars	37.97	March 1976
Compass	7.89	March 1976
Hand-held wind gauge	8.00	June 1976
U.S. Geological Survey maps, 7.5-min series	0.75 each	January 1976
Templates for 15 toxic gas hazard areas (included in *Accidental Episode Manual,* below), 7.5-min scale		
Thermometers, standard register and bimetallic	18.45	January 1976
MSA universal testing kit:		
Case	34.95	January 1976
Sampling pump	126.00	March 1976
Sample pyrolyzer	160.35	December 1976
Detector tubes	190.00	January 1976
Gas-sampling kit:		
Personal pump	79.50	January 1976
Sampling bags	8.50	January 1976
Check, on-off valves	2.75	January 1976
Sampling jars		
Weather station (wind speed, temperature, and direction)	245.00	January 1976
Pump, peristaltic action, 30-gal/h-capacity	179.10	January 1976
Tubing, vinyl plastic, 50-ft by ½-in inside diameter by ¾-in outside diameter	24.60	January 1976
Tubing, vinyl plastic, 50-ft by ⅝-in inside diameter by ⅝-in outside diameter	20.30	January 1976
Assorted containment materials including redwood plugs, lead wool, duct seal, duct tape, and heavy-duty garbage bags		
Flashlight(s), explosionproof		

Local Spill Team Equipment Reference Material Listing*   (Continued)

Description	Unit cost	Purchase date
Response Gear		
Absorbent pillows, 3M Type 240	67.00	January 1978
Absorbent boom, 3M Type 270	135.00	January 1978
J-W Model SS-P supersensitive indicator, explosion meter	592.00	March 1976
Carrying case with foam liner	40.65	March 1976
Extension probe	25.00	January 1978
Geiger counters and dosimeters (provided by disaster services authority)		
Oxygen meter		
Aqua ammonia (4 small bottles) for acid or chlorine vapor leak detection		
Soda ash, 300 lb, for small acid leak neutralization and containment		
Assorted tools including shovels, rubber mallet, wood mallet, cylinder wrench, screwdriver, crescent wrench, chlorine wrench, and monkey wrench		
Reference Materials		
N. Irving Sax, *Dangerous Properties of Industrial Materials,* 5th ed., Van Nostrand Reinhold Company, New York	$ 54.50	1979
G. G. Hawley, *Condensed Chemical Dictionary,* 8th ed., Van Nostrand Reinhold Company, New York	29.79	1976
CMA Chem-Card Manual, Chemical Manufacturers Association, Publications Service, 1825 Connecticut Avenue N.W., Washington, D.C. 20009		
DOT Emergency Action Guide for Selected Hazardous Materials, U.S. Department of Transportation, Office of Hazardous Materials, National Highway Traffic Safety Administration, Washington, D.C. 20590		
Laboratory Waste Disposal Manual, Chemical Manufacturers Association, 1825 Connecticut Avenue N.W., Washington, D.C. 20009	3.50	1975
CMA Chemical Safety Data Sheets, Chemical Manufacturers Association, 1825 Connecticut Avenue, N.W., Washington, D.C. 20009 (about 100 available)	.50	1976
RSMA Handling Guides for Hazardous Materials, Railway Systems and Management Association, P.O. Box 330, Ocean City, N.J. 08226	125.00	1976
RSMA Chemical Transportation Index, Railway Systems and Management Association, P.O. Box 330, Ocean City, N.J. 08226	3.50	1976
NSC Chemical Safety Slide Rule, National Safety Council, 425 North Michigan Avenue, Chicago, Illinois 60611	1.47	1976
Bureau of Explosives Pamphlets 1 and 2, Association of American Railroads, Washington, D.C. 20036	5.00	1976
Hazardous Materials, National Fire Protection Association, Batterymarch Park, Quincy, Massachusetts 02269	10.50	1976
Toxic and Hazardous Industrial Chemicals Safety Manual, International Technical Information Institute, Toronomon-Tachikawa Building, 6-5, 1 Chome Nishi-Shimbashi, Minato-Ku, Tokyo, Japan	80.00	1976

Description	Unit cost	Purchase date
Reference Materials		
James H. Meidl, *Flammable Hazardous Materials,* Glencoe Publishing Company, 17337 Ventura Boulevard, Encino, California 91316	5.00	1977
Charles J. Baker, *Firefighter's Handbook of Hazardous Materials,* Maltese Enterprises, Inc., P.O. Box 34048, Indianapolis, Indiana 46234	3.95	1978
Threshold Limit Values for Chemical Substances and Physical Agents in the Workroom Environment, American Conference of Governmental Industrial Hygienists, Secretary-Treasurer, P.O. Box 1937, Cincinnati, Ohio 45201	1.50	1977
Effects of Exposure to Toxic Gases, First Aid and Medical Treatment, Matheson Gas Products, Box E, Lyndhurst, New Jersey 07071	5.00	1977
Accidental Episode Manual, Document PB 210-814, U.S. Department of Commerce, National Technical Information Service, 5285 Port Royal Road, Springfield, Virginia 22161 (includes templates for vapor dispersion of 15 toxic gases scaled to U.S. Geological Survey 7.5-min maps)	15.00	1980

*This listing is based on the equipment inventory for the Dayton, Ohio, local response program (references 11 and 12) but is standardized to a unit cost basis to be adaptable to programs with varying numbers of team members.

†An alternative approach is to purchase pagers of the Plectron type used by many volunteer fire departments and to use tone-activated paging over an existing emergency radio channel.

REFERENCES

1. J. C. Zercher, "CHEMTREC—For Chemical Transportation Accident Assistance," in G. F. Bennett (ed.), *Proceedings of the 1976 National Conference on Control of Hazardous Material Spills,* New Orleans, April 1976, p. 110.

2. R. G. Sanders, S. R. Rich, and T. G. Pantazelos, "A Mobile Physical-Chemical Treatment System for Hazardous Material Contaminated Waters," p. 412, and R. W. Fullner and H. J. Crump-Wiesner, "Use of EPA's Environmental Emergency Response Unit in a Pesticide Spill," p. 345, in G. F. Bennett (ed.), *Proceedings of the 1976 National Conference on Control of Hazardous Material Spills,* New Orleans, April 1976; D. G. Mason, M. K. Gupta, and R. C. Scholz, "A Mobile Multipurpose Treatment System for Processing Hazardous Material Contaminated Waters," *Proceedings of the 1972 National Conference on Control of Hazardous Material Spills,* Houston, March 1972, p. 153.

3. T. W. Pearson, private communication, New York Department of Environmental Conservation, Avon, N.Y., June 1978.

4. *Local OSC Functions, Air Spills,* State of Ohio Environmental Protection Agency, Air Episode Plan, 361 East Broad Street, Columbus, Ohio 43216.

5. "National Oil and Hazardous Substances Pollution Contingency Plan," *Federal Register,* vol. 45, no. 55, part III, Mar. 19, 1980, p. 17838.

6. *Federal Water Pollution Control Act,* as amended, Pub. L. 92-500, 33 USC 466 et seq.

7. *Resource Conservation and Recovery Act,* Pub. L. 94-580.

8. *Comprehensive Environmental Response, Compensation, and Liability Act of 1980 (Superfund),* Pub. L. 96-510.

9. T. W. Pearson, "Response Capability—The First Man In," in G. F. Bennett (ed.), *Proceedings of the 1978 National Conference on Control of Hazardous Material Spills,* Miami Beach, Fla., April 1978, p. 447.

10. *A Hazardous Material Spill Response System,* University of Dayton Design of Systems Report, April 1975. Copies of this report may be obtained for $4.50 from Prof. Robert L. Mott, chairman of the Department of Mechanical Engineering Technology, University of Dayton, Dayton, Ohio 45469.

11. L. R. Froebe, "The Organization of a Local Environmental Emergency Response Team," in G. F. Bennett (ed.), *Proceedings of the 1976 National Conference on Control of Hazardous Material Spills,* New Orleans, April 1976, p. 156.

12. L. R. Froebe, "Montgomery County, Ohio, Responds to Spills," *Journal of Environmental Health,* vol. 40, no. 4, 1978, p. 184; id., "Environmental Emergency Response," *Ohio Journal of Environmental Health,* December 1976.

13. S. Keck, "Hazardous Materials Part I: The Spreading Threat," *Emergency,* September 1978, p. 42; id., "Hazardous Materials Part II: Meeting the Challenge," *Emergency,* October 1978, p. 36. Additional information may be obtained from Mr. R. C. Watts, radiological defense officer, Louisville and Jefferson County Civil Defense, Room 113, City Hall, Louisville, Ky., 40202.

14. H. N. Morse, "Compensation for 'Servants Only,'" *Fire Command,* December 1977, p. 38.

15. W. B. Katz, "A New Pair of Eyes," in G. F. Bennett (ed.), *Proceedings of the 1976 National Conference on Control of Hazardous Material Spills,* New Orleans, April 1976, p. 1.

16. R. G. Gore, "Hazardous Materials Training: A Present-Day Necessity," *Fire Command,* May 1978, p. 40; R. Yarbrough, Jacksonville Zeroes In on Hazardous Materials Incidents," *Fire Command,* December 1977, p. 22.

17. *Handling Hazardous Materials Transportation Emergencies,* Instructional Course No. SL-29, and others available from the National Fire Protection Association, Publications Sales Department, Batterymarch Park, Quincy, Mass. 02269.

18. *Pesticide Fire Safety,* slide presentation and narrative, 1974; available through regional EPA offices, some state EPA offices, and state fire marshals' offices.

PART 3
European Governmental Plans

A. J. Fairclough
U.K. Department of Transport

Introduction / 7-29
General Background in the United Kingdom / 7-30
Spills in Static Installations / 7-31
Spills during Transport / 7-31
Protection of Water Supplies / 7-38
Arrangements in Several Other European Countries / 7-39
Conclusion / 7-40

INTRODUCTION

In the nature of things a spill must be dealt with where it occurs. The role of governments thus tends to involve guidance, coordination, and backup rather than direct participation in cleanup.

Normally, response to a spill will be almost entirely at the local level (or occasionally at the national level) rather than at the international level. It is true, of course, that if there is a major disaster (e.g., that of Seveso, Italy, and some of the recent major oil spills), there is active cooperation between countries to assist in the cleanup; but such arrangements are usually made ad hoc. Moreover, even within a single country only in relation to the most serious spills of hazardous materials (amounting in effect to national disasters) will national governments be directly involved in the action to deal with spills. In the vast majority of cases action will be taken satisfactorily at the local level by the emergency services, backed up if necessary by whatever arrangements for reinforcement exist.

A spill of a hazardous material may well be significant (quite apart from its immediate threat to life, health, and the environment) for a number of reasons that will involve a variety of government ministries. For example, there is the obvious immediate potential threat to water supplies, special hospital arrangements to deal with injured people may be needed, agriculture or foodstuffs may be affected, and wildlife may be harmed.

In many cases, all these problems can be handled satisfactorily at the local level, provided adequate plans or arrangements exist for coordination of information and action between the emergency services, the local authorities, and the various other interested bodies. But in major spills there is a need for coordination between ministries at the national level, and it is important that arrangements to this end should exist; such arrangements have in fact been made throughout Europe. The United Kingdom position will be considered in detail in this part, and differences in a few other European countries will be indicated at the end of the part.

GENERAL BACKGROUND IN THE UNITED KINGDOM

In the United Kingdom, central responsibility for coordinating government advice to local authorities on dealing with "major accidents and natural disasters" (which include spills of toxic or corrosive substances) rests with the Home Office. The key document controlling response actions is a comprehensive circular, *Major Accidents and Natural Disasters*,[1] addressed in 1975 to all local authorities (and sent also to all chief officers of police, chief fire officers, regional and area health authorities, regional water authorities, and regional and divisional officers of the Ministry of Agriculture, Fisheries, and Food).

This circular addresses, in particular, major accidents and stresses that it is not concerned with the minor accidents with which the three emergency services of police, fire, and ambulance deal every hour of the day. It makes plain that the accidents with which it is concerned are those which by the nature of the hazard or the number and seriousness of the casualties are likely to create problems far beyond what it is reasonable to expect those three emergency services to clear up unaided. In outlining the types of accidents involved, the circular makes clear that they include accidents that may occur during manufacturing processes, in the storage or conveyance of dangerous substances, or in transit by road, rail, or air.

The purpose of the circular is to give guidance on the handling of major peacetime disasters and accidents. It notes that some local authorities have completed their plans to coordinate a rapid response in the event of an emergency and to ensure that responsibility for action is clearly defined and understood. The circular also emphasizes that its object is to encourage all local authorities to bring their contingency planning to a similar state of preparedness. It gives general guidelines, stresses the importance of coordination between all the various services involved, and underlines the need for coordination of action and resources between county and district authorities.

In addressing the subject of those major accidents which involve spills of hazardous materials, the circular focuses in particular on toxic or corrosive substances in the first instance and on escapes of poisonous gases or of substances which give off toxic fumes when involved in a fire in the second. Accidents involving corrosive substances which present no other hazard or substance which are toxic only if ingested will usually be of localized effect, involving only action by the emergency services and not implementation of the local authority's emergency plan. In accidents of the second kind, however, in which toxic gas or fumes could be widely dispersed and people thought to be at risk might have to be moved, coordination between the police, the fire service, and the local authorities would be vital and meteorological or other scientific advice would be needed to assess the extent of the areas at risk; all these are matters which would normally be covered in the emergency plans.

Other points stressed in the circular as being of importance in connection with spills of hazardous materials are:

1. The potentially serious hazard that could be presented if significant quantities of toxic substances (or even of nontoxic pollutants) found their way into water supplies or watercourses (or sewers or drains), since human or animal drinking water could be affected.

2. The special handling of incidents involving radioactive substances [*National Arrangements for Dealing with Incidents Involving Radioactivity (the NAIR Scheme)*].[2]

3. The fact that inland transport accidents involving toxic fumes and gases, the leakage of toxic substances or pollutants, etc., pose substantially the same problems as accidents arising at static installations, subject only to the fact that an accident during conveyance might present the additional difficulty that the risk in question could not reasonably have been foreseen, so that any special countermeasures are not readily available.

4. That although the majority of accidents will be dealt with by the police, fire, and ambulance services without invoking the emergency plans of the local authorities, it is desirable (without affecting the arrangements of the emergency services or the coordination between them) that local authorities should have a comprehensive plan designed to make the best possible use of all available resources in an emergency which cannot be cleared up solely by the three services usually providing the initial response. These plans (which are to cover a very wide range of matters such as accommodation, feeding, and financial provisions) must also cover the handling of spills of hazardous materials which are beyond the resources of the emergency services.

Such is the general emergency planning background in the United Kingdom for the handling of spills of hazardous materials that constitute "major accidents." But as the 1975 circular stresses, most accidents are not of this kind. Therefore, for most spills of hazardous substances (which are going to be dealt with directly by the emergency services) the main requirement, over and above adequate labeling and information on the substances in question, is specific and readily available guidance of a practical character on to how to deal with spills of commonly occurring dangerous substances.

SPILLS IN STATIC INSTALLATIONS

In a static installation, the first responsibility for dealing with a spill, whether of major of minor effect, must rest with the management of the plant in question. Advice setting out recommended procedures for handling major emergencies has been published by the Chemical Industries Association (CIA), based on the experience and practices of a number of chemical and petrochemical manufacturers, augmented by expert advice from the Health and Safety Executive (HSE) and the emergency services.[3]

The CIA publication stresses the statutory duty of management under the Health and Safety at Work Act of 1974 to protect employees and other persons against risks to health and safety arising out of or in connection with the activities of persons at work and spells out in some detail the procedures through which management should go to assess risks, communicate as necessary with the emergency services, and handle emergencies on site. For spills of hazardous materials, the publication stresses that in these circumstances arrangements for people to reach safe assembly points may be necessary; that where gas escapes occur, effective isolation of the source of emission and the dispersion of gas clouds are essential; and that when a large release of toxic vapors is involved, collaboration with the emergency services in downwind areas will be essential, together with meteorological advice.

The publication also mentions the importance of safety audits and lists the statutory duties of the emergency services and of the HSE. In fixed installations, the problem is simplified because the substances involved in an accident are usually known, the hazards are known, and emergency plans normally exist. Nonetheless, there will be cases of spills, especially those that the emergency services must be called in to deal with, in which guidance of a more detailed or technical character might be of assistance.

SPILLS DURING TRANSPORT

The most unpredictable spill situations, and thus the most difficult to deal with, are likely to arise during transport operations. The arrangements for dealing with such spills and the availability of effective practical advice to the emergency services are therefore particularly important. The nature of such advice must clearly vary with the nature of the hazardous material spilled and the circumstances of and mode of transport involved in the spill.

Spills at Sea

In 1976 the report[4] of an interdepartmental study aimed at assessing the risks of oil spills incidents at sea in the early 1980s was published. The report also drew attention to work then in hand on the problems likely to arise from major chemical spillages at sea and stressed that for such substances direct hazards to health could arise. Having concluded that emergency action might well be needed to deal with pollution resulting from accidents involving hazardous materials, the report recommended that consideration be given to coordinating arrangements for dealing with all types of marine emergency.

Subsequently several steps were taken to increase preparedness for dealing with incidents around the coasts of the United Kingdom, notable among which was the establishment of the Marine Pollution Control Unit under the Department of Trade's Marine Division.[5] This unit, although initially concerned with oil spills, is now increasingly involved with spills of hazardous materials. The coordination recommended by the interdepartmental committee is taking placing under the aegis of the Marine Division.

Spills at sea may involve response both at sea and on land (if materials spilled at sea are washed ashore). A spill more than 1 mi (1.6 km) offshore is one of the few spill situations in the United Kingdom in which the central government is directly involved immediately. As soon as a spill that is beyond the capability of the ship to deal with occurs, the Department of Trade is involved. The department's organization and contingency plans are essentially regional, being based on the principal marine officers in the nine marine survey districts covering the United Kingdom; these officers have the duty of preparing contingency plans that include a list of vessels which can be utilized at short notice to assist in dealing with spills.

Hitherto, the department's organization has focused on improving its capability for dealing with oil spills (by far the most common type of spill around the United Kingdom's coasts), but it is clear that it will increasingly need to be able to respond also to spills at sea of hazardous materials, and this is clearly a more complex task than is dealing with oil pollution.

There will doubtless arise many incidents in which the hazardous materials concerned have dispersed before emergency action can effectively be taken. In such cases, there is no option but to rely upon the vast dilution potential of the sea to render the materials harmless. There are, however, two sets of circumstances in which emergency action may be feasible: incidents involving bulk shipments of hazardous materials and incidents involving packages. In the latter case, the obvious course is to recover the packages, but this may only rarely be possible. In extreme cases (of which the *Cavtat,* which sank off Italy in 1970 with a substantial cargo of tetraethyl lead in drums, was an example)[6] it may be necessary to mount special recovery operations; in other cases packaged material may in due course come ashore.

For bulk shipments of chemicals and gases which in the event of an accident are more likely to pose serious hazards to human and environmental well-being, the Marine Division of the Department of Trade has recently developed a new scheme called Spillages Emergency Action for Chemicals (SEACHEM) to provide guidance in incidents at sea.[7]

Data sheets have been prepared for all substances contained in Chap. 6 of the Intergovernmental Maritime Consultative Organization (IMCO) Bulk Chemicals Code and Chap. 19 of the IMCO Gas Carrier Code.[8,9,10] The sheets include advice on both pollution and remedial action and hazard follow-up action appropriate for the substance in question. An example of a SEACHEM data sheet is given in Fig. 3-1, while Fig. 3-2 is a

ISOBUTYRALDEHYDE

Extremely high risk of violent explosion even at a distance from the ship

Threat to human life even at a distance from the ship

UN No: 2045

IMCO Class: 3

May be carried in ship types:	1, 2 or 3
Characteristics:	Liquid soluble volatile
Threat to life by fire or explosion:	On or near ship and at a distance
Threat to health:	On or near ship and at a distance

RESCUE SERVICES PRECAUTIONS AND REMEDIAL ACTION:

(a) Inhalation:	BA2:	Dangerous. Use self-contained breathing apparatus or trolley unit. Take anyone affected by fumes to hospital.
(b) Contact with skin:	C2:	Dangerous. Wear protective clothing, gloves and goggles and take reasonable care. Remove contaminated clothing and wash skin immediately.
Pollution rating:	Minor	
Pollution threat to marine life:	Minor	
Pollution threat to beaches:	Minor	
Pollution remedial action:	Nil	
Hazard follow-up action:	J:	Use water sprays to reduce local gas concentrations from volatile products.
	K:	Deliberate ignition of flammable vapour clouds or floating flammable liquids where this is deemed the safest course at a distance from the shore.
	M:	Loud hailing warnings and local radio broadcasts ahead of vapour clouds warning populations to take cover in closed rooms.
Dispersants:	Nil	

Fig. 3-1   Sample of SEACHEM data sheet.

Working Manual for Rail Staff

BR 30054/3
Apr 69

F. Fires and Accidents Involving Dangerous Goods

6. Inflammable Solids or Substances Liable to Spontaneous Combustion (Class 4(b))

FIRE

F6/1.   Water may be used to control fires involving these substances unless the packages bear the blue DANGEROUS WHEN WET label as well as the wagon label.

F6/2.   As the fire may break out again, the goods must be watched until instructions have been received for their disposal.

SPECIAL INSTRUCTIONS

(i)   Organo-metallic compounds

F6/3.   These substances—for example, alkyls of aluminium, lithium or zinc—although igniting spontaneously when exposed to air can also react violently with water. Such packages therefore bear the blue DANGEROUS WHEN WET label as well as the wagon label. However, the standard of packing is so high that it is unlikely that any of the contents would escape and cause a fire.

F6/4.   A fire in which intact packages are involved may be dealt with by water but if it is suspected that any of the contents have escaped, dry sand or dry fire-extinguishing powder must be used.

(ii)   Phosphorus (yellow or white)

F6/5.   This material will ignite spontaneously if exposed to air.

F6/6.   When conveyed in large quantities it is carried in water in tank vehicles which bear warning plates indicating the action to be taken in the event of accident or leakage.

F6/7.   If any receptacles become involved in a fire, water must be applied liberally to them until they can be moved to a safe place.

F6/8.   In dealing with phosphorus fires, water should be applied from the windward side because burning phosphorus gives off poisonous fumes.

F6/9.   If leakage occurs as a result of an accident, the leaking phosphorus will certainly take fire unless kept wet with water.

F6/10.   If phosphorus should come into contact with the skin, the affected area should be immediately flushed with water and kept wet until medical attention can be obtained.

Fig. 3-2 Sample of British Rail instructions to staff on action in the event of fires and accidents involving dangerouus goods.

copy of the evaluation sheet for incidents which is to be used under SEACHEM (and which details the various categories of action covered in the data sheets).

Hazardous Materials Washed Ashore

Advice has been given to local authorities in the United Kingdom on the preparation of appropriate plans to deal with hazardous materials that may be washed ashore following incidents at sea. In 1968, following the 1967 *Torrey Canyon* oil supertanker disaster in the English Channel, local authorities were advised[11] that they should prepare at the county level schemes of organization which would allocate executive responsibility for action against oil pollution on every part of the coast and make arrangements for mutual-aid reinforcement. The object was to bring into existence as soon as possible coastal organizations capable of dealing effectively both with routine minor pollution and with much larger-scale incidents. The circular also gave advice on the content of the schemes of organization as well as on such matters as early-warning arrangements, equipment, liaison with other organizations, and protection of water supplies.

In 1974, in another circular[12] dealing with emergencies arising from chemicals and other substances washed ashore, local authorities were advised to extend their oil pollution schemes to cover hazardous materials also. The circular gave further details regarding arrangements for early warning and for obtaining specialist advice (particular reference was made to a scheme called Chemsafe, operated by the CIA,[13] and to the availability of advice and assistance from the National Chemical Emergency Centre established at the Atomic Energy Research Establishment at Harwell[14]). There were a request for designated officers and outlines of schemes to be sent to the Department of the Environment.

The circular also stressed the importance of local authority emergency plans providing for the worst that could happen. The greatest risks identified were the large-scale release of dangerous gases or high-vapor-pressure liquids from bulk carriers near shore or in port (especially during loading and unloading) and the risk of collisions or strandings leading to the release of hazardous cargoes in enclosed waters such as estuaries. Evacuation plans, coordinated with police and port authorities, were to be prepared for such eventualities.

All coastal local authorities have prepared and deposited with the Department of the Environment contingency plans for dealing with the onshore effects of spills at sea. Such plans are closely coordinated with the Department of Trade's at-sea plans, and each contains a section specifically dealing with spills of hazardous chemicals. Also important (since many hazardous materials that are carried by sea are in drums or other packages that may be washed ashore either intact or damaged and leaking) will be ongoing work in the United Kingdom on durable labeling that will survive prolonged immersion in seawater.

Rail

The *Working Manual for Rail Staff* published by the British Railways Board[15] contains detailed instructions, with each of the categories of dangerous goods with which the *Manual* deals referenced separately, on the action to be taken in the event of fires and accidents involving dangerous goods; an example is shown in Fig. 3-3. The *Manual* also contains general instructions to railway staff for dealing with accidents (including spills and leakages). These require the prevention of injury to persons and, if possible, the isolation of the tank cars involved and immediate notification to the nearest signal box (which in turn will pass on the information via local and area offices to the emergency services) of a "rail dangerous goods emergency" with full details of the incident, including the emergency code (United Nations number and letters[16]) of the vehicles involved.

Road

For spills occurring as a result of road transport, there are two main sources of practical advice. The first is a Home Office handbook[17] with information on how to deal with fires and spills of dangerous substances; this book, orginally published in 1972, is being supplemented. To date, sections dealing with flammable liquids and corrosive substances have been published. The introduction to the document makes clear that it concerns dan-

EVALUATION SHEET

EXTREMELY HIGH RISK OF VIOLENT EXPLOSION EVEN AT A DISTANCE
   FROM THE SHIP
VAPOUR CAN BE IGNITED AND CAUSE FIRE OR EXPLOSION
THREAT TO HUMAN LIFE EVEN AT A DISTANCE FROM THE SHIP
THREAT TO HUMAN LIFE ON OR NEAR THE SHIP

RESCUE SERVICES PRECAUTIONS AND REMEDIAL ACTION

BA1 - ☐   Very Dangerous—Preferably use positive pressure self-contained breathing apparatus or trolley unit. Take anyone exposed to vapour to hospital even if no symptoms show.

BA2 - ☐   Dangerous—Use self-contained breathing apparatus or trolley unit. Take anyone affected by fumes to hospital.

BA3 - ☐   Moderately Dangerous—Avoid Breathing of Fumes.

C1* - ☐   Very Dangerous—Wear full protective PVC suit, gloves and boots. Remove any contaminated clothing immediately and wash exposed skin with soap and water. Take exposed person to hospital.

C2* - ☐   Dangerous—Wear protective clothing, gloves and goggles and take reasonable care. Remove contaminated clothing and wash skin immediately.

C3* - ☐   Wear gloves and goggles. Remove soaked clothes and wash splashed skin.

   * - ☐   Splashes to eyes should be washed with copious quantities of water.

F   - ☐   Danger of frostbite from contact with liquid or associated metal surfaces.

POLLUTION DATA

I   - ☐   Likely to give rise to major pollution hazard.
II   - ☐   Likely to give rise to a pollution hazard.
5   - ☐   Bioaccumulative.
6   - ☐   Moderately or highly toxic.
7   - ☐   Objectionable because of persistency, smell or poisonous or irritant characteristics—interference with use of beaches.

POLLUTION REMEDIAL ACTION

A   - ☐   Disposal and dilution by mechanical action.
B   - ☐   Containment of floating substances by booms.
C   - ☐   Dispersal by chemicals.
D   - ☐   Use of absorbents.
E   - ☐   Containment by surface chemicals.
F   - ☐   Prohibition on fishing by amateur fishermen. Contact MAFF [Ministry of Agriculture, Fisheries, and Food] to advise professional fishermen.
G   - ☐   Recovery of sunken vessel or pump from undamaged tanks of sunken vessel.
H   - ☐   Pump or dredge from seabed.
I   - ☐   In the case of spillage the chemical may be recoverable from the surface in solid form if sea and weather conditions so allow.

Fig. 3-3   Evaluation sheet for dangerous goods. [*"Handling and Conveyance of Dangerous Goods,"* Working Manual for Rail Staff, *British Railways Board, BR 30054, 1978.*]

HAZARD FOLLOW UP ACTION

J   - ☐   Use water sprays to reduce local gas concentrations from volatile products.
K   - ☐   Deliberate ignition of flammable vapour clouds or floating flammable liquids where this is deemed the safest course at a distance from the shore.
L   - ☐   Track visible vapour cloud or plume by boat or helicopter.
M   - ☐   Loudhailer warnings and local radio broadcasts ahead of vapour clouds warning populations to take cover in closed rooms.
N   - ☐   Prohibit bathing and use of beaches.

DISPERSANTS

A   - ☐   Hydrocarbon solvents.
B   - ☐   Concentrate dispersants.

Fig. 3-3   (*Continued*)

gerous substances that are conveyed by road and that the document's purpose is to present, in ready reference form for use in an emergency, information about special hazards presented by dangerous substances that are conveyed by road in commercial quantities, together with guidance on the personnel protection required and the fire-extinguishing media to be used.

Further guidance to response to spills of hazardous chemicals on the highways, in this instance addressed to local authorities, is contained in a Department of Transport circular of 1978[18] which draws together information on the requirements of the 1978 *Packaging and Labelling of Dangerous Substances Regulations*[19] and the 1978 *Hazardous Substances (Labeling of Road Tankers) Regulations*,[20] the Chemsafe scheme, and the NAIR scheme for incidents involving radioactive substances.

The circular makes three other points of importance:

1. The highway authority is likely to be involved in the removal of dangerous substances from the highway, once the immediate emergency is over; and, when necessary, technical advice regarding appropriate methods should be sought from either the company concerned or the National Chemical Emergency Centre. The circular also notes the names of companies (mainly members of the National Association of Waste Disposal Contractors) which are prepared to cooperate in the removal of spilled material from highways.

2. Washing away spilled chemicals may endanger water supplies or could cause water pollution. Therefore, before spilled chemicals are washed away, the advice of the water authority should be sought unless the delay involved is unacceptable on safety grounds. In any event, the water authority should be informed whenever spilled substances find their way into sewers, drains, or watercourses.

3. The Road Haulage Assocation has introduced a scheme whereby certain member companies may supply tank trucks and ancillary equipment for the transfer of a dangerous load from a damaged tank truck. The circular lists the companies participating in the scheme. Details of the scheme itself have been published separately by the Road Haulage Assocation.[21]

Air

The majority of spills of hazardous materials being carried by air occur either on or close to airports. The most probable causes are incidents during storage and handling at the airports, accidents during movement on airports, or air crashes during takeoff or landing.

In such cases, the airport authorities clearly have the initial responsibility, and they are required (under Civil Aviation Authority publication CAP 168,[22] which lays down criteria for airport licensing) to prepare plans for dealing with emergencies. In that sense, they are in much the same position as any other industrial installation. Spills resulting from accidents occurring far beyond the confines of airports would of course be dealt with by the emergency services in the locality in which they occur, under normal emergency plans.

The instructions[23] issued by airport authorities make it clear that the airport fire service would not normally be involved in dealing with spills of hazardous materials. For nonradioactive hazardous materials, the airport fire service will send a fire unit to the scene of the spill for such immediate action as is necessary, pending removal of the hazard, but the principal action is to alert the emergency services, which then deal with the incident. For incidents involving radioactivity, the action is to activate the NAIR scheme by informing the police.

Pipelines

The safety record of pipelines in Europe is good, and spills resulting from their failure are accordingly rare.[24] Although most pipelines carry oil, some hazardous materials other than oil are transported. Most European countries have legislation relating to the construction and operation of pipelines [a list that is relevant to *oil* pipelines, some of which relates also to other substances, is contained in a recent report[25] of the Oil Companies' International Study Group for Conservation of Clean Air and Water in Europe (CONCAWE)].

In the United Kingdom, the basic legislation is the Pipelines Act of 1962,[26] under which the government authorizes cross-country pipelines and is notified of local pipelines; under this act requirements relating to the safety of pipelines can be laid down, and provision is also made for accident notification. The act requires the owner of a pipeline, in the event of the accidental escape or the emission of anything in the line, to notify fire, police, and water authorities and to give them information to enable them efficiently to discharge their duties.

The other important provision of the act requires the responsible minister in exercising any of these powers under the act to have constant regard to the need to protect water against pollution. Technical bulletins are issued periodically by the Home Office and in appropriate cases will refer to pipeline incidents. For example, the bulletin[27] dealing with incidents involving bulk quantities of liquefied petroleum gas contains advice on the action to be taken in the event of pipeline leakage; emphasis is placed on pipeline operators being informed of incidents, since one of the most immediate actions that can be taken to limit the effects of any spill is to stop the flow by closing down the isolation valves on either side of the site of the incident.

PROTECTION OF WATER SUPPLIES

One of the most immediate and potentially troublesome consequences of large spills of hazardous materials is their possible impact on water supplies. This impact could be virtually direct (a spill into a watercourse not far above a water supply intake) or indirect (via groundwater or through disturbance of sewage treatment processes, leading to changes in the effluent from sewage works discharged to rivers).

It is an ever-present threat that a spill could cause poisonous substances to enter the water supply before action to control the situation could be taken, with serious consequences for human life and health on a potentially large scale. The importance of informing water authorities at the earliest possible moment of spills that take place and of making sure that emergency authorities dealing with spills are aware of the potential consequences for water quality and water supply, cannot be overemphasized. Emergency arrangements in the United Kingdom take full account of these considerations.

Nonetheless, spills will occur, and water supplies will be affected. The water industry therefore needs arrangements to deal with this situation. In the United Kingdom, response plans have in fact been thoroughly reviewed in recent years, following a number of serious incidents. A working party under the National Water Council has reviewed emergency procedures and reported with recommendations affecting the whole industry.[28] Most of these recommendations are concerned with internal arrangements, liaison, and joint planning with the emergency services, the need for familiarity with hazard information systems and for joint exercises, etc. Perhaps the most important of the recommendations are those which spell out clearly the need for all water authorities to maintain a state of preparedness so that they can respond appropriately in an emergency situation (with a communication center staffed on a 24-h basis to act as a focal point) and to maintain comprehensive manuals of emergency procedures. With arrangements of this kind in force, the risk arising from spills of hazardous materials can be reduced.

In addition to stressing the need to improve communication systems and liaison between the water authorities and the fire and police services, the water industry in the United Kingdom has argued that the basic policy in dealing with any spill of a hazardous substance should be to "contain and seek advice." This approach allows time for a water authority to consider whether, and if so how, the river and/or sewage works concerned can accept the spill if it is washed away; time is also gained to close off water supply intakes, if necessary, until the resultant pollution has passed. Moreover, if the substance concerned is one that cannot safely be washed away, such an approach avoids premature commitment to such a course of action.

These points only need be added:

1. The greatest risk to natural waters from spills of hazardous materials arises from transport by road.

2. All the regional water authorities in the United Kingdom now have regional emergency procedures manuals reflecting the recommendations of the report by the working party.

3. Arrangements have been made for all spills affecting water supplies to be reported and analyzed.[29]

ARRANGEMENTS IN SEVERAL OTHER EUROPEAN COUNTRIES

In other European countries, too, immediate responsibility for dealing with spills of hazardous materials is local.

Netherlands

In the Netherlands, each local fire brigade has a book describing how to recognize and deal with 800 dangerous chemicals. If in doubt or unable to recognize the substance involved in a spill, the emergency authorities can telephone a central alarm number in the Inspectorate of Fire Services in the Ministry of Internal Affairs.

The Dutch authorities say that they have never failed to recognize the chemical involved in a spill. The fact that chemicals which are a danger to the environment are sold only to companies licensed by the Ministry of Health and the Environment may be of assistance, since knowledge of the company involved in a spill may give some indication of the chemical involved.

Federal Republic of Germany

In the Federal Republic of Germany, chemical spills are dealt with either by squads belonging to the Catastrophe Control Service or by squads from the Civil Defense Service, depending upon the nature of the chemical concerned. The Catastrophe Control Service is operated by the states; its principal task is to protect the population against hazards caused by peacetime catastrophes. Each state operates its own catastrophe control unit (CCU) and has its own catastrophe control legislation. The responsibility for deploying the CCUs in the event of a spill (or of any other kind of incident) rests with the administrative districts within each state.

The Catastrophe Control Service is staffed by unpaid volunteers, mainly from local voluntary and professional fire brigades. Its equipment is provided by a publicly owned technical aid organization; employers are obliged by law to give CCU members time off for necessary training and actual catastrophe control work (they are reimbursed by the state governments). Most of the states operate special training schools financed by the federal government.

The involvement of the Civil Defense Service is of a backup character. In an actual chemical spill, the first responders on the scene would be the local CCU, which would either deal with the spill or report to the head of the local administrative district if it did not have the necessary equipment or expertise; in the latter case a squad from the local civil defense unit would be called in.

France

In France, too, initial responsibility for dealing with spills of hazardous materials is at the local level, responsibility being placed on the mayor of the commune (local authority) in which the spill occurs; it is the mayor's responsibility to decide how the spill should be treated initially. If in doubt, the mayor can contact the directeur départmentale de la protection civile (DDPC), who is the representative at the departmental level of the Ministry of the Interior.

If a spill is very serious, further levels in the central government machine can be involved: first, the prefect of the department, who in turn may contact the Ministry of the Interior in Paris. The ministry is the advisory body to which the local authorities may turn for guidance on how to deal with spills, and it regularly publishes material on how to cope with the most common spills.

CONCLUSION

In dealing with spills, action must inevitably take place on the spot—at the local level. The role of governments therefore is to encourage, through advice, guidance, instructions, and assistance, the organization at the local level of efficient and effective arrangements to enable spills and other accident and emergency situations to be satisfactorily dealt with and itself to ensure suitable backup and coordination arrangements so that the small proportion of cases that require a wider than local response are also properly catered for.

REFERENCES

1. *Major Accidents and Natural Disasters,* Circular No. ES 7/1975, Home Office, London, 1975.

2. *National Arrangements for Dealing with Incidents Involving Radioactivity (the NAIR Scheme),* Circular No. ES 7/1972, Home Office, London, 1972; Circular No. ES 3/1977, Home Office, London, 1977.

3. *Recommended Procedures for Handling Major Emergencies,* Chemical Industries Association, Ltd., 2d ed., reprinted, London, 1977.

4. Department of the Environment, Central Unit on Environmental Pollution, *Accidental Oil Pollution of the Sea: A Report by Officials on Oil Spills and Clean Up Measures,* Pollution Paper No. 8, H. M. Stationery Office, London, 1976.

5. *Improved Arrangements to Combat Pollution at Sea,* Department of Trade, London, 1979.

6. "Distant Drums," *Nature,* vol. 267, May 26, 1977, p. 301.

7. *SEACHEM—Spillages Emergency Action for Chemicals,* Department of Trade, London, 1980.

8. *Code for the Construction and Equipment of Ships Carrying Dangerous Chemicals in Bulk,* Intergovernmental Maritime Consultative Organization, London, 1977; Supplement No. 1.

9. *Code for the Construction and Equipment of Ships Carrying Liquified Gases in Bulk,* Intergovernmental Maritime Consultative Organization, London, 1976.

10. *Code for Existing Ships Carrying Liquified Gases in Bulk,* Intergovernmental Maritime Consultative Organization, London, 1976.

11. Ministry of Housing and Local Government, *Oil Pollution of Beaches,* Circular 34/68, H. M. Stationery Office, London, 1968.

12. Department of the Environment, *Emergencies Arising from Chemicals and Other Substances Washed Ashore,* Circular 123/74, H. M. Stationery Office, Londo, 1974.

13. *Chemsafe: A Manual of the Chemical Industry Scheme for Assistance in Freight Emergencies,* Chemical Industries Association, Ltd., 2d ed., London, 1976.

14. R. F. Cumberland, *The Harwell Chemical Emergency Centre,* AERE-R8079, U.K. Atomic Energy Authority, Harwell, England, 1975.

15. "Handling and Conveyance of Dangerous Goods," pink pages, *Working Manual for Rail Staff,* BR 30054, British Railways Board, 1978.

16. *Transport of Dangerous Goods: Recommendations Prepared by the Committee of Experts on the Transport of Dangerous Goods,* United Nations Publication GE.77-27512 (7351), rev. ed., United Nations, New York, 1977.

17. Home Office, *Dangerous Substances: Guidance on Dealing with Fires and Spillages,* H. M. Stationery Office, London, 1972.

18. *Spillages of Hazardous Chemicals on the Highway,* Circular Roads 27/78, Department of Transport, London, 1978.

19. *Packaging and Labelling of Dangerous Substances Regulations, 1978,* ST 1978 No. 209, H. M. Stationery Office, London, reprinted 1979.

20. *Hazardous Substances (Labelling of Road Tankers) Regulations, 1978,* ST No. 1702 1978, H. M. Stationery Office, London, 1978.

21. *Hazardous Liquids Emergency Load Transfer Scheme,* Road Haulage Association, Ltd., Tanker Functional Group, undated.

22. *Licensing of Aerodromes,* CAP 168, rev. ed., Civil Aviation Authority, London, 1979.

23. *Director's Operational Instructions, Heathrow Airport—London—Handling of Damaged Consignments Containing Hazardous Materials,* DOI/25/76, British Airports Authority, 1976; *Director's Operational Instructions, Heathrow Airport—London—Handling of Damaged Consignments Containing Radioactive Materials,* DOI/22/75, British Airports Authority, 1975.

24. *Spillages from Oil Industry Cross-Country Pipelines in Western Europe: Statistical Summary of Reported Incidents, 1975,* Report No. 7/76, CONCAWE, The Hague, 1976.

25. *Published Regulatory Guidelines of Environmental Concern to the Oil Industry in Western Europe,* Report No. 1/79, CONCAWE, The Hague, 1979.

26. *Pipelines Act, 1962,* 10 and 11 Eliz. 2 CH 58, H. M. Stationery Office, London, 1962.

27. *Fires and Other Occurrences Involving Bulk Quantities of Liquified Petroleum Gas,* Bulletin No. 1/77, Home Office, Fire Department, 1977.

28. *Emergency Procedures on Pollution of Inland Waters and Estuaries,* National Water Council, London.

29. Water Data Unit (A. V. Sheckley, N. C. Taylor, A. M. Kemp, and R. P. Donachie), *The Spillages System,* draft technical memorandum.

PART 4
United States Industrial Plans

Randolph A. Jensen, P.E.
President, Jensen Consultants

Definition of a Hazardous Spill / 7-43
Potential Spill Survey / 7-44
Spill Containment / 7-44
Planning for Spill Response / 7-47
Response to a Spill Event / 7-48

Spill response plans should be thought of as damage control methods. They are designed to prevent damage to the environment and reduce material loss. Since one-half to three-fourths of all spills are caused by human error, a program of education in spill control awareness can be an effective tool in spill prevention.

Commitment on the part of management is an important first step in achieving awareness on the part of all operating personnel. Long before a spill response plan can be put into action in an industrial plant, work must be done to define the problems and correct obvious deficiencies in the system. Materials that have serious spill potentials must be identified, and containment devices must be designed and constructed. Then a spill response plan can be devised to deal with a spill emergency when it occurs.

DEFINITION OF A HAZARDOUS SPILL

The question of what constitutes a spill that is hazardous was initially addressed by the U.S. Environmental Protection Agency (U.S. EPA) in the publication of the four-volume set *Determination of Harmful Quantities and Rates of Penalty for Hazardous Substances* in January 1975.[1] Four different methodologies were used to take into account such factors as aquatic toxicity, bioaccumulation, hazards to human health by oral intake or skin contact, and damage to the scenic or recreational environment. The degree of difficulty in removing or ameliorating a spill of material into a watercourse was brought into the evaluation by assigning values to significant physical and chemical properties, such as degradability and whether the material mixes with water or floats or sinks. Categories A through D were based on aquatic toxicity, which seemed to be the most severe constraint (Table 4-1).

Two of the methodologies defined a "critical volume" as that volume which when contaminated to the critical concentration resulted in significant damage. One of these methodologies started with an estimate of a static volume, while the other started with an estimate of a flowing condition. Since these two methodologies produced values that were, in general, neither extremely high nor extremely low, it seems reasonable to average the results to arrive at typical spill sizes that should cause concern.

TABLE 4-1 Hazardous Categories

Category	Assigned critical concentration (mg/L)	Aquatic toxicity, 96-h TL_M (mg/L)
A	0.5	less than 1.0
B	5.5	1.0–10
C	55	10–100
D	300	100–1000

Both methodologies recognized the difference in the size and type of the receiving water. A typical lake was considered to be about 45,600 m³ (37 acre-ft), and a typical small river to have a flow rate of 1 m³/s (36 ft³/s). A typical river capable of handling barge traffic was considered to have an average flow rate of 175 m³/s (6150 ft³/s). Estuaries were considered to have a critical volume of 48,000 m³ (39 acre-ft) in one method and 1,380,000 m³ (1120 acre-ft) in the other for an average of 714,000 m³ (580 acre-ft). Coastal waters were considered to have a critical volume of 1,380,000 to 3,350,000 m³ (1120 to 2717 acre-ft) for an average of 2,370,000 m³ (1918 acre-ft). The average critical spill quantity that is the product of the critical concentration and the critical volume is shown in Table 4-2.

POTENTIAL SPILL SURVEY

A list should be prepared of all materials within the plant that could possibly reach a watercourse. The typical shipment amount might well represent the size of a possible spill, but large amounts in bulk storage also should be noted. The hazard category should be determined by using Table 4-1 after referring to published data on the aquatic toxicity for the type of fish most likely to be representative of those in nearby waters. The previously mentioned EPA publications yield the data shown in Table 4-3 for several chemicals.

A typical spill candidate list is shown in Table 4-4, using the same chemicals as above. It is evident that a considerable potential for hazardous spills exists.

To respond properly to a spill emergency, one should know whether the material will sink or float or dissolve in water and whether it is a serious toxicant, very flammable or explosive, or a serious hazard to the public. These properties are summarized in Table 4-5 for the spill candidates. Such a summary is helpful in planning for a spill emergency. For example, all the materials listed in Table 4-5 will float on water and are flammable hazards. The first three, being monomers, must be protected against heat to avoid polymerization under confined conditions. However, since polymerization under nonconfining conditions is not usually hazardous, it may well be the most feasible way of decontaminating the spill area. Danger to the public would be slight for three of the four candidates, but if acrylonitrile is involved, the downwind area should be evacuated. Further, if acrylonitrile reaches public watercourses, livestock and humans must be prevented from using the water until it is proved safe.

SPILL CONTAINMENT

The containment capability of the entire manufacturing operation must be assessed with the view of keeping contamination out of storm sewers. Can curbs and drains be provided around equipment and pumps which handle material that may be hazardous if leaked to a watercourse? Can dikes be provided around storage tanks to confine the full tank con-

TABLE 4-2 Critical Spill Quantities (kg) of the Different Categories of Hazardous Materials as Originally Contemplated by the Battelle Reports

Category	Small rivers and lakes	Large rivers	Estuaries	Coastal waters
A	22.7	950	360	1,040
B	270	10,450	3,900	11,400
C	2,730	104,500	39,000	114,000
D	13,600	591,000	214,000	636,000

TABLE 4-3 Aquatic Toxicities and Hazard Categories

Material	96-h TL_M (mg/L)	Fish type	Hazard category
Methyl methacrylate	250	Bluegill	D
Acrylonitrile	12	Bluegill	C or B
Styrene	22	Bluegill	C
Toluene	24	Bluegill	C
Sodium dodecyl benzene sulfonate (50%)	19	Bluegill	C

TABLE 4-4 Typical Spill Candidates and Quantities within a Chemical Plant

Material	Typical container	Typical storage amount	Possible spill amount	Hazard category	Critical quantity (kg)*
Methyl methacrylate	76-m³ tank car	7,600 m³	76 m³	D	13,600
Acrylonitrile	15-m³ tank truck	None	15 m³	C	2,730
Styrene	15-m³ tank truck	76 m³	15 m³	C	2,730
Toluene	950-m³ barge	7,600 m³	38 m³	C	2,730
Sodium dodecyl benzene sulfonate (50%)	55-gal (0.2-m³) drum	100 drums	200 kg	C	2,730

*Amount is considered harmful if spilled to a lake or a river with a flow of at least 1 m³/s (36 ft³/s).

TABLE 4-5 Physical and Chemical Properties of Spill Candidates

	Methyl methacrylate	Acrylonitrile	Styrene	Toluene
Solubility in water	Moderate	Moderate	Very slight	Very slight
Density (water = 1)	0.94	0.82	0.91	0.87
Vapor toxicity	Moderate	High	Moderate	Moderate
Flammability	High	High	High	Extreme
Explosive?	If confined uner heat	If confined under heat	If confined under heat	No
Water-reactive?	No	No	No	No
Special precautions?	Polymerizes with heat	Polymerizes with heat	Polymerizes with heat	None
Danger to public?	Slight	Toxic by all routes	Slight	Slight
Cleanup equipment	Protective clothing, goggles	Self-contained breathing apparatus and impervious clothing	Protective clothing, goggles	Protective clothing, goggles

tents in the event of overfilling or a catastrophic tank failure? Can loading and unloading areas for materials considered hazardous be diked to trap loss by overfilling or equipment failure? Can drainage from production, storage, and loading areas be separated from storm drainage? Can a diversion unit be provided in the drainage system for trapping spills?

Unlikely combinations of events should be considered. For example, a storage tank may rupture during a heavy rainstorm so that released material would be immediately flushed to storm sewers. A chemical stored at an elevated temperature may not be considered to have spill potential if it is a solid at atmospheric temperature. Therefore, the storage tank on first consideration would not appear to require diking for spill control. However, if the material is appreciably soluble in water, it could escape in the rainwater runoff and create a problem before recovery of the cooled and solidified material could be completed.

Good spill control practice would dictate that tank car facilities for methyl methacrylate (see Table 4-4) should be provided with drainage to the chemical sewer or to a secure area within the plant, since the critical quantity, although relatively large, is still less than the possible spill amount. Similarly, tank truck facilities for both acrylonitrile and styrene should be provided with secure drainage within the plant, since the critical quantity of each is less than the possible spill amount. Furthermore, the styrene storage tank should be diked with sufficient containment volume to hold the contents of the tank in the event of catastrophic failure.

Toluene being received by barge is considered to be secure from spillage on the waterfront, but the in-plant storage tank must be provided with a leak-free dike that will be able to contain the entire contents of the tank. The possible spillage of one drum of the last chemical in Table 4-4 does not pose much of a hazard since the critical quantity is equivalent to 5450 kg (12,000 lb) of a 50% mixture.

A study by the U.S. EPA[2] over a 2-year period showed that more than 40 percent of hazardous material spills occurred in bulk-storage areas. Therefore, serious consideration should be given to the design of dikes around storage tanks. The diked area should be large enough to hold the entire contents of the storage tank and strong enough to withstand the force generated by a sudden collapse of the tank. Moreover, the dike wall should be designed to retain liquid that might spurt from punctures of the tank above the top elevation of the wall. It should be obvious that the dike should be free of cracks and holes and sealed with a material that is impervious to the material in storage.

Dikes built with lines to drain rainwater are not recommended since the valves could leak or the lines could be left open at a critical time, allowing the trapped material to escape. A better design provides drainage inside the dike to a sump in one corner. Here a pump suction line can be installed over the top of the dike to pump out the unwanted material after it has been established that it is safe to do so. If the material in the storage tank floats on water, fire fighters may protest that drains are needed to remove fire control water so that the floating material will not be displaced from the dike. However, removal of the lower layer via a suction line permanently installed over the top of the dike is preferable to use of an underground drain since the former eliminates the possibility of an accidental discharge from the dike.

A chemical sewer can handle minor day-to-day leaks into the normal wastewater flow. Spills, however, can create havoc in the downstream treatment system. Thus, a diversion system should be provided with a rapid response mechanism to confine the spill until it can be handled in a safe and acceptable way. It has been suggested that a minimum diversion tank or pond should hold 20 to 60 min of average flow. Of course, this means that retained material must be handled promptly to restore the diversion unit to its normally empty condition.

It may be advantageous to divide the plant into zones for specific control procedures, particularly if similar chemicals can be grouped in the same zone. For example, if the chemicals in one area are primarily inorganic and would normally require only pH adjustment if a spill occurred, it may be logical to keep the runoff from this area separate from that of the rest of the plant.

PLANNING FOR SPILL RESPONSE

A plot plan of the plant area should be available for ready reference. On this plan, the location of all process areas and storage areas should be shown, as well as the process-sewer layout. In addition, the location of critical catch basins leading to the storm sewer should be indicated. If storm drainage is only on the surface, these ditches should be shown with the direction of flow indicated. If storm drainage is underground, the location of these lines and all intervening manholes should be shown. The manholes should be labeled so that response can be made to a specific point if necessary to intercept a spill on emergency notice.

If surface drainage or the underground storm sewer from the plant area goes to a moderate or small stream, it would be desirable to have a topographic map of the area. Such a map is available from the U.S. Department of the Interior as the U.S. Geological Survey 7.5-min series (topographic) map. This map series is on a scale of 1:24,000, showing contour intervals of 10 ft (3 m) with most 5-ft (1.5-m) contours also shown by dotted lines. Study of the topographic map should indicate the likely path of a spill toward the receiving stream as well as possible sites for impoundment of the spilled material. Location of access roads for heavy equipment, bridges for possible survey sites, and the nature of the terrain gradients, whether gradual or severe, should be noted. Also of interest would be other drainage inputs to the receiving stream as well as likely water withdrawal points for either human or animal use.

Particularly when considering necessary response actions outside the plant area, the availability of heavy equipment such as bulldozers, backhoes, front-end loaders, dump trucks, tank trucks, vacuum trucks, cranes, winches, etc., from local contractors should be considered. Of particular concern should be probable cost, capabilities of the operators, and availability on short notice. Along with these considerations, various suppliers should be listed for the emergency supply of such materials as activated carbon, soda ash, peat moss, granulated limestone, laundry bleach, solvents such as acetone and alcohol, and sand for sandbags. Sand might be overlooked as a necessary material, but not after having faced the need for a sandbag dike in an area in which all the soil is fully frozen.

For rapid spill response a unit similar in design and concept to a fire truck is very desirable. This unit could be fully self-contained like a fire truck, or it could be a well-equipped trailer ready for quick hookup to any available plant vehicle. It should be supplied with full protective gear for at least two operators as well as several self-contained breathing units capable of being recharged on site. Thus, the unit should be equipped with its own engine-driven air compressor and sufficient air supply hose to supply a manifold while the unit itself is kept at a safe distance from the emergency.

The air supply also should be available to power diaphragm pumps, which can be operated in areas that would be hazardous for engine-driven pumps. At least one portable engine-driven pump should be included for use in nonhazardous and extremely remote locations. Of course, enough hose of suitable size and construction should be available to use these pumps effectively. Supplies such as several bags of oil absorbent, empty sandbags, tarpaulins, and tools such as shovels, rakes and brooms also should be included.

When the high cost and hopefully infrequent use of special-purpose equipment for spill control are considered, the advantages of joining or establishing a mutual-aid association become evident. An association might be established by a group of similar industries in an area or by an area group of diverse businesses. In some cases, joint operation of private and public units is feasible. This would be of particular interest in areas where receiving waters are extensive and requirements for protective gear are expensive. For example, if it is necessary to deploy floating booms to confine a floating spill for recovery, the investment in suitable motorboats and boom materials can be considerable. If each participant in the pact can contribute something compatible with the contributions of others, none needs to have a massive investment in equipment and personnel to provide the responsive task force.

RESPONSE TO A SPILL EVENT

The National Oil and Hazardous Substances Pollution Contingency Plan (see Chap. 7, Part 1) published in the March 19, 1980, *Federal Register*[3] lists several phases of response to a spill alert: Phase 1—discovery and notification; Phase 2—evaluation and initiation of action; Phase 3—containment and countermeasures; Phase 4—removal, mitigation, and disposal; and Phase 5—documentation and cost recovery. The same stages of reporting and response apply to industrial spills as well.

Discovery and Notification

The front-line supervisor is generally the first individual involved in a spill alert. In general, after completing the initial survey and notifying the person charged with responsibility for pollution control, the supervisor must return to normal duties.

Evaluation and Initiation of Action

In general, it will be the responsibility of the pollution control officer to evaluate immediately the spill report received from the front-line supervisor. The officer's first duty is to assess the dangers involved and evacuate the area if necessary. This evacuation may involve more than the immediate plant area and include downstream locations as well as downwind locations. Reference to Tables 4-4 and 4-5 will help determine whether it is necessary to take steps to prevent downstream water use by livestock or humans.

Study of the plot plan and the area map will aid the pollution control officer in deciding whether help is required from outside contractors and suppliers or whether the plant force is likely to be adequate. On-site inspection probably will be required to assess the extent of the problem. If the spill affects areas beyond the plant boundaries, state and federal pollution control authorities will have to be notified as soon as the problem can be described adequately.

Containment and Countermeasures

The first action in containment is to stop the discharge if possible. If the spill occurred on land, it may be possible to contain it by building a dike of sandbags or dirt or by use of a proprietary foamed-in-place dike.[4] Sometimes catch basins need to be covered by tarpaulins and weighted with sandbags to prevent escape of the spilled material. In some areas, excavation of a sumplike area will not only contain the spill but provide a site for pump suction.

If the material floats on water, floating booms and barriers[5] can be used to herd the material into an area for recovery or treatment. If the material is heavier than water and insoluble, sinks can be constructed in the bottom of the water body by excavation or by construction of underwater dikes. Spills of soluble material may require massive flow diversions or the containment of the entire water body for subsequent treatment. If the spill involves a hazardous volatile material, it may be necessary to attempt to knock down the vapor with water fog or a reactive material. However, no countermeasures that may produce secondary hazards nearly as bad as the original or worse should be undertaken. For example, use of strong caustic soda to neutralize a spill of sulfuric acid may be as bad or worse than no neutralization at all.

Removal, Mitigation, and Disposal

If it becomes necessary to treat a spill at the site, it would be a great help if the necessary prior work had been done to prepare a table similar to Table 4-6. In this table likely spill candidates can be listed, and possible treatment schemes can be outlined. Some laboratory development work could be done to confirm the suitability of the suggested scheme and develop time distribution information before a spill occurs.

TABLE 4-6   Possible On-Site Treatment Schemes for Spilled Chemicals

Material	Treatment process	Treatment methods	Remarks
Methyl methacrylate	1. Skim off to drums. 2. Dissolved material can go to biological treatment. *or* 3. Polymerize in place.	For No. 3, use redox emulsion polymerization procedure.	For No. 3, avoid excesses of redox agents.
Acrylonitrile	1. Skim off to drums. 2. Dissolved material should be treated with alkaline bleach.	Adjust pH to 8.5, then add bleach to Cl₂ residual. React at least 1 h.	Full protective gear is required. Keep drums closed.
Styrene	1. Skim off to drums. 2. Dissolved material can go to biological treatment slowly. *or* 3. Polymerize in place	For No. 3, use redox emulsion polymerization procedure.	For No. 3, avoid excesses of redox agents.
Toluene	1. Skim off to drums. 2. Dissolved material can be adsorbed on activated carbon.	Use 10 to 20 kg of activated carbon per kilogram of dissolved material.	
Sodium dodecyl benzene sulfonate (50%)	Adsorb on activated carbon.*	Use 10 to 100 kg of activated carbon per kilogram of dissolved material.	

*This has not been proved in actual use.

In all on-site treatment work, the first task is to remove as much as possible of the spilled chemical in as high a concentration as possible. This is desirable because it reduces the amount of on-site treatment required and may permit the recovery of some value from the waste. For example, methyl methacrylate, styrene, or toluene possibly could be recovered by distillation, or they could at least be used as sulfur-free fuel. If the chemical floats on water, recovery can be aided by the use of floating booms and skimming pumps or skimmers.

If the spill is insoluble and sinks to the bottom of the water body, vacuum dredges or submerged pumps would be required to remove the sunken material. Soluble material spilled into a water body would require more extensive treatment such as adsorption on carbon, adsorption by peat moss or some other adsorbent, ion exchange reaction with either synthetic or natural material, or a chemical reaction such as neutralization, oxidation or reduction, precipitation, or polymerization. Of course, if the material is reasonably volatile and does not constitute a vapor hazard, aeration by pumping or bubbling of compressed air may be a feasible way to mitigate the damage from the dissolved spilled material. Restoration of the spill site may involve washup with solvents, cleanup with solid adsorbents, or physical removal of contaminated soil and substitution of clean soil.

Disposal of the recovered spill material as well as contaminated cleanup materials such as adsorbents and contaminated soil from the spill area presents a serious problem. Often this disposal grows to be more of a problem than the original spill. Deposit in a secure landfill is usually suggested as the solution, but such areas are becoming increasingly difficult to find. Decontamination in a suitable incinerator probably would be the best general disposal technique, but this procedure is not easily implemented either. Future work in the field of solid waste disposal must be addressed to the development of generally available, environmentally acceptable versions of both these procedures. See also Chap. 14.

Documentation

As with all things, paperwork must be done. Reports to management and the civil authorities are required. The Federal Water Pollution Control Act requires that spills be reported to the National Response Center if subject to Sec. 311; spills to sewers normally must be reported to appropriate authority also. Moreover, detailed reports retained in department files may help avoid making mistakes in the handling of future emergencies.

Detailed Procedure for Front-Line Supervisor

When confronted with a spill emergency, the front-line supervisor usually is the first to face the questions of what to do and when to do it. Planning in advance of the spill can reduce the possibility of error, and if the plan can be reduced to the form of a visual aid, there is less possibility of misunderstanding. A logic diagram has been developed for use when a spill has been detected (Fig. 4-1). In developing this flow sheet it was assumed that the plant had separate chemical sewers and storm sewers in keeping with good hydraulic design practice. Furthermore, it was assumed that the chemical-sewer system had a diversion basin of some type for emergency use. By referring to Fig. 4-1, it can be seen how the supervisor can proceed in an orderly way to handle his or her spill responsibility and obtain assistance from other departments. This figure shows in computer format a typical logic diagram in which the overall problem is broken down into a series of question diamonds requiring YES or NO responses followed by action or DO blocks depending on the response chosen.

The first question that must be answered determines whether the spill can be contained and kept out of the sewers. If the answer is YES, then assistance should be requested from the fire brigade or an equivalent service group. Sometimes expert action

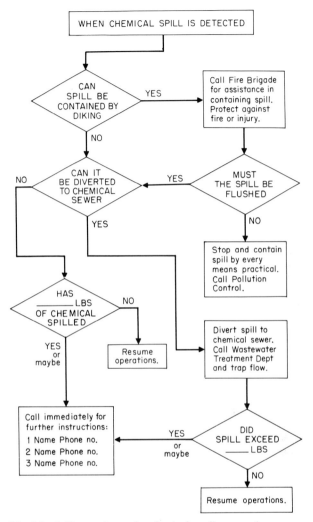

Fig. 4-1   Spill control procedure for the front-line supervisor.

with a backhoe or front-end loader can dike a spill quickly, while at other times the only procedure available is to dike the area manually and seal off the catch basins. Containment may result in a secondary hazard of possible fire or personal injury; thus the question of whether or not the spill must be flushed to a sewer to avoid these problems must be answered. If YES, one moves to the next question block, which also was the result of a NO answer to the original question.

Now one has a situation of whether or not a choice of sewer systems is available and, specifically, whether or not the spill can be directed to the chemical sewer. If YES, this action should be taken, and the department responsible for wastewater treatment should be notified of the action. Furthermore, the wastewater treatment department should be requested to activate the diversion basin as quickly as possible to trap the spill.

Now the size of the spill must be determined by asking the question of whether or not it exceeded a specific amount. The specified amount should be determined in advance by the pollution control officer as the level at which he or she *must* be notified immediately of the occurrence and the level below which the officer is satisfied that restoration of flows will not be hazardous. To determine this quantity, the spill candidates list of Table 4-4 and the hazard categories list of Table 4-3 should be consulted again to determine the most hazardous material listed. Then an estimate can be made of the tolerable concentration in the wastewater going to the treatment plant.

For example, if the most hazardous material has a 96-h median tolerance limit (TL_M) to fish of 12 mg/L and the wastewater must go to a biological treatment plant, the allowable concentration in the treatment plant probably should not exceed 10 percent of 12, or 1.2 mg/L. However, with an average holdup in the treatment plant of 8 to 10 h, a dilution of about 500 to 1 is available. Therefore, the allowable concentration in the sewer line would be 600 mg/L. If the average wastewater flow is 0.017 m^3/s, the treatment plant could presumably tolerate a leakage of 0.01 kg/s of the spilled material. Thus, some arbitrary time such as 1 h must be selected to establish a fixed amount of a spill. In this case, 36 kg (80 lb) would be the spill amount to be specified in the chemical-sewer part of Fig. 4-1.

Now one must return to the second level of the decision diagram and confront the NO response to the question of whether or not the spill could be diverted to the chemical sewer. Thus, if the spill did not go the chemical sewer, it must have gone to the clean-water sewer, and one must quantify the spill here. The question, therefore, is whether or not the spill exceeded a certain specified amount, which again must have been determined by the pollution control officer as the alarm point at which he or she *must* be notified immediately. To avoid a multiple entry at this point, an amount should be determined for the most hazardous material shown on the list in Table 4-4. This table already shows a critical quantity of 2730 kg (6000 lb) of this hazardous material that presumably would be harmful to the public domain. However, the pollution control officer surely would want to be aware of conditions before they would reach this level. Therefore, an arbitrary 10 percent of this quantity can be specified for the alarm point. For the example material considered earlier, this figure becomes 270 kg (600 lb), which is much larger than the figure for the chemical sewer. This is to be expected when it is recalled that the 2730-kg level given in Table 4-4 was derived for a watercourse flow of 1 m^3/s.

For other materials that form floating oil slicks or for materials that are serious fire or poison hazards, this alarm figure may be an extremely small amount. After the pollution control officer or a designated alternate has been notified, the line supervisor returns to the task of minimizing the effects of the spill and carrying out any supplementary instructions from the pollution control officer.

Detailed Procedure for Pollution Control Officer

After a spill is reported by the front-line supervisor, immediate decisions are again required, this time by the pollution control officer. A second logic diagram (Fig. 4-2) has been developed for this situation in the event that the second- or third-level backup person is required to assume the responsibilities of the pollution control officer. The sequence of decisions started by the report of a spill being received is shown in Fig. 4-2. The pollution control officer's first consideration is whether the spill is still contained on the surface or has escaped to underground drains. If the spill has been contained, the officer must consider several alternatives for recovery or destruction. Often it will be possible to recover spilled material by pumping to tank trucks or by vacuuming the spill into vacuum trucks. Although it may be impossible to recover the chemical for reuse, the disposal of the spilled material may be facilitated by trucking it to approved burning facilities or to an area where its decontamination can be readily controlled.

Occasionally it may be necessary to destroy the spilled material on site by using one

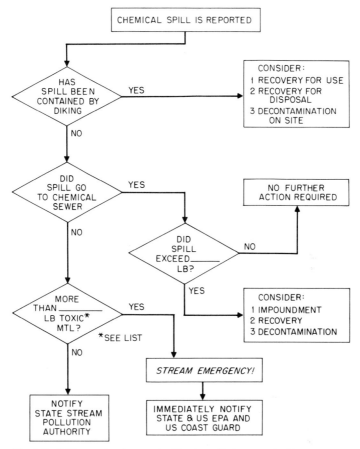

Fig. 4-2   Spill control action required by the pollution control officer.

of the methods mentioned earlier, depending on the nature of the material. Emulsion polymerization can be effective if the spill is a monomer in a substantial amount of water. Chlorination with laundry bleach or bleaching powder may be effective in converting the spill to harmless by-products. This procedure may be most effective if it is carried out in a dilute aqueous system. Other chemical agents could be used as long as it is recognized that the possibility of high reaction rates should be avoided. Acetone cyanohydrin, for example, can be effectively destroyed by chlorination; however, it can be readily ignited in undiluted form when sprinkled with bleaching powder.

If the chemical has gone underground, the question then becomes "Where did it go?" If the answer to the question "Did it go to the chemical sewer?" is YES, then it must be established whether or not the spill exceeded a certain amount. The quantity entered in this question diamond is that set by the pollution control officer; below this amount no further action is required. Previously an amount that would not upset the downstream treatment unit had been determined, so this figure can be used here also. If the spill exceeded this figure, the pollution control officer must decide what to do with the trapped material so that the impoundment area can be restored to service.

It must be determined whether the impounded material should be decontaminated at the impoundment site or removed immediately for recovery or for destruction or controlled reaction at a remote location. Two actual spill events demonstrate how these diagrams were used to prevent discharge of spilled materials to the river.

During the loading of a 45-m³ (12,000-gal) tank car of distilled monomer product, it was discovered that the transfer papers called for the loading of a 75-m³ (20,000-gal) shipment. By the time that the error was discovered, about 2730 kg (6000 lb) had overflowed the dome of the car. Fortunately, the entire loading area drained to the chemical sewer, and the monomer was confined to this system.

The shift foreman called the wastewater treatment department, advised it of the spill, and requested impoundment of the flow. As a result of this action, the spill was trapped in a 1500-m³ (400,000-gal) basin of water that now required action by the pollution control officer. Since the monomer concentration was low, recovery was out of the question, and it was decided to try polymerization to remove the dissolved material. This procedure was successful. Much of the resulting polymer was removed as a floating skin, and the balance was odorless and harmless, so the flows were returned to normal.

On another occasion, a similar amount of monomer was collected by the chemical sewer, but in this case the wastewater treatment department was called early enough so that the spill was trapped in a smaller quantity of water. Under these conditions, an organic layer formed on the surface. At this point, this material was pumped into drums, the water layer separated, and the monomer layer sent to the boiler house for fuel.

Now the NO response to the question of whether or not the spill went to the chemical sewer in the second level of the logic diagram must be considered. If the spilled material went to the underground sewers but not to the chemical sewer, it must have gone to the clean-water sewer.

Sometimes it is possible to trap a spill in an underground storm sewer if the spill occurs in dry weather. This operation may be successful if (1) the underground lines and manhole locations are well known and (2) quick action is taken to seal off the downstream location with an inflatable balloon in the line or a dike in a manhole. If such a stopper can be placed in time, the spill can be pumped out for recovery or treatment.

However, if the spill occurs in wet weather or if there is cooling water or natural flow in the clean-water sewer, time is of the essence and the amount of the spill must be gauged quickly so that the answer to the next logic step can be determined. The amount specified in the logic block at this point is that at which a stream emergency must be declared. Here the information assembled in the list referred to in Table 4-3 will be of help to the pollution control officer in determining the amount of chemical that constitutes a serious hazard to the receiving waters. The dilution available is of the utmost importance: a discharge of a spill into a receiving stream where the plant effluent constitutes the major element of flow can have much more serious consequences than a spill of the same amount into a river the size of the lower Ohio or the lower Mississippi. An attempt was made to include this factor in the U.S. EPA publications mentioned earlier.[1] A flow of 1 m³/s (36 ft³/s) was defined as sufficient to qualify a stream as a river. This was not an arbitrary choice; EPA's contractor determined that 95 percent of all identifiable streams in the United States had a flow of 1 m³/s or greater. This would be equivalent to a stream 7.6 m (25 ft) wide and about 1.5 m (5 ft) deep on the average. Thus, if the discharge from the plant goes to a stream smaller than the river as defined, the spill amount that should require an alarm will be less than the critical quantity as listed.

Guidelines for amounts of spilled material that should be considered hazardous must depend on the possible uses of the downstream water. If all streams are to be considered for public water supply and water-contact recreation, then such other uses as stock watering and industrial water supply are automatically included. However, probably the most sensitive indicators in the receiving stream are the fish and other aquatic life that are native to the area.

Consider, for example, a typical effluent standard that limits cyanide discharge to 0.2 mg/L. Suppose that the effluent flow rate is approximately 0.75 m^3/s (12,000 gal/min); then any cyanide spill of more than 1.5 $\times$ 10^{-4} kg/s would exceed the allowable discharge concentration. Studies with goldfish have indicated that this concentration can be tolerated without serious consequences, but it is close to the TL$_M$ of 0.45 mg/L. With such a small factor of safety of 2 with this test species, there very likely would be damage to other aquatic life at this concentration. For discharge into a stream where this effluent is a high percentage of the total flow, this amount of spilled material should be considered an emergency, for it would very likely result in a fish kill if it persisted more than an hour. On the other hand, a spill of this magnitude into an effluent that enters a stream with a flow of 280 m^3/s (10,000 ft^3/s) would be of slight consequence since the cyanide concentration would be approximately 5 $\times$ 10^{-4} mg/L.

The spill was expressed as a rate, whereas many spills are discrete events. To determine what the weight figure in the diagram should be, some flow time such as 1 h should be selected. Here again, judgment is required to balance between the hazards involved downstream as a result of the spill and the activation of the full emergency procedure.

If it is even remotely possible that the amount spilled is at or above the critical, the stream emergency should be declared immediately to minimize downstream damage. It is better to err on the safe side with an alert later to be canceled if, in fact, the anticipated emergency does not develop than to find that action was not taken in time because the alarm was not given in time. If the spill was less than the critical amount, the state stream pollution authority should be notified that the spill occurred and that the amount was less than that which would be a hazard to the watercourse.

REFERENCES

1. G. W. Dawson, M. W. Stradley, and A. J. Shuckrow, *Determination of Harmful Quantities and Rates of Penalty for Hazardous Substances,* vol. 1: *Executive Summary,* vol. 2: *Technical Documentation,* vol. 3: Appendices, vol. 4: *Comparative Analysis,* EPA-440/9-75-005, U.S. Environmental Protection Agency, January 1975.

2. J. L. Buckley and S. A. Wiener, "Documentation and Analysis of Historical Spill Data to Determine Hazardous Material Spill Prevention Research Priorities," in G. F. Bennett (ed.), *Proceedings of the 1974 National Conference on Control of Hazardous Material Spills,* San Francisco, August 1974, pp. 85–89.

3. Council on Environmental Quality, "National Oil and Hazardous Substances Pollution Contingency Plan; Final Revision," *Federal Register,* vol. 45, Mar. 19, 1980, pp. 17832–17860.

4. J. V. Friel, R. H. Hiltz, and M. D. Marshall, *Control of Hazardous Chemical Spills by Physical Barriers,* EPA-R2-73-185, U.S. Environmental Protection Agency, March 1973.

5. F. J. Freestone, W. E. McCracken, and J. P. Lafornara, "Performance Testing of Several Oil Spill Control Devices on Selected Flotable Hazardous Materials," in G. F. Bennett (ed.), *Proceedings of the 1976 National Conference on Control of Hazardous Material Spills,* New Orleans, April 1976, pp. 326–331.

PART 5
European Industrial Plans

A. H. Smith, M.C.I.T.
Distribution Manager
*Laporte Industries, Ltd.**

Introduction / 7-56
Basic Response Information / 7-56
Additional Response Arrangements / 7-57
Chemsafe / 7-57
Single-Product Consortia Arrangements / 7-60
Local Consortia of Manufacturers and Traders / 7-60
Organization of a Manufacturer's Response Team / 7-60
Future Developments / 7-63

INTRODUCTION

In most countries the emergency services (fire brigades and police) have the primary responsibility for dealing with incidents which occur during the transportation of hazardous chemicals. Legal responsibility for the carriage of chemicals rests with the carrier. This can be onerous if the carrier is a general carrier not specializing in the movement of chemicals.

Within Europe the chemical industry is increasingly accepting the need to establish suitable arrangements for the safe movement of hazardous loads in order to minimize the damage or injury that could arise should any incident occur. Information is being provided about the potential hazards of chemicals being transported, and the industry also accepts the further need to play a supporting role to the emergency authorities when requested by them.

BASIC RESPONSE INFORMATION

Mandatory requirements to provide drivers with instructions setting forth in detail the hazards of chemicals loaded on their vehicles are part of ADR regulations, which also require that road vehicles be labeled with the United Nations reference number of the chemicals being transported. Similar regulations apply to the labeling of consignments of chemicals moved by rail transport (RID). In the transport of chemicals within the United Kingdom, it is mandatory to label vehicles when they carry certain classes of chemicals: explosives, flammables, corrosives, and organic peroxides.

---

*Retired October 1981.

The European chemical industry has, via existing regulations, an immediate response to the emergency services by means of an indication that a vehicle load is hazardous and by identification of the hazards. The United Kingdom chemical industry has extended this supply of information for bulk tank trucks by the voluntary introduction of Hazchem signs (see Fig. 5-1), which indictate to the emergency services via a code reference the action to be taken immediately. This method of marking tank trucks forms part of new United Kingdom national regulations.

ADDITIONAL RESPONSE ARRANGEMENTS

Response activities beyond the initial information available on packages and vehicle labels vary throughout Europe. As a general rule, most manufacturers will respond to a call from the emergency services for advice on how to deal with chemicals involved in a traffic incident. Requests can be made directly to known manufacturers or to nearby manufacturers with which the emergency services have a liaison.

CHEMSAFE

The British chemical industry and the National Chemical Emergency Centre at Harwell have combined to develop a response system called Chemsafe. The system has four main elements as follows:

Action to Be Taken by the Chemical Manufacturer or Trader

The manufacturer should ensure that all bulk vehicles, freight containers, and packages containing hazardous materials are marked with (1) the statutory symbols and familiar

Fig. 5-1   Hazchem sign.

chemical name or with any agreed alternative such as a composite label including the United Nations number and the Hazchem code and (2) the manufacturer's or trader's emergency telephone number. The manufacturer should also ensure that the Tremcard (see Fig. 5-2) for the material being carried (or similar instructions in writing) is readily available in the vehicle.

UN No. 1830	TRANSPORT EMERGENCY CARD (Road)	CEFIC TEC (R)-10b MARCH 1976, Rev. 1 Class 8 ADR items 1a + b

Cargo	**SULPHURIC ACID (above 75%, except Oleum)** Often odourless, colourless - yellowish liquid.
Nature of Hazard	Corrosive. Causes severe damage to eyes and skin. Attacks many materials, clothing. Contact with a relatively small quantity of water creates violent reaction generating much heat and spattering of hot acid. Attacks many metals with liberation of hydrogen which is inflammable and forms explosive mixture with air.
Protective Devices	Goggles giving complete protection to eyes. Apron or other light protective clothing, boots and plastic or rubber gloves. Eyewash bottle with clean water.

EMERGENCY ACTION   Notify police and fire brigade immediately

* If possible move vehicle to open ground and stop the engine.
* Mark roads and warn other road users.
* Keep public away from danger area.

Spillage
* Contain leaking liquid with sand or earth, consult an expert.
* If substance has entered a water course or sewer or contaminated soil or vegetation, advise police.
* Do not absorb in sawdust or other combustible materials.
* Do not use water jet on a leak of the tank.

Fire
* Keep containers cool by spraying with water if exposed to fire.

First aid
* If the substance has got into the eyes, immediately wash out with plenty of water for at least 15 minutes.
* Remove contaminated clothing immediately and wash affected skin with plenty of water.
* Seek medical treatment when anyone has symptoms apparently due to inhalation or contact with skin or eyes.

Additional information provided by manufacturer or sender.

Telephone: In emergency telephone: 051-424 5555

Prepared by CEFIC (CONSEIL EUROPEEN DES FEDERATIONS DE L'INDUSTRIE CHIMIQUE, EUROPEAN COUNCIL OF CHEMICAL MANUFACTURERS' FEDERATIONS) Zurich, from the best knowledge available; no responsibility is accepted that the information is sufficient or correct in all cases.

Acknowledgement is made to V.N.C.I. and E.V.O. of the Netherlands for their help in the preparation of this card.

Applies only during road transport   ENGLISH

Fig. 5-2   Tremcard.

In addition, the manufacturer should ensure that a company procedure exists for dealing with incidents involving its materials in transit. Key elements in this procedure are:

1. An emergency telephone number staffed throughout the period when the materials are in transit

2. Availability of a person able to give advice to the emergency services on the best way to deal with an incident involving the company's materials

3. Facilities for a competent person to attend the scene of the incident to provide on-the-spot advice to the emergency services

Chemsafe Mutual-Aid Arrangement

Chemical Industries Association (CIA) members who are participants in the mutual-aid arrangement pool information about their freight emergency facilities: locations, telephone numbers, and range of products for which expertise is available. They undertake to respond favorably to requests from other participants for assistance at the scene of an incident, provided that this is practicable and within their own range of expertise and capability. The principal objective is to achieve better geographical cover and speedier effective response to requests for attendance and help at the scene of an incident.

Chemsafe Backup Procedures

The further development of company procedures and of intercompany arrangements, reinforced, it is hoped, by those of other industries, should result in the great majority of incidents being dealt with under the Chemsafe standard procedure. It must be recognized, however, that the circumstances of an incident may be such that the emergency information provided on the vehicle or packages cannot be read. For this reason a backup procedure whereby the authorities can obtain advice and/or assistance is needed even though the manufacturer or trader involved is either unknown or unobtainable or the product itself is unidentified.

The Chemsafe backup procedure involves a Chemsafe center operated in collaboration with the National Chemical Emergency Centre at Harwell, which was established by the Department of the Environment as part of the Toxic and Hazardous Materials Group of the Atomic Energy Research Establishment. The center has a continuously staffed emergency telephone through which the public emergency authorities can ask for technical advice in the case of any freight emergency in which the nature of the chemical hazards cannot be ascertained from other sources (viz., manufacturer, trader, or Tremcard). Qualified staff with practical experience over a broad range of chemicals is available on a call-out rota, and the center's chemical data bank is being continuously expanded with product information supplied from CIA member firms.

Chemsafe Advice on Action to Be Taken by Public Emergency Services

Chemsafe offers the following advice:

1. Look for the emergency telephone number on the vehicle, freight container, or package and for the Tremcard (or similar written instructions) giving product details.

2. If you need further assistance, first telephone the manufacturer's or trader's emergency telephone number and ask for advice.

3. If no emergency telephone number is available or if the manufacturer's or trader's emergency procedure fails to operate, telephone the National Chemical Emergency Centre at Harwell and ask for advice. Do not telephone Harwell before trying the manufacturer's or trader's emergency number; this procedure will only cause delay.

4. When the manufacturer or trader cannot be contacted, the emergency authorities have the option of contacting a local firm directly if they judge that this will result in the required help being obtained more rapidly.

SINGLE-PRODUCT CONSORTIA ARRANGEMENTS

Some manufacturers have organized themselves to deal with incidents in which a particular chemical is involved and a large tonnage is produced and transported. A typical example is chlorine. Producers in the United Kingdom have developed a mutual-aid scheme in association with the CIA. A booklet giving details of the scheme has been distributed to emergency services and other organizations likely to be involved in any transport incidents. Mutual-aid schemes covering the movement of chlorine also operate in France, Belgium, the Netherlands, and the Federal Republic of Germany.

Technical liaison exists in the United Kingdom between manufacturers of other hazardous chemicals (e.g., sulfuric and hydrofluoric acids) to improve safety in transport and to promote mutual aid when incidents occur during transportation, but these arrangements have not been developed as far as the arrangements for chlorine.

LOCAL CONSORTIA OF MANUFACTURERS AND TRADERS

In the United Kingdom local consortia have been developed to pool resources to enable a better response to calls for assistance from the emergency services. These consortia can be organized so that the individual companies provide technical advice on their own products directly in response to inquiries from the emergency services but any requirement for attendance at the incident plus provision of equipment is provided from a central control point, usually by one of the larger members of the consortium, the costs of transport and equipment being shared between the members.

ORGANIZATION OF A MANUFACTURER'S RESPONSE TEAM

The selection of a team to cope with emergency calls needs thought. It should not be just a question of choosing all technical personnel who have knowledge of the products but of selecting a mixture of personnel from different functions who have particular expertise to offer. A team, therefore, could comprise chemists, chemical engineers, packaging and distribution personnel, production management, safety officers, etc. There must be sufficient members to allow for the absence of personnel on other company work, sickness, holidays, etc. The total number in any team obviously must be relevant to the product distribution pattern. Large-tonnage movements from a manufacturing site probably could require from 6 to 10 team members.

Personnel in an emergency team must have available to them the following information in duplicate, one set to be available at their homes and the other at their place of work:

1. Full data on products: hazards, protective-clothing requirements, action in case of fire, spillage, first-aid treatment, any special medical treatment

2. Details of backup service available from the base site and any other company sites; names of technical specialists not in the team; availability of vehicles and types (e.g., flatbed and dump trucks, tank trucks); home and work telephone numbers of key personnel: site management, engineering, distribution

3. Availability of stocks of neutralizing agents

The team must have transport available to travel to any incident as and when required. In many cases the need to convey a member to give advice is all that is required in the way of transport. However, it is recommended that when there is a large movement of hazardous products, a truck or landrover or similar vehicle be promptly available to the team and that it be permanently loaded with essential basic equipment: full protective clothing for two persons, hand tools, drum-handling equipment, empty packages, small transfer pumps, and other more specialized items considered necessary (e.g., breathing apparatus, tank transfer pump).

Quick communication is vital in dealing with emergency calls, and companies should seriously consider the installation of special emergency telephone extensions for all team members and instruction to switchboard personnel to give priority to all emergency calls. The emergency extension numbers must not be made available to other personnel. The availability of team personnel should be known to switchboard operators and out of hours to security or other personnel who will receive calls. A weekly list giving this information needs to be provided. Contact between team members going to an incident and members at base can be valuable; it can be provided by air call equipment in the emergency vehicle.

The team has been selected, data on products are available, there is full information on backup services, and communications systems have been checked. In theory, then, the team can cope with emergency calls.

How the Team Operates

It is not possible to anticipate the nature of future emergencies. The team leader can, however, regularly call the team together to go through hypothetical incidents and to decide what advice or action would be required to assist the emergency services in such an eventuality.

The function of the team is to assist, not replace, government agency response. Emergency services in most European countries are responsible for dealing with incidents, and there is no question of a company emergency team taking over that responsibility. The requirement from a company team will be to give accurate technical advice as promptly as possible by telephone and to proceed to the incident to give further assistance if requested or if the company considers it advisable in the general public interest to do so. The provision of vehicles, neutralizing agents, special pumps, etc., may be part of the assistance offered to or requested by the emergency services as a follow-up to the technical advice.

The circumstances surrounding a particular incident will have some bearing on the advice given. This does not mean that the basic hazards of the chemical (toxic, corrosive, flammable, etc.) alter but that factors such as location or type of accident may require additional advice. The presence of toxic fumes is a more difficult problem in built-up locations than in the open countryside. Spillage of flammable products also is a greater hazard in built-up locations. The type of accident affects the advice and any requirement to attend the scene. A multivehicle collision may result in a mixture of chemicals and other substances which could increase the hazard substantially.

The vital ingredient toward giving an accurate response to the emergency services is to complete the checklist (see Fig. 5-3) accurately when receiving the first emergency call. The name of the chemical should be spelled out to ensure that there is no doubt: for example, receipt of telephone messages about incidents involving hydrochloric or, alternatively, hydrofluoric acid could lead to wrong or insufficient advice if care is not taken

---

EMERGENCY TELEPHONE CALLS: CHECKLIST

The following is suggested for persons receiving emergency telephone calls:

PART A: ESSENTIAL INFORMATION

DETAILS OF CALL

DATE	TIME	CALLER'S NAME	TELEPHONE NO. FOR RETURN CALL

STATUS OF CALLER (Indicate by tick)

Police_____   Fire_____   Ambulance_____   Vehicle driver_____   Public_____

PURPOSE OF CALL Is there any particular advice required immediately?	
NAME OF CHEMICALS To be spelt out clearly	
BRIEF DESCRIPTION OF INCIDENT nb Fire/Spillage Quantity involved Packaging details	
Location of incident	

PART B: INFORMATION TO BE OBTAINED IF READILY AVAILABLE

Has anyone been injured? Affected by chemicals?	Yes/No Yes/No	If so how many? If so how many?	
What first aid has been given?			
Has anyone been taken to hospital?	Yes/No		
Address of Hospital (If applicable)			
Is the road blocked?	Yes/No	Closed to traffic?	Yes/No
Who owns the Chemicals? Has the owner been informed?	Yes/No		
Vehicle registration number Vehicle owner Has the owner been informed?	Yes/No		
To whom was the load consigned?			

Fig. 5-3   Checklist for an emergency response duty officer at XYZ Chemicals Limited.

to check the actual product involved. An additional safeguard is to have readily available in all traffic departments details of consignments sent out over the past 48 h so that confirmation of the load can be obtained quickly.

Following any incident in which a response has been requested, the emergency team personnel involved should submit a report promptly. Important points to be reported are (1) the cause of the incident, (2) recommendations toward avoiding future incidents, and (3) an assessment of the team's response.

FUTURE DEVELOPMENTS

There are two areas in which there will be positive development and expansion of activity related to the supply of information and technical advice to the emergency services. First, there will be an expansion of pooling of technical information by manufacturers of a particular chemical. In a similar vein, there will be further development of consortia arrangements in particular geographical areas to share the costs and technical expertise in order to deal with emergencies in relation to the products manufactured by the members.

Second, there will be an expansion of the storage of relevant data similar to that of the Harwell activity as part of the Chemsafe scheme. This is likely to be followed by a world-wide linkup to obtain and supply chemical hazard data and emergency action information, which will be of great assistance in dealing with incidents during the transportation of chemicals. Such a linkup would cover not only domestic incidents involving road and rail but incidents on vessels at sea, with air freight, and in ports and harbors. The speed of response to calls for advice and technical assistance would then be improved.

PART 6
Land Transportation Plans

G. Stapleton
Motorway Maintenance Department
Worcester County Council

Protecting Oneself and the Scene / 7-65
Identification / 7-67
Specialist Advice and Specialist Response Plans / 7-70
Decontamination / 7-70

Response plans are implemented following an incident or a potential incident. On land, this involves the municipal or highway engineer for roads and the railway engineer for railways, in addition to the emergency services. The highway engineer is concerned mainly with and responsible for the infrastructure of the road network and, although interested in traffic volumes and weights of vehicles and loads, is generally surprisingly ignorant of the contents of vehicle loads, in particular, loads which are hazardous. This is in contrast to the railway authority, which, in addition to its responsibility for the infrastructure of the railway system is also responsible for and has jurisdiction over the rolling stock within that system and the material content, hazardous or otherwise, which constitutes a train load. The highway authority has no such control or jurisdiction over the hazardous loads which will travel the roads for which it is responsible, and, indeed, no one is even obliged to inform the police of the movement of radioactive material. Labeling and careful packaging are considered to be adequate protection.

It will therefore be appreciated that the highway or railway engineer is closely involved with incidents of all types on road and rail in assisting the emergency services and restoring the safety of the damaged road and rail infrastructure. What is not appreciated is that the civil engineer or the engineer's staff may be the first to arrive at the scene of an incident or indeed may be carrying out normal work on a site where an incident occurs. In this respect engineers or their employers are now legally obliged to protect their employees and the traveling public.[1] The highway authority also has a duty[2] and a right[3] to clear the highway of obstruction and debris. The U.K. Department of Transport has also given advice to highway authorities on the procedure to follow in dealing with hazardous chemicals on the highway.[4]

Many accidents cause structural damage, and it may be necessary for the highway or structural engineer to give an expert opinion or take action before the emergency services and other disciplines can do their work safely. Furthermore, if a multiple accident involves, say, molasses in addition to hazardous chemicals, the expertise of the beekeeper and the entomologist may be just as pertinent as the expertise of the chemist or the fire fighter.

The conclusion to be drawn from this brief introduction is that the problem of dealing with incidents of all types is multidisciplinary and that it is important for the key services of police, fire, ambulance, highway, and rail authorities to develop their own response plans until expert advice arrives in whatever field it is required.

The professional and industrial chemists have their own response plans,[5] which are described in Part 5 of this chapter, but if the conclusion previously drawn is accepted, the question arises as to how the basic services of police, fire, ambulance, and transport authorities can develop their response plans so as to protect, deal with, and survive the initial stages of an incident involving hazardous materials when they are amateurs in the field of chemicals and before representatives of other expert disciplines arrive. The way in which these response plans are developed may be different in an area away from a concentrated chemical center where expert advice may be some hours distant, as opposed to an area where expert chemical advice is at hand.

Having established the need to develop response plans for the amateur in terms of chemical expertise, the rest of this part will be devoted to developing such a response plan with as much safety as possible. This will include practical case histories.

The basic requirements for reaction in any response plans to an incident involving hazardous material being transported on land are to:

1. Protect oneself and the scene

2. Try to identify the materials involved in the incident or which may become involved

3. Obtain specialist advice and instigate specialist response plans

4. Decontaminate

Steps 1 and 2 can be reversed, for identification of the materials may dictate additional protection, isolation, evacuation, and, in extreme cases such as boiling-liquid–expanding-vapor explosions (BLEVEs), evacuation of the emergency services.

PROTECTING ONESELF AND THE SCENE

This requirement may sound simple in the first instance, but the first person to appear on the scene of an incident may be a total amateur in the field of hazardous materials. With the speed of modern transport the incident will in all probability be approached quickly and appear suddenly. Many people may approach the incident simultaneously, with a tendency to confusion and panic.

Even if the first arrival at the scene is a police officer, fire fighter, or highway or railway maintenance worker who may be used to dealing with incidents, such a person is relatively amateur in dealing with complex hazardous materials transported in a modern society. It may sound like stating the obvious, but an incident has to be approached before the implications of what has occurred are even partly realized, and the initial approach can be all-important. Whenever possible, the approach should be from upwind and uphill.

In the following incident in which the mixed contents of a waste tanker were reacting violently, the approaching nonchemist could only assume that the contents might be flammable, explosive, toxic, corrosive, or radioactive, or a combination thereof. The only faculties on which the nonchemist depends are the basic senses of sight, smell, sound, heat, taste, and touch. It may seem obvious that the incident is visible and there may well be flames, noise, or escaping liquid, but the incident may occur in darkness, when the first person to approach may detect it only by smell or noise. It may also be raining, there may be snow on the ground, or it may be foggy. The wind may be blowing the fumes away from the approacher, and a spill may be flowing in any direction, depending on the

Fig. 6-1 Wind currents caused by traffic and affecting the fumes from a spill of material by a chemical tank truck.

topography of the site. Wind currents caused by passing traffic can also cause a hazard to personnel working at the scene. Figure 6-1 shows swirling fumes caused by traffic. Taste and touch should never be used by an amateur. There is little future in tasting or touching toxic or caustic materials to see if they are palatable.

The second incident shows a tank truck (Fig. 6-2) of ethylene oxide which turned over after striking another vehicle. In this case, the potential dangers are not so obvious to the nonexpert. Although it was not known if the tank truck was leaking, much activity took place at the scene before expert advice was received by telephone to extinguish all sources

Fig. 6-2 Overturned ethylene oxide tank truck.

Fig. 6-3 Leaking sulfuric acid tank truck.

of ignition for 0.4 km (0.25 mi). Prior to receiving that advice, vehicles had been passing the tanker with their engines running, and personnel in the vicinity had been smoking.

Also, it was easy to overlook the most obvious sources of ignition: flashing beacons on police, ambulance, and other emergency vehicles, radiotelephone sets, highway telephones, highway hazard-warning signals, nylon clothing, etc. In addition, standing traffic trapped on the highway for 5 km (3 mi) on either side of the tank truck had to be extracted safely without spark or panic.

In this case, protecting the scene and isolating the stricken tank truck was time-consuming. Although the emergency services and highway department responded within minutes, assembling all the disciplines on site, together with the necessary recovery equipment, took 6 h. The highway was not reopened for over 10 h.

The incident shown in Fig. 6-3 depicts a ruptured tank truck carrying sulfuric acid which crashed over a small highway embankment in darkness. Acid leaked from the truck and splashed over several hundred square meters. Protecting the scene and amateur personnel working at the site was essential throughout the recovery operation.

The lack of expertise is apparent. The crane was not capable of recovering a tanker which was half full of acid, and during the recovery the crane operator with no protective clothing other than rubber boots picked up the towing cable, which had been lying in a pool of acid, and was burned. The recovery of this tanker and acid took 2 days, and decontamination of the area took many more days.

IDENTIFICATION

Up to the present time, the identification of hazardous material involved in many incidents has been something of a nightmare. The situation has, however, been improving gradually since the introduction in the United Kingdom of a voluntary tank truck marking scheme, and it is hoped that it will improve further now that the United Kingdom Hazard Information System (UKHIS) is mandatory.

The acid waste tank truck shown in Fig. 6-1 carried no placards. The driver's papers indicated that he was transporting sulfuric acid. It was eventually admitted that industrial waste from two different sources had been mixed together. In this type of incident, it is extremely difficult to determine the exact contents of a load, and they will certainly not be fully analyzed until after the incident has been dealt with. The final analysis of this particular load revealed nitric acid, hydrofluoric acid, iron, cadmium, chromium, copper, lead, nickel, and zinc.

Although the overturned tank truck of ethylene oxide (Fig. 6-2) was well labeled, the first reports given to the police described it as a tank truck of caustic soda. Even during the recovery of the vehicle conflicting reports expressing doubt as to its contents were still being received on site.

Identification of hazardous material in loads made up of smaller containers can sometimes be more difficult and hazardous than those in large tank trucks. The box container in Fig. 6-4 was filled with drums of formaldehyde, one of which was leaking badly. There were no labels on the vehicle or the container, and the only indication of trouble was that it was impossible to approach the vehicle from downwind because of acrid fumes.

In another incident, a trailer carrying a mixed load caught fire and left dangerous remains on the hard shoulder of the highway. The load consisted of carboys of concentrated nitric and sulfuric acids and separate containers of hydrogen peroxide, all of which were unlabeled. The only indication of the contents of the load was on the driver's papers.

Small drums and packages which fall off vehicles and are deposited on the highway or its shoulder often become a hazard to maintenance workers who are clearing the highway of debris as part of their normal duties. These packages are sometimes unlabeled, but even if they are labeled, it is only by studying the small print that the presence of constituents like hydrofluoric acid is revealed; in any case, this means very little to lay persons as to the true hazard of such a material. Again the only instruction which can be included in an amateur's response plan is to protect and not to investigate too closely such debris until expert advice is obtained.

Fig. 6-4   Cargo container containing a leaking drum of formaldehyde.

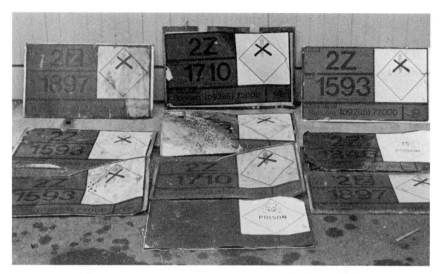

Fig. 6-5 Confusing set of labels removed from a single overturned vehicle.

An aspect of identification which receives little attention is a misunderstanding of how differently materials can behave in bulk compared with their behavior in small quantities. This together with a wrong association of names can be fatal. A particular example is aqueous ammonia solution. A lay person who is familiar with how unpleasant a small bottle of ammonia in the kitchen can be if mishandled would have no conception of the catastrophic results which could ensue from a bursting tank truck load of anhydrous ammonia. Similarly, the innocuous nature of phenolic soap could lead the lay person into assuming that a spilled load of phenolic acid might be relatively harmless.

Earlier in this part, it was stated that in any response plans the priorities of protection and identification could be reversed, depending on the circumstances of a particular incident. This is very important. It is also vital to continue the protection aspect throughout the identification and recovery process. There is little point in walking through a pool of acid to read an identification label which may tell the would-be identifier, if he or she is lucky enough to arrive at the label, that acid is dangerous.

When the Duke of Wellington was infected with the Malibar itch, he was advised to plunge into baths of nitric acid so strong that the towels on which he dried were burned. Not all rescuers are Iron Dukes and would be well advised to resist plunging into nitric acid even in a well-intentioned but ill-advised rescue attempt.

A parallel incident occurred when a young policeman attempting to read the label on an overturned tanker of methylene chloride walked through the spilled chemical only to find his shoes dissolving. Fortunately, the chemical was not acid.

The danger of fumes as part of a spill is often misunderstood, probably because fumes are more ethereal than liquids or solids. But fumes of sulfur trioxide emanating from a spill of oleum will form more sulfuric acid in combination with perspiration on the skin or moisture in the mouth, with the production of much heat and consequent burning. Fumes from hydrofluoric acid will etch eyeballs as well as glass.

Now that UKHIS labeling is a legal requirement and is fully enforceable, care will have to be exercised in the allocation and fixing of labels if the system is not to be discredited.

Figure 6-5 shows a selection of labels, some of which were printed on both sides, taken from one overturned tank truck. Readers can imagine the confusion which occurred at the scene while the true identity of the contents was being established.

SPECIALIST ADVICE AND SPECIALIST RESPONSE PLANS

Items under this heading are dealt with in other parts of this chapter, but all incident response personnel must remember the importance of the multidisciplinary approach to the problem of dealing with accidents involving hazardous material. The successful outcome of an emergency operation may depend just as much on the expert advice of the public utilities specialist as on the chemist.

DECONTAMINATION

Dealing safely with and decontaminating the scene of an incident together with thinking laterally through all the side effects is not always easy. For instance, it is easy to neutralize 10 tons of sulfuric acid with 10 tons of soda ash, but how many people, including members of emergency services, consider the end products of the subsequent chemical reaction? One product, 14 tons of soluble sodium sulfate, will attack concrete, and the 4 tons of carbon dioxide will inhibit breathing if the incident occurs in a confined space. A similar example is recorded in which three men died in a surface-water sewer which had become filled with carbon dioxide owing to acidic groundwater running over calcareous filter media in the drain.

If, in the case of the sulfuric acid spill, calcium hydroxide had been used to neutralize the spill, an insoluble calcium salt would have been produced and removed without so much risk of damaging the concrete.

Dangers also exist in the free use of emulsifiers. It is very easy to spread an emulsifier on a spilled chemical on the highway and wash the resulting mixture into a watercourse

TABLE 6-1   Checklist of Steps to Be Taken in Response to a Hazardous Materials Spills Incident

When approaching what appears to be a hazardous incident:
1. Stop smoking.
2. Assume that all loads and spills are flammable, corrosive, toxic, explosive, and radioactive.
3. Use the senses:
   Sight—fumes, flames, or escaping liquid
   Smell—fumes or spills that may not be visible
   Sound—explosion or chemical reaction   Consider position of load; consider
   Temperature—hot or very cold container   casualties
   Taste—do not taste
   Touch—do not touch
4. Wear protective clothing.
5. Protect the scene—send for emergency services.
6. Identify the load (if it can be safely done).
7. Determine the weather (obtain accurate meteorological forecast):
   Rain, snow, or fog
   Wind (weather)
   Wind (traffic)
   Temperature—_____, snow, frost
8. Ascertain aftereffects—condition of structures, road surfaces, drainage, and pollution.
9. Decontaminate—remove remains of spill, wash, clean clothes, tools, and vehicles, etc.

without being conscious of the spread of pollution or of the disrupting effect such an action can have on a sewage disposal works or a water supply source.

The washing down of a spill of china clay, although chemically inert, into a watercourse may drown fish because of clogging their gills with fine particles. Similarly, a spill of milk or beer entering a watercourse can play havoc with fish life by increasing the organic load with concomitant usage of the dissolved oxygen. Hence, in a spill of complex chemicals an expert chemist should be asked to give very careful consideration to the by-products of any decontaminants which are used.

In the case of the two acid spills depicted in this part, the full spread of contamination was not detected for several days. Isolated pockets of acid were later found in drainage catch basins and manholes. Acid was also splashed on fences and lay in small pools in the grass, all of which caused hazards to unwary workers. The pollution caused by the sulfuric acid tank truck spill was widespread and traveled several hundred meters in drains beneath the highway and across fields in which cattle were grazing. It is obviously very important for a full knowledge of the local drainage system near an incident to be made available to and considered by the chemist, water authority, and emergency services.

The decontamination of equipment, tools, personnel, clothing, and vehicles used at the scene of an incident is extremely important. It may be necessary to set up an area for decontamination through which personnel working at the incident must pass if contamination is not to be spread into ambulances and other vehicles leaving the site. Also it is very easy for workers to reuse tools which have been temporarily put on the ground and so become contaminated. The results can be unpleasant. Thorough personal washing by workers before handling food or drink is also important.

In one highway authority, the information given previously in this part is used for training purposes and is briefly summarized in Table 6-1. This table and the Hazchem scale card issued by Her Majesty's Stationery Office are useful aids to memory which are easy to carry.

REFERENCES

1. *Health and Safety at Work Act, 1974,* chap. 37, part 1, sec. 4.

2. *Highways Act, 1959,* part VII, sec. 129, as amended by the Highways (Amendment) Act, 1965, sec. 1, and the Control of Pollution Act, 1974, sec. 22.

3. *Highways Miscellaneous Provisions Act, 1961,* chap. 63, sec. 8.

4. *Spillages of Hazardous Chemicals on the Highway,* U.K. Department of Transport, Circular Roads 27/78, London, 1978.

5. *Road Transport of Hazardous Chemicals,* 1973, *The Chemical Industry Scheme for Assistance in Freight Emergencies (Chemsafe),* rev. 2d ed., 1976, Chemical Industries Association, Ltd., London.

PART 7
British Maritime Transportation Plans

H. P. Lunn
General Manager, Cargo and Operational Services Division
Overseas Containers Ltd.

Introduction / 7-72
Emergencies / 7-73
Repairs / 7-75
Training / 7-75

INTRODUCTION

Safety and the prevention of spills during the maritime transportation of dangerous substances in bulk or in packages can be divided into matters directly concerned with the dangerous substance and matters concerned with the circumstances of handling and transport. Concerning the dangerous substance, the following points are basic:

1. Information concerning the substance and its characteristics that are relevant to handling and transport. Substances should be identified by a United Nations number* as well as by name.

2. The proper containment of the substance in tanks, pipelines, compartments, or packages, compatible with the substance and at least sufficiently strong to withstand the stresses and strains that may be met during the normal course of handling and transport.

3. The proper handling and stowage aboard the ship, taking due account of segregation of a particular substance from an incompatible one, and other aspects.

4. The passage of proper information at the appropriate time to all those concerned about transport and handling, including persons at the port of discharge or in transit.

    The area or berth at which the dangerous substance is to be handled should be properly managed and well ordered. It should be provided with appropriate emergency equipment, taking account of the type of dangerous substance or substances to be handled.
    Personnel involved should be trained to understand the risks to a degree appropriate to their responsibility and involvement. Proper lines of communication with the appropriate emergency services should be established. The Intergovernmental Maritime Consultative Organization (IMCO) document *Recommendations on the Safe Transport,*

---

*United Nations numbers can be obtained from lists published by the United Nations itself or by various other organizations.

Handling and Storage of Dangerous Substances in Port Areas gives useful general guidance on these aspects.

It is important also that the ship be seaworthy in every sense. Guidance is available from a number of IMCO documents concerning ship structure and equipment and others dealing with operation, such as the Convention on Standards of Training, Certification and Watchkeeping for Seafarers, 1978 (not yet ratified by sufficient countries to be in force).

Guidance is also available from manuals published by the International Chamber of Shipping (ICS) and the Oil Companies International Marine Forum (OCIMF).

EMERGENCIES

The whole object of management and planning developed within national regulations and companies' routines is to minimize incidents, but some will inevitably happen. A ship at sea is isolated and cannot normally obtain help from others. The responsibility is therefore a matter for the immediate action of those present, and relevant information should be available to them. Much pressure arises for this information to be available in simple but precise terms for each substance carried, but the problem when related to ships is not easy for several reasons:

1. The quantity of different substances and mixture in stowage with other combustible goods (not necessarily hazardous).

2. The fact that the particular traits of the substances apply to the chemicals themselves and not the packaging. Thus, while the packaging remains intact, the goods are not different from general cargo, and fire spread perhaps can be prevented by the use of water.

3. The complicated nature of the ship's dangerous-goods manifest and/or plan which may list 400 or 500 different items.

IMCO has produced emergency information in somewhat broad terms for every scheduled chemical in the International Maritime Dangerous Goods Code. In addition, many companies organize their own forms of emergency action information. All these are good, particularly the IMCO work, which gives internationally agreed characteristics and emergency actions.

In many cases trouble may start in an area unconnected with dangerous goods. Thereafter, the appropriate action should take account of dangerous goods nearby, which could greatly worsen matters if they became involved, and probably do everything possible to prevent such involvement. In some cases, jettisoning is recommended.

While the foregoing gives some general background to the situation, it may be useful for those in authority who may be concerned with such an emergency to train themselves to think constructively along these lines:

• Emergency	What type?	Fire
		Spillage of liquid or powder
		Leakage of fumes
		Others
		Identification of substance and its effect
• Fire	Action	Inert gas
		Water—jet spray fog
		Foam
		Powder

- Spillage   Wash away or contain; effect of pollution? Fumes?
- Fumes   Wind effect, accommodation, ventilation
- Other matters   Side effects of the above and mixtures
- Personnel   Immediate safety
                    Breathing apparatus
                    Protective clothing
                    Minimizing risks
                    Proper backup

For emergencies involving dangerous goods to be dealt with properly, ships must be provided with adequate equipment. Such equipment should be in addition to the minimum required by regulations. It should be assessed in view of the ship's volume and the type of goods, but it should include at least two sets of self-contained breathing apparatus and four sets of appropriate protective clothing.* This equipment, together with other items, should be well maintained in a specially designated locker to which access is unlikely to be immediately closed by fire.

An essential element concerns shipboard routines additional to the normal good practices of fire drills, etc., which apply whether the ship carries dangerous goods or not.

As stated, the ship should be provided with a comprehensive manifest giving a detailed list of dangerous goods and their stowage position. The latter information is best shown in plan form by class of goods so that personnel can readily envisage the relative position of the dangerous goods.

Although the law usually requires one copy of this information to be given to a ship, it is preferable for two copies to be placed aboard and each to be kept in a separate place so as to minimize the likelihood of destruction in the initial stages of a fire. Suitable places would be the control center (probably the bridge) and, say, the forecastle area (if proper accommodation is available).

On receipt of the information concerning the cargo, ship routine should appoint a responsible officer to consider the emergency implications of the dangerous-goods stowage and to mark the plan accordingly. Some companies do this work ashore and give the ship the plan already marked wtih emergency information, perhaps by means of diamond-shaped stickers carrying a simple code for fire fighting, spills, and personnel protection. While this is useful for smaller cargoes, it may be preferable to consider the full implications of total stowage, which would better assist the master in deciding how best to deal with any emergency that may arise.

It is usual for ships at sea to be regularly patrolled. When dangerous goods are being carried on deck or in vehicle decks, patrols should be instructed to inspect stowage superficially and to report accordingly.

Should dangerous goods be leaking or otherwise present undue danger, the master may think it desirable to jettison. (As mentioned earlier, there are certain items for which IMCO's International Maritime Dangerous Goods Code recommends this course of action when emergency threatens.) This is not necessarily easy to do, and care should be taken to avoid danger from spillage or the possibly severe effects of wetting. It is also necessary to take account of pollution and environmental problems. Any jettisoning should be reported with precise position to competent authority of the flag country via the owners, with details of marks and labeling.

When a spill has occurred aboard a ship, decontamination is necessary. For most substances, copious water is useful, and ships at sea tend to be well provided with facilities,

---

*"Appropriate" means either specifically suited to what the ship is carrying or, for a general-cargo ship, suitable for as wide a range of chemicals as is practicable.

at least on deck. However, water may not be sufficient for all materials, and advice should usually be sought before its use. If the spill is below deck, care should be taken to avoid contaminating bilges and pumping systems.

As in so many matters, successful emergency action arises from previously carried out good routines and practices and the proper maintenance of the emergency equipment. It is most desirable that personnel be practiced in handling hoses and other equipment while wearing breathing apparatus and other protective equipment.

The prime object of all emergency procedures is, of course, to save lives and minimize injury. Therefore, when someone is overcome, rescue must be the first task, but care should always be taken to ensure that the way of doing so is properly thought about, albeit quickly, and the necessary precautions taken to ensure that the rescuers themselves are as safe and able to operate efficiently as possible. *Safe Working Practices aboard Merchant Ships,* published by the U.K. Department of Trade, and other publications tell how to do this; their advice should be followed. Recommendations include a backup team, proper fitting of protective equipment, a lifeline, and a water curtain.

REPAIRS

A general-cargo liner, particularly of a modern type, has packaged dangerous goods aboard much of the time, but it is nonetheless essential that routine repairs can be carried on while the ship is in port. It is also essential that careful regard be taken of the presence of dangerous goods, particularly those with flammable characteristics. Work involving burning or welding should not be carried out in any position where there is risk to cargo. However, since there is no likelihood of a major release of flammable vapor affecting the noncargo parts of the ship engine room or accommodation, repairs can usually proceed unhampered by the presence of the packages of dangerous goods aboard.

The reverse is usually the case when dangerous cargoes are carried in bulk in ships. In such ships repair work must be vigorously controlled and hot work probably prohibited.

TRAINING

It is customary for ships' officers and senior petty officers to undergo formal training in fire-fighting procedures. If they are to serve in dangerous-goods ships, they should also receive appropriate training so that they can properly understand and evaluate the risks involved and thus be better able to take proper action when an emergency arises.

Training should be in two basic areas:

1. For bulk liquefied gas or other specialist carriers, the training should be precise and enter into some detail of the cargoes involved, the emergency and other equipment carried, potential disasters, and ways to avoid them or, if the worst occurs, to save lives.

2. For packaged-cargo carriers, training should be of a more general nature designed to take trainees through the generality of the shipping process, including the basic elements of packaging, and to stress the need for individual decision on, say, fire fighting or spills according to the overall circumstances and the amount of dangerous goods involved. No attempt should be made to make trainees "little chemists" but rather to deal with the practical approach, using the information that should be available to them.

In both cases due emphasis should be placed on the need to:

- Follow through good routines
- Take advice as soon as possible (but not hesitate to take emergency actions designed to prevent spread of the incident)
- Call for help before it is too late

A typical curriculum for dangerous-goods training is issued by the U.K. Merchant Navy Training Board. The ICS gives guidance on bulk training.

PART 8
Coast Guard Organization for United States Coastal Areas

Gregory N. Yaroch, Lieutenant Commander
U.S. Coast Guard

Authority and Organization / 7-77
Research and Development / 7-80
Contingency Planning / 7-81
Response Methodology and Support / 7-82
Summary / 7-83

This chapter deals with the U.S. Coast Guard's approach to assure proper response to pollution discharges involving substances other than oil within the coastal areas of the United States.

AUTHORITY AND ORGANIZATION

An integral program component for controlling pollution in the marine environment is planning to contain, recover, and clean up rapidly any accidental discharges in order to minimize adverse effects. Specific legislative acts have established varying responsibilities for the Coast Guard to respond to emergency situations involving hazardous chemicals.

The National Oil and Hazardous Substances Pollution Contingency Plan[1] assigns the Coast Guard a leading role in working with various federal agencies to ensure a coordinated response to any polluting discharge. This plan is issued under the authority of the Federal Water Pollution Control Act (FWPCA),[2] as amended.

The Coast Guard was given authority to respond to pollution incidents involving oil and certain designated "hazardous substances" occurring in United States coastal areas, ports, harbors, and the Great Lakes. It also has authority to enforce regulations to prevent pollution from vessels and other transportation-related facilities.

The U.S. Environmental Protection Agency (U.S. EPA) shares this authority and has jurisdiction in the inland areas of the United States for response to and prevention of pollution incidents from non-transportation-related facilities. These Coast Guard and U.S. EPA obligations include the assignment of predesignated on-scene coordinators (OSCs) for all areas of jurisdiction, the development of regional and local response plans to identify potential problem areas, the identification of available pollution control resources, and the establishment of a means of responding rapidly and effectively to any pollution incident.

The Coast Guard maintains emergency port task forces at the Coast Guard Captain of the Port and Marine Safety Offices for carrying out these responsibilities. These forces

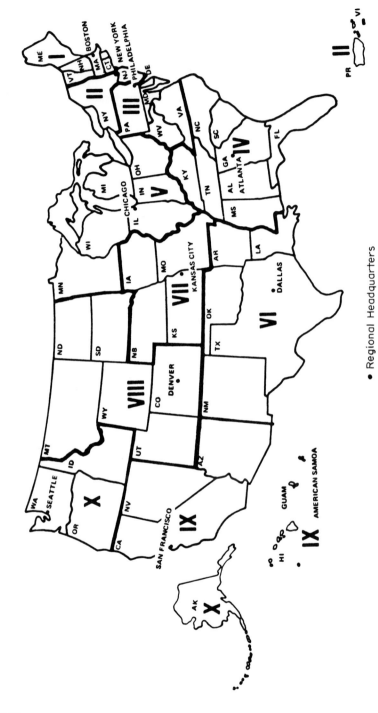

Fig. 8-1 Standard federal regions. [Federal Register, *vol. 45, no. 55, 1980.*]

• Regional Headquarters

consist of individuals specifically trained in techniques and methodologies for responding to pollution incidents.

The FWPCA and the national plan also charge the Coast Guard with maintaining the National Strike Force (NSF), which consists of three teams of specially trained and equipped pollution control experts. These strike teams provide technical advice, equipment, and personnel augmentation to OSCs during pollution incidents, particularly for incidents which exceed local pollution control capabilities.

In addition, the Coast Guard has been designated the appropriate federal agency to be notified when a pollution discharge occurs in the navigable waters of the United States, the contiguous zone, or waters beyond the contiguous zone to 370 km (200 nautical mi). A Coast Guard National Response Center (NRC),* established under the national plan, receives these discharge notifications and further disseminates pertinent information to the appropriate OSC. Once such a report is received, it is the OSC's responsibility to evaluate the situation and initiate whatever federal action may be required.

As a matter of policy, the party responsible for the discharge is encouraged to take appropriate cleanup actions. If the party refuses or its efforts are inadequate or untimely, the OSC will assume responsibility for the cleanup operation, using commercial contractors, emergency task force personnel and equipment, the National Strike Force, state or local response personnel, or any combination of these resources as the circumstances of the incident dictate. Cleanup actions continue until the pollutant has been removed to the extent considered feasible and necessary by the OSC.

If resources, advice, and/or assistance other than that available locally is required, the OSC can request the assistance of the regional response team (RRT). The RRT is the body responsible for coordinating pollution planning efforts in the region. Its membership consists of personnel from various federal agencies in the region (see Fig. 8-1). The RRT also actively seeks the participation of states and municipalities in planning and response efforts.

When assistance beyond regional capabilities is required, the RRT can turn to the national response team (NRT). The NRT, which consists of representatives of various federal agencies at the national level, can help to access needed expertise or assistance from various sources. It also maintains coordination between agencies. Thus, the federal response network can utilize many resources in the United States in a short time. The organization of the RRT and the NRT under the National Oil and Hazardous Substances Contingency Plan is shown in Fig. 8-2 and is discussed in greater detail in Part 1 of this chapter.

In addition to specific environmental mandates of the FWPCA, the Ports and Tanker Safety Act[3] directs the Coast Guard to safeguard the nation's ports, waterways, persons, and property from destruction, damage, loss, or injury and to protect navigable waters and the resources therein from environmental harm. The Coast Guard, in carrying out its responsibilities under the FWPCA, the National Oil and Hazardous Substances Pollution Contingency Plan, and the Ports and Tanker Safety Act, endeavors to respond to all reported chemical releases occurring within its jurisdiction.

The Coast Guard is also routinely called upon to assist local authorities in dealing with pollution emergencies which occur close to Coast Guard facilities even when the affected area is not within an area of normal Coast Guard jurisdiction. The National Oil and Hazardous Substances Pollution Contingency Plan recognizes the need to provide such expertise and encourages utilization of the response mechanism created by the plan under other authority when necessary.

---

*The NRC is operated by Coast Guard personnel (because it has the people and resources to do so) under the auspices of the national response team.

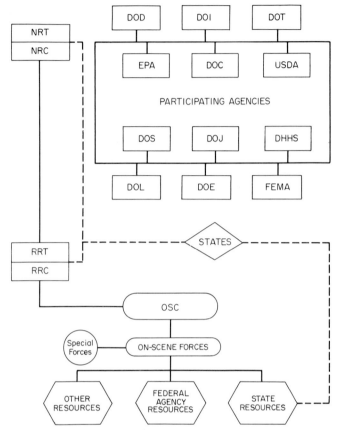

Fig. 8-2 Organization of the National Oil and Hazardous Substances Pollution Contingency Plan. [Federal Register, *vol. 45, no. 55, 1980.*]

RESEARCH AND DEVELOPMENT

In preparation for these responsibilities, research and development efforts have been under way since 1970 to develop and identify state-of-the-art equipment, information, and methodologies for responding to nonpetroleum spills. The problem areas addressed by the Coast Guard's Hazardous Chemical Discharge Amelioration Program are reflected in the following project areas:

1. *Prevention and reduction.*   This area encompasses the development of equipment and methods to prevent the discharge of a hazardous chemical from an endangered vessel and to stop or reduce an ongoing discharge originating from a vessel.

2. *Containment, treatment, and recovery.*   This area deals with the development of response techniques for a wide range of chemicals (i.e., mixers, floaters, sinkers, and vaporizers).

3. *Disposal.*   This area identifies disposal techniques for recovered hazardous chemicals.

4. *Detection, identification, and quantification.* This area encompasses the development of equipment and methods to:

   a. Locate a discharge and determine its boundaries
   b. Identify the chemical
   c. Determine the concentration of the chemical

5. *Personnel protection.* This area concentrates on the development of equipment and methods to protect personnel engaged in hazardous chemical spill response. Recent developmental efforts in personnel protection have concentrated on total body encapsulation employing a self-contained breathing system.

6. *Hazard assessment.* This area has resulted in the development of the Chemical Hazards Response Information System (CHRIS), of which the Hazard Assessment Computer System (HACS) is a part. These systems are designed to provide timely information essential for proper decision making by Coast Guard personnel during emergencies.

CHRIS has been utilized in the field for some time, and prototype encapsulating outfits are presently at each strike team location for testing and evaluation. The remaining project areas are in varying stages of research and development.

Although the response organization utilized for responding to the discharge of a hazardous chemical is similar to that for responding to an oil discharge as established by the National Oil and Hazardous Substances Pollution Contingency Plan, the state of the art in dealing physically with discharges of hazardous chemicals is limited. Thus, owing to these limitations and to the inherent risks involved in dealing with certain hazardous chemicals, certain available response actions will occasionally prove infeasible.

CONTINGENCY PLANNING

The task of removing or mitigating the effects of any pollution incident is affected by a combination of circumstances surrounding the discharge, including the source, type of pollutant, quantity, environmental conditions, and physical characteristics of the affected area. In each case, the particular set of circumstances must be rapidly assessed and an appropriate course of action determined. The responsibility for making these decisions rests with the OSC. The OSC must determine the nature, amount, and location of the substance discharged, the potential impact of the discharge on public health and welfare, and the resources which may be affected and priorities for protecting them. This can be accomplished rapidly only through contingency planning and the identification of accessible resources. The potential for harm to the environment and to public health and welfare from an accidental release of hazardous chemicals underscores the necessity for preplanning a response strategy. Such a strategy is effectively applied on a local level through the formulation of hazardous chemical spill response plans. The following points are representative of the considerations and actions that are required by the OSC in preparing to deal with the unique problems associated with emergencies involving hazardous chemicals:

1. *The development of a list of products carried within each OSC's area of responsibility.* This list is designed to identify the types and amounts of hazardous chemicals produced, utilized, or transported in the area; it includes a detailed list of commodities designated as "hazardous substances" by the U.S. EPA or as "hazardous materials" by Part 49 of the *Code of Federal Regulations.*

2. *The identification of points of production, utilization, and storage of these hazardous chemicals.*

3. *The determination of the modes of transporting hazardous chemicals in the area and an examination of the traffic patterns utilized.* There is always a potential for pollution entering navigable waters from nearby highways, railways, and pipelines as well as from the water mode of transport.

4. *The assessment of potential hazards which may exist upon the release of the hazardous chemicals which have been identified in local traffic patterns.* Worst-case situations and most likely cases are determined in making these assessments. The utilization of CHRIS and HACS is extensive in this phase of planning. In assessing hazards, OSCs consider such factors as exceptional volatility, reactivity, toxicity, proximity to vulnerable natural resources, and proximity to populated areas.

5. *The establishment of methods of chemical analysis to determine the identity of unknown pollutants.* Arrangements with local laboratory facilities capable of conducting chemical analyses are necessary to enable positive identification of the pollutant prior to active response operations. Until the identity of a suspected hazardous chemical is confirmed, the affected area is cordoned off to prevent entry. Once the identity has been established, additional steps may become necessary to monitor concentration levels continually in an affected area in order to ensure that human exposure is maintained at acceptable levels. This service can be performed by an industrial hygienist-toxicologist or by someone qualified to make such measurements.

6. *The determination of the capabilities of local commercial cleanup contractors to respond to a discharge of a hazardous chemical.* These capabilities may range from no preparedness at all to full capabilities for specific or diverse products. Chemical manufacturers are also contacted to determine whether or not they possess emergency response teams capable of responding to incidents involving their products. When response teams exist, their accessibility is ascertained.

7. *The establishment of liaison with local medical facilities to ensure ready access to medical support during a chemical spill incident.* Efforts are made to ensure that medical facilities are made aware of the types of chemical products moving in the area. The medical community is then better prepared for injuries caused by accidental releases of these products.

8. *The coordination with local officials in establishing procedures to be followed if the evacuation of civilian personnel should become necessary.* Although the OSC may determine that evacuation is advisable, responsibility to make such a decision rests with local civilian officials. Close liaison with local police and fire officials is an absolute necessity, especially since they are usually the first emergency response personnel to arrive on scene.

RESPONSE METHODOLOGY AND SUPPORT

To maintain a coordinated federal response to discharges of hazardous chemicals, the OSC proceeds autonomously according to a prescribed policy and safety guidelines. Subject to the availability of resources, every reported discharge involving hazardous chemicals is promptly investigated. Extreme care is taken during the initial investigation while determining the extent of the spill and the identity of the chemical or chemicals involved. Although the chemical identification process is time-consuming, it is essential to identify the pollutant before a clear course of action is established and removal and mitigatory efforts are undertaken. A "passive" response attitude, consisting of evacuation of the affected area and maintenance of a safe perimeter, is *always* undertaken whenever the identity of the chemical is unknown or uncertain.

The Coast Guard's detection, identification, and quantification efforts remain in varying stages of research and development. Therefore, local arrangements with laboratories and the chemical manufacturing industry facilitate the identification of unknown pollutants and the monitoring of human health hazards. Additional sources of assistance for chemical identification exist in other federal agencies. The accessibility of regional EPA laboratories is determined at the RRT level.

The use of outside resources to augment the OSC's response forces is paramount in emergency situations involving hazardous chemicals. Many chemical manufacturers have chemical response teams that have been trained to deal with accidents involving specific products made by their firms. Full use of these response teams is made whenever they are available and have the needed expertise. Normally, initial access to industrial emergency response assistance is gained by getting in touch with the Chemical Transportation Emergency Center (CHEMTREC) operated by the Chemical Manufacturers Association. A number of commercial cleanup contractors may also have expertise in chemical response. Such contractors are used in federal response to chemical discharges in preference to state and federal response personnel whenever they have the required expertise and equipment and are available.

The National Strike Force and the EPA environmental response team (ERT) also possess expertise in chemical spill response. The NSF is composed of three teams: the Atlantic, Gulf, and Pacific strike teams. Each is an operating unit of the Coast Guard with a commanding officer and is charged with providing special pollution control equipment, technical knowledge, and supervisory assistance to the Coast Guard, EPA, or other federal OSCs. The EPA's ERT operates in similar fashion (i.e., at the disposal of the OSC) and possesses specific technical expertise in such areas as biology, chemistry, toxicology, and waste disposal.

Although some risks may be taken to protect public health and welfare, risks to personnel are reduced to the minimum possible level consistent with the operational situation and *are not* incurred for purely environmental reasons.

Requirements have been established for the periodic medical evaluation of all Coast Guard personnel who are exposed to specified hazardous chemicals or physical agents. Medical evaluation guidelines have been established to monitor the health of personnel who are potentially exposed to hazardous chemicals during spill response operations. Personnel are examined annually and/or immediately following a response to a pollution incident involving a hazardous chemical during which the potential for harmful exposure existed or direct exposure occurred.[4] Recognizing the complexities of an effective medical monitoring system for response personnel, the Coast Guard will continue to seek improvements in this program as experience is gained in dealing with incidents involving hazardous chemicals.

SUMMARY

The *national response team (NRT)* consists of representatives from the various participating federal agencies (Fig. 8-2) and serves as the national body for planning and preparedness actions prior to a pollution discharge. It acts as an emergency response team to be activated in the event of a discharge involving oil or hazardous chemicals which

1. Exceeds the response capabilities of the region in which it occurs
2. Transects regional boundaries

3. Involves significant numbers of persons or significant amounts of property

4. Results in a request by any participating agency representative

The *National Response Center (NRC)* is located at U.S. Coast Guard headquarters in Washington (telephone: 800-424-8802). It receives notification of polluting discharges and emergency situations involving the uncontrolled release of hazardous chemicals and disseminates this information to the appropriate OSC for action. The NRC also provides physical communications facilities for the coordination and control of a pollution emergency if national-level involvement is necessary.

The *regional response team (RRT)* consists of representatives of the primary and advisory agencies located within the region. It performs functions similar to those carried out by the NRT at the national level. The RRT is activated automatically in the event of a major or potential major discharge [10,000 gal (37.9 m^3) of oil inland, 100,000 gal (378.5 m^3) of oil in coastal waters, or *any* discharge of a hazardous chemical threatening public health and welfare] or during any other emergency upon the request of any primary-agency representative.

The *emergency port task forces* consist of personnel trained to evaluate, monitor, and supervise pollution response. Additionally, they have limited "first aid" response capability to deploy equipment prior to the arrival of a cleanup contractor. These forces are established in major port areas; they are composed of Coast Guard personnel and may, in certain areas, have additional representation from other federal, state, and local agencies.

It is the policy of the Coast Guard to ensure that timely and effective response action is taken to control and remove discharges or the threat of discharges of oil and hazardous chemicals in port and coastal waters. This course of action proceeds unless such removal is being conducted properly by the party responsible for the discharge or if such removal will cause greater environmental damage or risk to personnel than the discharge alone. Coordination and direction of federal pollution control efforts at the scene of a discharge of hazardous chemicals is accomplished through the *on-scene coordinator (OSC)*. The OSC is the single federal official predesignated by the regional contingency plan as the responsible individual for initiating containment and countermeasures.

The OSC must quickly determine the nature, amount, and location of the material discharged, the probable distance and the time of travel of the pollutant, the potential impact on human health and welfare, and the resources and installations which may be affected and the priorities for their protection. If the predesignated OSC cannot respond immediately to a discharge, the first federal official on scene from any of the participating agencies shall assume the coordination role until the arrival of the predesignated OSC. Whether a discharge is removed by the federal government or by the party responsible for the discharge, the OSC must determine when removal is complete and operations may be terminated.

The broad base of legislative authority in this area provides Coast Guard personnel with the tools necessary for controlling potentially disastrous situations arising from the discharge of hazardous chemicals. These situations, if uncontrolled, may seriously threaten the safety and security of vessels, structures, facilities, and the general population. Success in mitigating such disasters lies in the state of preparedness of the entire response mechanism, from the OSC to the NRT. Preparedness to carry through an aggressive response action from notification of discharge to termination of removal is provided by *contingency planning*. The Coast Guard's experience in preplanning for emergency situations has contributed to a successful record in minimizing the potential for harm to the environment and public health and welfare within the coastal areas of the United States.

REFERENCES

1. "National Oil and Hazardous Substances Contingency Plan," *Code of Federal Regulations,* Title 40, Part 1510.

2. *Federal Water Pollution Control Act,* 33 USC 1321, as amended by the Clean Water Act, Pub. L. 95–217; 91 Stat. 1566.

3. *Ports and Tanker Safety Act,* Pub. L. 95–474, 33 USC 1221.

4. *U.S. Coast Guard Safety and Occupational Health Manual,* Commandant Instruction M5100.29, U.S. Coast Guard, November 1978.

PART 9
Fire Service Role

Charles L. Page
Training Specialist
Fire Protection Training Division
Texas Engineering Extension Service
The Texas A&M University System

Introduction / 7-86
Authority of the Fire Service / 7-86
Planning for Response to Hazardous Material Incidents / 7-87
Training / 7-93
Summary / 7-96

INTRODUCTION

Historically, the fire service has been the public's first line of defense in most emergencies. Today, fire departments respond not only to fires of every conceivable nature but also to drownings, bomb threats, attempted suicides, cave-ins, building collapse, and natural disasters such as earthquakes, tornadoes, hurricanes, and floods. Many fire departments also provide emergency medical services for their communities.

As technology has increased, so have the number and the amount of hazardous materials produced, stored, and transported. Fire and police departments and state troopers normally respond first to emergencies concerning hazardous materials. The actions taken by these groups in an emergency may make the difference between an incident causing minimal damage and a disaster.

AUTHORITY OF THE FIRE SERVICE

In most nations, the fire department has broad authority during emergencies. In the United States, for example, the Constitution delegates public safety to the several states. State and local governments are responsible for fire protection, law enforcement, and other aspects of public safety. In most states, statutory authority places the fire chief or the chief's representative at the scene of a hazardous material emergency in charge of operations concerning imminent fire-related hazards. If a clear and present fire hazard exists, the ranking fire officer present is generally acknowledged as the sole authority until that hazard is eliminated.

Precedents have held that such authority may be extended through common or customary law to include the release of certain hazardous materials (e.g., the dispersion of a toxic nonflammable gas such as chlorine). The authority may be extended even though there is no evidence of an immediate fire hazard.

Each fire chief should review state laws, local ordinances, and legal precedents with the city attorney. The chief should also understand national water pollution laws.

Equally important, the fire chief should know the duties and understand the role of the federal on-scene coordinator (OSC). Public safety and environmental protection demand that the fire chief and the federal OSC understand and respect each other's responsibilities. The public is best served when federal, state, and local authorities cooperate and respect the duties and expertise of each other.

PLANNING FOR RESPONSE TO HAZARDOUS MATERIAL INCIDENTS

It is essential that each fire department have a preplan or contingency plan for hazardous material incidents. The following may prove helpful in planning for response to such incidents.

The Hazardous Material Survey

This survey should determine what hazardous materials are stored within the community and in what quantity, where they are located, and in what kind of containers. As far as possible, the survey should determine the primary modes as well as the standard routes by which these materials are transported into and through the area. Railroads, highways, waterways, and pipelines should be included in the survey.

In the survey, owners of property on which hazardous materials are stored should be asked to mark specific locations of hazardous materials with National Fire Protection Association (NFPA) 704M placards. (The NFPA 704M placard system should be used in accordance with guidelines set forth in *NFPA Standard 704-M-75*.[1]) Fire department response, or "run," cards (cards upon which the dispatcher may find, among other pieces of information, the probable contents of a building and the special hazards that fire fighters might encounter) should be updated to reflect hazardous materials that might be involved should an emergency occur at a specific location.

The survey should assess the release potential of hazardous material associated with such facilities as chemical plants, agricultural supply warehouses, bulk-storage units, railroads, and highways.

If the survey is sufficiently detailed, the preplanning personnel should have a reasonably accurate estimate of the types of spills that might occur, the probable magnitude of the spill most likely to occur, and the probable limits of the maximum spill that could occur.

Preparing the Preplan or Contingency Plan

In a hazardous material incident, it is essential that fire officials be able to anticipate both the immediate real dangers and the potential for disaster. Critical parameters of the plan should include, as a minimum, procedures for evacuation, tactics for fire protection and suppression, and strategies for waterway protection.

Resources

The preplan should inventory materials such as aerial photographs, street and topographic maps, and plans of storm sewers as well as sanitary sewers and make sure that these materials are readily accessible in case of a hazardous material incident. The plan should provide both direction and information and should be flexible. It should list personnel, equipment, and supplies necessary to implement the plan, while also providing a mechanism for activating the plan.

The plan should also provide for the safety of personnel. Safety measures may include furnishing protective equipment such as acid suits and breathing apparatus, providing for medical examinations during or after the incident, and rotating personnel during weather extremes.

It should also outline steps for protecting the public. Any evacuation plan should provide both a means for alerting persons endangered by the incident and assembly points for emergency food, clothing, and shelter (Fig. 9-1). Disaster relief agencies such as the Salvation Army and the American Red Cross may be helpful with this portion of the plan.

The plan should include arrangements for aerial surveillance of the incident, should it be needed.

It should also allow for the worst possible conditions. Severe weather such as icing bridges, snowstorms, thunderstorms, and flooding contribute to the frequency of hazardous material incidents (Fig. 9-2).

It should similarly reflect that incidents often occur during rush-hour or holiday traffic, near a large gathering of people (such as a major sporting event), in inaccessible areas, in areas with limited or no water supply, or during a major fire. The plan should be designed so that it works even if telephone lines are jammed, radio problems occur, or key persons are out of town.

The plan should include a list of federal, state, and local agencies and others who should be notified in the event of a hazardous material incident. It should include sources of aid and assistance. A list of emergency phone numbers prepared by the Fire Protection Training Division, Texas Engineering Extension Service, Texas A&M University Sys-

Fig. 9-1   Fire fighters evacuate a hazardous material incident area. [*Photo: Bill Meeks,* The Eagle.]

Fig. 9-2 Weather conditions contributed to this transportation-related incident. [*Photo: Bill Meeks, The Eagle.*]

tem, to assist Texas fire departments in preparing lists of contacts for aid and assistance is shown in Table 9-1.

In the United States, the list should certainly include the Chemical Transportation Emergency Center, Chemical Manufacturers Association (CHEMTREC), toll-free number (800-424-9300) and the National Response Center (NRC) toll-free number (800-424-8802). In Canada, it should include the regional phone number for the Trans-

TABLE 9-1 Emergency Phone Number List Prepared for a Fire Department Contingency Plan

CHEMTREC (Chemical Transportation Emergency Center)
    Call toll-free when truck or train wreck involves hazardous materials.
    Give name of material, transportation line, cargo numbers, and other
    shipping information when possible. CHEMTREC will provide
    immediate information by phone and will notify shipper for you.
	1-800-424-9300

National Response Center
    Report any oil or hazardous substance spill that has reached or may
    reach a waterway.
	1-800-424-8802

Department of Transportation	202-426-1830
Bureau of Explosives	202-293-4048
DOT Office of Pipeline Safety Report pipeline gas leaks.	202-426-0700
Radiological Emergency hotline (Texas Department of Health)	512-458-7460
Nuclear Regulatory Commission Report accidents involving nuclear or radioactive materials in Texas.	505-264-4667

Table 9-1 Emergency Phone Number List Prepared for a Fire Department Contingency Plan (*Continued*)

Missouri Pacific Transportation Control Center		314-622-2119 314-622-2224
Southern Pacific Railway Main Office: Hazardous Material Control		713-223-6000 713-223-6304 or 713-223-6307
Southern Railways System Safety Office		404-688-0800
Santa Fe Railway: Operations Dispatcher		817-773-3451
Texas Railroad Commission: LPG Division Report LPG emergencies.		512-475-4351
Texas Department of Water Resources	Weekdays Nights and weekends	512-475-2786 512-475-2651
Report oil and hazardous substance spills.		
Poison Center (Southeast Texas)		713-654-1701

Department of Public Safety
  Highway Patrol
  District Office
		_____

Emergency Operations Center, Austin (Call local or district office *first*.)		512-452-0331 Extension 295
Police		_____
Constable		_____
Sheriff's Office		_____
City manager and/or mayor		_____
Ambulance(s)		_____ _____
Doctor(s)		_____ _____
Hospital(s)		_____ _____

Explosive Disposal
  East Texas and Louisiana, Fort Polk, Louisiana
  South Texas, Naval Air Station, Corpus Christi, Texas
  West Texas, Fort Bliss, Texas
		318-578-5505
		512-939-2991
		915-568-8905

	or	915-568-8703
North and Central Texas, Fort Hood, Texas If no answer, call		817-685-2309 817-685-2176
American Rescue Dog Association:	Texas alerting phone National alerting phone	214-271-0079 206-937-3460
Volunteers with search dogs for locating disaster victims (these are not bloodhounds).		
Mutual aid		_____ _____ _____
Fire Protection Training Division and		713-845-7641
Oil and Hazardous Material Control Training Division		713-845-3418

portation Emergency Assistance Program, Canadian Chemical Producers' Association (TEAP), and the regional number for Environment Canada.

These sources of assistance may be able to provide important information regarding what steps to take or not to take during a hazardous material incident. For example, in transportation emergencies CHEMTREC notifies the shipper, who in turn may send a trained industry response team or at least provide advice concerning ways to handle the product. See Chap. 3, Part 4.

The list should also include telephone numbers for state and provincial agencies, such as the state department of natural resources, state troopers (in Canada, the provincial police or the Royal Canadian Mounted Police), the health department, the poison center, and civil defense. Local telephone numbers should include police, hospitals, the city manager, mutual aid, the ambulance service, and local chemists or chemical engineers, among others.

The plan should list resources necessary to stabilize and control hazardous material incidents and the availability of resources. For example, the fire officer must be capable of accurately estimating personnel needs as well as determining the best means of meeting the needs. The plan should outline ways to obtain bulldozers, dump trucks, vacuum trucks, acid suits, cranes, and other special items such as chlorine tank and valve repair kits. A city's public works department should not be overlooked as a resource for dump trucks, barricades, earth-moving equipment, sand, and other resources. The *Disaster Checklist* prepared by the International Association of Fire Chiefs' Emergency Preparedness Committee and *Fire Officer's Guide to Disaster Control*[2] may prove helpful in preparing the resource list.

The U.S. Coast Guard and Environment Canada each maintain computer systems to assist in locating spill control equipment. The Coast Guard system is called SKIM (Spill Cleanup Inventory System). In an emergency, the fire officer may obtain information from SKIM by calling the NRC at 800-424-8802. In Canada, information from NEELS (National Emergency Equipment Locater System) may be obtained by calling Environment Canada at 613-997-3742 or any regional office of Environment Canada. The U.S. Environmental Protection Agency's (U.S. EPA's) regional contingency plans may also prove helpful in determining the location of specialized equipment or contractors.

Response

The plan should outline certain standard operating procedures for hazardous material incidents. It should include guidelines for the fire department dispatcher upon receiving notification of a hazardous material incident and for the fire officer in charge responding to the incident.

If the dispatcher is to take appropriate action, he or she must know (1) what the nature and the location of the incident are, (2) what hazardous material or materials are involved (if known), (3) whether or not any product is escaping, (4) whether or not there are injuries, and (5) whether or not there is an immediate danger to area population.

When dispatching fire department units, the dispatcher should alert them that the incident involves hazardous materials. Responding police and ambulance crews should also be advised of any known dangers. Using prepared references, the dispatcher may be able to provide characteristics of the hazardous material by radio to responding personnel. The dispatcher may relay information received from CHEMTREC to the fire officer responding to an incident. If radiotelephone patch equipment is available, the dispatchers may be able to connect or "patch through" the fire officer to CHEMTREC for direct communications. During initial communications, officials should spell the name of the hazardous material to avoid confusion because many materials have similar-sounding names.

The officer in charge responding to a hazardous material incident should, among other things, consider the protection of personnel, making sure that all persons are wearing full protective clothing and breathing apparatus. The dangers of specific hazardous materials may necessitate special protective clothing such as acid suits and rubber gloves. The normal response route to a location may need to be changed to allow apparatus to approach from upwind. If the hazardous material is heavier than air, the approach should be made from uphill where possible. Driving vehicles into flammable gas or liquid may cause a fire or explosion. If flammable gases such as propane or flammable vapors such as gasoline are drawn into the air intake of a diesel engine, they may cause the engine to overspeed and disintegrate.

Upon arrival at the incident, the fire officer must first assess the situation. As quickly as possible, the officer should determine or confirm what hazardous materials are involved. Binoculars are useful for safely checking placards from a distance. Shipping documents and labels may also identify the materials involved. Useful information may sometimes be obtained by talking to the driver, conductor, brakeman, nightwatchman, or others involved with the storage or transportation of the material. One should not guess what materials are involved.

Extreme caution should be exercised in attempting to identify the material. The potential hazards of the material may be established by using reference books such as the NFPA's *Fire Protection Guide on Hazardous Materials*[1] or the Association of American Railroads' *Emergency Handling of Hazardous Materials in Surface Transportation,*[3] by calling the NRC and requesting information from the Chemical Hazards Response Information System (CHRIS), and by contacting the manufacturer's and/or carrier's emergency response team.

The EPA maintains a computer system, the Oil and Hazardous Materials Technical Assistance Data System (OHM-TADS), which can be helpful in identifying unknown hazardous materials by their characteristics such as smell, color, solubility, and specific gravity. OHM-TADS can also provide technical data on over 1000 substances. The Hazardous Materials System (HAZMATS) of Environment Canada can provide technical data on some 3500 chemicals. Information may be obtained by telephone from Environment Canada's regional offices. For details on information systems, see Chap. 3.

The plan should include guidelines to assist the fire officer in establishing priorities for decisions. Human life must take first priority. Decisions concerning rescue and evacuation must be made as soon as possible. If evacuation is necessary, the officer must consider:

1. Geographic area to be evacuated
2. Time available for evacuation
3. Where to relocate evacuees, such as schools or Salvation Army or Red Cross shelters
4. Personnel available to assist with evacuation, i.e., fire and police personnel, state troopers, civil defense personnel, national or state guard units

Next, the officer should decide what steps are necessary to control the situation. If the material is burning, the officer should decide whether or not to extinguish the fire. In some cases it may be best to let the fire continue to burn, particularly if the incident is in an unpopulated area and there is danger of an explosion, or if the material itself is more dangerous than the by-products of combustion (for example, H_2S gas), or if a heavier-than-air flammable gas is involved and the fuel supply cannot be cut off.

The plan should also provide for establishment of a command post. In an incident such as a tank truck rollover, for example, a chief's car or other vehicle may serve the purpose (Fig. 9-3). However, during major incidents, the plan should provide for the

Fig. 9-3   Fire chief's automobile serving as the initial command post during a hazardous material incident. [*Photo: Bill Meeks,* The Eagle.]

establishment and operation of a command post near the location of the incident itself. The command post provides a focal point for the officer in charge to receive information and to issue instructions to personnel and to mutual-aid fire units responding to the incident. Additionally, the command post serves as a staging area for equipment and personnel necessary to control the incident and provides a central meeting place for all response personnel involved in control and cleanup. The command post coordinates the actions of local public safety personnel, industry response teams, and state and federal agency representatives to handle incidents safely and properly.

Plan Evaluation

After the plan has been drafted, it should be evaluated. Particularly suitable for such evaluation is the checklist in Table 9-2, adapted from the oil spill control course conducted by the Oil and Hazardous Material Control Training Division of the Texas Engineering Extension Service.

After the plan has been evaluated and the necessary changes made, the fire chief should distribute draft copies to all officers for their constructive criticism . At this point it may be advantageous to conduct classroom simulations using a tabletop or rear-screen simulator. After the simulations, the plan should be revised as necessary. It is advisable to conduct full-scale drills to test the plan. After the drills, reviews and critiques should be conducted to remove any remaining flaws from the plan. The information in the plan should be periodically checked and updated.

TRAINING

Experienced, well-trained fire-fighting staffs are essential for successful implementation of the plan. Training, orientation, and motivation of personnel are important factors. Fire

TABLE 9-2   Contingency Plan Evaluation Checklist

1. General
    a. Is the problem described?
    b. Is prevention emphasized?
    c. Does the plan specify actions in the event of an incident?
    d. Are there too many unnecessary details?

2. Equipment
    a. Does the plan include a list of resources such as earth-moving equipment, tank trucks, and specialized equipment?
    b. Does the plan include mutual aid when necessary?
    c. Does the plan include guidelines for the use of fire-fighting chemicals and dispersants?
    d. Does the plan provide for a command post and for adequate communications equipment?

3. Notification
    a. Are the dispatcher's duties outlined?
    b. Does the plan specify who will handle public information releases?
    c. Does the plan fail if certain officers are out of town?

4. Procedures
    a. Have fire and explosion prevention been emphasized?
    b. Does the plan provide for alternative action when necessary, such as during adverse weather conditions?
    c. Does the plan provide for personnel safety including breathing apparatus, protective clothing, specialized chemical-resistant suits, etc.?
    d. Are there plans for controlling the spill flow?
    e. Does the plan provide for adequate documentation such as a detailed written report, photographs, and videotapes?
    f. Are guidelines for final disposal outlined?
    g. Does the plan provide for protection of potential victims, their property, public water supplies, wildlife, etc.?
    h. Does the plan provide for security of the spill area?
    i. Does the plan outline communication procedures and specify the radio frequencies to be used for operations, mutual aid, logistics, etc.?

Fig. 9-4   Spill training tank at the British Fire Service Technical College. [*Photo: David White, College Station, Texas.*]

Fig. 9-5 Fire fighters experience hands-on training using this loading-terminal simulation at Texas A&M University. [*Photo: Fire Protection Training Division, Texas A&M University.*]

Fig. 9-6 This gasoline-processing-unit simulation is one of many training fires used by the Fire Protection Training Division of the Texas Engineering Extension Service at Texas A&M University. [*Photo: Fire Protection Training Division, Texas A&M University.*]

officers responsible for implementing the plan must be trained to evaluate effectively the changing situations at the spill incident site and to make the best decisions possible in order to protect the public, prevent fires and explosions, minimize environmental damage, and protect property. Training should be conducted at regular intervals in order to accommodate changes in fire-fighting personnel and technological advances and to introduce new equipment and strategies.

Personnel should be taught a basic understanding of hazardous materials. They should learn the characteristics of the most common hazardous materials and the way in which these react to fire, shock, and exposure to water or to air. A basic understanding of how gases react to heat and pressure (Boyle's law and Charles' law) should also be learned.

Personnel should learn to use standard sources of information such as CHEMTREC, CHRIS, OHM-TADS, *Fire Protection Guide on Hazardous Materials,* and *Emergency Handling of Hazardous Materials in Surface Transportation.* They should learn the basic construction and safety devices of tank trucks and rail tank cars. Fire fighters should learn to recognize Department of Transportation (DOT) placards and be alert to some shortcomings of the placarding system (e.g., the fact that the nonflammable gas placard used for ammonia does not reflect the dangers involved).

The fire department should conduct both classroom and full-scale drills. Whenever possible, the drills should involve representatives of other agencies or organizations who may participate in the control and cleanup of a hazardous material incident. Simulators, either rear-screen or tabletop, present hypothetical drills effectively during officer-training sessions.

In the United Kingdom the Fire Service Technical College conducts hazardous material training for British fire brigades. A special tank approximately 9.1 by 1.2 m (30 by 4 ft) simulates currents and waves in streams and rivers; in it trainees learn to use booms and other containment equipment (Fig. 9-4).

Large-scale training exercises should be conducted once or twice a year to determine the time required to mobilize personnel and equipment resources, obtain other specialized equipment, reach the incident area and take control measures, and estimate the time required to evacuate the public from the danger area. Fire fighters should also receive hands-on training in combating large-scale training fires (Fig. 9-5). Hands-on training is essential for fire fighters. They should receive such training involving hazardous material spills and fires before being sent to hazardous material incidents (Fig. 9-6).

SUMMARY

The fire service plays an essential role in confining, controlling, and containing hazardous material incidents. It performs this role best when it bases its comprehensive training program on careful preplanning.

REFERENCES

1. *Fire Protection Guide on Hazardous Materials,* 7th ed., National Fire Protection Association, Quincy, Mass., 1978.

2. *Fire Officer's Guide to Disaster Control,* National Fire Protection Association, Quincy, Mass., 1978.

3. *Emergency Handling of Hazardous Materials in Surface Transportation,* Association of American Railroads, Bureau of Explosives, Washington, 1977.

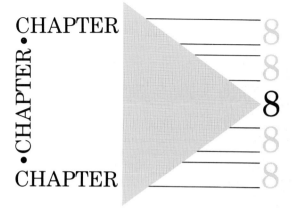

CHAPTER 8

Spills into Watercourses

PART 1
Sampling, Analysis, and Detection / 8-2

PART 2
Bioassay / 8-14

PART 3
Dispersion Modeling / 8-22

PART 1
Sampling, Analysis, and Detection

Joseph P. Lafornara, Ph.D.
Acting Chief, Chemical Evaluation and Safety Section
Environmental Response Team
U.S. Environmental Protection Agency

Introduction / 8-2
Sampling Spill-Contaminated Water Bodies / 8-2
Chemical Analysis / 8-5
Remote Sensing and Detection / 8-9

INTRODUCTION

The success of a response to a hazardous material spill into water often depends on the quality of the sampling, analysis, and detection program that is implemented to support it. Although the success of an oil spill response or cleanup is readily apparent, many hazardous materials dissolve in water. Therefore, sophisticated technology must be used to discover and monitor their presence. This chapter outlines some of the required technology and assesses constraints. Topics discussed are sampling methods and sample preservation techniques, chemical analysis, and remote-sensing and aerial reconnaissance. The section on sampling refers to definitive texts and manuals on the subject. The section on chemical analysis does not describe chemical methods in detail but addresses mechanisms that can be used to provide the needed analysis in various types of spill situations. Several sources are cited for actual detailed analytical procedures that can be used during a spill response activity. The section on remote sensing describes how this technique, including the mechanisms and sources of the technology, can be useful for spill response.

Superimposed on all aspects of environmental-emergency response is the need for strict attention to safety procedures. Although such procedures are not cited specifically in each section, the reader is advised that before any sampling and analysis or other on-site spill response activity is undertaken, a detailed safety program should be planned and implemented to ensure that there is no undue risk to personnel. This safety program should include an assessment of the hazard from the spilled materials, the selection of equipment and procedures to safeguard the workers, and the use of a standard operating sequence during very dangerous procedures.

SAMPLING SPILL-CONTAMINATED WATER BODIES

The methods employed for sampling spills are usually the same as those used for stream surveys and the monitoring of industrial effluents. Several publications describe these methods in detail and include equipment needs, container specifications, and procedures for obtaining samples.[1-6] Table 1-1, drawn from one of these publications, summarizes the sampling and preservation techniques needed for the determination of several common water quality parameters. Similar sampling requirements for other determinations are often provided by the analytical method. Close consultation with the analyst in the laboratory is recommended to ensure that valid handling procedures are followed. Prescribed sample labeling and documentation procedures should be followed, and chain-of-custody procedures are advisable if the analytical data are to be used as evidence in any future spill litigation.

Even if all equipment is in place and good sampling methods are practiced by trained, experienced personnel, the invisibility and transient nature of chemical spills can make it extremely difficult to obtain a sample that is truly representative of the spill plume. Sampling at a given point at the wrong time may yield a sample which contains little or none of the spilled material. For this reason, it is necessary to sample more frequently and in several locations to ensure that the spill plume is actually tapped. It may be possible to use field tests to detect the presence of a spill plume prior to taking a sample for laboratory analysis and thus to reduce the number of samples (see subsection "Field Analysis Kits" below). Visual observations and dispersion models can also be valuable in estimating the movement of a spill plume. For instance, a very toxic chemical may dissolve into the water column, killing fish and other animal life. Frenzied activity of fish can indicate that a spill plume is moving through an area. However, because it may also indicate that the plume has passed, frequent sampling at a point far enough downstream from this area may be necessary.

When no immediate biological effect is apparent, the problem of selecting a sampling location is more difficult. Estimates of time of travel may have to be based on on-site determinations of current speed and direction (for large lakes, estuaries, etc.). These determinations may be performed with sophisticated techniques such as current meters, dye studies, or marker buoys or with techniques as simple as timing the movement of a floating twig or leaf between two points in a stream.

For continuous spills (i.e., a slow leak from a tank car or from the runoff or leaching of spill-contaminated soil), samples should be taken upstream as well as downstream of the discharge point to ensure that diffusion or back eddying has not spread the plume upstream.

It should be emphasized that an analysis is only as valid as the sample on which it is performed. Therefore, the best techniques and equipment available to the responder should be used. In many past cases, however, grab samples taken in tin cans or mayonnaise jars were the only samples available, and the data derived from these samples were valuable to the response operation. Lack of sophisticated sampling equipment should not deter the responder from taking a sample. To wait until the proper equipment is available on scene often means that the spill plume will be missed. In urgent cases, a sample should be taken in any available container and later transferred to the proper one. The sample may receive some contamination from the original container, but at least a sample of the plume will be on hand. The original container should be saved so that the analyst can estimate the amount of any interfering substances that may have been introduced into the sample.

In summary, lack of proper equipment and uncertainty as to where and when to take a sample are the principal problems associated with implementation of a sampling program to support a spill response. Contingency planning prior to a spill can include the prepackaging of a spill response sampling kit. The use of visual observations, data on stream currents, frequent sampling, and common sense can help ensure that samples are taken at the proper time and place.

TABLE 1-1 Recommendation for Sampling and Preservation of Samples According to Measurement[a]

Measurement	Volume required (mL)	Container[b]	Preservative	Holding time[c]
Physical properties				
Color	50	P, G	Cool, 4°C	24 h
Conductance	100	P, G	Cool, 4°C	24 h[d]
Hardness	100	P, G	Cool, 4°C HNO_3 to pH < 2	6 months[e]
Odor	200	G only	Cool, 4°C	24 h
pH	25	P, G	Determine on site	6 h
Residue				
Filterable	100	P, G	Cool, 4°C	7 days
Nonfilterable	100	P, G	Cool, 4°C	7 days
Total	100	P, G	Cool, 4°C	7 days
Volatile	100	P, G	Cool, 4°C	7 days
Settleable matter	1000	P, G	None required	24 h
Temperature	1000	P, G	Determine on site	No holding
Turbidity	100	P, G	Cool, 4°C	7 days
Metals				
Dissolved	200	P, G	Filter on site HNO_3 to pH < 2	6 months[e]
Suspended	200		Filter on site	6 months
Total	100	P, G	HNO_3 to pH < 2	6 months[e]
Mercury dissolved	100	P, G	Filter on site HNO_3 to pH < 2	38 days (glass) 13 days (hard plastic)
Total	100	P, G	HNO_3 to pH < 2	38 days (glass) 13 days (hard plastic)
Inorganics; nonmetallics				
Acidity	100	P, G	None required	24 h
Alkalinity	100	P, G	Cool, 4°C	24 h
Bromide	100	P, G	Cool, 4°C	24 h
Chloride	50	P, G	None required	7 days
Chlorine	200	P, G	Determine on site	No holding
Cyanides	500	P, G	Cool, 4°C NaOH to pH 12	24 h
Fluoride	300	P, G	None required	7 days
Iodide	100	P, G	Cool, 4°C	24 h
Nitrogen				
Ammonia	400	P, G	Cool, 4°C H_2SO_4 to pH < 2	24 h
Kjeldahl method, total	500	P, G	Cool, 4°C H_2SO_4 to pH < 2	24 h[f]
Nitrate plus nitrite	100	P, G	Cool, 4°C H_2SO_4 to pH < 2	24 h[f]
Nitrate	100	P, G	Cool, 4°C	24 h
Nitrite	50	P, G	Cool, 4°C	48 h
Dissolved oxygen				
Probe	300	G only	Determine on site	No holding
Winkler titration	300	G only	Fix on site	4 to 8 h
Phosphorus				
Orthophosphate, dissolved	50	P, G	Filter on site Cool, 4°C	24 h
Hydrolyzable	50	P, G	Cool, 4°C H_2SO_4 to pH < 2	24 h[f]
Total	50	P, G	Cool, 4°C H_2SO_4 to pH < 2	24 h[f]
Total, dissolved	50	P, G	Filter on site Cool, 4°C H_2SO_4 to pH < 2	24 h[f]

Measurement	Vol. required (mL)	Container[b]	Preservative	Holding time[c]
Silica	50	P only	Cool, 4°C	7 days
Sulfate	50	P, G	Cool, 4°C	7 days
Sulfide	500	P, G	2 mL zinc acetate	24 h
Sulfite	50	P, G	Determine on site	No holding
Organics				
Biochemical oxygen demand	1000	P, G	Cool, 4°C	24 h
Chemical oxygen demand	50	P, G	H_2SO_4 to pH < 2	7 days[f]
Oil and grease	1000	G only	Cool, 4°C H_2SO_4 or HCl to pH < 2	24 h
Organic carbon	25	P, G	Cool, 4°C H_2SO_4 or HCl to pH < 2	24 h
Phenolics	500	G only	Cool, 4°C H_3PO_4 to pH < 4 1.0 g $CuSO_4$ per liter	24 h
MBAS	250	P, G	Cool, 4°C	24 h
Nitrilotriacetic acid	50	P, G	Cool, 4°C	24 h

[a]More specific instructions for preservation and sampling are found with each procedure as detailed in the source manual. A general discussion on sampling water and industrial wastewater may be found in ASTM handbooks.

[b]Plastic (P) or glass (G). For metals, polyethylene with polypropylene cap (no liner) is preferred.

[c]It should be pointed out that the holding times listed above are recommended for properly preserved samples on the basis of currently available data. It is recognized that for some sample types extension of these times may be possible, while for other types these times may be too long. If shipping regulations prevent the use of the proper preservation technique or the holding time is exceeded, as in the case of a 24-h composite, the final reported data for these samples should indicate the specific variance.

[d]If the sample is stabilized by cooling, it should be warmed to 25°C for reading, or a temperature correction made and results reported at 25°C.

[e]If HNO_3 cannot be used because of shipping restrictions, the sample may initially be preserved by icing and be shipped immediately to the laboratory. Upon receipt in the laboratory, the sample must be acidified to a pH < 2 with HNO_3 (normally 3 mL 1:1 HNO_3 per liter is sufficient). At the time of analysis, the sample container should be thoroughly rinsed with 1:1 HNO_3 and the washings added to the sample (volume correction may be required).

[f]Data obtained from the National Enforcement Investigations Center, Denver, Colo., support a 4-week holding time for this parameter in sewage systems (SIC 4952).

SOURCE: *Methods for Chemical Analysis of Water and Wastes,* EPA-600/4-79-020 (NTIS-PB-297686/AS), U.S. Environmental Protection Agency, Environmental Monitoring and Support Laboratory, Cincinnati, 1979.

CHEMICAL ANALYSIS

The role of chemical analysis in hazardous material spill response can vary from incident to incident. In one case, it may be necessary to perform a rapid analysis to characterize or identify the spilled substance before an effective countermeasure can be initiated. In another case, a detailed, time-consuming sampling and analysis plan may be necessary to define the spill-impacted area and thus make possible an adequate assessment of ecological damage. In other situations, water analysis is necessary to determine whether spill mitigation procedures have been effective. Therefore, the methods used and the apparatus and instrumentation required will vary in sophistication and complexity with the particular situation. This section discusses several levels of chemical analysis capability, including field analysis kits, remote (off-site) analytical laboratories, and mobile laboratories. Their limitations with regard to hazardous spill response are assessed. In addition, case histories on their use and recommendations for future applications are given.

Field Analysis Kits

Field kits for the on-site analysis of water have been available from many suppliers for several years. Their chief limitation is that they usually consist of a number of wet chemical tests, each highly selective for a specific compound or class of compounds, and therefore are useful only when the background concentration of interfering substances is low (i.e., in process samples and drinking water). Further, the kits are almost totally inadequate unless the identity of the polluting substance is already known.

However, if the hazardous substance has been identified and there are no background interferences, the field kits can be useful in delineating the portion of a water body that has been contaminated. For example, if a spill of phenol from a train wreck flows into a stream, a phenol test kit (if one is immediately available) can be used in the following manner by the first environmental or public health official on scene. A point should be selected sufficiently downstream from the spill site to ensure that the phenol slug* has not passed. At short intervals the kit should be used to test the water. When a positive indication is found, a dye can be added to the stream to mark the leading edge. Continued analyses can be conducted and dye added until positive test indications are no longer found. The presence of the dye has a dual function: (1) it serves as a warning to downstream users that the water is contaminated, and (2) it marks the spill plume so that effective countermeasures such as in-stream treatment can be instituted.

Commercially available water analysis kits have been of limited use because the identity of the substance had to be known and the proper kit had to be available to the responder. Since many of the tests are highly specific and sensitive to interferences and since hazardous substances are diverse, a response organization would need to keep a wide variety of test kits in its inventory to ensure availability of the proper kit when a spill occurs.

The Oil and Hazardous Materials Spills Branch (OHMSB) of the U.S. Environmental Protection Agency (U.S. EPA) contracted with the U.S. Army's Chemical Systems Laboratory (CSL) to develop a diversified kit which would have the capability of detecting most of the chemicals on the EPA's list of designated hazardous polluting substances.[7] A list of 33 chemicals representative of the EPA-designated substances was used to evaluate commercially available and specially designed tests. Following an initial screening with tap-water solutions, the most promising methods were further evaluated against samples of natural waters polluted in the laboratory with compounds from the model list. It was decided that a selection of 15 nonspecific detection systems (Table 1-2) could be organized into the desired portable kit. It was also concluded that more than 85 percent of the EPA-designated compounds would respond to at least 1 of the 15 tests. A final report entitled *Development of a Kit for Detecting Hazardous Materials Spills in Waterways*[8] describes the project in detail and contains sufficient information to duplicate the kit.

The prototype kits have been used several times by the Environmental Emergency Response Unit (EERU) and others during actual spills. During the Kepone cleanup at Hopewell, Virginia,[9] it was necessary to monitor the pH of the influent and effluent of the water being treated. The EPA kit functioned well in this mode and was used several times daily. The author has made several attempts to use the kit to monitor and detect spills of cholinesterase-inhibiting pesticides with only limited success. Outdated reagents and operator training have been problems with several of the tests in the kit (benzene and cyanide are the most pronounced examples), but a regular training and maintenance schedule would alleviate these shortcomings. An updated version of the EPA kit is now being marketed by the Hach Chemical Company.

---

*In a stream, the contaminant will flow almost in plug-flow fashion, so that one can expect that the major amount of chemical will be contained in a small volume of water (slug) and pass a given point in a short time.

TABLE 1-2 Summary of Methods Selected for the Detector Kit

Detection parameter	Type*	Process	Reagents†	Time (min)	Types of substances detected	Maximum sensitivity in natural water (mg/L)
Cholinesterase inhibitors	Qualitative	Enzyme ticket	M30A1, chemical agent detector kit	<8	Certain insecticides; very acid materials	<3
Benzene	Qualitative	Detector tube	Bendix/Gastec	<2	Benzene, styrene, xylene, toluene, and other organic materials	1
Heavy metals	Qualitative	Extraction/test tube	Hach/laboratory	<1	Many metals; also some organic materials	<10
Phenol	Qualitative	Extraction/cent. tube	Hach	<1	Phenolics	1
Cyanide	Qualitative	Detector tube	Kitagawa	<1	Cyanide ion; certain other organic and inorganic materials	10
pH	Quantitative	Meter	· · ·	<1	Acidic or basic organic and inorganic materials	<3
Conductivity	Quantitative	Meter	· · ·	<1	Inorganic compounds and other materials with high ionic strength	10
Nitrate nitrogen	Quantitative	Spectrophotometer	Hach	<6	Nitrate ion; certain other organic and inorganic substances	10
Color	Quantitative	Spectrophotometer	· · ·	<1	Highly colored organic or inorganic substances	<5
Sulfate	Quantitative	Spectrophotometer	Hach	<10	Sulfate ion	10
Phosphate	Quantitative	Spectrophotometer	Hach	<2	Phosphate ion; certain arsenates, arsenites, and bromates	<1
Ammonia nitrogen	Quantitative	Spectrophotometer	Hach	<10	Wide range of organic and inorganic ammonia-containing materials	<1
Chloride	Quantitative	Spectrophotometer	Laboratory	<10	Wide range of sulfides, bromides, cyanides, sulfides, thiosulfates, and chlorides	10
Fluoride	Quantitative	Spectrophotometer	Hach	<2	Fluoride; various inorganic and organic materials	10
Turbidity	Quantitative	Spectrophotometer	· · ·	<1	Low-solubility materials	Not determined

*In this context "quantitative" means that a numerical value is assigned to the data.
†Laboratory indicates specially prepared reagents.

In summary, field detection kits in general and the EPA kit in particular are useful analytical tools if (1) the identity of the chemical is known and one of the tests in the kit is applicable, (2) there are few background interferences, (3) the kit has been well maintained and restocked appropriately, and (4) a well-trained individual is available to conduct the tests.

Off-Site Analytical Laboratories

The use of a well-equipped chemical laboratory is recommended whenever circumstances permit. In this setting, standard methods can be used and adequate quality assurance maintained.[1,2,10-16] This method is particularly important when sophisticated sample preparation and instrumentation are required. It is imperative if the data are to be used in litigation.

The primary disadvantage of using an off-site laboratory is the time required to prepare a sample for transportation and to ship it. Often it is necessary to have the results of an analysis as rapidly as possible. For example, during a spill cleanup it may be necessary to treat contaminated water. The effluent from the treatment process must be analyzed before discharge to ensure that the contaminant has been suitably removed. The time available to obtain analytical results therefore is limited to the amount of time required to exhaust the available clean-water storage capacity. If there is unlimited storage, the time for sample transport is not critical, but if there is little or no storage capacity, it may be necessary to terminate the treatment operation until analysis of the effluent is complete. If a laboratory capable of performing the required analysis can be found near the treatment site, the time of sample transport may be short, but if the sample must be transported several hundred miles, the delay may render the entire treatment process infeasible.

At the Kepone cleanup at Hopewell, Virginia,[9] samples were flown by charter aircraft to the OHMSB laboratory at Edison, New Jersey, 480 km (300 mi) away. This time-consuming operation was made possible by the use of several large-capacity railroad tank cars as storage vessels for the effluent from EPA's mobile physical-chemical treatment system. However, at a pentachlorophenol incident at Haverford, Pennsylvania,[17] only 11.4 m^3 (3000 gal) of effluent storage was available, and rapid analyses were performed on site in a mobile laboratory.[18] Constant effluent monitoring is not required in all spill treatment operations. Intermittent analysis of effluent can be sufficient if no sudden increases are expected.

In the first phase of a PCB cleanup of the Duwamish River in Seattle, Washington,[19] the water was treated on a batch basis. The U.S. EPA Region X Laboratory, 48 km (30 mi) from the site, was used to analyze each batch prior to discharge. On the basis of experience from this first phase, analyses during the second phase were performed daily at the regional laboratory.

In summary, a well-equipped off-site chemical laboratory should be used whenever data are to be developed for litigation purposes. This laboratory should be proficient in analyzing refereed or documented reference samples. Whenever possible, reference standards should be analyzed concomitantly with spill samples to establish quality control data. It is also recommended that whenever conditions permit such a laboratory be used to ensure maintenance of quality control.

Mobile Laboratories and Field Laboratories

Time constraints and logistical considerations often do not permit analysis of spill samples in off-site laboratories, and it may be necessary to deploy a laboratory at the spill site.

The first documented use of an on-site field laboratory was the 1971 endrin cleanup at Pond Lick Lake, Ohio.[20] A gas chromatography laboratory was set up in a nearby motel room to monitor the effluent from a field-improvised activated-carbon adsorption water treatment system. The use of motel rooms as field laboratories, however, is generally impractical and unsafe without adequate modifications (i.e., fume hood or glove box).

Accordingly, mobile laboratories equipped with the latest chemical instrumentation have recently been developed. The first of these in the United States was fabricated by the EPA OHMSB at Edison to provide on-site analytical services in conjunction with the EPA's mobile physical-chemical treatment system.[18,21,22] Constructed in a 10.7-m (35-ft) semitrailer equipped with heating and air conditioning, the EPA mobile laboratory is fitted with laboratory benches and a gas chromatograph–mass spectrometer (GC-MS), two gas chromatographs, an atomic absorption spectrophotometer, an infrared spectrophotometer, a fluorescence spectrophotometer, and an emission spectrograph. The laboratory is equipped with a fume hood, a vented solvent locker, an explosionproof refrigerator, running water, a safety shower, an eyewash, a fire blanket, fire extinguishers, and "once-through" heating and air conditioning. In addition, a vented glove box is included to permit safe handling of concentrated hazardous waste samples. Support equipment on board includes a telefacsimile, a pH meter, balances, hot plates, magnetic stirrers, desiccators, a steam bath, and all glassware, solvents, reagents, and other supplies necessary for fully independent operation in remote field locations.

The EPA mobile laboratory has been used several times both in conjunction with the physical-chemical treatment system and independently of it. The laboratory was used at Haverford, Pennsylvania, to monitor effluent,[17] and it was also used for analysis during the first phase of the Love Canal cleanup in Niagara Falls, New York,[21] and at the cleanup of abandoned hazardous waste disposal sites at Dittmer, Missouri,[23] and Oswego, New York.[24] It was used independently of EPA's mobile physical-chemical treatment system at Kernersville, North Carolina,[25] in conjunction with pilot-plant treatment studies and toxicity investigations resulting from the spill of an unknown chemical into a water supply reservoir. Other operations included the screening of samples from the defunct Chemical Control Company,[21] at Elizabeth, New Jersey, to determine disposal options and the monitoring of Susquehanna River water for dichlorobenzene at Pittston, Pennsylvania, to assess the extent of contamination from an illegal chemical waste spill into an abandoned coal mine.

Other mobile laboratories available from third-party spill cleanup contractors specializing in hazardous material incidents and from state and federal agencies do not have the variety of instrumentation of the EPA laboratory but can be adequate for a given spill. In the majority of cases, only one type of chemical analysis is required, and a laboratory equipped to handle that type will function satisfactorily. However, mobile laboratories are not recommended when time constraints and logistics permit the analysis of samples by a fully equipped off-site permanent laboratory.

REMOTE SENSING AND DETECTION

Remote sensing and detection during spills have several functions. Robot monitors allow continuous surveillance over a given water body without requiring on-site personnel. Aerial surveillance allows the monitoring of areas that are vast in comparison with the capability of ground-based observations, and it can accurately define the full extent of environmental impact. Both techniques can often be instrumental in discovering the source of a mystery spill. This section is devoted to outlining the advantages and shortcomings of each of these remote-sensing techniques and presenting case histories of their use.

Robot Monitors

Various water quality monitoring systems have been in operation in the United States for several years. Noteworthy among them is the system operated by the Ohio River Valley Water Sanitation Commission (ORSANCO). Another, the Raritan Bay–New York Harbor system, was operated and maintained by the EPA and its predecessor agencies until the mid-1970s, when the system was turned over to the states of New York and New Jersey.

Routine monitoring allows the detailed profiling of water quality parameters (pH, dissolved oxygen, conductivity, water temperature, etc.) in a water body and correlation of these parameters with flow, air temperature, and other environmental conditions. This detailed information is necessary for water basin planning and environmental-impact studies.

Most robot monitors are of limited utility in the detection of spills, however, because the background levels of most of the measured parameters fluctuate greatly. (For example, dissolved-oxygen concentration, pH, and temperature vary with the time of day and other natural conditions.) Thus it is nearly impossible to distinguish a spill-induced fluctuation from one induced by a natural event. Another major weakness of these systems for spill detection has been the lack of a sensor for dissolved or emulsified organic chemicals.

In an effort to overcome these shortcomings, the EPA OHMSB undertook a series of studies[26–28] which culminated in the fabrication of a prototype mobile spill alarm system. This system was designed to have the following characteristics:

1. Off-the-shelf detection components ready for use with a minimum of modification

2. Detection components rugged enough for use in an untended remote station and capable of continuous functioning

3. Components resistant to fouling, scaling, etc., to such a degree that cleaning is necessary only once every 14 days or longer

4. Detection components which produce an electrical response that can be transmitted by wire or radio to a distant receiver

5. Detection components which can function under adverse conditions such as high flow or rapid current in a watercourse

6. System capable of distinguishing between a slug of a given hazardous material or materials and concentration levels of the substances normally present in a watercourse

7. System that does not degrade any of the beneficial uses of a watercourse

The EPA mobile spill alarm system is capable of sensing both organic and inorganic hazardous materials. The system does not contain components which selectively detect the presence of any given substance. Rather, several detectors capable of sensing wide classes of these substances are included. A pH, oxidation-reduction potential, and conductivity sensor package, an ultraviolet absorptiometer, and a total organic carbon analyzer were integrated into a detection package with provision for automated recalibration and audio and telephone alarm systems.

The entire package was mounted in a heated and air-conditioned 8.2-m (27-ft) recreational-type trailer so that it could be conveniently moved from one site to another. To date this system has been tested at Rockwell International's Santa Susanna field laboratory and other sites on the Los Angeles River in Los Angeles and on a tributary to the Raritan River in Edison. These tests have shown that the system is capable of long-term field deployment with a minimum of personnel, but as yet only one release, an ultraviolet-absorbing substance found at Los Angeles, has been detected.

The system was used at Pittston, Pennsylvania, to continuously monitor the discharge of chemical wastes that had been dumped into an abandoned coal mine. The total organic carbon analyzer and ultraviolet absorptiometer proved to be most applicable in this instance, which involved waste oil and aromatic and halogenated organic chemicals being released intermittently from a mine drainage tunnel into the Susquehanna River.

Other OHMSB research prototype detectors soon to be incorporated into the mobile spill alarm system include a cyclic colorimeter for heavy metal ion detection[27] and a continuous aqueous monitor (CAM) for the detection of cholinesterase-inhibiting materials (organophosphate and carbamate pesticides).[29]

Aerial Reconnaissance

The use of aerial reconnaissance to detect and monitor spills of petroleum is widespread. Photographic, electronic, and visual methods have been useful in estimating the spreading and movement of waterborne oil slicks and the extent of shoreline or bank contamination. The same methods will also be useful for floatable insoluble hazardous materials.

However, caution is urged when any aerial or satellite surveillance method is used because imagery keys[30,31] must be developed in conjunction with ground observations (ground truth) to make a meaningful interpretation of what is observed. For example, naturally occurring phenomena such as algae blooms, fish oils, and wave or current movement may appear to the untrained eye as "pollutant sheen." Even experienced observers and interpreters can feel uncomfortable unless ground-based data are available to aid them. It is prudent, therefore, whenever possible to use aerial reconnaissance as a supplement and guide rather than as the sole source of data on which decisions must be made.

This is not to imply that aerial reconnaissance is of limited value. In the course of a spill response, it can be a vital tool. Visual spotting from a light plane or a helicopter is often the only method which will give an overall picture of the impact of an incident. An overflight of a chemical spill caused by a train wreck in a congested area may reveal critical watercourses that may be contaminated or even a hospital, nursing home, or school that must be evacuated. These facilities may not be apparent from the ground.

Aerial photographs of an area can save hours and even days of work in locating water intakes and sensitive environmental areas such as marshlands and other critical habitats. These areas have the highest priority for protective and mitigative action. Aerial photography and visual overflights are also invaluable in deploying control and removal equipment or decontamination systems. Precious hours can be lost in attempting to locate these countermeasures by using maps only to find during the implementation phase that the maps do not show fences, ditches, and buildings which inhibit response.

Varying degrees of sophistication are available in this area of aerial reconnaissance, and a case-by-case determination must be made to select the method to be employed. For example, aerial photographs and detailed interpretation may be justified for a response to a spill which mainly impacts land, where the situation is stable, but for response to a spill into a moving stream the situation often changes too rapidly for aerial photographs to be used except as a historical record. In the latter cases, helicopter overflights will probably be much more desirable.

Firms that can provide either aerial photography or light planes or helicopters are located throughout the industrialized world, and provision for rapidly obtaining these services in the event of a spill should be addressed in an organization's spill contingency plan.

REFERENCES

1. *1979 Annual Book of ASTM Standards,* part 31: *Water,* American Society for Testing and Materials, Philadelphia, 1979.

2. *Methods for Chemical Analysis of Water and Wastes,* EPA-600/4-79-020 (NTIS-PB-297686/ AS), U.S. Environmental Protection Agency, Environmental Monitoring and Support Laboratory, Cincinnati, 1979.

3. E. R. DeVera, B. P. Simmons, R. D. Stevens, and D. L. Storm, *Samplers and Sampling Procedures for Hazardous Waste Streams,* EPA-600/2-80-018, U.S. Environmental Protection Agency, Municipal Environmental Research Laboratory, Cincinnati, 1980.

4. *NPDES Compliance Sampling Manual,* U.S. Environmental Protection Agency, Office of Water Enforcement, Washington, undated.

5. *Handbook for Sampling and Sample Preservation of Waters and Wastewaters,* EPA-600/4-76-049 (NTIS-PB-259946/AS), U.S. Environmental Protection Agency, Environmental Monitoring and Support Laboratory, Cincinnati, 1976.

6. D. J. Harris and W. J. Keffer, *Wastewater Sampling Methodologies and Flow Measurement Techniques,* EPA-907/9-74-005, U.S. Environmental Protection Agency, Surveillance and Analysis Division, Kansas City, Mo., 1974.

7. "Designation of Hazardous Substances," *Federal Register,* vol. 40, no. 250, Dec. 30, 1975, p. 59961.

8. A. Silvestri, A. Goodman, L. M. McCormack, M. Razulis, A. R. Jones, Jr., and M. E. P. Davis, *Development of a Kit for Detecting Hazardous Material Spills in Waterways,* EPA-600/2-78-055, U.S. Environmental Protection Agency, Industrial Environmental Research Laboratory, Oil and Hazardous Materials Spills Branch, Edison, N.J., 1978.

9. J. P. Lafornara, U. Frank, and I. Wilder, "Environmental Emergency Response Unit (EERU) Operations at the Hopewell, Virginia, Kepone Incident," in G. F. Bennett (ed.), *Proceedings of the 1978 National Conference on Control of Hazardous Material Spills,* Miami Beach, Fla., April 1978, pp. 236–239.

10. J. F. Thompson (ed.), *Analysis of Pesticide Residues in Human and Environmental Samples,* U.S. Environmental Protection Agency, Health Effects Research Laboratory, Research Triangle Park, N.C., 1977.

11. *AMEPA Manual for Organic Analysis Using Gas Chromatography Mass Spectrometry,* EPA-600/8-79-006 (NTIS-PB-297164/AS), U.S. Environmental Protection Agency, Environmental Monitoring and Support Laboratory, Cincinnati, 1979.

12. *Handbook for Analytical Quality Control in Water and Wastewater Laboratories,* EPA-600/4-79-019 (NTIS-PB-297451/AS), U.S. Environmental Protection Agency, Environmental Monitoring and Support Laboratory, Cincinnati, 1979.

13. *Publications on the Analysis of Spilled Hazardous and Toxic Chemicals and Petroleum Oils,* U.S. Environmental Protection Agency, Oil and Hazardous Materials Spills Branch, Edison, N.J. 1978.

14. J. R. Simons, *Bibliography on Hazardous Materials Analysis Methods,* U.S. Environmental Protection Agency, Oil and Hazardous Materials Spills Branch, Edison, N.J. (In preparation.)

15. "Guidelines Establishing Test Procedures for the Analysis of Pollutants; Proposed Regulations," *Federal Register,* Dec. 3, 1979, pp. 69464–69575; Dec. 18, 1979, pp. 75028–75052.

16. "Guidelines Establishing Test Procedures for the Analysis of Pollutants," *Code of Federal Regulations,* part 136.

17. H. L. Lamp'l, T. Massey, and F. J. Freestone, "Assessment and Control of a Ground and Surface Water Contamination, Haverford, Pennsylvania, November–December, 1976," in G. F. Bennett (ed.), *Proceedings of the 1978 National Conference on Control of Hazardous Material Spills,* Miami Beach, Fla., April 1978, pp. 145–147.

18. M. Urban and R. Losche, "Development and Use of a Mobile Chemical Laboratory for Hazardous Material Spill Response Activities," in G. F. Bennett (ed.), *Proceedings of the 1978 National Conference on Control of Hazardous Material Spills,* Miami Beach, Fla., April 1978, pp. 311–314.

19. J. C. Williams, J. Blazevich, and H. Snyder, "PCB in the Duwanish," in G. F. Bennett (ed.), *Proceedings of the 1976 National Conference on Control of Hazardous Material Spills,* New Orleans, April 1976, pp. 351–355.

20. *Pesticide Poisoning of Pond Lick Lake, Ohio, Investigation and Resolution,* OHM-71-06-002, U.S. Environmental Protection Agency, Oil and Special Materials Control Division, Washington, 1971.

21. U. Frank, M. Gruenfeld, R. Losche, and J. P. Lafornara, "Mobile Laboratory Safety and Analysis Protocols Used on Site at Abandoned Chemical Waste Dumpsites and at Oil and Hazardous Chemical Spills," *Proceedings of the 1980 National Conference on Control of Hazardous Material Spills,* Louisville, Ky., May 1980, pp. 259-263.

22. M. Gruenfeld, F. J. Freestone, and I. Wilder, "EPA's Mobile Lab and Treatment System Responds to Hazardous Spills," *Industrial Water Engineering,* September 1978, pp. 18-23.

23. H. Gilmer and F. J. Freestone, "Clean-Up of an Oil and Mixed Chemical Spill at Dittmer, Missouri, April-May, 1979," in G. F. Bennett (ed.), *Proceedings of the 1978 National Conference on Control of Hazardous Material Spills,* Miami Beach, Fla., April 1978, pp. 131-134.

24. J. P. Lafornara, F. J. Freestone, and M. Polito, "Spill Clean-Up at a Defunct Industrial Waste Disposal Site," in G. F. Bennett (ed.), *Proceedings of the 1978 National Conference on Control of Hazardous Material Spills,* Miami Beach, Fla., April 1978, pp. 152-155.

25. J. Stonebreaker, F. J. Freestone, and W. H. Peltier, "Clean-Up of an Oil and Mixed Chemical Waste Spill at Kernersville, North Carolina, June 1977," in G. F. Bennett (ed.), *Proceedings of the 1978 National Conference on Control of Hazardous Material Spills,* Miami Beach, Fla., 1978, pp. 182-186.

26. R. J. Pilie, R. E. Baier, R. C. Ziegler, R. P. Leonard, J. G. Michalovic, S. L. Pek, and D. H. Bock, *Methods to Treat, Control and Monitor Spilled Hazardous Materials,* EPA-670/2-75-042, U.S. Environmental Protection Agency, Oil and Hazardous Materials Spills Branch, Edison, N.J., 1975.

27. D. Bock and P. Sullivan, *Selected Methods for Detecting and Tracing Hazardous Materials Spills,* EPA-600/2-79-0064, U.S. Environmental Protection Agency, Oil and Hazardous Materials Spills Branch, Edison, N.J., 1979.

28. M. Kirsch, R. W. Melvold, and J. J. Vrolyk, *A Hazardous Materials Spill Warning System,* U.S. Environmental Protection Agency, Oil and Hazardous Materials Spills Branch, Edison, N.J. (in preparation.)

29. L. Goodson, W. B. Jacobs, and A. W. Davis, *A Rapid Detection System for Organophosphates and Carbamate Insecticides in Water,* EPA-R2-72-010, U.S. Environmental Protection Agency, Oil and Hazardous Materials Spills Branch, Edison, N.J., 1972.

30. H. V. Johnson, *Aerial Reconnaissance of Hazardous Substances Spills and Spill Threat Conditions,* EPA-600/4-79-029, U.S. Environmental Protection Agency, Environmental Monitoring and Support Laboratory, Las Vegas, Nev., 1979.

31. R. Landers and H. V. Johnson, "Photo Interpretation Keys for Hazardous Substances Spill Conditions," in G. F. Bennett (ed.), *Proceedings of the 1978 National Conference on Control of Hazardous Material Spills,* Miami Beach Fla., April 1978, pp. 124-127.

PART 2
Bioassay

Royal J. Nadeau, Ph.D.
Acting Chief, Environmental Impact Section
Environmental Response Team
U.S. Environmental Protection Agency

Introduction / 8-14
On-Scene Toxicity Testing / 8-16
Conclusion / 8-20

INTRODUCTION

Hazardous material spills often cause environmental perturbations that require specialized techniques for detection, containment, and mitigative measures. Aquatic toxicology frequently provides assistance in evaluating the environmental hazards of spills and spill cleanup operations.

Bioassays are the most commonly used procedures in aquatic toxicology. Various types of bioassays (i.e., static acute, flow-through) exist and can be designed for spill use. Among the uses to which bioassays can be applied are the following:

- *Determining the most environmentally acceptable mitigative measure to counteract a spill.* If more than one means of treating a particular hazardous material spill exists, the most environmentally acceptable approach should be utilized. If the selected method involves treating the material with another chemical to neutralize or convert the spilled material to a different form or state, a bioassay can be conducted to determine the toxicity of the new compounds formed during the treatment process.

- *Determining the environmental acceptability of the final product from the spill treatment process.* With the development of mobile treatment systems for chemically treating contaminated water bodies, the question of environmental acceptability of the effluent from these systems often arises. For most organic material spills, applicable effluent standards that can serve as guidelines for establishing treatment levels do not exist. In lieu of effluent standards for these treatment systems, an attempt should be made to monitor the effluent by performing acute toxicity tests.

- *Determining the biological "zone of influence" resulting from the spill.* The extent of biological impact can be estimated through on-scene bioassays. The premise used in predicting the impact is that the spilled material initially will be discharged in a concentrated form in a large volume (slug), followed by the discharge of a smaller volume (drainage).

To plot the movement of the slug in the receiving water, all the physical factors that affect dispersion, diffusion, and mixing must be considered. The effect that the material will have on the biological populations of the receiving waters depends upon the concentration (dose) and the length of time during which these populations are exposed to the slug (period of exposure).

Toxicity is a measurable, quantifiable water quality parameter. It is defined here as any observable response that is considered irregular and detrimental to the test organism. Toxicity is usually a response that is part of a syndrome which ultimately detracts from the survivability of the organisms. However, standard bioassay data such as the median tolerance limit and the median lethal concentration often have little utility for hazardous material spill management without adequate assessment by toxicologists. Sublethal responses may be more significant than mortality in denoting observable effects since mortality is insensitive and represents the ultimate no-return point. In implementing counteractive measures it is more useful to know when the organisms are under stress. Responses observed in test chambers must be compared with the controlled experiments to determine whether a true effect or response is resulting from the spilled material or from a preexisting stress condition in the test organisms.

Recommendations for conducting bioassays for evaluating oil spills were published in 1977 by the American Petroleum Institute.[1] A critical need identified by the investigating panel was the necessity for promoting standardized toxicity testing and bioassay procedures. Other critical points were identified by the panel for conducting toxicity tests to assess the effects of spilled oil on plankton. Advice on these points is as follows:

- Use the shortest possible exposure time to obtain a measurable response. To generate meaningful toxicity information for spill management decisions, it is especially important to minimize the time lag.

- Use test species that have regional economic or ecological importance. Although this practice may provide environmentally relevant data, economic and ecologically important species may not always be available for testing when needed. Most aquatic toxicologists use a species that is laboratory-cultured and disease-free as a test organism in bioassays. Organisms collected from nature are often disease-ridden and cannot easily be acclimated to the laboratory.

- Rapidity in obtaining a specific reproducible response to the toxic material is another important feature for determining a test organism. Sensitivity is equally important. Organisms that are insensitive will not respond to the toxic material at low enough concentrations for making spill management decisions.

- Bioassays should be designed with specific statistical tests, and the number of observations should vary according to the null hypothesis being tested, the desired significance level, and the variability encountered in the measured response.

Priority should be given to addressing the following types of questions in managing hazardous material spill operations. Sound decisions can then be made in choosing the most environmentally safe and effective cleanup procedures.

- How much toxicity has resulted from the spilled material?
- How is this toxicity distributed within the affected water body?
- What is happening to this toxicity over time? Is there any natural attenuation of the toxicity from interaction with aquatic chemical and biological factors (i.e., pH, natural ionic constituents, hydrolysis, hardness, alkalinity, and biodegradation)?
- What is the bioaccumulation potential of the spilled material?
- What is the potential for residual toxicity after the initial impact?

ON-SCENE TOXICITY TESTING

Portable Toxicity-Measuring Devices

Hazardous material spill incidents have produced innovative methodology to address the toxicity problems created by these incidents. Persons involved in hazardous material spills should be aware of the technology that has recently been developed for assessing toxicity. It is particularly valuable for technical support personnel who may be on scene at a spill incident to be aware of this technology.

Toxicity-measuring apparatus (e.g., mobile bioassay units) that has been used at hazardous material spills was never intended for that specific function. However, it has been made available for the purpose. This apparatus has great utility, but it may not always be available since it is being used to a greater extent for biomonitoring studies implemented by water resource management agencies.

The instruments mentioned below were designed to address the kinds of toxicity problems associated with hazardous material spills. They can be easily transported and set up within a short time to generate data for spill management purposes.

CAM-4

The cholinesterase antagonist monitor (CAM) was designed and fabricated by the Midwest Research Institute under a U.S. Environmental Protection Agency (U.S. EPA) contract.[2] The latest unit, the CAM-4 (Fig. 2-1), is a battery-powered (12V dc) portable version of the CAM-1. It can be operated continuously for 8 h by using a 12-V automobile battery. The CAM-4 is particularly well suited to monitoring water bodies that have been affected by spills of organophosphate and carbamate insecticide compounds. The CAM-4 operator can determine when toxic or subtoxic levels of cholinesterase-antagonistic substances are present in the water. Both CAM systems are generally more sensitive for organophosphate pesticides that have phosphate $(-O-P=O)$ compounds than for those that have phosphorothioate $(-O-P=S)$ or phosphorodithioate $(-S-P=S)$ compounds.

The CAM system is considered a toxicity-measuring instrument because of its capacity to measure cholinesterase inhibition. Cholinesterase is an essential enzyme that is

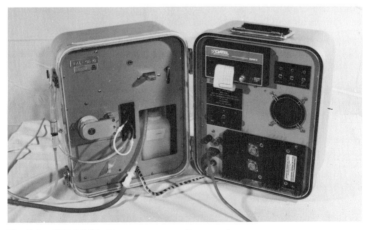

Fig. 2-1 CAM-4 (cholinesterase antagonist system).

Fig. 2-2   Luminescent-bacteria toxicity-monitoring system (Microtox).

required to deactivate acetylcholine produced at the nerve muscle activation site. In the CAM system, immobilized horse-serum cholinesterase collects the organophosphate and carbamate molecules at its active sites, resulting in a decrease in enzyme activity. This decrease is measured by the enzyme's capacity to hydrolyze a substrate (butyralthiocholine iodide) presented to it. One of the hydrolysis products is thiocholine iodide, which when present causes a low voltage potential to occur between two electrodes. When cholinesterase inhibition occurs, enzyme activity is impaired, the quantity of thiocholine iodide formed is small, and the potential between the electrodes is greater.

The operation and thus the reactivity of the enzyme are temperature-dependent, with low temperatures requiring a longer period for hydrolytic action to occur. Although the CAM-4 system does become less sensitive as the temperature decreases, it retains sufficient sensitivity at 10°C to detect subtoxic levels of the compounds tested in field evaluations of the instrument.

The CAM-4 is encased in a watertight fiberglass case occupying a space of approximately 0.03 m³ (1 ft³) and weighs less than 13.6 kg (30 lb), exclusive of the battery.

Luminescent-Bacteria Toxicity-Monitoring System

Beckman Instruments, Inc., has developed a toxicity-monitoring and -measuring apparatus, Microtox, that utilizes the light output from luminescent bacteria as an indicator of the presence and concentration of toxic materials in water (Fig. 2-2). When exposed to a waterborne toxicant, the light intensity is altered proportionally to the concentration of the toxicant.

Freeze-dried luminescent bacteria are hydrated and mixed in small vials with a buffer solution. A baseline light production level is measured in a specially designed photometer which is equipped with a rotary shutter containing a temperature-controlled cuvet holder. About 5 min is required for the rate of light output to be stabilized.

A total of 500 μL of the test material (toxicant) which has been adjusted to a salinity of 20 parts per thousand and precooled to 15°C is added to the luminescent-bacteria-cell suspension. A toxic response is the decrease in light production as measured by the photometer at 2, 5, and 10 min after addition. The initial light output level minus the final light level indicates the light reduction caused by the toxicant. This value divided by the initial light level and multiplied by 100 gives the percentage decrease at each time interval:

$$\frac{\text{Initial light level } - \text{ final light level}}{\text{Initial light level}} \times 100 = \text{percentage decrease in light level}$$

The analyzer expresses the percentage of light lost (% Δ) on a digital display.

The Microtox is a compact, self-contained bench-top unit with temperature control, preincubation, and two sensitivity levels built into the unit. Supplies include lyophilized cells, disposable glass cuvets, and dilution preparation glassware.

The compactness and simplicity of the Microtox are positive features when it is being considered for use as a portable toxicity-measuring device at hazardous material spills. It can be easily and quickly set up in a mobile trailer-laboratory requiring a few square feet of bench space and 120 V ac. Reagents and glassware are minimal. The procedure is simple and rapid, and many samples can thus be processed in a short period. Quick turn-around in analyses is imperative for any analytical procedure that is to be used for hazardous material spill decisions.

The analyst–data interpreter using the Microtox system on scene must be aware that the data it produces are relative. The inhibited-light response does not have a high ecological-impact significance, particularly for freshwater systems. The Microtox is valuable to show the presence of toxicity in the affected water body and/or the effluent of the treatment system being used to detoxify the spill. However, it does not measure the impact of the spilled material.

Portable Fish and Macroinvertebrate Toxicity-Monitoring System

A portable toxicity-measuring system that is simple and easy to operate and is easily transported was developed some time ago. A prototype system (Fig. 2-3) was designed and fabricated by an EPA employee at a laboratory in Gulf Breeze, Florida. Essentially this system is a gravity-feed proportional diluter system for small-volume exposures. It was originally intended to be used with mysid shrimp, which do not require as large a water volume as fish.

The dilution water and toxicant solutions are stored in large vessels on top of the unit. The smaller-volume vessels are calibrated and arranged to provide the appropriate proportions of toxicant to dilution water for the various dilutions in the exposure-chamber array at the bottom of the unit. The flow is regulated by a bar clamp which closes against flexible rubber tubing that leads from the proportioners to the mixing chambers. This clamp is operated by a solenoid that is controlled by a multiple-cycle timing device located on the side. The timer is adjusted to activate the solenoid to open the bar clamp long enough to empty the proportioners into the mixing chambers, which then empty into the exposure chambers.

The advantage of this system is its compactness and simplicity. The EPA environmental response team has acquired this prototype and is modifying it to fit inside especially designed wooden footlockers that comply with airline size specifications for luggage. The modified diluter system can be easily transported by air to a hazardous spill scene and be set up and operating within an hour of arrival. It is designed to use small fish (e.g., fry stage) or small macroinvertebrates (e.g., daphniad cladocerans). Use of a cultured species is preferable, and the test organisms therefore must be transported with the diluter system. Cultures of fathead minnows *(Pimephales promelas)* and cladocerans *(Daphnia magna)* or similiar acceptable-for-bioassay species must be maintained ready for use in the portable diluter system.

Static bioassays are simpler and require the transport of less hardware, and in some instances static tests may be appropriate. However, the advantages of the intermittent-flow system are that with its use optimal dissolved-oxygen levels are easier to maintain and metabolic wastes are flushed out of the exposure chambers. The organisms are exposed to consistent levels of toxicants in the flow-through system, in contrast to the static system, in which concentrations are altered during the exposure period.[3]

The portable toxicity-measuring system is being modified to accommodate the expo-

Fig. 2-3 Portable fish and macroinvertebrate toxicity-monitoring system (Allison diluter).

sure of multiple species simultaneously. Optimally, these species will be *Daphnia* and the fry of fathead minnows. However, if a spill incident were to occur in an area where a species of high interest (e.g., a salmonid) was a major concern, the fry of this species plus *Daphnia* would be used in lieu of the fathead minnow.

In addition to compactness and simplicity, an advantage of this system is the small volume of dilution and test water required. No more than 30 L (7.9 gal) of each is required to conduct a 24-h bioassay on two species with at least five exchanges in the exposure chambers. In this manner, effluent from a treatment system could be tested every 24 h with a single grab sample from the effluent.

The modified Allison proportional diluter will be capable of testing four distinct toxicant levels and will have a control for generating a TL_{50} on a single water sample. In this procedure, death will probably not be the measured response. It is more useful to denote the presence of toxicity by other toxic observations. Such effects as erratic swimming, loss of reflex, discoloration, excessive mucus production, opaque eyes, hemorrhaging, or abnormal behavior should be documented.

Four distinct water samples (possibly collected from distinct sampling locations within the affected water body), plus a control (unaffected water body), could be tested in the modified system by shunting the proportioner device and going directly into the mixing chambers from the sample reservoirs on top. For these tests, smaller reservoirs (6-L capacity) are required for 24-h bioassay.

In Situ Field Bioassays

Cages with live fish have been utilized to assess the condition of receiving waters that have been affected by hazardous material spills. This approach measures stress response on test organisms that are exposed to ambient background levels of toxic materials. This bioassay technique is best used as a monitoring tool for determining changes in toxicity patterns associated with a hazardous material spill. The response indicates the total effect being elicited from the spilled material alone or an additive toxic effect from other materials in the water that react with the spilled material. A control area of similar water quality characteristics but uncontaminated must be tested to evaluate possible stresses not related to the spill.

An *in situ* bioassay approach was used to evaluate a hazardous material spill in 1974 at Clarksburg, New Jersey, where a small pond was contaminated by the herbicide 2,4-dinitro-6-sec-butylphenol (DNBP), resulting in a massive fish kill. A total of 6.9 mg/L of DNBP was found in the pond at the site of entry 24 h after the incident had occurred. The *in situ* bioassay, using indigenous fish species of similar physical and biological characteristics collected in a nearby pond, was employed to monitor continually the changes in toxicity of DNBP and its degradation products.

A variety of juvenile fish species were seined from the control pond and placed in the exposure cages. These cages were plastic minnow traps ($\sim$ 45-mm mesh) which were placed in the littoral zone of each pond. The pond was completely impounded with sandbags while a trailer-mounted carbon column treatment system which cycled the contaminated pond water through the system was installed. Caged fish were placed at two sites in the contaminated pond. One site was near the treatment effluent hose, and the other was in the area of the influent line. An additional site was located in the control pond.

There was excessive mortality in the cages at the test and control sites during the first 48 h of exposure, possibly owing to the shock and injury sustained during the seining and transfer operation. In general, fish are in their lowest health condition during warm weather, immediately after spawning, and during long periods of food scarcity. Additional cages were set out with healthy laboratory-acclimated fish *(Rhinichthys atratulus)* at the same sites 48 h after the initial *in situ* cages had been set out.

The *in situ* bioassays were monitored at 24-h intervals for 184 h for mortality, abnormal behavior, pH, dissolved oxygen, and water temperature. The *in situ* bioassay results did indicate that there were materials in the contaminated pond that were toxic to fish. On the basis of the results from the *in situ* bioassay, it was recommended to the on-scene spill coordinator that release of the contaminated pond water be delayed until the acute toxicity had been removed. The problem of chronic effects to downstream species would thus be minimized, and restoration of fish populations by natural means or by restocking would most likely succeed.

CONCLUSION

A most important aspect of toxicity testing conducted at a hazardous material spill incident is that the technical adviser and spill managers realize that each test provides specific data that have limited application. In many incidents, it may be desirable (although not always feasible) to conduct a battery of tests utilizing more than one toxicity-testing system. The important point is to be familiar with each system's optimal application, information applicability, and operation requirements so as to be able to decide when and where the tests should be used (Table 2-1). The real challenge lies in interpreting the environmental significance of the data being generated from on-scene bioassays. This requires a comprehensive understanding of basic ecological, toxicological, and pharma-

TABLE 2-1   Toxicity Measurement Systems for Hazardous Material Spills

Equipment	Mobility	Number of responses measured	Number of species	Type of material	Commercial availability
CAM-4	Airline luggage	Single; cholinesterase inhibition	Not applicable	Organophosphates and carbamate compounds	Pending
Microtox	By air	Single; luminescent inhibition or enhancement	Single	Water-soluble or miscible compounds	Beckman Instruments
Portable proportional diluter (Allison diluter)	By air	Multiple; depending on test species	Multiple	Water-soluble or miscible compounds	No
Mobile bioassay vans	Must be driven or towed to scene	Multiple; depending on test species	Single (multiple with modification)	Water-soluble or miscible compounds	Service available from environmental laboratories

cological principles integrated with an awareness of the spill management scene. On-scene biologists-toxicologists can offer an invaluable service by providing this knowledge to hazardous material spill managers.

REFERENCES

1. *Oil Spill Studies: Strategies of Techniques,* API Publication No. 4286, American Petroleum Institute, Washington, 1977.
2. Louis H. Goodson and B. R. Cage, *CAM-4: A Portable Warning Device for Organophosphate Hazardous Materials Spills,* EPA-600/2-80-033, U.S. Environmental Protection Agency, 1980.
3. W. Peltier, *Methods for Measuring the Acute Toxicity of Effluents to Aquatic Organisms,* EPA-600/4-78-012, U.S. Environmental Protection Agency, 1978.

PART 3
Dispersion Modeling

Michael T. Kontaxis
Project Engineer
Hydroqual, Inc.

Joseph A. Nusser
President, Nusser and Associates

Introduction / 8-22
General Comments / 8-23
Solutions / 8-24
Application / 8-28
Conclusion / 8-28

INTRODUCTION

A number of time-variable water quality models have been utilized for analysis of the dispersion of pollutant discharges. In general, these models are complex and computer-oriented. Although their conceptual basis is quite general, their applications are site-specific, require considerable time to apply, and need extensive information for their calibration and execution. Finally, their application requires someone quite familiar with modeling.

While these models are invaluable tools for detailed risk analysis of a specific location, there are in the literature simpler models which can be used in the screening of sites before application of a detailed model or, more important, in a post-spill-event situation to gauge the temporal and spatial scale of the spill impact. Because these simpler models are analytical, they facilitate computation; they are in fact ideal for programmable calculators. They can be formulated so that less detailed information and a lower level of expertise are required for their application.

The bulk of this chapter considers the development of such a simple model for application to a one-dimensional water body such as a stream, a tidal river, or an estuary. In the interest of simplicity and utility of the resulting model, certain simplifying assumptions have been made. Whenever such assumptions have been introduced, ample consideration has been given so that they do not improperly affect the validity of the model. Furthermore, an effort has been made to provide guidance for selecting and applying modifications to the resulting model equation when more general situations are to be considered.

GENERAL COMMENTS

The general differential equation for the concentration of a miscible material in a body of water is

$$\frac{\partial c}{\partial t} = - E_x \frac{\partial^2 c}{\partial x^2} + E_y \frac{\partial^2 c}{\partial y^2} + E_z \frac{\partial^2 c}{\partial z^2} - u \frac{\partial c}{\partial x} - v \frac{\partial c}{\partial y} - w \frac{\partial c}{\partial z} \pm \Sigma S_0 - \Sigma S_i \quad (1)$$

where $c(x, y, z, t)$ is the concentration of the material at a point and time within the water body, E_x, E_y, E_z are the dispersion coefficients in the longitudinal, lateral, and vertical directions respectively, and u, v, w are the net average transport velocity components in the x, y, z direction. The term ΣS_0 represents all the causes that increase the concentration, and the term ΣS_i represents all the causes that decrease the concentration. These terms are known as the "source" and "sink" terms.

Certain observations about Eq. (1) should be made before detailed solutions are considered. It is important to notice that in Eq. (1) the transport of the substance is due to advection, $-\left(u \frac{\partial c}{\partial x} + v \frac{\partial c}{\partial y} + w \frac{\partial c}{\partial z} \right)$, and fickian diffusion $\left(E_x \frac{\partial^2 c}{\partial x^2} + E_y \frac{\partial^2 c}{\partial y^2} + E_z \frac{\partial^2 c}{\partial z^2} \right)$. Thus, an observer on a reference system moving with the advective velocity component u, v, w would only have to consider the fickian diffusion. Such an observer could therefore adopt directly all the solutions developed for the description of heat conduction in solids.[1,2,3]

It is also important to notice that Eq. (1) is a linear equation. This linearity allows the use of the principle of superposition, enabling one to write solutions of complex cases when solutions of simpler cases are known.

The detailed forms of the sink and source terms require separate attention. There are two feasible alternatives for the sink term, one for conservative and one for reactive substances. For conservative substances the sink term is zero, and solutions to Eq. (1) can be considerably simplified. For reactive substances, to consider anything more complicated than a first-order decay would be to render the solution of Eq. (1) intractable with analytic tools.

The form of the source term is usually determined by the type of discharge. In general, that term represents the spatial and temporal form of the introduction of the material to the system. Here it describes the spatial and temporal characteristics of the discharge and not the enrichment of the material by any means of physiochemical transformation.

Of the possible functional dependencies of the source term on time t, three forms seem to be simple but quite useful: (1) the instantaneous release of mass, (2) the release of mass at a constant rate for a finite amount of time, and (3) the release of mass at a constant rate for all time. The first two types are usually associated with dye data analyses for the determination of dispersion coefficients, whereas the third is associated with steady-state water quality modeling of point discharges. One may also consider periodically varying source terms that could be used when the time record of a discharge is Fourier-analyzed.

Various spatial configurations for the sources have been considered in the literature.[4,5,6] These include point, line, and plane sources. The solution techniques for these types of sources enable one to write solutions for source terms when analytic expressions can reasonably approximate the surface of submerged bodies.

From this discussion it follows that the source term alone can introduce complications which can make the solution of the problem very difficult to handle even though the simplest geometries and boundary conditions may be assumed. It is therefore our judgment that the treatment of a spill is best represented by the case of a point source discharging mass at a constant rate for a finite period of time. In the following discussions the term "spill" will carry that connotation.

SOLUTIONS

The solution of the conservation-of-mass equation for a spill in a one-dimensional body of water is constructed by superposition of solutions for an instantaneous release. This method is also helpful for the introduction of a sink term as a first-order decay. The governing differential equation for an instantaneous release of a conservative substance of mass M at the origin of the space and time coordinates is

$$\frac{\partial c}{\partial t} + u\frac{\partial c}{\partial x} = E_x\frac{\partial^2 c}{\partial x^2} + M\delta(x, t) \tag{2}$$

where c is the areally averaged concentration over the cross section perpendicular to the advective velocity u, E_x the longitudinal dispersion coefficient, and $\delta(x, t)$ the well-known Dirac function. The solution to Eq. (2) is well known:

$$c_I^c(x, t) = \frac{M\,u}{Q\,\sqrt{4\pi E_x t}}\,e^{-(x-ut)^2/4E_x t} \tag{3}$$

where Q is the net freshwater flow, the superscript c indicating that the substance is conservative and the subscript I indicating the instantaneous nature of the discharge. If the substance undergoes a decay that can be approximated by a first-order term, then Eq. (2) becomes

$$\frac{\partial c}{\partial t} + u\frac{\partial c}{\partial x} = E_x\frac{\partial^2 c}{\partial x^2} + M\delta(x, t) - Kc \tag{4}$$

where K is the decay rate. It can be shown by substitution that if $c_I^d(x, t)$ is the solution of Eq. (4), then

$$c_I^d(x, t) = c_I^c(x, t)e^{-Kt} \tag{5}$$

$$c_I^d(x, t) = \frac{M\,u}{Q\,\sqrt{4\pi E_x t}}\,e^{[-(x-ut)^2/4E_x t] - Kt} \tag{6}$$

For the remainder of this discussion a first-order decay will always be included, and therefore the superscript d will be omitted. To calculate the concentration c_f (finite-duration spill) at any point for a constant-rate discharge over a time interval τ, one has to create a superposition of Eq. (6) over time, thus:

$$c_f(x, t > \tau) = \int_0^\tau c_I(x, t - t^*)dt^* \tag{7}$$

Physically, Eq. (7) indicates that the discharge can be subdivided into a number of δ-like functions or square pulses of finite height and width. The effect at any location x of each pulse is given by Eq. (6), where the time variable has been adjusted to keep the correct sequence of releases. For the net effect at x after the last release has arrived, sum the effects of the previous releases at x. Stated mathematically, if the mass M is released at the constant rate of μ over a time interval τ and the release is subdivided into N-many square pulses, the mass of the ith pulse is $\dfrac{M}{N} = \dfrac{\mu\tau}{N} = \dfrac{\mu\,(\text{duration of spill})}{N}$.

$$c(x, t) = \frac{\mu \tau u}{N} \frac{1}{Q} \frac{1}{\sqrt{4\pi E_x}} \sum_{i=1}^{N} \frac{1}{\sqrt{t^*}} e^{-\{[(x - ut^*)^2/4E_x t^*] + Kt^*\}} \tag{8}$$

where $t^* = t - (i - 1)\dfrac{\tau}{N}$.

Equations (7) and (8) yield the same result as $N \to \infty$. Equation (7) can be integrated to yield a closed-form solution. The result[7] has also been stated as

$$
\begin{aligned}
c(x, t, \tau) = &\frac{\mu u}{2QD} e^{(u-D)x/2E_x} \left\{ \mathrm{erf}\, \frac{x - D(t - \tau)}{\sqrt{4E_x (t - \tau)}} - \mathrm{erf}\, \frac{x - Dt}{\sqrt{4E_x t}} \right\} \\
&- \frac{\mu u}{2QD} e^{(u+D)x/2E_x} \left\{ \mathrm{erf}\, \frac{x + D(t - \tau)}{\sqrt{4E_x(t - \tau}} - \mathrm{erf}\, \frac{x + Dt}{\sqrt{4E_x t}} \right\}
\end{aligned} \tag{9}
$$

where $\qquad\qquad D = \sqrt{u^2 + 4KE}$

Let $\qquad\qquad R_{+D} = \dfrac{e}{D}^{(u-D)x/2E_x} \left\{ \mathrm{erf}\, \dfrac{x - D(t - \tau)}{\sqrt{4E_x(t - \tau)}} - \mathrm{erf}\, \dfrac{x - Dt}{\sqrt{4E_x t}} \right\} \tag{10}$

Then $\qquad\qquad c(x, t > \tau) = \dfrac{\mu u}{ZQ} \{R_{+D} + R_{-D}\}$

In Eqs. (8), (9), and (10) the discharge point is located at the origin. Should the discharge be at another point, say, x_0, then the same equations hold but with the simple substitution $x \to x - x_0$.

Figure 3-1 is a graph of the error function, which is defined as

$$\mathrm{erf}\,(Z) = \frac{2}{\sqrt{\pi}} \int_0^Z e^{-t^2}\, dt$$

The results of a sensitivity analysis of the maximum concentration observed at a point to (a) dispersion coefficient E, (b) spill duration τ, and (c) decay coefficient K are shown in Fig. 3-2. In all these cases, 454 kg (1000 lb) of material is spilled into a river whose flow is 283 m³/s (10,000 ft³/s) with a velocity of 0.3 m/s (1 ft/s). Figure 3-3 contains the results of a sensitivity analysis of the temporal concentration at selected mile points to (a) the dispersion coefficient E and (b) spill duration τ. From an examination of Figs. 3-2a

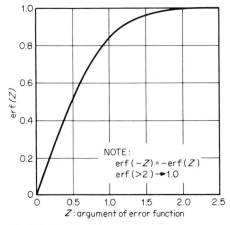

Fig. 3-1 Graph of the error function.

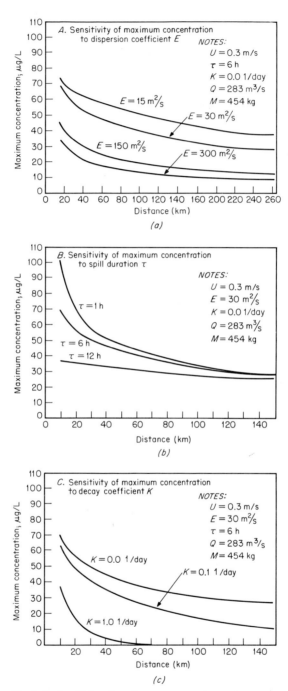

Fig. 3-2 Sensitivity of maximum concentration to dispersion, spill duration, and decay coefficient.

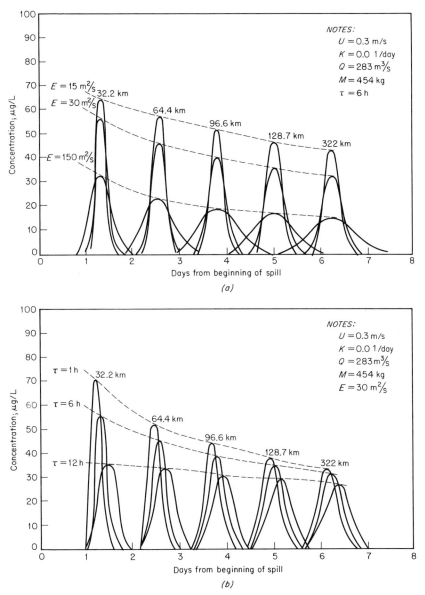

Fig. 3-3 Temporal concentrations at various distances as a function of dispersion and spill duration.

and 3-2b, it becomes apparent that the reduction of the maximum concentration is due to the spreading of the temporal concentration, which in turn implies that low limiting concentrations may be exceeded for longer periods of time as the dispersion coefficient increases. An inspection of Fig. 3-3b reveals that an increase in the spill duration causes a spreading in the temporal concentrations and a delay in the time of arrival of the maximum concentration.

APPLICATION

The success with which either Eq. (8) or Eq. (9) can be applied to an actual spill situation depends on (1) how closely the actual situation resembles the conditions under which these equations were derived and (2) knowledge of the parameters involved in Eq. (8) or Eq. (9).

The assumption of vertical and lateral homogeneity requires that ideally the concentration at the point of discharge be uniform. Since this can hardly be achieved, Eqs. (8) and (9) will apply only after a point at which discharged substances become well mixed across the cross section. This length[8] is taken as

$$L_m = \frac{0.445}{\alpha} \frac{W^2}{H}$$

where L_m = distance for complete lateral mixing
W = average width of river
H = depth of the river
α = dimensionless constant relating velocity, depth, and tranverse dispersion coefficient; usual range is 0.02 to 0.04

The longitudinal dispersion coefficient will, in general, be the least-known system parameter of the parameters involved in the evaluation of Eqs. (8) and (10). The literature on this term is quite extensive.[9] The relationship[10]

$$D_L = \frac{\beta Q^2}{\mu R^3} \tag{11}$$

with

$$\beta = 0.18 \left(\frac{u^*}{u} \right)^{1.5} \tag{12}$$

may be used in estimating longitudinal dispersion coefficients. The parameters are

Q = streamflow
u^* = shear velocity = $\sqrt{gRS}$
R = hydraulic radius, or mean depth in large, wide rivers
g = acceleration due to gravity
μ = mean velocity

When Eqs. (11) and (12) furnish the only means of estimation, sensitivity analysis to E within one order of magnitude is recommended.

CONCLUSION

This modeling of spill events has been intentionally simplified so as to be readily applicable with a minimum of expertise and data input. Thus results should be cautiously interpreted as a first approximation. Should a spill event require more extensive evaluation, verified computer-oreinted models with finer hydraulic and geometric definition should be used.

REFERENCES

1. H. S. Carslaw and J. C. Jaeger, *Conduction of Heat in Solids*, Clarendon Press, Oxford, 1959.

2. Necati M. Ozisik, *Boundary Value Problems of Heat Conduction*, International Textbook Company, Scranton, Pa., 1968.

3. J. Crank, *The Mathematics of Diffusion*, Clarendon Press, Oxford, 1956.

4. R. W. Cleary and D. D. Adrian, "New Analytical Solutions for Dye Diffusion Equations," *Journal of the Environmental Engineering Division*, American Society of Civil Engineers, vol. 99, EE no. 3, June 1973, pp. 213–228.

5. G. T. Yeh and Y. J. Tsai, "Analytical Three Dimensional Transient Modeling of Effluent Discharges," *Water Resources Research*, vol. 12, no. 3, June 1976.

6. B. A. Benedict, "Analytical Models for Toxic Spills," in G. F. Bennett (ed.), *Proceedings of the 1978 National Conference on Control of Hazardous Material Spills*, Miami Beach, Fla., April 1978, pp. 439–443.

7. D. Tsahalis, "Mitigation of Chemical Spills; RIDIS: A River Dispersion Model," in G. F. Bennett (ed.), *Proceedings of the 1978 National Conference on Control of Hazardous Material Spills*, Miami Beach, Fla., pp. 427–431.

8. E. R. Holley, *Transverse Mixing in Rivers*, Delft Hydraulics Laboratory, December 1971.

9. M. K. Bansal, "Dispersion in Natural Streams," *Journal of the Hydraulics Division*, American Society of Civil Engineers, vol. 97, no. HY11, November 1971.

10. H. Liu, "Predicting Dispersion Coefficients of Streams," *Journal of the Environmental Engineering Division*, American Society of Civil Engineers, vol. 103, EE no. 1, February 1977.

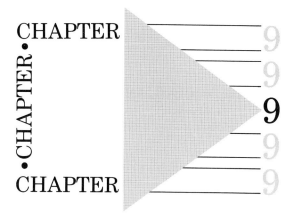

CHAPTER 9

Spill Cleanup

PART 1
Field-Implemented Measures / 9-2

PART 2
Technology Development / 9-24

PART 3
Biological Measures / 9-40

PART 4
Chemical and Physical Measures / 9-50

PART 1
Field-Implemented Measures

Robert C. Scholz
Environmental Research Center
Rexnord Inc.

Introduction / 9-2
Spill Summaries / 9-3
Termination of Discharge / 9-4
Containment / 9-4
Recovery and Treatment / 9-9
Summary / 9-21

INTRODUCTION

The purpose of this part is to classify and describe some of the available methods for responding to spills of hazardous chemicals, particularly those for which experience has been reported. These experiences will be summarized, and the lessons learned from them concerning the applicability and effectiveness of control methods will be discussed. In addition, new technology which has been demonstrated as suitable for spill response will be identified.

The three primary aspects of spill response that will be described are (1) control at the source, (2) containment, and (3) recovery and/or treatment. Alternative forms of ameliorating hazards by dilution and dispersion are not discussed here. In many spill events the quantity and toxicity of spilled chemicals are great enough to cause catastrophic and enduring biotic damage. For this reason it is essential that effective methodology to minimize spill damage be readily available. Other important aspects of spill response such as ultimate disposal, control of volatile substances, fire and explosion control, sampling and analysis, and water treatment methods for specific chemicals are discussed elsewhere in the *Handbook*.

Control at the Source

If the first line of defense against spills of hazardous chemicals is spill prevention, then the second line of defense is control at the source. Control at the source includes all actions taken to terminate discharge of hazardous materials and to isolate already spilled materials to the immediate vicinity of the discharge point, thus preventing dispersion of the chemicals over and into soils and into groundwater and surface-water supplies.

Termination of discharge usually includes actions such as righting an overturned and spilling container, pumping the contents of a leaking container into a different container, plugging a leak, or closing a valve, but it may also include stopping improper disposal activities. If the discharge point is very localized, isolation of spilled materials to the immediate vicinity of the discharge point can be accomplished with such measures as impermeable barriers and dikes and gelling agents. Isolation can also be facilitated by selectively employing emergency response procedures which do not act to disperse contaminants (e.g., the use of chemical foam rather than water to fight chemical fires).

The containment, recovery, and treatment of hazardous chemicals dispersed in groundwater and surface-water supplies are extremely expensive and time-consuming and usually are incomplete. However, control at the source can be effective and nearly complete, and it may lead to product recovery.

Containment

Containment includes the action of any physical barrier, natural or artificial, to limit the dispersion and permit the recovery and treatment of hazardous chemicals which have reached groundwater and surface-water supplies. Typical measures that have been reported on spills of hazardous chemicals into small streams and ponds include floating booms, underflow or impermeable dams, and bypasses for isolating sections of flowing streams. The use of containment measures on actual spills into soils and groundwater has not been reported.

The effectiveness of containment measures depends on the nature of the spill, the mode of dispersion, the size of the contaminated water body, and the rapidity of response. In most cases, the flow patterns of the groundwater or surface-water supply dictate the speed and extent of dispersion. These flow patterns can be quite variable, with peaks occurring during and immediately following storms.

The efficiency and cost-effectiveness of recovery and treatment measures have depended primarily on the ability to contain spills successfully.

Recovery and Treatment

Whether spilled hazardous materials sink, float, solubilize, and/or become suspended in the water column, recovery of the contaminant from groundwater and surface water almost always involves some form of water treatment. Rarely is the contaminated water volume low enough that it can be transported and disposed of. Because sinking substances can be resuspended or dissolved by dredging, solid-liquid separation must usually be accompanied by treatment of the transport liquid, which may range from 2 to 10 times the amount of settled material recovered. Only in the recovery of relatively thick layers of floating substances will water treatment be unnecessary.

Field treatment methods can be classified broadly as one of two types: (1) those which require removal of the contaminated fluid to a treatment system located at the spill site where contaminants are detoxified or reduced to safe levels and (2) those in which treatment is applied directly to contaminated fluid in place (*in situ* treatment). Both types have been used in reported spill case histories. When contaminants were removed to a treatment system adjacent to the water body, containment of the contaminated water body was necessary so that the total volume to be treated did not become prohibitively large. *In situ* methods have been employed in fast-moving streams when containment was not feasible. Several other treatment methods may include one or more of the following: chemical reaction, aeration, sedimentation, filtration, carbon adsorption, and ion exchange.

Treatment of contaminated soils has been infrequently reported, as contaminated soils are usually excavated and transported to a disposal site.

SPILL SUMMARIES

Anyone who has participated in the development of a spill contingency plan or the design and specification of spill response equipment and materials must have experienced the feeling of being overwhelmed by the limitless variety of possible spill situations. Selecting a proper response method demands an understanding of technical feasibility, costs of alternative control measures, and technical application problems. There is no substitute for direct experience in such evaluations. Therefore, the discussion of control measures in this part centers in historical data as presented in reported evaluations of spill response measures.

Spill response efforts have been limited, and detailed reports of these responses have been even more limited. Of the cases reported in the literature, 12 are cited in the following discussions of spill response technology. To provide the reader with a better understanding of the conditions surrounding each of these spill responses, information concerning the type and quantities of materials spilled, the ways in which water was contaminated, and the response methods used is presented in Table 1-1.

TERMINATION OF DISCHARGE

Transportation

Actual experience has shown that temporary leak-patching methods can permit derailed and leaking tank cars to be righted and moved to a nearby location where they can be pumped out. In one reported case, the amount of sulfuric acid loss had been low because the tank car ended up on its side with the punctured end elevated. This allowed enough time to construct a bolted patch plate for sealing the leak. A most difficult and dangerous situation arises when the leak is low on the tank as it lies and when for safety reasons or lack of accessibility the car cannot be quickly turned to put the leak at the top.

The U.S. Environmental Protection Agency (U.S. EPA) has developed a foam plugging method which allows a person to insert a probe into a puncture from a few feet away and to inject an expanding urethane foam sealant.[13] The unit is self-contained and includes provisions for storing and mixing foam components and transferring them to the applicator tip. The sealant material is compatible with most hazardous chemicals and can be used to stop leaks up to 100 mm in diameter underwater as well as above water.

The process of pumping out leaking tank cars can be difficult because of safety and accessibility reasons. In one reported derailment (Incident 7) the railroad shipped an explosionproof rail-mounted pumping system to evacuate a tank car partially filled with a highly flammable chemical.

Drum Leaks and Punctures during Transit and Handling

Drums rubbing on sharp objects are a major cause of leaks during transit, and improper handling with such devices as forklift trucks is a major cause during handling. A large chemical corporation has developed an "overpack drum" with which it can encapsulate leaking drums.[14] It consists of a stainless-steel drum body large enough to enclose any size of 0.21-m^3 container, a lid, a formfitting gasket, and a ring-type closure.

CONTAINMENT

Isolation of Sections of Flowing Streams

Spills into small streams and reservoirs can sometimes be contained and prevented from contaminating downstream water supplies by isolating a stream section and treating it. Water bodies that fall into this category generally have too low a flow to produce rapid dispersion and dilution of contaminants, yet too high a flow to allow flow stoppage by damming even for short periods. Flow in these water bodies is usually quite variable and subject to large surges during storms.

TABLE 1-1 Spill Response Summaries

Incident	Hazardous chemicals	Mode of discharge	Dispersion mechanisms	Termination of discharge	Containment	Recovery or treatment	Reference number
1	Oils and solvents containing 2% of a polychlorinated biphenyl (PCB), Aroclor 1260	Illegal disposal site, which owner filled with soil when legal action was pending	Leaching through soils and into nearby creek	Legal action	Creek isolation using dams and a gravity bypass pipe	Excavation and removal of soils from pit area; leachate treated with field-constructed carbon columns using 2.8 m^3 of carbon preceded by a disposable filter	1
2	Endrin (chlorinated hydrocarbon), 18.6% in 3.8 L of pesticide	Willful dumping into reservoir	Overflow of reservoir spillway at 1 L/s into stream (dry weather)	Instantaneous spill	Stream isolation using an earthen dam at inlet to reservoir, a sandbag dam at reservoir outlet, and a stream bypass pumping system (Fig. 1-1)	First attempt—broadcasting 300 g of granular activated carbon on surface (failed); second attempt–reservoir fluids pumped through a field-constructed carbon column (Fig. 1-1); succeeded	2
3	Phenol (80,000 kg)	Ruptured railroad tank cars at a derailment	Flow of 5 percent downhill into stream; seepage into soil of 95 percent, which began to leach into stream during rainfalls	Almost instantaneous spill that emptied contents of tankers	Collection channel built to act as a reservoir to prevent runoff and leachate from entering stream	Flow from collection channel directed to a field-constructed carbon column using 28.3 m^3 of carbon; clean underflow from carbon column discharged to the stream	3
4	Transformer fluids containing PCBs in waste oil pits	Rupture of a pit which spilled 740,000 L of oil and acidic water	Runoff of escaping fluid into a creek	Loss of entire contents of pit	Booms used in creek to contain oil	Oil removed from the creek; fluid content of remaining five pits treated by using powdered activated carbon, sedimentation, and sand filtration (Fig. 1-2); after dewatering, oils remaining in pits immobilized by adding earth and dry solid waste	4

TABLE 1-1 Spill Response Summaries (*Continued*)

Incident	Hazardous chemicals	Mode of discharge	Dispersion mechanisms	Termination of discharge	Containment	Recovery or treatment	Reference number
5	Oil and 1% pentachlorophenol (PCP; 3700 m³)	Injection into shallow disposal well	Leaching into groundwater, pentration into joints of culvert, and discharge into stream	Shallow-well disposal stopped by company 9 years prior to reports of stream contamination	None	Three 200-mm recovery wells near the disposal well and trenches at culvert entry points; fluid pumped into vacuum trucks where oil-water separation occurred (Fig. 1-10); water portion pumped to U.S. EPA mobile physical-chemical treatment system: sedimentation tank with oil skimming, filtration, and carbon adsorption	5
6	Anhydrous ammonia (12,000 kg)	Hose failure during transfer from tank truck	Materials washed into creek by water sprays used to disperse them	At conclusion of hose spraying of spilled materials	None; unrestricted runoff	Peat-moss dams used to adsorb and buffer the ammonia; hydrochloric acid (8000 L) added to stream to reduce pH (ammonia is most toxic at elevated pH)	6
7	Acrylonitrile (32,000 L)	Tanker rupture during train derailment	Material washed by fire hoses into pools in area adjacent to tracks	Leakage of contents of tanker out to the level of the puncture; remainder of tanker contents pumped out	Natural containment by terrain; containment temporary because of seepage into soils	pH of affected area raised above 10 with 4100 kg of lime; area then sprayed with sodium hypochlorite solution (1100 kg)	7
8	Creosote	Creosote wastes from a manufacturing facility discharged into a small stream over many years	Material caused by currents to permeate a long length of river bottom and to form large pockets where stream widened at bridges	Termination of discharge practices after people wading in river developed severe rashes	None	Dredging of stream bottom accomplished with specially designed dredge (Fig. 1-7) and hand-held devices (Fig. 1-6); froth flotation and skimming used to	8

remove some of the creosote from the mud; remainder of treatment similar to that used in Incident 6 (EPA physical-chemical treatment trailer)

No.	Contaminant	Source	Mechanism of spread	Action to stop discharge	Emergency containment	Treatment
9	Polychlorinated biphenyls (PCBs; 980 L)	Fall of a transformer during loading onto a ship; unit split open and fluid spilled into bay	Settling of material on river bottom where tidal flows began to spread it	Transformer moved to location where leakage did not reach river before all the material had leaked out (time frame unknown)	None	Top layer of bottom mud dredged with hand-held devices by skin divers; settling tanks and flocculants used to remove particulate; fluid treated (Fig. 1-8) by the EPA's physical-chemical treatment trailer using sedimentation, filtration, and activated-carbon adsorption before being discharged back to river
10	Hydrocarbons (80% gasoline; 1850 m³)	Source not defined	Leaching of contaminants into ground, migration along top of aquifer, and discharge into stream, domestic wells, and sewer system	Not discussed	None	One 500-mm recovery well at a point of substantial accumulation; two pumps used in well, one to pump underflow water to storm sewer and the other to pump oil to storage tank (Fig. 1-9)
11	Pesticide for termites (9.2 L); chlordane, 39%; heptachlor, 19%; related compounds, 7%; petroleum derivatives, 29%	Pesticide injected into soils around house	Material leached into open drainpipe under house which discharged into a stream	Discharge stopped when exterminator completed work	Streamflow (2 to 4 m³/h) stopped by downstream earthen dam	Streamflow treated with EPA mobile physical-chemical treatment unit; water injected into the same places into which pesticide had been injected around house to speed leaching process into the drainpipe

TABLE 1-1 Spill Response Summaries *(Continued)*

Incident	Hazardous chemicals	Mode of discharge	Dispersion mechanisms	Termination of discharge	Containment	Recovery or treatment	Reference number
12	Dinitrobutylphenol (DNBP)	Herbicide improperly applied at full strength around parking lot on a hill	Pesticide washed by storm runoff into an artificial lake, a tributary of a river	No special action taken; applicator unaware that he was applying material improperly	Spillway at outlet of lake dammed with sandbags	Parking lot washed down with hoses and runoff collected in a ditch and treated with mixed-media filtration and activated-carbon adsorption, using EPA's mobile physical-chemical treatment system; lake fluid treated in similar manner	12

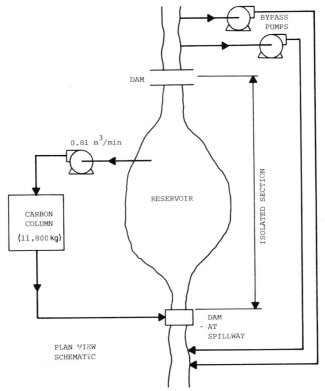

Fig. 1-1 Stream bypass and field-constructed carbon adsorption system to treat reservoir (Incident 2).

Two cases in which stream diversion was successfully used during spill cleanup have been reported. In one case (Incident 1), the normal flow in a small creek was so low (0.22 m^3/h) that stream diversion could be accomplished with dams and a long length of flexible tubing laid in the streambed. This approach eliminated the necessity of using a pump, but the presence of the tube in the streambed interfered with treatment of the isolated section.

In another case (Incident 2 and Fig. 1-1) a reservoir was isolated from the stream that fed it by using dams and two 9.5-m^3/min pumping systems to bypass the flow through 280-mm aluminum pipe.

The EPA has developed a portable stream diversion system with the capability of bypassing 0.35 m^3/s a distance of 0.3 km, or a lesser flow a distance of 0.91 km.[15]

RECOVERY AND TREATMENT

Physical-Chemical Treatment Systems

The most widely used field treatment method for recovery of chemicals from surface water and groundwater has been activated-carbon adsorption. It has been used to treat contaminated streams, ponds, and reservoirs (Incidents 2, 4, and 12), dredging fluids (Incidents

8 and 9), and leachate and runoff water (Incidents 1, 3, 5, and 11). A laboratory study conducted by the EPA concluded that many of the chemicals regarded as very hazardous and with a high likelihood of being spilled, e.g., acetone, cyanohydrin, acrylonitrile, and phenol, could be treated by this method.[16] It was found that most pesticides and herbicides could be removed through coagulation, filtration, and carbon adsorption.

Field-Constructed Carbon Columns

Field-constructed carbon adsorption systems utilizing both granular and powdered carbon have been successfully applied in several spill response efforts. The major limitation to the use of such systems is that they are time-consuming to construct and thus require containment of the spill. In addition, unless there is access to information concerning proper design, the ultimate degree of effectiveness could be severely diminished by problems of plugging with solids, channeling, or air-binding. The two last-named types of failures permit contaminants to pass through the system untreated.

A major problem in constructing a granular carbon column in the field, besides acquiring the carbon media, is providing a suitable container to house the treatment process. At a railroad accident site, a granular carbon container was constructed of railroad ties lined with plywood and sealed with tar and okum (Incident 3). In another case, an available cypress tank was shipped to the spill site from another state (Incident 2). In a third case (Incident 1), a long section of locally available concrete culvert was used in a design proposed by an EPA research publication on field-improvised water treatment systems.[17] After being stood vertically on end, this tank design provided a favorable height-to-diameter ratio of 3, which helped in preventing the channeling to which squat tanks such as were used in Incidents 1 and 2 are subject.

One common feature of all three responses was that the spills were sufficiently contained to allow the necessary time (several days) to purchase and construct the carbon columns and make the systems fully operational.

In two of the cases (Incidents 1 and 2), dams and stream bypasses were used to isolate a reservoir and a stream. In the third case (Incident 3), a collection channel was built as a reservoir for contaminated runoff and leachate.

Field construction of systems in the three reported cases was simplified because the solids content of the water was low enough that filtration pretreatment was unnecessary except in Incident 1, in which 3- to 5-μ disposable bag filters were used ahead of the carbon column.

Powdered Activated Carbon

In two reported spill response efforts, powdered activated carbon was injected into contaminated water and subsequently recovered. In one of the cases (Incident 4 and Fig. 1-2), a powdered-carbon treatment system was implemented in four units:

1. Collection and equalization tank
2. Carbon-mixing chamber
3. Sedimentation tank
4. Sand filter

The initial carbon injection rate, 27 mg/L, was adjusted as necessary to maintain the effluent concentration of the contaminant at an acceptable value.

A powdered-carbon system was well suited for this application because it is not subject to blinding by oil, as a carbon column is. Nor is it subject to breakthrough because of exhaustion of the carbon column. A drawback to the use of this method is the size of the carbon recovery system (sedimentation and filtration tanks).

The use of large excavated sumps in containing three of the four treatment operations

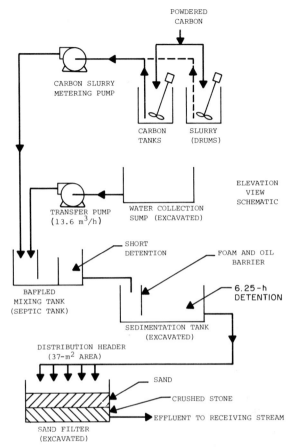

Fig. 1-2 Field-constructed powdered-carbon adsorption system including carbon injection and recovery (Incident 4).

expedited system construction. However, the excavations were not sealed with impervious liners. Although in this case a groundwater survey revealed no contamination from PCBs, in many cases leaching from excavated treatment tanks could cause serious groundwater contamination problems.

In the second of the reported cases, an *in situ* treatment approach in which powdered carbon was mixed into a small stream where it contacted water contaminated with acrylonitrile was employed.[18] Farther downstream, the carbon was collected on a filter dam constructed of rock and covered with peat moss. Unfortunately, the report lacked details which could shed some light on the problem of injecting and recovering powdered carbon in flowing streams.

Unless carbon is slurried and distributed throughout a stream cross section or mixed with the stream by mechanical action of some sort (outboard motors, obstacles, etc.), the job of providing uniform mixing and proper contact must be accomplished by the turbulence of the stream itself. The problem of providing sufficient mixing and contact time is directly coupled to the recovery problem. The farther the recovery dam is downstream from the injection point, the longer the contact time but also the greater the percentage

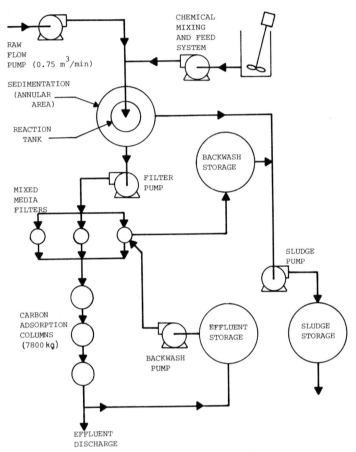

Fig. 1-3 Simplified process flow sheet for the EPA mobile physical-chemical treatment system.

of carbon which settles on the stream bottom and eludes collection unless dredging is performed.

Mobile Physical-Chemical Treatment Unit

In 1972 the EPA constructed, and has since continuously operated, a mobile spill response unit employing chemical reaction, sedimentation, granular media filtration, and carbon adsorption of water contaminated with hazardous chemicals. The purpose of constructing and operating the mobile system is to demonstrate that it is feasible to apply treatment processes at a spill site and that a quick, safe, effective, and versatile way to do this is to employ preconstructed treatment components strategically located and on call.

A simplified process flow sheet for the mobile system is shown in Fig. 1-3. Filters and carbon columns and their piping, valving, and control systems are mounted in fixed locations on the trailer (Fig. 1-4). Other large fluid treatment and storage systems are contained in collapsible tanks adjacent to the trailer. One such tank for chemical reaction and sedimentation is shown in Fig. 1-5. The entire system is self-contained and self-powered (with an engine-generator set).

Fig. 1-4 The EPA mobile physical-chemical treatment system in operation at its first spill response in 1972 (Incident 8).

The features which make this system very suitable for field response are:

1. *Speed of application.* The items which are the most time-consuming to acquire, assemble, and make operational in the field, i.e., tanks, pumps, piping, valves, and underdrain systems (filters and carbon columns), are all preassembled, thus requiring a minimum of setup time for the mobile unit.

Fig. 1-5 A large portable sedimentation tank constructed of rubber-covered fabric provided rugged service with a minimum of site preparation (Incident 8). This figure appears in the following report: C. A. Hansen and R. G. Sanders, *Removal of Hazardous Material Spills from Bottoms of Flowing Waterbodies,* EPA 600/2-81-137, U.S. Environmental Protection Agency, July 1981.

2. *Minimization of site preparation.*   The use of heavy-duty collapsible tanks elim-inates the need for the special site preparations which are required when either swim-ming pools or excavated and lined tanks are used to provide large fluid storage and treatment capacities.

3. *Safety.*   Instead of attaching and reattaching hoses during the various phases of treat-ment (e.g., backwashing of filters), the system is manifolded with valves that permit flow redirection without disconnections. Hose changes can cause splashing of contam-inated fluids. Workers on field-implemented systems should be protected from contact by contaminated fluids on their eyes and skin by using protective clothing and work practices which prevent splashing.

Chemical mixing and metering equipment eliminates manual handling and close contact with treatment chemicals.

A carbon slurry injection and removal system provides for quick media changes without the need to enter the carbon column and manually shovel out the media.

4. *Protection against carbon blinding.*   Construction of a filtration system as a pri-mary treatment step or as a pretreatment step to prevent carbon blinding is one of the most difficult field processes to construct and operate because of the need to backwash the filters effectively.

5. *Monitoring.*   Water samples taken to monitor the performance of a treatment system will most often be transported to a distant laboratory for analysis. In such a case it is necessary to provide protection against a particular form of treatment system failure termed "carbon column breakthrough." Depending on the amount of chemical to be treated and the volume of carbon in the system, the adsorptive capacity of the carbon column could be reached during the treatment process, thus allowing untreated chem-icals to be discharged. One method to prevent this failure is to utilize sufficient carbon in the system so that the adsorptive capacity can never be reached. However, this may not be feasible. The mobile system utilizes a more practical approach termed "revolv-ing-series flow." There are three carbon columns in series with provisions for moni-toring after each of them. When a breakthrough of the first column is detected, this column can be recharged and brought back into service as the last column in the series. In this way there is assurance that, at the time of the first column breakthrough, the adsorptive capacity of the system will still be adequate to provide effective treatment during the analytical lag time.

Although revolving-series operation provides a high degree of reliability for monitor-ing carbon adsorption treatment, the most reliable monitoring approach is one in which an effluent reservoir is provided with a storage capacity given by the relationship

$$C \geq Q \times t \tag{1}$$

where C = storage capacity in cubic meters
   Q = treatment flow rate in cubic meters per hour
   t = analytical lag time in hours

In a treatment system with a high flow rate located at a substantial distance from an analytical laboratory, this amount of storage may not be practical. In Incident 5, a mobile on-site laboratory developed and operated by the EPA decreased the analytical lag time for gas chromatography to 3 h, making it feasible to store cleaned effluent from a 22.7-m^3/h treatment system by using 68 m^3 of tank capacity. (Information on the mobile lab-oratory is presented in Reference 19.)

Manual for Hazardous Spills Control

As field experience in treating hazardous materials spills has increased, two needs have become apparent:

1. The need for people who are properly trained in the selection and use of field response procedures
2. The availability of design information which can be used to construct field-improvised treatment systems

Thus the EPA devloped a manual specifically for field response personnel.[17] This manual presents a logical approach to determining a suitable spill-handling method for a particular situation, presents treatment schemes for 303 hazardous chemicals, and details design, construction, and operational procedures for filtration, carbon adsorption, ion exchange, gravity separation, and chemical reaction.

Porous Peat-Moss Dams in Flowing Streams

Substances such as peat moss are known to have adsorptive ability, although to a much less extent than activated carbon. Peat-moss dams are documented in a number of spill response case histories involving spills of ammonia (Incident 6), acrylonitrile, paint, herbicides, pesticides, and oil. Usually these porous dams were among the first countermeasures employed, probably because materials for their construction were all readily available and the small size of the contaminated streams simplified erection of the dams. Fence posts driven into the streambed, chicken-wire fencing, bales of straw, crushed limestone, and loose peat moss served as construction materials in various configurations. Dams were employed for their adsorbing and filtering capabilities, and in one case (Incident 6) the peat moss was considered an auxiliary to the use of acid in lowering stream pH.

In Situ Neutralization

Neutralization has been used in a number of spill response efforts to minimize the environmental damage from contaminated runoff, particularly owing to tank punctures in tank truck and railroad tanker derailments and hose failures during chemical transfer. In situ treatment has been applied to pools of materials in the vicinity of an accident, in trenches dug to intercept contaminated runoff, and in small streams which have received a runoff. Unnecessary dispersion of a contaminant because of water used to put out fires or to suppress fuming was a disruptive factor in applying this treatment during several incidents.

Acid spills have been neutralized with crushed lime, sodium carbonate, and soda ash. Lime has also been used as part of a two-part treatment procedure to neutralize a drainage area soaked with acrylonitrile (Incident 7). Hydrochloric acid has been used to neutralize pools of sodium hydroxide and to lower the hazard from anhydrous ammonia in a flowing stream by lowering the pH of the stream (Incident 6). The latter experience provided the lesson that personal protection from skin contact and inhalation is absolutely necessary to prevent severe irritation because of acute exposure of response personnel to acid.

Recovery of Sinking Hazardous Substances from the Bottom of Watercourses

Dredging can be used to recover hazardous materials of a specific gravity greater than 1.0 which are discharged into surface waters. The extensive experience and available equipment typically used for underwater excavation to deepen watercourses for navigation or for underwater mining can also be applied in spill response efforts.[20] Hydraulic dredging is much better suited to spill response than mechanical dredging (dippers, clamshells, etc.) because it can recover liquid as well as solid contaminants and releases the smallest amount of hazardous substances back into the water column. Hydraulic dredging is most suitable for removing intact masses of liquid or granular chemicals or unconsolidated sand, silt, or clay. The primary concerns in using hydraulic dredging are fluid-solid separation and the treatment of a quantity of water 2 to 10 times the quantity of soil recovered.

Two documented cases of small-scale dredging operations to remove contaminated bottom muds (Incidents 8 and 9) have provided some insight into the unique problems of this response method.

In Incident 8, both hand-held devices (Fig. 1-6) and a specially designed hydraulic-powered pontoon dredge (Fig. 1-7) were employed during the cleaning of a small stream by using self-priming trash pumps located onshore. The small size (76 mm) and long length of the suction hose, plus a lack of augers on the suction heads, caused frequent clogging of the suction lines, which had to be cleaned by backflushing.

Another problem in dredging is control of the suction head. The pontoon dredge had a power-assisted suction line, but the hand-held suction devices relied on the operator to control penetration. These devices had an effect similar to that of a magnet; i.e., little force could be felt by the operator until the suction head came in close proximity to the mud, at which time the suction device would uncontrollably bury itself in the mud and deadhead the pump. For this reason, the operator typically positioned the suction head in a suitable location to prevent burying it, resulting in a high ratio of liquid-to-solid pickup.

Whereas in Incident 8 the operators had to rely on their "feel" to determine the best way to move the suction head, the skin divers in Incident 9 could witness visually the hazardous material pools lying on the silt. The experience gained on Incident 9 revealed the problems unique to using swimming pools in field-constructed solids separation systems. A schematic diagram of the system is provided in Fig. 1-8. Originally, the dredged material was pumped directly to a portable swimming pool used as a flocculation and sedimentation basin. However, the pool liner was repeatedly torn by surging hoses, which

Fig. 1-6   Hand-held suction hoses frequently became clogged with debris. Here one worker raked up contaminated bottom muds and fed the material in a uniform pattern to the suction head held by the other worker (Incident 8).

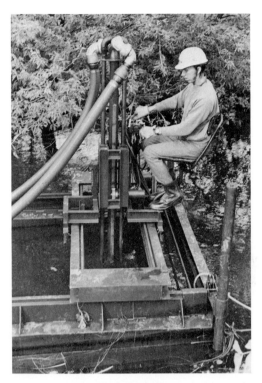

Fig. 1-7 A hydraulic-powered pontoon dredge permitted scouring of bottom muds contaminated with creosote from a narrow stream (Incident 8). This figure appears in the following report: C. A. Hansen and R. G. Sanders, *Removal of Hazardous Material Spills from Bottoms of Flowing Waterbodies,* EPA 600/2-81-137, U.S. Environmental Protection Agency, July 1981.

periodically broke loose from their anchorings within the tank. Moreover, the pump surges caused unwanted resuspension of settled solids. A separate equalization tank solved the problem and permitted the use of a single nonpulsing pump to transfer the dredgings into a stilling well in the flocculation tank. Periodically, the settled solids were vacuumed from this tank by an operator who entered the pool wearing hip waders.

Before being discharged back into the river, the supernatant of the flocculation-sedimentation tank underwent further sedimentation, mixed-media filtration, and carbon adsorption to remove the hazardous chemicals which had been put into solution by the pumps or were bound to very fine particles which had not been removed by the flocculation process.

In both of the above recovery operations, the contaminated and partially dewatered dredging solids were removed from the site for disposal.

Recovery and Treatment of Contaminated Groundwater

Conditions under the ground are very complex hydrogeologically and are not easily scrutinized. The problem of migration of contaminants into groundwater aquifers is one of the most unmanageable spill situations. Test borings and the services of a hydrogeologist

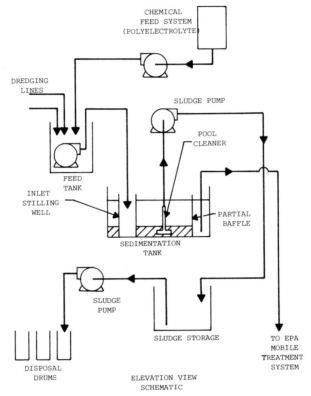

Fig. 1-8   System for removal of dredging solids from carrier water (Incident 9).

are needed to trace the movement of leachate through soils and to develop a strategy for recovery and treatment of contaminated groundwater.

Most groundwater contamination occurs because of surface spills into zones of recharge of aquifers. However, leakage of underground tanks and pipelines is also common, and numerous improperly sealed landfill disposal sites also pose severe problems. In addition, shallow-well discharge of toxic wastes, although this practice was discontinued years ago, still poses a substantial problem because the materials remain in place.

Rainfall and melting snow are prime movers of hazardous materials into subsurface aquifers. The velocity of groundwater can be so low that migrating contaminants may take months or years to reach freshwater supply wells or receiving surface waters. The time factor of leaching is perhaps the most difficult aspect of recovery efforts. Any potential recovery and treatment system must be fairly automated and designed for long-term use if such cleanup is to be economically feasible.

Several cases of recovery and treatment of groundwater, all involving the use of interceptors, have been reported. In two cases (Incidents 5 and 10) the quantities of contaminants were very large, 3700 and 1850 m^3 respectively, and recovery wells were employed after hydrogeological surveys had established the zones of maximum accumulation of the contaminants. Owing to the permeability of the groundwater aquifer, the action of pumping out a recovery well located at the center of a zone of maximum accumulation creates what is termed a "cone of depression." The groundwater level lowers as it approaches the well, and this lowering provides the gravity head to permit flow into the well from all directions. Incidents 5 and 10 both involved the recovery of lighter-than-water hydrocarbons from groundwater, and oil-water separation was therefore an integral part of the recovery process.

In Incident 10 a single 510-mm-diameter steel-cased well was dug, and two pumps were installed, a lower one for water and an upper one for oil (Fig. 1-9). The oil was permitted to thicken to a layer of about 0.6 to 1.0 m before it was pumped in order to prevent drawing water out with the oil. The system was designed to operate continuously because of the required 3-day start-up period before steady-state conditions could be achieved. Operation was essentially automatic, although the system required close attention during abrupt changes to the water table after major storms. During 4 months of operation, an estimated 15 percent of the contaminants was recovered. A 2-year program was envisioned, with a second well to be added.

The other case (Incident 5) involved both interceptor wells and trenches and employed an oil-water separation system located outside the wells and trenches. All the recovered fluid was pumped to a tank truck to allow time for oil-water separation (Fig. 1-10). The separated water required treatment to recover a soluble toxic fraction before discharge. This mode of operation was maintained for 1 month, after which treatment was terminated. Because substantial amounts of contaminant were still in the soil and groundwater, alternative long-range recovery-treatment options were still being sought at the time of writing this case history.

Four other incidents in which runoff and leachate were recovered from excavated trenches or isolated stream sections have been reported. Some notable techniques applied

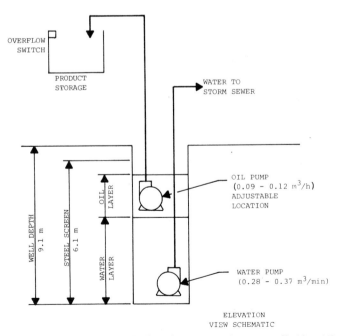

Fig. 1-9 Interceptor well with oil- and water-pumping systems (Incident 10).

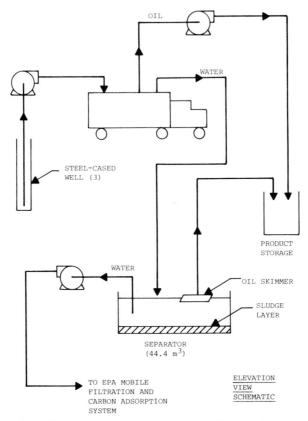

Fig. 1-10 Interceptor well, fluid pumping, oil separation, and water treatment (Incident 5).

in these response efforts either to automate or to speed up the recovery process were as follows:

Incident 1. A float switch was used to activate a pump to treat stream water in a carbon column when leachate and runoff caused the water level in a dammed stream section to rise above a preset level.

Incident 3. The slope of the terrain was used to provide gravity flow of contaminated leachate and storm runoff into a carbon column without the use of pumps.

Incident 11. Clean water was injected into the soil in the same spots where soil contamination had occurred, thus speeding the recovery process.

Incident 12. Pond water was used to flush a contaminated parking area into an excavated ditch. The collected runoff was then treated.

Recovery of Floating Materials

Oil spill control technology has been developed and used extensively to contain and recover floating oil from water bodies. This equipment includes:[21]

1. **Booms.** Floating barriers which can be either moored or towed. This method is limited to currents of less than about 1.3 m/s and to moderate wave conditions.

2. **Sorbents.** Oleophilic materials in common use include straw, plastic foams, and granular organic materials. Sorbents may be contained in a porous barrier, broadcast upon the spill and recovered with mechanical harvesters, or applied in the form of a continuous belt from which the oil is removed by pressure.

3. **Skimmers.** The oil layer is removed from the top of the water principally by using suction, a weir, or vortex action.

The EPA has conducted a study to determine whether booms and skimmers could be used effectively on floating hazardous materials as well as on oil.[22] In general, the study found that the performance of this equipment on hazardous materials did not vary substantially from the performance in recovering oil when the materials were similar to oil.

Recovery and Treatment of Contaminated Soils

The principal method for treating spills of chemicals into soils has been excavation, usually followed by containerizing and shipment for disposal. Such an operation was detailed in Incident 1, stressing the need for workers to be clothed in protective suits with air-supplied respiratory protection.

The EPA has developed a new mobile system for treating soils that utilizes grout to seal off a section of soil and chemical treatment, including recovery, to treat the soils in the sealed-off area.[23]

SUMMARY

It is apparent that a substantial amount of technology is available for ameliorating the damage caused by spills of hazardous chemicals. This part has provided an overview of a number of incidents in which technology has been applied successfully under difficult field conditions.

In many situations it seems to be not so much a question of whether or not containment, recovery, and treatment can be accomplished, but rather whether these steps are economically feasible and can be carried out quickly. The various demonstration projects undertaken by the EPA have shown that quick, safe, and effective recovery and treatment are best performed with preconstructed portable systems kept on ready alert for spill response.

Unfortunately, the few case histories provided in the literature describe the complicated situations under which the cited spills happened, and only a small section of each account is devoted to technical details. With minor exceptions (e.g., the 12 incidents discussed in this part), only sketchy details are presented in the literature. Unfortunately, these details do not provide sufficient information to allow others to adapt the reported experience to similar situations. Having the technology to solve problems is one thing; understanding the methodology to apply it effectively under adverse field conditions is quite another.

REFERENCES

1. H. Gilmer and F. J. Freestone, "Cleanup of an Oil and Mixed Chemical Spill at Dittmer, Missouri, April–May, 1977," in G. F. Bennett (ed.), *Proceedings of the 1978 National Conference on Control of Hazardous Material Spills,* Miami Beach, Fla., April 1978, pp. 181–134.

2. Ryckman, Edgerley, Tomlimson and Associates, Inc., *Pesticide Poisoning of Pond Lick Lake, Ohio: Investigation and Resolution,* U.S. Environmental Protection Agency, Office of Water Programs, Oil and Hazardous Materials Program Series, OHM 71-06-002, 1971.

3. W. L. Ramsey and J. M. MacCrum, "How to Cope with Hazardous Spills: A Case History," in G. F. Bennett (ed.), *Proceedings of the 1974 National Conference on Control of Hazardous Material Spills,* San Francisco, August 1974, pp. 234–237.

4. F. B. Stroud, R. T. Wilkerson, and A. Smith, "Treatment of PCB Contaminated Water and Waste Oil: A Case Study," in G. F. Bennett (ed.), *Proceedings of the 1978 National Conference on Control of Hazardous Material Spills,* Miami Beach, Fla., April 1978, pp. 135–144.

5. H. J. Lamp'l, T. Massey, and F. J. Freestone, "Assessment and Control of a Ground and Surface Water Contamination Incident, Haverford, PA, November–December, 1976," in G. F. Bennett (ed.), *Proceedings of the 1978 National Conference on Control of Hazardous Material Spills,* Miami Beach, Fla., April 1978, pp. 145–147.

6. K. M. Harsh, "Toxicity Modification of an Anhydrous Ammonia Spill," in G. F. Bennett (ed.), *Proceedings of the 1978 National Conference on Control of Hazardous Material Spills,* Miami Beach, Fla., April 1978, pp. 148–151.

7. K. M. Harsh, "In-Situ Neutralization of an Acrylonitrile Spill," in G. F. Bennett (ed.), *Proceedings of the 1978 National Conference on Control of Hazardous Material Spills,* Miami Beach, Fla., April 1978, pp. 187–189.

8. J. P. Lafornara and I. Wilder, "Solution of the Hazardous Material Spill Problem in the Little Menomonee River," in G. F. Bennett (ed.), *Proceedings of the 1974 National Conference on Control of Hazardous Material Spills,* San Francisco, August 1974, pp. 202–207.

9. J. C. Willmann, J. Blazevich, and H. J. Snyder, "PCB Spill in the Duwamish—Seattle, WA," in G. F. Bennett (ed.), *Proceedings of the 1976 National Conference on Control of Hazardous Material Spills,* New Orleans, April 1976, pp. 351–355.

10. H. B. Eagon, Jr., J. F. Howard, A. C. Smith, and R. E. Diefenbach, "Removal of Hazardous Fluid from the Ground Water in a Congested Area: A Case History," in G. F. Bennett (ed.), *Proceedings of the 1976 National Conference on Control of Hazardous Material Spills,* New Orleans, April 1976, pp. 373–377.

11. R. W. Fullner and H. J. Crump-Wiesner, "Use of EPA's Environmental Emergency Response Unit in a Pesticide Spill," in G. F. Bennett (ed.), *Proceedings of the 1976 National Conference on Control of Hazardous Material Spills,* New Orleans, April 1976, pp. 345–350.

12. J. P. Lafornara, M. Polito, and R. Scholz, "Removal of Spilled Herbicide from a New Jersey Lake," in G. F. Bennett (ed.), *Proceedings of the 1976 National Conference on Control of Hazardous Material Spills,* New Orleans, April 1976, pp. 378–381.

13. R. C. Mitchell, J. J. Vrolyk, R. L. Cook, and I. Wilder, "System for Plugging Leaks of Hazardous Materials," in G. F. Bennett (ed.), *Proceedings of the 1978 National Conference on Control of Hazardous Material Spills,* Miami Beach, Fla., April 1978, pp. 332–337.

14. P. W. Pontius, "Containment and Disposal of Product from Leaking Drums in Transit," in G. F. Bennett (ed.), *Proceedings of the 1974 National Conference on Control of Hazardous Material Spills,* San Francisco, August 1974, pp. 217–218.

15. F. J. Freestone and J. Zaccor, "Design, Fabrication, and Demonstration of a Mobile Stream Diversion System for Hazardous Material Spill Containment," in G. F. Bennett (ed.), *Proceedings of the 1978 National Conference on Control of Hazardous Material Spills,* Miami Beach, Fla., April 1978, pp. 371–377.

16. M. K. Gupta, *Development of a Mobile Treatment System for Handling Spilled Hazardous Materials,* EPA 600/2-76-109, U.S. Environmental Protection Agency, July 1976.

17. K. R. Huibregtse, R. C. Scholz, R. E. Wullschleger, J. H. Moser, E. R. Bollinger, and C. A. Hansen, *Manual for the Control of Hazardous Material Spills,* vol. 1: *Spill Assessment and Water Treatment Techniques,* EPA 600/2-77-227, U.S. Environmental Protection Agency, 1977.

18. J. C. Fizer, "Transporting Hazardous Materials by Rail," in G. F. Bennett (ed.), *Proceedings of the 1976 National Conference on Control of Hazardous Material Spills,* New Orleans, April 1976, pp. 332–341.

19. M. Urban and R. Losche, "Development and Use of a Mobile Chemical Laboratory for Hazardous Material Spill Response Activities," in G. F. Bennett (ed.), *Proceedings of the 1978 National Conference on Control of Hazardous Material Spills,* Miami Beach, Fla., April 1978, pp. 311–314.

20. T. D. Hand and A. W. Ford, "The Feasibility of Dredging for Bottom Recovery of Spills of Dense, Hazardous Chemicals," in G. F. Bennett (ed.), *Proceedings of the 1978 National Conference on Control of Hazardous Material Spills,* Miami Beach, Fla., April 1978, pp. 315–324.

21. Arthur D. Little, Inc., *Chemical Hazard Response Information System Response Methods Handbook,* U.S. Coast Guard, Washington, November 1974.

22. W. E. McCracken and S. H. Schwartz, *Performance Testing of Spill Control Devices on Floatable Hazardous Materials,* EPA 600/2-77-222, U.S. Environmental Protection Agency, November 1977.

23. K. R. Huibregtse, J. P. Lafornara, and K. H. Kastman, "In Place Detoxification of Hazardous Material Spills in Soil," in G. F. Bennett (ed.), *Proceedings of the 1978 National Conference on Control of Hazardous Material Spills,* Miami Beach, Fla., April 1978, pp. 362–370.

PART 2
Technology Development

Frank J. Freestone, Chief
Hazardous Spills Staff

John E. Brugger, Ph.D.
Physical Scientist

Oil and Hazardous Materials Spills Branch
U.S. Environmental Protection Agency

Introduction / 9-24
Organization of the Program / 9-25
Prevention of Hazardous Material Spills / 9-26
Preresponse Planning / 9-27
Cleanup Site Safety / 9-27
Situation Assessment and Analytical Support / 9-29
Containment and Confinement Techniques / 9-29
Concentration and Separation Techniques / 9-32
Ultimate Disposal / 9-33
Restoration of Damaged Areas / 9-35
Information Transfer and Processing / 9-36
Conclusion / 9-38

INTRODUCTION

For the past 10 years, the U.S. Environmental Protection Agency (U.S. EPA) has sponsored a research and development program on the control of environmental emergencies. This part describes the progress of that program to date and outlines current goals and objectives.

An environmental emergency involving a hazardous material is defined as the acute release or threat of acute release of a chemical substance into the environment such that there is an actual or a potential effect on the public health and/or the ecosystem. Included in this definition are spills and other accidental releases as well as acute releases from uncontrolled hazardous waste sites. An "acute" release is intended to be distinct from a "chronic," or long-term, release.

Environmental-emergency response means those activities conducted *primarily* for the protection of human health and the environment from hazardous material exposure situations. Environmental response to a spill typically occurs after response by first-on-scene personnel such as fire fighters. While the EPA program does concern itself with certain decisions that are made by fire fighters regarding hazardous material emergencies—such as the toxic effects of a smoke plume from a pesticide warehouse fire, a burn–no-burn decision for such a fire, and first-on-scene responses to releases not involving fires—the predominant thrust of the program relates to activities that take place after fire fighters have completed their initial response and the immediate hazard to life, limb, and property has been eliminated. Environmental response refers to the actions that are taken during a situation that has been essentially stabilized with respect to fire and explosion but still presents hazards to human health and the environment.

Hazardous materials normally include toxic chemicals, mixtures, and wastes as well as explosives, etiological agents, and radiological agents. The hazards associated with these materials are varied. While some consideration has been given to certain types of radiological hazards, the primary thrust of the EPA program has been to ameliorate chemical hazards. No activity has been conducted in the areas of etiological agents or explosives. The EPA has conducted a program on the control of oil spills, but that program is not included in the following discussion.

The program emphasis to date has been on the development of hardware and techniques to control environmental emergencies. Development is carried from the concept stage through the prototype stage to field testing. The final objective of field tests and demonstrations is to encourage commercialization of the devices and techniques developed. Indeed, one of the primary measures of success is the commercial adoption or adaptation of a prototypical device developed through federal funding. Examples of EPA-developed devices that have already been commercialized include an instrument to detect instability in an earthen dike, a mobile physical-chemical water treatment system, and a mobile laboratory for analytical support in the field.

The program is being conducted by the Hazardous Spills Staff of the Oil and Hazardous Materials Spills Branch, Solid and Hazardous Waste Research Division, Municipal Environmental Research Laboratory, Cincinnati, Ohio. The branch, which is located at Edison, New Jersey, has become an internationally recognized center of expertise for the control of environmental emergencies.

ORGANIZATION OF THE PROGRAM

The program is divided into nine areas. Eight of these involve technology development, in which procedures or hardware are taken to experimental or prototype stages. The eight technology areas are selected to represent the major technical elements of environmental-emergency response; they are listed in the order in which this information is typically needed: before, during, and after an emergency situation. These areas are:

- Prevention of hazardous material spills
- Preresponse planning
- Cleanup site safety
- Situation assessment and analytical support
- Containment and confinement techniques
- Concentration and separation techniques
- Ultimate disposal
- Restoration of damaged areas

The ninth area, information transfer and processing, is an activity to organize and disseminate, by various means, information on the technology developed in the other eight areas. Further, prototypical equipment is tested and field-proved as part of the information transfer category.

For each of the eight technology development areas, the two principal outputs are prototypical hardware for and manuals on the cleanup of released materials. Hardware is developed on a systematic basis from a conceptual stage, through state-of-the-art and feasibility studies, to design and fabrication of prototypes. Once a prototype has been fabricated, it is turned over to the Environmental Emergency Response Unit (EERU) for shakedown, field testing, and adaptation by the commercial community. In all cases of hardware development, every effort is made to adapt existing equipment from related technology areas. Emphasis is on practical use of equipment in the field at lowest cost by labor with minimum training.

PREVENTION OF HAZARDOUS MATERIAL SPILLS

The objective of this program area is to generate technological approaches to reduce the frequency and severity of releases. Several projects have been undertaken to develop information and systems useful for spill prevention.

An acoustical system has been developed to detect instability in earthen dikes and thus prevent dike failures.[1,2] This system, which has been commercialized in at least three versions, has received wide recognition as a simple, portable, inexpensive tool ($2500; Fig. 2-1). The device can be used to assess the stability of impoundments in which haz-

Fig. 2-1 Acoustical system for detecting instability in earthen dikes.

ardous chemicals or wastes are treated or stored. Many such impoundments are not well engineered and are prone to failure, with devastating environmental consequences.

Documentation and analysis of historical data relating to hazardous material spills have been ongoing for several years. The Buckley and Wiener study[3] has analyzed such spill report data as are available in government files, newspaper accounts, and other sources. The data include the substances spilled and the associated quantities, frequencies, damage, hazards, and other factors. The information is particularly useful from the viewpoint of spill prevention planning.

Currently under way is a project to develop a microwave-based system for the nondestructive assessment of impoundment stability. This system is intended to be an adjunct to the acoustical technique. The microwave system uses an antenna and a receiver that can be transported in a vehicle along the top of a dike. Water pockets, voids, and other subsurface discontinuities in the dike can be located rapidly and flagged for subsequent investigation with the acoustical system.

A project currently being planned is the preparation of a manual on the prevention of hazardous material spills at small fixed facilities. This manual is intended to be a practical guide for use by small-facility operators. Such operators are typically less able to develop spill prevention techniques and planning than are the operators of large facilities.

PRERESPONSE PLANNING

The objective of this program area is to prepare planning materials for use by organizations that respond to hazardous chemical incidents. These materials will assist in maximizing the effectiveness of responses through coordination and planning.

Contingency planning guidelines for environmental emergencies were recently completed. A manual and a training course were prepared to assist communities and state-level response personnel with the upgrading of existing contingency plans or the preparation of new ones. These planning materials, now available through the Federal Emergency Management Agency (FEMA),[4] stress the technical and resource hierarchy of response personnel, from those first on scene, through state health and environmental officials, to the federal level, and ending with the national response team (NRT). The NRT can bring the considerable resources of the federal government to bear on a hazardous material release situation.

Ongoing is the development of guidelines on contingency planning techniques and program requirements for preparedness at the state and local levels in response to transportation-related radiological incidents. This project is being undertaken for the Nuclear Regulatory Commission.

A project currently in the planning stages is the automation of multimedia environmental-emergency contingency plans at the EPA regional level. These plans constitute the link between state and local planning activities and the National Oil and Hazardous Substances Pollution Contingency Plan. One of the major problems associated with contingency planning is maintaining plans in an updated status. This project will explore the application of automatic data-processing techniques for maintaining these rapidly evolving plans in a state of readiness.

CLEANUP SITE SAFETY

The objective of this program area is to develop products to provide a safe working environment for personnel engaged in spill cleanup operations. In addition to the EPA, many federal organizations such as the National Institute for Occupational Safety and Health,

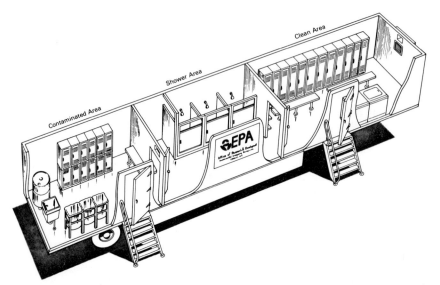

Fig. 2-2 Mobile personnel decontamination station for use at cleanup activities involving toxic materials.

the Occupational Safety and Health Administration, the U.S. Coast Guard, and the FEMA are concerned with safety for response personnel. EPA R&D activities in this program area are limited to specific technology development requirements for cleanup site safety that are not being addressed by other organizations.[5]

A project recently completed was the development of a mobile decontamination station for use by personnel responding to spills. The system provides for a shower and a clothing change for personnel involved in the cleanup of highly toxic substances. In the past, when spill response activities were undertaken at remote locations, personnel involved in cleanup were often forced to leave the site without adequate opportunity for a clothing change and washup. Such a deficiency is not acceptable when people are dealing with highly hazardous substances. The new system consists of a self-contained semitrailer having a "clean" locker room, a "contaminated" locker room, and a shower room in between (Fig. 2-2). The system is equipped with freshwater supply and waste-holding tanks and a washer and dryer for spill response clothing. The system is used in conjunction with an external arrangement for the washdown and removal of exterior protective clothing.

Ongoing is the development of specialized systems and techniques for the decontamination of equipment used for waste site cleanup operations. Since many cleanups involve the rental of locally available construction equipment such as backhoes and drilling apparatus, this equipment must be properly decontaminated before being released for subsequent use in normal construction activities. The project also includes improvements to specifications for protective-clothing use and safe distances for command posts, evacuations, etc., around a cleanup activity.

A project currently being planned is the development of specialized safety equipment for personnel exposure monitoring. This project will evaluate the current state of the art in personal exposure monitors for use in hazardous or toxic chemical situations and will develop new or novel systems for ambient monitoring as well as for monitoring the exposure of individual cleanup site workers.

SITUATION ASSESSMENT AND ANALYTICAL SUPPORT

The objective of this program area is to provide decision-making information to spill control managers. Sampling and analysis information is critical for spill control decision making. Analytical information is needed to define the type of material spilled, the extent of environmental contamination, the progress of cleanup, and the concentration of any residual after cleanup is "complete." Decisions include such first-response considerations as burn–no-burn choices for pesticide fires and evacuation decisions for the release of gases or vapors or toxic smoke from chemical fires. Decisions to be made after the situation is more stable include the determination of priorities of actions in a situation in which resources are insufficient to address all aspects of the cleanup problem simultaneously, the selection of alternative cleanup techniques for cost-effectiveness and least overall environmental impact, and, finally, the rational determination as to when cleanup is complete, or "how clean is clean."

A completed project in the subcategory of decision making is the *Manual for the Control of Hazardous Material Spills,* Volume I: *Spill Assessment and Water Treatment Techniques.*[6] This manual provides guidance for the logical and systematic evaluation of a spill situation to determine the media affected and the extent of the spill.

A key ongoing project is the production of a decision guide for a systematic and analytical approach to spill cleanup decision making. This guide is intended to address three of the questions noted above, namely, what to do first, including the consequences of deferral of some aspects of cleanup, how to choose among alternative cleanup procedures from a cost-effectiveness viewpoint, and how to make a first-cut rational determination of the required extent of cleanup.

A currently planned project is the preparation of a manual that will furnish a clear and logical approach to the evaluation of alternative cleanup procedures. This manual will provide detailed information on the application of various technologies. It will be aimed specifically at release situations that involve the use of complex treatment technologies for air, water, sediments, soil, and groundwater.

In the subcategory of analytical support, the development of a field kit to detect hazardous materials in waterways has been completed.[7] This kit, which is currently available commercially from the Hach Chemical Company, can be used to determine the extent of a spill by detecting many different types of known contaminants in water (Fig. 2-3).

An ongoing project is the development of rapid field-use analytical methods that can be applied to spill situations. These methods can be used to provide information on the presence of spilled material, the nature of the material present, and sufficient identification to enable cleanup-related judgments to be made. The methods are specifically tailored for use in an on-site or mobile laboratory and are designed to be both space- and time-efficient (see Fig. 2-4).

A planned project in the analytical category is the development and demonstration of portable kits and instrumentation to assist in determining the identification and concentration of contaminants in various environmental media.[8] These systems are intended predominantly for use at uncontrolled hazardous waste sites, where much analytical information is necessary to determine the nature of a particular release problem.

CONTAINMENT AND CONFINEMENT TECHNIQUES

The objective of this program area is to limit the spread of a spill in all environmental media, including air, soil, groundwater, surface water, and sediments. Indeed, first-

Fig. 2-3 Field kit for detecting a variety of hazardous materials in water.

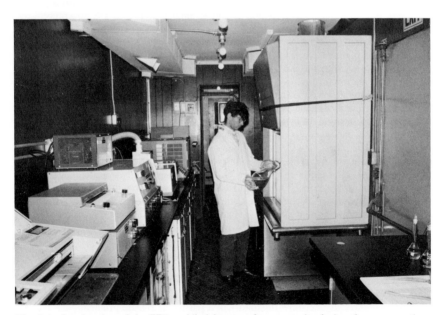

Fig. 2-4 Interior view of the EPA mobile laboratory for use on site during cleanup operations. Instrumentation includes computerized gas chromatographs, a gas chromatograph–mass spectrometer, an atomic absorption spectrophotometer, and other highly sensitive analytical tools. Special sample-handling techniques and glove boxes permit safe handling of high concentrations of highly toxic chemicals.

Fig. 2-5   Portable foam diking system capable of forming a barrier to restrict the flow of liquids over asphalt or concrete or of plugging drains to prevent spilled material from entering.

response personnel, after an initial analysis of a situation, must take stabilizing actions, including those that limit the spread of the spill. This task becomes challenging when all environmental media must be addressed and the wide range of solubility and dispersibility of hazardous materials must be considered.

A completed key product in this area is a portable polyurethane-foam diking system for use when a small temporary impoundment can be employed to halt the spread of a spilled liquid (Fig. 2-5). An example is found in the common urban situation when a spill from a gasoline tank trailer runs onto a concrete or asphalt roadway. The foam dike can be used to stop the flow of the fuel along a gutter or across a roadway, and it can also be used to plug storm drains to prevent entry of the contaminant. This system is now commercially available from the Mine Safety Appliances Co.[9,10]

An ongoing project is the development of a trailer-mounted system for injecting decontaminating agents or grout into soil to neutralize contaminants or to form migration barriers for hazardous materials percolating into and through the soil (Fig. 2-6). This system can also be used to flush contaminants from soils by injection and withdrawal of water, to decompose organic substances in soils by injection of microorganisms, or to neutralize or detoxify contaminated soils by injection of chemicals. Preliminary testing of the system has been completed, and more thorough tests were scheduled for the summer of 1982.[11]

An ongoing project that has engendered widespread interest is an evaluation of the use of fire-fighting foams to reduce the rate of evaporation of volatile spilled liquids. Several commercially available foams have been examined for their ability to form and maintain a blanket over pools of selected hazardous materials. Technical effort on this project is now complete; the final report will soon be available.[12]

A planned project is the preparation of a manual on the control of air pollution from hazardous material releases. This manual will categorize frequently occurring situations of air pollution at hazardous material incidents and will define practical, environmentally appropriate alternatives for controlling releases of gases and vapors. Previous efforts pertaining to the use of foams, gelation agents, coolants, and sorbents for vapor control will be consolidated in this manual.

Fig. 2-6 Mobile soils decontamination system for injecting treatment agents into contaminated soil or for forming a grout curtain to modify the horizontal permeability of the soil in order to limit the spread of contaminants.

CONCENTRATION AND SEPARATION TECHNIQUES

The objective of this program area is to develop techniques and equipment to collect and concentrate a dispersed released material to a minimum practical volume and to separate the material from collected air, water, soil, or sediments. This program area has received considerable emphasis over the first several years of the life of the hazardous incidents program, particular attention being given to the development of water treatment technologies usable in field situations. With the current authority provided to the EPA under the Comprehensive Environmental Response, Compensation, and Liability Act (Superfund) to give consideration to multimedia releases, greater attention can now be turned toward controlling releases into the air, groundwater, soils, and sediments.

A completed device which has been used widely around the United States and adapted for commercial use is the mobile physical-chemical treatment system[13] (Fig. 2-7). This

Fig. 2-7 Mobile physical-chemical treatment system, which, when set up with equipment not shown, provides for gross settling, coagulation-sedimentation, and mixed-media filtration for the removal of suspended solids and inorganic substances and carbon adsorption for the removal of dissolved organic materials.

system is designed to remove many inorganic and organic substances from water and consists of a series of physical-chemical treatment steps employed in the field:

- Settling to remove suspended solids and allow flotation of oily material

- Coagulation-sedimentation with chemical addition to remove finely divided suspended solids and dissolved inorganic materials

- Mixed-media filtration to capture any suspended-solids carry-over from the preceding steps

- Carbon adsorption to remove many dissolved organic materials

This system has been used at more than 30 cleanups of uncontrolled hazardous waste sites and spills of hazardous materials around the United States.[14-17] Commercial units patterned after the EPA system are now routinely employed.

An ongoing project of particular interest is the development of a mobile separation system for stripping spilled hazardous materials from excavated soils by using a water-based cleaning agent. The system consists of a high-energy water scrubber and a lower-energy multistage separator, all mounted on semitrailers. The prototype system will be capable of processing 2 to 14 m^3 (3 to 18 yd^3) of soils per hour, depending upon the composition of the feed. The intent of the soil-scrubbing process is to reduce the contamination level in a given soil volume to a level suitable to allow return of the soil to the location from which it was removed or to permit disposal of the soil into a lower-grade landfill than would otherwise be acceptable. The system was scheduled for delivery to the Edison laboratory in the summer of 1982.

A planned project is the development of advanced techniques for the control of sediments contaminated with toxic substances. This project will develop new or novel procedures and equipment for the removal and treatment, or for in-place treatment, of contaminated sediments. Particular attention will be focused on sediments in ponds and small rivers. Large water bodies can be dredged with hydraulic equipment that is relatively available and well developed. Small streams can have their flows bypassed to allow drying of the contaminated segment so that the dried sediments can be removed mechanically. However, many rivers and ponds that are exposed to sediment contamination are too small for hydraulic dredging and also too large for bypassing. Cost-effective techniques of sediment treatment, both in place and after removal, will be examined.

ULTIMATE DISPOSAL

The objective of this program area is to provide permanent destruction or encapsulation of residuals from spill cleanup operations, with emphasis on technologies usable on site. The elements that constitute hazardous substances cannot be destroyed. Ultimate disposal by "destruction" means a process of rearranging structures that impart a hazardous quality (such as by thermal conversion of PCBs into CO_2, H_2O, and salt) or of incorporating toxic elements (e.g., mercury or arsenic) into less hazardous low-leachability glasses, rocks, or other matrices. The goal is to make the products of the disposal process compatible with the receiving environment's assimilative capacity.

A key completed project is the manual *Guidelines for the Disposal of Small Quantities of Unused Pesticides*. This manual provides specific recommendations for disposing of washings and diluted formulations through burial in appropriate trenches and through other techniques. Recommended sources to contact are provided to facilitate the disposal of heavy metal-containing or chlorinated pesticides.[18]

The use of molten sodium metal at 220°C has been demonstrated at the pilot scale to

Fig. 2-8   Mobile incineration system for the on-site destruction of organic hazardous wastes accumulated during cleanup operations.

be an effective method for reductively disintegrating even refractory (PCB-type) organics. Carbon appears as graphite that can be filtered out; hydrogen is flared; halides form sodium salts. The existing system can be used to destroy small lots of chlorinated organics; the design and operating data are available for commercialization on a larger scale.[19]

The most significant ongoing project in the entire program is the development of a mobile incineration system.[20,21,22] The unit is mounted on three interconnected semitrailers (Figs. 2-8 and 2-9). Liquid, sludge, and solids feed equipment and a 1.321-m- (52-in-) inside-diameter by 4.88-m- (16-ft-) long refractory-lined kiln are mounted on the first trailer; a 1.32-m-inside-diameter by 11-m- (36-ft-) long secondary combustion chamber,

MOBILE INCINERATOR SITE LAYOUT

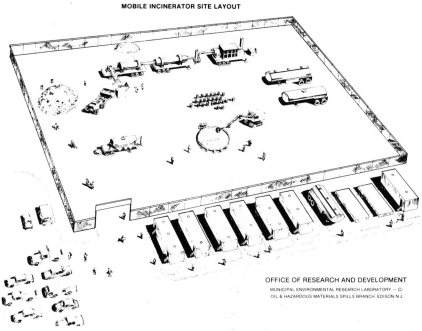

OFFICE OF RESEARCH AND DEVELOPMENT

MUNICIPAL ENVIRONMENTAL RESEARCH LABORATORY — Ci

OIL & HAZARDOUS MATERIALS SPILLS BRANCH, EDISON N.J.

Fig. 2-9   Artist's conception of a site layout of the mobile incinerator. Depicted are numerous support vehicles including office spaces, a mobile laboratory, a personnel decontamination station, maintenance support, and other personnel support.

on the second; and air pollution control equipment (particulate and mass transfer scrubbers), an induced-draft fan, and a stack, on the third. The system is designed to provide the EPA-recommended 1200°C (2200°F) and 2-s dwell time for PCB incineration. The heat release of the system is 1.58×10^{10} J/h (15 million Btu/h), which is provided either by the material to be incinerated or by fuel oil. Capacities are 4080 kg/h (9000 lb/h) of dry sand, 680 kg/h (1500 lb/h) of slightly contaminated water, the equivalent of 284 L/h (75 gal/h) of fuel oil, or any combination of these three components that does not overload the system. The unit is sized to handle 11,000 m^3 (STP)/h of flue gas. Field trials were initiated in the fall of 1981.

Future efforts in ultimate disposal will focus on novel systems for *in situ* treatment of contaminated soils and novel, relatively low-energy systems for treating refractory compounds. Thermal destruction is very energy-intensive and costly. Techniques must be found for altering the nature of synthetic refractory organics by using low-energy, low-cost technology.

RESTORATION OF DAMAGED AREAS

The objective of this program area is to develop technology and equipment to provide for the natural or accelerated assimilation of spilled materials into the environment and for the return of the local area to its prespill condition. Not all spills nor all portions of every spill are cleaned up with mechanical equipment. Further, the extent of cleanup in a given situation may depend upon an estimate of the sensitivity of the local environment to a given concentration of released material and the corresponding ability of the local environment to assimilate the contaminant. Ultimately, the natural processes of biodegradation, photooxidation, and dissipation have an effect on the concentration of the contaminant remaining in the local environment after the cleanup is "complete." When these processes are studied and understood to some extent, they can potentially be used as a planned technique for the "last phase" of a cleanup or in place of mechanical cleanup.

Land farming of oily wastes and *in situ* decomposition of gasoline spills have been shown to be successful when indigenous microorganisms are supplied with food, nutrients, and oxygen. Ideally, biochemical processing of organic materials should yield carbon dioxide and water along with nitrates, phosphates, and the like. At best, heavy metals will be bioaccumulated but may not be in harvestable form. Aerobic biodegradation processes are akin to thermal and chemical oxidation methods, but they operate for a longer time at a low-energy requirement.

A completed project evaluated the use of selected pure-culture microorganisms for degrading pentachlorophenol, hexachlorocyclopentadiene, and methyl parathion in aqueous solutions. While all chemicals studied exhibited some removal, it is not known to what extent the removal mechanism involved complete metabolism and degradation of the chemicals. A reduction of 80 to 90 percent in the concentration of pentachlorophenol was observed during a 2-day pilot study using a fungal culture.[23]

In a related project, the effects of seeding a simulated spill with sewage sludge were examined. Several test plots were dosed with spilled materials and treated with sludge. The plots subsequently redeveloped their vegetative cover. The experiments did not show conclusively that all hazardous materials were fully biodegraded to nontoxic materials, specifically, carbon dioxide and water.[24]

Future projects in this area will concentrate on the applications of natural processes. The ultimate objective includes the possibility of genetically engineered materials that have the desirable properties of several organisms and can provide a potent and efficient microbiological technique for the degradation of toxic materials. A potential application is to contaminated deep soils that cannot be economically excavated.

INFORMATION TRANSFER AND PROCESSING

The objective of this program area is to deliver information about products of technology development activities in forms acceptable to end users. These products include software, technical assistance, national conferences, and hardware.

Software includes user manuals for field activities, technical reports and papers describing the development of various processes, and visual materials such as slides, photographs, slide-tape presentations, videotapes, and 16-mm films. The software products are consolidated at the EPA Oil and Hazardous Materials Spills Branch (OHMSB) into an in-house activity called the Technical Information Center (TIC). The TIC is organized in parallel with the eight major program areas and has extensive technical holdings that collectively represent the state of the art of control technology for environmental emergencies. It responds to many hundreds of requests per year received by the OHMSB for information on spills technology developments.

OHMSB professional personnel are frequently called upon to provide technical assistance to EPA on-scene coordinators. Requests include specific questions regarding cleanup problems, treatability studies, treatment scheme designs, and recommendations regarding the use of specific apparatus or techniques for control of emergency situations.

The OHMSB supports national conferences on control of hazardous material spills and on management of uncontrolled hazardous waste sites. Papers on OHMSB projects are typically presented at these conferences. The proceedings of the conferences form an excellent cross section of information on current control technologies as well as on interesting case histories.[25-30]

Information transfer regarding hardware developments is a function of both the TIC and the EERU. The EERU is an on-site contractor-operated activity at Edison, New Jersey, that consolidates testing and demonstration functions for spills control hardware developed under the eight main program categories.[31]

As prototypical hardware is developed, the equipment is provided to the EERU for a specific series of steps in the completion of the development process. These steps include shakedown to assure field readiness, field trials to demonstrate the applicability of the equipment, and activities to encourage commercialization of the products developed at public expense. The EERU has used the mobile physical-chemical treatment system and other prototypical devices around the United States at a wide variety of situations involving spills and uncontrolled hazardous waste sites. Information pertaining to each item of hardware in the EERU inventory, including in most cases plans, specifications, and field-use reports, is available from the TIC.

Major items of equipment in the EERU inventory include:

- The mobile incineration system, which was tested during the fall of 1981.

- The mobile carbon reactivator, designed to thermally regenerate granular carbon used for the concentration of refractory, politically sensitive substances such as dioxin and PCBs (Fig. 2-10). Activated carbon used to treat such substances cannot now be reactivated commercially.

- The mobile stream diversion system, designed to bypass the flow [up to 0.37 m^3/s (13 ft^3/s)] of a small stream for distances up to 915 m (3000 ft) when the bottom sediments of the stream have become contaminated (Fig. 2-11).

Once an item of equipment has been through the development and shakedown process, it is designated field-ready and is made available for field response by the EERU at the direction of the EPA environmental response team. Thus, prototypical hardware is avail-

Fig. 2-10 Mobile reactivator system for regenerating granular activated carbon used in the mobile physical-chemical treatment system. The reactivator is designed for on-site destruction of compounds that cannot be treated in commercial reactivators, such as PCBs or Kepone.

able for cleanup situations when the use of government-owned state-of-the-art equipment is desirable. The EERU does not routinely respond to spills, however, and all efforts are made by the EPA to use commercial capabilities when available.

Use of government-developed equipment at many cleanup activities has served to stimulate commercialization of complex systems such as the mobile physical-chemical treatment system and the mobile spills laboratory.

Fig. 2-11 Mobile stream diversion system for isolating segments of small streams so that contaminated sediments can be removed easily with mechanical equipment. This technique is an alternative to dredging, which typically requires extensive water treatment to remove contaminants that become suspended or dissolved during the pumping operation.

CONCLUSION

The state of the art of control technologies for environmental emergencies is evolving rapidly. Previous emphasis of government-sponsored research has been in the area of advanced mobile water treatment technologies. With the authorities provided to EPA under Superfund, a greater thrust can be given to the control of multimedia releases. Therefore, future R&D efforts will be focused on removing or treating contamination in groundwater, sediments, soil, and air. With the incentives provided by Superfund for the cleanup of releases of hazardous substances, private industry is expected to be more interested than ever before in the application of the products of technology development.

REFERENCES

1. *Acoustic Monitoring to Determine the Integrity of Hazardous Waste Dams,* Centec Corporation, Reston, Va., 1979.
2. R. M. Koerner and A. E. Lord, Jr., *Spill Alert Device for Earth Dam Safety Warning Systems,* U.S. Environmental Protection Agency, Cincinnati, 1982. (In preparation.)
3. J. L. Buckley and S. A. Wiener, *Hazardous Material Spills: A Documentation and Analysis of Historical Data,* EPA-600/2-78-066, U.S. Environmental Protection Agency, Cincinnati, 1978.
4. F. C. Gunderloy and W. L. Stone, *Planning Guide and Checklist for Hazardous Materials Contingency Planning,* FEMA-10, Federal Emergency Management Agency, Washington.
5. M. L. Sproul, "Safety Considerations at Spills and Dump Sites," *Proceedings of the 1980 National Conference on Control of Hazardous Material Spills,* Louisville, Ky., May 1980, pp. 255–258.
6. K. R. Huibregtse, R. C. Scholz, R. E. Wullschleger, J. H. Moser, E. R. Bollinger, and C. A. Hansen, *Manual for the Control of Hazardous Material Spills,* vol. I, *Spill Assessment and Water Treatment Techniques,* EPA-600/2-77-227, U.S. Environmental Protection Agency, Cincinnati, 1977.
7. A. Silvestri, A. Goodman, L. M. McCormack, M. Razulis, A. R. Jones, Jr., and M. E. P. Davis, *Development of a Kit for Detecting Hazardous Material Spills in Waterways,* EPA-600/2-78-055, U.S. Environmental Protection Agency, Cincinnati, 1978.
8. A. Silvestri, M. Razulis, A. Goodman, A. Vasquez, and A. R. Jones, Jr., *Development of an Identification Kit for Spilled Hazardous Materials,* EPA-600/2-81-194, U.S. Environmental Protection Agency, Cincinnati, 1981.
9. J. V. Friel, R. H. Hiltz, and M. D. Marshall, *Control of Hazardous Chemical Spills by Physical Barriers,* EPA-R2-73-185, U.S. Environmental Protection Agency, Washington, 1973.
10. R. H. Hiltz and F. Roehlich, Jr., *Emergency Collection System for Spilled Hazardous Materials,* sec. 10, EPA-600/2/77-162, U.S. Environmental Protection Agency, Cincinnati, 1977.
11. K. R. Huibregtse and K. H. Kastman, *Development of a System to Protect Groundwater Threatened by Hazardous Spills on Land,* EPA-600/2-81-085, U.S. Environmental Protection Agency, Cincinnati, 1981.
12. S. S. Gross and R. H. Hiltz, *Evaluation/Development of Foams for Mitigating Air Pollution from Hazardous Spills,* U.S. Environmental Protection Agency, Cincinnati, 1982. (In preparation.)
13. M. K. Gupta, *Development of a Mobile Treatment System for Handling Spilled Hazardous Materials,* EPA-600/2-76-109, U.S. Environmental Protection Agency, Cincinnati, 1976.
14. G. J. Moein, "Containment, Treatment, Removal, Disposal and Restoration of Large Volumes of Oil and Hazardous Substances in a Land Site in Chattanooga, Tennessee," *Proceedings of the 1980 National Conference on Control of Hazardous Material Spills,* Louisville, Ky., May 1980, pp. 46–48.

15. H. Gilmer and F. J. Freestone, "Cleanup of an Oil and Mixed Chemical Spill at Dittmer, Missouri, April–May, 1977, in G. F. Bennett (ed.), *Proceedings of the 1978 National Conference on Control of Hazardous Material Spills,* Miami Beach, Fla., April 1978, pp. 131–134.

16. H. J. Lamp'l, T. Massey, and F. J. Freestone, "Assessment and Control of a Ground and Surface Water Contamination Incident, Haverford, Pennsylvania, November–December 1976," in G. F. Bennett (ed.), *Proceedings of the 1978 National Conference on Control of Hazardous Material Spills,* Miami Beach, Fla., April 1978, pp. 145–147.

17. J. P. Lafornara, F. J. Freestone, and M. Polito, "Spill Clean-Up at a Defunct Industrial Waste Disposal Site," in G. F. Bennett (ed.), *Proceedings of the 1978 National Conference on Control of Hazardous Material Spills,* Miami Beach, Fla., April 1978, pp. 152–155.

18. E. W. Lawless, T. L. Ferguson, and A. F. Meiners, *Guidelines for the Disposal of Small Quantities of Unused Pesticides,* EPA-670/2-75-057, U.S. Environmental Protection Agency, Cincinnati, 1975.

19. J. S. Greer, G. H. Griwtaz, S. S. Gross, and R. H. Hiltz, *Laboratory Feasibility and Pilot Plant Studies on Novel Sodium Metal Fluxing Processes and Glassification Methods for the Ultimate Disposal of Spilled Hazardous Materials,* U.S. Environmental Protection Agency, Cincinnati, 1982. (In preparation.)

20. R. Tenzer, B. Ford, Jr., W. Mattox, and J. E. Brugger, "Characteristics of the Mobile Field Use System for the Detoxification/Incineration of Residuals from Oil and Hazardous Material Spill Clean-Up Operations," *Journal of Hazardous Materials,* vol. 3, 1979, pp. 61–75.

21. J. E. Brugger, R. E. Tenzer, W. A. Mattox, and F. J. Freestone, "Design and Testing of Mobile Incineration System for Spilled or Waste Hazardous and Toxic Material," *Proceedings of the 1980 National Conference on Control of Hazardous Material Spills,* Louisville, Ky., May 1980, pp. 467–475.

22. R. Tenzer, B. Ford, Jr., W. Mattox, and J. E. Brugger, "Mobile System for the Detoxification/ Incineration of Cleanup Residuals from Hazardous Material Spills," *Disposal of Oil and Debris Resulting from a Spill Cleanup Operation,* American Society for Testing and Materials, Special Technical Publication No. 703, 1980, pp. 118–136.

23. N. K. Thuma, P. E. O'Neill, S. G. Browlee, and R. S. Valentine, *Novel Biodegradation Processes for Disposal of Spilled Hazardous Materials—Laboratory Feasibility and Pilot-Plant Studies,* U.S. Environmental Protection Agency, Cincinnati, 1982. (In preparation.)

24. R. Wentsel, R. H. Foutch, W. E. Harward III, and W. E. Jones III, *Restoring Hazardous Spill-Damaged Areas: Technique Identification/Assessment,* EPA-600/2-81-208, U.S. Environmental Protection Agency, Cincinnati, 1981.

25. *Proceedings of the 1980 National Conference on Control of Hazardous Material Spills,* Louisville, Ky., May 1980.

26. *Proceedings of the 1980 National Conference on Management of Uncontrolled Hazardous Waste Sites,* Hazardous Materials Control Research Institute, Silver Spring, Md., 1980.

27. G. F. Bennett (ed.), *Proceedings of the 1978 National Conference on Control of Hazardous Material Spills,* Miami Beach, Fla., April 1978.

28. G. F. Bennett (ed.), *Proceedings of the 1976 National Conference on Control of Hazardous Material Spills,* New Orleans, April 1976.

29. G. F. Bennett (ed.), *Proceedings of the 1974 National Conference on Control of Hazardous Material Spills,* San Francisco, August 1974.

30. *Proceedings of the 1972 National Conference on Control of Hazardous Material Spills,* Houston, March 1972.

31. *Environmental Emergency Response Unit (EERU) Capability,* U.S. Environmental Protection Agency, Municipal Environmental Research Laboratory, Oil and Hazardous Materials Spills Branch, Edison, N.J., 1980.

PART 3
Biological Measures

Neal E. Armstrong, Ph.D.
Professor of Civil Engineering
University of Texas at Austin

Elements of Spill Control / 9-40
Requirements of a Countermeasure / 9-40
Biological Countermeasure Rationale / 9-41
Cleanup Methods / 9-44
Storage / 9-47
Deployment / 9-47
Undesirable Bacteria / 9-47
Conclusions / 9-48

ELEMENTS OF SPILL CONTROL

The National Oil and Hazardous Substances Pollution Contingency Plan (40 CFR 1510) delineates five classes of actions that comprise the elements of spill control. The actions are Phase I—discovery and notification (discovery of a spill by the discharger, patrol vessels, or incidental observation and the reporting of that discovery to the proper agency); Phase II—evaluation and initiation of action (evaluation of the magnitude and severity of the spill, the feasibility of removing it, and the effectiveness of removal actions); Phase III—containment and countermeasures (actions taken to restrain the movement of the spilled material and to minimize its effects on water-related resources); Phase IV— cleanup, mitigation, and disposal (actions taken to recover the spilled material and to monitor the scope and effectiveness of removal actions); and Phase V—documentation and cost recovery. The time to implement any of these phases will depend on the location of the spill, the material spilled, the magnitude of the spill, and so forth. Employment of a biological countermeasure imposes special constraints on the activities in Phases III and IV and requires that its use be carefully considered in Phase II. To understand these special constraints, the requirements of a general countermeasure and the information needed to judge the suitability of a biological countermeasure must be discussed.

REQUIREMENTS OF A COUNTERMEASURE

Dawson et al.[1] suggested the following criteria for evaluating potential countermeasures:

1. Countermeasures should be highly effective.
2. Countermeasures should be applicable to a large number of substances.

3. Countermeasures should be amenable to rapid, easy deployment. (Highly specialized equipment and/or chemicals which require wide deployment and stockpiling prior to a pollution incident or which cannot be rapidly conveyed to the scene of an accident are undesirable.)

4. Countermeasures should be free from potentially harmful secondary effects in the aquatic environment, including noxious sludges.

5. Countermeasures developed to combat spills of hazardous polluting substances should take advantage of technology, particularly technology developed to combat oil spills, to the maximum possible extent.

They discussed and evaluated several physical and chemical countermeasures and pointed to the difficulties of biological countermeasures. They also discussed the dynamics of a spill and the problems of containment and mitigation, given the type of material spilled and the nature of the receiving water. The most important parameter, they concluded, was the time lag between the spill and the initiation of treatment because effects on organisms and in many cases process removal efficiencies are functions of the concentration of the spilled material. One could add to this response time the time for removal of the spilled material to safe levels. Huibregtse et al.[2] incorporated such requirements in their user's manual for hazardous material spills but did not include biological countermeasures. For the biological countermeasure to be considered feasible, the first four criteria should be satisfied to the extent possible and be competitive with or at least complementary to the physical-chemical countermeasures available.

BIOLOGICAL COUNTERMEASURE RATIONALE

The attractiveness of a biological countermeasure for hazardous material spills is twofold: (1) bacteria are natural components of ecological systems, and their use as a countermeasure will not constitute the introduction of a "foreign" material; and (2) bacteria will metabolize organic hazardous materials to the principal end products carbon dioxide and water according to the general equation

$$C_xH_yO_z + O_2 \xrightarrow{\text{enzyme}} CO_2 + H_2O \tag{1}$$

The microbial utilization of a hazardous material in a finite-volume mixed reactor is described by the following equation:[3]

$$\frac{dS_1}{dt} V = QS_0 - QS_1 - qX_aV \tag{2}$$

Change in hazardous material mass in system = input − output − removed by cells

where S_0 = influent hazardous material concentration (mg/L)
S_1 = effluent hazardous material concentration (mg/L)
Q = flow into and out of the reactor (L/day)
X_a = average microorganism concentration in reactor (mg/L)
V = reactor volume (L)
t = time
q = hazardous material removal rate = $\dfrac{\text{mg hazardous material removed/day}}{\text{mg microbes}}$

For a system with no flow (e.g., a batch reactor), Eq. (2) reduces to

$$\frac{dS}{dt} = -qX_a \tag{3}$$

which has the solution

$$S = - qX_a t \tag{4}$$

where

$$q = \frac{S_0 - S}{X_a t} \tag{5}$$

Eckenfelder[4] has described this same process by the equation

$$\frac{dS}{dt} = -kX_a S \tag{6}$$

where S = hazardous material concentration (mg/L)
$\quad X_a$ = average microorganism cell concentration in reactor (mg/L)
$\quad k$ = removal rate (mg S removed/day/mg S/mg X_a)
In Eq. (6), the product kS is equivalent to the term q in Eq. (3); therefore,

$$\frac{S_0 - S}{X_a t} = kS \tag{7}$$

These equations represent the overall biodegradation process; usually a number of bio-chemical reactions take place in the microorganism as the hazardous material is reduced to elemental forms.

Likewise, the growth of bacteria in a reactor may be expressed by the following equation:[3]

$$\frac{dX_a}{dt} V = QX_0 - QX_1 + \mu X_a V - k_d X_a V \tag{8}$$

Δ cell mass = input − output + growth − decay in system

where X_0 = influent microorganism cell mass concentration (mg/L)
$\quad X_1$ = effluent cell mass concentration (mg/L)
$\quad Q$ = flow into and out of system (L/day)
$\quad X_a$ = average cell mass concentration in system (mg/L)
$\quad u$ = microorganism growth rate (mg cells produced/day/mg cells)
$\quad k_d$ = microorganism death rate (mg cells removed by death/day/mg cells)
$\quad t$ = time
$\quad V$ = volume
For a system with no flow, Eq. (8) reduces to

$$\frac{dX_a}{dt} = (\mu - k_d)X_a \tag{9}$$

It is important to note here the nature of the relationship between the hazardous material concentration and bacterial growth rate. This relationship has been shown[3] to be very similar to the Michaelis-Menten kinetic model for enzymatic action and may be expressed as

$$\mu = \hat{\mu} \left(\frac{S}{K_s + S} \right) \tag{10}$$

where $\hat{\mu}$ = maximum growth rate (mg cells produced/day/mg cells)
$\quad S$ = hazardous material concentration (mg/L)
$\quad K_s$ = hazardous material concentration at one-half the maximum growth rate (mg/L)

The techniques for deriving the maximum growth rate and the Michaelis constant K_s have been given by Pearson.[3]

The substrate removal rate q may be transformed to the growth rate μ by multiplying by the yield coefficient Y as follows:

$$qY = \mu = \frac{\hat{\mu}S}{(K_s + S)} \tag{11}$$

$$\frac{\text{mg hazardous material removed}}{\text{mg cells} - \text{day}}$$

$$\times \frac{\text{mg cells produced}}{\text{mg hazardous material removed}} = \frac{\text{mg cells produced}}{\text{mg cells} - \text{day}}$$

Thus:

$$q = \frac{\hat{\mu}S}{Y(K_s + S)} \tag{12}$$

Under steady-state conditions, it may be shown from Eq. (2) that q may be determined in a continuously stirred reactor by

$$q = \frac{Q(S_0 - S_1)}{VX_a} \tag{13}$$

In a batch reactor, q may be calculated from Eq. (5).

It is well known in microbiological research that the introduction of a small inoculum of bacteria into a medium with usable substrate results initially in growth of the bacteria at a maximum rate $\hat{\mu}$ with concurrent reduction of the substrate concentration (see Fig. 3-1) according to Eq. (3). This period is termed the maximum-growth phase. After a

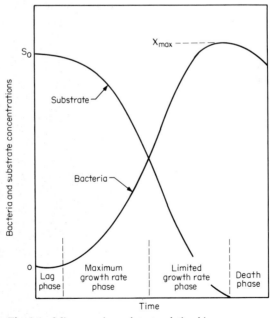

Fig. 3-1 Microorganism-substrate relationship.

short time, the substrate concentration is reduced to a level which becomes limiting to the growth of bacteria, and the bacterial concentration quickly reaches a maximum concentration. This period is termed the declining-growth phase. Following the peak concentration a decline in concentration occurs due to autooxidation and death; this is the death phase or the often-called endogeneous-respiration phase.

A delay in growth, called the lag phase, may occur initially. The extent of the lag phase is a function of the physiological condition of the bacteria, the size of the inoculum, the state of acclimation of the bacteria to the substrate, and perhaps other effects. Once growth begins, the maximum rate reached is characteristic of the bacterium, the other nutrients required by the bacteria for growth, the temperature of the medium, and the toxicity of the substrate. If the lag phase is very long, the rate at which the substrate is consumed also lags. Since one of the most important requirements of a countermeasure[1] is that it be capable of immediate use and application, the lag time must be minimized and the bacterial and substrate characteristics which influence the time lag must be defined.

Equations (3) and (9) may be used to describe the mitigation of a hazardous material and the increase in bacterial concentration respectively in a batch system or in a spill situation in which the spill occurred instanteously (or over a very short time) and onto which bacteria were deployed. The reduction of the hazardous material and the growth of the bacteria would follow approximately the curves shown in Fig. 3-1.

Equations (2) and (8), on the other hand, would be applicable to a continuous-flow biological treatment system which is operated so that high removal of the hazardous material and high bacterial retention in the system are achieved. Operational parameters for these treatment systems have been developed in practice for wastes containing hazardous substances.[4] Equations may also be developed to describe the transport of hazardous material in flowing systems including biodegradation and other sink terms.

To apply these equations, information must be obtained by experimental means for the terms of the equations. The constants for the growth rate–substrate relationship are especially important, as is determination of the substrate removal rate.

CLEANUP METHODS

Biological countermeasures may be employed in one of two ways: they may be applied *in situ* to a spill of hazardous material, that is, in the receiving water itself; or by pumping the spilled hazardous material to a biological treatment system brought to the site. For both treatment techniques, the nature and amount of the hazardous material spilled should be determined so that the appropriate bacteria may be used and the proper amount of bacterial culture applied. Containment of the spilled hazardous material is also desirable to avoid the diluting effects of the natural system and to provide a controlled environment for the bacteria. Once the nature and amount of the hazardous material spilled have been determined and once containment has been achieved or at least ambient concentrations determined, the bacterial countermeasures may be deployed.

If the spills occur in rivers and the use of portable treatment facilities is restricted, the bulk sludge application method may be a solution for the treatment of the spills. As usually practiced, the dispersion in rivers may be considered one-dimensional.

The one-dimensional dispersion equation is usually expressed as

$$\frac{\partial S}{\partial t} + u \frac{\partial S}{\partial x} = D \frac{\partial^2 S}{\partial x^2} \tag{14}$$

where S = concentration at a distance x and time t (M/L^3)
  u = mean flow velocity (L/T)
  D = dispersion coefficient (L^2/T)

For a point source injection of a substance, the concentration distribution has been solved to be

$$S = \frac{M_s}{A_r \sqrt{4\pi Dt}} \exp\left\{ -\frac{(s - ut)^2}{4Dt} \right\}$$ (15)

where M_s = total mass of substance injected (M)
  A_r = cross-sectional area of channel (L^2)

Fischer[5] reported that the mechanism of initial diffusion, characterized by tail effects, was markedly different from that described by Eq. (15), so that Eq. (14) was the only effective formulation after the initial period.

If dispersive acclimated activated sludge is deployed over the spills according to the pollutants dispersion model, the distribution function of both sludge and pollutants may be approximated as Eq. (15). Then, the total mass removal rate can be expressed as

$$\frac{dM_s}{dt} = -\int_{-\infty}^{\infty} \frac{A_r k X S}{K_s + S}\, dx - \int_{-\infty}^{\infty} A_r k_r S\, dx$$ (16)

where k_r is the pollutant removal rate coefficient (time^{-1}) contributed by volatilization and bottom sorption, etc., other than biological reactions.

The usefulness of acclimated activated sludge can be increased by reducing the amount of idle sludge which is a by-product of sludge settling and return. If sludge-containing cloth bags are introduced into the reactors, this process is not necessary. Floating cloth bags can also be used for *in situ* treatment to prevent sludge settling if the mixing intensity is not great enough for sludge suspension. The cloth pores are rapidly clogged with bacterial floc when the bags are filled with concentrated sludge. Through the cloth, liquid exchange is free while the transport of bacterial floc is greatly inhibited.

Containment of a spill in a concentrated form provides better treatment efficiency and, in addition, reduces damage to the total environment. Brown[6] reported that an inflatable plastic barrier was highly effective in confining phenol, methanol, and other soluble hazardous substances. The barrier was constructed of a highly flexible fiber-reinforced plastic with an air-inflated seal, which sealed the barrier to the bottom of the waterway. The barrier was maintained in position by a mooring system.

In confining-barrier application tests carried out in a model river, Kim[7] used the sludge-containing cloth bags as the acclimated-bacteria source with phenol and methanol as substrate, with groundwater flowing in the model river, and without any chemical aids at 28°C. Oxygen was supplied at the cloth bag sites by means of aeration. The average pH for phenol removal in the groundwater was 7.0. Therefore, k_r at 28°C is estimated to be 0.0724 h^{-1} for phenol. The average pH for methanol removal in the groundwater was 7.8. Therefore k_r for methanol at 28°C is estimated to be 0.1143 h^{-1} ($= 0.2656 \times 0.91 \times 0.473$). Detailed experimental conditions and information necessary for the prediction of S and X aeration time are available in Kim.[7]

At the end of 2 days of aeration, about 2 percent of the phenol sludge and about 5 percent of the methanol sludge applied initially leaked from the cloth bags. Theoretically computed substrate removal patterns of phenol and methanol, on the assumption that all the sludge stayed within the cloth bags, are compared with the observed substrate concentrations in Figs. 3-2 and 3-3.

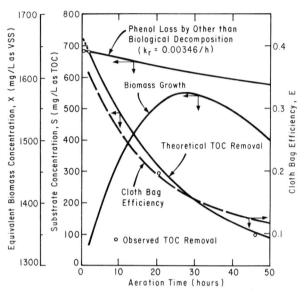

Fig. 3-2 Phenol removal using a fixed barrier with sludge-containing cloth bags.

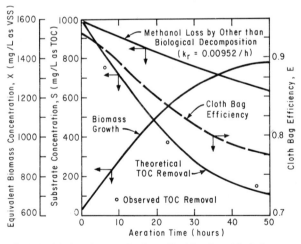

Fig. 3-3 Methanol removal using a fixed barrier with sludge-containing cloth bags.

STORAGE

The principal constraints of the biological countermeasure system emerge in the production, storage, and deployment system and in the introduction of the undesirable bacteria into a natural system. Having to produce a single bacterial species in large quantities for use on a specific kind of hazardous material would impose a serious constraint on the feasibility of biological countermeasures. Thus, it is desirable to find bacteria which may break down a wide variety of organic substrates and which may be produced easily in large volumes. The pseudomonads are such a versatile group. *Pseudomonas fluorescens,* for example, may be grown on sugars, amino acids, organic acids, alcohols, aromatic compounds, and other cyclic organic compounds.[8] Several species of the genus *Pseudomonas* other than *P. fluorescens* have been shown to use a variety of aromatic compounds.[9,10]

Storage of this large quantity of bacterial material may potentially be in several forms: as a liquid culture in which substrate is supplied continually and bacteria are produced continually, in a frozen form in which large quantities of bacterial culture must be continually refrigerated, or in a lyophilized form in which the bacteria are stored in large quantities in airtight containers in a powdered form. Storage in the liquid form is desirable in the sense that the bacteria are ready for application immediately, but the expense of maintaining these cultures at a number of points around the United States near potential spill areas may be prohibitive. Storage in the frozen form is more desirable because less volume is required, but continual refrigeration is required and a lag time is needed to reconstitute the bacteria for application. The lyophilized form may be the most desirable in terms of storage because the bacteria are in a powdered form, may be stored at room temperature, and may be maintained at many points around the country or even shipped with the hazardous material for which bacteria have been cultured. One possible disadvantage of this storage form is the time, a few hours, for reconstitution of the bacteria to an active state. Because the storage method to be used must be determined for each culture developed, investigation of this phase of deployment is required.

DEPLOYMENT

The methods for *in situ* deployment of the bacteria would depend on the physical-chemical state that the hazardous material takes in the receiving water and the storage mode of the bacterial culture. Spraying from a helicopter or a boat or from shore should be adequate. Physical deployment of the portable treatment plant and start-up of the bacterial culture appear to be the critical steps of a biological waste treatment countermeasure.

UNDESIRABLE BACTERIA

Use of biological countermeasures may result in the addition of undesirable bacteria to an aquatic system. Application of pathogenic bacteria, for example, to a spill of hazardous material could result in the proliferation of those bacteria as long as the hazardous material remains present. Also, use of activated sludge from treatment plants for *in situ* application or for the portable treatment plant may result in the application of undesirable bacteria. This potential hazard may not be serious, however, since the bacterial groups which may break down hazardous materials will not likely be pathogenic, and disinfection of the effluent from the portable treatment system would remove the undesirable bacteria from the waste stream before discharge to the aquatic system.

TABLE 3-1 Products Presently Accepted by the U.S. Environmental Protection Agency under Annex X of the National Oil and Hazardous Substances Pollution Contingency Plan

NOSCUM	Natural Hydrocarbon Elimination Company 5400 Memorial Drive Houston, Texas 77007
Petrodeg 100 Petrodeg 200	Bioteknika International, Inc. 7835 Greeley Boulevard Springfield, Virginia 22152
Petrobac R Phenobac R Hyrdrobac	Polybac Corporation 1215 South Cedar Crest Boulevard Allentown, Pennsylvania 18103

Another undesirable by-product of biological countermeasures is the consumption of dissolved oxygen by the bacteria during the breakdown of the hazardous material. Low levels of dissolved oxygen are typically found downstream of domestic and industrial waste discharges. This problem may be avoided by aerating the receiving water artificially after applying the bacteria or by using small amounts of bacteria so that excessive oxygen consumption does not occur.

Cautions about biological countermeasures are expressed in Annex X of the National Oil and Hazardous Substances Pollution Contingency Plan. The plan states that biological countermeasures

> may be used only when such use is the most desirable technique for removing oils or hazardous substances and only after obtaining approval from the appropriate state and local public health and water pollution control officials. Biological agents may be used only when a listing of organisms or other ingredients contained in the agent is provided to EPA in sufficient time for review before its use.

Under the provisions of Annex X of the plan the U.S. Environmental Protection Agency (U.S. EPA) has currently accepted five biological agents. This entitles them to be considered for use by federal on-scene coordinators, as prescribed in Annex X, on a case-by-case basis. Acceptance by the EPA does not constitute approval of the spill cleanup agent or imply compliance with any EPA criteria or minimum standards for such agents, but it means that the manufacturer has submitted to the EPA technical product data which meet Annex X requirements and are maintained on file by the agency (Table 3-1).

CONCLUSIONS

It may be concluded that:

1. Microorganisms are highly effective in removing certain hazardous materials, as research investigations of some of the most significant hazardous materials show that the majority are biodegradable.
2. There are microorganisms such as the pseudomonads which attack a variety of hazardous materials.

3. It should be possible to deploy microorganisms *in situ* or in a portable treatment system easily and rapidly in a fresh, liquid state, a powdered state, or a freshly reconstituted state.

4. Potentially harmful secondary effects should be minor since microorganisms are a natural part of the aquatic environment, pathogenic bacterial will not likely constitute a significant part, if any, of the countermeasure, and noxious sludges should not persist because the microorganisms should oxidize themselves following consumption of the hazardous material.

REFERENCES

1. G. W. Dawson, A. J. Shuckrow, and B. W. Mercer, "Strategy for Treatment of Waters Contaminated by Hazardous Materials," *Proceedings of the 1972 National Conference on Control of Hazardous Material Spills,* Houston, March 1972, pp. 141–144.

2. K. R. Huibregtse, R. C. Scholz, R. E. Wullschleger, C. A. Hansen, and I. Wilder, "User's Manual for Treating Hazardous Spills," in G. F. Bennett (ed.), *Proceedings of the 1976 National Conference on Control of Hazardous Material Spills,* New Orleans, April 1976, pp. 249–253.

3. E. A. Pearson, "Kinetics of Biological Treatment," in E. F. Gloyna and W. W. Eckenfelder, Jr. (eds.), *Advances in Water Quality Improvement,* University of Texas Press, Austin. 1968.

4. W. W. Eckenfelder, Jr., *Water Quality Engineering for Practicing Engineers,* Barnes & Noble, Inc., New York, 1970.

5. H. B. Fischer, "A Note on the One-Dimensional Dispersion Model," *Air and Water Pollution International Journal,* vol. 10, 1966, p. 443.

6. L. S. Brown, "A Physical Barrier System for Control of Hazardous Material Spills in Waterways," *Proceedings of the 1972 National Conference on Control of Hazardous Material Spills,* Houston, March 1972, pp. 93–102.

7. J. W. Kim, "Biological Counter Measures for the Removal of Phenol and Methanol Spills in Waters," Ph.D. dissertation, University of Texas, Austin, 1977.

8. R. Y. Stanier, M. Doudoroff, and E. A. Adelberg, *The Microbial World,* Prentice-Hall, Inc., Englewood Cliffs, N.J., 1957.

9. R. J. Chapman, "An Outline of Reaction Sequences Used for the Bacterial Degradation of Phenolic Compounds," *Degradation of Synthetic Organic Molecules in the Biosphere,* National Academy of Sciences, Washington, 1972.

10 D. T. Gibson, "Initial Reactions in the Degradation of Aromatic Hydrocarbon," *Degradation of Synthetic Organic Molecules in the Biosphere,* National Academy of Sciences, Washington, 1972.

PART 4
Chemical and Physical Measures

W. W. Eckenfelder, Jr.
Distinguished Professor of Environmental and Water Resources Engineering
Vanderbilt University

Alternative Technologies / 9-50
Neutralization / 9-50
Sedimentation / 9-51
Coagulation / 9-52
Precipitation / 9-53
Ion Exchange / 9-54
Carbon Adsorption / 9-54
Oxidation-Reduction / 9-56
Biodegradation / 9-57

Water treatment technology can be applied to the control of hazardous waste spills either *in situ,* during the transfer of the spilled material to a portable treatment system, or when the material is being discharged to a wastewater treatment plant. Selection of treatment protocols will depend on available resources, the magnitude and geographic location of the spill, and the physical and chemical properties of the spilled materials.

ALTERNATIVE TECHNOLOGIES

There are several technologies which can be used to detoxify hazardous waste spills. Some of these may be employed for *in situ* treatment, some for external treatment, and some for either alternative. The primary spill treatment alternatives are provided in Fig. 4-1 and Table 4-1.

The choice of treatment technology in Fig. 4-1 would be influenced by the characteristics of the spilled material. Depending on the material, it could be discharged either to a watercourse or to a wastewater treatment plant at any stage in the treatment sequence. Further discussion of these treatment alternatives follows.

NEUTRALIZATION

Theoretically, acid or alkaline spills can usually be neutralized *in situ*. Acid spills can be neutralized by using lime, limestone, or soda ash. These weak bases have the advantage that an overdose will not result in an excessively high pH. Alkaline spills can be neu-

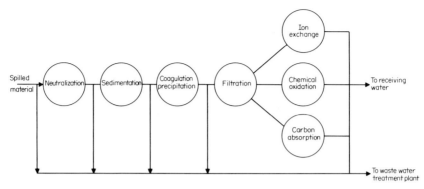

Fig. 4-1 Alternative treatment technologies.

tralized with HCl (in some cases, with H_2SO_4). For example, the amount of 90% lime required to neutralize 95% H_2SO_4 is

$$H_2SO_4 + Ca(OH)_2 \rightarrow CaSO_4 + 2\,H_2O$$
$$ 98 74$$
$$\frac{74}{98} \times \frac{0.95}{0.90} = 0.80 \text{ kg lime/kg } H_2SO_4$$

If the acidity of the material is not known, a sample of the spilled material should be titrated with a standard lime solution (or other alkali if it is to be used) to pH 7.0. The quantity of lime per cubic meter of spilled acid can then be directly calculated. When neutralizing spills in *actual emergency field situations,* the limiting factors will often be the magnitude of the spill, the violence of the reaction kinetics, and the evolution of toxic gases.

Ammonia requires special consideration since un-ionized ammonia (NH_3) is extremely toxic to aquatic life. Because the percentage of ammonia as NH_3 increases with increasing pH, it is important to reduce the pH below 7.0 before discharge. Caution must be exercised in neutralization since the reaction generates considerable heat.

SEDIMENTATION

Sedimentation, i.e., the removal of suspended particles by gravity separation, can be accomplished *in situ* or by external treatment in a gravity separation basin or tank. The time required for separation *in situ* can be estimated by observing subsidence in a beaker.

TABLE 4-1 Treatment Alternatives for Hazardous Waste Spills

Material	Technology	*In situ*	External
Acids or alkalies	Neutralization	Yes	Yes
Ammonia	Neutralization	Yes	No
Suspensions	Sedimentation	Yes	Yes
Heavy metals	Precipitation	Yes	Yes
	Ion exchange	No	Yes
Colloidal dispersions	Coagulation-filtration	No	Yes
Organics	Adsorption	No	Yes
Organics	Chemical oxidation	No	Yes

External treatment in a continuous-flow basin is related to the overflow rate in the basin expressed as cubic meters per square meter per day. This can be roughly estimated by observing the time required for the particles to settle 1.2 m (4 ft) in a cylinder. The settling rate in meters per hour can be computed:

$$\frac{1.2 \text{ m}}{\text{Hours required to settle}} = \frac{\text{meters}}{\text{hour}}$$

The overflow rate in cubic meters per square meter per day then is

$$\frac{\text{Meters}}{\text{Hour}} \times 24$$

To compensate for turbulence and short-circuiting, for design purposes the overflow rate estimated above should be divided by 2.

COAGULATION

Spilled materials containing inorganic or organic colloidal suspensions can be treated by chemical coagulation. Coagulation can be defined as the addition of a chemical to a colloidal dispersion that results in particle destabilization and the formation of complex hydrous oxides which form flocculent suspensions. The flocculent suspensions are subsequently removed from the liquid by sedimentation.

The most common coagulants in use today are alum, iron salts, and lime. In some situations, organic polyelectrolytes (cationic, anionic, or nonionic) can be effectively used as a primary coagulant or in conjunction with alum, iron, or lime.

When using iron or alum, the charge on the colloidal particle is neutralized by the Al^{+++} or Fe^{+++} ion and by positively charged microflocs, which are rapidly produced when the coagulant is added to water. Flocculation of the mixture for 20 to 30 min will result in the production of large flocs which can subsequently be removed by sedimentation, floctation, or filtration.

Effective coagulation is a function of the dosage of the coagulant and the pH of the solution. The alkalinity must be sufficient to effect a reaction with the added coagulant. The coagulation sequence is shown in Fig. 4-2.

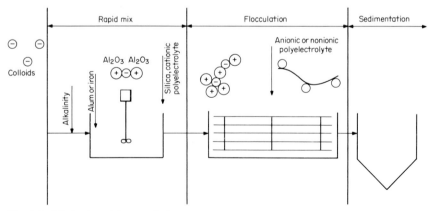

Fig. 4-2 Mechanism of the coagulation process.

TABLE 4-2   Removal of Heavy Metals by Precipitation

Metal	Process	Effluent level	Constraint
Arsenic	Precipitation with S^-	0.05 mg/L	pH 6–7
	Carbon adsorption (low levels)	0.06 mg/L	
	$Fe(OH)_3$ coprecipitation	0.05 mg/L	
Barium	$BaSO_4$	0.5 mg/L	
Cadmium	$Cd(OH)_2$	0.1 mg/L	pH 10.0
	$Fe(OH_3)$ coprecipitation	None	Complexing ions, e.g., CN^-, requires pretreatment; pH 8.5
	H_2O_2 oxidation	. . . . .	Oxidizes CN^- and Cd^{++} to oxide
Copper	$Cu(OH)_2$	0.2 mg/L	pH 9.0–10.3
	$Fe(OH)_3$ coprecipitation	0.3 mg/L	pH 8.5
Lead	$Pb(OH)_2$	0.5 mg/L	pH 10.0
	$Pb(OH)_3$	0.001 mg/L	pH 8.0–9.0
	PbS	. . . . .	pH 7.5–8.5
Mercury	$Fe(OH)_3$, $Al(OH)_3$ coprecipitation	0.1 mg/L	Na_2S added
Nickel	$Ni(OH)_2$	0.15 mg/L	pH 10.0
Selenium	SeS	0.05 mg/L	pH 6.5
Zinc	$Zn(OH)_2$	. . . . .	pH 8.5

A series of jar tests which can readily be performed in the field should be conducted to determine the optimum pH and coagulant dosage. The test procedure involves varying the pH (usually over a range of pH 4 to pH 10) with a constant coagulant dosage which will produce a floc. Once the optimum is established, the coagulant dosage is varied to determine a coagulant dose which yields optimal removal. Adams et al.[1] present test procedure details on this subject.

PRECIPITATION

Most heavy metals can be precipitated as the hydroxide $(Me(OH)_x)$ by the addition of caustic soda (NaOH) or lime $(Ca(OH)_2)$. The reaction which occurs is

$$Me^{+x} + Ca(OH)_2 \rightarrow Me(OH)_x + Ca^{++}$$

Most metals can also be precipitated as a sulfide. Precipitation of the metal to an insoluble form eliminates the problem of seepage into the soil or the surrounding area. To define the alkali requirements, the following procedure is suggested:

1. Titrate a sample with a standard lime solution to the terminal pH of minimum solubility (see Table 4-2). Depending on the volume of spilled material, the quantity of lime required can be calculated.

2. Spread the lime over the spill area, ensuring contact between the spilled material and the lime for precipitation.

3. If external treatment is employed, lime in slurry form should be added and mixed with the waste at a rate determined from Par. 1.

4. It should be noted that for concentrated metal solutions, the quantity of sludge produced may be equal to the quantity of spilled material. This will require on-site dewatering or disposal as a wet slurry.

Extreme caution should be exercised when using lime or other caustic materials. Direct exposure to these materials by vapor inhalation or skin contact must be avoided. Exposure of treatment equipment to these materials can result in corrosion problems as well.

ION EXCHANGE

Heavy metals and other salts can also be removed by ion exchange in which the metal is exchanged for sodium ion:

$$Me^{++} + Na_2Z \rightarrow MeZ + 2\ Na^+$$

Ion exchange is a process in which ions held by electrostatic forces to functional groups on the surface of a solid are exchanged for ions of a different species in solution. This exchange takes place on a synthetic resin. Various kinds of resins, including weakly and strongly acidic cationic exchangers and weakly and strongly basic anion exchangers, are available. The ions are exchanged until the resin is exhausted, at which time the resin is regenerated. The capacity of resins varies, and pertinent data on the resin must be obtained by the manufacturer. Since a resin usually will not be regenerated on site, the necessary quantity of resin should be predetermined.

Ion exchange is an external treatment in which the spilled material is pumped through a column containing the appropriate resin. The system can be considered a detoxification process, although the resulting concentrated salt solution will require a controlled discharge to a receiving water or a municipal sewer or removal to a suitable disposal site.

CARBON ADSORPTION

Many organics can be removed by carbon adsorption. In the adsorption process molecules attach themselves to the solid surface through attractive forces between the adsorbent and the molecules in solution. Adsorption continues until equilibrium is established with the concentration in solution. The ability of organics to be adsorbed on carbon depends upon such factors as molecular structure, solubility, and the substitute groups in the molecule. Extensive adsorption studies have recently been conducted by Dobbs et al.[2] on a wide variety of toxic organics and priority pollutants. Organic chemicals that are adsorbed by carbon are shown in Table 4-3. Some organics which are not removed on carbon are given in Table 4-4. The relative capacity (milligrams per gram) for adsorption of the listed organics on carbon at 1.0 mg/L influent concentration is also given in Table 4-3.

Organic removal using carbon generally employs external treatment through granular carbon columns brought to the site. The spilled material is pumped through multiple columns in series (Fig. 4-3). A breakthrough curve of the type shown in Fig. 4-4 should be prepared. As illustrated in Fig. 4-3, when breakthrough occurs in a three-column series system, the carbon in the first column (Col. 1) is replaced, and the first column now

TABLE 4-3   Selected Organics Removed by Activated Carbon

Compound	Carbon capacity (mg/g)*	Compound	Carbon capacity (mg/g)*
Hexachlorobutadiene	360	1,2,3,4-Tetrahydronaphthalene	74
Anethole	300	Adenine	71
Phenyl mercuric acetate	270	Nitrobenzene	68
p-Nonyl phenol	250	Dibromochloromethane	63
Acridine yellow	230	Ethyl benzene	53
Benzidine dihydrochloride	220	o-Anisidine	50
n-Butyl phthalate	200†	5-Bromouracil	44
N-Nitrosodiphenylamine	220	Carbon tetrachloride	40
Dimethylphenyl carbinol	210	Ethylene chloride	36
Bromoform	200	2,4-Dinitrophenol	33
β-Naphthol	100	Thymine	27
Acridine orange	180	5-Chlorouracil	25
α-Naphthol	180	Phenol	21
α-Naphthylamine	160	Trichloroethylene	21
Pentachlorophenol	150	Adipic acid	20†
p-Nitroaniline	140	Bromodichloromethane	19
1-Chloro-2-nitrobenzene	130	bis-2-Chloroethyl ether	11
Benzothiazole	120	Chloroform	11
Diphenylamine	120	Uracil	11
Guanine	120	Cyclohexanone	6.2
Styrene	120	5-Fluorouracil	5.5
Dimethyl phthalate	97	Cytosine	1.1
Chlorobenzene	93	Ethylenediaminetetraacetic acid	0.86
Hydroquinone	90	Benzoic acid	0.80
p-Xylene	85	Benzene	0.70
Acetophenone	74		

*Capacity at C_0 = 1 mg/L; milligrams of chemical adsorbed per gram of carbon.
†Adsorption capacities at pH 3.

becomes the final column in the series (Col. 4). This procedure is continued until all the spilled material has been treated.

When the applicability of using carbon for the removal of the spilled organic is unknown, a laboratory batch test can rapidly define carbon effectiveness. In this test, various quantities of powdered carbon are mixed with the spilled material in a shaker assembly and mixed for 1 h. The mixture is then filtered to remove the carbon, and the concentration of the organic remaining is measured. The results are then plotted (Fig. 4-5). Since the exhausted carbon removed will be in equilibrium with the influent, the quantity of carbon required for the treatment can be estimated from Fig. 4-5 by extrapolation to the influent concentration of the organic and determining the kilograms of organic removed per kilogram of carbon. The total carbon requirement can then be computed from a knowledge of the volume of spilled chemical to be treated.

TABLE 4-4   Examples of Chemicals Not Adsorbed by Activated Carbon

1. Acetone cyanohydrin	6. Ethylenediamine
2. Butylamine	7. Hexamethylenediamine
3. Choline chloride	8. Morpholine
4. Cyclohexylamine	9. Triethanolamine
5. Diethylene glycol	

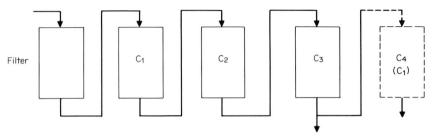

Fig. 4-3 Carbon column configuration.

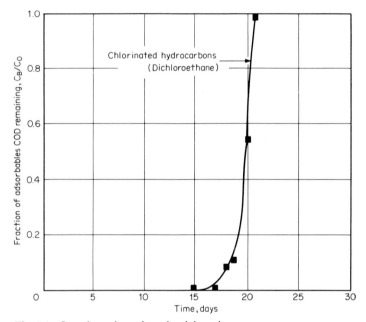

Fig. 4-4 Granular carbon column breakthrough curve.

OXIDATION-REDUCTION

Chemical oxidizing or reducing agents can be applied to a variety of spilled materials, although the magnitude of the spill, the type of spilled material, etc., can limit the choice of these agents. Hexavalent chromium (Cr^{+6}), for example, can be reduced to the trivalent state (Cr^{+3}) by the addition of a reducing agent such as SO_2 or sodium metabisulfite (NaS_2O_5) under acid conditions (pH 2.0). The trivalent chromium can then be precipitated by the addition of lime at pH 8.3 as the insoluble $Cr(OH)_3$. The reactions are

$$Cr^{+6} + \begin{cases} SO_2 \\ Na_2S_2O_5 \end{cases} \xrightarrow{\quad pH = 2.0 \quad} Cr^{+3} + SO_4^{=} \rightarrow Cr^{+3} + Ca(OH)_2 \rightarrow Cr(OH)_3 + Ca^{++}$$

Sulfur dioxide (SO_2) is introduced as a gas, while metabisulfite is a dry powder. The advantage of SO_2 is that it hydrolyzes in water to form the acid H_2SO_3, eliminating the

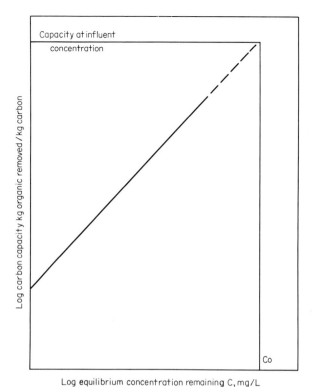

Capacity at influent concentration

Log carbon capacity kg organic removed / kg carbon

Co

Log equilibrium concentration remaining C, mg/L

Fig. 4-5 Carbon capacity estimation from laboratory data.

need for additional acid for pH adjustment. However, removal of the chromium hydroxide sludge is necessary.

Theoretically, spills of chlorine can be reduced with sulfite to the chloride ion, while cyanide can be oxidized to harmless end products (N_2 and CO_2) by oxidation with chlorine under alkaline conditions (pH 8.5). The reactions for the latter case are

$$CN^- + 2OH^- + Cl_2 \rightarrow CNO^- + 2Cl^- + H_2O$$
$$2CNO^- + 4OH^- + 3Cl_2 \rightarrow 2CO_2 + N_2 + 6CL^- + 2H_2O$$

It is important to maintain the pH in the alkaline range to avoid producing toxic by-products.

Other oxidants such as hydrogen peroxide (H_2O_2) and ozone (O_3) offer future promise, but available data on their application are currently insufficient.

BIODEGRADATION

A number of factors should be evaluated when considering the use of biological means for treating hazardous waste spills. Several of the hazardous chemicals are complex organics which require long periods of acclimation before effective biodegradation will occur. In

TABLE 4-5   Relative Biodegradability of Certain Organic Compounds

Biodegradable organic compounds*	Compounds generally resistant to biological degradation
Acrylic acid	Ethers
Aliphatic acids	Ethylene chlorohydrin
Aliphatic alcohols (normal, iso, secondary)	Isoprene
Aliphatic aldehydes	Methyl vinyl ketone
Aliphatic esters	Morpholine
Alkyl benzene sulfonates (with the exception of propylene-based benzaldehyde)	Oil
Aromatic amines	Polymeric compounds
	Polypropylene benzene sulfonates
Dichlorophenols	Selected hydrocarbons:
Ethanolamines	Aliphatics
Glycols	Aromatics
Ketones	Alkyl-aryl groups
Methacrylic acid	Tertiary aliphatic alcohols
	Trichlorophenols
Monochlorophenols	
Nitriles	
Phenols	
Primary aliphatic amines	
Styrene	
Vinyl acetate	

*Some compounds can be degraded biologically only after extended periods of seed acclimation.

many cases concentration limits exist, and large dilution may be needed. Excessively long periods of aeration may also be required to reduce the contaminant to a level suitable for discharge to a watercourse or a sewer. All these factors would usually tend to discourage the use of biological treatment at the spill site.

The possibility exists, however, of discharging the spilled chemical to a municipal sewer at a controlled rate which will avoid shock loading of the wastewater treatment plant to ensure degradation of the chemical in the biological treatment plant. The biodegradability of various organic compounds is shown in Table 4-5. Information on some specific compounds is available in various published sources, e.g., Verschueren.[3] If the chemical is biodegradable, the suggested procedure is to control the discharge rate so that the organic loading rate (F/M)* as kilograms BOD per day per kilogram of mixed liquor volatile suspended solids (MLVSS) does not exceed 0.2 (on the basis of the chemical discharged). This will therefore be in addition to the organic loading normally received by the plant. (It should be confirmed that the total loading to the biological plant does not exceed 0.5.) The following example will illustrate.

A spilled chemical, phenol, is to be treated in a biological treatment plant:

Aeration volume: 7.57×10^3 m^3

Concentration of volatile suspended solids: 3000 mg/L

$$\text{Allowable F/M} = 0.2 = \frac{\text{kg BOD applied/day}}{\text{kg MLVSS}}$$

*F/M is the food/microorganisms ratio in kilograms of chemical BOD per kilogram of mixed liquor volatile suspended solids.

$$= \frac{\text{kg BOD applied/day}}{7.57 \times 10^3 \text{ m}^3 \times 3000 \text{ mg/L} \times 10^3 \text{ L/m}^3 \times 10^{-6} \text{ kg/mg}}$$

Thus the maximum BOD_5 that can be applied each day is 4542 kg, and the allowable kilograms of phenol application (based on 1.87 mg BOD/mg phenol) = 4542/1.87, or 2429 kg/day = 101 kg/h. Hence, for adequate treatment the phenol should be fed to the wastewater treatment plant at a rate of less than 101 kg/h.

The possibility exists that preacclimated cultures in dry form could be employed in some treatment plants or that biological sludge from an industrial wastewater treatment plant treating the same or similar chemicals could be used. Generally, the organic loading of the spilled material fed to the biological process should be adjusted to 0.1 kg BOD per day per kg MLVSS to ensure a low effluent concentration of the pollutant.

REFERENCES

1. C. Adams, D. Ford, and W. W. Eckenfelder, Jr., *Development of Process Design Criteria for Wastewater Treatment Processes,* Enviropress, Inc., 1979.
2. R. A. Dobbs, R. J. Middendorf, and J. M. Cohen, *Carbon Adsorption Isotherms for Toxic Organics,* U.S. Environmental Protection Agency, Municipal Environment Research Laboratory, Cincinnati, May 1978.
3. K. Verschueren, *Handbook of Environmental Data on Organic Chemicals,* Van Nostrand Reinhold Company, New York, 1977.

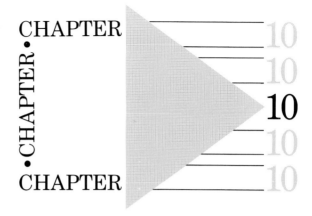

CHAPTER · CHAPTER · CHAPTER · CHAPTER

10
10
10
10
10
10

Volatile Materials

PART 1
Commercially Available Monitors for Airborne
Hazardous Chemicals / 10-3

PART 2
Atmospheric Dispersion / 10-13

PART 3
Vapor Hazard Control / 10-21

PART 4
Ammonia / 10-34

PART 5
Chlorine / 10-52

PART 6
Flammable Liquid Gases / 10-65

PART 1
Commercially Available Monitors for Airborne Hazardous Chemicals

Sheridan J. Rodgers
MSA Research Corporation

Introduction / 10-3
Hazardous Chemical Spills / 10-3
Monitoring Equipment / 10-4
Conclusion / 10-12

INTRODUCTION

The purpose of this section is to present an overview of monitoring devices which can be used to measure the concentration of airborne hazardous chemicals under emergency conditions. The monitors described are applicable to the hazardous chemicals most likely to be involved in a spill. Two major hazards are considered: toxicity and flammability.

Basic assumptions are required to provide a meaningful overview of monitors which can be used for hazardous spills. First, it must be assumed that normal line voltage may not be available to provide power for the monitor. Second, it must be assumed that the individual using the monitor may not be a highly trained instrumental analyst. Therefore, the monitors which are described are either manually operated or have self-contained battery packs, and the readout is either in concentration if the concern is toxicity or in percentage of the lower explosive limit (LEL) if the concern is flammability.

HAZARDOUS CHEMICAL SPILLS

A discussion of monitoring systems must include an assessment of chemicals which might be involved in hazardous chemical spills. A good source of information is the U.S. Coast Guard's *Hazardous Chemical Data*,[1] which provides information needed for decision making during spill emergencies. Specific chemical, physical, and biological data are presented for 985 hazardous substances. Information important to in-field monitoring includes data on fire and toxicology hazards (Table 1-1).

A spill requires prompt reaction by local fire departments, with possible follow-up by an emergency team skilled in handling hazardous chemical spills. Spills resulting from truck accidents will likely involve only one chemical. Conversely, an accident involving trains or river barges could result in the spill of a number of different chemicals. In the case of a single spill, i.e., the release of only one chemical to the atmosphere, the problem of monitoring the environment is minimized since the type of chemical released and the threshold-limit value (TLV) and LEL may be known. Multicomponent spills will com-

TABLE 1-1   Fire and Health Hazard Data Available in *Chemical Hazards Response Information System, Volume 2: Hazardous Chemical Data*

Fire hazards	Health hazards
Flash point	Threshold-limit values
Flammable limits	Short-term inhalation limits
Ignition temperature	Irritant characteristics
Burning rate	Odor threshold

TABLE 1-2   Monitoring Problems Presented by Different Types of Spills

Type of transport	Spill characteristics	Monitoring requirements
Truck	Single chemical	TLV and LEL probably known
Barge	Single or multichemical	TLV and LEL either known or unknown
Rail	Multichemical	TLV and LEL probably not known

plicate monitoring since the TLVs and LELs for the various chemicals will be different. The problems associated with the different types of spills are summarized in Table 1-2.

The type of spill which occurs (single-component or multicomponent) will require some subjective judgment on the part of personnel assigned to the scene. Current monitoring equipment enables the hazards of single-component spills to be accurately assessed. Multicomponent spills pose a more demanding assessment of hazards owing to the mix of TLV and LEL values which result from the spill. Methods for calculating the TLV of mixtures are presented in App. C of an American Conference of Governmental Industrial Hygienists text.[2]

MONITORING EQUIPMENT

The monitoring devices which will be described are capable of providing information on the hazards resulting from spills. This discussion is limited to portable monitors: detector tubes, combustible-gas analyzers, gas chromatographs, and infrared analyzers. The first two types of monitors require no technical training, but the last two need some expertise in instrument operation and calibration.

Detector Tubes

Detector tubes are used to measure toxic levels of hazardous chemicals. The American Conference of Governmental Industrial Hygienists provides a listing of threshold-limit values for hazardous chemicals.[2] This list is upgraded annually. Detector tubes are designed to provide information on these airborne chemicals, the range generally being from one-fifth to 5 times the TLV. They are not meant to be precise measuring devices, but they do provide an accuracy of about ±25 percent. With this accuracy and when used properly, detector tubes can provide guidance on respiratory protection requirements for emergency personnel and decision-making information on whether an area should be evacuated.

Industry has developed detector tubes for approximately 150 different hazardous chemicals. The tubes are filled with a reactive chemical which produces a color change and/or stain as a sample of contaminated air is drawn through the tube. In most cases, the length of the stain rather than the color change is used as a measure of the gas concentration.

A typical detector tube kit is shown in Fig. 1-1. The dimensions of the carrying case are 508 by 356 by 102 mm (20 by 14 by 4 in); it contains a pump, spare parts, an instruction card, a data sheet, and up to 20 packages of detector tubes or reagent kits.

The variable-orifice–variable-volume pump, which draws an accurate volume of ambient air through the detector tube, is illustrated in Fig. 1-2. The volume of aspirated air is controlled by the distance to which the handle is withdrawn. The variable orifice is located at the inlet to the piston pump. The ability to vary sample size and sampling rate results in optimum sample collection efficiency.

Two types of calibration scales which are employed with the detector tubes are shown in Fig. 1-3. In the case of the CO_2 calibration scale, the tube is placed on the calibration scale corresponding to the number of pump strokes, and the concentration is read directly as a function of the length of stain in the tube. The bromine and chlorine calibration technique employs a standard millimeter scale in which the length of stain is measured and the concentration is derived from the table of stain length and pump strokes. For example, a stain 10 mm long resulting from two pump strokes would indicate a chlorine concentration of 10 ppm. Some tubes have the concentration scale inscribed directly on the glass surface.

The detector tube kit can be made more versatile with various accessories. A special adapter for chemically treated filter papers provides a means of colorimetrically reacting

Fig. 1-1   Detector tube kit. [*MSA.*]

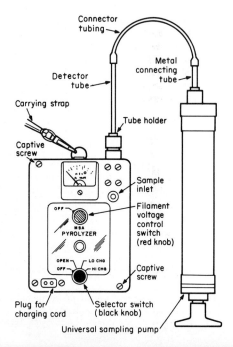

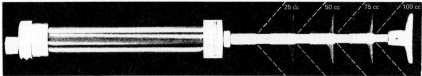

Fig. 1-2   Sample pump and pyrolyzer for use with a detector tube. [*MSA.*]

a sample after collection. A battery-operated pyrolyzer unit can be used to convert halogenated hydrocarbons to free halogen. The liberated halogen produces a stain in the tube proportional to the contaminant concentration (Fig. 1-3).

The simplicity of the detector tube unit provides a means of rapidly assessing the hazards of a chemical spill. With little training, emergency personnel can quickly determine potential hazards. This method of hazard assessment is particularly applicable to spills of a single chemical. Multicomponent spills complicate the survey and require subjective judgments by emergency personnel.

Combustible-Gas Detectors

Nearly all portable combustible-gas analyzers are based on the hot-wire–Wheatstone bridge principle. The simplest form employs a single catalytic filament as one of the four arms of the circuit. The use of two filaments in the Wheatstone bridge circuit provides a greater degree of sensitivity and a higher degree of accuracy. The change in resistance produced by the heat of combustion of the combustible gas is a measure of the gas concentration. Some detectors use a thermal conductivity principle in which the presence of a gas other than air causes the removal of heat from the circuit and a change in conductivity in the electrical circuit is proportional to the gas concentration.

One type of combustible-gas analyzer employing a single heated platinum filament is shown in Fig. 1-4. The monitor weighs approximately 2 kg (4 lb) and measures 102 by 152 by 127 mm (4 by 6 by 5 in). The sample is drawn through the instrument with an aspirator bulb. The monitor has an on-off switch, a zero-adjust knob, and a meter with a range of 0 to 100 percent of the LEL. The inlet coupling accommodates filters and varying-length sampling lines. Power is supplied by six standard D cell batteries. The instrument is generally calibrated for pentane (LEL = 1.5 percent), although it can be calibrated for other hazardous gases.

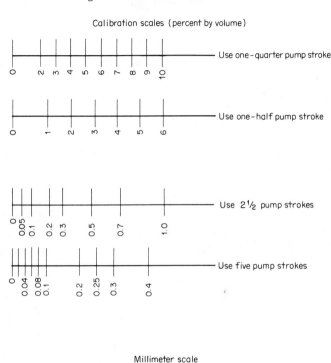

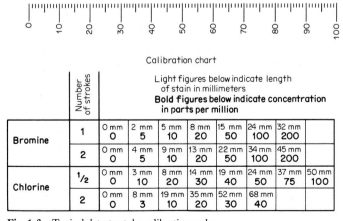

Fig. 1-3 Typical detector tube calibration scales.

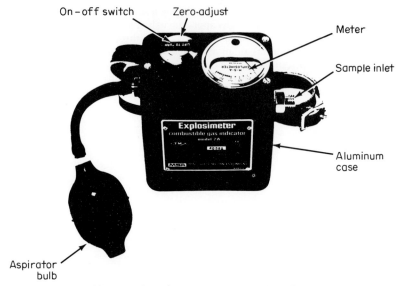

Fig. 1-4   Combustible-gas analyzer (0–100 percent LEL). [*MSA.*]

Figure 1-5 is a photograph of a portable monitor which employs both a heated filament circuit and a thermal conductivity filament. The monitor weighs approximately 3 kg (6 lb) and measures 102 by 152 mm by 152 mm (4 by 6 by 6 in). The gas sample is aspirated through the monitor with a squeeze bulb. A selector switch provides a choice of either the LEL or the concentration of the gas. Power is supplied by eight standard D cells. This instrument is normally calibrated with methane gas (LEL = 5.3 percent) since

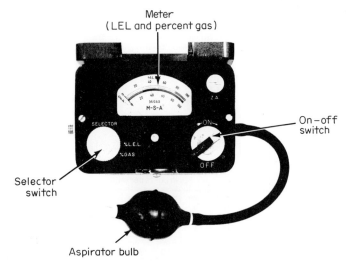

Fig. 1-5   Combustible-gas analyzer (0–100 percent LEL, percent gas). [*MSA.*]

it is frequently used to monitor natural gas distribution systems. However, the monitor can be calibrated with other gases as well.

A combustible-gas monitor which employs two heated filaments in the Wheatstone bridge circuit is shown in Fig. 1-6. The monitor weighs 3 kg (6 lb) and measures 127 by 102 by 152 mm (5 by 4 by 6 in). The instrument is powered with eight standard D cell

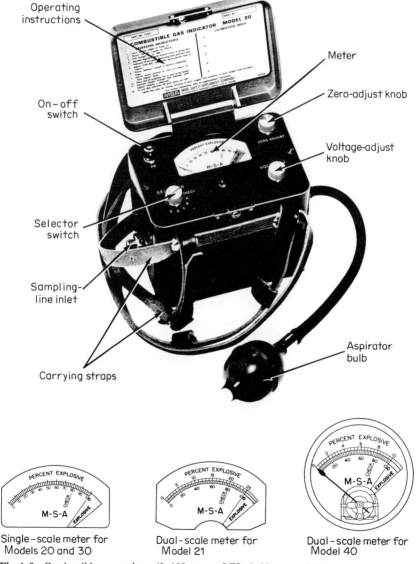

Single-scale meter for Models 20 and 30

Dual-scale meter for Model 21

Dual-scale meter for Model 40

Fig. 1-6   Combustible-gas analyzer (0–100 percent LEL, 0–25 percent LEL, 0–10 percent LEL). [*MSA.*]

batteries. The face of the monitor has an on-off switch, a voltage-adjust knob, a zero-adjust knob, and a selector switch.

Since this instrument is more accurate than those previously described, a choice of scales is available (0–100 percent LEL; 0–25, 0–100 percent LEL; and 0–10, 0–100 percent LEL). This monitor can be calibrated for five separate combustible gases. A response curve manual, which covers over 50 combustible gases and can be used to correlate meter readings to the actual percentage of LEL, is available with one model of this instrument. A reference chart, which also converts the 0 to 10 percent LEL readings to concentration (parts per million) for many combustible gases, is supplied.

Figure 1-7 shows a combustible-gas monitor which employs a hot Pelement detector. An inactive hot Pelement in the bridge circuit compensates for varying conditions of flow, temperature, and humidity. The unit measures 254 by 178 by 102 mm (10 by 7 by 4 in) and is powered by a rechargeable Ni-Cd battery pack. The instrument has both a visual and an audible alarm circuit. The monitor is factory-calibrated using pentane, but it can be calibrated for any number of combustible gases.

The instruments described are typical of commercially available equipment. Prices range from $150 to $600, depending upon the type of instrument and desired accessories.

Infrared Analyzers and Chromatographs

Monitoring devices to determine the presence and concentration (TLV or LEL) of hazardous chemicals are commercially available. They perform quite well when a single known airborne contaminant is present. However, area monitoring to determine toxic or explosive levels requires some judgment on the part of the emergency team. Since all spills will differ, it is impossible to present a specific set of rules for monitoring, but certain influencing factors should be considered with any spill: physical and chemical characteristics of the chemical, wind direction, wind velocity, size of the spill, and temperature.

Multicomponent spills severely complicate the problem of monitoring for both toxic and explosive levels. Grouping chemicals according to TLVs and/or LELs appears to be

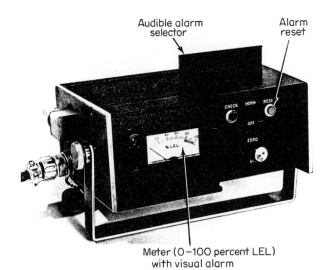

Fig. 1-7   Combustible-gas monitor (0–100 percent LEL). [*MSA.*]

an approach which will provide decision-making information for emergency crews using detector tubes or combustible-gas analyzers. Other portable instruments such as infrared analyzers and gas chromatographs deserve some comment, although these instruments are expensive and require certain skills in the use and interpretation of the data.

Nearly all gases and vapors absorb infrared radiation at wavelengths from 2 to 15 μ. There are two main types of infrared instruments that use this principle for quantitative analysis: (1) the nondispersive instrument that operates at a fixed wavelength and (2) the dispersive instrument that operates at user-selected wavelengths.

The nondispersive infrared analyzer has proved to be an accurate and reliable instrument for a single component for concentrations ranging from the TLV to 100 percent composition. The instrument is sensitized at the time of manufacture and is limited to the analysis of one substance. This type of infrared analyzer would not be practical for monitoring spills.

A portable infrared analyzer that overcomes this disadvantage by using a variable filter to select a wavelength band suitable for the compound of interest is available. The manufacturers of this instrument have compiled a list of maximum tolerable limits set by the U.S. Occupational Safety and Health Administration (OSHA) for more than 400 gases and vapors (*Federal Register,* vol. 40, May 28, 1975, p. 23072).[3]

Many of the gases and vapors have characteristic infrared absorption and can be detected by infrared. The infrared analytical data contained in the compiled list provides calibration information necessary for testing with the portable infrared analyzer. These data include the minimum concentration that the analyzer will detect, the appropriate instrument settings, and the absorbance value of the safe concentration limit. A 10 percent accuracy is claimed for concentrations close to the safe limit. Pure air is required to set the baseline before an analysis is made, but it can be obtained by using a suitable filter cartridge. The cost of the standard instrument is approximately $4000.

Portable gas chromatographs that can quantitatively analyze gases and vaporized liquids in air are commercially available. A full range of detectors is available, but those most commonly used for gas analysis are the thermal conductivity detector and the flame ionization detector. A gas chromatograph offers the advantage of individual monitoring of each compound in a multichemical spill. Its main disadvantage is that a skilled technician is required to operate the instrument and interpret the results. Also, present portable gas chromatographs, unlike detector tubes and combustion analyzers, are not designed to be explosionproof and could be hazardous in an already explosive atmosphere. Since flame ionization detectors use a hydrogen flame to burn the pollutants, it is unlikely that these units could be made explosionproof. Costs range to $4500 depending on the type of detector required.

Calibration

A discussion of hazardous airborne-gas detectors would be incomplete without some discussion of calibration techniques. Instruments should be calibrated at the frequency recommended by the manufacturer. From the foregoing discussion, it is obvious that a single calibration will not cover all the combustible gases which might be encountered. Practically, it may be necessary to group combustible gases according to their LEL, using the gas with the lowest LEL in a given grouping for calibration. In this way the readings will always be on the "safe" side.

Calibration gases are available from manufacturers of combustible-gas monitors. These may be packaged in aerosol cans or compressed-gas cylinders. In Fig. 1-8, a typical calibration configuration is shown for the combustible-gas monitors which have been described. The calibration gas is delivered through a needle into the bladder. The standard gas is then aspirated through the monitor.

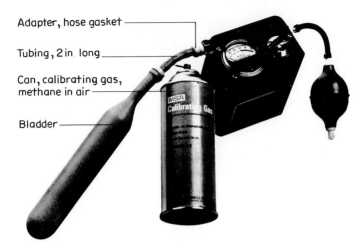

Adapter, hose gasket

Tubing, 2 in long

Can, calibrating gas, methane in air

Bladder

Fig. 1-8   Typical calibration configuration for an airborne-gas detector. [*MSA.*]

CONCLUSION

Current monitoring equipment permits reliable surveys of single-component spills to be made, provided the specific chemical is known. Multicomponent spills result in a mixture of airborne hazardous chemicals which cannot be discriminated by detector tubes or combustible-gas analyzers. Portable gas chromatographs and infrared analyzers are commercially available but require personnel skilled in the use of the instruments and interpretation of the data. The user of any instrument for monitoring hazardous airborne chemicals should be familiar with the capabilities and limitations of the device.

REFERENCES

1. *Chemical Hazards Response Information System,* vol. 1: *Hazardous Chemical Data,* M16465.12, U.S. Coast Guard, 1978.

2. *Threshold Limit Values for Chemical Substances and Physical Agents in the Workroom Environment with Intended Changes for 1978,* American Conference of Governmental Industrial Hygienists, Cincinnati.

3. *Pocket Guide to Chemical Hazards,* U.S. Department of Health, Education, and Welfare and U.S. Department of Labor, NIOSH-OSHA, September 1978.

PART 2
Atmospheric Dispersion

Richard H. Schulze
President, Trinity Consultants

Types of Spills / 10-13
Danger Area / 10-14
Calculation of Hazard Area / 10-17
Comparison of Rule 36 with the More Generalized Equation / 10-17

Whenever there is an accidental spill of a liquid or a gas, some or all of the material volatizes. As the cloud of volatile material and air moves gradually downwind, the natural process of atmospheric turbulence reduces the concentration of the contaminants. If the spill is of a hazardous material, the concentration at some point downwind will fall below the toxic level. Regardless of the nature of the spill, atmospheric turbulence is the major factor in diluting the plume as it moves downwind. The initial turbulence of the actual release generally has only a minor effect on concentrations at distances beyond 250 m or so from the spill.

TYPES OF SPILLS

There are three types of spills: liquids, gases stored as liquids, and gases. In no case is a buoyant plume created, unless of course there is a fire. Observations indicate that plumes not created by burning hug the ground and do not become elevated above the surface. A more detailed description of the three types of spills follows.

Liquid Spills

The spill of a material which is normally a liquid at atmospheric pressure and temperature causes a pool to form. Examples of such liquids are benzene, acrylonitrile, and vinyl chloride. The liquid will evaporate from the pool at a rate that can be calculated on the basis of the vapor pressure of the liquid which governs the rate of liquid-to-gas transfer, the area of the pool, the heat of evaporation, the air temperature, the ground temperature, and the heat transfer of convection and radiation. The viscosity of the liquid will determine the rate of growth of the pool and, hence, its area.

For most liquids the initial rate of emission will be greater per unit area than the equilibrium rate. As evaporation takes place, the pool will be cooled, reducing the temperature of the adjacent air into which the liquid is evaporating. Wu and Schroy have provided a good summary of the thermodynamics and heat transfer associated with liquid spills.[1]

10-13

If the spill of a liquid occurs on water, the release rate of the material will be greater than if the spill occurs on land. The rate of release of toxic material decreases as the temperature of the pool decreases. If the pool is floating on water, circulation patterns in the water will help keep the pool relatively warm. On land there is no circulation, only conduction, which is a much less efficient means of heat transfer.

Liquid Gas Spills

Many materials are stored in pressurized and/or refrigerated containers. These include ammonia, chlorine, and liquefied natural gas (LNG). If the release rate is slow enough, the material will enter the atmosphere as a gas. In a sudden release such as a tank rupture, the dispersion process can be described as four phases according to Kaiser and Walker:[2]

1. There is the turbulence, called "flash-off," caused by the rapid escape of the liquid or gas. For a 20-ton ammonia tank rupture, tests suggest that about 20 percent becomes vapor, while 80 percent remains as a liquid aerosol. The liquid aerosol appears to observers as a cloud. The cloud at this point resembles a column or a vertical cylinder. The gases within this cylinder are quite cold owing to the evaporation of part of the ammonia.

2. This cylinder starts to slump, much in the way that a column of water starts to spread out. During slumping, air is entrained fairly slowly, and this second stage will typically last 30 to 40 s for a 20-ton ammonia spill.

3. The gas cloud now enters the ground-hugging phase, in which the entrainment of air takes place at a rate less than that expected for a passively diffuse plume. The area of the top surface is much greater than the area of the sides, and therefore most of the entrainment of air takes place through the top surface. Depending on the ambient temperature, the gases in the cloud will reach ambient-air density in 3 to 5 min after slumping starts from a 20-ton ammonia spill.

4. The natural process of atmospheric turbulence takes over, and the cloud diffuses like any other material in the atmosphere.

The boundaries between the phases are rather blurred and indistinct. The description of the four phases, however, is useful when discussing the atmospheric dispersion of dense or cold gases.

Gas Spills

The spill of a material which is normally a gas both at atmospheric conditions and at storage and/or transportation conditions results in the formation of an air-gas mixture at a temperature somewhat below ambient temperature. The reduced temperature is the result of the pressure reduction between storage and/or transportation conditions and atmospheric pressure. An analysis of such a release using hydrogen sulfide as an example is provided by Echols.[3]

DANGER AREA

The most frequent question asked about hazardous spills is "If there is an accident, how large an area will receive a hazardous dose?" The answer to this question usually has two parts. First, the maximum radius of exposure is determined on the basis of a worst-case scenario. Second, a more refined analysis is made to limit the area of evacuation when a hazardous spill actually occurs.

Regulatory agencies are sometimes concerned with the first part. One of the best examples is Rule 36 of the Texas Railroad Commission, the organization that regulates oil and gas production in Texas. This rule was adopted in 1976, 2 years after a tragic accident. In 1974, a pipeline carrying hydrogen sulfide had ruptured, and as a result several oil company employees and a ranching family died. Rule 36 seeks to protect the public from concentrations under worst-case meteorology, and it has withstood the scrutiny of public hearings.

Four steps are followed in the development of the rule:

1. *Levels not to be exceeded.* The first step is to determine the levels which should not be exceeded. For example, Rule 36 is based on a 5-min concentration of 500 ppm and a 30-min concentration of 100 ppm of hydrogen sulfide. The desired levels for other pollutants are normally given as a concentration for a period of time and are contained in several references such as Sax's *Dangerous Properties of Industrial Materials*[4] and works by Sittig[5] and the International Technical Information Institute.[6]

2. *Meteorological conditions.* The second step is to define appropriate meteorological conditions. Atmospheric turbulence is greatest on bright, sunny days and least on still, cloudless nights. Thus dilution is least on clear nights with very light winds. Rule 36 is based on F stability (clear night with a radiation inversion) and a 1-mi/h (0.446-m/s) wind. If the winds are lighter, they generally meander over broad sectors of the compass. If the dispersion is over water, the plume width may approximate F stability under cloudy conditions during either day or night.

3. *Initial dispersion.* All spills result in an initial dispersion that has some width. This is certainly the case of rupture of an ammonia tank. It is also true of the evaporation of material from the pool of a spill of liquid; the pool has some width. The jet effect of an escaping gas stream also creates an initial dispersion.

The initial dispersion results in larger values for plume width, thus producing lower concentrations. If the initial width of the plume is w and the height is h, the adjusted dispersion coefficients are calculated as follows:

$$\sigma_{yadj} = \sqrt{\sigma_y^2 + (w/4.3)^2} \tag{1}$$

$$\sigma_{zadj} = \sqrt{\sigma_z^2 + (h/2.15)^2} \tag{2}$$

where σ_z = vertical dispersion coefficient
σ_{zadj} = adjusted vertical dispersion coefficient
σ_y = horizontal dispersion coefficient
σ_{yadj} = adjusted horizontal dispersion coefficient

Rule 36 assumes, conservatively, no adjustment for initial dispersion.

4. *Dispersion coefficients and averaging time.* The rate of atmospheric diffusion has been studied by numerous investigators. The values most commonly used are called the Pasquill-Gifford (P-G) values. Among all sets of commonly used values, the P-G curves forecast the slowest rate of dilution. Rule 36 uses the P-G values and assumes that they represent 1-h values. It then adjusts the concentrations to the shorter time periods of 30 min and 5 min. In reality, the P-G values represent 3- to 10-min values, and thus the adjustment in Rule 36 tends to increase the distance at which levels not to be exceeded will occur.

In addition to knowing the maximum downwind distance, it is also necessary to know the horizontal sector that will be affected by the spill. The wind direction measured every minute usually meanders over a 20 to 40° arc centered on the average wind direction.

Viewed from above, the plume resembles a snake as its direction meanders downwind. It is generally safe to consider the plume width to be 45° centered on the average wind direction. During an accidental release, however, the wind direction should be carefully monitored by watching the plume. If the wind shifts, new areas may be endangered by the accidental spill.

For a more detailed analysis, the article by Echols[3] provides a procedure for estimating the area affected by a release of hydrogen sulfide. An article by Simmons et al.[7] describes a method for defining the area affected by a chlorine spill. Both of these articles presume that the rate of release, atmospheric stability, and wind speed are known.

Often it is desirable to have a quick reference guide to aid nontechnical on-site personnel to take proper action. The U.S. Department of Transportation supplies a guide from which the information in Table 2-1 is extracted.[8] The guide assumes that a spill from a tank car or a tank truck is about 10 m in diameter and that the wind speed is 2.7 to 5.4 m/s. For lighter winds the evacuation distance is greater, but fortunately one also has more time to notify individuals in the affected area because of the slow speed of the traveling hazardous cloud.

TABLE 2-1   Evacuation Table Based on Prevailing Wind of 6 to 12 mph (2.7 to 5.4 m/s)

Material	Radius of immediate danger area (km)	Dimension of evacuation area	
		Downwind (km)	Crosswind (km)
Acrolein	.69	8.05	4.83
Acrylonitrile	.03	.32	.16
Ammonia	.08	.64	.48
Carbon disulfide	.04	.32	.16
Chlorine	.31	3.22	2.41
Dimethylamine	.14	1.13	1.29
Epichlorohydrin	.05	.32	.32
Ethylene oxide	.04	.32	.16
Ethyleneamine	.35	3.22	2.41
Fluorine	.20	1.61	1.61
Hydrogen chloride	.24	2.41	1.61
Hydrogen cyanide	.12	1.13	.64
Hydrogen fluoride	.30	3.22	1.61
Hydrogen sulfide	.15	1.61	.81
Methyl mercaptan	.09	1.29	.48
Methylamine	.14	1.13	1.29
Monomethylamine	.14	1.13	1.29
Nitric acid	.13	1.13	.64
Nitrogen tetroxide	.14	1.13	1.29
Oleum	.35	3.22	1.61
Phosgene	.75	8.05	4.83
Phosphorus trichloride	.14	1.21	.81
Sulfur dioxide	.13	1.21	.81
Sulfur trioxide	.35	3.22	1.61
Sulfuric acid	.35	3.22	1.61
Trimethylamine	.35	3.22	2.41

SOURCE: *Emergency Action Guide for Selected Hazardous Materials*, U.S. Department of Transportation, 1978.

Zajic and Himmelman discuss hazardous spills in Chap. 3 of their book.[9] The work of Simmons et al.[7] is summarized, and an abbreviated version of Table 2-1 is included.

The same four steps are used whether one is concerned with a worst-case scenario or with analyzing downwind concentrations that are occurring as a result of an actual spill. The remainder of this section describes the detailed development of Rule 36 of the Texas Railroad Commission.

CALCULATION OF HAZARD AREA

Rule 36 uses the following equations to determine the radius of an area in which concentrations exceeding 100 ppm and 500 ppm could be found under a worst-case scenario:

$$100 \text{ ppm radius of exposure } X = [(1.589)(\text{mol fraction } H_2S)(Q)]^{.6258} \quad (3)$$
$$500 \text{ ppm radius of exposure } X = [(0.4546)(\text{mol fraction } H_2S)(Q)]^{.6258} \quad (4)$$

where X = radius of exposure in feet
Q = maximum volume determined to be available for escape in cubic feet per day
H_2S = mol fraction of hydrogen sulfide in the gaseous mixture available for escape
The value (mol fraction H_2S)(Q) is the volume of the hydrogen sulfide gas that is emitted per day. Since concentrations are expressed in parts per million by volume, the daily volumetric rate of escape of any gas can be used in this equation. Thus this equation can be used for any compound, not just hydrogen sulfide.

COMPARISON OF RULE 36 WITH THE MORE GENERALIZED EQUATION

The basic dispersion equation for concentrations from ground-level releases is

$$\chi = \frac{Q}{\pi \sigma_y \sigma_z u} \quad (5)$$

where χ = concentration in grams per cubic meter
Q = emission rate in grams per second
σ_y = horizontal dispersion coefficient in meters
σ_z = vertical dispersion coefficient in meters
u = wind speed in meters per second
The equivalency between this general equation and the rules of the Texas Railroad Commission can be demonstrated:

1. The expression (mol fraction H_2S)(Q) is the cubic feet of H_2S released per day.

1 g mol = 22.414 L at 0°C = 273.15 K
1000 L = 1 m^3
Standard conditions = 60°F = 15.6°C = 288.7 K
n = molecular weight
1 m^3 gas of mol weight n = $\dfrac{1000 \text{ L}}{m^3} \cdot \dfrac{1 \text{ g mol}}{22.414 \text{ L}} \cdot \dfrac{273.15}{288.70} = 42.21 \, n$ g
1 ft^3 of gas of mol weight n = 1.195 n g
1 ft^3 of gas of mol weight n/day = 13.83 $n \cdot 10^{-6}$ g/s
100 ppm = .004221 n g/m^3
500 ppm = .02111 n g/m^3

TABLE 2-2   Comparison of Dispersion Coefficients

Stability	Model	Downwind distance (meters)	σ_y (meters)	σ_z (meters)	$\sigma_y\sigma_z$ (meters)	(meters)
F	RAM	1000–2000	$.0629x^{.9105}$	$.1770x^{.63227}$	$.01113x^{1.543}$	$.00178w^{1.543}$
		500–1000	$.0610x^{.9149}$	$.1232x^{.6847}$	$.00752x^{1.600}$	$.00112w^{1.600}$
		200–500	$.0590x^{.9203}$	$.0642x^{.7841}$	$.00379x^{1.704}$	$.00050w^{1.704}$
		100–200	$.0571x^{.9264}$	$.0544x^{.8156}$	$.00311x^{1.742}$	$.00039w^{1.742}$
F	TEM	500–5000	$.0625x^{.911}$	$.1930x^{.6072}$	$.01206x^{1.518}$	$.00199w^{1.518}$
		100–500	$.0625x^{.911}$	$.05645x^{.805}$	$.00353x^{1.716}$	$.00046w^{1.716}$
F	Railroad Commission		$.0640x^{.922}$	$.176x^{.676}$	$.01126x^{1.598}$	$.00169w^{1.598}$
D	RAM	100–500	$.1175x^{.9218}$	$.0923x^{.8512}$	$.01085x^{1.773}$	$.00132w^{1.773}$
C	RAM	100–500	$0.1801x^{.9200}$	$.1102x^{.9147}$	$.01985x^{1.835}$	$.00224w^{1.835}$

NOTE:
F stability = clear and partly cloudy nights, light winds.
D stability = overcast conditions, day or night.
C stability = clear to partly cloudy daylight conditions.
x = downwind distance in meters.
w = downwind distance in feet.

2. The expressions for σ_y and σ_z were developed by measuring the slope of the curves for F stability. These can be compared with the values in the RAM model developed by the U.S. Environmental Protection Agency and the TEM model developed by the Texas Air Control Board, as shown in Table 2-2.

3. The time-averaging method used with the regulation is based on the suggestion of Gifford that concentrations for any time period can be calculated from values for a base time period as follows:

$$\chi_t = \chi_o(t_o/t_t)^{0.2} \tag{6}$$

where χ_t) = concentration for time period t
χ_o = concentration for base time period
t_o = base time period
t_t = time period of interest

If the calculated concentrations by using the Pasquill-Gifford equation are assumed to be 60-min values,

$$\chi_{30} = \chi_{60}(60/30)^{0.2} = 1.149\chi_{60} \tag{7}$$
$$\chi_5 = \chi_{60}(60/5)^{0.2} = 1.644\chi_{60} \tag{8}$$

4. The assumed wind speed is 1 mi/h, or 0.4464 m/s. Restating the general equation,

$$\chi = \frac{Q}{\pi\sigma_y\sigma_z u}$$

$$\sigma_y\sigma_z = \frac{Q}{\pi\chi u}$$

$$.00169\,w^{1.598} = \frac{Q}{\pi\chi u}$$

Restating for w;

$$w^{1.598} = \frac{591.72Q}{\pi \chi u}$$

Calculate Q/χ for 500 and 100 ppm:

$$Q/\chi \text{ for 500 ppm for 5 min} = \frac{(13.83 \; n \cdot 10^{-6} \; c)(1.644)}{(.02111 \; n)} = .001077 \; c$$

$$Q/\chi \text{ for 100 ppm for 30 min} = \frac{13.83 \; n \cdot 10^{-6} \; c)(1.149)}{(.004221 \; n)} = .00376 \; c$$

c = number of cubic feet per day of pollutant

Substitute Q/χ in the expression for $w^{1.598}$.

$$u = .4464 \text{ m/s}$$

For 500 ppm: $w^{1.598} = \dfrac{(591.72)(001077 \; c)}{\pi \cdot 0.4464} = .4544 \; c$

$$w = (.4544 \; c)^{.6258}$$

This compares with the rule where $X = (.4546 \; c)^{.6258}$

For 100 ppm: $w^{1.598} = \dfrac{(591.72)(.00376 \; n)}{\pi \cdot 0.4464} = 1.586 \; n$

$$w = (1.586 \; n)^{.6258}$$

This compares with the rule where $X = (1.589 \; n)^{.6258}$.

REFERENCES

1. J. M. Wu and J. M. Schroy, "Emissions from Spills," paper presented at a conference sponsored by the Air Pollution Control Association at Gainesville, Fla., February 1979. *Proceedings* available from the Air Pollution Control Association, Pittsburgh.

2. G. D. Kaiser and B. C. Walker, "Release of Anhydrous Ammonia from Pressurized Containers—The Importance of Denser-than-Air Mixtures," *Atmospheric Environment*, vol. 12, 1978, pp. 2289–2300.

3. W. T. Echols, "Estimating the Hazardous Radius of Exposure from Accidental Release of Hydrogen Sulfide Gas," in G. F. Bennett (ed.), *Proceedings of the 1976 National Conference on Control of Hazardous Material Spills*, New Orleans, April 1976, pp. 201–207.

4. N. I. Sax (ed.), *Dangerous Properties of Industrial Materials*, 5th ed., Van Nostrand Reinhold Company, New York, 1979.

5. M. Sittig, *Hazardous and Toxic Effects of Industrial Chemicals*, Noyes Data Corp., Park Ridge, N.J., 1979.

6. *Toxic and Industrial Chemicals Safety Manual for Handing and Disposal with Toxicity and Hazard Data*, International Technical Information Institute, Tokyo, 1976.

7. A. Simmons, C. Erdmann, and N. Naft, "Risk Assessment of Large Spills of Toxic Material," in G. F. Bennett (ed.), *Proceedings of the 1974 National Conference on Control of Hazardous Material Spills*, San Francisco, August 1974, pp. 149-156.

8. *Emergency Guide for Selected Hazardous Materials*, U.S. Department of Transportation, 1978.

9. J. E. Zajic and W. A. Himmelman, *Highly Hazardous Material Spills and Emergency Planning*, Marcel Dekker, Inc., New York, 1978.

PART 3
Vapor Hazard Control

Ralph Hiltz
Senior Scientist
MSA Research Corporation

Mechanical Covers / 10-22
Induced Air Movement / 10-22
Sorption and Vapor-Phase Reaction / 10-23
Liquid-Phase Modification / 10-23
Surface Cooling / 10-24
Film and Foam Covers / 10-25

Many of the chemicals appearing in the various hazardous substances lists pose a significant vapor hazard. Although the released vapors may ultimately disperse into the environment with little chronic effect, they do pose an immediate hazard to life and property downwind of the spill. Additionally, there is a hazard to those responding to the spill who must remain in the area for the duration of the incident.

The vapor hazard from spilled chemicals takes several forms. Toxic fumes which pose a life hazard even at low (parts-per-million) concentrations may be released, or flammable vapors for which minimum dangerous concentrations are generally above 1 percent may be released. Some chemicals may exhibit both hazards, but toxicity with its lower allowable concentration will be the controlling factor.

The great difference in minimum hazard levels creates two distinctly separate problems. In the case of flammable vapors, small increments of reduction may be meaningful. For toxic materials, the ability to provide mitigation of the hazard may lie with reduction of the equilibrium vapor pressure. For compounds whose threshold-limit value (TLV) approximates the anticipated vapor concentration above the liquid surface, the potential exists to mitigate the hazard at least for short-term exposure. When vapor concentrations can be expected to be greater than the TLV, significant reduction of the toxic hazard does not appear possible within existing technology.

Since the advent of concern for vapor hazards associated with spilled volatile chemicals, a number of possible vapor amelioration techniques have been considered. Only a few have exhibited significant potential. The most recent review of techniques was conducted under U.S. Coast Guard sponsorship.[1] The techniques which received consideration were:

1. Mechanical covers
2. Foam covers
3. Surface cooling
4. Induced air movement

5. Vapor scrubbing (adsorption and absorption)

6. Reaction–vapor phase

7. Liquid-phase modification to change the vapor species released or to dilute the surface concentration

Of these, only surface cooling, foam covers, and liquid-phase modification can currently pass criteria of cost, availability, deployment, and application. Only foam covers and liquid-phase modification have had any practical demonstration.

MECHANICAL COVERS

Placing a lid over a spilled chemical is a direct approach for containing the toxic vapors with nearly 100 percent efficiency. Three basic techniques have been considered: (1) total cover of the spill area by cloth or other continuous material, (2) spray of a continuous cover such as urethane, and (3) buoyant particles (either spheres or polygonal shapes). Theoretically, such covers would contain essentially all vapor release, but in practice some leakage is to be expected. However, leakage is not a problem. Being able to use such materials requires acquisition in advance of the spill and storage until needed, and in all but small spills deployment may be problematical.

Floating-cover assemblies are presently being fabricated by a number of suppliers for such purposes as protecting drinking-water reservoirs, collecting methane gas from sewage waste lagoons, and sealing a broad range of liquid chemicals for atmospheric isolation. Implementing such covers for spills in the field would appear to be a difficult if not an impossible task. Attempts are being made to integrate such covers with floating booms, but demonstrable equipment is not yet available.

The distribution of light particulates over a liquid surface is another means of developing a physical barrier to encapsulate liquids and reduce evaporation losses. Evaporation is decreased by the presence of the densely packed layer of particles, which reduce convection currents and insulate the liquid surface. This technique is presently used for open storage tanks, ponds, and reaction vessels to restrict evaporative losses. Effective materials, in the form of hollow spheres or closed-cell plastic foams, include glass, polypropylene, and polyurethane. Polyurethane has the best combination of chemical resistance and mechanical integrity.

An evaporation control system which incorporates the geometrical ordering of dodecahedrons to produce a close-packed array has been reported;[2] 1000 particles will cover an area of 0.38 m² (4.12 ft²). The dodecahedron geometry permits a single layer to cover more than 99 percent of a surface when in a close-packed arrangement.

Although particulate covers are potentially effective, cost is a deterrent to their use. The technique would require the dispersal of a minimum of 3000 particles per square meter of spill surface. At present prices, material costs would be approximately $1000 per square meter, with up to 100 times this cost for equipment to disperse the particles.

INDUCED AIR MOVEMENT

Simple dilution provides a direct approach to vapor hazard amelioration. The dilution technique involves the transport and mixing of uncontaminated air with the vapors released from a chemical spill. The volume of uncontaminated air must be large enough to maintain the concentration of hazardous chemical vapors below their TLV or lower flammability limit (LFL).

Performance specifications can be calculated from data generated with natural dispersion models. These models predict evaporation rates for spills of volatile hazardous chemicals to range from 1 to 3.5 m³ of vapor released per hour per square meter of spill surface. If an average TLV of 10 ppm is assumed, 1.0×10^5 to 3.4×10^6 m³ of uncontaminated air per square meter of spill surface must be added hourly to keep the concentration of the hazardous chemical vapor at this limit. If an average LFL of 1 percent by volume is assumed, 1.0×10^2 to 3.4×10^5 m³ of uncontaminated air must be delivered hourly to keep the vapor concentration at this limit. These are achievable but difficult quantities to obtain.

Such air movement requires the availability of high-velocity fans, jet air movers, etc., on a major scale. If these are available, their use appears to be justified.

Water aerosols which displace and dilute the vapor offer a similar action at a reduced efficiency. However, the large quantities of water required will dilute the spill and possibly spread it, thus complicating the situation. If the hazardous material floats on water, spreading will increase the surface area and may increase, rather than decrease, the vapor hazard.

SORPTION AND VAPOR-PHASE REACTION

The use of sorbents to remove spilled hazardous chemicals is described in Chap. 9. With respect to vapor release, sorbents can decrease the vapor hazard as long as they are unsaturated. To be effective, they must be continuously replenished. If they are allowed to become saturated, the vapor hazard increases owing to the large increase in surface area for vapor release.

Sorption of vapor from the air, using devices to draw the air through a purifier much like a dust scrubber, is a possibility. No commercial equipment exists, however, and there are no data on removal efficiencies.

A vapor-scrubbing mechanism which combines an intake fan with a high-expansion foam generator is under consideration. The contaminated air is used to generate foam and is encapsulated by it. The contaminant is trapped by the bubble wall. Scrubbing efficiencies of 20 to 80 percent have been achieved, depending on the solubility of the contaminant, the residence time in the foam, and the bubble size.[3,4] For the system to be cost-effective, it should be totally contained, enabling the solution from the collapsed foam to be recycled. The use of foam blankets is discussed below in the section "Film and Foam Covers."

LIQUID-PHASE MODIFICATION

The use of water or other chemical agent to reduce the vapor release through surface dilution or to change the species being released can be a useful, practical approach to vapor-phase hazard mitigation. The basic use of dilution, neutralization, and chemical reaction to ameliorate the hazard of a liquid spill is described in Chap. 9. Such procedures should be used when practical to maintain the vapor reduction.

Certain chemicals such as ammonia, sulfur trioxide (including oleum), and silicon tetrachloride can exhibit beneficial changes in the nature of the released vapor through reaction with water. Owing to their high heats of hydration, the addition of water is hazardous.

Gel formation involves interaction between a high-molecular-weight molecule (macromolecule) and a liquid. It is one form of liquid-phase modification which has seen extensive work, but for liquid immobilization rather than for vapor hazard control. The gel structure is a combination of physical and chemical interaction that generally results in the formation of a two- or three-dimensional network of macromolecular cages, entrapping the liquid phase. Irzhak[5] has reported experimental work and theoretical interpretations of gelation reactions.

The formation of a gel has some influence on the evaporation rate and therefore on the vapor concentration of a spilled chemical. Evaporation rate reduction is achieved by forming a continuous cover of gelled material to encapsulate the more volatile spilled liquid. However, the primary benefit obtained from gelation is immobilization or confinement. The ability to limit or stop the spread of the spilled chemical reduces evaporation since the amount of vapor released is proportional to the exposed surface area.

The time required for the gelling reactions to be completed is a limiting factor for this technique. There are a few gels which may be formed within minutes, but many gelling reactions can take hours.

A Universal Gelling Agent, developed under U.S. Environmental Protection Agency (U.S. EPA) sponsorship by the Calspan Corporation,[6] has a specific blend of gelling agents combining the flexibility and rapid reaction rates required for treating spills of hazardous chemicals. A listing of the hazardous chemicals used to test the performance of this agent is presented in Table 3-1, which is taken from the report of Pilie et al.[7]

In addition to the Calspan work, Fuller[8] and Goldstein[9] have reported on gelling techniques developed specifically for oil spills. Development programs conducted by Bannister et al.[10] and Weaver et al.[11] have potential applications in the gelation of spills. However, there are no data to indicate any vapor concentration reduction.

SURFACE COOLING

Reduction of the surface temperature of a spilled chemical can suppress vapor release. Reduction to the ice point ($0°C$) would reduce equilibrium vapor pressure by at least 40

TABLE 3-1   Compounds for Which Universal Gelling Agent Has Been Shown to Be Effective*

Acetone	Formaldehyde
Acetone cyanohydrin	Gasoline
Acrylonitrile	Isoprene
Ammonium hydroxide	Isopropyl alcohol
Aniline	Kerosine
Benzaldehyde	Methanol
Benzene	Methyl ethyl ketone
Butanol	Octane (2,2,4-trimethylpentane)
Carbon disulfide	Orthodichlorobenzene
Carbon tetrachloride	Petroleum ether
Chlorine water, saturated	Phenol (89%)
Chloroform	Pyridine
Cyclohexane	Sulfuric acid
Cyclohexanone	Tetrahydrofuran
Ethanol	Trichloroethylene
Ethyl acetate	Water
Ethylene dichloride	Xylene
Ethylene glycol	

*As reported by R. J. Pilie, R. E. Baier, R. C. Ziegler, R. P. Leonard, J. G. Michalovic, S. L. Pele, and D. Bock, *Methods to Treat, Control and Monitor Spilled Hazardous Materials,* Contract No. 68-01-0110, EPA-670/2-75-042, U.S. Environmental Protection Agency, June 1975.

TABLE 3-2   Theoretical Refrigeration Values and Costs of Coolants

Coolant	State	Net refrigeration (cal/g)	Cost ($/kg)
Nitrogen (N_2)	l, 77.4 K (195.8°C)	97	7.27×10^{-2}
Carbon dioxide (CO_2)	l + g, 274 K (2068 kPa, −16.8°C)	71	8.81×10^{-2}
	S, 194,7 K (−78.5°C)	153	2.09×10^{-1}
Water ice	S, 372 K (0°C)	80	3.33×10^{-2}

NOTE: l = liquid; g = gas; S = solid.

percent and for some chemicals up to 95 percent. Greater reductions can be achieved by employing coolants with lower boiling or freezing points.

Although a number of refrigerants exist, only wet ice, dry ice, and liquid CO_2 appear to be practical. Liquid nitrogen may be considered, but mechanical problems in application and economic problems in storage limit its efficiency of use. The theoretical refrigeration values and estimated costs of these three coolants are shown in Table 3-2. These represent a theoretical limit. Real-time reductions will almost certainly be smaller.

Liquid carbon dioxide stored at 2068 kPa (300 psi) will yield about 43 cal/g of cooling in the spilled material after allowing for transfer inefficiencies. Solid carbon dioxide will absorb approximately 153 cal/g and solid water (ice) will absorb about 80 cal/g. To cool to the ice point the heat removal requirements are conservatively estimated at 136 kcal/m^2 for initial spill material cooling, plus a continuing 41 kcal/($m^2 \cdot$h), including heat losses to the environs for the duration of the spill.

The efficiency of cryogenic cooling to ameliorate spills of hazardous chemicals is controlled by several factors. Physical properties such as melting point, freezing point, vapor pressure, specific heat, viscosity, surface tension, and heat of fusion of both the spilled material and the cryogen determine the theoretical efficiency of this response technique. However, the potential of the technique remains to be measured.

Delivery of the cryogenic material affects the ultimate efficiency in much the same way as the other techniques. It is imperative that the cryogen be spread evenly, over as large an area as possible, very quickly. Vaporization rate reductions will be directly proportional to the efficiency with which a homogeneous cooling blanket of cryogen is placed over the spill. A program to define the basic efficiencies and the mechanisms of application has recently been completed.[12]

FILM AND FOAM COVERS

Of all the available techniques for mitigating the vapor hazard from spilled chemicals, foam systems are the only ones which have been widely used in the field. The ability of a foam cover to block vapor release was recognized as an outgrowth of its use in controlling chemical fires. Even though foams are used in this manner and both governmental and independent specifications for these materials incorporate a requirement for vapor control, it has been only recently that any definitive measure of capabilities under field conditions has been made.[13]

In a similar category are surfactant films. Foams block vapor release by means of a three-dimensional blanket of continuous bubbles. Films block release by means of a spatially oriented layer floating on the spill surface.

Surfactant Films

Early use of monomolecular films to reduce the evaporation rates of water and aqueous solutions were carried out by Hedestrand,[14] Rideal,[15] and Langmuir and Langmuir.[16] More recent developments and applications were conducted by La Mer,[17] and the use of fluorinated surfactants for the suppression of fuel evaporation was studied by Tuve and Jablonski,[18] Bernett et al.,[19] and Moran et al.[20]

Two mechanisms contribute to vaporization rate reductions obtained from monomolecular films: film resistance and surface quiescence. Langmuir and Langmuir,[16] Sebba and Rideal,[21] Archer and La Mer,[22] and Rosano and La Mer[23] have found that the organic chain length and stereochemical nature of the surfactant affect the rate of evaporation of liquids through monomolecular films. Additionally, surface turbulence and convection currents are reduced by the generation of a surface pressure when the monomolecular film is established.

Tuve et al.[24] proposed a unique method of forming this thin layer, i.e., the use of fluorochemical surfactant foams [aqueous film-forming foams (AFFF)], which collapse to form a thin film. Bernett et al.[19] and Moran et al.[20] showed the feasibility of this method for hydrocarbon fuels.

Fluorochemical surfactant films were found to reduce vaporization of fuels by 90 to 98 percent when present at a thickness of about 10 μ. This material works best for hydrocarbon fuels having relatively low vapor pressures, but it may be applied to most chemicals having a surface tension equal to or greater than 0.20 N/m.

This technique in its current state of development utilizes fluorochemical surfactants in water solution to form a thin film over the surface of a spilled chemical. The surfactant-water film should be between 10 and 20 μ thick to seal the surface of the hazardous chemical and reduce evaporation.

Two distinct methods may be used to form the film: (1) an aerosol or mist of the surfactant-water solution sprayed over the surface of the spill or (2) the generation of a foam which is formulated for rapid collapse. Systems are commercially available for both foam generation and fluorocarbon surfactant application.

The amount of solution required for this response technique can be calculated from the size and volume estimates at the spill and the required film thickness.

Fire-fighting experience with aqueous film formers has shown the need for 100 to 200 times as much solution as is theoretically necessary to form the surfactant-water film. Using approximately 100 times the minimum surfactant-water solution required for film formation permits visual verification of the foam cover and a realistic safety factor for maintaining the film.

The safety factor is necessary since relatively minor discontinuities in the film result in large changes in effectiveness. The influence of film continuity was shown by Archer and La Mer,[22] who reported that 1 percent film area breakage (holes) reduced the efficiency of vaporization suppression by more than 90 percent. This difficulty affects the suitability of surfactant films under field conditions where wind, rain, or other factors causing spill movement could be expected to cause breaches in the film.

Commercial Foam Covers

Water-based foams have been effective in controlling vapor release from volatile hydrocarbons. A significant amount of data has been developed as a result of the extensive use of foams in the control of fires of such materials. Extended tests have been conducted with foam in the control of fires and spills of liquefied natural gas. In addition to fairly well detailed tests, restricted testing has been conducted with a number of other materials.

Foam blankets act to isolate surfaces from ignition sources and radiant energy. The blocking of radiant energy slows the vaporization rate, but vapor release is also reduced by the limited permeability of the foam, its capacity to absorb the chemical, or dilution of the surface layer. Aggressive water-reactive chemicals can be treated by foam. Foam allows the gentle application of water to the surface to effect dilution and/or conversion to less hazardous foams without violent reaction.

Water-based foams are not necessarily applicable to all volatile chemicals. Although the reasons are not very clear, the available data show emphatically that commercially available foams are degraded by materials whose dielectric constant is greater than 3. Several investigations have attempted to find mechanisms to stabilize foam against polar compounds.

Two basic foam types are currently available and in wide use by fire services: (1) protein-derived material and (2) surfactant-based concentrates. Within the surfactant category are two foams: hydrocarbon and fluorocarbon foams. The latter, marketed under the generic name aqueous film-forming foams (AFFF), tend to be film formers rather than persistent foam formers. Protein-base systems also have modifications. The fluoro-proteins are combinations of fluorocarbon surfactant materials with protein-base agents. The "alcohol foams" designed for use with polar compounds are normally protein-base, although polar AFFF foams are now being made available.

As the applicability of foams to spill control has developed, certain novel foam formulations have evolved. These formulations have been devised to be compatible with a wide variety of hazardous chemical classes, to provide extended life and/or resistance to atmospheric effects, or in some cases to be specific to one class (e.g., acids or bases) or to a particular material (e.g., ammonia). These materials are not readily available to fire services because of cost or restricted application. As their utility is shown, they can be expected to appear in the arsenal of industry and government response teams or third-party cleanup contractors.

Protein foam systems are widely used in the control of hydrocarbon fuel fires. In this application their ability to restrict vapor release has long been recognized. A measure of this capability is contained within both governmental[25] and independent[26] foam specifications and test procedures.

Surfactant foam systems are also in wide use in the control of Class B fires. Unlike protein-base materials, which are limited to low-expansion applications (usually 10:1 or less), surfactants are suitable for both low and high expansion defined as greater than 100:1. They are not as well regulated as protein materials, and there is a wide variety of surfactant-based concentrates on the market. Some approach the capability of protein foams in controlling hydrocarbon vapors, but many are of dubious quality. The AFFF materials can be quite capable in this application because of their film-forming ability.

The evolution of data on the use of foams on nonreactive materials has been reported by the Maritime Administration,[27] Hiltz and Friel,[28] Gross,[29] and Normal and Dowell.[30] To define the applicability of foam for vapor mitigation, the EPA[13] sponsored a program to review existing data and conduct tests necessary to document the technique fully. That program is in the final stages of evaluation and analysis of data derived from field tests. In earlier stages of the EPA program, a foam use matrix was prepared on the basis of existing data including data from laboratory tests. This matrix is presented in Table 3-3. The field test data now available are not expected to change the basic premises used in preparing the matrix. They will define the influence of atmospheric factors, principally wind.

In using the matrix, it must be understood that the nature of the spill and the conditions at the scene will exert an influence on the selection of the most applicable foam type. Low-expansion foams by their nature are less influenced by the atmospheric effects of wind, rain, or high temperature. For vapor mitigation, they will provide the longest time delay before vapor breakthrough occurs. However, they require larger volumes of water

TABLE 3-3 Matrix of Foam Capabilities to Suppress or Otherwise Minimize the Release of Toxic or Flammable Vapors from the Spilled Hazardous Chemicals Used

		Recommen-dation	1	2	3	4	5	6	7
Aliphatic Organics:									
Acids:	Acetic acid	R	A+	B+	ND	ND	ND	ND	U
	Caproic acid	ND	U	ND	ND	ND	ND	ND	U
Alcohols:	Amyl alcohol	ND	U	U	ND	ND	ND	ND	ND
	Butanol	R	E−	E−	E−	E−	A+	E−	−
	Butyl cellosolve	ND	ND	ND	U	ND	ND	ND	U
	Methanol	R	E−	E−	E−	E−	A+	E−	−
	Octanol	R	B+	A+	B+	B+	U	ND	U
	Propanol	R	E−	E−	−	E−	A+	E−	−
Aldehydes	Acetone	R	E−	E−	E−	E−	A+	E−	ND
and	Methyl butyl ketone	R	B+	A+	U	ND	ND	ND	ND
ketones:	Methyl ethyl ketone	ND	U	U	U	ND	ND	ND	ND
Esters:	Butyl acetate	ND	U	−	−	U	ND	ND	ND
	Ethyl acetate	ND	U	−	−	U	ND	ND	ND
	Methyl acrylate	R	B+	A+	ND	U	ND	ND	ND
	Methyl methacrylate	R	B+	A+	ND	U	ND	ND	ND
	Propyl acetate	ND	U	−	.	U	ND	ND	ND
Halogen-	Butyl bromide	R	A+	U	ND	ND	ND	ND	ND
ated:	Methyl bromide	ND	U	U	ND	ND	ND	ND	ND
	Tetrachloroethane	R	B+	A+	ND	ND	ND	ND	ND
Hydrocar-	Heptane	R	B+	A+	B+	B+	E−	C+	U
bons:	Hexane	R	B+	A+	B+	B+	E−	C+	U
	Octane	R	B+	A+	B+	B+	E−	C+	U
Nitrogen-bearing:	Dimethyl formamide	ND	U	E−	ND	ND	ND	ND	ND
Aromatic Organics:									
Hydrocar-	Benzene	R	B+	A+	−	−	E−	ND	ND
bons:	Tetrahydronaphthalene	R	B+	A+	E−	ND	ND	U	ND
	Toluene	R	B+	A+	−	−	E−	ND	ND
Organics— Alicyclic:									
	Cyclohexane	R	B+	A+	B+	B+	E−	C+	U
Industrial Organics:									
	Gasoline	R	B+	A+	B+	B+	E−	C+	U
	Kerosine	R	B+	A+	B+	.B+	E−	C+	C+
	Naphtha	R	B+	A+	B+	B+	E−	C+	U
	Paint thinner	ND	U	ND	ND	ND	ND	U	ND
Organics—Cryogens:									
	Liquefied natural gas	R	E−	A+	F−	F−	F−	F−	ND
Inorganics:									
	Silicon tetrachloride	R	C+	C+	C+	C+	ND	F−	ND
	Sulfur trioxide	R	C+	C+	C+	C+	ND	F−	ND
Inorganics—Cryogens:									
	Ammonia	R	A+	A+	B+	B+	ND	E−	ND
	Chlorine	R	B+	A+	B+	B+	ND	E−	ND

NOTE:
1	= low-expansion surfactant.
2	= high-expansion surfactant.
3	= protein.
4	= fluoroprotein.
5	= alcohol.
6	= AFFF.
7	= mechanical.
+	= data available indicating vapor pollution control.
−	= data available indicating no vapor pollution control.
U	= limited data available; capabilities uncertain.

ND	= no data.
R	= foam use recommended over spill.
N	= foam use not recommended over spill.
P	= foam use prohibited over spill.
A	= best foam formulation.
B	= next-best foam formulation.
C	= acceptable in some situations.
E	= unsuitable foam formulation.
F	= deleterious foam formulation.

TABLE 3-4   Foam Mitigation Times in Minutes: Time to Reestablish a Vapor Concentration of 1 Percent by Volume

Foam agent	Cyclohexane	Benzene	Toluene	Ethyl ether	Acetone	n-Butyl acetate	n-Octane
Regular protein	60	60	60	Collapsed	Collapsed	Collapsed	60
Fluoroprotein	60	60	60	Collapsed	Collapsed	60	NT
Alcohol	60	60	60	20	60	60	60
Low-expansion surfactant	60	35	55	Collapsed	Collapsed	25	NT
AFFF 3 percent	60	13	60	15	Collapsed	60	60
AFFF 6 percent	NT	6	NT	16	Collapsed	NT	NT
High-expansion surfactant	60	25	60	Collapsed	Collapsed	Collapsed	60
High-expansion AFFF 3 percent	NT	10	NT	NT	NT	NT	NT

NOTE:
NT = not tested.
Foam depths (millimeters):
  Alcohol: 57
  Fluoroprotein: 57
  Protein: 57
  Low-expansion surfactant: 102
  Low-expansion AFFF: 102
  High-expansion surfactant: 457
  High-expansion AFFF: 457

than high-expansion systems to effect a foam cover, and this can spread the spill or cause an overflow of impoundments.

In Table 3-4, typical mitigation times are shown for several chemicals as a function of foam types. The times indicated can be extended by adding foam to compensate for breakdown, drainage, and vapor saturation of the initial blanket. It is in this operation that expansion (the amount of water used) becomes important.

High-expansion foams use about one-half of the water volume required for the equivalent cover of low-expansion foams. Their use will be controlled by wind conditions, the containment of the spill, and the nature of the spilled material. With winds below 4.5 m/ s, maintenance of an adequate high-expansion blanket should be possible. At higher wind speeds, some method of downwind containment of the foam mass by fencing or other structure will be necessary.

In addition to using less water, high-expansion foam can exhibit a lower vapor concentration above the blanket than low-expansion foam. In any given free spill situation, a vapor profile is developed above the liquid. Under stagnant conditions, this profile is a reproducible state of equilibrium. At any given height above the liquid, the vapor concentration will be a constant with or without a foam blanket.

A foam blanket acts to slow vapor release, but ultimately the equilibrium profile will redevelop. Since greater foam depth can be built with high expansions, the maximum vapor concentration above the foam level will be less than for a thinner low-expansion blanket. With a minimum of 457 mm (18 in) of foam (for the majority of volatile flammable liquids), the range of flammable concentrations will occur within the foam mass. Thus, for well-contained spills for which response and cleanup times are short, low-expansion foams will be best. When spill duration is long, the benefits of high-expansion foam will predominate.

However, foam blankets become saturated with the material they cover, and flammable vapor concentrations will develop within the bubbles. Ignition sources penetrating the foam can result in fire. The higher the expansion, the greater the possibility of deflagration of the foam.

Most of the discussion to this point has concerned flammable materials. The degree of vapor mitigation will be the same regardless of the volatile chemical involved. Unfortunately, since toxic levels are small, it is doubtful that foam of any kind could maintain vapor concentrations below the TLV in the immediate spill area. Some reduction may occur in the downwind extension of the minimum toxic level, but no data have been developed in this regard.

The discussion of foam systems has concerned generic foam types. A number of different companies currently are providing foam concentrates in each category. In absolute terms all these concentrates should provide equal capabilities. The bubble walls are the principal barrier. These consist basically of water, but the surfactant does have some influence. Relating the properties of the concentrates and the developed foam to the vapor-mitigating capability within each group shows only a relationship to foam drainage. Within each class of foams the slower the foam drainage, the better the vapor hazard mitigation and the longer the duration of control without foam makeup.

New Foam Systems

Most work on foams for vapor mitigation has been carried out with commercial foams in use by fire services. These foams were developed for fire protection, not for vapor hazard control. On the basis of data which have been developed, new foam systems with better adaptability to vapor control have been devised.

Three foam systems with demonstrated superiority over commercial materials are now available. These are the Universal foam of National Foam System, Inc., Type L foam of Mine Safety Appliances Co. (MSA), and Light Water ATC of the Minnesota Mining and Manufacturing Company (3M). The first two materials are applicable to a wide range of volatile chemicals, are useful in both low and high expansion, and exhibit extended life. This latter property is based upon much earlier work for the containment of radioactive gases.[31] The 3M material is a fluorocarbon surfactant base (AFFF) which is suitable for polar compounds. Full capabilities of these three agents have not yet been documented in the literature.

Water-Soluble and Water-Reactive Materials

A significant number of the volatile chemicals on the various hazardous substances lists are water-soluble, and one approach to vapor suppression is through direct dilution. This approach presupposes that the water addition does not spread the spill. When such an effect is possible, foams can provide effective vapor control through a combination of dilution and vapor scrubbing.

Some volatile materials can be considered water-reactive, exhibiting a violent response to water additions. For other materials, the addition of water to either dilute or convert them to less hazardous substances may be the best approach to vapor hazard mitigation; however, this approach is usually precluded by potential exaggeration of the hazard.

Foams have been demonstrated to be a means of adding water to such materials in a sufficiently gentle manner that the desired end point is reached with a minimum of reaction violence. The data on this process currently available are limited to five volatile materials, but what is available provides a base to consider the use of foam with spills of like materials.

Foams have been successfully demonstrated in the control of spills of ammonia[13,30,32] and of sulfur trioxide (SO_3) and oleum,[33] and there are data to support applicability to silicon tetrachloride.[34,35] Foams have not shown benefits with anhydrous hydrogen fluoride,[13] while results with chlorine have been mixed.[13,36]

With SO_3 and oleum, the objective is to convert the SO_3 to a sulfuric acid solution with a minimum release of SO₃ vapor. A test program conducted by the principal manufacturers of sulfur trioxide shows that foam, principally high-expansion foam, is the best method to achieve the objective. Even with foam, the SO_3-water reaction is quite energetic, and an extensive vapor cloud is formed. This tends to be a sulfuric acid mist rather than SO_3 vapor. Sulfuric acid is itself a hazardous chemical but not of the same magnitude as SO_3. In addition, mists tend to coalesce and drop out of the atmosphere, reducing the magnitude of the affected zone downwind.

Ammonia follows a similar pattern. Dilution and conversion to ammonium hydroxide (NH_4OH) solution is a preferred approach. Vapor release becomes an NH_4OH mist rather than an NH_3 vapor. The initial work on ammonia[32] has been carried further by MSA[13] and National Foam [30] with the development of foams which do not collapse as readily as normal fire-fighting grades. They, therefore, provide effective vapor scrubbing along with dilution. A mechanism to gel foam in the presence of alkaline materials which has effectively controlled ammonia vapors in laboratory tests has been evolved;[37] however, no field-scale testing has been accomplished.

The use of foam on materials which decompose is restricted to silicon tetrachloride ($SiCl_4$), and evidence of beneficial effects is limited.[35] In an actual spill situation involving $SiCl_4$[34] foam was one of the control mechanisms attempted. Reaction of water with $SiCl_4$ results in the formation of SiO_2 and HCl, with significant amounts of the latter being released as mist and vapor. As with SO_3 and H_2SO_4, HCl is less toxic than $SiCl_4$, and such conversion can be beneficial. Visual observation, however, would indicate that foam provides a worse condition than uncontrolled boil-off from the $SiCl_4$. This illustrates the need to base spill responses on sound technical precepts rather than on visual or other uncertain observations.

Initial work on chlorine[36] showed foam to have a beneficial effect on vapor release, but subsequent work was not as positive. Chlorine causes accelerated collapse of foams, and initial applications may exaggerate the vapor hazard. The available evidence indicates that continued application will result in a measurable reduction in the vapor release rate.[13] If a better response mechanism is lacking, foam should be considered in the control of spilled chlorine. However, larger-scale and more extended test work is needed.

REFERENCES

1. U.S. Coast Guard Contract No. DOT-CG-51870-A, June 30, 1975.

2. Plastic Systems, Santa Ana, Calif.

3. T. E. Ctvrtnicek, C. M. Moscowitz, S. J. Rusek, and L. N. Cash, *Application of Foam Scrubbing to Fine Particle Control,* EPA-600/2-76-125, U.S. Environmental Protection Agency, May 1976.

4. T. E. Ctvrtnicek, C. M. Walburg, C. M. Moscowitz, and H. H. S. Yu, *Application of Foam Scrubbing to Fine Particle Control, Phase II,* EPA 68-02-1453, U.S. Environmental Protection Agency, December 1976.

5. V. I. Irzhak, "Statistical Theory of Gelatinization Sol Fraction," *Vysokomolekuliarnye Soedinennia,* Series A, vol. 17, no. 3, 1975, p. 535.

6. R. E. Baier, J. G. Michalovic, V. A. De Palma, and R. J. Pilie, *Universal Gelling Agent for the Control of Hazardous Liquid Spills,* reprint, Calspan Corporation, Buffalo, N.Y., August 1974.

7. R. J. Pilie, R. E. Baier, R. C. Ziegler, R. P. Leonard, J. G. Michalovic, S. L. Pele, and D. Bock, *Methods to Treat, Control and Monitor Spilled Hazardous Materials,* Contract No. 68-01-0110, EPA-670/2-75-042, U.S. Environmental Protection Agency, June 1975.

8. H. L. Fuller, "Use of Floating Absorbents and Gelling Techniques for Combating Oil Spills on Water," *Journal of the Institute of Petroleum,* vol. 57, no. 553, 1971, p. 35.

9. A. M. Goldstein, R. M. Koros, and B. L. Tarmy, "Engineering Study of an Oil Gelation Technique to Control Spills from Distressed Tankers," paper presented at the Fourth Joint Conference on Prevention and Control of Oil Spills, Washington, 1973.

10. W. W. Bannister, J. R. Pennace, H. H. Reynolds, and W. A. Corby, "Gelation of Oil by Amine Carbamates as a Means of Removal and Recovery," paper presented at the American Chemical Society (Southeastern Regional) Meeting, Norfolk, Va., October 1974.

11. M. O. Weaver, E. B. Bagley, G. E. Tanda, and W. M. Doane, "Gel Sheets Produced by Hydration of Films from the Potassium Salt of Hydrolyzed Starch—Polyacrylonitrile Graft Polymer," *Applied Polymer Symposia*, no. 25, 1974, pp. 97–105.

12. Rockwell International Corporation, Contract No. N8520025SP, Sept. 21, 1978.

13. U.S. Environmental Protection Agency, Contract No. 68-03-2478, Oct. 29, 1976.

14. G. Hedestrand, "Influence of Thin Surface Films on the Evaporation of Water," *Journal of Physical Chemistry*, vol. 28, 1924, p. 1245.

15. E. K. Rideal, "Influence of Thin Surface Films on the Evaporation of Water," *Journal of Physical Chemistry*, vol. 29, 1925, p. 1585.

16. I. Langmuir and D. C. Langmuir, "The Effect of Monomolecular Films on the Evaporation of Ether Solutions," *Journal of Physical Chemistry*, vol. 31, 1927, p. 1719.

17. V. K. La Mer, *Retardation of Evaporation by Monolayers*, Symposium on Transport Properties, 1960, Academic Press, Inc., New York, 1962.

18. R. L. Tuve and E. J. Jablonski, *Compositions and Methods for Fire Extinguishment and Prevention of Flammable Vapor Release*, U.S. Patent No. 3,258,423, June 1966.

19. M. K. Bernett, L. W. Halper, N. L. Jarvis, and T. M. Thomas, "Effect of Adsorbed Monomolecular Films on the Evaporation of Volatile Organic Liquids," *Industrial and Engineering Chemistry Fundamentals*, vol. 9, 1970, p. 150.

20. H. E. Moran, J. C. Bernett, and J. T. Leonard, *Suppression of Fuel Evaporation by Aqueous Films of Fluorocarbon Surfactant Solutions*, National Research Laboratory Report No. 7247 AD 723 189, April 1971.

21. F. Sebba and E. K. Rideal, "Permeability in Monolayers," *Transactions of the Faraday Society*, vol. 37, 1941, p. 273.

22. R. J. Archer and V. K. La Mer, "The Rate of Evaporation of Water through Fatty Acid Monolayers," *Journal of Physical Chemistry*, vol. 59, 1955, p. 200.

23. H. Rosano and V. K. La Mer, "The Rate of Evaporation of Water through Monolayers of Esters, Acids and Alcohols," *Journal of Physical Chemistry*, vol. 60, 1956, p. 348.

24. R. L. Tuve, H. B. Peterson, E. J. Jablonski, and R. R. Neill, *A New Vapor-Securing Agent for Flammable-Liquid Fire Extinguishment*, Naval Research Laboratory Report No. 6057, March 1964.

25. *Foam Liquid, Fire Extinguishing, Mechanical*, Federal Specification No. O-F-555C, Jan. 3, 1969.

26. *Standard for Air Foam Equipment and Liquid Concentrates*, UL 162, Underwriters Laboratories, Northbrook, Ill., June 26, 1975.

27. *Tanker Tank Cleaning Research Program, Phase I*, Report No. MARD 900-7401-7, U.S. Department of Commerce, Maritime Administration, March 1974.

28. R. H. Hiltz and J. V. Friel, "Application of Foams to the Control of Hazardous Chemical Spills," in G. F. Bennett (ed.), *Proceedings of the 1976 National Conference on Control of Hazardous Material Spills*, New Orleans, April 1976, pp. 293–302.

29. S. S. Gross, "Evaluation of Foams Mitigating Air Pollution from Hazardous Material Spills," in G. F. Bennett (ed.), *Proceedings of the 1978 National Conference on Control of Hazardous Material Spills*, Miami Beach, Fla., April 1978, pp. 394–398.

30. E. C. Norman and H. A. Dowell, "The Use of Foams to Control Vapor Emissions from Hazardous Material Spills," in G. F. Bennett (ed.), *Proceedings of the 1978 National Conference on Control of Hazardous Material Spills*, Miami Beach, Fla., April 1978, pp. 399–405.

31. J. E. Mecca and J. D. Ludwick, *The Development of a Long Lived High Expansion Foam for Entrapping Air Bearing Noble Gas,* Douglas United Nuclear Report No. DUN-SA-41, August 1970.

32. W. D. Clark, "Use of Fire Fighting Foam on Ammonia Spills," paper presented at the Eightieth National Meeting of the American Institute of Chemical Engineers, Boston, September 1975.

33. J. R. Pfann and C.A. Sumner, "Sulfur Trioxide Spill Control," in G. F. Bennett (ed.), *Proceedings of the 1976 National Conference on Control of Hazardous Material Spills,* New Orleans, April 1976, pp. 192–195.

34. T. R. Hampson, "Chemical Leak at a Bulk Terminal Tank Farm," in G. F. Bennett (ed.), *Proceedings of the 1976 National Conference on Control of Hazardous Material Spills,* New Orleans, April 1976, pp. 214–218.

35. W. C. Hoyle and G. L. Melvin, "A Toxic Substance Leak in Retrospect: Prevention and Response," in G. F. Bennett (ed.), *Proceedings of the 1976 National Conference on Control of Hazardous Material Spills,* New Orleans, April 1976, pp. 187–191.

36. C. H. Buschman, "Experiments on the Dispersion of Heavy Gases and Abatement of Chlorine Clouds," paper presented at the Fourth International Symposium on Transport of Hazardous Cargoes by Sea and Inland Waterways, National Academy of Sciences, 1975.

37. J. V. Friel, R. H. Hiltz, and M. D. Marshall, *Control of Hazardous Chemical Spills by Physical Barriers,* EPA Report No. R2-73-185, U.S. Environmental Protection Agency, March 1973.

PART 4
Ammonia

Phani Raj
President, Technology and Management Systems

Introduction / 10-34
Properties of Anhydrous Ammonia / 10-35
Production, Storage, and Transportation / 10-39
Behavior of Releases in the Environment / 10-41
Nomenclature / 10-49
Conclusion / 10-50

INTRODUCTION

Anhydrous ammonia is a compound of nitrogen and hydrogen (NH_3) which is used in a variety of industrial processes and in agriculture. At normal temperatures and atmospheric pressure, it is a gas, but it can be liquefied by application of modest pressures or by cooling to low temperatures (240 K). Ammonia is produced by the catalytic reaction of hydrogen and nitrogen at high temperature and high pressure. In most ammonia-producing facilities in the United States, natural gas (methane) forms the feedstock to produce the hydrogen. The nitrogen for the process is obtained from the atmosphere. Ammonia is also produced in small quantities as a by-product in the production of coke.

Ammonia ranks very high in the worldwide use of chemicals. Modern agriculture is heavily dependent on nitrogenous fertilizers, and ammonia is the basic product in their manufacture. In 1968 approximately 89 percent of the United States production of ammonia was utilized in the manufacture of fertilizers such as ammonium phosphate, ammonium nitrate, urea, and ammonium sulfate.[1] Recent statistics indicate that an increased proportion is going to industrial uses, with a concomitant decline in agricultural use (now at 75 percent). Ammonia is also applied directly to the soil in gaseous form or as a solution in water. In the United States, there is an increasing use of ammonia as a fertilizer for direct application under the soil surface. Because of the toxic nature of ammonia vapor and the handling of pressurized liquid that this application procedure involves, there is a potential health hazard.

The other uses of ammonia are found in the manufacture of explosives, dyes, and artificial fibers, as a chemical reagent (e.g., in the formation of amines and ammonium compounds), for nitriding special steels, for preventing corrosion in oil refineries, for preventing flue-duct, economizer, and air-heater corrosion in boiler plants, for preventing acid smut emission from oil-fired boilers, as a preservative in rubber latex, in high-speed diazo printing, in metallurgical processes such as bright annealing, for flameproofing textiles, for water sterilization, for neutralizing acid effluents, and as a nonaqueous ionizing solvent for many chemical compounds.[2] Ammonia is also used in some refrigeration systems. Currently ammonia is the prime contender for the working fluid in the Rankine cycle to be utilized for extracting ocean thermal energy.[3] A potential large-scale use of ammonia is in power plants for reducing the emission of the oxides of nitrogen from the stack.[4]

While the benefits from the uses of ammonia are significant, improper handling or accidental releases of liquid or vapors pose potential public health hazards because of the toxic nature of the material. Human health can be affected by either acute exposure to high concentrations in accidental releases or chronic exposure to low concentrations as in the workplace or as an air pollutant.

Ammonia is an irritant that most commonly affects the skin, eyes, mucous membrane, upper respiratory tract, and lungs. When ingested, it has corrosive effects on the mouth, esophagus, and stomach.[5] The level of effects and the degree of damage suffered by human beings depend on a number of factors including ammonia concentration, duration of exposure, and type of exposure (external contact versus respiration or ingestion). Therefore, understanding the properties of ammonia, its proper handling procedures, and the nature of the dispersion of vapors in the atmosphere is essential to prevent or minimize ammonia hazards.

The objective of this part is to provide basic facts about the properties of ammonia and its behavior in the environment when released accidentally and to indicate data on production and transportation, modes of shipping, and magnitude of problems caused by accidental spills. No attempt is made to describe in detail the technology and economics of ammonia production, storage and handling, or transportation.

PROPERTIES OF ANHYDROUS AMMONIA

Physical Properties

Anhydrous ammonia is a colorless gas with a pungent odor. Its important thermodynamic properties are listed in Table 4-1. A comprehensive nomograph of the physical properties of saturated ammonia is shown in Fig. 4-1. Gaseous ammonia at normal temperature and at atmospheric pressure can be liquefied either by increasing the pressure or by cooling to 239.8 K at atmospheric pressure. Empirical equations have been developed for the thermodynamic and transport properties of anhydrous ammonia for liquid and vapor at both saturated and superheated conditions.[3]

Ammonia is highly soluble in water and forms a strongly alkaline solution. Its solubility is limited by the liberation of heat during the dissolution, which causes boiling and prevents further dissolution. Removal of heat solution by a cooling surface permits the formation of a solution up to 35% (by weight) at atmospheric pressure. The enthalpy-concentration equilibrium diagram for ammonia-water mixtures is shown in Fig. 4-2.

Potentially Hazardous Properties

Ammonia forms a combustible mixture with air when the concentration is in the range of 15.8 to 25.7% by volume and with oxygen in the range of 14.3 to 79%.[6] Combustible ammonia-air mixtures also explode under partial or complete confinement. The magnitude of pressure rise observed for ammonia is approximately three-quarters of that of common hydrocarbon fuels under similar conditions.[6] There are many documented cases of burning and/or explosion of ammonia. In one case, excess ammonia vapor released on top of a 30,000-ton storage tank is reported to have burned. The fire caused extensive damage to the insulation on the top of the tank and melted the aluminum valve bodies.[7] It is also reported that ammonia vapor ignitions have been experienced during the welding

TABLE 4-1 Thermodynamic Properties of Anhydrous Ammonia

Molecular symbol	NH_3
Molecular weight	17.032 kg/kmol
Boiling temperature at atmospheric pressure*	239.8 K
Melting point at atmospheric pressure	195.4 K
Critical temperature	406.2 K
Critical pressure	114.25×10^5 N/m^2
Liquid density at 239.8 K and atmospheric pressure	682.2 kg/m^3
Vapor density at 239.8 K and atmospheric pressure	0.89 kg/m^3
Specific heat of vapor at 288 K:	
At constant pressure (C_p)	2190.1 J/kg K
At constant volume (C_v)	1672.3 J/kg K
Heat of vaporization at 239.8 K, 1 atmosphere	1.3707×10^6 J/kg
Ignition temperature (in a quartz container)	1123.2 K
Flammable limits in air by volume	16 to 25 percent
Thermal conductivity† at 273.15 K of:	
Saturated liquid	0.502 W/m·K
Saturated vapor	0.022 W/m·K
Viscosity† at 258 K of:	
Saturated liquid	2.5×10^{-4} N s/m^2
Saturated vapor	8.5×10^{-6} N s/m^2

*Physical properties based on data in *Safety Requirements for the Storage and Handling of Anhydrous Ammonia*, ANSI K61.1-1972, American National Standards Institute, New York, 1972.

†*Thermodynamic Properties of Refrigerants*, American Society of Heating, Refrigerating, and Air-Conditioning Engineers, Inc., New York, 1969.

of pipes in ammonia plants. Unfortunately, notwithstanding these experiences there is a widespread misconception that ammonia is nonflammable. The U.S. Department of Transportation (DOT) and the U.S. Coast Guard have classified ammonia as a nonflammable gas for the purpose of transportation.

Ammonia forms explosive compounds with mercury, the halogens, and hypochlorites. It explodes when mixed with iodine and produces violently explosive and extremely sensitive compounds when contacted by any of the azides.[8] Moist ammonia attacks copper and zinc. These metals or their alloys, such as brass and bronze, should never be used in equipment to handle ammonia. Steel is normally resistant to such corrosion.

Physiological Properties

The deleterious effect of ammonia on human beings is caused by direct external contact and by ingestion. Low concentrations of ammonia in air cause irritation of the nose, throat, and eyes. Generally, the human nose can detect ammonia in the range of 10 to 20 ppm.[2] A high degree of correlation is said to exist between the ammonia concentration just outside a plant boundary and telephone complaints from near-plant residents.[9] On the basis of this correlation, the smell threshold is estimated to be 5 ppm. Low concentrations of ammonia, even at the limits of detectable odor, are irritating to the skin. Contact of cryogenic liquid anhydrous ammonia produces freeze burns, and skin exposure to liquid or gas under pressure results in second-degree burns with the formation of blisters. Exposure of skin to 10,000-ppm vapor concentration causes mild irritation, to 20,000-ppm concentration causes increased irritation, and to 30,000-ppm concentration produces blisters in a few minutes.[5]

Ammonia in high concentrations is toxic to both human beings and animals. The metabolic toxicity of ammonia in man, the toxicity symptoms, and the research to date on this subject are treated exhaustively in a 1979 publication.[5] At low concentrations, acute

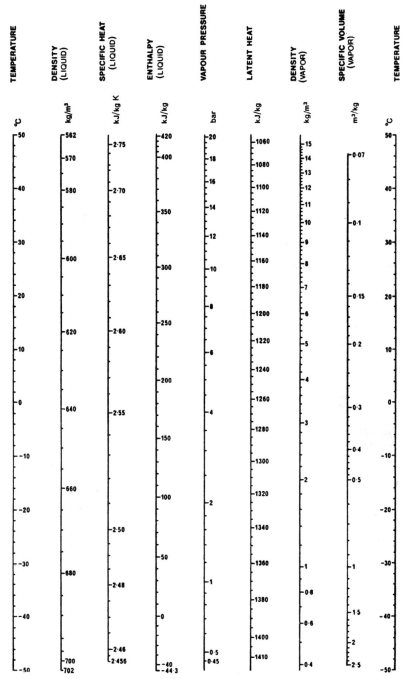

Fig. 4-1 Nomograph indicating the thermodynamic properties of anhydrous ammonia. [*Reprinted with permission from ICI.*]

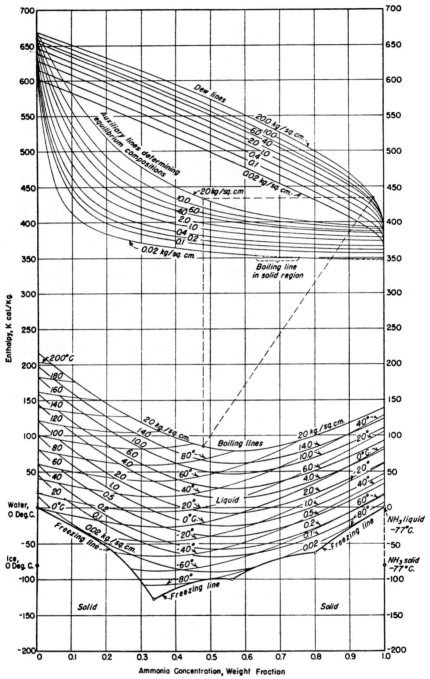

Fig. 4-2 Enthalpy-concentration diagram for ammonia-water mixtures. [*Reprinted from R. H. Perry and C. H. Chilton,* The Chemical Engineers' Handbook, *with permission from McGraw-Hill, Inc.*]

TABLE 4-2   Physiologic Response to Ammonia Concentrations in Air

| Physiologic response | Concentration equal to or greater than | | Remarks |
	In ppm	In mg/m³	
Odor detectable	0.7	0.5	Most sensitive people
	5.0	3.5	Average threshold value
	20.0	14.0	Complaint level
	50.0	35.0	Least sensitive people
Immediate throat irritation	400	280	
Irritation of eyes	700	490	
Coughing	1700	1200	
Threat to life after 30-min exposure	2400	1700	
Rapidly fatal for short exposure	5000–10,000	3500–7000	

SOURCE: *Ammonia*, by the Subcommittee on Ammonia, Committee on Medical and Biologic Effects of Environmental Pollutants, of the National Research Council. Copyright © 1979 by University Park Press, Baltimore, Md. Reprinted with permission.

exposures can cause mild irritation of the respiratory tract. High concentrations produce severe symptoms involving laryngeal spasm, bronchopneumonia, and fatality. The physiologic response of human beings to various ammonia concentrations in air is shown in Table 4-2. Chronic exposure to low levels of ammonia concentrations (50 ppm), as in a work environment, results in upper-respiratory-tract diseases. A few inconclusive cases of ammonia exposure causing malignancy in the lung and urinary tract and gastric and lymphatic neoplasia have also been reported in persons exposed to 2 to 3 times the minimum allowable concentration of 35 ppm or 25 mg/m³.[5]

Exposure Standards

The current United States federal standard for exposure to ammonia in an 8-h time-weighted average (TWA) is 50 ppm or 35 mg/m³. The American Conference of Governmental Industrial Hygienists (ACGIH)[10] has recommended a TWA of 25 ppm (18 mg/m³). More recently the U.S. Department of Labor published a proposed rule[11] that would limit the ceiling concentration of exposure to ammonia in work environments to 50 ppm for any 5 min. Generally, an average person is extremely uncomfortable in an atmosphere of ammonia above 25 ppm. The U.S. Navy set 25 ppm as the limit for continuous exposure and 400 ppm as the maximum concentration for 1 h in a submarine.[12] The standards in several industrialized countries differ, varying from a low of 30 ppm in the U.S.S.R., Hungary, and Poland to a high of 100 ppm in Yugoslavia and England for maximum allowable concentration.[5]

PRODUCTION, STORAGE, AND TRANSPORTATION

Manufacturing Process and Capacity

Large-scale commercial production of anhydrous ammonia is based on synthesizing atmospheric nitrogen and hydrogen from a hydrocarbon. The process involves four major steps, including (1) the reaction of water (steam) and hydrocarbon, producing a "synthesis gas," a mixture of hydrogen and carbon monoxide (CO); (2) "shift conversion," in which the hydrogen fraction is enriched by the additional reaction of CO and steam; (3) the

TABLE 4-3   Ammonia Production (Millions of Metric Tons)

	1955	1965	1975	1977	1978	1979	1980
World production*	11.8	28.0	64.0	76.6	81.9	90.0‡	N.A.
United States production†	2.9	8.1	14.9	16.1	15.6	16.4	15.9–16.1‡
United States imports†	....	0.2	0.7	1.0	1.4	1.8	2.3–3.2‡
United States exports†	....	0.17	0.17	0.30	0.47	0.71	N.A.

*Based on data from O. J. Quartulli and L. J. Buividas, "Some Current and Future Trends in Ammonia Production Technology," *Nitrogen*, March–April 1976, p. 6.
†Based on data from *Chemical Products Synopsis*, reporting service of Mansville Chemical Products, Cortland, N. Y., March 1980.
‡Projected estimate.
NOTE: N. A. = not available.

purification of hydrogen; and (4) the final synthesis of nitrogen and hydrogen in the presence of a catalyst. Modern plants employ single-train units, and the largest ones produce up to 1700 metric tons per day. The history of the ammonia process, the chemical reactions, and the process design are discussed by Slack and James,[1] while the current technology and the operating experience are enumerated by Turner.[13] Quartulli and Buividas[14] discuss the trends in the industry and analyze the outlook for the future.

Ammonia production statistics, both worldwide and in the United States, are given in Table 4-3. Worldwide production of ammonia almost tripled between 1963 and 1976, and United States production doubled in the same period. United States demand is rising at about 5 percent per year. At present, the United States is suffering from underutilization of capacity.[15] This may be due to competition from low-cost imports and the rising cost of natural gas. The current at-plant price is below $100 per metric ton in the United States, whereas import and spot prices may be lower.[15]

Storage and Handling

Large quantities of ammonia are stored in three different ways: (1) in pressurized hortonspheres, (2) as aqua ammonia in low-pressure storage, and (3) as a refrigerated liquid at atmospheric pressure. Receiving tanks at transportation and import terminals have up to 27,000-metric-ton storage capacity, while dealers' tanks may have as much as 80-metric-ton capacity. Detailed specifications, standards, and safety requirements for the storage and handling of anhydrous ammonia have been published by the American National Standards Institute (ANSI)[16] in the United States and by the Chemical Industries Association, Ltd. (CIA)[17] in the United Kingdom. The ANSI code deals with refrigerated storage, portable containers, and systems mounted on trucks, trailers, and farm wagons. The CIA code deals with similar issues for large-scale storage and, in addition, specifies criteria for inspection and maintenance, employee training and safety, emergency planning, etc. The U.S. Department of Labor[18] has also established a code for storage and handling of ammonia. The eight sections of this code specify requirements for the safe handling and movement of liquid and gaseous ammonia. The criteria for siting multiple storage facilities have been prescribed by the Department of the Environment, Alberta.[19]

Recommended distances between storage tanks and other public areas are shown in Table 4-4. A recent publication by Reed[20] deals with current practices in tank storage, construction materials, types of tanks, economics, safety precautions, and emergency spill plans related to the handling, storage, and shipping of ammonia. An operational safety manual published by the Fertilizer Institute[21] and a publication by Imperial Chemical Industries, Ltd.,[22] are also useful references.

Transportation

Ammonia is transported in bulk by road, rail, and sea and is pumped through pipelines. In road and rail transport, it is carried as a liquid at or about ambient temperature but is pressurized. In barge and ship transport, it is normally refrigerated and is at atmospheric pressure. Unit shipments vary from about 10 metric tons in road tankers, 25 to 80 metric tons in rail tank cars, and 2700 metric tons in barges to as high as 20,000 metric tons in specially constructed oceangoing ships. In the United States the marine transport of anhydrous ammonia is regulated by the U.S. Coast Guard[23] and road and rail transport by the DOT.[24] In recent years in the United States approximately 250,000 metric tons have been transported annually on rails and 1.5 million metric tons by barge (principally on the Mississippi and Ohio rivers). Among the principal ports at which imported ammonia is off-loaded from ships are Tampa, Savannah, and New Orleans.

There are two major ammonia pipelines in the United States, and a 2500-km pipeline is to be constructed in the U.S.S.R.[14] In the United States the Gulf Central Pipeline runs north from New Orleans and branches at St. Louis. The east-running line terminates in Indiana, while the western leg ends in Nebraska. The second major line runs between the Texas Panhandle and Iowa. Annually, these two pipelines carry approximately 1 million metric tons and 400,000 metric tons respectively. The pipeline pressure varies between 21.7 and 100 bar (between 2170 and 10,000 kPa), and the ammonia temperature ranges from 275 to 300 K.[25]

BEHAVIOR OF RELEASES IN THE ENVIRONMENT

Principal Causes of Releases

Accidental releases of ammonia into the environment are likely to occur at any of the stages of production, storage and handling, transportation, and end use. The releases may be very small, as in the case of a leaking pipe gland, or catastrophic, as from an accidentally damaged rail tank car. Other types of accidental releases may arise from (1) overfilling storage tanks during loading, (2) structural failure of storage tanks, (3) venting of

TABLE 4-4   Minimum Recommended Distances between Storage Facilities and Public Structures

Range of nominal capacity of ammonia storage container (m^3)	Minimum distance (m) from storage containers to:		
	Line of adjoining property which may be built upon, highways, and mainline railroad	Place of public assembly	Permanently occupied residential buildings
2–7.5*	8	45	75
7.5–75*	15	90	150
75–115†	..	...	230
115–380	15	140	230*
Over 380	15	185	305,* 380†

*Data converted from *Safety Requirements for the Storage and Handling of Anhydrous Ammonia*, ANSI K61.1-1972, American National Standards Institute, New York, 1972.

†Data converted from *Guidelines for the Location of Stationary Bulk Ammonia Storage Facilities, Standards and Approvals Division*, Department of the Environment, Edmonton, Alberta, 1977, p. 9.

excess vapor, (4) transportation accidents such as derailed and punctured rail tankers or overturned road trucks, (5) failure of loading or unloading pipelines at docks, (6) damage to interstate pipelines, and (7) rollover in storage tanks.

The phenomenon of rollover is caused in atmospheric-pressure storage tanks because of a very small thermal gradient established owing to heat leaks. When a warm liquid, at a certain depth, with its vapor pressure lower than the local hydrostatic pressure moves up (owing to buoyancy) to a higher level where its vapor pressure exceeds or equals the hydrostatic head, the liquid flashes, releasing very large quantities of vapor in a very short time (several minutes). In such cases, the excess vapor would be vented at enormous rates by the customary pressure relief system on storage tanks.

While the rollover phenomenon is unlikely to occur in pure-ammonia storage tanks, the addition of impurities such as water results in a change in the temperature and density gradients needed to impart free convective circulation. In addition, if concentration gradients are also created by imperfect mixing or by stratification, the potential for rollover increases. Unstable conditions of this nature are expected to result when the moisture content approaches 1 mol percent.[26] To reduce stress corrosion cracking about 0.2 mol percent water is generally added to the storage anhydrous ammonia. Germeles[27] and Drake[28] have discussed the rollover phenomenon in detail with regard to other energy liquids.

Cases of Accidental Releases

During the period 1971–1975 a total of 239 incidents involving anhydrous ammonia transportation accidents were reported to the Office of Hazardous Materials Operation of the DOT.[5] Injury and death related to handling or transportation of anhydrous ammonia are reported to have occurred in 61 incidents from 1971 to April 1977.

A number of small-scale incidents in which careless handling resulted in serious injuries and fatalities have been reviewed by Kamin.[5] These have included spills during the unloading of a tractor trailer, bursting of hoses, tank truck unloading, and the transport of air-conditioning equipment containing ammonia. There have also been accidents involving large-scale releases which have resulted in injury to and evacuation of people unconnected with the ammonia operations. Some of the major accidents are indicated in Table 4-5. These accidents have involved releases of ammonia from both storage tanks and transports (including pipelines). Pressurized and refrigerated ammonia have been released. Except for one incident, the safety record in water transport of anhydrous ammonia has been exceptionally good.

Dispersion in the Environment

The behavior of ammonia in the environment depends on the type of release, the location, and the atmospheric and environmental conditions. Ammonia released over land from a pressurized container behaves quite differently from cryogenic ammonia released onto water. The vapors generated from these spills disperse quite differently in the atmosphere.

Pressurized-Ammonia Releases

When pressurized liquid ammonia is released into the atmosphere, a certain fraction flashes. This mass fraction of flash (f) can be calculated by

$$f = \frac{h_L(P_T) - h_L(P_a)}{h_v(P_a) - h_L(P_a)} \tag{1}$$

TABLE 4-5 Some Major Ammonia Accidents and Releases

Place, time, date, and type of release	Cause of release	Mass of released ammonia (metric tons)	Release duration (minutes)	Container gauge pressure before accident (bars g)	Atmospheric temperature (K)	Weather conditions	Result of the accident; extent of dispersion and other observations	Reference number
Houston, Texas 11:15 A.M. May 11, 1976 (P)	Road accident: tanker crash, tank burst open	19	Almost instantaneous		300	Light wind, bright sun	30-m-high cloud formed; slow dispersion; danger persisting for 2½ h; 5 deaths, 178 injuries	29
Potchefstroom, South Africa 4:15 P.M. July 13, 1973 (P)	Spontaneous structural failure of dished end of horizontal bullet-type tank	38	Instantaneous	6.2	292	Initially still air; a few minutes later, a slight breeze	Initial cloud 150 m in diameter, 20 m deep; later 300 m wide at 450 m downwind; 18 deaths, 65 injured	30
Pensacola, Florida 6:06 P.M. Nov. 19, 1977 (P)	Train derailment resulting in 8-cm by 1-m tear in tank head of the car	~40	About half of the contents quickly released		293	Overcast and light rain; wind about 1.5 m/s	Cloud that seemed to have dispersed at ground level for many kilometers; also tracked on radar	31
Glen Ellyn, Illinois May 1976 (P)	Railcar rupture due to collision	51.5	Almost instantaneous				3000 residents evacuated for 16 h; no deaths, 15 nonserious injuries	29
Crete, Nebraska 6:30 A.M. Feb. 18, 1969 (P)	Train derailment, collision, and tank car rupture	76	Instantaneous	1.4	258	Temperature inversion; practically no wind	Cloud that spread at ground, blanketing a wide area	31
Blair, Nebraska 5:45 A.M. Nov. 16, 1970 (R)	Overfilling of a 40,000 metric-ton refrigerated tank	160	2.5 h	0		Stable atmosphere initially; light wind later	Spill in remote area; no casualties; covered 2.7 km downwind and 3.6 km²	32

TABLE 4-5 Some Major Ammonia Accidents and Releases (*Continued*)

Place, time, date, and type of release	Cause of release	Mass of released ammonia (metric tons)	Release duration (minutes)	Container gauge pressure before accident (bars g)	Atmospheric temperature (K)	Weather conditions	Result of the accident; extent of dispersion and other observations	Reference number
Conway, Kansas 4:30 A.M. Dec. 6, 1973 (P)	Rupture of overpressurized pipeline	230	>0.5 h	83	267	4.5-m/s wind, clear sky, stable atmosphere	Foglike vapor visible for 800 m; irritating to nose and eyes for 6 km; two people hospitalized; no fatalities	31
Hutchinson, Kansas 1:00 P.M. Aug. 13, 1976 (P)	Rupture of overpressurized pipeline	360	Few hours	100		1.8-m/s wind	200 people evacuated; several treated in hospital; no fatalities; photographs indicating dispersion of vapor close to ground level	31
Kekur Peninsula Alaska October 1974 (R)	Sinking of a barge	8160	Unknown	0	260	Stormy weather	Escape of entire cargo to marine environment and atmosphere; no humans exposed; 2.6 km² of nearby forest laid waste by NH₃ fumes	5

NOTE: (P) = pressurized ammonia release; (R) = refrigerated ammonia release.

where the h's represent the enthalpy at saturated condition of the liquid L and vapor v at the pressures indicated, tank T and ambient a. The percentage of liquid flashing when released from different tank pressures is shown in Fig. 4-3. In the case of liquid ammonia stored at ambient temperature (293 K) and released to ambient pressure, about 18 percent of the mass flashes. The remaining liquid may drain onto the ground (in the case of small leaks) or be released in the form of aerosols.

The evaporation of a liquid pool from the ground depends on the soil thermal properties, soil temperature, and rate of heat transfer from the air to the pool by convection. Experiments with 1-metric-ton and 10-metric-ton LNH_3 spill pools indicate a measured maximum liquid regression rate of 1.35×10^{-5} m/s [$\simeq 0.01$ kg/(m^2·s)] and an apparent evaporative loss rate of 7.35×10^{-6} m/s over a long time.[26] A fivefold increase in wind speed over the pool from about 2.5 m/s to 12 m/s increases the steady-state evaporative loss rate by a factor of about 1.5.

It is speculated that in many large accidental releases from pressurized containers the liquid released after the flash is in the form of fine aerosols. Eyewitness accounts and photographic records of accidental releases indicate the formation of a fog which disperses close to the ground. The fog is visible owing to the presence of ammonia aerosols and the condensed atmospheric moisture. The aerosols in the cloud would react with the moisture in the atmosphere, producing heat and aqua ammonia aerosols. The rate of evaporation of these aerosols is a function of dilution of the vapor cloud by air as well as of the relative humidity of the atmosphere.

Controlled experiments involving the release and dispersion of pressurized ammonia (0.1 metric ton to 1 metric ton) have been conducted by Resplandy.[33] Theoretical analysis of the dispersion phenomenon of vapor-containing aerosols has been discussed by Kaiser and Walker.[34] More recently, a detailed analysis[35] has been made by using a numerical model to account for the reaction between ammonia aerosols and the atmospheric moisture and its effects on cloud density variation.

Basically there are three different stages of dispersion of a heavier-than-air ammonia vapor cloud containing aerosols. Let us consider the case of continuous release of such a vapor cloud. Initially, because of its higher-than-air density, the cloud spreads laterally,

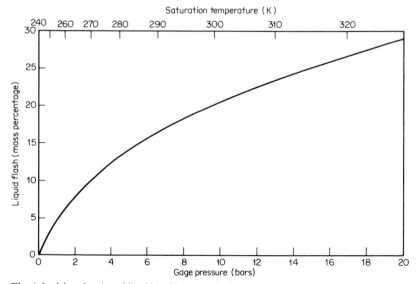

Fig. 4-3   Mass fraction of liquid flashing versus initial pressure.

owing to gravity, as it travels downwind. The lateral spread velocity in this initial regime is given by

$$\frac{dy}{dt} = \sqrt{\left(\frac{\rho - \rho_a}{\rho}\right)gH} \tag{2}$$

where y is the lateral extent of cloud, ρ is the combined vapor-aerosol density, and H is the cloud height. This lateral spread results in the entrainment of air into the cloud and its dilution. The aerosols react with moisture in the air, evaporate partly, and form ammonium hydroxide droplets. Because of cloud dilution, the density decreases, and hence the tendency to gravity spread decreases. In the second stage of dispersion, mixing of air and vapor is affected both by gravitational spreading and by atmospheric turbulence. Several criteria for the termination of gravity spread are discussed by Kaiser and Walker.[34] These criteria include (1) equating the turbulent energy density and the mean potential energy density of the cloud, (2) terminating gravity spread when the fractional density deviation is less than 0.1 percent, and (3) equating the rate of spread of cloud for atmospheric turbulence to the rate of cloud growth by gravitational spread.

The third and final stage of dispersion (far-field dispersion) is dominated by atmospheric turbulence. The concentration of vapor in the cloud varies both laterally and vertically and depends on the stability of the atmosphere. Exhaustive discussion and details of models used in the calculation of atmospheric dispersion of pollutant vapors are given by Slade.[36]

It can be safely assumed that by the time that the dispersion is dominated by atmospheric turbulence alone, the ammonia cloud is so dilute that no ammonia aerosols will be present. It is not known, however, whether heating from the ground and solar heating will cause this vapor to rise in the atmosphere. Conservative calculations should be based on the assumption that all the vapor disperses essentially close to ground level. Reasonable accuracy of ammonia vapor concentrations can be obtained by using the equations based on gaussian models as given by Slade.[36] These are, for continuous release,

$$c(x,y,z) = \frac{\dot{M}_c}{\pi U \sigma_y \sigma_z} e^{-y^2/2\sigma_y^2} \left[\frac{e^{-(x-H)^2/2\sigma_y^2} + e^{-(z+H)^2/2\sigma_z^2}}{2}\right] \tag{3}$$

and for instantaneous release,

$$c(x,y,z,t) = \frac{M_i}{(2\pi)^{3/2}\sigma_y^2\sigma_z} e^{-((x-Ut)^2 + y^2)/2\sigma_y^2} \left[e^{-(z-H)^2/2\sigma_z^2} + e^{-(z+H)^2/2\sigma_z^2}\right] \tag{4}$$

where c is the time-averaged vapor concentration, M_i and $\dot{M}_c$ are respectively the total mass and the mass rate of release of ammonia, and H is the height of cloud center above ground at the downwind location x; σ_y and σ_z are respectively the crosswind and vertical dispersion coefficients which are dependent on the stability of the atmosphere and on the downwind distance. The correlations between σ_y, σ_z, and x are given by Slade.[36] H is generally taken as zero (ground level) for pressurized-ammonia releases.

Cryogenic Ammonia Behavior

Anhydrous ammonia reacts with water, releasing heat and resulting in the formation of ammonium hydroxide solution (NH_4OH). Furthermore, ammonia dissolves in water in all proportions. Because of these properties, a spill of liquid anhydrous ammonia into water is expected to behave quite differently from a spill on land. Scenarios of cryogenic ammonia spills into water are of interest because of large-scale transportation of refrigerated ammonia in ships and barges.

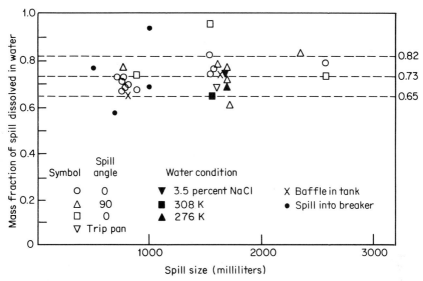

Fig. 4-4 Mass fraction dissolved in water versus spill size for laboratory ammonia spill experiments.

A laboratory investigation of the behavior of liquid (cryogenic) ammonia (LNH_3) on water was conducted by Lima and Reid.[37] They concluded that the process of dissolution was violent and very quick. Nominally, about 50 percent of the liquid spilled dissolved; the remainder was released as vapor into the atmosphere.

A three-stage experimental investigation of the release of LNH_3 on water has been conducted by Arthur D. Little, Inc.[38] The quantities of LNH_3 released varied from 10^{-3} m³ in the laboratory to 0.19 m³ (50 gal) in a lake. Small-scale tests indicated that regardless of conditions existing at the time of the spill, approximately 27 percent of the liquid was vaporized. The test results are shown in Fig. 4-4. A thermodynamic model to explain these results has been developed by Raj and Reid.[39]

In larger spills, approximately 50 percent of the spilled liquid vaporized, indicating the effects of the dynamics of a spill on vaporization. The NH_4OH solution formed is warm and disperses in the water as a stably stratified fluid. The water dispersion is discussed by Raj et al.[38]

The experimental results for the maximum boiling zone radius $R_e(m)$ and duration of boiling $t_e(s)$ were correlated with volume $V(m^3)$ of LNH_3 spilled instantaneously by

$$R_e = 6.17 \ V^{0.375} \tag{5}$$

and

$$t_e = 14.76 \ V^{0.25} \tag{6}$$

The validity of these correlations for LNH_3 spill volumes greater than 0.2 m³ is unproved.

The experiments reported above showed that the vapor cloud produced by the spill of LNH_3 on water rises and disperses in the atmosphere as a buoyant gas. The rising vapor puff generated by an instantaneous spill of 0.19 m³ of LNH_3 onto the water surface is shown in Fig. 4-5. The rate of rise of the vapor cloud center depends on the wind speed

Fig. 4-5 Rising vapor cloud generated by the spill of 0.19 m³ of LNH_3 onto water under a low-wind-speed condition. [*Reprinted with permission from the Department of Transportation, U.S. Coast Guard.*]

and the stability of the atmosphere. For most spills, the height of the cloud center with downwind distance can be calculated by

$$\frac{H}{l} \sim 3.43 \left(\frac{x}{l}\right)^{2/3} \tag{7}$$

where x is the downwind distance from the spill center, H the height of the cloud center above the water surface, and l the buoyancy length, which in turn is calculated:

$$l = \frac{Vg\Delta_v}{\pi t_e} \tag{8}$$

where V is the volume of LNH_3 spilled, g is the acceleration due to gravity, Δ_v is the fractional density defect of saturated ammonia vapor with respect to air, and t_e is the boiling time.

In a *stable atmosphere*, the center of the cloud reaches a maximum height given by

$$H_\infty = l \, S^{2/3} \tag{9}$$

The stratification parameter S is related to the potential temperature gradient $\left(\dfrac{\partial \theta}{\partial z}\right)$ in the atmosphere and wind speed U and atmospheric temperature T by

$$S = \frac{U}{l \sqrt{\dfrac{g}{T}\dfrac{\partial \theta}{\partial z}}} \tag{10}$$

The vapor concentration at ground level at any downwind distance can be obtained by using Eq. (4), in which the value of H obtained from Eq. (7) is substituted. The above equations and results have been used to predict the ammonia hazard distances from potential LNH_3 spills of different quantities from barges onto the water.[38] The results indicate that the hazard distance varies with the quantity spilled and the atmospheric condition. For a 100 metric-ton (150-m³) spill with a stable atmosphere and about 2 m/s wind, the downwind distance to 1000-ppm ammonia concentration at ground level is about 2.2 km, whereas for a 3000-metric-ton (4500-m³) spill under the same conditions the distance is about 10 km. These calculations do not account for the dilution of vapor by additional mixing caused by dispersion over urban areas or for the loss of vapor by absorption in water if the dispersion occurs over a large body of water.

The amount of reliable experimental information available to make any meaningful predictions of potential hazard distances is inadequate. The effects of sizes of spills are poorly understood. The spill sizes of concern involving releases of LNH_3 from potential marine accidents (barges and ships) are in the range of 2000 to 3000 metric tons (3000 to 4400 m^3). In evaluating the hazard zones from such spills, extrapolation of the 0.2-m^3 spill experiment results have been utilized.[38] It is questionable whether an extrapolation by four orders of magnitude is valid. Similar uncertainty exists in predicting the behavior of pressurized-ammonia releases. Noting these deficiencies, a joint industry-government group[35] has sponsored a comprehensive experimental study. The objective of the study is to determine the dispersion characteristics of ammonia vapors generated from spills of 6 m^3 of LNH_3 onto water and of 6-m^3 and 125-m^3 pressurized (liquid) releases on land.

NOMENCLATURE

Symbol	Significance	Unit of measure
c	Ammonia vapor concentration in air	kg/m^3
f	Mass fraction of vapor produced by the release of pressurized liquid ammonia	
g	Acceleration due to gravity	m/s^2
h	Saturated enthalpies of vapor and liquid ammonia	J/kg
H	Height of center of vapor cloud above ground level; also used for depth of cloud	m
l	Buoyancy length $= \dfrac{Vg\Delta_v}{\pi t_e}$	m
P	Pressure	bars
R_e	Maximum radius of spread on water of instantaneously released LNH_3	m
S	Stratification parameter [Eq. (10)]	
t	Time	s
t_e	Time for complete evaporation of a spill	s
T	Atmospheric temperature	K
U	Mean wind speed	m/s
V	Volume of liquid spilled	m^3
x	Downwind distance from the source	m
y	Crosswind distance from the source	m
z	Vertical distance above ground	m

Greek		
Δ_L	Fractional density deviation of liquid $= \left(1 - \dfrac{\rho_L}{\rho_w}\right)$.	
Δ_v	Fractional density deviation of vapor $= \left(1 - \dfrac{\rho_v}{\rho_a}\right)$.	
ρ	Density	kg/m^3
σ	Dispersion parameter	m
$\dfrac{\delta\theta}{\delta z}$	Potential temperature gradient in the atmosphere	K/m

Subscripts	
a	Air, ambient
L	Liquid
T	Tank condition
v	Vapor
w	Water

CONCLUSION

Ammonia is a useful industrial and agricultural chemical. However, it has to be handled very carefully because of its hazardous properties. The hazards vary from minor discomfort to mild burns to toxic poisoning. Exercise of great care is therefore a requisite in handling anhydrous ammonia in all phases from production to end use. For those who handle ammonia as a part of the job as well as for those in emergency response, a knowledge of the properties and behavior of ammonia in the environment is extremely important.

REFERENCES

1. A. V. Slack and G. R. James (eds.), *Ammonia*, Marcel Dekker, Inc., New York, 1973.

2. *Anhydrous Ammonia*, Pamphlet No. 1415, Imperial Chemical Industries, Ltd., Agricultural Division, Billingham, England, 1978.

3. W. J. Rowan, R. W. Zub, W. P. Goss, and J. G. McGowan, *Thermophysical and Physical Properties and Handling Characteristics of Ammonia for Application as a Working Fluid in OTEC Cycles*, technical report to the U.S. Energy Research and Development Administration, Contract No. EG-77-S-02-4238, University of Massachusetts, Amherst, Mass., August 1977.

4. *Control Techniques for Nitrogen Oxide Emissions from Stationary Sources*, 2d ed., EPA-450/1-78-001, U.S. Environmental Protection Agency, Office of Air Quality Planning and Standards, Research Triangle Park, North Carolina, January 1978.

5. H. Kamin, Chairman, Subcommittee on Ammonia, *Ammonia*, National Research Council, Division of Medical Sciences, Committee on Medical and Biological Effects of Environmental Pollutants, University Park Press, Baltimore, 1979.

6. W. L. Buckley and H. W. Husa, "Combustion Properties of Ammonia," *Chemical Engineering Progress*, vol. 58, no. 2, 1962, pp. 81–84.

7. R. E. Gustin and D. A. Novacek, *Ammonia Storage Vent Accident*, Paper No. 43f, Symposium on Safety in Ammonia Plants and Related Facilities, National Meeting of the American Institute of Chemical Engineers, Miami Beach, Fla., November 1978.

8. O. Kirk, *Encyclopedia of Chemical Technology*, vol. 2, Interscience Publishers, New York, 1963, pp. 288–292.

9. H. Mayo, "Experiences with Ammonia," personal communication, Farmland Industries, Lawrence, Kans., January 1979.

10. *Threshold Limit Values for Chemical Substances and Physical Agents in the Workroom Environment*, American Conference of Governmental Industrial Hygienists, Cincinnati, 1980.

11. Occupational Safety and Health Administration, "Ammonia—Proposed Standards for Exposure," Part II, *Federal Register*, vol. 40, no. 220, Nov. 25, 1975, pp. 54684–54693.

12. *Submarine Atmosphere Habitability Data Book*, NAVSHIPS 250-649-1, rev. 1, U.S. Department of the Navy, Bureau of Ships, 1962, p. 181.

13. W. Turner, *Ten Years of Single Train Ammonia Plants*, AMPO 74, Paper No. 1, ICI Operating Symposium, 1974, p. 11.

14. O. J. Quartulli and L. J. Buividas, "Some Current and Future Trends in Ammonia Production Technology," *Nitrogen*, March–April 1976, p. 6.

15. "Key Chemicals—Ammonia," *Chemical and Engineering News*, Jan. 22, 1979, p. 11.

16. *Safety Requirements for the Storage and Handling of Anhydrous Ammonia*, ANSI K61.1-1972, American National Standards Institute, Inc., New York, 1972.

17. *Code of Practice for the Large-Scale Storage of Fully Refrigerated Anhydrous Ammonia in the United Kingdom*, Chemical Industries Association, Ltd., London, May 1975.

18. U.S. Department of Labor, Occupational Safety and Health Administration, *Storage and Handling of Anhydrous Ammonia*, 29 CFR 1910, p. 111.

19. *Guidelines for the Location of Stationary Bulk Ammonia Storage Facilities, Standards and Approvals Division*, Department of the Environment, Edmonton, Alberta, 1977, p. 9.

20. J. D. Reed, "Handling, Storage and Shipping of Ammonia," in A. V. Slack and G. R. James (eds.), *Ammonia*, Marcel Dekker, Inc., New York, 1979, chap. 12.

21. *Operational Safety Manual for Anhydrous Ammonia*, Fertilizer Institute, Washington, 1971, p. 21.

22. *Anhydrous Ammonia—Safe Storage and Handling of Bulk Liquid*, Pamphlet No. 1453, Imperial Chemical Industries, Ltd., Agricultural Division, Billingham, England, 1978.

23. U.S. Department of Transportation, Coast Guard, 46 CFR, *Shipping*, chap. I, subpart 98.25, "Anhydrous Ammonia in Bulk."

24. U.S. Department of Transportation, 49 CFR, *Transportation*, chap. I, "Material Transportation Bureau," sec. 179.

25. W. A. Inkofer, "Ammonia Transport via Pipeline," *Chemical Engineering Progress*, vol. 65, March 1969, pp. 64–68.

26. W. L. Ball, " A Review of Atmospheric Ammonia Research Study,"*Ammonia Plant Safety Manual*, vol. 12, American Institute of Chemical Engineers, New York, 1970, pp. 1–7.

27. A. E. Germeles, *A New Model for LNG Tank Rollover*, Paper No. H-2, Cryogenic Engineering Conference, Kingston, Ont., July 1975.

28. E. M. Drake, "LNG Rollover—Update," *Hydrocarbon Processing*, January 1976, pp. 119–122.

29. *Report on Accident at Houston*, Report No. 6050911, *Report on Accident at Glen Ellyn*, Report No. 606004, U.S. Department of Transportation, Office of Hazardous Materials Operations, Hazardous Materials Incident Reporting System, Washington, 1976.

30. H. Lonsdale, "Ammonia Tank Failure—South Africa," *Ammonia Plant Safety*, vol. 17, 1975, pp. 126–131.

31. National Transportation Safety Board, *Report on Crete Accident*, Report No. NTSB-PAR-71-2, 1971; *Report on Conway and Hutchinson Accidents*, Report No. NTSB-PAR-74-6, 1974; *Report on Pensacola Accident*, Report No. NTSB-PAR-78-4, 1978.

32. J. G. MacArthur, "Ammonia Storage Tank Repair," *Ammonia Plant Safety Symposium*, 1971, pp. 1–3.

33. A. Resplandy, "Étude experimentale des propriétés de l'ammoniac," *Chimie et industrie—Géne chimique*, vol. 102, 1969, pp. 691–702.

34. G. D. Kaiser and B. C. Walker, "Releases of Anhydrous Ammonia from Pressurized Containers—The Importance of Denser than Air Mixtures," *Atmospheric Environment*, vol. 12, 1978, pp. 2289–2300.

35. *Joint International Industry and Government Spill Study on Ammonia*, organized by the Fertilizer Institute and the U.S. Coast Guard, Washington, 1978.

36. David H. Slade (ed.), *Meteorology and Atomic Energy*, U.S. Atomic Energy Commission, Office of Information Services, Washington, July 1968.

37. U. M. R. Lima and R. C. Reid, *Liquid Ammonia Spills on Water*, Massachusetts Institute of Technology, Department of Chemical Engineering, Cambridge, Mass., 1972.

38. P. K. Raj, J. H. Hagopian, and A. S. Kalelkar, *Prediction of Hazards in Spills of Anhydrous Ammonia on Water*, NTIS AD-779-400, report to U.S. Coast Guard, March 1974.

39. P. K. Raj and R. C. Reid, "Fate of Liquid Ammonia Spilled onto Water," *Environmental Science and Technology*, vol. 12, no. 13, December 1978, pp. 1422–1475.

PART 5
Chlorine

Robert L. Mitchell, Jr.
Executive Director
The Chlorine Institute

Introduction / 10-52
Properties of Chlorine / 10-53
Chlorine Incidents / 10-53
Respiratory Protection from Chlorine / 10-56
Investigating Leaks / 10-56
Emergency Assistance / 10-57
Controlling Leaks / 10-57
Emergency Kits / 10-58
Punctured Containers / 10-59
Area Affected by a Chlorine Release / 10-61
Disposal of Chlorine / 10-63
Planning for Emergencies / 10-64

INTRODUCTION

One of the world's basic chemicals, chlorine has been an industrial chemical for more than 80 years. It has been shipped both in small containers and in bulk for most of that period. Tank car shipments began in the United States in 1909. Today, the United States produces approximately 11 million metric tons of chlorine annually out of a world production estimated at 28 million metric tons. In the United States annual shipments of chlorine by all modes amount to about 4 million metric tons. It is difficult to estimate world shipments; 9 million metric tons per year seems a reasonable total.

When the volume of chlorine production is considered along with its toxicity, the safety record over the years has been extremely good. In United States rail transportation between 1909 and 1978, there were only two incidents that resulted in fatalities. The history of transportation by barge and tank truck is much shorter and the tonnage significantly less; here no United States transportation fatalities have been recorded. The record of incidents at producing and consuming plants throughout the world is not as good, although fatalities rarely occur.

There are many reasons for chlorine's good accident history; two are especially important:

- Chlorine's strong, acrid odor makes it readily detectable at low concentrations: people do not stay in a dangerous concentration if they can help it.

- Chlorine's toxic properties are well known to producers, consumers, regulatory authorities, and, for that matter, most of the public. Chlorine demands respect from all who handle it.

No one knows the need for such respect better than the chlorine producer. The need for safety in all its aspects has been a concern for many years. Company programs in safety and industrial hygiene were in place long before any regulatory requirements. In addition, the industry has taken steps to work on this problem cooperatively. The Chlorine Institute has been concerned since 1926 in developing and promoting the safe handling of this basic chemical. Similar organizations were later established in other countries.

Although the safety record over the years has been a good one by any standard, it can always be improved. It is hoped that the following information will contribute to this improvement.

PROPERTIES OF CHLORINE

Before people can take effective action in controlling a chlorine accident, they must have a working knowledge of chlorine's properties. The important properties are as follows:

- Chlorine when unconfined at normal ambient temperatures is a gas, but when confined in a shipping container or storage tank under pressure it has both a liquid and a gas phase.
- Chlorine boils at $-34°C$ ($-29°F$) and has a vapor pressure (gauge) of approximately 265 kPa at $0°C$, 568 kPa at $2°C$, and 1025 kPa at $40°C$.
- Chlorine is neither flammable nor explosive. However, like oxygen, it will support combustion of many substances. Many organic chemicals react readily with chlorine, sometimes with explosive violence. At higher temperatures, most metals will burn in chlorine; for example, carbon steel ignites at about $250°C$ ($480°F$).
- Chlorine gas is greenish yellow in color. With a molecular weight of 70.9, it is about 2½ times as heavy as air.
- Liquid chlorine is amber-colored and is about 1½ times as heavy as water.
- When water-free, chlorine is usually handled in carbon steel containers and piping. All chlorine shipping containers are of carbon steel. However, when water is present, the steel can rapidly corrode.
- Most people can readily smell chlorine in air at ½ ppm or less.
- The current American National Standards Institute (ANSI) consensus standard[1] acceptable 8-h time-weighted average concentration for chlorine is 1 ppm with an acceptable ceiling concentration of 2 ppm.

Among the above properties, the significant hazard of chlorine is that of inhalation by persons or animals. Emergency handling must focus on this hazard.

CHLORINE INCIDENTS

Experience suggests that chlorine incidents should be considered under two categories: transportation incidents and incidents at producer or consumer plants.

Transportation Incidents

With rare exceptions, transportation incidents involve leakage from valves and other fittings on chlorine shipping containers. The much more serious leakage from, or failure of, the container itself is far less common. The predominant transportation containers are the railway tank car, the highway tank trailer, and small containers like the ton container and the cylinder. In the United States chlorine is also moved in tank barges. North American rail tank cars have net contents as high as 81 metric tons, while in other parts of the world tank cars holding more than 50 metric tons are exceptional. (See Fig. 5-1.) Chlorine is moved in bulk on highways in trailers of about 16-metric-ton capacity, but some have been authorized with contents as high as 27 metric tons. Although not usual in North America, highway transportation is more common in Europe. In the United States the smaller containers are the ton container (907-kg capacity; see Fig. 5-2) and the cylinder of either 45.4- or 68.1-kg capacity. Similar containers are used in Canada and Mexico. In Europe, the "ton" container may be as large as 1200 kg, and a small number of intermodal tanks of 15 to 18 metric tons are also in service.

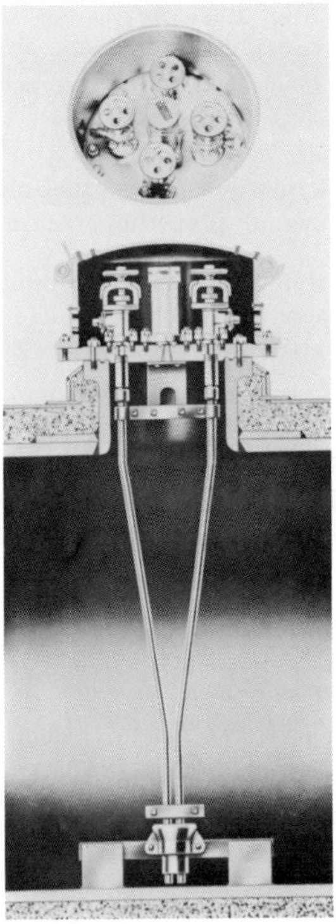

Fig. 5-1 Valve arrangement and manway for a single-unit tank car.

In North America, tank cars, barge tanks, and highway tanks must have safety relief valves, while cylinders and ton containers must be fitted with fusible plugs. In some countries in Europe, safety relief valves are prohibited on chlorine tanks.

The chlorine industry has always felt that the first line of defense in transportation safety is a safe shipping container. For example, the American chlorine tank car is designed with a tank tested at 3448 kPa, or about twice the pressure required by its vapor pressure. The manway is the only opening in the tank, and all tanks are insulated. Barge tanks are independent uninsulated pressure vessels, and no piping may be connected to them during transportation. These and many less obvious design factors along with careful maintenance provide a safe shipping container. However, even the best container cannot survive every incident. The forces at hand in train derailments or shipping collisions can exceed the capabilities of any practical container.

In many transportation incidents, identification of the involved commodities is a difficult matter; this is especially true of train derailments. Chlorine shipping containers are marked with the word CHLORINE in North America and, in appropriate languages, in most other parts of the world. Bulk shipments are also identified with appropriate placards (see Fig. 5-3). When the presence of a chlorine tank car or container has been established or when there is an indication of chlorine in the area, inspection for leakage must be carried out.

Plant Incidents

In contrast to transportation incidents, accidents that occur at a chlorine production plant are characterized by (1) experienced personnel on the scene, (2) ready availability of

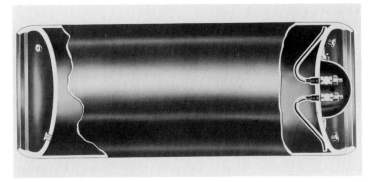

Fig. 5-2 United States chlorine ton container.

appropriate countermeasure equipment, and (3) the practicality of a prearranged emergency plan.

Small releases of chlorine are of little significance to the public as a result of large plant operations. On the other hand, most chlorine plants handle significant quantities of chlorine during processing, in stationary storage tanks, and in loaded transportation tanks. The possibility of a major release is always present, but because industry is well aware of this potential, there is a constant effort to improve equipment and emergency handling techniques. Major releases are a rarity. Big releases which permit chlorine to escape from the plant property are especially well handled by the joint efforts of the plant and the appropriate emergency services. In addition, in areas where there is a concentration of manufacturing facilities, industrial mutual-aid plans greatly help to minimize the effect of serious incidents.

The situation at chlorine-consuming points ranges all the way from a very high level of competency to almost none. A large chemical plant or paper mill with a chlorine-consuming rate of 100 or more metric tons a day can have nearly the capability of a manufacturing plant, while a village water treatment facility or a municipal swimming pool may very often have little or no capability.

It is recommended that any location using chlorine have, at the minimum, a simple emergency plan. The plan must include (1) means of notifying emergency services, (2) means of evacuation of the facility, and (3) means of obtaining appropriate specialized advice and, if necessary, summoning help to the scene. The key to this planning is the very fact that it is done *before* the emergency. Arrangements can be made with the responding fire department or other emergency service unit. Appropriate respiratory equipment and emergency kits can be ordered. There is time to plan for everyone or at the least for two persons on every shift to be competent in the use of respiratory equipment, and all the many other details can be considered in an orderly fashion.

Fig. 5-3 Chlorine placard. (USA)

RESPIRATORY PROTECTION FROM CHLORINE

Severe exposure to chlorine can occur whenever chlorine is handled or used. A person making or breaking a chlorine connection should have on his or her person a suitable escape-type respirator. In addition, suitable protective equipment for emergency use should be available outside chlorine rooms near the entrance, away from areas of likely contamination. If chlorine is used in widely separated locations, protective equipment should be available at each use point. All respirator equipment must be of a type approved by the designated authority.

Respiratory protective equipment (as distinguished from escape respirators) should be available for handling emergencies. Respirators should be located outside the probable location of any leak. They should be routinely inspected and maintained in good condition. They should be cleaned after each use and at regular intervals. Equipment used by more than one person should be sanitized after each use. All equipment should be used and maintained in accordance with the manufacturer's instructions.

Equipment may be of the following types:

1. Self-contained breathing apparatus is suitable for high concentrations of chlorine and is the preferred means of respiratory protection for the usual chlorine consumer. It provides protection for a period which varies with the amount of air, oxygen, or oxygen-producing chemicals carried. Oxygen masks should not be used in a tank or other closely confined area where there may be danger of sparks or fire.

2. The hose mask with blower, with a full facepiece and with air supplied through a hose from a remote hand-operated blower, is suitable for high concentrations of chlorine. The blower air supply must be free of air contaminants.

3. The industrial-canister-type mask, with a full facepiece and a chlorine canister, is suitable for moderate concentrations of chlorine, provided sufficient oxygen is present. The mask should be used only for a relatively short exposure period. It may not be suitable for use in any emergency since at that time the actual chlorine concentration may exceed the safe limit (1 percent by volume) and the oxygen content may be less than 16 percent (by volume). The wearer must leave the contaminated area immediately on detecting the odor of chlorine or on experiencing dizziness or difficulty in breathing; these are indications that the mask is not functioning properly, that the chlorine concentration is too high, or that sufficient oxygen is not available. Unless the presence of other gases requires the use of an all-purpose canister, the chlorine canister (which has a large capacity) should be used. Exceeding manufacturer's recommended limits on maximum nonuse shelf life might be hazardous. Regular replacement of overage canisters, though unused, is recommended.

For further details on respiratory protection, see Chap. 13.

INVESTIGATING LEAKS

Aqua ammonia, preferably commercial strength (26° Be), is used to test for chlorine, but a newly opened bottle of household ammonia will do if commerical-strength ammonia is unavailable. Put the ammonia in a plastic squeeze bottle with a spout so that its vapor, not liquid, can be directed in the vicinity of the leak. If a squeeze bottle is unavailable, tie a rag to a stick and saturate the rag with ammonia. A white cloud of ammonium chloride will result if there is any chlorine leak. Do not let the liquid ammonia come in contact with the brass parts of valves.

Those investigating the leak must have on their persons and know how to use appropriate respiratory protection. For highest safety, self-contained breathing apparatus should be used. However, canister-type respirators have been used successfully on many occasions for cylinders and ton containers if chlorine concentrations are low.

Impervious protective clothing is not mandatory and is rarely used by experienced chlorine emergency teams. Irritation of the moist parts of the body can sometimes occur; conventional fire-fighting clothing is helpful. Protective gloves of neoprene or similar material should be worn to avoid frostbite from possible liquid chlorine leakage. The buddy system should be observed; working alone is not good practice. On the other hand, unnecessary persons should be kept away from the affected area until the cause of the leak has been discovered and the trouble corrected. If the leak is extensive, all persons in the path of the fumes must be warned to leave the area. When possible, workers should keep upwind, to the side or above the leak. Because chlorine gas is 2½ times as heavy as air, it tends to lie close to the ground; however, this may not be true under all conditions, especially inside buildings or where local air currents are present.

EMERGENCY ASSISTANCE

If a chlorine emergency cannot be handled promptly by personnel at the site, the nearest office or plant of the supplier should be called for assistance. If the supplier cannot be reached, call CHLOREP, the 24-h, 7-day-a-week *CHLOR*ine *E*mergency *P*lan, organized in the United States and Canada by the Chlorine Institute[2] to advise and assist in resolving chlorine emergencies. In the United States, calls for help go through the Chemical Transportation Emergency Center (CHEMTREC) in Washington. In Canada calls are currently made to the nearest Transport Emergency Assistance Program (TEAP) control center. Write the Chlorine Institute for an information card giving the current telephone numbers. Similar plans have been organized in the United Kingdom and some continental European countries. The telephone numbers of one's supplier and of CHEMTREC, TEAP, or other plan should be posted in suitable places so that they will be quickly available if needed. When telephoning for assistance, the following information should be given:

1. Nature, location, and extent of the emergency
2. Your company name, address, telephone number, and the person or persons to contact for further information
3. Type and size of the container or other equipment involved (with a tank car or barge, the car or barge number will be helpful)
4. Corrective measures being applied
5. Name of the chlorine supplier
6. Travel directions to the emergency site

CONTROLLING LEAKS

If a leak occurs in equipment or piping, the chlorine supply should be shut off to relieve the pressure and permit necessary repairs. If welding is unavoidable, the system must be purged with dry air (nitrogen or carbon dioxide also may be used) before proceeding. Welding should comply with all applicable codes.

Valve Leaks

Leaks around container valve stems usually can be stopped by tightening the packing gland or packing-gland nut. If this does not stop the leak, the container valve should be closed; if the valve does not shut off tight, the outlet cap or plug should be applied.

Leaks in Transit

If a chlorine leak develops in a container being transported through a populated area, it is generally advisable to keep the vehicle or tank car moving until open country is reached in order to disperse the gas and minimize the hazards of its escape. Appropriate emergency measures should then be taken as quickly as possible. If a vehicle transporting chlorine containers is wrecked and there is any possibility of fire, the containers should be removed from the vehicle. Tank cars or barges should be disconnected and pulled out of the danger area. If no chlorine is escaping, water should be applied to cool containers which are exposed to fire heat and cannot be moved. If any container is leaking, appropriate measures should be taken to stop or minimize the leak. If the area is congested, the container should be moved if possible to an area of reduced hazard.

If a tank car or a tank truck is wrecked and chlorine is leaking, the danger area should be evacuated and clearing of track or highway should not be started until safe working conditions are restored. All unauthorized persons should be kept at a safe distance.

Container Leaks

If a container is leaking chlorine, it should be positioned so that the leak is uppermost, allowing gas instead of liquid to escape; this procedure is not often practical with tank cars or trucks. The quantity of chlorine that escapes from a gas leak is about one-fifteenth of the amount that escapes from a liquid leak through a hole of the same size. As noted earlier, it may be desirable to move the container to an isolated spot where it will do the least harm. In plant incidents it may be practical to reduce pressure in the container by removing the chlorine as gas (not as liquid) to be processed or taken to a disposal system. This will reduce the temperature of the chlorine and thus reduce the vapor pressure. In some incidents it may be preferable to remove the chlorine as liquid; this will not reduce the pressure, but transfer may take less time. If available, an appropriate device from an emergency kit should be applied.

To estimate the danger and evacuation area for major leaks, see Part 2, "Atmospheric Dispersion."

EMERGENCY KITS

About 30 years ago, a study of chlorine industry incidents revealed that a great majority of chlorine container leaks involved leaking valves, valve packings, and gaskets and similar causes. In recognition of this situation, the chlorine industry conceived the idea of emergency kits. Kits for cylinders, ton containers, and tank cars were first made available in the late 1940s by the Solvay Process Division and some other manufacturers. In the early 1960s an industry committee concluded that a standardized kit would be desirable, and development by the Chlorine Institute was begun. Today, there are three standardized emergency kits for use with chlorine containers. These are the A kit for 45-kg (100-lb) and 68-kg (150-lb) cylinders, the B kit for 908-kg (ton) containers, and the C kit for railroad tank cars and highway trailers.

Currently more than 6000 of these kits are kept at thousands of locations throughout

Fig. 5-4   Chlorine Institute emergency kit A.

the United States and Canada and a small number in other parts of the world where United States containers are commonly used. The success of these kits depends in no small part on the early realization by the chlorine industry of the safety advantages of standardized containers. Currently the manways of all North American tank cars and tank trucks are identical. The valves on cylinders and ton containers also have been standardized. A practical kit is much simpler to develop when only one variety of valve is to be encountered.

Several large fire departments own emergency kits for cylinders and ton containers, and some have all three kits. Members of these departments receive periodic training in their use.

The emergency kit A for cylinders (Fig. 5-4) contains a clamp to control fusible-plug leaks, a hood to cover a leaking valve with means to hold it in place, and a patch to control leaks from small holes in the side of the cylinder. The emergency kit B for ton containers (Fig. 5-5) contains a hood to cover a leaking valve and a beam to hold it in place, a hood to cover leaks in a fusible plug, and a patch to control small leaks in the side of the container. The Emergency kit C for tank cars and tank trucks (Fig. 5-6) contains an angle-valve hood, a safety-valve hood, and a beam to hold either hood in place against the manway cover, thus preventing leakage through the valve or the joint between the valve and the cover. There are no parts to handle leaks in the tank itself. All kits contain wrenches and other tools, but no respiratory equipment is included.

All but a very few incidents involving leakage from small containers can be handled with the kits, provided that they are applied by experienced and trained technicians. Instruction books and training slide sets are available to aid in this training.

PUNCTURED CONTAINERS

In most situations, a large puncture in a cylinder or ton container will permit the entire contents to escape before any effective emergency action can be taken. Occasionally, with a moderate-size puncture, some of the chlorine will remain in the container, and its rate of escape can be reduced by turning the container so that the hole is in the gas phase. In

Fig. 5-5 Chlorine Institute emergency kit B.

Fig. 5-6 Chlorine Institute emergency kit C.

the case of a tank car or a tank trailer, a significant quantity of chlorine may remain in the container. On two occasions, contents of punctured tank cars have been successfully transferred to another car, and on two occasions the remaining contents were transferred to a pit containing dilute caustic soda, thus neutralizing the chlorine. Since preparations for such actions involve getting specialized equipment and an empty chlorine tank or appropriate neutralizing chemicals to the scene, there is also time to get thoroughly experienced personnel to handle the problem. The transfer of loads or neutralizing by fire fighters or railroad employees definitely is not recommended.

In the period before actual handling of the chlorine, the people at the scene must take what steps they can to minimize the effects of release. To do this effectively, they must estimate the extent or quantity of the release and evaluate conditions such as the terrain, the weather, the site of the leak (whether indoors or outdoors), the location of the nearest residences and businesses, and related matters. Without this knowledge, effective handling of the incident cannot take place. If there has been a catastrophic failure of a chlorine tank, the responding emergency services must determine the wind direction and initiate an appropriate evacuation of persons who could be affected. If there is actual spillage of liquid chlorine from the damaged tank, it is possible to construct dikes or other means to limit the ground spread of the liquid chlorine remaining after the initial flash. In some situations, it may be possible to reduce the evaporation rate by applying special foam to minimize the heat input to the evaporating pool. Experiments have been undertaken in England, Scandinavia, and the United States in an effort to determine the best foam to apply and the best means to apply it. Further work is planned, but specific recommendations cannot be made at this time.

AREA AFFECTED BY A CHLORINE RELEASE

Chlorine releases must be considered under two headings: (1) the continuous release of gas or liquid, usually due to the failure of a valve or other fitting or a small leak; and (2) the essentially instantaneous release of a large quantity of chlorine, usually due to the failure of a tank, piping, or process equipment. The more common continuous release results in a downwind toxic hazard which continues as long as the emission continues. The flash release, fortunately rare, is so named because about one-quarter of the tank contents immediately evaporates. This flash removes heat from the remaining liquid, thus cooling it to the atmospheric boiling point. At ordinary storage temperatures about 25 percent flashes; the exact portion which will flash depends on the temperature (see Fig. 5-7). The remaining liquid will boil off at a slower rate determined by the heat flow into the boiling liquid. If liquid chlorine is spilled on the ground, the evaporation rate will be governed by the size of the pool and the ground temperature. The estimated evaporation rate is shown in Fig. 5-8.

Cloud Travel

The manner in which chlorine or other vapor is released will influence its concentration within the affected downwind area.[3] A continuous release of chlorine from a point source results in a cloud which diffuses both horizontally and vertically as it moves downwind. The shape of the cloud is roughly a cone with a semicircular cross section. The concentration in the cloud, while decreasing in the downwind direction from the source, persists as long as the emission continues. Eddy diffusion currents set up by variations in the terrain and by wind fluctuations can cause variations in these patterns. A sudden release of chlorine, as from a ruptured vessel, will form a single hemispherical cloud which will have a much higher concentration at any point downwind than would the same quantity of chlorine released over a finite time period. However, the danger area is limited to the area where the concentration is great enough to be injurious for the length of time required for the cloud to pass over. The isopleths of uniform gas concentrations within the cloud are shaped roughly like a cigar. Weather, especially wind, conditions have an important effect. Under stable conditions, the cloud will persist for a long time and will travel a considerable distance. However, it should also be noted that the length of time during which an area is exposed to the cloud is relatively short. Where unstable conditions exist, the cloud dilutes and the concentration drops quite rapidly.

Following an instantaneous release, depending on whether the liquid spills on the ground or is contained in the vessel, a much lesser amount of chlorine will be emitted continuously as the liquid evaporates because of the heat picked up from its surroundings. This cloud will then follow along the same path taken by the initial cloud. It will persist as long as the liquid is exposed.

At producing plants, estimates of the area which would be affected by a potential leak can be effectively studied. Equipment can be a complete weather station tied to a com-

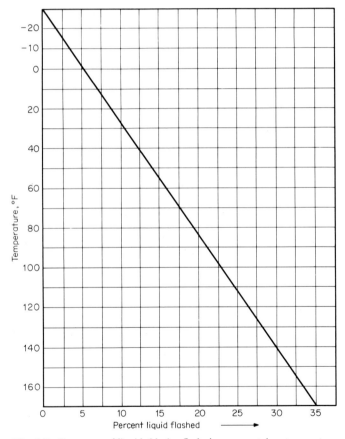

Fig. 5-7   Percentage of liquid chlorine flashed versus container temperature.

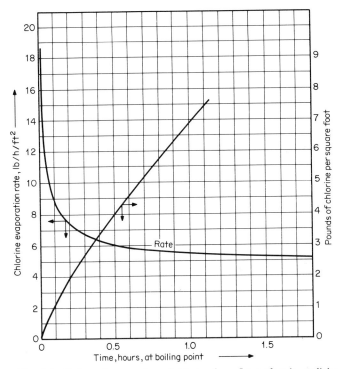

Fig. 5-8 Estimated evaporation of chlorine from flat surface in sunlight versus time (not including adiabatic flash of warm liquid).

puter which, with other input data, can calculate the downwind cloud travel. Smaller plants or large consuming plants may use a wind sock and a map with a plastic overlay.

DISPOSAL OF CHLORINE

After the emergency is under control, there often remains the problem of disposal of the chlorine container. In the case of transportation incidents in which an emergency kit has been applied to a leaking container, it is necessary to transport the defective container to another location for emptying. It is advisable to take the time for careful inspection of the container to determine whether it is safe to transport it and to determine the nearest possible disposal point before actually shipping the defective container. Regulations do not permit the shipment of a leaking chlorine container or a container which has been exposed to fire, whether full or partially full. However, it may be desirable under some circumstances to ship an otherwise defective chlorine container (including tank cars) to which an emergency device has been applied. Normally, these decisions are reached jointly by the chlorine shipper and the carrier. If the chlorine shipper facility is not extremely remote, particularly with cylinders and ton containers, it is often the conclusion that the damaged container will be picked up and removed from the scene by the shipper.

At consumer locations it is often best to discharge the contents of a leaking container into the normal consuming process. Gas should be removed if it is desirable to reduce the pressure and minimize a leak. Liquid can be removed if the process will handle it safely. In the case of tank cars and trucks, a serious effort should be made to use the chlorine at the site.

In most situations, including those where an emergency kit C has been applied, it is possible to unload the tank about as usual. Unloading lines can be connected to one of the liquid valves, and the tank can be unloaded at the usual rates which apply at the location. If this is not possible, an investigation of alternative solutions to the problem must be undertaken and a decision reached jointly by the parties involved.

Alkali Absorption

If the process cannot be expected to handle chlorine under emergency conditions, consideration should be given to using an alkali absorption system. This is practical with small containers. A tank large enough to hold the required alkaline solution is necessary. Alkaline solutions can be prepared in accordance with the following table:

Chlorine (kg)	Caustic soda (kg)	Water (liters)	Soda ash (kg)	Water (liters)	Hydrated lime (kg)	Water (liters)
100	125	334	300	834	125	1043

The quantities shown are chemical equivalents, and it is desirable to provide excess over these quantities in order to facilitate absorption. Hydrated-lime solution must be continuously and vigorously agitated when chlorine is to be absorbed. Chlorine should be passed into the solution through a suitable connection properly submerged and weighted to hold it under the surface. The chlorine container must not be immersed in the solution.

PLANNING FOR EMERGENCIES

As has been indicated, planning for chlorine emergencies is recommended for all locations which manufacture, store, or consume chlorine. In addition, carriers and emergency services must include in their general emergency planning some consideration of emergencies involving chlorine and other toxic compressed gases. When these emergency plans involve calls for help, it is essential that appropriate telephone numbers be readily available to those who will need to use them in an emergency. Training in the use of respiratory protective equipment is an important part of any emergency planning so that the emergency forces will be engaged in controlling the emergency and not in becoming a part of the emergency. Development of emergency plans for any location must be an important function of the location management.

Help in developing portions of the plan relating to chlorine is available through most chlorine suppliers. Chlorine consumers should take advantage of this help whenever possible.

REFERENCES

1. *Accepted Concentrations of Chlorine,* ANSI Standard Z37.25-1974, American National Standards Institute, New York, 1974.
2. *Chlorine Manual,* The Chlorine Institute, Inc., New York, 1969.
3. *Estimating Area Affected by a Chlorine Release,* The Chlorine Institute, Inc., New York, 1969.

PART 6
Flammable Liquid Gases

L. Edward Brown
President

Larry M. Romine
Senior Engineer
Energy Analysts, Inc.

Introduction / 10-65
Properties of Liquid Gases / 10-66
Hazards of Liquid Gases / 10-66
Design Spill Considerations / 10-70
Downwind Travel of Flammable Gas Clouds / 10-71
Prediction of Fire Hazards / 10-73
Hazard Control Systems / 10-74
Manufacturers of Hazard Detection Systems / 10-78

INTRODUCTION

The present and projected worldwide demands for liquid gases as fuels and chemical building blocks indicate that the processing and transport of liquid gases will continue to increase for some years. Release of liquid gases in processing plants and transportation accidents has resulted in a number of serious fire and explosion incidents.[1,2,3,4] As a consequence, new and conservative governmental regulations are being proposed for facilities for handling liquid gases (both processing and transportation) in many parts of the world. Additionally, a number of the world's major processors of liquid gases are designing their new liquid gas facilities to handle a variety of potential liquid gas spill scenarios.

The hazards associated with liquid gas releases are reasonably well understood. Considerable effort has been expended by many organizations to model mathematically the hazards produced by releases of liquid gas. In broad terms, these models are designed to describe the development and dispersion of liquid gas clouds subsequent to a liquid gas spill, the deflagration (explosion) hazards of liquid gas clouds, and the plume and pool fire hazards of liquid gas spills.

Some experimental liquid gas spill test work has been done under controlled conditions. This work has yielded valuable data for use in hazard modeling. Experimental studies have also been conducted to identify practical techniques that can be used to limit the extent of the hazard generated by a liquid gas spill.

The hazard potentials of all liquid gases are not identical but depend on the properties of each liquid gas, the atmospheric conditions at the time of a spill, the spill scenario, and available spill hazard mitigation systems.

10-65

PROPERTIES OF LIQUID GASES

The term "liquid gas" is generally used to denote pure-component fluids whose boiling point at atmospheric pressure is below 20°C (68°F). A broad range of fluids can be classified as liquid gases, ranging from liquid hydrogen at −217°C (−423°F) and atmospheric pressure to butane stored as a pressurized liquid at 20°C and 220 kPa (Table 6-1).

Two of the most common liquid gases are liquefied natural gas (LNG) and liquefied petroleum gas (LPG). Both LNG and LPG are mixtures of various liquid gases. The major constituents of LNG and LPG are methane and propane respectively.

Liquid gases are stored as either pressurized liquids or refrigerated liquids. In pressurized storage, the liquid gas is stored at atmospheric temperature under its vapor pressure. As the atmospheric temperature changes, the vapor pressure of the liquid gas changes, and correspondingly the pressure on the storage container changes. As an example, for propane stored at 0°C (32°F), the pressure on the storage container would be 483 kPa, while at 35°C (95°F) the pressure in the propane storage tank would be 1213 kPa.

The critical temperatures of methane and hydrogen are well below ambient temperature. Usable quantities of these two fluids can be stored at ambient temperature in either high-pressure cylinders or large-volume atmospheric-pressure gas holders. An alternative storage technique for these two gases is to chill and liquefy them and store the liquid at or near atmospheric pressure. This type of storage is commonly called refrigerated storage. At atmospheric pressure, the temperature of LNG is about −162°C. One volume of liquid-phase LNG contains the equivalent of about 600 volumes of gas at atmospheric pressure and 20°C. This volume reduction makes refrigerated storage very attractive for both LNG and hydrogen.

The other liquid gases presented in Table 6-1 are also stored as refrigerated liquids, especially when very large volumes of gas are to be stored. The atmospheric-pressure storage temperatures for some liquid gases are also listed in Table 6-1.

HAZARDS OF LIQUID GASES

A pressure-enthalpy diagram for a liquid stored at ambient temperature T_a and vapor pressure P_i (pressurized storage) is shown in Fig. 6-1. If the liquid gas is released to the atmosphere, the pressure drops to P_a and the temperature of the residual liquid drops to T_f. As the pressure decreases, adequate liquid is vaporized to chill the remaining liquid to T_f. The vapor released forms a gas cloud with the surrounding air and is carried downwind. The hazards of both the residual spilled liquid and the vapor cloud are described below.

Cryogenic Hazard

At atmospheric pressure, the liquid-phase temperature of many liquid gases is below −45°C (Table 6-1). At this temperature, the carbon steel used in most construction becomes brittle and if stressed can fracture. Storage tanks designed to contain liquids at temperatures near or below −45°C must be constructed of alloy steels and/or concrete having suitable strength at the design operating temperature of the storage tank. Nickel steel alloys, stainless steels, aluminum, and prestressed concrete can be used in low-temperature service. Critical steel structures in liquid gas facilities should either be constructed of cryogenically compatible metals or provided with cold protection if the structures could be contacted by cold liquid.

TABLE 6-1 Properties of Some Liquid Gases

	Pressure (in kPa at 21°C)	Temperature (in °C at 101.3 kPa)	Specific gravity at atmospheric pressure (boiling point/4°C)	Autoignition (°C)	Pilot ignition (°C)	Flammability limits in air (percent)
Hydrogen	...*	−259.9	0.071	609	...†	4.1–74.2
Methane	...*	−161.5	0.43	573	−188	5.3–14.0
Ethane	3885	−88.9	0.54	551	−135	3.0–12.5
Etheylene	...*	−103.9	0.57	579	−136	3.1–32.0
Ethylene oxide	149	10.7	0.9	464	−29	3.6–100.0
Propane	860	−43.2	0.57	503	−104	2.3–9.5
n-Butane	215	−0.6	0.60	440	−60	1.9–8.5
i-Butane	311	−11.8	0.60	...†	...†	...†
Butene-1	474	−6.1	0.63‡	419	−80	1.6–10.0
trans-Butene-2	217	1.1	0.615‡	359	−73	1.8–9.7
cis-Butene-2	192	3.7	0.642‡	359	−73	1.7–9.0
i-Butene	264	...†	0.63‡	...†	...†	...†
1,3-Butadiene	251	−4.4	0.65‡	464	−76	2.0–11.5

*Critical temperature is less than atmospheric temperature.
†Not available.
‡Specific gravity (16°C/16°C).
NOTE: This table was compiled from the following references:

1. K. E. Starling, *Fluid Thermodynamic Properties for Light Petroleum Systems*, Gulf Publishing Co., Houston, 1973.
2. N. I. Sax, *Dangerous Properties of Industrial Materials*, 4th ed., Van Nostrand Reinhold Company, New York, 1975.
3. *Toxic and Hazardous Industrial Chemicals Safety Manual for Handling and Disposal with Toxicity and Hazard Data*, International Technical Information Institute, Tokyo, 1976.
4. *Fire Protection Guide on Hazardous Materials*, 6th ed., National Fire Protection Association, Boston, 1975.
5. R. H. Perry and C. H. Chilton (eds.), *Chemical Engineers' Handbook*, 5th ed., McGraw-Hill Book Company, New York, 1973.
6. J. B. Maxwell, *Data Book on Hydrocarbons Application to Process Engineering*, D. Van Nostrand Company, New York, 1950.

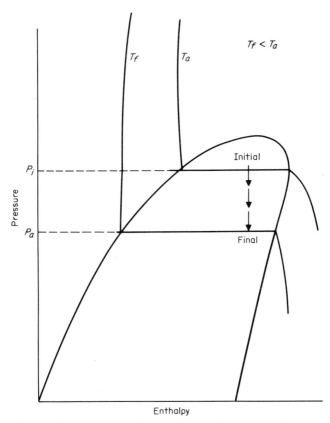

Fig. 6-1 Pressure-enthalpy diagram for a liquid stored at ambient temperature T_a and vapor pressure P_i, i.e., in pressurized storage.

TABLE 6-2   Toxicity Information for Some Liquid Gases

Hydrogen	High concentration causes asphyxiation.
Methane	High concentration causes asphyxiation.
Ethane	High concentration causes asphyxiation.
Ethylene	High concentration causes paralysis.
Ethylene oxide	TLV = 50 ppm; causes pulmonary edema paralysis and possible death by inhalation of large quantities of gas.
Propane	TLV = 1000 ppm; high concentration causes paralysis.
N-Butane	TLV = 500 ppm; high concentration causes asphyxiation.
i-Butane	High concentration causes dizziness, difficult breathing, or loss of consciousness.
Butene-1	High concentration causes asphyxiation.
cis-Butene-2	High concentration cuases asphyxiation.
trans-Butene-2	High concentration causes asphyxiation.
1-3, Butadiene	TLV = 1000 ppm; high concentration causes anesthesia.

Vapor Cloud Hazard

The vapor cloud produced by a liquid gas spill can be toxic and flammable. Toxicity characteristics of the liquid gases considered here are shown in Table 6-2, while Table 6-1 contains data on the limits of flammability for each of the liquid gases in air. Most hydrocarbons are simple asphyxiants; that is, they act by exclusion of air. As a consequence, the toxic hazard of hydrocarbon clouds in the open air occurs only when the hydrocarbon concentration is at levels in excess of 30 percent in air.

If a gas cloud concentration in air is within its flammable concentration limits and is ignited, the resulting flame can flash back to the spill source and under some conditions may deflagrate. At hydrocarbon concentrations in air below the lower flammable limit (LFL), the gas cloud cannot be ignited. Similarly, at gas concentrations in excess of the upper flammable limit (UFL), the gas-air mixture cannot be ignited. Even if the gas concentration over a significant portion of an open-air hydrocarbon cloud exceeds the UFL, at least a portion of the cloud will be within the flammable zone. If the portion of the cloud within the flammable concentration range is ignited, the flame will burn along the edge of the gas cloud, draw in combustion air, and consume the entire gas cloud.

Unprotected persons exposed to a burning gas cloud can sustain serious burns. Combustible and heat-sensitive objects exposed to a burning gas cloud can be ignited or damaged. Typical process equipment exposed to a burning gas cloud would sustain nominal fire damage.

Except for LNG gas clouds,[1] the vapor clouds produced by the liquid gases shown in Table 6-1 can deflagrate (explode) in the open air.[5] There are a number of theoretical and empirical papers[6,7,8,9] that discuss the conditions required to produce an open-air explosion of a flammable gas cloud. Until recently, however, little experimental research work has been done to identify the conditions required to produce deflagration in an open-air vapor cloud. Parameters which are known to be important include the concentration of gas in the cloud, atmospheric conditions, flame velocity in the cloud, objects in the path of the flame front, and the strength of the ignition source.

Fire Hazard

Both the flammable vapor cloud and any residual spilled liquid gas in liquid form present a fire hazard. If the residual spill liquid is ignited, objects within the flame zone can be damaged by direct flame contact, while objects outside the flame zone can be heated and possibly damaged by thermal (heat) radiation released by the fire.

Overpressurization

When heat is added to a liquid gas, a portion of the liquid gas vaporizes, thereby increasing the gas pressure in the storage container. The container can be a tank, a section of pipe, a pump, etc. If heat is continuously added without release of vapor or inadequate release of vapor, catastrophic failure of the container can occur. The term "boiling-liquid–escaping-vapor explosion (BLEVE)" has been used as an acronym for this phenomenon. BLEVEs have resulted in major property damage and the loss of a number of lives.

Reid[10] has proposed that under some conditions the vapor clouds associated with a BLEVE expand at a rate significantly higher than predicted from equilibrium thermodynamics. He has pointed out that under some conditions confined liquid gases which are being heated can become supersaturated in energy. However, once a maximum liquid-phase supersaturation energy level has been exceeded, the liquid phase evaporates at a very high rate. The resulting gases increase the pressure on the container rapidly and generally beyond the venting capabilities of relief valves.

DESIGN SPILL CONSIDERATIONS

The best spill control system for liquefied gas is to prevent spills from occurring through facility design, good construction, construction inspection, careful facility operations, and preventive maintenance. However, liquefied gas spills do occur, and practical steps can be taken to restrict the extent of spill hazard if potential spill conditions are included as design criteria for the facility.

Codes for liquefied gas tanker ships and land-based facilities are undergoing review and modification to reflect accident experience within the industry and spill scenarios conceived by industry, regulators, and consensus code bodies. Among the potential liquid gas spill conditions being considered in liquefied gas facility design are the following:

1. It is assumed that pump packing, valve packing, and gaskets in liquefied gas service may fail. The design spill rate and volume are then computed on the basis of system operating pressure, size of the potential release orifice, the time to detect the spill, and the time to isolate the flow. The resulting spill volume is usually multiplied by a safety factor of 2 or 3 to obtain the design spill volume for these equipment components.

2. A rupture in liquid gas transfer piping is assumed, and the design spill rate and volume are computed on the basis of maximum design flow rate through the line, the time to detect the release, and the time to terminate the flow and drain the line. Most liquid gas pumps have low-pressure discharge switches which will automatically shut the pumps down if pressure drops below a preset value. For spills just downstream from a pump, the low-pressure switch will detect the spill and shut down the pump. The design spill volumes calculated for piping failures also are multiplied by a safety factor of 2 or 3 to establish the design spill volume.

3. For liquid gas tanks with bottom connections, some codes require that the design spill be determined by assuming that the largest connection is severed and liquid spills under the hydrostatic and pressure head in the storage tank. It is further proposed that if either excess flow or remote-actuated foot valves are provided at the bottom of the tank, a spill time of up to 60 min should be considered in design.[11] If tanks are equipped with bottom connections and no internal shutoff valves, it is generally held that one should design for total loss of tank contents through the tank bottom connections. Some regulatory bodies are proposing to eliminate future construction of such tanks with bottom connections.

4. For marine operations of liquefied gas tankers, many industrial organizations are examining the potential consequence of the loss of liquefied gas contents from one cargo tank aboard a tanker.

Utilizing these design spills, the facility operator is generally required to:

- Limit the travel of the potential flammable gas cloud to within the facility fence line or to areas where the population is small

- Minimize the potential for liquid gas cloud ignition within a facility

- Limit the potential fire hazard at the facility fence boundary

- Limit the potential fire hazard risk experienced by personnel of the facility

- Limit the potential fire-heating effects on critical facility components

DOWNWIND TRAVEL OF FLAMMABLE GAS CLOUDS

Historically, the downwind travel of flammable vapor clouds has been modeled by using the standard gaussian-type dispersion equation.[12] More recently, gas cloud dispersion models have been developed on the basis of transport phenomena theory. It is not within the scope of the *Handbook* to discuss the relative merits of these two model types. Havens[13] has, however, compared the results obtained by using the two types for LNG spills onto water.

The impact of atmospheric-stability class and wind speed on the downwind travel of pollutants has been discussed in Part 2 of this chapter. The principal effects of atmospheric parameters on the downwind travel discussed there are applicable to all pollutants, including flammable liquid gas clouds.

Regardless of the pollutant dispersion modeling technique, all dispersion models require the input of a source strength Q (the rate at which the pollutant is released into the air). The source strength term for many pollution problems can be assumed to be constant. However, for liquid gas spills the source strength term is generally time-dependent.

Four driving forces can produce vapor from a liquid spill: (1) flashing of the liquid, (2) heat transfer from the subsurface to the spilled liquid gas, (3) heat transfer from the atmosphere to the liquid surface, and (4) ullage gas. Each of these terms must be considered in developing downwind concentration profiles for liquid gas dispersion.

1. **Release of gas from gas phase (ullage gas).** The quantity and rate of gas released from the gas phase of a liquid gas storage tank can be computed if a release orifice size is assumed, the fraction of the storage vessel containing gas is known, and the pressure on the container is available. The most common release source from the gas phase of liquid gas containers is the relief valve or valves. Flow characteristics of relief valves are valve-specific; thus, flow performance data on a given relief valve type should be obtained from the valve manufacturer.

2. **Flashing of spilled liquid gas.** When liquid gas from pressurized storage is spilled, a fraction of the liquid gas flashes so that the temperature of the residual liquid achieves its equilibrium value at atmospheric pressure. The fraction of liquid flashed to vapor depends on both the pressure-enthalpy characteristics of the liquid gas and the initial storage temperature of the liquid.

   The percentage of liquid gas that would flash for two different liquid gases is shown in Table 6-3. The dependence of percentage flashed on initial liquid spill temperature is also shown. No butane flashes at liquid temperatures below 0°C (32°F).

3. **Heat transfer from the spill surface.** In Table 6-1, the equilibrium temperature of various liquid gases at atmospheric pressure is shown. Hydrogen, methane (LNG), ethane, ethylene, and propane have liquid temperatures well below common ambient

TABLE 6-3 Percentage of Liquid
Flashed When Released to the Atmosphere

Liquid gas	Percentage flashed, ambient temperature, °C			
	−18	4	16	27
Butane	0	3.1	9.4	21.0
LPG (propane)	12	25.4	32.7	40.4

temperatures at atmospheric pressure. Thus, if these liquids are spilled onto ambient-temperature surfaces, the liquid will be heated and vaporized while the spill surface is chilled. As the spill surface chills, the rate of heat transfer from the spill surface to the spilled liquid gas decreases with time.

For all the hydrocarbons discussed, the initial mode of heat transfer from a spill subsurface to the liquid gases is in the nucleate boiling regime ($\Delta T \gtrsim 3°C$) and in some cases in the film boiling regime ($\Delta T \gtrsim 60°C$). As the spill subsurface cools, the rate of heat transfer from the subsurface to the spill declines in a complex manner. LNG evaporation rate data for spills onto water, typical soil, and insulating concrete are shown in Fig. 6-2.[14]

The source strength used in the dispersion model for LNG boil-off is strongly dependent on the spill subsurface type (Fig 6-2). Similar experimental boil-off data for other liquid gases have not been published in the open literature. Boil-off rates can be estimated by using boiling-heat transfer data in the literature and standard head conduction models.[15]

4. *Heat transfer from the atmosphere.*   Convective and radiative heat transfer from the wind and the sun respectively will add to the vaporization rate of spilled liquid gas. However, for most liquid gas this heat input is of minor importance when com-

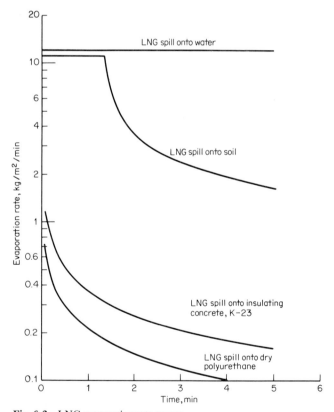

Fig. 6-2   LNG evaporation rate curves.

pared with other sources of heat input for vaporization. This input source may become dominant after the spill has chilled all the spill subsurfaces.

The total amount of vapor generated subsequent to a spill depends on the thermal properties of the spill surface, the area over which the spill spreads, and the temperature of the spill surface. After a spill each of these properties will change as a function of time. Thus, the computation of vapor generation rate from a spill surface requires an area source dispersion model; this is a complex calculation.

Spills from liquefied gas storage tanks, piping, and process areas may be confined to dikes with significant vapor holdup capability. If vapor holdup is provided, the source term for vapor cloud dispersion analysis is not the rate at which vapor is generated from the spill but the rate at which vapor is released from the dike. When adequate vapor hold is available to confine flashed vapor plus vapor generated from the spill surface during the initial high-rate boil-off period, the subsequent downwind travel of the flammable vapor cloud can be significantly reduced.

The vapor clouds produced following spills from most hydrocarbon liquid gases are heavier than air. The difference in density between the hydrocarbon gas cloud and the surrounding air causes the gas cloud to sink and spread laterally along the ground. This phenomenon is called gravity dispersion and is not accounted for in the classical gaussian-type atmospheric-dispersion models. When gravity is not included in the vapor dispersion modeling procedures, the predicted travel distances are somewhat longer than when gravity spreading is considered.

The gas concentrations computed by using most vapor dispersion models are at grade and represent average gas cloud concentrations at a point. Owing to wind gusts, actual gas cloud concentrations at a point downwind of a liquefied gas spill will vary above the average. The amount of variation in gas cloud concentrations depends on gustiness, and the variation in concentration is usually expressed as a ratio of the peak gas concentration at a point divided by the average concentration. Peak to average concentration ratios for Pasquill F conditions range from 1.2 to 1.5, while for Pasquill D the range is from 2 to 3.

A peak gas concentration is generally preceded and followed by a gas concentration below the average. As a consequence, if the average gas concentration at a point is below the LFL and the potential peak concentration is above the LFL, a discrete pocket of flammable gas can be formed, but if it is ignited, it will not usually propagate the ignition back to the spill source.

Generally speaking, the lower the wind speed and the more stable the atmosphere, the farther downwind a flammable vapor cloud can travel. However, the frequency of low wind speed conditions combined with stable atmospheres is site-dependent. Specific atmospheric-stability classifications and wind speeds used in a vapor dispersion model for a given facility should be based on site weather data. If these data are not available, realistic conservative assumptions relative to wind speed and atmospheric stability must be made.

PREDICTION OF FIRE HAZARDS

The hazard zones produced by a liquid hydrocarbon fuel fire can be computed. The steps in the computation process include the following:

1. Predict the potential spill area.
2. Compute the potential flame height by using the method of Thomas.[16]
3. Compute the effect of wind on lighting of the flame.[17]

4. Compute the geometric view factor between the flame and objects near the flame.[18]
5. Compute the amount of heat[12] received by personnel and objects by using the input from Steps 1 to 4.
6. Correct for local humidity conditions.[12]

The following heat radiation criteria can be used to determine the consequences of heat radiation on facilities.

Damaged condition	Heat radiation load (kW/m^2)
Piloted ignition of wood in 1 min	37.8
Personnel skin burns in 30 s	5.0
Storage tank damage	11.7
Wiring damage	12.6

HAZARD CONTROL SYSTEMS

During the design of any facility handling liquefied gases, the prevention and control of liquid spills and gas releases should be carefully considered. The design objectives of the hazard control systems incorporated into plant design should include the following:

1. Prevent liquefied gas spills and releases.
2. Detect spills and fires in their early stages if releases occur.
3. Limit the amount of liquefied gas that can be spilled by providing emergency shutdown systems.
4. Limit the spread of a spill by providing controlled drainage systems and dikes.
5. Control the hazards associated with a spill by using both passive and active spill control systems.

The following subsection discusses various hazard control systems and how they can be utilized.

Hazard Detection Systems*

Process Detection

Process detectors may be used to detect abnormal facility operating conditions, sound alarms, and shut down equipment if an equipment-operating parameter approaches the design limit for that equipment component. Process detectors prevent spills and releases from occurring.

Low-Temperature Detection

Low-temperature detectors are useful for detecting liquefied gas releases from process equipment and piping if the liquefied gas in question has a boiling point sufficiently below the ambient temperature. Generally, they are used in conjunction with diking systems and are located beneath process equipment and piping where spills might occur. The diking systems help assure that the spill will come into contact with the detector. The

*For the names and addresses of manufacturers that supply this equipment, see the list at the end of this part.

temperature alarm point will depend on the specific liquefied gas or gases in question, but it must be low enough to prevent false alarms in cold climates. Detector elements are available in single-point types (sensing temperature at only one location) and in a continuous-strip type.

Combustible-Gas Detection

Combustible-gas detectors can be used to detect combustible hydrocarbon gas vapors in the air. They are available in three main types: infrared analyzer, catalytic bead, and solid-state electrolytic cell. Infrared analyzer types use a pump to draw in atmospheric samples from the various plant locations to a central point where an infrared analyzer is located. These sample streams are sequentially injected into the infrared analyzer to determine the combustible-gas concentrations at each sample point.

Solid-state electrolytic cell detector systems (thermal conductivity) operate on the principle of allowing the combustible-gas molecules to diffuse into a semiconductor, thereby decreasing its electrical resistivity. The magnitude of the resultant current flow is related to the concentration of combustible-gas molecules in the semiconductor, which in turn depends on their concentration in the atmosphere. The current flow is sensed by a control-indicator module and is displayed on a meter in terms of percent LFL.

Catalytic bead sensors are the type usually chosen for use in petroleum facilities, but they do have limitations. They will not work in inert atmospheres because they need oxygen to support combustion on the catalyst. They are inaccurate when the combustible-gas concentration exceeds 100 percent LFL (5 percent methane in air), and they can be very misleading if the gas-air mixture exceeds the stoichiometric ratio.

Flame Detection

Flame detector types for possible use in liquefied gas plants include ultraviolet sensors, rate-of-temperature-rise sensors, and high-temperature sensors. Ultraviolet sensors detect the presence of larger-than-normal amounts of ultraviolet radiation when a flame is present. Ultraviolet flame detectors are very sensitive and cannot generally distinguish between fires, lightning, and arc welding.

Rate-of-temperature-rise and high-temperature detectors are now being used in conjunction with ultraviolet fire detectors in many facilities. This detection combination provides the fast response of the ultraviolet detectors as well as an alarm confirmation using a detector that is not generally prone to false alarms. Rate-of-temperature-rise and high-temperature detectors must be carefully located near the expected fire location to assure proper response.

Emergency Shutdown Systems

All piping handling liquefied gas or its vapor should be equipped with fail-safe emergency isolation valves. These valves should be located so that all major equipment and piping sections can be isolated. Valve closure times for the emergency shutdown system should be less than 30 s if the fluid dynamics of the piping system permits. Emergency shutdown system actuation switches should be located in the main plant control room and at key locations in the plant. Additionally, signals from process detectors discussed previously should be used to actuate emergency shutdown automatically.

Spill Confinement

Diking is an effective passive means for liquefied gas spill hazard control. Liquefied gas spill confinement provides the following hazard control benefits:

1. Spill confinement prevents the uncontrolled spreading of a spill. This decreases the potential pool size, thus reducing the downwind travel of the flammable vapor cloud, and decreases the distance at a given radiant-heat flux level.

2. Spill confinement permits more effective use of spill hazard mitigation systems by confining a spill to a definite, predetermined area.

Details of spill confinement should be evaluated on a plant-by-plant basis to accomplish optimum spill control.

Passive Spill Protection

Passive spill protection is protection built into a facility such as separation between equipment components, fire-protective coatings, cryogenic protection, and other systems that require no actuation to perform their design functions.

Equipment Spacing

Initial plant layout can be an effective hazard control method if enough area is available to allow critical equipment items to be spaced sufficiently distant from one another so that an accidental release of liquid gas in one critical area will not endanger other plant areas. Since the large land areas required to separate all plant areas totally from each other are rarely available to the plant designer, other hazard control systems must be used in conjunction with plant layout.

Surface Insulation

The effect of using insulation as dike floors on the rate of vapor generation from LNG spills is shown in Fig. 6-2. The use of insulating material such as lower-density concrete can in some cases significantly reduce the vapor generated per unit time subsequent to a spill. As a consequence, the downwind travel of the flammable vapor cloud is significantly reduced.

Fire-Protective Materials

Critical piping and pipe supports should be provided with fire-protective coatings with at least a 2-h rating. Specific attention should be given to piping and supports near spill confinement areas where liquefied gas fires will be confined. If fire control equipment must be located near the hazard to be protected, it should be provided with a fire-protective coating.

Cryogenic Protection

Piping supports, the outer storage tank shell, and other equipment components that could be contacted by a cryogenic liquefied gas should be provided with protection to prevent low-temperature embrittlement and failure.

Active Spill Control

Active spill and fire control systems should be provided when hazard control cannot be achieved by facility design or passive spill control systems. These systems provide rapid spill and fire control for spills which can be expected to occur in liquefied gas facilities.

Dry Chemical Systems

Dry chemical fire-fighting agents are effective in extinguishing fires in exposed locations.[19,20] Tests to evaluate the effectiveness of dry chemical agents on other liquefied gas fires have been performed. However, since the data from these tests have not as yet been published, the effectiveness of these agents on liquefied gas fires other than LNG is unknown. The time during which a fire burns before dry chemical application has a significant effect on whether the fire can be extinguished with a dry chemical agent. Firefighting systems employing dry chemical agents include hand-held portable wheeled engines and nonportable units using hose lines, monitor nozzles, or fixed-pipe–nozzle systems for dispersing the dry chemical powder.

High-Expansion Foam Systems

High-expansion foam (HEF) will not extinguish an LNG fire, but it can be used in three types of hazard protection systems within an LNG plant: vapor dispersion,[21,22] fire control, and exposure control. High-expansion foam has also been evaluated for its effectiveness when applied to spills of LPG, ethylene, and butadiene.

Applying HEF to an unignited LNG spill has been shown to be an effective method for aiding the dispersion of the flammable gas vapors, thereby reducing the downwind travel of the vapor cloud. This reduces the extent of the flammable zone, which in turn decreases the probability of ignition.

Applying HEF to an ignited LNG pool is an effective fire control method for reducing the damage potential of the fire. The foam blanket reduces the feedback of heat energy from the flame to the LNG pool. This causes a much lower vaporization rate of LNG, thereby decreasing the size of the flame column. The flame size reduction is often great enough to decrease the thermal radiation from the fire by 95 percent, drastically reducing the hazard potential. The ability of HEF to accomplish vapor and fire control depends on its quality and the foam application rate. Ability to control fire burn-back is another important characteristic which should be evaluated in foam selection.

The vapors generated from spills of hydrocarbon liquefied gases other than LNG will be denser than air and will spread laterally along the ground under the influence of the earth's gravitational force. The application of foam may warm these vapors to a point at which the cloud will not be visible but the vapor will still be denser than air so that dispersion will not be significantly improved by the application of foam.

Evaluation of the effectiveness of HEF on some liquefied gas fires revealed that butadiene and LPG fires can be extinguished and the intensity of an ethylene fire reduced by about 76 percent. Extinguishing fires involving liquefied gases generating vapors heavier than air must be carefully evaluated on a case-by-case basis. Unless adequate water spray systems are available to control the flammable vapor cloud, it may be better to control the fire rather than extinguish it.

HEF can also be used in exposure control on critical pieces of equipment. By applying foam on a structure that is near a fire (but not involved in the fire), it is possible to keep the structure cool since the radiant heat from the fire is removed by heating and subsequently vaporizing the water in the foam. The temperature of the structure will thus remain below the boiling point of water as long as sufficient foam is being applied.

Low-Expansion Foam Systems

Some low-expansion foams have been evaluated for their effectiveness in vapor and fire control on selected liquefied gases. Low-expansion foams show positive results only on butadiene fires, have little effect on LPG fires, and worsen LNG and ethylene fires. They are useful for controlling fires for fuels having a boiling temperature at or above ambient temperatures at atmospheric pressure. Test results show that low-expansion foams are useful for fires involving liquefied gases whose boiling point is only slightly below ambient temperature.[23]

Fire Water Systems

Water will not extinguish a liquefied gas fire, but it is useful for vapor dispersion and exposure control. For vapor dispersion, the water is generally applied as vertical sprays around the liquefied gas spill. The water sprays increase the mixing of the vapors with air, thereby decreasing the downwind travel of the flammable vapor cloud. These sprays can also be useful in exposure protection since they will "intercept" some of the radiant-heat energy from a fire. Normally, however, the water is sprayed directly on the equipment for which thermal protection is desired. Application of water to a liquefied gas spill will increase the spill evaporation rate. Since this increases both the vapor cloud and the fire hazard associated with the spill, it is *not* recommended.

Inerting Systems

Halogenated hydrocarbons and carbon dioxide systems can be used for extinguishing liquefied gas fires in closed spaces and for explosion prevention. Halogenated hydrocarbons are now replacing carbon dioxide for inerting applications owing to the limited toxic effects of these chemicals at their normal use concentrations in air. Halogenated hydrocarbon systems are installed in compressor buildings, in central control rooms, aboard ships, and on offshore platforms for fighting fires and inerting closed spaces.

Ventilation

Ventilation of enclosed process areas can be used to prevent fires or explosions. For areas in which no flammable gas or liquid is handled (e.g., control rooms), flammable-gas detectors can be used to stop the ventilation system if the gas concentration outside the room reaches the LFL. For compressor buildings, etc., where flammable gases or liquids are handled, gas detectors within the compartment can be used to increase the ventilation rate if the gas concentration approaches the LFL. If the ventilation rate is sufficiently high, the atmosphere in the compartment can be prevented from reaching the LFL.

MANUFACTURERS OF HAZARD DETECTION SYSTEMS

Process, High-Temperature, and Low-Temperature Detection

Automatic Switch Company
50–56 Hanover Road
Florham Park, New Jersey 07932

Fenwal Inc.
Post Office Box 309
Ashland, Massachusetts 01721

Honeywell Commercial Division
Honeywell Plaza
Minneapolis, Minnesota 55408

United Electric Controls Co.
85 School Street
Watertown, Massachusetts 02172

Combustible-Gas Detection

Bacharach Instrument Co.
625 Alpha Drive
RIDC Industrial Park
Pittsburgh, Pennsylvania 15238

General Monitors, Inc.
3019 Enterprise Street
Costa Mesa, California 92626

Mine Safety Appliances Co.
600 Penn Center Boulevard
Pittsburgh, Pennsylvania 15235

Seiger Gasalarm Division
The Condit Company, Inc.
4615 Southwest Freeway
Houston, Texas 77027

Flame Detection

Detector Electronics Corp.
7351 Washington Avenue South
Minneapolis, Minnesota 55435

McGraw-Edison Co.
Grenier Field Municipal Airport
Manchester, New Hampshire 03103

REFERENCES

1. C. D. Lind and J. C. Whitson, *Explosion Hazards Associated with Spill of Large Quantities of Hazardous Materials, Phase II,* Report No. CG-0-85-77, U.S. Coast Guard, November 1977.

2. M. A. Elliott, C. W. Siebel, F. W. Brown, R. T. Arts, and L. B. Berger, *Report on the Investigation of the Fire at the Liquefaction Storage and Regasification Plant of the East Ohio Gas Co., Cleveland, Ohio, October 20, 1944,* R.I. 3867, U.S. Department of the Interior, Bureau of Mines, February 1946.

3. Lee N. Davis, *Frozen Fire,* Friends of the Earth, San Francisco, 1979.

4. J. R. Welker, L. E. Brown, J. N. Ice, W. E. Martinsen, and H. H. West, *Fire Safety aboard LNG Vessels,* Report No. DOT-CG-42355-A, U.S. Coast Guard, January 1976.

5. K. Gugan, *Unconfined Vapor Cloud Explosions,* Gulf Publishing Co., Houston, 1978.

6. D. C. Bull and J. A. Martin, "Explosion of Unconfined Clouds of Natural Gas," paper presented at the American Gas Association Transmission Conference, St. Louis, May 1977.

7. A. A. Boni, M. Chapman, J. L. Cook, and G. P. Schnyer, "On Combustion Generated Turbulence and Transition to Detonation," paper presented at the Fifth Aerospace Meeting, American Institute of Aeronautics and Astronautics, Los Angeles, 1977.

8. J. H. Lee, "Initiation of Gaseous Detonation," *Annual Review of Physical Chemistry,* vol. 28, 1977.

9. S. M. Kogarko, V. V. Adoshkin, and A. G. Lyamin, "An Investigation of Spherical Detonation of Gas Mixture," *International Chemical Engineering,* vol. 6, p. 393, 1966.

10. R. C. Reid, "Superheated Liquids," *American Scientist,* vol. 64, March–April 1976, pp. 146–156.

11. *Standard for the Production, Storage and Handling of Liquefied Natural Gas (LNG),* Standard 59A, National Fire Protection Association, Quincy, Mass., 1979.

12. *LNG Safety Programs Interim Report on Phase II Work,* Project IS-3-1, American Gas Association, Arlington, Va., 1974.

13. J. A. Havens, *Predictability of LNG Vapor Dispersion from Catastrophic Spills onto Water: An Assessment,* Report No. CG-M-09-77, U.S. Coast Guard, April 1977.

14. R. C. Reid and R. Wang, "The Boiling Rates of LNG on Typical Dike Floor Materials," *Cryogenics,* July 1978.

15. L. E. Brown and C. P. Colver, "Nucleate and Film Boiling Heat Transfer to Liquefied Natural Gas," *Advances in Cryogenic Engineering,* vol. 12, 1968, p. 647.

16. P. H. Thomas, "The Size of Flames from Natural Fires," *Ninth International Symposium on Combustion,* Academic Press, Inc., New York, 1963.

17. J. R. Welker and C. M. Sliepcevich, *Susceptibility of Potential Target Components to Defeat by Thermal Action,* AD 875-925L, Edgewood Arsenal, Md., 1970.

18. R. G. Rein, C. M. Sliepcevich, and J. R. Welker, "Radiation View Factors for Tilted Cylinders," *Journal of Fire and Flammability,* vol. 1, 1970, p. 140.

19. H. R. Wesson, J. R. Welker, and L. E. Brown, "Control of LNG Spill Fires on Land," *Advances in Cryogenic Engineering,* vol. 20, 1974, p. 151.

20. H. R. Wesson, L. E. Brown, and J. R. Welker, "Vapor Dispersion, Fire Control, and Fire Extinguishment of High Evaporation Rate LNG Spills," *American Gas Association Operating Section Proceedings D-1974,* 1974.

21. H. H. West, L. E. Brown, and J. R. Welker, "Vapor Dispersion, Fire Control, and Fire Extinguishment for LNG Spills," paper presented at the Combustion Institute 1972 spring meeting, San Antonio, 1974.

22. J. R. Welker, H. R. Wesson, and L. E. Brown, "Use Foam to Disperse LNG Vapors," *Hydrocarbon Processing,* vol. 52, February 1974, p. 119.

23. L. E. Brown and L. M. Romine, "Liquefied Gas Fires: Which Foam?" *Hydrocarbon Processing,* vol. 58, September 1979, p. 321.

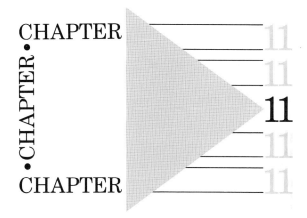

CHAPTER
•CHAPTER•
CHAPTER

11

Case Histories: Volatile Materials

PART 1
Pesticide Fires / 11-2

PART 2
Bulk Terminals: Silicon Tetrachloride
Incident / 11-11

PART 3
Seveso Accident: Dioxin / 11-18

PART 1
Pesticide Fires

Russell E. Diefenbach
Environmental Emergency Section
Region V, U.S. Environmental Protection Agency

Introduction / 11-2
Plant Location / 11-3
Action Levels / 11-6
Fire Units: Preplanning / 11-7
Conclusions / 11-10

INTRODUCTION

The most difficult and potentially dangerous fire with respect to a threat to the environment and human life is a pesticide plant fire. A facility that manufactures or blends or stores pesticides (and herbicides) is a potentially dangerous fire hazard because it contains both petroleum products and nerve gas. The formulation that kills unwanted insects in many cases is nerve gas.

If these products are not controlled, damage to the environment can be extensive. Such is the case when pesticide manufacturing or blending plants or pesticide warehouses catch fire. The pesticides are released by incompletely combusted smoke-carrying particles and by water used to put out the fire. Because of environmental contamination the effects of the aftermath of a pesticide fire are often as severe as the fire itself.

Pesticide fires are occurring with increasing frequency, and the fire fighters who respond to them are often in a dangerous situation that poses a threat to their health. A threat also exists to persons who inhabit an affected area. Within the six states of U.S. Environmental Protection Agency (U.S. EPA) Region V (Illinois, Indiana, Michigan, Minnesota, Ohio, and Wisconsin), at least two "problem" pesticide fires occur annually. These incidents pose severe threats to the environment and result in costly postfire cleanup and disposal problems. Other pesticide incidents, usually associated with transportation and handling, expose fire fighters to contamination, resulting in hospitalization.

In this part the term "pesticide" is used to include insecticides, herbicides, rodenticides, poisons, fumigants, bactericides, defoliants, nematocides, and fungicides. There are in existence approximately 1000 pesticide compounds, and these are formulated and/or blended in nearly 40,000 different products. Many are identical in their basic chemical composition but are packaged under different labels.

The problems that occur during a fire at a pesticide facility usually are the result of an inadequate fire code and a lack of location restrictions. Many pesticide facilities are encouraged by well-meaning city administrators and politicians to locate within a community because of jobs that are created, tax base generation, etc. These well-intentioned persons probably know only that complex chemicals will be in the new plant. The danger and potency of these materials is not apparent to them.

Packaged pesticide products are labeled as follows:

1. Mildly toxic or low-toxicity materials bear the label CAUTION with directions.

2. Moderately toxic chemicals are identified on the label with the word WARNING.

3. Highly toxic materials are labeled with a skull and crossbones and the word POISON.

All these designations are on the final-sale package label. The salable products are in a blended state with inert products or petroleum products as a carrier and in some cases with a propellant. But in a manufacturing or blending plant where mildly toxic products are packaged the source material is probably more nearly lethal. Fire fighters and emergency crews who must respond should consider all pesticide facilities to contain highly toxic, highly flammable, and explosive materials.

Governmental units and pesticide manufacturers must work together for a safe location, a safe facility, and a plan of action that minimizes danger to inhabitants and the environment. After a bad pesticide fire, the facility is usually unwanted in the community. This rejection may be justified, but the facility operator is not always a culprit; the operator, too, may be an innocent victim.

PLANT LOCATION

At the request of a local fire chief in a community on the Fox River in Illinois, approximately 56 km (35 mi) northwest of Chicago, a pesticide plant was inspected. The inspection revealed the following situation.

The plant was located 1.2 km (¾ mi) from a community hospital. The prevailing winds were such that should a fire occur, the smoke would most likely blow toward the hospital and cause an emergency evacuation of the patients. The plant was also located at a high point above one of the city's sewage treatment plants. In the pesticide plant, numerous drains went to the sanitary system. On the property adjoining the building were storm drains. Fire runoff water contaminated with pesticides would reach both the Fox River and the sewage treatment plant. The result of pesticide-contaminated runoff would be a serious pollution problem in the river and possible temporary loss of sewage treatment owing to the killing of all the bacteria.

In the building it was found that blending operations were isolated but that source materials, blended and packaged materials, and petroleum products all were stored in a common area. Further, few workers knew of the dangers and potency of the materials with which they were working; few of them could read, and some did not speak English. The storage area was protected by a sprinkler system using only water. The building was also occupied by other tenants and was bordered on two sides by other industries and on the other two sides by residential areas. The city's elected officials had helped the plant locate at this site after the company was forced to vacate its building in a larger nearby city.

No fire codes or building codes existed to ensure a safe pesticide facility. Indeed, the dangers and problems of a pesticide facility are too specialized and restrictive to be covered by the general codes of a community. The technical recommendations for a pesticide facility to reduce its potential hazards should a fire occur are discussed later in this part.

The problems that occur during and after a pesticide plant fire are exhibited by two pesticide fires discussed in the following subsections.

Alliance, Ohio, Fire

At approximately 8:30 P.M. on July 14, 1974, during a severe thunderstorm, lightning struck a power line leading to a pesticide and herbicide plant in Alliance, Ohio.[1] The resulting fire caused an estimated $1 million damage to the plant, which contained quantities of paint, pesticides, herbicides, carrier distillates, and raw materials.

About 2:00 A.M. on July 15, the U.S. EPA was alerted because of air and water pollution problems caused by the fire and runoff water from fire-fighting efforts. The EPA district oil and hazardous materials coordinator left for the site to function as the federal on-scene coordinator. The Ohio EPA was also contacted for assistance. At 3:00 A.M., evacuation procedures began, the Alliance police being joined by reserves and Stark County sheriff's deputies and auxiliaries, who went from door to door in the area of the fire and suggested that residents leave their homes and seek shelter at the local YMCA. At this point, the air was nearly stagnant, with a slight breeze pushing the smoke in a northeasterly direction toward the Alliance water treatment facilities. Both particulates and gases were in the plume formed by the fire, which was raging out of control. Alliance fire fighters were joined by off-duty members of their department as well as by units from the towns of Sebring and Lexington and the township of Washington in battling the blaze. Constant exploding of 5-gal (19-L) paint and pesticide cans and aerosol bombs hurled debris from the flames and hampered fire fighting.

The fumes from the fire caused fire personnel and residents to be sent to the hospital suffering from smoke inhalation and exhibiting symptoms that included nausea, burning eyes and throat, and dizziness. Treatment consisted of administering atropine sulfate and oxygen; following treatment the patients were released. The treatment with atropine sulfate (1 cm³) was discontinued when it was established that a report of parathion in the fire had been incorrect.

To minimize runoff and extinguish the fire, chemical foam was flown in and used. Several steps were taken to minimize environmental damage from the runoff. Street crews dumped bags of lime into gutters and storm-sewer openings in an attempt to neutralize acidity. Several filter-fence dams were erected across a small creek, at the east side of a cemetery, into which the storm-sewer outfall from the plant flowed.

The dams, composed of peat moss and fine limestone, filtered some of the toxicants from the water before it reached the Mahoning River and, subsequently, the Berlin Reservoir. An EPA team of persons knowledgeable in air monitoring and pesticides arrived by charter flight in the early morning of July 15. At the request of Alliance hospital officials, the EPA also brought a medical expert in pesticide poisoning; he was Donald P. Morgan, M.D., Ph.D., project director of the Iowa Community Pesticides Study.

The National Weather Service was consulted; the winds at the site were predicted to shift to a southeasterly direction as a new cool air system moved in at approximately 5:00 A.M. A new area of concentration for the evacuation efforts was determined; a hospital located approximately 2.4 km (1.5 mi) from the scene, alerted of the possible air pollution danger, prepared evacuation procedures in case the plume should reach it.

One incident, initially reported as a fire-related casualty, but later proved to be unrelated, was the death of a 76-year-old woman at 9:15 A.M. on July 15 because of heart failure complicated by an enlarged liver, according to the coroner's report. The victim had had a history of heart trouble.

The wind shifted as expected and forced relocation of the temporary shelter for evacuees set up at the YMCA to a church farther from the scene. National Guard personnel and police cooperated in maintaining security in the evacuated area; as a consequence, there was only one reported burglary.

Upon the arrival of air and pesticide experts from the U.S. EPA, Region V, on the morning of July 15, air samples were taken from the plume near the still-smoldering fire site. Hydrogen chloride was found to be the main constituent in the fumes at a concentration of 6.5 ppm. On scene at this time were representatives of the following agencies: the Ohio EPA, the U.S. EPA, the Ohio State Wildlife Commission, state and local fire departments, local police, the National Guard, and municipal officials.

During the afternoon of July 15, the mayor called a meeting that included himself, the chiefs of the fire and police departments, and the U.S. EPA emergency response team. Topics discussed were the evacuation situation, the health effects of the hydrochloric acid, possible pesticide residues left by the fumes, and procedures which needed to be implemented to alleviate the hazards posed. Recommendations included the washing of automobiles, clothing, dishes, silverware, and all other materials which could possibly come in contact with the residue. Thorough washing of fruits and vegetables and disposal of all open foods were also recommended. At 5:00 P.M., after further discussion with U.S. EPA representatives, the mayor ended the evacuation order, as the fire was now extinguished and the air pollution hazard eliminated.

At 9:30 P.M., in a meeting with the U.S. Army Corps of Engineers, it was decided to set up a sample schedule coordinating state and federal efforts to determine the water pollution hazard to the Berlin Reservoir, a recreation center and infrequent source of water for several towns. A 2-day ban was placed on all activities in the lake; additionally, there was a recommendation that, once the lake was reopened, fish caught there not be consumed.

On the morning of July 16, a meeting was held at Alliance police headquarters to determine what action to recommend to the company. At another meeting that afternoon, the federal EPA, the state EPA, the Army Corps, and local police and fire personnel assessed the situation and discussed all aspects of the cleanup.

In the afternoon, at the company's offices, cleanup procedures as well as the role of each agency and the company in the total operation were discussed. It was determined that several large storage chemical holding tanks could be salvaged but that all other material in the two affected buildings had to be disposed of in an approved landfill. The site of the land disposal was to be agreed upon by all agencies involved; hauling to the landfill would be in enclosed trucks to prevent any further contamination of the area. In addition, approximately 0.3 m (1 ft) of topsoil was to be removed from the plant site and disposed of because of contamination from the fire runoff. During the cleanup operation, the storm drains on the site were to be covered to prevent additional amounts of hazardous substances from entering the Mahoning River through the city storm-sewer system.

On July 18, the U.S. EPA had Dr. Morgan advise the local medical profession and the local hospital on treatment and diagnosis of pesticide poisoning. Dr. Morgan also met with local police and fire fighters to answer questions and to allay fears arising from rumors.

Follow-up action, including the monitoring of the cleanup and hauling operation of the debris to the approved landfill site in Smith Township, was implemented. The landfill was an old strip mine in a basin formed of 8-m- (26-ft-) thick limestone and clay, which ensured that nothing would seep to the surrounding area. A hotline for citizen and press inquiries was established and maintained in the Alliance City Hall by Ohio EPA personnel, and an extensive water-sampling program was set up to monitor water quality in the Mahoning River and the Berlin Reservoir.

Federal sampling was carried out on July 17, 18, and 20; all chemical concentrations were within acceptable drinking-water standards for the materials monitored. It was determined that the ban on activities in the Berlin Reservoir could be lifted and the water in the area declared safe.

In conclusion, two buildings at the chemical plant had been completely destroyed by the flames, resulting in a loss of nearly $1 million. A total of 500 residents had been evacuated from their homes in the area of the fire, and 180 persons had received treatment at the emergency room of the local hospital.

The response effort required the coordination of the U.S. EPA, the Ohio EPA, the Ohio State Wildlife Commission, the Army Corps of Engineers, the state fire marshal, local fire officials, civil defense personnel, the National Guard, local police, hospital personnel, municipal workers, the Red Cross, the Salvation Army, the YMCA, and company officials. Total costs incurred by city, county, state, and federal governments amounted to $800,000; the added costs borne by the hospital and relief agencies are unknown.

Later, the plant was relocated in a rural area, and today only executive offices remain within the city of Alliance.

Brooklyn Center, Minnesota, Fire

On January 6, 1979, at about 9:30 A.M., fire struck a pesticide warehouse in Brooklyn Center, Minnesota. The building contained an estimated 80 metric tons of active ingredients for 100 different pesticides and some herbicides and fertilizers. Fire fighters used an estimated 3000 m^3 (800,000 gal) of water in fighting the fire.

The runoff water was contained in a nearby stream before it could reach the Mississippi River. Further, the weather was a comforting ally; owing to extreme cold, the runoff water froze almost immediately and reduced the imminent threat of contamination.

Laboratory analysis of the runoff revealed high levels of pesticides, particularly atrazine with concentrations ranging from 100 to 1000 mg/L, indicating even insoluble substances. There was no doubt that the runoff was a potential problem. Although frozen temporarily, when the runoff thawed, water supplies could be endangered. The runoff had to be cleaned up and disposed of properly.

Removal and disposal costs were estimated at $260,000. The owner of the facility would pay only for cleanup costs on his property. He felt that the fire departments were responsible for the contaminated water. All the involved governmental agencies looked to each other to assume the expense. Finally, the governor of Minnesota appropriated special funds.

However, Minnesota had no identified disposal sites for such material. Final disposal was not accomplished until March 17, with monitoring of wells for contamination continuing after this date.

Problems Identified

Incidents such as those described above occur about twice a year. Each one results in high cleanup costs, and, worse, in almost every case some fire fighters suffer from exposure to the pesticide. These ill effects are not due to careless acts, falling debris, and other fire-associated types of injuries but solely to the chemicals and the lack of prior warning of the dangers from plant operators. Additionally, pesticide fires require expensive after-the-incident cleanup actions. The question is "How can the response from both an economic and a safety standpoint be improved?"

ACTION LEVELS

The acute toxicological danger of pesticides occurs in three modes:

1. Dermal

2. Respiratory

3. Oral ingestion

In the first method of entry into the human body, many pesticides can be absorbed through the skin, thereby entering the bloodstream and attacking the nervous system. The mildest results can be headaches and vomiting, progressing to the spitting of blood; the most advanced result can be paralysis and/or brain damage. Respiratory intake is faster than dermal intake but just as severe in its effect. Oral ingestion occurs usually by accidental swallowing and is the least probable means of entry into the body during a fire. However, it can happen at a fire when people gulp to relieve body stress in an atmosphere containing pesticides.

The best method of avoiding the poisoning of pesticide fires is to reduce the hazard of fire and minimize the exposure of fire fighters.

Location and Plant Design

The first step to be taken is to locate the pesticide plant in an area of low population density; the more remote the area, the better. The plant should also be situated in a location that minimizes drainage to streams. In addition, there should be no in-plant floor drains that connect to the sanitary system. Drains to the sanitary system should be limited to toilet facilities, washbasins, showers, etc. Storm drains on the surrounding plant property should be capable of being closed if a fire should occur.

All stocks of petroleum products used as a carrying medium for the pesticides should be stored in a fireproof-vaulted area equipped with an automatic foam extinguishing system. Blended products in bottles, cans, and aerosol cans, boxed, or otherwise packaged should be warehoused independently from petroleum stocks. Source materials stocked for blending can be in the same building but should be isolated. These areas should be protected by a foam system in preference to water sprinklers. The blending, formulation, and packaging area should be separate from storage areas. This area should be well ventilated and fire-protected, and all employees should be trained in emergency procedures should a fire occur.

Consideration should be given to diking the plant during its design or to implementation of an action plan for emergency diking during a fire incident. Full consideration and forethought should be given to fully containing or minimizing fire water runoff.

FIRE UNITS: PREPLANNING

Fire units having pesticide facilities within their areas of protection should give serious consideration to the use of foam in fighting a pesticide fire. They should also consider specialized training for their personnel, directed toward personal safety at the fire site and decontamination of equipment and gear after the blaze has been extinguished.

Preplanning can be the most efficient means of making a safe response to a pesticide fire. Preplanning activities should be under the direction and active pursuit of the commanding officer of the fire unit, but they should include other elements of government and industry. The recommended steps of preplanning are as follows:

1. Establish a relationship with the facility owner for plant familiarity. Responding units should have a floor plan and knowledge of the chemicals stored. Fire fighters should know what products are in the plant, and their location should be established on the floor plan.

2. The medical director of the local hospital should be informed of materials in the plant in case of poison admissions from fire-fighting activities. Fire units, medics, and hospitals should be provided with copies of *Recognition and Management of Pesticide Poisonings.*[2]

3. As in all major fires, the method of fighting the fire should be preplanned and well thought out, and all fire fighters should be well versed in the various options. Consideration should be given the following fire-fighting techniques:

   a. Plan to attack the fire with foams exclusively; use water only when the foam supply is exhausted. If the fire can be successfully extinguished with foam, contamination of streams is avoided or reduced and danger to sewage treatment plants is reduced. Also, the carrying of pesticide particulate matter from a plume is greatly reduced.

   b. If water must be used, either exclusively or after foam supplies have been exhausted, arrangements should be made for drains to be plugged, drainage systems isolated and used as traps, the sewage treatment plant notified, and runoff water from the fire isolated at the plant if possible. Emergency activated-carbon filter dams should be constructed at drainage points and at sewer entries to reduce the contamination reaching both storm and sanitary systems (for field techniques, see the U.S. EPA report[3]). Preplanning sessions with sewage and street department officials are necessary, and the employees of these departments should have responsibility for the necessary actions to be taken in an emergency.

   c. Preplanning will also involve the establishment of evacuation areas and evacuation routes. Parameters should be established to determine when to evacuate. Alternative temporary quarters for evacuees should be selected. This selection process should involve police and civil defense departments and relief agencies.

   d. If the pesticide facility is isolated or the fire itself is not a threat to other property, *consideration should be given to letting the fire burn rather than extinguishing it.* This may be the safest and most economical process when the cost of postfire cleanup is considered; complete or nearly complete combustion will reduce the amount of pesticide particulates in the plume. Firefighters can concentrate on protecting surrounding property without subjecting themselves to pesticide poisoning by entry into the facility. Other hazards for firefighters to consider, in addition to pesticides and petroleum products, are associated fine dusts that are flammable and explosive and aerosol cans that can become incendiary propellants or flying bombs. All such fires give off toxic fumes, and the plume should be monitored by grab sampling, if conditions permit, to predict the need for evacuation. An important safety point should be noted: a really hot fire will decompose most pesticides, but after the fire fumes can still be given off by hot undecomposed material. Thus, the cleanup crew can be in danger in a period normally thought to be safe.

4. The plan must extend beyond the fire emergency. After the fire all runoff water should be checked to see if the concentration of toxic materials is dangerous; if so, the contaminated water must be treated. Further, any food crops in the area and affected residences should be checked for contamination caused by the fire. All fire equipment and fire fighters' personal safety gear should be decontaminated immediately upon return to the fire station. All those who have worked in close proximity to the fire should be monitored for symptoms of pesticide poisoning.

Recommended Safety Actions during the Incident

Fire fighters and other responding officials should be protected by self-contained air packs and acid-resistant rubberized clothing including gloves. The fire fighters should be encapsulated to avoid completely skin and respiratory contact with the pesticides. Persons in fringe areas can use canister masks with organic vapor elements, but self-contained air should be the only consideration for fire fighters in close proximity to the fire.

Upon return to home stations, all should shower with copious amounts of soap and water (the water should be as hot as possible). All clothing should be thoroughly washed. All outer gear (fire coats, boots, helmets, gloves, etc.), spanner wrenches, tools, and fire-fighting equipment including trucks should be washed with solutions of soda ash or sodium hypochlorite and detergent. Use 1 L of sodium hypochlorite, 5% sodium carbonate, or 5% trisodium phosphate with one cup of detergent to each 4 L of water. After washing, all equipment should be rinsed thoroughly, making sure that the water is directed to the sanitary sewer.

All persons involved should be monitored for the following symptoms:

• Headaches

• Giddiness

• Nausea

• Blurred vision

• Chest pains

Any of these symptoms may indicate pesticide poisoning; suspect persons should report to medical authorities. Such persons should be watched closely for additional symptoms such as sweating, pinpoint pupils, drooling, and/or vomiting.

Should fire runoff water not be fully contained according to the prearranged plan with the street and sewer departments, all possible efforts to dam or treat with activated-carbon filter dams should have a secondary priority to fire efforts. Evacuation procedures should be followed without hesitation if the content of the smoke and vapors is unknown.

Monitoring of the area after the fire is necessary to remove all doubts concerning contamination. Contaminated runoff water should be treated with activated carbon to a safe level before entering a sewage treatment plant or a natural stream. Portable activated-carbon units can be used to treat contamination runoff water that has been captured. Also, activated-carbon beds may be used at the sewage treatment plant to treat runoff water. If streams used for water supply are contaminated, an activated-carbon bed may be necessary at the water treatment plant. The adsorption potential of activated carbon for selected pesticides is shown in Table 1-1. Adsorption plots can be obtained for other compounds from carbon suppliers, but intake and discharge monitoring is recommended when a carbon filter unit is used.

TABLE 1-1*   Pesticide Adsorption Isotherms on Activated Carbon

Compound	Molecular weight	Initial concentration (mg/L)	Weight, percent adsorbed†	Type of carbon
Aldrin	365	0.048	2.45	FS-300
Chlordane	409.57	0.057	6.5	FS-300
DDD	320.1	0.056	0.01	FS-300
DDE	. . . . . .	0.038	0.47	FS-300
DDT	354.5	0.041	0.82	FS-300
Dieldrin	381	0.019	1.4	FS-300
Endrin	. . . . . .	0.062	7.6	FS-300
Lindane	290.84	0.100	2.5	FS-300
Naphthalene	128.16	9.7	2.77	FS-300

*Courtesy of Calgon Corporation, Pittsburgh.
†Kilograms of pesticide per kilogram of activated carbon.

The area of the pesticide fire should be secured against public or any unauthorized entry because of the danger from residual chemicals. Again, all workers removing the debris should wear proper protective clothing and organic-vapor-dust canister face masks. All debris should be removed to proper approved disposal sites by closed hopper trucks. The use of open trucks would result in blown dust and dangerous distribution of pesticide-contaminated debris.

The manufacturer of the pesticide base products should always be requested to send a representative for technical consultation and advice. Local, state, or federal environmental authorities should also be called in to assist local authorities both during the fire and for postfire monitoring.

Rural fire units should also be aware that they may be exposed to pesticides when fighting barn or other farm utility building fires. In addition, hardware, garden supply, and other retail outlets usually have stocks of pesticides.

CONCLUSIONS

The potential dangers from a pesticide fire can be reduced by proper location, plant facility design, preplanning of coordinated actions of all involved units, and radical departure from conventional fire-fighting techniques. The most difficult decision in the plan will be to allow a pesticide facility to burn. This decision must consider location, surrounding elements, weather conditions, and contents of the smoke plume. To ask a fire unit commanding officer to let a pesticide facility burn without forethought and evaluation of these factors places the officer in the awkward position of being unprepared. However, these factors should be given full consideration by the fire unit's commanding officer for maximum safety considerations for both fire fighters and the community. As society develops and uses a greater variety of complex chemicals, the fire fighters' job becomes increasingly difficult, complex, and dangerous. Specialized training needs should be developed and offered to fire fighters.

REFERENCES

1. R. E. Diefenbach, "On-Scene Decisions—The Resultants in Effects—Impact," in G. F. Bennett (ed.), *Proceedings of the 1976 National Conference on Control of Hazardous Material Spills,* New Orleans, April 1976, pp. 356–367.

2. D. P. Morgan, *Recognition and Management of Pesticide Poisonings,* U.S. Environmental Protection Agency, 1976.

3. K. R. Huibregtse, R. C. Scholz, R. E. Wullschleger, J. H. Moser, E. R. Bollinger, and C. A. Hansen, *Manual for the Control of Hazardous Material Spills,* vol. 1: *Spill Assessment and Water Treatment Techniques,* EPA-600/2-77-227, U.S. Environmental Protection Agency, November 1977.

PART 2
Bulk Terminals: Silicon Tetrachloride Incident

William C. Hoyle, Ph.D.
Corporate Analytical Services
The Continental Group, Inc.

Friday, April 26, 1974 / 11-12
Saturday, April 27, 1974 / 11-14
Sunday, April 28, 1974 / 11-15
Monday, April 29, 1974 / 11-16
Tuesday, April 30, 1974 / 11-16
Analysis / 11-16

One of the largest and most disruptive hazardous material incidents recorded to date occurred in 1974 at the Bulk Terminals storage facility in Chicago, Illinois. This incident is noteworthy for several reasons including the quantity of material involved (3300 m^3, or 876,000 gal), the location (a large metropolitan area), the duration (leakage for 8 days, emissions for 20 days), and the nature of the chemical (highly toxic, strongly irritating, and highly reactive, yielding volatile, strongly irritating toxic products). All these features, when combined, required a response effort second only to the responses required by oil tanker disasters. Hence it is worthwhile to review the chain of events following the initial release of silicon tetrachloride (siltet) and to examine the cleanup effort, with the idea that an approach may be developed which will help prevent future incidents, or if an incident does occur, that the system developed will minimize the effects on the public and the environment.

Bulk Terminals is a storage farm located in the Calumet Harbor area of Chicago's South Side. Its 78 storage tanks, which have capacities ranging up to 4900 m^3 (1.3 million gal), contain large quantities of liquid animal fats, vegetable oils, and chemicals. Shortly after 12:30 P.M. on April 26, 1974, several Bulk Terminals employees heard a sudden thud and "noticed fumes rising out of the dike area surrounding Tank 1502."

Much later, the cause was attributed to events occurring earlier that year when a pressure release valve on a 152-mm (6-in) line leading to the tank had been inadvertently closed. A flexible coupling on the line had burst under the pressure, and the entire piping system had shifted, cracking a 76-mm (3-in) line at the tank wall.

Within hours the ruptured tank affected the lives of tens of thousands of Chicagoans. As the escaping silicon tetrachloride reacted with the moisture-laden air, an enormous, breathtaking, eye-watering, nausea-inducing acid cloud spewed forth. At times the acid cloud measured from 8 to 16 km (from 5 to 10 mi) long as it moved across the city, disrupting traffic, work, and normal living activities.

No fewer than nine governmental agencies and departments became involved in the incident. More than 16,000 area residents were evacuated, as state, city, and federal agencies worked around the clock for 8 days to stop the leak, neutralize the spill, and transfer the material to another tank.

Someone not actually involved in the response will find it difficult, if not impossible, to imagine how one leaking tank could withstand the efforts of so many people for such a long time. However, a chronology of events will provide a dramatic picture.

FRIDAY, APRIL 26, 1974

A U.S. Coast Guard crew, during the course of routine observations in the Lake Calumet area, was the first governmental unit to notice the leak at Bulk Terminals. The Coast Guard notified the Chicago regional office of the Illinois Environmental Protection Agency (EPA), which in turn dispatched to the scene two engineers who had been trained in incident response procedures and at 1:05 P.M. also notified the Emergency Action Center (EAC) of the Illinois EPA. At this time preliminary information about the chemical, its reactivity, and possible countermeasures was relayed to the regional office and the response engineers.

By 3:00 P.M. the leak was getting progressively worse; the acid cloud was now about 400 m (¼ mi) wide, about 300 to 450 m (1000 to 1500 ft) high, and about 1600 m (1 mi) in length. The corporate response efforts were minimal, while employees awaited the arrival of chemical experts from Cabot Corporation, owners of the silicon tetrachloride. The fire department had not arrived; since there was no fire, the chief in command did not think the department should become involved. The exchange of information between the scene and the EAC continued via the Coast Guard telephone.

At this time, the first good estimate of the extraordinary quantity of material (3300 m³) was obtained. The EAC, which is staffed with chemists, engineers, and meteorologists, calculated the hazard area from information gathered at the scene. Information regarding the hazard area and weather conditions was updated for the on-scene personnel and the state civil defense agency. At this time, Bulk Terminals officials were advised by the EAC to use lime to neutralize the spilled material.

At 3:40 P.M. the first report of people becoming sick was received. The Illinois EPA diverted lime trucks from a nearby facility to the scene, turned in a fire alarm, and requested additional equipment such as acid suits and breathing gear from the metropolitan sanitary district. The lime trucks were refused admittance to the scene by the Bulk Terminals officials.

By 4:45 P.M. the chemical rescue unit of the Chicago Fire Department was in charge of the emergency operations. The acid concentration in the air was increasing as the spill worsened and the silicon tetrachloride began to pool within the dike area. EAC personnel discussed the ramifications of evacuating the Altgeld Gardens area, which housed from 20,000 to 30,000 persons approximately 2.8 km (1¾ mi) from the tank. Three National Guard battalions were put on alert by the governor of Illinois.

Shortly after 5 P.M. the chemical experts arrived at the scene; two fire fighters had been overcome by the fumes, and visibility was greatly reduced. The Friday-evening rush-hour traffic leaving the city on the nearby interstate highway was using headlights because of the dense acid cloud. The local and national television news media were now giving live on-the-scene reports of the incident and interrupting normal programming with updated bulletins.

Shortly after 6 P.M., the Illinois EPA received the first complaints from the Altgeld Gardens area. Chicago Police Department units began evacuation of the area almost immediately. Altgeld Gardens residents were taken to Carver High School, where first aid was administered by the fire department. Approximately 38 persons were treated in nearby hospitals. At about the same time, an 8-km (5-mi) section of the Calumet Express-way was closed because of almost-zero visibility.

Additional air-monitoring equipment was being adapted for continuous monitoring of hydrochloric acid (see Fig. 2-1 for the area being monitored). The meteorological conditions and acid concentrations in the residential and industrial areas were now being constantly monitored by Illinois EPA personnel. The initial attempt to "hot-tap" the 152-mm pipe at the base of the tank was unsuccessful because of the adverse conditions. Meanwhile, evacuation from Altgeld Gardens was made mandatory; the shift in the wind necessitated that those in Carver High School be relocated by bus to Fenger High School. Approximately 16,000 persons were evacuated.

Fig. 2-1 Bulk Terminals aerial view. Note the nearby residential and industrial areas which had to be constantly monitored. [*Photo courtesy of Illinois Department of Transportation.*]

As nightfall approached, the acid plume was still worsening. The diked area was now flooded to an approximate depth of 100 mm by the 110 m³ (29,000 gal) of silicon tetra-cloride which had leaked. Skids were now required to bridge the moat to the leak area. The concentrations of HCl in the residential areas were reported to be in excess of 35 ppm.

During this time EAC personnel were in contact with the safety directors and chemists of more than 15 chemical corporations discussing various approaches to solving the situation. Suggestions ranged from neutralizing the pool with lime, to flooding the dike area with water, to covering the pool with table tennis balls to reduce the emissions. Because the Chicago Department of Environmental Control could not locate a sufficient supply of table tennis balls, that plan was abandoned.

The best approach was believed to be the covering of the pool with some material to reduce the emissions first, thereby providing time to remove the material in the tank. Two methods for reducing the emissions appeared to be most viable: (1) a nonprotein high-expansion foaming agent and (2) No. 6 fuel oil.

Fig. 2-2   Addition of lime to the moat area.

SATURDAY, APRIL 27, 1974

Shortly after midnight the foam and experts from Dow Corning Corp. were helicoptered to the scene. Simultaneously, a shift in wind direction and speed moved the acid cloud away from the residential area toward Lake Calumet. The EAC meteorologist predicted that this condition would persist for the next 24 to 36 h; on this basis, the police allowed the people who had been evacuated to return to their homes. During the night the plume shifted periodically, causing problems in isolated areas. Fortunately, the plume stayed either over Lake Calumet or in industrial areas which were closed for the weekend. Because of the size of the plume, air traffic patterns in the area were sporadically disrupted.

At 4:10 A.M. foam was applied, but it was ineffective in reducing the emissions. A decision was made to apply the foam after the fuel oil had coated the surface. The transfer pumps still were not working.

By morning an estimated 230 m³ (60,000 gal) had leaked, and the rate was increasing. At 8 A.M. approximately 4 m³ (1000 gal) of No. 6 fuel oil was added to the moat area; simultaneously eight truckloads of lime were added at the opposite side (Fig. 2-2). The effects were dramatic: emissions were drastically reduced 1 h later. The HCl concentration was only 12 ppm approximately 3.2 km (2 mi) from the scene. More oil and lime were blown on throughout the morning as pockets of fumes erupted.

The 16-member chemical warfare unit of the U.S. Army Technical Escort Center (TEC) arrived at the site about 10:30 A.M. with its 8200 kg (18,000 lb) of equipment. The first pumping started at 10:40 A.M. at a rate of 45 m³ (12,000 gal)/h. About noon, 24 h after the leak had begun, the first calm appeared in the area of Tank 1502. The pump was working, and with the second pump being installed, the Bulk Terminals officials said that it would require 30 to 40 h to pump the remaining 2800 m³ (750,000 gal) out of the tank. Emissions were low; continuous monitors showed the source emissions were 1900 ppm of HCl at noon, but this figure was reduced to 635 ppm by 5:30 P.M. As night approached, little vapor could be seen at the site, and most fire and police department personnel went home. At 10:40 P.M. the EAC reported that there was a 30 percent probability of rain, with the percentage increasing as a front approached.

SUNDAY, APRIL 28, 1974

By 1 A.M. Sunday morning only one police car was on the scene, and all fire equipment was removed. Emissions were negligible 180 m (200 yd) downwind from the tank. By 5 A.M. all pumps had stopped because of acid corrosion.

The frontal system entered the Chicago area at approximately 8:00 A.M. At 7:30 A.M. a light rain began to fall at the site. Within a short time heavy rains were falling, and the wind was gusting to 48 km (30 mi)/h. At the onset of heavy rains, the EAC personnel predicted much heavier emissions in the immediate vicinity of the tank, with the HCl being effectively scrubbed by the rain before reaching the residential areas. The fuming was so severe that at 8:15 A.M. the electric power lines were being corroded by the HCl: "The lines and transformers were sparkling and exploding like a Fourth of July fireworks display."

By 9:30 A.M. Sunday morning, the situation was worse than at any previous time. Four pumps were rendered inoperable by the HCl corrosion, 300 to 400 m³ (80,000 to 100,000 gal) had leaked, and area residents were put on alert for possible evacuation since the rain had stopped and the gusty winds were pushing the 14.5-km- (9-mi-) long cloud to the ground. The fire department ordered the resumption of adding lime and asphalt. Meanwhile, the electrical power failure halted all pumping until emergency generators could be obtained.

As the day progressed, the situation worsened. The probability of the worst-case situation was increasing: the main pipe leading to the tank (Fig. 2-3) where the leak had occurred had been severely weakened by the acid. It appeared that this pipe might break,

Fig. 2-3   Tank 1502 on Sunday, April 28, 1974. [*Photo courtesy of Illinois Department of Transportation.*]

allowing the remaining 2300 m³ (600,000 gal) of siltet to flow into the moat, which had already been severely reduced in capacity by the addition of lime, sand, and fuel oil.

At the peak of this new crisis, communications were faltering: civil defense radios were ineffective because of the storm, phone lines were rendered inoperable because of the HCl, and high, gusty winds were curtailing the use of helicopters for surveillance and communication purposes. Illinois Bell emergency mobile units were rushed to the scene for Illinois EPA use. As the afternoon progressed, a large pit was dug and lined with lime to prevent the overflow from the dike area from reaching the city sewer lines and Lake Calumet. Simultaneously, the U.S. Army and private contractors prepared to seal the area of the leak with quick-drying cement. Wearing M-3 toxicological suits, the Army TEC personnel went into the leak area to help guide the flexible hose lines. The pouring of 110 m³ (150 yd³) of cement required nearly 8 h in near-zero visibility. Mixed reports indicated that either the leak was sealed or was continuing to leak at a reduced rate. Because of the low visibility it was impossible to tell.

MONDAY, APRIL 29, 1974

The rain, which restarted just before midnight, continued throughout Monday morning. Additional fuel oil and plastic sheeting were placed on the spill area to reduce the fuming. An early-morning reconnaissance of Tank 1502 was inconclusive; visibility of less than 457 mm (18 in) did not permit accurate appraisal. By late afternoon, inspection teams reported that the concrete seal was deteriorating and siltet was leaking from the tank; a rumor reporting that the concrete had been poured on the wrong pipe was never confirmed. A second concrete pour was completed at 11:30 P.M.

TUESDAY, APRIL 30, 1974

The emissions were greatly reduced by this time, and the EAC recommended and the Army and the fire department agreed that a water curtain would be discharged into the acid cloud in an attempt to scrub the acid from the cloud, which was now blowing toward nearby residential areas. Marine Engine 58 was positioned, and with a discharge of 0.4 m³/s (6500 gal/min) of water the HCl concentrations were dramatically reduced.

ANALYSIS

It would take 3 more days to pump the remaining material to the adjacent tanks. It would take until May 15 finally to reduce emissions from the moat area to tolerable levels. Storage Tank 1502 sat there like a frowning green monster, puffing clouds of noxious fumes until it literally ran out of gas: 1800 m³ (480,000 gal) of silicon tetrachloride was transferred; 1100 m³ (284,000 gal) was spilled; 60 m³ (15,000 gal) remained in the tank. One person was killed; approximately 160 persons were hospitalized.

The response actions which were carried out during this incident can be divided into three major categories: (1) actions which monitored the concentration of the HCl emissions, (2) actions which attempted to reduce the HCl emissions from the moat, and (3) actions to remove the silicon tetrachloride to other tanks. A reduction of the emissions would, of course, reduce the risk to the general public by reducing exposure to the hazardous acid cloud. The constant monitoring of the HCl concentrations in both residential and industrial areas helped ascertain the degree of risk at any given time. Because of the size of the leak and the duration of the emissions from Tank 1502, the Illinois EPA monitored the HCl by length-of-stain measurements and continuous monitors. During the response effort, occasions arose which required the use of predictive models in addition to direct monitoring. The duration of the incident presented a unique opportunity to calculate ambient ground-level HCl concentrations and to compare these results with actual measured concentrations.

The modeling resulted in very good estimates of the hydrochloric acid concentrations. For example, on the afternoon of Sunday, April 27, during the period when the 76-mm pipe was suspected of breaking, dispersion models were used to predict the HCl concentration for a period of time before, during, and after the break. Fortunately, the pipe never broke; however, the results of the model for the period before the break correspond well to the actual concentrations measured. Generally the deviation was well within 20 percent.

GENERAL REFERENCES

1. W. C. Hoyle, G. L. Melvin, and J. G. Coblenz, "A Statewide Hazardous Materials Program," in G. F. Bennett (ed.), *Proceedings of the 1974 National Conference on Control of Hazardous Material Spills,* San Francisco, August 1974, p. 149.

2. *Chemical Leak at the Bulk Terminals Tank Farm,* Illinois Legislative Investigating Commission, June 1975.

3. W. C. Hoyle and G. L. Melvin, "A Toxic Substance Leak in Retrospect," in G. F. Bennett (ed.), *Proceedings of the 1976 National Conference on Control of Hazardous Material Spills,* New Orleans, April 1976, p. 187.

4. W. C. Hoyle, *A Practice Response to an Air Pollution Emergency,* Illinois Environmental Protection Agency, 1973.

5. *Evacuation Risks—An Evaluation,* EPA-520/6-74-002, U.S. Environmental Protection Agency, National Environmental Research Center, Las Vegas, Nev., June 1974.

PART 3
Seveso Accident: Dioxin

Alex P. Rice
Director of Client Services
Cremer and Warner, Ltd.

Introduction / 11-18
Occurrence of the Accident on July 10, 1976 / 11-19
First Responses to the Accident / 11-22
Decontamination and Rehabilitation: General Considerations / 11-23
TCDD Analysis and Monitoring / 11-24
Theoretical Analysis of the Discharge, Its Dispersion, and Deposition / 11-29
Definition of Criteria for Contamination Levels / 11-33
Destruction of TCDD / 11-34
Decontamination Operations / 11-37
Appendix / 11-40

INTRODUCTION

At approximately 12:37 P.M. on Saturday, July 10, 1976, at the ICMESA factory some 20 km north of Milan, a bursting disk on a reflux column ruptured. Through the vent pipe, which projected through the factory roof, a discharge took place at sonic velocity, forming a cone-shaped cloud which drifted downwind in a south-southeast direction. The ambient temperature at the time was about 35°C, and much of the cloud was carried upward by thermal currents rising from the hot land surface, but there was also deposition as the cloud moved away from the source.

Providentially, the cloud was carried along a corridor of open fields before it reached any built-up areas. Apart from one isolated pocket of residential properties, inhabited areas were spared the heaviest fallout. In no other direction in which the cloud might have been carried by the wind was there a comparable "window" in the relatively urban environs of the factory (Fig. 3-1).

The discharge originated from a reactor being used in the manufacture of 2,4,5-trichlorophenol (TCP) by alkaline hydrolysis of 1,2,4,5-tetrachlorobenzene. The particular process in use, with ethylene glycol as the solvent, normally proceeded at atmospheric pressure. Following completion of the hydrolysis, the solvent was removed by distillation at reduced pressure. The fact that a bursting disk, rated at approximately 355 kPa, actually ruptured could only be accounted for by some abnormal rise in temperature in the reactor, which was normally controlled at a maximum of 185°C. It is known that when sodium trichlorophenate is heated above about 180°C, the highly toxic compound 2,3,7,8-tetrachlorodibenzo-p-dioxin (TCDD) begins to form according to the reaction

The amount formed varies according to the temperature excess and the time for which this is maintained.[1] Thus, it became apparent that the discharge from ICMESA had released a mixture of chemicals, containing an unknown amount of TCDD, later estimated as about 2 kg, which dispersed as contamination over a wide area downwind of the works. Hay[2] has given an account of the aftermath of the accident and its immediate effect on the local communities.

Previously there had been accidents at factories in other countries where 2,4,5-TCP was being made, with releases of reactor contents contaminated with TCDD.[3-10] But in all the other accidents the escapes were confined to the interior of the building housing the plant. The ICMESA accident was unprecedented in the dispersion of TCDD beyond the factory site over a considerable area in which a large number of people lived and worked.

There was a considerable number of animal mortalities, largely attributable to ingestion of contaminated fodder, but no deaths attributed to TCDD poisoning in the human population, at least within 2 years of the accident. A report prepared by workers from the Istituto Superiore di Sanità in Rome[11] summarized the results of the epidemiological survey over the first 2 years. There was increased frequency (about 0.6 percent in a screening of over 32,000 people) of chloracne mainly in children and young persons, and the condition of those affected was improving. There was some evidence of subclinical neurological damage as well as cases of clinically detectable polyneuropathy in adults, based on the screening of about 700 individuals including 446 from Zone A (Fig. 3-2). The pathological signs were most frequently found in the peripheral nervous system. There was no evidence for a positive correlation between incidence of neurological symptoms and chloracne.

A limited percentage of cases of liver enlargement had been reported, without giving any criteria for determining the diagnosis. Some alterations were observed in some exposed people in one or more liver tests (mainly transaminases and γ-GT). Immunological investigations, cytogenetic examination, and embryomorphology analysis in cases of therapeutic or spontaneous abortions gave no abnormal results.

Although the clinical surveillance of TCDD-exposed persons was planned to continue for several more years, the results of the first 2 years' work indicated that the alarming predictions voiced by some experts around the world immediately after the accident were ill-considered. Nevertheless, taking account of the disruption of people's lives, including the loss of homes and possessions, the personal anxieties caused on health and economic grounds, the social and ethical problems created, and the costs of emergency and remedial measures, it is justified to describe the accident and its consequences as a disaster.

OCCURRENCE OF THE ACCIDENT ON JULY 10, 1976

According to the *Final Report of the Italian Parliamentary Commission of Inquiry into the Seveso Disaster (FRIPC Report)*,[12] the process being used at ICMESA was a variation from the original process patented by Givaudin in 1947.[13] In the original process, the reaction mixture in the first stage comprised 1,2,4,5-tetrachlorobenzene (TCB), caustic soda, and ethylene glycol in the molar ratio of 1:2:11.5, and the reaction was carried out at a temperature between 160 and 200°C, preferably in the range 170 to 180°C. This

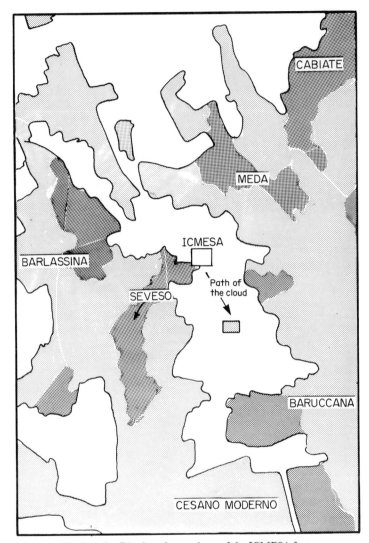

Fig. 3-1 The "window" in the urban environs of the ICMESA factory.

was followed by acidulation with hydrochloric acid in a second stage. Next the ethylene glycol was distilled off and recovered, and finally the 2,4,5-trichlorophenol was recovered and purified.

The variations in the ICMESA procedure are identified by the *FRIPC Report* as the following:

Stage 1 reaction mixture. TCB: NaOH: ethylene glycol—1:3:5.5 molar plus an unspecified quantity of xylene and a maximum temperature of 185°C.

Stage 2. Distillation of solvents under reduced pressure.

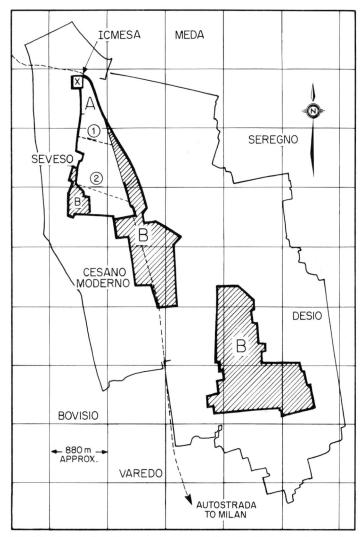

Fig. 3-2 Boundaries of contaminated zones—Zone A with intermediate southern limits and the original Zone B—and Zone R (Zone of Respect).

Stage 3. Acidulation with hydrochloric acid.

Stage 4. Recovery and purification of 2,4,5-TCP.

The batch quantity was about 6000 kg, and the process times for each stage were 8,
6, 1, and 8 h respectively. These times allowed for one complete batch to be finished in
a 24-h day involving three shifts.

A new batch commenced production at 4 P.M. on Friday, July 9, 1976, and the first
stage would have been completed by 4 A.M. the next morning. The evidence is that at 5
A.M. on July 10, when distillation of the solvent was proceeding, the production cycle was

interrupted. Exactly what this entailed is not described, apart from stopping the reaction agitator and leaving the mixture to cool down naturally. No shift was due to take over at that point, the day being a Saturday.

At some time later in the morning, the temperature ceased dropping and then began to rise again. The reason for this is not discussed. The report refers to Milnes[14] with regard to a possible exothermic reaction in mixtures of ethylene glycol and caustic soda. This is said to start at about 230°C and to proceed rapidly up to about 410°C. The heating of the reactor, in the first stage, was with steam at 1216 kPa from the factory steam system. No evidence is cited in the report that the normal maximum reaction temperature of 185°C for the first stage was exceeded, and apparently the system had cooled naturally to a somewhat lower temperature. There remains the problem of what caused the initial rise in temperature up to 230°C, at which the exothermic reaction referred to above is said to start. Subsequent efforts by Givaudin in the laboratory to cause a repetition of the event deliberately were not successful.

At 12:37 P.M. on July 10, a bursting disk, fitted to the top of the reflux column attached to the reactor, gave way. From the report, it appears that the purpose of the bursting disk was not to provide relief of overpressure due to abnormal reaction conditions. After the removal of the solvents by distillation under vacuum, the procedure was to transfer the sodium trichlorophenate to another vessel in which the acidulation was carried out. This transfer was performed by the application of compressed air, and the bursting disk's function was to protect the equipment in the event of faulty control of the air pressure during this operation. This accounts for the bursting disk's being vented to the outside through a short stack.

The bursting disk is reported as rated at approximately 355 kPa, which if related to the temperature-pressure characteristics of the reaction mixture, would not have prevented a dangerous temperature rise, to about 240°C at least.

Eyewitness accounts, reported in the press, describe the discharge as lasting for about 2 or 3 min and forming an inverted cone-shaped cloud of a height variously estimated at between 20 and 50 m. Initially the cloud started to drift toward the northeast, but after a few seconds the wind changed and carried the cloud in a south-southeast direction. The discharge was audible, the sound being described as a whistle. This attracted the attention of the ICMESA department manager, who happened to be near the works, and he turned on the cooling water and informed one of the senior technical staff, substituting for the technical manager, who was on vacation.

FIRST RESPONSES TO THE ACCIDENT

The first samples for analysis were collected on July 11, 1976. These were taken from foliage which had been discolored by contact with the cloud released from the vent. The samples were taken the same day to the Givaudin Research Co. laboratory in Switzerland for analysis.[15] Results of the analyses, obtained on July 14, showed the presence of TCDD, and this was confirmed by further tests during the next few days.

On July 15, the mayors of Meda and Seveso decided to surround the zone thought to have been affected by the cloud with notices warning of contamination. In the evening of the same day dermatological symptoms were observed, especially on some children. There were also mortalities of animals. The local authorities of Seveso and Meda ordered the burning of the animal corpses and affected crops, but this was not done.

During the next few days more cases, including four children with severe chemical burns, were hospitalized. Meanwhile, ICMESA personnel continued to collect samples of various kinds from the area downwind from the works. The first evidence for the distribution of the contamination began to accumulate, but because of the variety of samples—leaves, grass, mixtures of grass and soil, and wipe samples from hard surfaces—drawing significant conclusions from the results was difficult.

The weather remained hot and dry. At the time of the accident the temperature was about 35°C, and soil conditions were very dusty. Most of the open fields were covered in thick herbage standing ½ to 1 m tall, and it seemed most likely that the heaviest contamination remained deposited on the vegetation. It was suggested that spread of contamination might be reduced by sealing it onto the vegetation, bare earth, and other surfaces with soil-stabilizing resin emulsions. Arrangements were made to have stocks of such resins in readiness, together with spraying equipment.

This idea had the additional attraction that once the contamination was firmly attached to the vegetation, reaping and collection of the latter would effectively collect up and bring under control at least a major proportion of the TCDD which had been released.

Unfortunately, Givaudin did not receive permission to proceed, and in mid-August the dry weather abruptly ended in a series of violent thunderstorms and heavy rains. These transferred contamination from the vegetation to the soil, thereby considerably increasing the problems of decontamination.

The first definition of the most heavily contaminated area was made on July 24 on the basis of analytical results and the incidence and location of animal mortality. This area, called Zone A, originally extended about 750 m south-southeast from the ICMESA works and covered about 15 ha. It was decided to evacuate the population and prohibit access. By July 26, a total of 225 people had been evacuated.

As the flow of results was increased, the southern boundary of Zone A was moved to more than twice its original distance from ICMESA within a few more days. There was yet another extension a little later, and by the end of August 1976, Zone A extended some 2.2 km from the works and covered 108 ha. The number of people evacuated exceeded 730.

The continuing analysis of samples indicated contamination by TCDD, at smaller concentrations than in Zone A, in areas farther removed from the ICMESA factory. This led to the first definition of Zone B. General evacuation from this zone was not proposed, but other special measures were imposed on August 2. All children under the age of 12 were removed from the area during the daylight hours, as were women in the first 3 months of pregnancy, and there were rigorous controls on food and water supplies. The initial area of Zone B (which consisted of two separate parts at that stage) was 269.4 ha in all.

Finally, the authorities defined a "buffer zone," designated Zone R, isolating Zones A and B from the areas considered to be unaffected by TCDD contamination. Zone R (Zone of Respect) was also sampled, and results were either below the detectable level for TCDD or at least very low.

These first definitions of the zones considered to form the area of concern were mapped provisionally in August 1976 and became definitive when approved by the Lombardy Regional Council on October 7, 1976.

The boundaries of the three zones are shown in Fig. 3-2. The interim southern boundaries of Zone A are also shown.

DECONTAMINATION AND REHABILITATION:
GENERAL CONSIDERATIONS

The unprecedented release of TCDD contamination over a wide area in the open meant, among other things, that the relevance of the experience and information relating to previous accidents involving TCDD needed reevaluation. Regrettably, excessive publicity was given early to views or suggestions proffered by people, including scientists, unappreciative of the particular situation at Seveso. This inhibited the implementing of practical programs for mitigating the disruption and economic loss suffered by the communities affected.

The key questions involved in formulating a strategy for decontamination were as follows:

1. What was released from the ICMESA reactor? How much in total? How much TCDD was associated with the release? How much of what was released came down to ground level? What was the distribution of ground-level concentration of TCDD? Could the initial ground-level contamination spread?

2. Assuming that the windward side of buildings was the most exposed, what was the general probability of contamination having penetrated to the interiors of the buildings?

3. Since, on first assessments, the greater part of the contamination had been deposited in the open, on vegetation, what would be the effect of exposure, e.g., to rainfall?

4. Since laboratory toxicological studies with animals were designed to ensure that known doses were administered in the most assimilable form, how would this compare with the kind of environmental exposure to be expected in the Seveso area?

5. Was it reasonably possible to propose a tolerance limit of TCDD concentration as a criterion of acceptable contamination?

6. Would it be practicable to provide analytical services with the necessary techniques and capacity to handle large numbers of samples with reliable results?

7. What techniques were available for decontamination?

8. What methods could be used for the destruction of TCDD?

The account which follows covers the development of the answers to such questions and outlines the program developed for the cleanup and rehabilitation of the affected areas.

TCDD ANALYSIS AND MONITORING

As a result of earlier industrial incidents and with the realization that the herbicide 2,4,5-trichlorophenoxyacetic acid (2,4,5-T) could contain significant traces of TCDD, analytical procedures for determination of very small amounts of TCDD had already been developed. By the use of gas chromatography linked with mass spectrography it had been found possible to determine as little as 5 pg under the most favorable conditions.[16,17-21]

In the immediate aftermath of the release at Seveso, ICMESA personnel were able to collect a limited range of samples in the vicinity of the factory. These samples consisted of vegetation, soil, mixtures of the two, and wipe tests from suitable surfaces, and they were examined in Switzerland* with positive but highly variable results for TCDD content. Insufficient information was provided by this early work to give any indication of the distribution of the contamination or of the amount of TCDD deposited in the affected areas.

*Initial analyses were done in the Givaudin Research Co. laboratory at Dübendorf, Switzerland.[15]

Fortunately there existed in Milan* both equipment and trained personnel to permit the rapid establishing of local analytical facilities. These were commissioned in an emergency program which included examining pathological samples from dead animals collected from affected areas and samples of soil, vegetation, swabs from wipe tests of building surfaces, and samples of drinking water collected from a wide variety of sites in the communes considered to be affected by the release. The results of some of these early surveys are summarized in Tables 3-1, 3-2, and 3-3.

These results illustrate three points of practical importance. First, the high values found on vegetation and especially on foliage taken from trees indicated that a substantial part of the contamination had been deposited on upstanding materials in the path of the cloud. Most of the open fields immediately downwind of the ICMESA factory were not under horticulture but were covered in wild plant growth standing ½ to 1 m tall. Much of the contamination could have been collected by harvesting this surface growth, but unfortunately the opportunity was lost.

The second factor was the low incidence and level of TCDD contamination found inside buildings. This was accounted for by the fact that the accident occurred on a Saturday, when factories, workshops, schools, and other public premises were closed. In addition, because of the heat most houses had almost all their windows closed and shuttered, a normal Mediterranean custom.

The third factor was the absence of any evidence of contamination of well water and other drinking-water supplies, not only by TCDD but also by 2,4,5-TCD sodium salt, which is far more water-soluble. This confirmed that there was little penetration in depth of contamination in the soil.

The analytical results of the first "emergency" phase of sampling, together with the distribution of animal mortalities and dermatological symptoms in the population, comprised the evidence on which the health authorities had to make their first definitions of the boundaries of the controlled zones (Fig. 3-2). However, a more systematic survey was necessary to improve the definition of zone boundaries, particularly those of Zone B, in relation to specific criteria for the degree of contamination and for the purposes of planning a logical decontamination program. In the period between August 28 and September 6, 1976, some 213 samples of soil were collected in Zone B. The sampling points were arranged on a 150-m by 150-m square grid following cartesian coordinates. Samples were collected by removing surface vegetation and taking cores 6.5 cm diameter and 7 cm deep. Three cores at the corners of an equilateral triangle with a side length of 1 m were collected at each sampling point. Two of the cores were combined for determination of TCDD, and the third was kept as a reference sample. The combined cross-sectional areas of the two cores was 66 cm². Total TCDD was determined and the result, when multiplied by 150, was expressed as a soil surface concentration in micrograms per square meter.

To check the efficiency of recovery of TCDD from contaminated soil, separate tests were made in which uncontaminated soil was spiked with known amounts of TCDD and then processed by the standard extraction and cleanup procedure. These tests showed that when the TCDD concentration was 0.01 μg/500 g soil or more, average recovery was 86

---

*In Italy the analytical survey work was coordinated through the Laboratory of Hygiene and Prophylaxis, Province of Milan. Analytical work there was under the direction of Dr. Aldo Cavallero. TCDD analysis was also carried out in the Institute of Pharmacology and Pharmacognosy, School of Pharmacy, University of Milan. Most of the analyses of animal tissues from veterinary autopsies were done there before the facilities noted above were commissioned. Those involved were Drs. Flaminio Cattabeni and Giovanni Galli, who subsequently with Dr. Cavallaro were joint editors of the book *Dioxin: Toxicological and Chemical Aspects*.[17] Analytical facilities also existed at the Mario Negri Institute (Dr. S. Garrattini). These, together with the others mentioned above, were deployed in an overall program coordinated through Dr. Cavallaro's laboratory.

TABLE 3-1 Vegetation: Results of Analysis of Various Samples Collected from Late July to Early August 1976

Commune	Total number of samples	Number of positive results	Maximum TCDD, μg/100g	Notes
Meda	11	4	2.7	
Seveso	43	19	5130.0	Foliage from trees 150–200 m from source
Cesano Moderno	20	9	1.25	Foliage from trees
Desio	25	4	0.40	Garden crops
Bovisio, Masciago, and Seregno	5	2	0.19	Corn

NOTE: Limit of detection, 0.01 μg/100g.

TABLE 3-2 Buildings: Results of Analysis of Various Samples Collected in August 1976

Commune	Exterior samples			Interior samples		
	Total number of samples	Number of positive results	Maximum TCDD, μg/m^2	Total number of samples	Number of positive results	Maximum TCDD, μg/m^2
Meda	6	..	....	..	..	....
Seveso	6	5	32.8	21	9	0.064
Cesano Moderno	16	11	10.6	73	20	0.73
Desio	7	1	<0.75	26	4	0.013

NOTE: The buildings included some residences, industrial premises, and craft business premises. The highest internal sample result was from a building under construction in Cesano Moderno. Wipe test samples were usually of a 1-m^2 area, occasionally larger.

TABLE 3-3 Drinking Water: Results of Analysis of Samples Collected in Early August 1976

Commune	Number of samples	Source	Number of positive results	
			For TCDD	For 2,4,5-TCP
Meda	5	Wells	...	
	8	From ICMESA works	...	
Seveso	6	Wells	...	
	4	Wells		...
Cesano Moderno	8	Wells	...	
	5	Wells	...	...
Desio	8	Wells	...	
	5	Wells	...	...
Barlassina	2	Wells	...	
Bovisio	2	Wells	...	
Masciago	2	Wells	...	...
Seregno	2	Wells	...	
Varedo	4	Wells, drinking fountains	...	

NOTE: Limits of detection: for TCDD, 0.005 μg/L; for 2,4,5-TCP, 0.01 μg/L.

TABLE 3-4 Results of Analysis of Soil Samples Collected in a Systematic Survey in Zones A and B in August-September 1976

Range of TCDD concentration, $\mu g/m^2$	Zone A		Zone B	
	Number of samples	Maximum, $\mu g/m^2$	Number of samples	Maximum, $\mu g/m^2$
N.V.–0.75 ⎱	32		98	
0.75–4.99 ⎰			91	
5.0 –9.99 ⎱	6	13.0	14	
10.0 –14.99 ⎰			1	
15.0 –29.99	9	24.7	4	
30.0 –49.99	8	49.5	1	43.8
50.0 –199.99	25	155		
200.0 –499.99	7	474		
500.0 –999.99	7	942		
1000 –4999	10	4256		
Over 5000	4	21212		
Not determined	1		2	
Total number of samples	109		211*	

NOTE: N.V. = no value. Limit of sensitivity, 0.75 $\mu g/m^2$, equivalent to absolute amount of 5 ng of TCDD in total soil sample.

*Two further samples were collected but not analyzed.

percent. The recovery decreased to 66 percent when the soil concentration was as low as 0.005 $\mu g/500$ g. All analytical results from the soil samples were corrected for these recovery efficiencies, the lower efficiency factor being applied when the total TCDD content found did not exceed 9 ng.

A description of the extraction and cleanup procedure is given in the Appendix to Part 3. The gas chromatography–mass spectrometry (GC-MS) results were cross-checked between four different facilities in Milan and one in Rome with good agreement.

The survey in Zone B was immediately followed by a new one in Zone A. In this case the sampling grid was based on a 100-m pitch in the north-south direction and a radial grid originating from a point just north of the ICMESA factory, yielding 109 soil samples collected in the same way as for Zone B.

The results of these surveys are reported in Table 3-4. The four highest results in Zone A all came from within 450 m of the source, as did the four in the next highest group. Another five in this group came from within the 450- to 600-m range. However, within the same distance and from adjacent sampling points, samples with no detectable TCDD were collected, illustrating the extreme nonuniformity of distribution of the contamination close to the source. The very high readings were probably attributable to deposition of flakes of reactor products formed in the reflux condenser as the system cooled and released in a series of low-energy puffs after the main discharge pressure was spent.

A new map of Zone B, plotting the distribution of TCDD concentrations in the soil, was issued on September 18, 1976, and a corresponding map of Zone A followed on September 21. Sampling of soil in the 150-m by 150-m grid continued beyond the boundaries of Zone B, in Zone R, until the systematic survey had covered practically the whole area contained within Zone R. Then, having established the tolerance level of 5 $\mu g/m^2$ for soil contamination, it was eventually possible to redefine the boundaries of Zone B, as shown in Fig. 3.5.

Another aspect of the TCDD contamination problem investigated during this phase was to study the distribution of contamination in depth in the soil. Samples were taken on various sites in both Zone A and Zone B. Two sampling methods were used: (1) by

TABLE 3-5 Sampling of Education Premises for TCDD Contamination: Results of Analysis of Samples Collected between August 1976 and March 1977

Commune	Number of premises sampled	Soil samples					Interior surface samples				
		Total	N.V.	N.D.	Positive	Maximum, $\mu g/m^2$	Total	N.V.	N.D.	Positive	Maximum, $\mu g/m^2$
Meda	18	..	..	..	..	....	74	73	..	1	0.02
Seveso	15	30	26	4	..	....	157	146	8	3	0.53
Cesano Moderno	23	16	16	..	..	....	112	103	5	4	0.09
Desio	14	15	12	..	3	3.66	115	89	13	13	0.57
Bovisio Masciago	1	..	..	..	..	....	4	4	..	..	....
Barlassini	2	..	..	..	..	....	14	7	7	..	....
Totals	73	61	54	4	3		476	422	33	21	

NOTE: N.V. = soil samples <0.75 $\mu g/m^2$, interior samples <0.01 $\mu g/m^2$.
N.D. = not determined owing to analytical difficulties.

core sampling down to depths of 50 cm from the surface and (2) by digging a trench to a depth of 50 cm and then taking core samples horizontally in the trench wall at various depths.

Results on samples taken by the trench method tended to show that virtually all the TCDD contamination was confined to the top 2 cm of soil. With the core-sampling technique, TCDD remained just detectable beyond a depth of 14 cm but never below 25 cm. There remained some doubt whether the sampling method had itself carried these traces down to the lower levels, since the significant concentrations still remained in the top few centimeters of soil. Repetitions of sampling at the same locations showed no significant change of distribution with time even after intervening heavy rainfall.

As stated in the section "Introduction," the portion of the population most affected by the incident consisted of children, among whom there was an increased frequency of chloracne, mostly mild. Once the health authorities had adopted tolerance levels for the interior surfaces of buildings and for the ground in their precincts, a systematic sampling program covering all educational premises both within and close to the periphery of Zone R was started. The results of this survey, which continued in phases from August 1976 to March 1977, are summarized in Table 3-5. Most premises were visited more than once, and it was a matter for concern when, in February 1977, positive results started to come up where previously there had been no detectable TCDD. Eventually it appeared likely that at least some of these results were false positives originating from the cleanup apparatus. A complete new apparatus was installed, and a further series of checks confirmed the earlier results.

If we bear in mind that within 6 months of the accident's occurrence on July 10, 1976, the samples of all kinds collected and analyzed for TCDD by the GC-MS method already numbered in thousands, it was a remarkable achievement that such a high standard of reliability was maintained.

THEORETICAL ANALYSIS OF THE DISCHARGE, ITS DISPERSION, AND DEPOSITION

At a very early stage, it was evident that it was going to take several weeks to analyze enough samples collected systematically to begin to map the distribution of TCDD contamination. The consultants considered that it would be valuable to make a theoretical study to arrive at a predicted distribution of the deposition. This could be done quite quickly by using modeling techniques and computer programs which they had already developed. The predicted distribution provided a useful reference picture against which the analytical results from the sampling program could be compared. In this way it was possible to build up the general overall picture without having to wait until many analytical results were available.

The first requirement in the theoretical study was to estimate the distance of rise of the vapor cloud above the point of emission in order to determine an "effective height." For the computation the following unauthenticated data were used:

Composition of Reactor Contents (in Kilograms, Estimated)

Ethylene glycol	2800
Trichlorophenols (sodium salts)	2030
Sodium chloride	542
Caustic soda	562
Total	5934

Bursting-disk rupture pressure   376 kPa
Vent pipe internal diameter   127 mm
Discharge height above ground   8.0 m

As a first approximation the bursting-disk rupture pressure was assumed to be due to the vapor pressure of ethylene glycol at about 250°C. From these data the exit velocity was calculated as 274 m/s.

It was necessary to consider two extreme cases for the state of the issuing plume:

Case A. An equilibrium mixture of vapor and liquid to give a reasonable estimate of entrainment, resulting in a bulk density of 8.99 kg/m^3.

Case B. Pure vapor with no entrainment; vapor density, 1.61 kg/m^3.

The estimation of plume rise considered two components, the vertical rise followed by the horizontal displacement to give a bent-over plume. For the calculations the transition point was taken at the point where the vertical velocity of the plume equaled the wind speed, assumed to be 4 m/s. The results obtained were:

	Case A (m)	Case B (m)
Total plume rise	82.75	54.9
Downwind distance of maximum height	103	95.4

The calculated plume trajectories of the two cases are shown in Fig. 3-3.

The dispersion and deposition computations had to take account of the finite release, which, on the basis of estimates available at the time, indicated a total release of about 500 kg of reactor materials with a TCDD content of about 0.35 percent by weight. This would amount to 1.75 kg of TCDD in all. For the calculation a figure of 2 kg was used. The relevant meteorological conditions assumed were a wind speed of 4 m/s and Pasquill Category B for atmospheric stability. Since the interest was in the quantity of material deposited at the ground surface, rates of deposition were calculated by using published data deemed most appropriate and allowing for the progressive depletion of the particulate burden of the cloud as it traveled downwind.

The early calculations were based only on the mean wind direction and resulted in a series of isopleths of ground-level deposition as functions of downwind distance from the source and crosswind distance from the axis of the cloud (Fig. 3-4).

Three valuable conclusions were immediately obtained from this study. First, at distances beyond about 1.5 km there was no practical difference between the results for the two extreme cases. Since eyewitness accounts and other such evidence showed that the actual discharge conformed to some intermediate case, the results would be valid for that also, certainly beyond the boundary of Zone A. The second important conclusion derived from the good agreement found between predicted values and those found by analysis of samples. This showed that the assumed total release of 2 kg of TCDD could not be far wrong. As some publicity had been given, on television and in the press, to totally unfounded statements by some scientists that the release could have been as high as 130 kg of TCDD, this finding was of great value to those dealing with problems in Seveso. The third conclusion was that there was no reason to expect to find high levels of contamination farther afield.

The studies outlined above were continued and refined when more detailed information became available. The wind speed and direction traces from the nearest anemograph station (Carate Brianza) were analyzed to give 1-min average values of wind direction

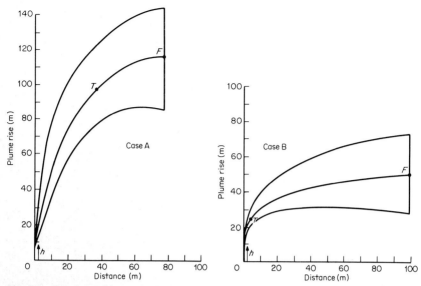

Fig. 3-3 Calculated plume trajectories. Point T is the transition point from the vertical plume to the bent-over plume. Point F is the point of final rise of the plume.

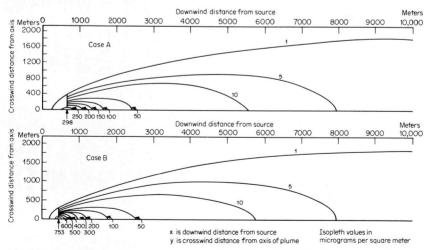

Fig. 3-4 Deposition isopleths tor a release containing 2 kg of TCDD; Case A, vapor discharge only; Case B, vapor with entrained liquid.

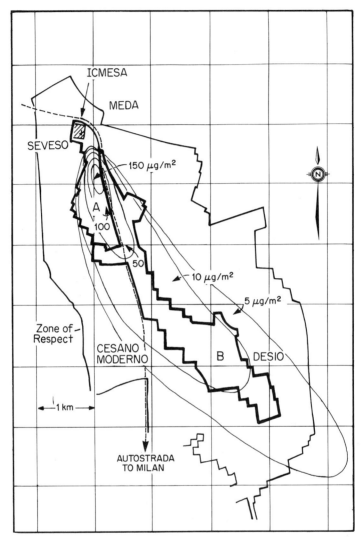

Fig. 3-5 Deposition isopleths (recalculated to allow for minute-by-minute variations of average wind direction and speed, as recorded at Carate Brianza) shown in relation to Zone A and redefined Zone B.

and speed over the period of the incident. Recalculation, using these data, resulted in a more precise positioning of isopleths on a map which was found to coincide remarkably well with the final mapping of Zones A and B, a process which continued up to the end of 1976 (Fig. 3-5).

The first phase of these studies was reported in detail in the final report prepared by the consultants for the Lombardy Regional Council in October 1976.[22] Comer has described the refined treatment and results.[23]

DEFINITION OF CRITERIA FOR
CONTAMINATION LEVELS

This was the most controversial of all the problems confronting the Italian authorities responsible for emergency and decontamination policy. By the time of the Seveso accident, there was a considerable and growing body of toxicological research data on TCDD which left no doubt that it was a very potent poison.[24-28] All these data related to experiments with animals. There was also a substantial body of experience of the effects on humans, as a result of previous accidents or industrial exposure in the manufacture of 2,4,5-TCP.[3-10,29-38] However, the latter did not include any quantitative data on the amounts of TCDD absorbed or ingested by individuals to correlate with observed clinical symptoms. In the absence of such quantitative data for humans, there were naturally those who advised against any attempt to reclaim the contaminated areas. Some said that the only permissible tolerance level was zero, overlooking the logical consequences of that philosophy.

The practical requirement was a criterion for defining "contaminated" areas where decontamination activities would have to be applied and tolerance levels as the criteria for assessing the effectiveness of decontamination. This required judgments with the attendant risk of errors. The first proposals on this basis were made by the Medicoepidemiological Commission in response to a request made on September 16, 1976.

The commission assessed the risk of oral ingestion of TCDD as a result of incidental contact with contaminated soil, surfaces, objects, etc., inside or adjacent to houses, public buildings, and workplaces. They used known toxicity data for the most sensitive animal (the guinea pig), as found by Schwetz et al.[39] Taking the toxic dose as 1 μg/kg body weight, the equivalent absolute dose for a young child of 15 kg was set at 15 μg. Thus, if a surface concentration of 5 μg/m^2 were considered, the implication was that the child would have ingested all the TCDD from an area of 3 m^2 of ground. If we bear in mind the convention adopted in expressing the analytical results as a concentration per unit area, in practice the child would have to ingest all the soil to a depth of 70 mm in a total surface area of 3 m^2, i.e., 0.21 m^3 (7.4 ft^3), a highly improbable incidental occurrence. The commission concluded that a tolerance limit of 5 μg/m^2 for the soil represented a significant safety margin.

Quite independently, the consultants were asked to propose a criterion level of soil contamination as a basis for redefining the boundaries of Zone B. In this case, rather than ingestion, the primary risk of absorption through the skin was considered. Schwetz et al.[39] had published an LD_{50} = 275 μg/kg body weight for topical applications of TCDD in acetone solution to the inside of a rabbit's ear. The absolute dose for an adult rabbit was taken as 550 μg, with a probability of 100 percent for receipt of the administered dose. If the probability of reaching this absolute dose by incidental contact were taken as 1 percent, 100 contacts would be required to accumulate the full dose, and the average incremental dose per contact would be 5.5 μg. By distributing this over 1 m^2 of surface, there would evidently be a substantial margin of safety against acquiring a toxic dose by incidental contact. Here again, a tolerance limit of about 5 μg/m^2 for the soil was indicated.

The only documented example of TCDD poisoning from incidental contact with a soil surface occurred at a horse-breeding farm in Missouri in 1973.[37] The training arena was sprayed with TCDD-contaminated oil wastes to suppress the dust, and a 6-year-old child who frequently played there developed hemorrhagic cystitis and signs of focal pyelonephritis. Two out of three other children developed chloracne. The TCDD content of the soil was found to be as high as 33 μg/g (equivalent to about 3000 mg/m^2 in Seveso terms). Furthermore, the TCDD was in solution in still residues mixed with oil, thus increasing the probability of adhesion to the skin and absorption. Since the affected children in this case all apparently recovered,[40] the poisoning was disabling only. Thus a tolerance limit of 5 $\mu g/m^2$ for the soil represented a concentration only 1.5×10^{-6} of a level demonstrably disabling.

Tables of the probabilities of deaths (per person per year) due to exposure to a variety of risks have been published.[41] It is generally considered that the risk of death due to a given exposure is of an order of magnitude smaller than the risk of disabling injury. In the case of poisoning the figures would be 1 in 100,000 and 1 in 10,000 respectively. Such considerations as these reinforced the conclusion that the proposed tolerance limit of 5 $\mu g/m^2$ for the soil would indeed reduce the risk of serious toxic effects from TCDD to a level, if anything, lower than the everyday risk of poisoning.

When it came to indoor areas such as schools, the probability of contact would obviously be greater than with the soil surface in the open. A lower tolerance limit was logically required, and the Medicoepidemiological Commission proposed an additional safety factor of 500. This gave a tolerance limit of 0.1 $\mu g/m^2$ for interior surfaces, which also applied to stocks, products, and machines located outside factories or workshops.

For the ground or soil in the immediate vicinity of schools, public buildings, and workplaces, a tolerance limit of 0.75 $\mu g/m^2$ was adopted, with defoliation and removal of grass from the areas concerned.

These tolerance limits were at least practical. They did not go too close to the sensitivity limit of the analytical procedures, so that reasonable confidence in results could be preserved. This was of the first importance. They did make it possible to draw up a plan of campaign for systematic decontamination and restoration in Zones A and B. From a purely scientific point of view, there was a case for keeping the tolerance limits under review, since raising them by a factor of 2 or 3 would be of little significance in terms of the indicated safety margins but would have very real practical effects in the decontamination program. However, it was not possible for the Italian authorities to consider this under the circumstances of the Seveso disaster.

DESTRUCTION OF TCDD

Incineration

The only data available on the thermal behavior of TCDD prior to the incident were reported by R. H. Stehl and coworkers.[42] These data indicated complete decomposition at 800°C with an overall residence time of 2½ s in the pyrolysis tube. There was nothing to indicate for how much of that residence time the TCDD was at the stated temperature. There was also some doubt whether or not higher temperatures would be necessary for pure TCDD, compared with TCDD in the presence of trichlorophenol.

Stehl's data were considered kinetically[43] and suggested that in the interval between 700 and 800°C there was a change in reaction mechanism accompanied by a greatly accelerated rate of decomposition. Consideration of the energy changes involved indicated that at lower temperatures a sluggish second-order oxidation reaction took place, but at approximately 700°C it underwent a transition into a first-order thermal cracking reaction. Since reliable values of the Arrhenius constant A for molecules of the TCDD type were known, with orders of 10^{-13} to 10^{-14}, a "more probable" model for the reaction system at very high temperatures was proposed (Fig. 3-6). The indicated activation energy was 258 kJ/mol, which was deemed credible for a dioxin-type molecule.[44]

From this model, it was possible to predict the half-life of TCDD at higher temperatures. For example, at 1200°C the indicated half-life was only 9×10^{-5} s. It was per-

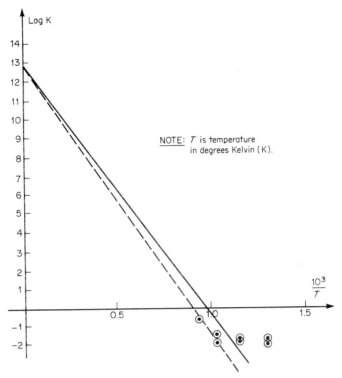

Fig. 3-6   Arrhenius diagram of a "more probable" model for thermal crack-ing of TCDD at higher temperatures; assumed activation energy of 258 kJ-mol and reaction constant A of 10^{13}. Points in circles plot kinetic data derived from R. H. Stehl et al., in Etcyl H. Blair (ed.), *Chlorodioxins—Origin and Fate,* American Chemical Society, Washington, 1973, pp. 119–125.

fectly possible to design an incinerator with an afterburner system to operate at this tem-perature, and virtually total destruction of TCDD could be expected.

The consultants proposed that such an incinerator be installed, sized for the ultimate destruction of contaminated combustible materials such as vegetation, dismantled wooden structures (rabbit hutches, chicken houses, sheds, etc.), and discarded decontamination materials (solvent swabs, vacuum cleaner contents, used protective clothing, etc.), but not for contaminated soil. However, local feelings against any incinerator reached a point that made it politically impossible to adopt this method.

Photolysis

Apart from high-temperature incineration, photolysis was the only other known practic-able method for the decomposition of TCDD. Plimmer et al.[45] had reported the photo-decomposition of TCDD in methanol solution (5 mg/L) under irradiation by light with an intensity of about 100 W/cm² at 307 nm and also on exposure to natural sunlight. The role of the methanol, in addition to being the solvent, was that of a hydrogen donor.

Givaudin workers[46] investigated the possibility of using an alternative hydrogen donor that could be employed on contaminated vegetation, etc., *in situ*. They selected olive oil and worked out two formulations: (1) 80% olive oil and 20% cyclohexanone to reduce viscosity; and (2) a 40% aqueous emulsion with 4% biodegradable emulsifying agent. Experiment showed that 350 L/ha of (1) or 400 L/ha of (2) were sufficient to produce a practically continuous film on vegetation and smooth surfaces. Both formulations had reasonably good resistance to rain and gave good prospects for rendering much of the TCDD in the environment accessible to accelerated ultraviolet degradation.

Both formulations were tried out in early September 1976 on test plots located in Zone A, with a control plot of equal size which was untreated. The experiments were carried out jointly by Givaudin scientists and an Italian team from the Istituto Superiore di Sanità in Rome. Unfortunately, after the first 2 days the experiment was interrupted by bad weather with very heavy rain, after which it was continued for several more days. The overall results for all three plots showed substantial reductions in the TCDD content of the vegetation. In the case of the control plot, this could only be attributed to the contamination being washed off by the heavy rain. In the two test plots there were positive indications of reductions in TCDD content in the period before the rain, in sharp contrast to the control plot. After the rain, there was less distinction in the behavior of the three plots, and this brought controversy in the interpretation of the overall results. In the end, the chances of using the method to advantage faded with the arrival of autumn at Seveso and with it a period of very heavy rainfall.

Liberti et al.[47] reported on experiments carried out in a closed room, substituting an ultraviolet source for the central light bulb, to give an intensity of about 20 $\mu W/cm^2$. As the hydrogen donor they used a 1:1 solution of ethyl oleate and xylene and demonstrated photodegradation. They claimed complete decay of TCDD in about 1 h at 2 mW/cm^2 or in 72 h at 20 $\mu W/cm^2$.

The consultants suggested the use of photolysis for items such as decorative conifers and evergreens in gardens, patios, and verandas as well as for a final treatment for door and window frames and other less accessible contaminated surfaces such as the outside walls of houses. Instead of olive oil, it was suggested that corn oil could be used in some areas, and even linseed oil on or in buildings.

Microbiological Degradation in Soil

Kearney, Woolson, and others[48,49] have studied the behavior of TCDD in various soils and looked for evidence of it in soils which had been subjected to heavy applications of 2,4,5-trichlorophenoxyacetic acid, which before 1969 could have contained up to 40 ppm of TCDD. Their findings indicated that TCDD had a half-life of 1 year.

In October 1976, A. L. Young and coworkers reported on studies on biodegradation plots and field test areas in three climatically different regions of the United States.[50] In all cases, the residual concentrations of both herbicide and associated TCDD showed a progressive decrease with time, and the results indicated that there was degradation of TCDD by soil microorganisms, especially when in the presence of other chlorinated hydrocarbons, and that leaching of TCDD by water alone did not appear to occur. There was no clear evidence that the process could be speeded up, for example, by the use of additives which would normally intensify microbiological activity; in these experiments lime, organic matter, and NPK fertilizer were uniformly mixed in known ratios in the top 30 cm of soil in the test plot, but the rate of degradation was not significantly different from that of untreated soil.

The lengthy time involved in obtaining results from further field trials could not be accommodated in the plans for decontamination of residential areas. Here the only acceptable procedure where soil surface levels exceeded the "tolerance level" of 5 $\mu g/m^2$ was the physical removal of the top 10-cm layer in which virtually all the TCDD was contained. Therefore, the extent to which microbiological degradation would ultimately contribute to the destruction of TCDD in the soil could only be a matter for long-term observation rather than part of a scheduled plan.

DECONTAMINATION OPERATIONS

Because of an overlapping of administrative functions of regional, provincial, and communal authorities, special enabling legislation by the central government was necessary to define responsibilities for dealing with the problems of the Seveso incident. The Lombardy Regional Council was assigned the task of drawing up the plans for decontamination and reclamation of the affected areas. The detail work was coordinated by the Commissione Bonifica, led by Prof. A. Giovanardi.

International scientific consultation was organized. On September 30 and October 1, 1976, there was a meeting of experts in Milan, organized with assistance from the Commission of the European Communities.[51] This was followed, on October 23–24, 1976, by an international workshop on dioxin attended by workers with special knowledge of the toxicology and chemistry of TCDD.[17−21,27,28,46,47] The Lombardy Region also retained a London-based firm of consulting engineers and scientists to make an independent study and report, with recommendations for remedial work. Its final report was submitted on October 25, 1976.[22]

It became policy, at an early stage, to incorporate into all planning the installation of a suitably designed incinerator as the only practicable means for the destruction of TCDD which had been deposited on combustible materials. There were divergencies of opinion about the desirability and practicability of any large-scale incineration of contaminated soil. Thus, the general concept was that stockpiles of combustible contaminated materials would accumulate while the incinerator was being constructed and commissioned. Thereafter, collection and incineration would proceed in parallel. The selection of a suitable site for the incinerator and stockpiles, with controlled access routes for bringing in contaminated materials and disposal of ash, was given much consideration.

All operations within the boundary of Zone A had to be carried out under strictest control of access and with maximum protection of personnel. Procedures were modeled on those adopted in "hot" areas of nuclear installations, and an existing sports building was modified to provide the necessary changing accommodation. All personal clothing remained in lockers, and operators changed into new undergarments, one-piece lightweight weatherproof overalls with integral hoods, cleansed knee-length rubber boots, and new gloves. They were also required to wear dust respirators and goggles. All items except boots were worn once only and then discarded for ultimate incineration. Return from the field side to the locker rooms was by way of a shower room where used towels were also collected for disposal. This control station was conveniently situated just south of the midpoint of the western boundary of Zone A and was fully operational for the use of workers collecting soil and other samples in the autumn of 1976.

Similar "filter stations" were recommended, sited together with local depots where contaminated materials could be transferred from small in-field trailers to larger trucks for conveyance to main stockpiles. All such working areas required sealing with bitumen-based surfacing laid to drain down to sumps. Activated carbon was recommended as a trap for any contamination washed into the sumps. High-pressure water jets were called for to wash down the outsides of vehicles, especially the wheels and tires, before leaving the area.

Apart from the organizational problems, deterioration of the weather caused delays. In October and November 1976, rainfall was extremely heavy and there was considerable flooding in the immediately adjacent areas, where the Certesa stream overflowed. This

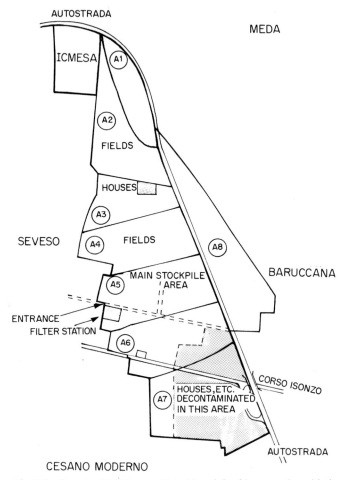

Fig. 3-7   Zone A with Subzones A1 to A7 as defined in connection with the decontamination program.

stream actually traverses the northern part of Zone A. But there were also increasing signs of popular opposition to the installation of an incinerator. This appeared to be due to the belief that incineration could not destroy the TCDD but would only redisperse it into places which had previously escaped contamination. Nevertheless, in January 1977 the Lombardy Region had not yet abandoned plans based on the use of incineration.

By this time, the interior decontamination of the evacuated houses in the south of Zone A (Zones A6 and A7; Fig. 3-7) was going ahead under the supervision of Swiss personnel from Givaudin. The main methods employed were high-intensity vacuum cleaning and washing with detergent and/or solvents. The vacuum cleaners were fitted with special filters to prevent redispersion. These methods were appropriate for relatively impermeable surfaces of various kinds. Other materials such as soft furnishings, fabric-upholstered furniture, and the like had to be removed and replaced with new materials.

The collection of vegetation from the open fields in the same two subzones was also virtually complete. All materials removed from these areas were transferred to protective storage at a site in the middle of Zone A5. However, by about March 1977, it was clear that a final decision on the method to be adopted for the destruction of TCDD in contaminated materials was not imminent. Accordingly, the Lombardy Region had to restrict its plans to the collection of contaminated materials and their removal and storage under conditions which would prevent redispersion and public access. It was also necessary to safeguard against the spread of contamination from stored materials by the movements of rats and other scavengers.

The most immediate objectives of the program, adopted with the approval of the Technicoscientific Commission in Rome, were (1) to dispose of any "hot spots" of TCDD contamination with Zone B so that the situation there could be stabilized as far as the health risks to people were concerned, and (2) to complete the decontamination of the residential area in Zones A6 and A7, together with other measures necessary to permit the displaced occupants to return to their homes.

As a criterion for physical removal of soil, it was decided that in open fields a level below 15 $\mu g/m^2$ would not be disturbed but that access would be restricted by fencing. Above that level, the top 10 cm was scraped off and removed to the stockpile area in Zone A5. Within the curtilage of premises, e.g., in private gardens, topsoil was to be removed to a depth of 10 cm wherever the surface concentration exceeded 5 $\mu g/m^2$. This applied in Zones A6 and A7 and in Zone B.

Similarly, the collection of vegetation from gardens was required. This was collected into plastic sacks which, at the long-term storage site, were further enclosed in plastic silos. Most of this contaminated vegetation was stored in Zone A5, but some from Zone B had to be taken to local depots in the Cesano Moderno and Desio districts, owing to the absence of suitable access routes to Zone A except through uncontaminated areas. The subzones A1 to A5 were closed to all unauthorized access by the erection of a tall fence.

Decontamination of the exteriors of properties in Zones A6 and A7 was completed. This was achieved by a combination of high-intensity vacuum cleaning followed by washing with high-pressure water jets. All decontamination work was checked for effectiveness by the GC-MS of samples.

Improved arrangements were completed for the safe storage of the 40,000 or so carcasses of small farmyard animals (rabbits, chickens, etc.), most of which were destroyed by the veterinary services when their economic value disappeared after the TCDD release. There were about 700 large plastic containers in which the carcasses had been deposited in a strong caustic soda solution. These containers were transported to the ICMESA works, where they were placed in the concrete decantation tanks of the effluent treatment plant under construction at the time of the accident. The containers were then buried under earth filled to give 1 m of cover.

For the longer term, the program was addressed to the decontamination and removal of the TCP plant from the ICMESA works, the possible decontamination and recovery of the houses in Zone A2, and the replacement of fresh topsoil in gardens and, above all, to seeking a solution to the problem of the ultimate destruction of TCDD in the accumulated stockpiles.

In June 1977, the *FRIPC Report* was able to refer to partial but significant progress of the work but recognized the importance of finding an acceptable solution to the problem of the ultimate disposal of the contamination. Incineration under correct conditions still seemed the only practicable answer, and there was a proposal to install an experimental unit at the ICMESA site. Nevertheless, by late 1977 most of the people evacuated from Zones A6 and A7 were able to return to their houses. In March 1978, a report issued by the department of dermatology of the University of Milan concluded that the incidence of chloracne in the TCDD-contaminated areas had decreased, that the condition of previously affected patients was improving, and that the numbers of skin cases in those areas was getting closer to those found in pollution-free areas.

Thus, amid recurring alarms, there have also been reassuring developments to justify the decontamination efforts carried on in the face of exceptional difficulties. But the story was far from complete when this part was written.

Cost of Cleanup

According to a report in the December 22, 1980, issue of *Lloyd's List,* Givaudin agreed to pay 103 billion lire (about $115 million) in compensation. Under this agreement, the Italian government and the Lombardy Region withdrew all civil actions against Givaudin in Italy and other countries. The company was to make an immediate payment of 40.5 billion lire to the region and 7.5 billion lire to the Italian government for the expenses it had incurred. The total payment was to include expenses already paid by Givaudin and estimated expenses for future decontamination of the affected area.

APPENDIX

STANDARDIZED ANALYTICAL PROCEDURE FOR A SYSTEMATIC SURVEY OF SOIL CONCENTRATIONS OF TCDD

1. *Extraction of samples*
   a. The sample of soil was weighed and placed in a 2-L flask fitted with a ground-glass stopper.
   b. To this was added 400 mL of hexane-acetone mixture (4:1 by volume), and the flask was thoroughly shaken for 5 min. The solvent was decanted through a filter.
   c. Then 300 mL of the same solvent mixture was added, and the flask was again shaken for 5 min, followed by decantation and filtration as before.
   d. Step *c* was repeated.
   e. The three extracted fractions were combined and carefully evaporated to dryness in a rotating evaporator. NOTE: The total volume of extraction solvent was kept constant and sufficient to deal with the heaviest samples. Actual sample weights ranged between 350 and 1150 g approximately.

2. *Purification of extracts*
   a. A chromatographic column 20 mm in diameter and 200 mm long was made up with successive layers as follows, starting from the bottom:

	Quartz wool plug
5 mm	Na_2SO_4, anhydrous (bottom layer)
15 mm	Silica gel, unactivated (Fisher 60–200 mesh)
15 mm	Na_2SO_4 + $NaHCO_3$ mixture (9:1 by weight)
50 mm	Celite + concentrated H_2SO_4 (prepared by mixing 6 g of Celite 545 with 3.7 mL concentrated H_2SO_4 in a mortar)
15 mm	Na_2SO_4, anhydrous (top layer)

b. The column made up as above was washed with 40 to 60° petroleum ether.

c. The dried extract from Step 1*e* was extracted three times with 5-mL portions of 40–60° petroleum ether, and the extracts were transferred quantitatively onto the column.

d. The column was then eluted with 40–60° petroleum ether until 40 mL of eluate had been collected in a 50-mL flask.

e. The eluate was evaporated to dryness and then reextracted with three portions of hexane (1 mL + 0.5 mL + 0.5 mL).

f. The hexane extract was transferred quantitatively onto a second column of 5-mm diameter containing 1.2 g of neutral alumina (Brockmann activity = 1; activation by heating for 12 h at 130°C).

g. This column was first eluted with 7 mL of a hexane–methylene chloride mixture (99:1 by volume) and the first eluate discarded.

h. The column was eluted again with 5 mL of hexane–methylene chloride mixture (4:1 by volume) and the eluate collected and evaporated to dryness. This residue would contain any TCDD present in the original sample and was finally analyzed by gas chromatography–mass spectrometry.

3. Gas chromatography–mass spectrometry.

Ultimately four separate facilities existed at three different laboratories in Milan, and a further facility was used in Rome. There were differences in the detailed instrument, column, and operating specifications. The following data relate to the new installation specially set up at the LPIP laboratory in Milan following the Seveso accident.

Column.   OV 101 2% on Chromosorb W 80–200 mesh; glass 2 m long
              Temperature, 220°C
              Injection temperature, 270°C
              Separation temperature, 210°C
              Transport gas: helium at 165 kPa

M.S.   Finnigan 3200 with processor
          Electron energy, 60 eV

REFERENCES

1. *Evaluation of the Carcinogenic Risk of Chemicals to Man,* International Agency for Research on Cancer Monographs, vol. 15, 1977, p. 54.

2. A. W. M. Hay, *Disasters,* vol. 1, no. 4, 1977, pp. 289–308.

3. R. R. Suskind, International Agency for Research on Cancer meeting, Lyon, January 1978.

4. A. W. M. Hay, "Toxic Cloud over Seveso," *Nature,* vol. 262, 1976, p. 636.

5. P. Dugois and L. Colomb, "Acné chlorique au 2-4-5 trichlorophénol," *Bulletin de la société française de dermatologie et de syphiligraphie,* vol. 63, 1956, p. 262.

6. P. Dugois, P. Amblard, M. Aimard, and G. Deshors, "Acné chlorique collective et accidentelle d'un type nouveau," *Bulletin de la société française de dermatologie et de syphiligraphie,* vol. 75, 1968, p. 260.

7. M. F. Hofmann and C. L. Meneghini, *Giornale italiano de dermatologia,* vol. 103, 1962, p. 427.

8. J. Bleiberg, M. Wallen, R. Brodkin, and I. L. Applebaum, "Industrially Acquired Porphyria," *Archives of Dermatology,* vol. 89, 1964, p. 793.

9. G. May, "Chloracne from the Accidental Production of Tetrachlorodibenzodioxin," *British Journal of Industrial Medicine,* vol. 30, 1973, p. 276.

10. L. Jirasek, J. Kalensky, and K. Kubec, "Acne chlorina a porphyria cutanea tarda při výrobě herbidic," *Československa dermatologie*, vol. 48, 1973, p. 306.

11. F. Pocchiari, V. Silano, and A. Zampieri, *Human Health Effects from Accidental Release of TCDD at Seveso [Italy]*, Istituto Superiore di Sanità, Rome, 1978.

12. *Final Report of the Italian Parliamentary Commission of Inquiry into the Seveso Disaster (FRIPC Report)*, in Italian, part 2, Rome, June 1977, chap. 1.5.

13. U.S. Patent No. 2,509,245, Givaudin, 1947.

14. M. H. Milnes, "Formation of 2,3,7,8-Tetrachlorobenzodioxin by Thermal Decomposition of Sodium 2,4,5,-Trichlorophenate, *Nature*, vol. 232, 1971, p. 395.

15. E. Homberger, G. Reggiani, J. Sambeth, and K. H. Wipf, "The Seveso Accident: Its Nature, Extent and Consequences," *Annals of Occupational Hygiene*, vol. 22, 1979, pp. 327–367.

16. R. Baughman and M. Meselson, "Improved Analysis for Tetrachlorodibenzo-*p*-dioxins," in E. H. Blair (ed.), *Chlorodioxins—Origin and Fate,* American Chemical Society, Washington, 1973, pp. 92–104.

17. Bo Holmstedt, "Mass Fragmentography of TCDD and Related Compounds," in F. Cattabeni, A. Cavallero, and G. Galli (eds.), *Dioxin: Toxicological and Chemical Aspects,* Spectrum Publications, Inc., New York, 1978, pp. 13–25.

18. H.-R. Buser, "Analysis of TCDD's by Gas Chromatography–Mass Spectrometry Using Glass Capillary Columns," in F. Cattabeni, A. Cavallero, and G. Galli (eds.), *Dioxin: Toxicological and Chemical Aspects,* Spectrum Publications, Inc., New York, 1978, pp. 27–41.

19. J. Freudenthal, "The Quantitative Determination of TCDD with Different Mass Spectrometric Methods," in F. Cattabeni, A. Cavallero, and G. Galli (eds.), *Dioxin: Toxicological and Chemical Aspects,* Spectrum Publications, Inc., New York, 1978, pp. 43–50.

20, R. L. Harless and E. O. Oswald, "Low- and High-Resolution Gas Chromatography–Mass Spectrometry (GS-MS) Method of Analysis for the Presence of 2,3,7,8-Tetrachlorodibenzo-*p*-dioxin (TCDD) in Environmental Samples," in F. Cattabeni, A. Cavallero, and G. Galli (eds.), *Dioxin: Toxicological and Chemical Aspects,* Spectrum Publications, Inc., New York, 1978, pp. 51–57.

21. P. W. O'Keefe, "A Neutral Cleanup Procedure for TCDD Residues in Environmental Samples," in F. Cattabeni, A. Cavallero, and G. Galli (eds.), *Dioxin: Toxicological and Chemical Aspects,* Spectrum Publications, Inc., New York, 1978, pp. 59–78.

22. Cremer and Warner, Ltd., *The ICMESA Accident (Seveso)—Final Report on the Accident, Its Consequences, and Recommended Remedial Action,* in Italian, Lombardy Regional Council, October 1976, Apps. I and II.

23. P. J. Comer, "The Dispersion of Large Scale Accidental Releases, Such as at Seveso," paper presented to the Royal Meteorological Discussion Meeting, London, June 15, 1977.

24. B. A. Schwetz, J. M. Norris, G. L. Sparschu, V. K. Rowe, and P. J. Gehring, "Toxicology of Chlorinated Dibenzo-*p*-dioxins," in E. H. Blair (ed.), *Chlorodioxins—Origin and Fate,* American Chemical Society, Washington, 1973, pp. 55–69.

25. K. S. Khera and J. A. Ruddick, "Poly(chloro)dibenzo-*p*-dioxins: Perinatal Effects and the Dominant Lethal Test in Wistar Rats," in E. H. Blair (ed.), *Chlorodioxins—Origin and Fate,* American Chemical Society, Washington, 1973, pp. 70–84.

26. W. N. Piper, J. Q. Rose, and P. J. Gehring, "Excretion on Tissue Distribution of 2,3,7,8-Tetrachlorodibenzo-*p*-dioxin in the Rat," in E. H. Blair (ed.), *Chlorodioxins—Origin and Fate,* American Chemical Society, Washington, 1973, pp. 85–91.

27. G. Matthiaschk, "Survey about Toxicological Data of 2,3,7,8-Tetrachlorodibenzo-*p*-dioxin (TCDD)," in F. Cattabeni, A. Cavallero, and G. Galli (eds.), *Dioxin: Toxicological and Chemical Aspects,* Spectrum Publications, Inc., New York, 1978, pp. 123–136.

28. E. E. McConnell and J. A. Moore, "The Toxicopathology of TCDD," in F. Cattabeni, A. Cavallero, and G. Galli (eds.), *Dioxin: Toxicological and Chemical Aspects,* Spectrum Publications, Inc., New York, 1978, pp. 137–142.

29. R. H. Schulze, *Archiv für klinische und experimentelle Dermatologie,* vol. 206, 1957, p. 589.

30. J. Kimmig and K. H. Schulz, "Berufliche Akne durch chlorierte aromatische zyklische Äther," *Dermatologica,* vol. 115, 1957, p. 540.

31. H. Bauer, K. H. Schulz, V. Schultz, and X. X. Spiegelberg, *Archiv für Gewerbepathologie und Gewerbehygiene,* vol. 18, 1961, p. 538.

32. P. Dugois and L. Colomb, "Remarques sur l'acné chlorique," *Journal de médecine de Lyon,* vol. 38, 1957, p. 899.

33. P. Dugois, J. Maréchal, and L. Colomb, "Acné chlorique au 2,4,5-trichlorophénol," *Archives des maladies professionelles,* vol. 19, 1958, p. 626.

34. A. P. Poland and D. Smith, "A Health Survey of Workers in a 2,4-D and 2,4,5-T Plant with Special Attention to Chloracne, Porphyria Cutanea Tarda, and Psychologic Parameters," *Archives of Environmental Health,* vol. 22, 1971, p. 316.

35. K. A. Telegina and L. I. Bikbulativa, "Sostoianie kozhi u kontaktiruiushchikh s metoksonom v usloviiakh ego promyshlennogo proizvodstva," *Vestnik dermatologie i venerologie,* vol. 44, 1970, p. 35.

36. L. Jirasek, J. Kalensky, J. Pazderova, and E. Lubas, "Acne chlorina, porphyria cutanea tarda a jiné projevy celkové intoxikace při výrobě herbidic," *Československa dermatologie,* vol 49, 1974, p. 145.

37. C. D. Carter, R. D. Kimbrough, J. A. Liddle, R. E. Cline, M. W. Zack, and W. F. Barthel, "Tetrachlorodibenzodioxin: An Accidental Poisoning Episode in Horse Arenas," *Science,* vol. 188, 1975, p. 738.

38. R. M. Oliver, "Toxic Effects of 2,3,7,8-Tetrachlorodibenzo-1,4-dioxin in Laboratory Workers," *British Journal of Industrial Medicine,* vol. 32, 1975, p. 49.

39. B. A. Schwetz, J. M. Norris, G. L. Sparschu, V. K. Rowe, P. J. Gehring, J. L. Emerson, and C. G. Gerbig, "Toxicology of Chlorinated Dibenzo-*p*-dioxins," *Environmental Health Perspectives,* experimental issue no. 5, 1973, p. 87.

40. M. G. Beale, W. T. Shearer, M. M. Karl, and A. M. Robson, "Long-Term Effects of Dioxin Exposure," *Lancet,* Apr. 2, 1977, p. 748.

41. Royal Commission on Environmental Pollution, *Sixth Report,* H. M. Stationery Office, London, September 1976.

42. R. H. Stehl, R. R. Papenfuss, R. A. Bredeweg, and R. W. Roberts, "Stability of Pentachlorophenol and Chlorinated Dioxins to Sunlight, Heat, and Combustion," in E. H. Blair (ed.), *Chlorodioxins—Origin and Fate,* American Chemical Society, Washington, 1973, pp. 119-125.

43. Cremer and Warner, Ltd., *The ICMESA Accident (Seveso)—Final Report on the Accident, Its Consequences, and Recommended Remedial Action,* in Italian, Lombardy Regional Council, October 1976, App. VI.

44. A. F. Trotman-Dickinson, *Gas Kinetics,* Butterworth & Co., London, 1955, p. 131.

45. J. R. Plimmer and U. I. Klingebiel, "Photochemistry of Dibenzo-*p*-dioxins," in E. H. Blair (ed.), *Chlorodioxins—Origin and Fate,* American Chemical Society, Washington, 1973, pp. 44-54.

46. H. K. Wipf, E. Homberger, N. Neuner, and F. Schenker, "Field Trials on Photodegradation of TCDD on Vegetation after Spraying with Vegetable Oil," in F. Cattabeni, A. Cavallero, and G. Galli (eds.), *Dioxin: Toxicological and Chemical Aspects,* Spectrum Publications, Inc., New York, 1978, pp. 201-217.

47. A. Liberti, D. Brocco, I. Allegrini, and G. Bertoni, "Field Photodegradation of TCDD by Ultra-Violet Radiations," in F. Cattabeni, A. Cavallero, and G. Galli (eds.), *Dioxin: Toxicological and Chemical Aspects,* Spectrum Publications, Inc., New York, 1978, pp. 195-200.

48. P. C. Kearney, A. R. Isensee, C. S. Helling, E. A. Woolson, and J. R. Plimmer, "Environment Significance of Chlorodioxins," in E. H. Blair (ed.), *Chlorodioxins—Origin and Fate,* American Chemical Society, Washington, 1973, pp. 105-111.

49. E. A. Woolson and P. D. J. Ensor, "Dioxin Residues in Lakeland Sand and Bald Eagle Samples," in E. H. Blair (ed.), *Chlorodioxins—Origin and Fate,* American Chemical Society, Washington, 1973, pp. 112-118.

50. A. L. Young, C. E. Thalken, E. L. Arnold, J. M. Cupello, and L. G. Cockerham, USAFA-TR-18, U.S. Air Force Academy, Colorado Springs, Colo., October 1976.

51. A. Berlin, A. Buratta, and M.-T. Van der Venne (eds.), *Proceedings of the Expert Meeting on the Problems Raised by TCDD,* Milan, Sept. 30 and Oct. 1, 1976.

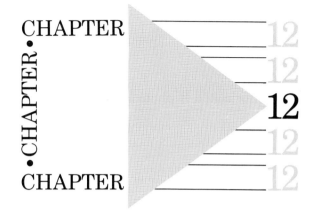

Case Histories: Land and Water Spills

PART 1
Polychlorinated Biphenyls / 12-2

PART 2
Phenol / 12-12

PART 3
Pesticides / 12-18

PART 1
Polychlorinated Biphenyls

Al J. Smith, P.E.
Chief, Environmental Emergency Branch
Air and Waste Management Division
U.S. Environmental Protection Agency

Historical Background / 12-2
Characteristics / 12-2
Case Histories / 12-5

HISTORICAL BACKGROUND

Polychlorinated biphenyls (PCBs) were first utilized in 1929 as nonflammable nonconductors of electric current. During the next 40 years, PCBs were widely used for this purpose. These chlorinated hydrocarbons were also used as plasticizers, lubricants, and valuable ingredients in inks, putty, waterproofing, sealers, adhesives, and waxes. Some estimates indicate that more than 362,000 metric tons have been produced.

In the late 1960s concentrations of PCBs were discovered in various components of the environment in the United States. National regulation began in the summer of 1970, to the extent that production was substantially limited to the single use of nonconducting insulating fluid for electrical transformers. The obvious need for PCBs was for safety purposes (i.e., to ensure nonexplosive transformers for hospitals, schools, etc.). However, large quantities of PCBs still remain in use. Therefore, these materials continue to be involved in spills and environmental emergencies.

CHARACTERISTICS

The information in this section was adapted from the U.S. Environmental Protection Agency's (U.S. EPA's) Technical Assistance Data System (TADS), Office of Emergency and Remedial Response, Washington, and is a condensation of worldwide information.

Basic Chemistry

Polychlorinated biphenyls are mixtures of the compounds formed by the chemical bond of two benzene molecules into a biphenyl molecule with varying numbers of chlorine atoms attached to the biphenyl molecule. The degree of chlorination greatly affects the toxicity and persistence of the chemical. A general list of the characteristics of PCBs follows:

12-2

1. Trade names: Aroclor, Dykanol, Noflanol, Pyranol, Chlorentol, Therminol, Inerteen.
2. Nonflammable and stable (nonexplosive).
3. Specific gravity, 1.182.
4. Boiling point, 340°C.
5. Flash point, 200°C.
6. Combustion characteristic: PCBs emit highly toxic vapors.
7. Persistence, very high.

Physical, Logistical, and General Health Characteristics

Chronic Hazard Level
PCBs may cause liver and skin ailments in humans and thin eggshells in birds.

Degree of Hazard to Public Health
PCBs are some of the most toxic materials known. They are considered to be strong irritants and are highly toxic when inhaled or ingested. Chronically toxic with inhalation or skin absorption, they rapidly accumulate in the food chain. Some of the hazards of PCBs can be attributed to polychlorodibenzofuran contaminants.

Air Pollution
PCBs are toxic in the air.

Action Levels
The air authority should be notified. Access to affected areas should be restricted.

Disposal Methods
High-temperature incineration should be used.

Water Pollution
PCBs are toxic in water with industrial fouling potential. They are not acceptable in food-processing waters.

Threat to Major Water Uses
Fisheries, potable supplies, and recreational waters all are threatened.

Physical State
PCBs are liquid; they will sink and dissolve only slightly in water.

Color in Water
PCBs are colorless.

Potential for Accumulation
The potential is high in liver and fatty tissues.

Food Chain Concentration Potential
PCBs display the same accumulative characteristics as DDT and other chlorinated pesticides.

Emergencies

Allowable Concentration   The allowable concentration of PCBs remaining in clay-type soils after the emergency cleanup of a spill is 50 mg/L and 1.0 μg/L in the leachate. An allowable concentration in totally porous soils (sand) or lime rock, in which solution channels are possible, is 25.0 μg/L. NOTE: The latter is the value recommended for drinking water and represents the leachate from the unrestricting soil and body; it reflects no migration inhibition or absorption.

Drinking-Water Recommended Limit   The recommended limit is 25.0 μg/L.

Biodegradation after 3 Years   There is little to no biodegradation (actual *in situ* study). NOTE: The allowable concentration of 50 mg/L is referenced in the *Federal Register*[1] and is based on long-term experiences with cleanup of PCB incidents. It is an attainment-oriented standard that is now being reviewed (for the purpose of lowering it) by the EPA. The 25 μg/L is an attainment-type recommended limit for emergency situations. The EPA currently uses 1.0 μg/L as a "concern level" and chronic discharge limit. In some locations in the southeast, background levels of PCBs in the order of 1.0 to 1.5 μg/L have been discovered. Generally, these are areas near old dumps or heavily industrialized areas. The migration and biodegradation statements are based on the only known long-term study of spilled PCBs.[2]

Inhalation Limit

The inhalation limit is 0.5 mg/m^3.

Toxicological Characteristics

Fresh Water

The 96-h TL_M (blue gill) is 0.278 mg/L.

Salt Water

Lethal: pink shrimp, young—48-h exposure = 0.100 mg/L

Lethal: oyster (100 percent decrease in shell growth)—96-h exposure = 0.100 mg/L

LC_{50} aerated: shrimp—48-h exposure = 0.3–1.0 mg/L

LC_{50} aerated: cockle—48-h exposure = 3.0–10.00 mg/L

LC_{50} aerated: pogge—48-h exposure = +10.0 mg/L

Inhibited growth: diatoms—+ 0.100 mg/L

Animal

Lethal: rat (oral)—500 mg/kg body weight

Lethal: guinea pig (oral)—170 mg/kg body weight

Lethal: mallard (oral)—2000–3000 mg/kg body weight

Lethal: pheasant (oral)—1000–3000 mg/kg body weight

Lethal: bobwhite (oral)—600–3000 mg/kg body weight

Human Trauma and Long-Term Effects

Human clinical manifestations of PCB poisoning are vividly portrayed in *PCB Poisoning and Pollution*.[3] This well-documented publication emphasizes the 1968 case study of a contaminated-rice-oil episode in northern Kyushu, Japan. In this incident, one of the worst mass poisonings ever recorded, PCB in the form of Kanechlor 400 (a mixture of chlorinated diphenyls and chlorobiphenyls used as an agent in the manufacture of rice oil) contaminated a large quantity of edible rice oil. The maximum concentration of PCBs in edible portions that was believed to have been consumed was 2000 mg/L. A total of 1200 patients were affected during a 2-year period. The results in terms of acute and subacute effects (including 22 deaths) were disastrous.

While nothing like this has happened in the United States, the realization of this potential is an adequate stimulus for rigid regulatory control to include both prevention of accidents and proper response to such accidents when they occur. A brief review of the state of the art of cleanup during the early and middle 1970s will offer the reader an opportunity to fathom the opportunities for innovation in this area. Obviously, much research remains to be done.

CASE HISTORIES

Kingston, Tennessee

On March 5, 1973, a transformer loaded with Askarel being transported along a mountainous highway 8 km (5 mi) south of Kingston, Tennessee, began leaking. The truck driver stopped and, after being warned by a local law enforcement official that "his vehicle was leaking something that was eating up the asphalt road," proceeded to dump the liquid onto an adjacent mountainside. Thus began an incident which was to:

1. Impact the public over a 52-km^2 (20-mi^2) area and become of national concern.
2. Cost well over $1 million in cleanup costs and legal damages.
3. Precipitate claims of clinical manifestations of various PCB diseases (chickens with no feathers, chickens laying eggs with no shells). There was even one claim that a young person had become mentally deficient from drinking well water contaminated with the chemical.
4. Force regulatory officials to become directly involved at the local, state, and federal levels in a situation which for years had been mostly ignored.

Early in the incident, heavy rains caused excessive soil depth penetration owing to a combination of slow response by regulatory and industrial officials and a lack of committal for cleanup. In addition, there was a general lack of understanding of how serious the problem was to become. Actually, this was to be one of the first major cleanup operations of PCBs in the United States. The spilled Askarel was a mixture of PCBs and the solvent trichlorobenzene (TCB). While TCB is both toxic and a strong irritant, its environmental and public health impact is of a lesser degree than that of PCBs. For example, spilled TCB is naturally degraded by more than 90 percent in 2 to 3 years as compared with PCBs, which exhibit only a 50 percent degradation in 20 years.

Experts from all over the world were questioned about the toxicology, drinking-water limits, and cleanup criteria for the spilled PCBs. Even though many conflicts developed throughout the 6-month $1.5 million cleanup effort, a scenario of cleanup and damage mitigation began to emerge. Much of this experience has found its way into state and federal regulations.

Although an elaborate sampling program was used to define *in situ* concentrations of PCBs, waste disposal decisions were difficult. Should the contaminated material be incinerated? What was the technology? Should the PCBs be buried? If so, where? What were the problems to be encountered?

Finally, crude though it was, a decision was made to ship 11,000 drums (55 gal, or

0.21 m³, each) of contaminated soil to Texas for subsequent burial in special concrete vaults. Sufficient incineration capacity was simply not available in the United States.

Cleanup concentrations were set initially at 5 mg/L PCBs remaining in the earth. After only a few days, it was obvious that to meet this limit a mountain of contaminated soil would have to be moved from Tennessee to Texas. Much discussion and on-scene testing finally established that an attainable concentration in plastic soils would have to be 500 mg/L. A 3-year study of the mobility of PCBs at this site later validated the 500-mg/L limit. To protect the water supplies (mostly private wells) a tight perimeter of deep well testing and a monitoring program for all private wells were undertaken. The more than 70 wells existing in a 16-km (10-mi) radius were tested; only 1 well had PCB concentrations greater than 2.0 μg/L. However, many of the wells as far as 16 km away revealed concentrations of 0.8 to 1.5 μg/L. Since no known spill had occurred in this area and the pattern was random, it was concluded that this concentration was "background" for the locale.

Public concern became very emotional but began to subside when the local people became convinced that the scenario was rational and was working. Well testing continued into September 1973. There was evidence of at least two cases of human disease derived from PCBs.

Edible portions of a number of farm animals were tested and found to be contaminated; the animals were destroyed. Clinical evidence of the impact of the spilled PCBs on public health was available to the author, and in subsequent litigation substantial awards were made to the plaintiffs for personal injuries caused by ingestion and inhalation of the chemical. During the cleanup, there were no reported cases of illness among the workers handling the PCBs.

During these operations, PCB "pockets" having concentrations of over 3000 mg/L were found in the soil. Also found in isolated cases were pools of free liquid Askarel. In view of the 200-mg/L ingestion limits at Kyushu, this was certainly not an incident to be ignored. A startling revelation from this incident, after a significant amount of testing in a 26-km² (10-mi²) area, was the discovery that background concentrations of 0.8 to 1.5 μg/L PCBs typified the groundwater of this rural area. The Japanese found in their country that 1.0 mg/L PCBs is not a surprising background in certain areas. These findings, however, should not be extrapolated beyond the context in which they are given.

Cedar Bluff, Alabama

In December 1974, a truck carrying a transformer loaded with 5.3 m³ (1400 gal) of Pyranol was wrecked in a rural area near Cedar Bluff, Alabama. Most of the Pyranol (a mixture of PCBs and TCB) ran into a dry ditch along the road.

As in Kingston, excavation was chosen as the method of cleanup (see Fig. 1-1). Once again, the ideal limit of 500 mg/L residual PCBs in the soil was attempted. However, the operation was halted when the *in situ* concentration of PCBs was reduced to less than 1000 mg/L. The reason for this change in target concentrations was the fact that excavation was destroying a long stretch of Alabama highway. A perimeter of wells was established 0.8 km (½ mi) around the spill area and monitored for 1 year. Today a water main is being constructed through the spill site, and no ill effects are anticipated. The price for the cleanup effort was $300,000.

Trion, Georgia

In November 1974, a truck carrying an electrical transformer loaded with Pyranol overturned, spilling approximately 1.1 m³ (280 gal) of the mixture in Trion, Georgia (Figs. 1-2 and 1-3). Once again, hundreds of drums loaded with a soil-PCB mixture were shipped to Texas for entombment.

Fig. 1-1 Excavation at Cedar Bluff, Alabama.

Fig. 1-2 Precipitation protection of spill area in Trion, Georgia.

Fig. 1-3 Excavation and drumming operation, Trion, Georgia.

Of paramount concern was the protection of the Trion water supply and the reduction of the concentration of PCBs in the soil to less than 500 mg/L. Protection of the water supply was achieved by monitoring the spill source for 1 year. Again, a perimeter of sampling wells was placed to ensure that the PCBs did not travel any significant distance from the spill site. As was the case in the Cedar Bluff incident, the 500-mg/L residual level in the soil was not achieved. The cleanup cost for this spill amounted to $200,000.

Whitehouse, Florida

In June 1976, the EPA responded to what was thought to be an oil spill at Whitehouse, near Jacksonville, Florida. Strong antiseptic orders alerted cleanup personnel to what was the beginning of a nightmare.

PCBs were found in oil, water, and sludge in several old dump pits that had been used as waste reservoirs for transformer liquids for approximately 30 years. A total of 26,500 m^3 (7 million gal) of PCB-contaminated water (1.0 to 1.5 mg/L) was immediately threatening the St. Johns River as well as oil and sludges with PCB concentrations approaching 10,000 mg/L. The problem here was failing dike walls, porous soil, close proximity to local residents, and the St. Johns River.

In contrast to Kingston, Cedar Bluff, and Trion, there was no logistical way to excavate this significantly large quantity of solids and liquids for subsequent landfilling. An alternative solution,[4] which consisted of draining all the liquids from the dump pits into a field-improvised treatment system, was initiated. The system comprised several processes including pH adjustment with lime, powdered activated-carbon adsorption, sedimentation, and filtration. The effluent from this operation, which had a PCB concentration of less than 1.0 μg/L, was released to the St. Johns River. The residual carbon-PCB sludges were transferred back to the dewatered pits, which were then backfilled and stabilized with fuller's earth topped with a contoured blanket of compacted clay. Following this stabilization technique, monitoring wells were placed around the area to check for migration of PCBs. Those wells are still being continuously monitored by the Jacksonville Health Department. The price of PCB cleanup at Whitehouse (see Figs. 1-4, 1-5, 1-6, and 1-7) was approximately $200,000.

Discussion

Experiences at the incidents described above have substantiated the fact that PCBs are very persistent and are not subject to significant microbial degradation in nonacclimated environments.[2] These incidents have also shown PCBs to migrate deeply into the soils following considerable precipitation. At Kingston, for example, 1200-mg/L concentrations of PCBs were found at a 4.6-m (15-ft) depth in clay. At Cedar Bluff, 300 mg/L was measured at 1.8 m (6 ft) beneath the surface in hard-compacted clay. Admittedly, these were "hot spots" and did not typify the entire spill site, but they do negate the assumption that contamination is limited to upper soil layers. Also, Kingston, Cedar Bluff, Trion, and Whitehouse have taught the profound lesson that as "dilution is not the solution to pollution," excavation is not always the answer to hazardous material land spills.

Research is needed for PCBs and other hazardous materials in the area of cleanup limits, disposal methodology, and emergency environmental limits for drinking water, edibles, etc. One fact is clear, however; PCB is harmful to biosystems and cannot be left in the environment for consumption by an unsuspecting public.

It must be recognized that a PCB spill is an emergency situation. A complete cure

Fig. 1-4 Dewatered PCB pit in Whitehouse, Florida, showing contaminated sludges.

Fig. 1-5 A pH adjustment pit in Whitehouse, Florida, showing perforated piping and lime filter.

Fig. 1-6 Activated-carbon batching shed and final settling pond in Whitehouse, Florida.

Fig. 1-7 Whitehouse after final cleanup, "crowning," and drainage.

with pristine restoration and absolutely fail-safe environmental and public protection may never be achieved. Knowing this, regulatory and public health officials must effect the best long-term public health protection possible.

REFERENCES

1. "Polychlorinated Biphenyls (PCBs)—Manufacturing, Processing, Distribution in Commerce, and Use Prohibitions," *Federal Register,* vol. 44, No. 106, May 31, 1979, p. 31516.

2. *Study of the Distribution and Fate of Polychlorinated Biphenyls and Benzenes after Spill of Transformer Fluid,* EPA 904/9-76-014, U.S. Environmental Protection Agency, Region IV, January 1976.

3. T. Higuchi (ed.), *PCB Poisoning and Pollution,* Kodansha, Ltd., Tokyo, 1976.

4. F. B. Stroud, R. T. Wilkerson, and A. Smith, "Treatment and Stabilization of PCB Contaminated Water and Waste Oil: A Case Study," in G. F. Bennett (ed.), *Proceedings of the 1978 National Conference on Control of Hazardous Material Spills,* Miami Beach, Fla., April 1978, pp. 135–144.

PART 2
Phenol

Austin Shepherd
Manager of Environmental Control
Singer Company

Introduction / 12-12
The Accident / 12-12
State Action / 12-13
Alternative Solutions / 12-14
Treatment Action / 12-14
System Start-Up / 12-16
Summary / 12-16

INTRODUCTION

The use of phenol and phenolic compounds has expanded rapidly in the United States over the last half century, as the markets for plastics, wood-bonding resins, and various pesticide and herbicide products have grown. In 1980 a total of 1,120 million kg (2.46 billion pounds) of phenol were produced. Most of this product finds its way from the manufacturer to the user by bulk rail or truck tankers. Therefore, the opportunity for an accident and subsequent environmental emergency is relatively high.

THE ACCIDENT

On June 27, 1972, one such accident occurred as a Western Maryland Railway freight train rounded a curve near Slabtown, Maryland. A shift of several heavy steel castings in a gondola car caused that car to derail, pulling from the track 10 other cars including 3 tankers carrying liquid carbolic acid.

Two of the three tank cars ruptured, spilling approximately 95 m³ (25,000 gal) of the 90% phenol solution. The tank cars were insulated to prevent crystallization of the phenol solution, and when they were ruptured, hot acid spilled on the railroad right-of-way and then ran down a small hillside toward Jennings Run, a tributary of the Potomac River (see Fig. 2-1). Vapors immediately engulfed the scene of the accident, driving away all forms of wildlife and chemically burning all vegetation within the immediate area.

Emergency railway crews, local police, and fire department officials were summoned. Upon arrival at the site, they surveyed the problem and decided to drench the area with water to cool the acid solution and prevent further drift of phenol vapors. This eliminated the immediate danger, but once the hot acid had been cooled, a giant cleanup problem remained.

STATE ACTION

State officials were notified of the spill, and within several hours of the accident representatives of the Maryland Water Resources Administration, located in Annapolis, arrived at the site. The first order of concern was for the extent of contamination of Jennings Run and for the possible impact on drinking-water supplies downstream. Water samples were collected at the spill site and at intervals downstream on Jennings Run, Wills Creek, and the Potomac River.

Initially the phenol concentration in the stream at the point of the spill was determined to be approximately 1000 mg/L. However, the levels approximately 16 km (10 mi) downstream had decreased to about 50 μg/L. Water utilities downstream were notified of this spill and temporarily halted pumping until the initial slug of phenol had passed. The amount of phenol entering the stream dropped rapidly from that point, and a second health emergency was averted.

Attention then focused on the spill itself, where concern was expressed for the possible leach-out of large quantities of phenol during subsequent rainfalls. At that point state officials requested the railway company to take action to prevent further contaminated discharge.

To aid in selecting a course of action to prevent further leaching of the phenol, the Cooperative Extension Service of the University of Maryland was enlisted. Core samples of the soil were taken at varying depths over the site of the spill. The samples were shipped to Maryland State Laboratories in Cumberland, where the phenol was extracted, distilled, and analyzed by colorimetric determination. From these analyses it was estimated that 74,800 kg (165,000 lb) of phenol remained in the upper layers of soil in the area immediately adjacent to the site. This meant that approximately 4500 kg (10,000 lb), or only 5 percent of the liquid, had entered the system. While the core samples were

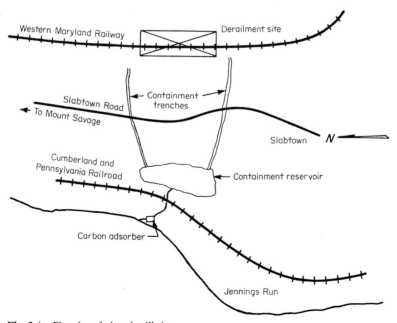

Fig. 2-1   Plot plan of phenol spill site.

being taken and analyzed, the railroad initiated action to prevent further runoff from entering Jennings Run. Trenches were dug to collect and store the runoff until a suitable disposal technique could be developed.

ALTERNATIVE SOLUTIONS

The Cooperative Extension Service offered three alternative solutions for handling the site contamination problem. The first involved removal of the upper 152 mm (6 in) of soil in the contaminated area and subsequently adding fresh topsoil and reseeding. The contaminated soil would then be distributed over a large land area to a thickness of 3.175 mm (⅛ in) or less and plowed into the ground. Bacterial action would then degrade the phenol and render it harmless. This approach was not pursued, however, owing to the prohibitive cost. Also of concern was the soil erosion potential on private grounds and roads in the areas.

A second alternative involved the use of flame throwers to incinerate the top layers of soil and thermally degrade the phenol. This idea was rejected by the Maryland State Forestry and Water Resources Departments primarily because of the potential fire hazard.

The third alternative involved the containment of the contaminated runoff and subsequent treatment of the water with granular activated carbon before discharge to the stream. This treatment technique had been proved for phenol removal in both industrial and municipal wastewater treatment applications and was highly recommended by the U.S. Environmental Protection Agency.

TREATMENT ACTION

Following further consultations with industrial carbon adsorption specialists, the railway company and state personnel elected to install a granular activated-carbon adsorption system. At the toe of the slope between the railroad right-of-way and Jennings Run, a collection channel was dug to collect all runoff from the contaminated site area. A small reservoir was also dug to provide storage for the runoff during exceptionally heavy rains. From the reservoir a 305-mm (12-in) corrugated pipe was installed to direct the contaminated runoff to a carbon adsorber situated adjacent to the stream (see Fig. 2-2 for a profile of the treatment system).

While the earth work was in progress, adsorption isotherm tests were conducted on the phenol-contaminated water (see Fig. 2-3 and Appendix to Part 2). These tests indicated that the average phenol contamination of about 1000 mg/L could be reduced to less than 1.0 mg/L at a carbon adsorption capacity of about 25 percent. Industrial adsorption experts indicated that a minimum contact time of about 30 min would be necessary to achieve these objectives. On the basis of this information, the carbon adsorber was designed to hold 28 m³ (1000 ft³) of Calgon Filtrasorb-300 granular carbon.

Since time was of the essence in getting the treatment system on line, it was decided to construct an adsorber out of easily obtainable railroad ties (Fig. 2-4). Heavy bridge ties 3.7 m (12 ft) long were hauled to the site and used to construct a container 3.3 by 3.7 by 3.7 m (11 by 12 by 12 ft) high. This was done by stacking and interlocking the tie ends. The tie walls were further reinforced with buttresses. The adsorber was then lined with exterior-grade plywood and sealed with okum and tar.

When the adsorber box was completed, an underdrain and lateral pipe assembly was designed and fabricated in Cumberland and delivered by rail. The underdrain consisted of a 152-mm (6-in) steel header pipe 3.3 m (11 ft) long with 18 laterals measuring 76

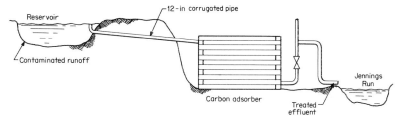

Fig. 2-2 Profile of treatment system.

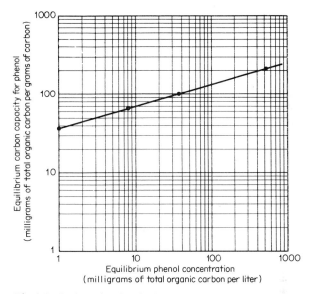

Fig. 2-3 Isotherm for phenol adsorption expressed in terms of total organic carbon.

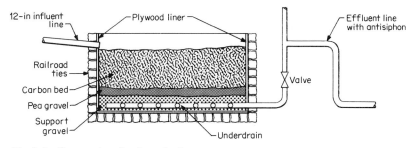

Fig. 2-4 Cross section of carbon adsorber.

mm (3 in) in diameter by 1.5 m (5 ft) long welded to it. Holes measuring 9.5 mm (⅜ in) and spaced on 102-mm (4-in) centers were drilled in the underside of the lateral pipes for collection of the water.

Following installation of the underdrain in the wooden adsorber, 152-mm (6-in) header pipe was connected to an antisiphon loop that exited at the bottom and rose to a height just above the top of the adsorber. A drain valve was installed in the bottom of the exit line to ensure against overflow of the top of the adsorber during heavy rainfall.

Prior to filling the adsorber with carbon, 6350 kg of Type B-6 gravel with a partial size from 13 to 25 mm (½ to 1 in) was placed around and over the underdrain to a depth of 305 mm (12 in). This bottom gravel layer was then covered with 76 mm (3 in) of pea gravel for the carbon support.

The adsorber was then filled with water and tested for leaks. Because of the manner in which the adsorber had been constructed, it was difficult to stop all leaks completely. They were minimized, however, and the adsorber was then filled with 13,600 kg (30,000 lb) of Calgon Filtrasorb-300 carbon. This provided a bed depth of approximately 3 m (10 ft).

SYSTEM START-UP

During the week of July 21, the treatment facility was started up. The system was operated in a downflow mode to enable gravity flow from the collection reservoir through the adsorber and into the stream. This also allowed any silt carried into the adsorber to be filtered on the surface of the carbon. Periodically, the top 0.3 m (1 ft) of carbon could be removed to prevent any hydraulic problems.

From start-up in late July through October the phenol in the influent to the system ranged from 100 to 1000 mg/L depending upon weather conditions. Effluent concentrations of phenol from the system during this period ranged from 0.5 to 13 mg/L. Some minor channeling of water through the carbon bed was noticed at several low-flow periods, but effluent concentrations of phenol were still within an acceptable level for discharge.

The initial fill of carbon lasted approximately 90 days and began to produce unacceptable results only shortly before the winter freeze. It was thus decided to postpone further treatment activities until the spring thaw.

The following April the spent carbon was removed from the adsorber, and minor damage to the system caused by winter freezing was repaired. The adsorber was then refilled with carbon and treatment reinitiated.

The system was operated successfully through the summer with a minimum of operating problems. By September 1973 it became apparent that most of the phenol had been leached from the soil. At that time the influent levels of phenol had dropped to less than 10 mg/L. Soil samples were taken over the area of the spill, and these showed only trace amounts of phenol left at the site. On October 3, 1973, the state of Maryland had sent a letter of compliance to Western Maryland Railway, and the treatment unit was permanently shut down. The spent carbon was removed, drummed, and shipped to a secured fill for disposal.

SUMMARY

The course of action in the handling of this spill of phenol followed a classic example. The immediate environmental dangers were eliminated by preventing the drift of noxious vapors and by notifying downstream water utilities of the spill and thus preventing contamination of drinking-water supplies.

Once the immediate dangers had been eliminated, action was taken to contain the spill and prevent further runoff and subsequent contamination of the receiving stream. Treatment alternatives were then evaluated and the selected alternative implemented. Carbon adsorption proved to be a technically viable option, providing a level of treatment permitting direct discharge of the contaminated runoff.

Treatment was initiated and continued until the concentration of phenol leaching from the soil was reduced to acceptable levels.

APPENDIX

ADSORPTION ISOTHERMS

An adsorption isotherm is a batch test designed to demonstrate the degree to which a particular dissolved organic compound (adsorbate) is adsorbed on activated carbon (adsorbent). The data generated show the distribution of adsorbate between the adsorbent and solution phases at various adsorbate concentrations. From the data a plot of the amount of impurity remaining in solution at constant temperature can be generated. For a single adsorbate, a straight-line plot can be obtained when using the empirical Freundlich equation:

$$\frac{x}{m} = kc^{1}/n \text{ or } \log \frac{x}{m} = \log k + \frac{1}{n} \log c$$

where x = amount of contaminant adsorbed
   m = weight of carbon
   c = equilibrium concentration in solution after adsorption
k and n = constants

For mixtures of adsorbates, a series of straight lines can be obtained. The presence of a nonadsorbable component will result in a curvature of the line when in combination with an adsorbable component and in a vertical line when alone.

Data for generating this type of isotherm are obtained by treating fixed volumes of the water sample with a series of known weights of carbon. The carbon-liquid mixture is agitated for a fixed time at a constant temperature. After the carbon has been removed by filtration, the residual adsorbate concentration is then determined. The amount of organic adsorbed by the carbon x is divided by the weight of carbon in the sample m to give one value of x/m for the isotherm.

The adsorption isotherm interpretation yields a theoretical equilibrium carbon dosage.

REFERENCES

1. W. L. Ramsey and J. M. MacCrum, "How to Cope with Hazardous Spills," in G. F. Bennett (ed.), *Proceedings of the 1974 National Conference on Control of Hazardous Material Spills,* San Francisco, August 1974, pp. 234–237.

2. James A. Dent, *Handbook of Industrial Chemistry,* Van Nostrand Reinhold Company, New York, 1974.

PART 3
Pesticides

Lee Frisbie
President, Environment Plus

Truck Trailer / 12-19
Railroad Car / 12-19
Ship / 12-21
Water Supply Contamination / 12-25
Tank Trailer Leakage of Liquid Heavier than Water into a Stream / 12-26

Although dictionaries define pesticide simply as an agent used to destroy pests, this section will be concerned only with chemicals used to control insects, weeds, and nematodes. However, within this limitation there are both relatively innocuous and highly toxic chemicals and a full range between them.

In this part actual case histories will be used to illustrate the various factors involved in properly handling a pesticide spill. Names, dates, and places are sometimes omitted because of possible legal complications. Reference to "the company" will mean the author's former employer, the Chemagro Division of Mobay Chemical Corp.

The types of pesticide spills to be discussed are as follows:

1. Container leaking in a truck trailer
2. Container leaking in a railroad car
3. Container leaking in a ship
4. Pesticide accidentally introduced into the drinking-water system of a home or a small town
5. Leakage of a liquid heavier than water into a stream

The pesticides involved in the incidents described here are primarily of the organophosphate family, but the principles of decontamination apply to most pesticides. The toxicity hazard presented by these materials is nearly all due to contact with liquid or powder. Although most of the materials have strong odors and the public tends to equate a high odor level with high toxicity, the odorous vapors are usually nontoxic.

One of the primary ingredients of a pesticide spill is the adverse psychological reaction of the public. The term "pesticide" has a very bad connotation to most people. Therefore, the person handling a pesticide spill must always keep this foremost in mind, using every reasonable means to avoid exploitation of the spill by the media. Remaining very calm and avoiding overstatement of the problem are essential.

In any pesticide spill, the important points for the person in charge to remember are:

1. Protect the public and the environment.
2. Contain the spill; stop the leakage.
3. Remove all contamination (if feasible).
4. Properly dispose of contaminated material.

TRUCK TRAILER

The most common type of spill is from a container leaking into a truck trailer. Such incidents, usually involving small quantities of material, may result in some contamination of adjoining cargo and the trailer floor. Routine cleanup involves sweeping up powder or granules in the case of solid material or of picking up liquid pesticide with absorbent material. The contaminated area is then scrubbed with detergent and water to remove any remaining thin film. Since strong oxidants will break down many pesticides, the last step is to soak the area for several hours with household bleach, followed by a final water rinse.

In some cases of severe leakage, it has been necessary to remove and replace a portion of a wooden trailer floor. In every situation, a check should be made to ensure that no food, clothing, or similar material was near the leaking container. Any of these materials adjacent to the spill should be destroyed under supervision to prevent recovery and use. If there is any evidence of contamination on the outside of the carton containing these materials, both the carton and the materials contained in it should be destroyed.

The spill described here was a relatively simple case involving solid material which presents very little danger of exposure of response personnel. A phone call concerning a broken bag of granular insecticide was received from a truck terminal. When the trailer was opened, one bag was noted to have broken, evidently from improper loading followed by a slight load shift in transit. About 4 kg of the 110-kg bag had spilled.

The terminal manager was instructed to have the employee assigned to cleanup wear gloves and use care not to cause dusting of the material. The broken bag was first placed in a heavy plastic bag. A broom and dustpan were used to transfer the loose material from the floor to the same plastic bag. The floor was washed with detergent and water. The broom, dustpan, gloves, and the employee's shoes were also washed with soap and water. The consignee (a distributor) was nearby, and the plastic bag containing the damaged bag and spilled material was delivered with the rest of the load. This material would be sold to a customer at a reduced price and used normally. This procedure avoided return shipment, additional handling, possible disposal, etc., and thus was most efficient from both environmental and practical standpoints.

RAILROAD CAR

This type of spill is similar to a truck trailer spill, except that movement of the car is more restricted. The case described here involved a piggyback railcar, so the leakage involved a trailer plus a railcar.

The call came from a yardmaster, whose people had noticed an odor and then observed liquid dripping from the railcar. The railcar had been placed on an isolated siding and a 24-h guard posted to prevent anyone from approaching it. This situation obviously required a company representative to go to the scene. The call came on a weekend, so that the company agent could not obtain detailed information on the contents before leaving home.

The spilled material was a liquid Class B poison insecticide, from a 55-gal. (0.21-m³) drum, loaded in a shipboard container for export. This and other drums had been loaded so that the doors could only be partly opened in order to avoid theft and discourage entrance.

The location was about 800 km (500 mi) away, but the site and air travel times made driving seem the best alternative; there was no immediate hazard to require a charter flight. By driving, it was possible to take more equipment than would have been reasonable by commercial flight. The following standard items were kept in a prepacked case ready to go for such incidents:

1. Rubber boots

2. Rubber apron

3. Rubber gloves (for most pesticides, latex gloves are the best protection against absorption into the skin)

4. Control clothing: white uniform shirt and pants

5. Respirator and extra cartridges (for organic vapors and acid gases, plus several others to cover various needs)

6. Plastic trash bags

7. Paper towels

8. Container closures and closure tools

9. Scoop

10. Isopropyl alcohol (solvent for cleaning gloves and contaminated surfaces)

11. Household detergent

12. Household bleach (it can be purchased locally when air travel is involved)

The author arrived at this small town about 5:30 P.M. on a hot Sunday evening. The drip had stopped, and after protective clothing was donned, an attempt was made to locate the leaking drum. It was extremely hot inside the railcar, limiting the work period to approximately 10 min. Bungs were removed and levels in the drums examined by flashlight (the restricted door opening prevented light from entering as well as restricting ventilation). The smell inside was quite strong but was from nontoxic ingredients in the mixture. The pesticide itself had a low vapor pressure and thus was not a problem.

There was no crane available to move the container from the railcar and thus no way to unload it to get at the leaking drum. It appeared that approximately 60 L (15 gal) had escaped, and the drip began again after the bung was removed. The author decided that a material transfer was the best approach, but no supplies were available from the railroad. A 55-gal (0.21-m³) drum was essential.

The proprietor of a service station obliged with an overpriced but clean oil drum which had been rinsed by hauling gasoline. This was more than might have been expected in a small town at 7:00 P.M. on a Sunday evening. A convenience grocery store provided 15 m (50 ft) of garden hose.

The drum was placed on the ground beside the railcar and a siphon established through the hose. Approximately 130 L (35 gal) of pesticide was recovered.

After the leaking drum had been emptied, cleanup began. There was barely enough room under the container to allow reaching under it with a paper towel to absorb the liquid. The latex gloves did not cover the forearms completely, and several times work was stopped to cleanse the forearms of liquid pesticide. Jugs of water were carried into the railcar for washing. The cleanup site was about 1.6 km (1 mi) from any building. The cleanup was being done after dark, and only the author was involved. The guard was available in case of a problem.

The wet paper towels were collected in a plastic bag for disposal as hazardous material. After the liquid had been absorbed, the contaminated gravel and soil from beneath the railcar were scooped up and placed in the same plastic bag. There was no way to wash the interior of the container because the drums were all tightly blocked in place.

The affected flatcar surface was washed with detergent and water, and then rinsed with water. Household bleach was poured into the container at the site of the leaking drum so that the bleach would follow the flow pattern of the pesticide that had leaked out. Bleach was also poured on the flatcar surface, with the excess following the flow pattern to the ground. A railroad foreman brought two of his strongest men to assist in getting the partially full drum (approximately 250 kg) from the ground into the container. There was only a 3-cm clearance on each side of the drum going into the slightly open container door, but the drum's loading was accomplished.

The men also brought wood to brace this drum in place. The plastic bags of contaminated trash were also placed in the container, and it was resealed. The yardmaster agreed that a full cleanup could not be accomplished until the container could be unloaded in the nearest major city, 130 km (80 mi) away. When the author arrived home, a check was made with the railroad, and the car had not been shipped. The cleanup had been completed about 2:00 A.M., and the day-shift people saw the bleach still dripping, noticed some odor, and would not move the car. A call got the car moving, and a better cleanup was made after the container was unloaded. However, it was necessary to ship the container back to the company's plant for a final decision, and the floor was replaced.

SHIP

This case history was presented by the author at the 1974 National Conference on Control of Hazardous Material Spills[1] to illustrate the responsible attitude of the company toward protection of the public and the environment even in 1968, when this incident occurred. Since the pesticide involved had been sold to Pakistan through the U.S. Agency for International Development program, the company had no legal responsibility. This made the assignment even more sensitive for the author, as he was instructed to prevent any handling problems but without taking any legal responsibility; in other words, he was an adviser without authority.

The USS President Hayes freighter left San Francisco in December 1968 en route to Japan and around the world to New York. About 5 days out, an extremely violent storm caused the cargo to shift. Part of the freighter's cargo consisted of Guthion, an organophosphate insecticide in 25-L cans, stored in two holds. In the full hold, there was no leakage, but the 100- by 100-mm (4- by 4-in) shoring in the partially filled hold broke loose. The Guthion cans then slid back and forth across the floor as the ship rolled heavily, and some ruptured from contact with 55-gal (0.21-m³) drums of tallow and 135-kg (300-lb) bales of paper pulp, which also were sliding. Since some of the drums of tallow were flattened, it was surprising that only 68 of the 785 cans of the pesticide in this hold were damaged to the point of leakage.

The Guthion odor was detected by the crew, and when the cargo was noted to be a Class B poison, the captain radioed for help. He went to Honolulu as a port of refuge, and the author arrived there in time to assure port and fire officials that this was not an explosion or fire hazard. There was almost no pesticide usage in Hawaii, and thus there were no experienced personnel to draw on.

One of the difficulties in this problem was that 19 agencies had some degree of control. It was hard to brief and reassure agency personnel concerning the scope of the problem and the minimum danger involved. The printed booklets carried by the author were invaluable in reinforcing his advice, as they described proper handling procedures. These booklets had been prepared to train personnel involved in the product distribution system for handling spills.

The following conditions were outlined by the author as necessary before work could begin:

1. Wearing of rubber boots and gloves and use of isopropyl alcohol to wash them.

2. Wearing of overalls as controlled clothing (no personal clothing allowed).

3. Availability of shower facilities in case of skin contact with pesticide by a worker.

4. Adequate medical coverage for accidental exposure. This item included the local firm's physician, the emergency room of a nearby hospital, and atropine sulfate, the antidote for the pesticide, with adequate knowledge by the medical facilities of atropine therapy (a balance must be maintained, as excess atropine can be a problem).

5. Instruction in safe handling for stevedores and any others involved in the cleanup. The stevedores were from the Philippines and spoke little English, presenting a formidable communication barrier.

6. A means of either storing or disposing of the hazardous material.

The most difficult question at this meeting was the insistence by the U.S. Department of Labor representative that the stevedores wear respirators in the hold, as he was convinced that there would be Guthion in the air. The writer objected strenuously, for two reasons. First, Guthion is a waxy solid after the solvent evaporates, with no vapor or dusting tendencies. It was mixed with the tallow in the hold, forming a greasy, gelatinous mass. Second, the stevedores had never worn respirators, so they would unconsciously and consciously be touching and pulling the respirators away from their mouths. With greasy gloves, the grease would be transferred to the faces of the stevedores, providing a possibility of adsorption through the skin, the primary hazard in handling Guthion. It required great effort to prevail at this point. A faulty conclusion 3 days later (described below) caused the wearing of respirators temporarily, but all the grease had been removed before that, and no problems resulted.

Operational Control

The operations managers from the local firm had charge of the work crews, the analytical laboratory people, and, later, the cleanup crews. The general-average surveyor (a marine insurance adjuster) soon arrived and assumed total charge of the operation, since all parties involved in shipping share equally in problems as a joint venture. After the first week, it was determined that the ship would have to be legally certified as clean. The author could not take this responsibility, so a toxicologist was brought in.

Disposal

The author preferred incineration or burial in a secure landfill, but a commitment for this procedure could not be obtained initially. Thus, an open barge was leased and tied alongside the ship for the storage of hazardous material. When the cleanup had been completed, approval was received for burial in a secure U.S. Navy landfill.

Removal of Contaminated Material

The first obstacle to beginning work was concern over fire or explosion when removing the hatch cover on the hold, because of possible solvent vapors. Using great care, the hatch was safely removed. Then it was discovered that the ship had spent the 3 days in getting to Honolulu with the hatch cover off to reduce the odor, and the solvent had evaporated.

The stevedores began work but were stopped after 3 h on the order of the local firm's physician. A conversation with another physician at the University of Hawaii had convinced him that there was an inhalation hazard. The other physician was checking for cholinesterase inhibition as part of a U.S. Public Health Service grant and wanted to become involved in this situation. However, the analytical work concerned was in a fairly primitive state, and few organizations outside organophosphate producers had the necessary expertise. This made it seem desirable to avoid the entry of an unknown capability into an already difficult situation, so the university group was resisted for several days. However, it was brought in on about the fourth day and provided excellent gas chromatographic analysis and cholinesterase testing.

A barge was not available until the third day after the ship had entered the harbor, so a false floor was installed and only uncontaminated material handled for the first 2 days.

On the third day, unknown to the operations control group, a seaman went to the United States public health clinic, was diagnosed as having "organic phosphate poisoning," and sent to a nearby Army hospital. No seaman had been allowed in the contaminated area, and so any exposure would have occurred 3 to 6 days before. Since symptoms from exposure show up within hours, this was not a legitimate complaint; this was confirmed by a follow-up.

Over the author's objections, air sampling was done on the third day. Exactly as predicted, the simple sampling method (air sparge through sodium hydroxide solution) provided a positive test for phosphorus. This was not strange in view of the phosphate fertilizer being unloaded from a nearby ship. The positive test for phosphorus came back on the fourth day, and respirators were ordered. The primary hazard (greasy material) had been removed, so the main problem with the respirators was a work slowdown from fear and unfamiliarity.

At this point, the author went to the university group, checked its capabilities, and asked for its help in resolving the problem. The staff worked on the weekend, and provided evidence on Sunday (the fifth day) of only a minute amount of Guthion in the air. This satisfied everyone but the stevedores, who insisted on continuing to wear the respirators until the toxicologist and the stevedore supervisor spent 4 h without respirators in the hold with them and showed no ill effects.

The university group also did blood sampling followed by cholinesterase testing, and on the sixth day provided the information that the average result of the stevedores' tests was about equal to the population average. This was very helpful in alleviating many fears.

On the fourth day, the writer carefully entered the lower levels of the affected hold. Guthion was found on every level. Before this, it had been hoped that the hatch seals had held the contamination to a single level and that the ship could sail in a week or less. The lower levels contained automobiles, forklift trucks, cases of evaporated milk, and containers (vans) of household goods. There were also refrigerated compartments with citrus fruits and frozen compartments with chickens and other meat. This was the only hold with food, and the wrong place for a pesticide.

Cleaning of Contaminated Surfaces

On the fifth day, a local tank-cleaning firm was found and employed. Its workers were given careful instructions. The tan crystallized Guthion was easy to see, and the first step was to scrape off and collect all the visible material. Then the surfaces were scrubbed with rags soaked in alcohol to dissolve the remainder. Any residue on the surfaces was destroyed by a thorough washing with sodium hydroxide solution, as Guthion hydrolyzes rapidly in alkaline solution. The caustic was removed by a hot water-steam rinse, and the final rinse was made with water from a fire hose.

By this time, the toxicologist had arrived. The sampling method involved masking tape 30 cm (12 in) long and 5 cm (2 in) wide. Each piece of tape was pressed against a surface to be sampled, then removed and analyzed for traces of Guthion. The smooth surfaces were clean. However, many of the steel surfaces were rusty and porous, making cleaning difficult. The masking tape removed some of the rust from such surfaces and indicated traces of Guthion. Because of the rust problem, the above treatment was repeated twice more before the toxicologist certified the ship as clean. This took 14 days because of the large area and the limited number of workers.

Personnel Exposure during Cleanup

The entire operation was completed with very minimum exposure of the personnel involved. There was no real evidence of symptoms of organophosphate exposure.

On the seventh day, at a time when the author was busy elsewhere, two cleaners were sent to the hospital with reported phosphate poisoning, including eye symptoms (pinpoint pupils). There was about an hour's delay in notifying the author, and when he arrived at the hospital, the men had already been released. One reported to the emergency room physician that he was not really sick, he just did not like the smell. The other man was emotionally upset by working around a toxic material. No symptoms were observed by physicians or other qualified observers.

This circumstance upset the stevedores, who were necessary to unload each new layer after cleaning. There were some "hangover" problems, and when two stevedores became nauseated, these problems appeared to have been a factor, combined with heat and odor.

Questions of Cargo Contamination

About 600 drums of tallow that were only slightly contaminated were taken to a U.S. Coast Guard facility for cleaning. They were loaded onto cradle-type skids and cleaned on the skids. After any visible material had been scraped off, the drums were sprayed with a caustic solution and allowed to bake in the sun for 24 h. Then a fire hose was used to remove the caustic solution. This procedure was very effective.

The Guthion cans were small enough to be easily cleaned on the deck and stored on the pier for reloading. The battered but nonleaking cans were stored for repackaging before shipment. Only 68 of the 785 cans in the problem hold leaked, a remarkably low 9 percent, in view of the battering by the much larger drums. There were approximately the same number of cans in the other fully packed hold, in which there was no leakage and little damage. The cases of evaporated milk were separated according to degree of contamination. If a case had been soaked, it was destroyed. If there were only a few surface spots, the cans were removed from the carton, caustic-washed, and recased in a new carton.

There was some circulation of outside air to the refrigerated box containing citrus fruits, and it had been entered for inspection during the time when the hold was contaminated. Analysis indicated minute traces of Guthion on the surface of some fruit. The toxicologist certified the fruit as safe for human consumption, but use of the food was rejected because of a recent poisoned-food episode in Hong Kong; it was destroyed.

The frozen-food compartment had not been entered, and there was no circulation of outside air. Thus, no Guthion was detected, but the meat was also destroyed.

Damages and Legal Aspects

The losses in this incident totaled approximately $1 million (1968). American President Lines claimed that the Mobay containers were not satisfactory, but excellent quality control reports plus actual results refuted this contention. It also considered the responsibilities of the ship captain, but he had detailed records of course and speed changes made to reduce hazards in the storm. The real cause was improper loading of the Guthion in a hold with foodstuffs and inadequate shoring to stabilize the containers.

WATER SUPPLY CONTAMINATION

Contamination of a water supply is not a typical pesticide spill, and the author has been involved in only one such case. However, spills of this kind have a significant potential for bad public relations.

A report was received about 3:00 P.M. from a state official of a company pesticide being accidentally introduced into a water system of a home (and possibly of a small town) approximately 400 km (250 mi) away. Although there appeared to be no company responsibility, the author left by car at about 5:00 P.M. to provide technical assistance. An attempt was made to obtain highway patrol escort, but it was denied.

The basic information available was that at about 5 P.M. on the preceding evening a farmer living on the edge of town was filling an 0.75-m³ (200-gal) spray tank, containing several liters of a liquid emulsifiable concentrate of an organic phosphate pesticide, from a garden hose attached to the city water system. The end of the hose was below the water surface. A water shutoff for maintenance of a water tower caused siphoning of an unknown amount of the spray mixture into the water pipe. The town had grown out to the farmer, and he was on the end of the line.

After the water pressure returned, the spray tank was filled and the spraying done. The farmer later remembered the long fill time, and estimated that the shutoff had come at about 0.26 m³ (70 gal). When he came in from spraying, he drank some water from the faucet, observing some milkiness. Since he had been spraying this material for several hours, the odor in the water was not recognized. He then took a shower and, when rinsing off, noted the milkiness of the water and mentioned it to his wife. He then went to sleep. His wife saw a television program concerning pesticides, woke him up, and had him shower at his son's house several blocks away.

The farmer called his insurance agent about 10:00 A.M. the next day and found that his insurance would not be effective unless he notified the city water department. After his notification, word spread, and school was dismissed about noon. State authorities were notified and arrived at about 4:00 P.M. They sampled the water at the farmer's house and told the residents of five houses apparently on the same line with the farmer not to use their water until it was released by the state. They left about 6:00 P.M. after advising the city water department to flush the line with detergent and then bleach.

Response

When the author arrived at about 8:30 P.M., his welcome was very friendly, as no one else knowledgeable in pesticides was on the scene. The author sought out the farmer and discovered that the exposure had occurred 24 h before. It was explained to the farmer that symptoms usually appear within 6 to 8 h, so that his danger period was over. He mentioned a slight nervous feeling, but further questioning revealed a trip to his doctor and some atropine sulfate therapy. This explained the nervous feeling. The farmer's apparent good health, plus the rural environment and the author's presence with technical knowledge, combined to allow a complete absence of media coverage. This was a pleasant surprise in view of money offers for news tips by a radio station in a larger town approximately 48 km (30 mi) away.

The water department personnel were fairly certain that they had disconnected the 910-m (300-yd) extension from the main to the farmer's house about 10 years before, when the five houses were built between the end of the main and the farmhouse. They thought that the farmer's water came from the end of the stub line serving these houses but were digging to be sure.

Digging was completed the next morning and revealed that the extension was still the supply line; no milky water was found at the connection to the main. The state official had suggested running a lot of water out of the end of the line and had not restricted use of the water by the rest of the town.

The author checked the farmer's water system and found a line to a mobile home not currently in use that contained odorous milky water, indicating that some pesticide was present. The farmer's hogs had been watered from the system the night before, but the only symptoms were that a few hogs had had dysentery and now were all right.

After extensive flushing of the farmer's system with detergent, plus several hours' standing full of bleach, the author took samples and left. He could not recommend any action but left that to the state authorities. He did advise the farmer at least to replace his hot-water heater and water softener. This led to the disclosure that the water softener company had encountered several instances in which this type of incident had occurred in a one-home water system and the softeners were replaced.

Summary

It was of benefit to both the community and the company to send a representative even though there was no liability. This service helped protect the public, prevent hysteria, and avoid bad publicity for the company.

The state authorities sent their samples to the nearby regional EPA laboratory, and personnel there communicated with company specialists concerning analytical techniques. Results from the two laboratories were compared, and the water systems were cleared for use by the state within a few days.

TANK TRAILER LEAKAGE OF LIQUID HEAVIER THAN WATER INTO A STREAM

In this incident the organic liquid involved was lighter than water when spilled but became heavier than water quickly as the solvent (toluene) evaporated. Hence the expertise and equipment available for oil spills were of no value.

The material spilled was a hazardous waste from the production of pesticide intermediates but initially was thought to be free of pesticide contamination. As a result of this spill, waste material shipments were brought under the same level of control as product shipments, with formal specifications, analysis, etc. This was the most significant effect of the spill (if the $750,000 cost for cleanup is not considered).

The waste disposal firm involved in the cleanup provided its own tank trailers to haul the bulk liquids to an incinerator. Since there were several specific types of waste liquids, each was handled separately; they were estimated to comprise 14 m^3 (3700 gal) of water and 3 m^3 (800 gal) of organic material. The water was to be "incinerated" because of the odor associated with it.

The trailer involved in the spill was loaded normally and was on an interstate highway east of Fort Smith, Arkansas, when a section of discharge pipe under the trailer, ahead of the first valve, opened owing to corrosion that had occurred because by mistake a section of aluminum pipe had been installed in contact with carbon steel. The waste being hauled at the time of repair was not corrosive to aluminum, but various wastes from many companies had been hauled following the repair, and corrosion occurred.

A state air pollution employee was behind the truck when the failure occurred, and he signaled the driver to pull over. Since it was obvious that the whole load would be lost, the driver said that he was asked to back up to a concrete spillway built to carry off rainwater, but others involved have not confirmed this request. If the truck had traveled only a short distance farther, the material would have soaked into the soil without reaching a creek, and cleanup would have been easy.

The spillway went down a steep embankment, where it joined the runoff ditch under the highway coming from the hills above. The ditch then crossed a slight downslope and entered a creek. The creek bank was about 2 m (6 ft) above the creek surface at this point. Most of the runoff water percolated through the rocky soil into the creek bottom. High flows caused some water to run over the top of the creek bank.

The incident happened at about 4:30 P.M. on Tuesday (Day 1). The first contact with the company generating the waste was a call the next morning from an Arkansas air pollution control official concerned with odor complaints from the area. Although the disposal firm's rules called for the driver to notify the generator immediately of a spill, this did not occur.

Since neither the pesticide contamination of the water nor the creek was known when the author left Kansas City at about 11:30 A.M. on Wednesday, the projected action was to use earth-moving equipment to cover the spill area with soil, thus eliminating the odor. Since the generating company was very spill-conscious, a sample was taken from the waste tank for analysis. When the author called in at 4 P.M., he found that there were indications of pesticide contamination. This drastically changed the complexion of the whole incident, and the plant production manager was to be sent to take charge of the situation. (The author was manager of environmental protection at that time.)

Initial Response

Some of the waste liquid had run over the creek bank, and some was still percolating into the creek, as evidenced by the oily sheen. At normal flow, the creek was 150 to 300 mm (6 to 12 in) deep and 3 to 6 m (10 to 20 ft) wide. The channel was approximately 1.8 m (6 ft) deep and 12 m (40 ft) wide at the spill point, with evidence of very high flow from the surrounding hills after a rain.

Odor permeated the area. There were a number of small dead fish in the creek. Their deaths could have been caused either by immersion in the solvent (toluene) or by its concentrated vapors (gasoline could have caused a similar effect). Since an analysis of the spilled waste material had not yet been run, the author could say only that there might be some pesticide in the material. This naturally made the state official suspicious and impaired communication.

Later that evening, the presence of pesticide was determined. A cold, steady rain began the next morning (Day 3) and continued all day.

Topographical maps revealed that the creek joined two other larger creeks before reaching (in 29 km, or 18 mi) a river-reservoir that was an important fishing area (see Fig. 3-1). Access to the creek was limited, as the area was hilly and sparsely populated, with rocky soil and many scrub trees and brush. Roads crossing the creek at points approximately 5, 10, and 26 km (3, 6, and 16 mi) below the spill point became prime sampling points. It was established that no one used the creek for potable water but that cattle did. Access to the creek at the spill site from two farmers' homes was via a steep, crude road. The creek was in a valley approximately 370 m (400 yd) wide at this point.

No further fish or animal kills were observed along the creek during Day 3. The organic material, whose volume was estimated to be 0.6 to 0.8 m³ (150 to 200 gal), had settled in puddles of various sizes along the creek bottom. The normal streamflow was estimated at 4.5 m³/min. There appeared to be no organic material below the natural dam. The primary problem seemed to be removal of the organic material, so a work crew was organized for Day 4.

Since odor was a significant problem, 140 kg (300 lb) of calcium hypochlorite,

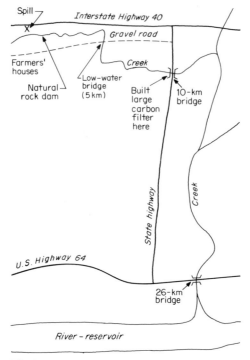

Fig. 3-1.

obtained from a pool service, was put on the contaminated soil. Water was applied to leach the hypochlorite into the soil, and a bulldozer was sent to scrape additional soil over the area to decrease the odor during the hypochlorite-waste reaction. Several small fires were started before the workers began to exercise the caution necessary with the handling and use of hypochlorite.

The creek cleanup operation was abandoned by early afternoon, as the hand-operated pumps could not retrieve the organic material from the creek bottom. A call was made to the company plant for a crew equipped with a gasoline-powered compressor, air-operated diaphragm pumps, and a four-wheel-drive vehicle.

At about 9:00 P.M., dead fish were found at the bridge, following about 48 h when no effects on fish had been noted. The following day, an investigation was made into the fish toxicity of the hydrolysis products of the pesticide. These appeared to be more toxic to fish than the pesticide, with some effects reported on one species at only 6 μg/mL and others in the 20- to 50-μg/mL range.

Secondary Response

On Day 5, an activated-carbon dam was built at a low-water bridge 5 km (3 mi) below the spill point. Bags of carbon (carbon was obtained from a local water treatment plant and placed in pillowcases) were placed in front of the culvert openings. This restricted the flow so that about one-half of the water went over the top of the bridge. A second carbon dam was placed upstream at a natural rock dam 0.8 km (½ mi) below the spill point. Here much of the water flowed over, instead of through, the carbon also.

Samples were taken twice a day and flown to the company laboratories. The morning sample results were usually available by the time of the nightly reporting and strategy meeting (11 P.M. to 1 A.M.).

On Day 6, the top management of the company inspected the scene to understand better the problems of cleanup and the need for a large commitment of resources. The chief recommendation of the cleanup team at that time was that the spill area had to be isolated to allow effective removal of the organic materials and prevent further pickup of hydrolysis products.

This concept was adopted, and irrigation pumps and piping were used to bypass the entire streamflow around the spill area. As a result, 0.8 km (½ mi) of stream was isolated with dams and the water bypassed.

Straw was placed before the carbon, but it proved ineffective in adsorption and attracted a cow into the water; the cow died. Hence a decision was made to contact all residents with cattle grazing near the creek all the way to the reservoir and ask them to keep their cattle away from the creek if possible, even though there appeared to be no hazard except in the spill area. An electric fence was built on the spill site farm to keep cattle on the high ground, away from the spill area.

To obtain a quick biological monitoring response, fish cages were placed in the creek at several sample points. Fish were now observed to be dying below the low-water bridge.

On Day 7, the creek was "posted" to the low-water bridge, indicating the danger of contamination. A large carbon filter was being constructed at the next access point, 10 km (6 mi) below the spill site. In this way, only the smaller of the two creeks joining the spill creek before the reservoir would have to be treated.

The creek was diverted to one side while a gravel bed was prepared. The carbon was placed on the gravel, having been delivered in a bulk tank truck, and was slurried into place in an area approximately 3 by 6 by 1.2 m (10 by 20 by 4 ft) deep. Gravel was placed on top of the carbon, followed by a plastic sheet, and covered with soil. The creek was channeled through the filter on Day 8.

On that day, the creek bypass became operational with two 3-m^3/min pumps running continuously and delivering water through two 200-mm (8-in) lines. A third pump was on standby.

The state provided radio communication cars at the command post and at the spill site. CB radios were used to communicate between the scattered work groups.

The threat of rain causing more flow than the pumps could handle resulted in consideration of a bypass (possibly permanent) around the spill area. A permanent bypass was considered because it appeared that any illness of cattle or other animals downstream of the spill would be blamed on the spill as long as the old streambed was involved.

Discussion of permits and approvals from agencies and the farmers, plus agreement on location of the bypass, took 2 days. A final decision was made on Day 10 to construct the bypass, and work was begun on the following day.

Meanwhile, Day 9 was spent in refining the operation of the pipeline bypass and the carbon dam. The carbon dam was found to be effective in reducing the concentration of the pesticide and hydrolysis products. A rain washed out the dam during its fourth day of operation, but it had served its primary purpose in absorbing the hazardous materials during its operational period. Even though the carbon was lost downstream, the residue would slowly be extracted by fresh water, but the breakdown of the pesticide and hydrolysis products would be rapid enough to prevent this from being a problem.

On Day 10, representatives of a consulting firm arrived; they were well prepared to clean up organic materials lighter than water, but their absorption pads and other methods were not effective in this spill, and they left the next day.

Since the fish in the cages were found to be dead at the 13-km (8-mi) point, it appeared that there was plug flow of the water from the spill without the hoped-for mixing. The concentration had been greatly reduced by the carbon dam but was still a problem in the creek at a concentration of 20- to 100-μg/L range. However, it would not cause a serious impact at the reservoir.

A final plan of action for the spill area was formulated with approval of the permanent bypass. Holding basins would be constructed for the water in the isolated section of the creek. After the creek was pumped dry, clay granules would be added to absorb any residual organic liquid. Since decomposition of the pesticide and its hydrolysis products was much quicker in the presence of high pH and high temperature, unslaked lime would be added. The whole creek bed would be filled with dirt. Protection from drainage from the hills and interstate highway was necessary, so a drainage ditch would be dug to join the creek bypass below the spill area.

Construction was begun on Day 11 on the bypass and holding lagoons. Bulk shipments of clay and lime were ordered. Permission was obtained to build a temporary road from the interstate highway to the spill area to expedite these shipments, and the road was built the next day.

Rain caused a washout of the carbon filter at the 10-km (6-mi) point on Day 12. The heavy flow was not in the spill creek; otherwise the spill might have become more dispersed.

On Day 13, the bypass was completed; it was a ditch 4 m (12 ft) wide and 2 m (6 ft) deep; the pumps were shut down. Pesticide concentrations decreased by 50 percent at the 10-km (6-mi) point, indicating that the slug had passed.

Holding Lagoons

Two holding lagoons were built so that the water stored in one could be passed through a carbon filter into the other one. After analysis, the water could be released. A carbon column with associated pumps and piping became operational on Day 23. There were more than 3800 m³ (1 million gal) of water to be treated [rains and runoff brought the total to nearly 7600 m³ (2 million gal) before treatment was completed]. One pass through the carbon was not enough (80 percent removal), so a third lagoon was constructed, and all the water passed through the column again (additional 90 percent removal). The resultant liquid was added to the creek at a carefully controlled rate to keep the concentration of the contaminant well below the fish toxicity level. The treatment process required almost 3 weeks.

Plug Flow

The slug of contaminated water associated with the original spill moved downstream more slowly and with less mixing than had been anticipated. However, the concentration was greatly reduced by the carbon filter at the 10-km (6-mi) point, so that it was toxic to some fish and not to others. Samples above and below the slug showed very low concentrations of pesticide and decomposition products. Arriving at the reservoir 32 days after the spill, the slug created no problems.

Creek Bed

The spill site was treated with clay and lime as planned and then covered with soil. The new creek bed (bypass) was expanded to accommodate a 100-year rain, to a depth of 4 m (12 ft) and a width of 12 to 15 m (40 to 50 ft). Riprap was used for erosion control.

Follow-Up

The site was inspected and the creek sampled many times in the next 18 months. There were a few pesticide determinations above the detectable limit (0.1 $\mu g/mL$), evidently caused by runoff percolating through the hillside beside the interstate highway. Since some of these "hot spots" were found in stagnant pools in the runoff ditch during dry weather, evaporation could have concentrated the chemical.

Conclusion

This case history had some interesting aspects that are worthy of being repeated:

- Heavier-than-water material
- Creek contamination that would never have occurred if the truck had not been moved
- Surprising presence of pesticide contamination
- Field construction of activated-carbon treatment
- *In situ* neutralization
- Stream bypass construction
- Cleanup cost of $750,000 for a spill of only 14 m^3

REFERENCE

1. L. H. Frisbie, "Environmental Problems in Pesticide Decontamination of a Freighter Hold," in G. F. Bennett (ed.), *Proceedings of the 1974 National Conference on Control of Hazardous Material Spills,* San Francisco, August 1974, pp. 325–328.

Personnel Safety Equipment*

*The opinions or assertions contained herein are the private ones of the author and are not to be construed as official or as reflecting the views of the commandant or of the Coast Guard at large.

Dennis Rome
Lieutenant Commander
U.S. Coast Guard

Introduction / 13-2
Protective Clothing / 13-4
Respiratory Protection / 13-6
Portable Personnel Monitors / 13-14
System Integration / 13-17
Conclusion / 13-20

INTRODUCTION

Whenever personnel must respond to a hazardous material spill, a number of different situations can develop. A response team can expect to encounter any or all of the following types of hazards:

1. *Toxicity.* Hazards involving the toxic or irritating effects of hazardous chemicals through inhalation, ingestion, or skin contact.
2. *Fire.* Hazards involving extreme heat.
3. *Oxygen deficiency.* Hazards involving an oxygen content insufficient to sustain consciousness or life.
4. *Explosion.* Hazards involving shock waves.

These hazards require that each spill situation be carefully analyzed to determine the approach to be used in spill cleanup. The factors which must be determined are the identity of the chemical or chemicals, the location of the spill (land or water), the likely chemical reaction products, the quantity spilled, environmental conditions, the type or types of hazard likely to be present, etc. This information permits the on-scene coordinator (OSC) to evaluate the consequences of either entering or not entering a contaminated area.

If the decision is made to enter the area for spill cleanup, the persons entering the area must receive the best protection possible. The types of personnel safety equipment which a response team must have available are protective clothing, respiratory protective devices, and personnel monitors to warn of toxic conditions and exposure levels. Combinations of this equipment can be assembled to maintain protection from each type of chemical hazard. As shown in Table 1, levels of protection categories which vary for each type of hazard can be set up. With the general outline of the table, a number of distinct alternatives for protective clothing, respiratory devices, and personnel monitors can be explored to correlate a desired level of protection with a level of hazard.

TABLE 1 Personnel Safety Equipment Requirements

Level of protection	Toxicity	Fire	Oxygen deficiency	Explosion
Minimum	Protection from hazardous chemicals capable of producing changes in the body which are reversible and which will disappear following termination of exposure	Protection from conditions normally encountered while combating a fire [no flame contact, ambient heat not greatly exceeding 93.3°C (200°F) at any time]	Protection from an oxygen concentration in the ambient air that is near the lower level required to sustain life (i.e., 19.5 percent as set by NIOSH)	Protection from an explosion caused by a small quantity of low explosive, such as black powder
Moderate	Protection from hazardous chemicals capable of producing irreversible as well as reversible changes in the body but changes which do not threaten life or cause serious permanent physical impairment	Protection from conditions encountered in the vicinity of a fire [minimum contact with flame; ambient heat not exceeding 260°C (500°F); radiant heat not exceeding 815.5°C (1500°F)]	Protection from an oxygen concentration in the ambient air that is near the lower level at which the average person will retain consciousness (i.e., approximately 16 to 19.5 percent)	Protection from an explosion caused by a low explosive, such as black powder in moderate quantities, or a small low-pressure vessel
Maximum	Protection from hazardous chemicals which are capable of threatening life or which can cause permanent physical impairment or disfigurement	Protection from conditions encountered in a fire [total contact with flame; ambient heat not exceeding 815.5°C (1500°F); radiant heat not exceeding 1371°C (2500°F)]	Protection from an oxygen concentration in the ambient air that is below the lower level at which the average person will retain consciousness (i.e., less than 16 percent)	Protection from an explosion caused by a low explosive in large quantities, any amount of high explosive such as TNT, or a high-pressure or large low-pressure vessel

SOURCE: W. M. Hammer and K. R. Nicholson, *Survey of Personnel Protective Clothing and Respiratory Apparatus for Use by Coast Guard in Response to Discharges of Hazardous Chemicals*, U.S. Coast Guard Report No. CG-D-89-75, NTIS Access No. AD A010 110, September 1974.

PROTECTIVE CLOTHING

Selection of items of protective clothing depends on the level of protection desired for response personnel. A number of sources of information on the hazard presented by a chemical can be consulted in making this decision.[1-17]

The specific items which are available for spill response personnel are boots, gloves, safety glasses, goggles, face shields, hard hats, aprons, splash suits, and fully encapsulated suits. Depending on the identity of the hazardous material spilled, the level of protection to prevent physical harm can vary from a minimum of boots, gloves, glasses, and hard hats to a maximum of a totally encapsulated suit. Selection and purchase of protective clothing depend on the behavior of the clothing's material when challenged by a spilled chemical.

Most of the information on chemical protective clothing is in the form of technical data which delineate the physical properties of a particular compound, coating, or synthetic fiber. Any of the major elastomer manufacturers[1,2,3] will provide literature on volume change, tensile strength retained, elongation retained, hardness, and surface condition and will rate their materials against common chemicals with general qualifying expressions such as excellent, good, fair, unacceptable, unchanged, attacked, and not visibly affected. In many cases, the test conditions or procedures used are not provided, and a comparison with expected use conditions therefore cannot be made.

Laboratory investigations have been performed at the Chambers works of E. I. du Pont de Nemours and Co. to determine the chemical resistance of samples of its elastomers. The general test procedure is outlined by Adrian L. Linch,[4] together with recommendations of other properties which must be evaluated:

1. Flame resistance
    a. Ignition
    b. Sustained combustion
    c. Melting range
    d. Drip properties
    e. Flame-retardant coatings

2. Static properties
    a. Permanent staticproofing
    b. Temporary staticproofing applied after laundering

3. Substantiality
    a. Material not removed in laundering operation

4. Durability
    a. Abrasion resistance
    b. Tenacity of coated fabrics; tendency to peel off
    c. Flex resistance
    d. Resistance to degradation by heat and light
    e. Tear and puncture resistance

5. Workability
    a. Elasticity (three-dimensional)
    b. Seam strength
    c. Sealing properties
        (1) Cement
        (2) Heat

   d. Weight
      (1) Density of coating
      (2) Thickness versus resistance
   e. Temperature effects
6. Comfort
   a. Thermal conductivity
   b. Surface roughness
   c. Drape and hand; softness versus stiffness
   d. Moisture absorption and evaporation
7. Safety
   a. Wet coefficient of friction (i.e., slippery when wet) for gloves or shoe soles
   b. Flushing with water before garment removal
   c. Decontamination
   d. Heat transfer and resistance; reflective coatings; air layer
   e. Interference with body movements; bulk
   f. Proper fit; adequate fabrication design
   g. Hidden design flaws: inadequate fasteners, exposed threads, improper seals, etc.

Until a standard test and evaluation procedure is established by the American Society for Testing and Materials for chemical protective clothing, particularly for chemical permeation, much of the input for using a particular material will come from elastomer manufacturers. For general usage of various materials, the following guidelines are offered for common elastomers:[5]

Natural rubber.   Natural rubber is attacked by mineral and vegetable oils, benzene, toluene, and chlorinated hydrocarbons. It is severely attacked by strong oxidizing agents (nitric acid, concentrated sulfuric acid, dichromates, permanganates, sodium hypochlorites, chlorine dioxide). Natural rubber is not affected by most inorganic salt solutions, alkalies, and nonoxidizing acids. It has high resiliency and tensile strength and retains its mechanical properties at both high and low temperatures.

Acrylonitrile-butadiene (nitrile rubber or Buna N).   Nitrile rubbers have good resistance to oils, solvents, alkalies, and aqueous salt solutions. They swell slightly in aliphatic hydrocarbons, fatty acids, alcohols, and glycols. Their physical properties are not greatly affected by the swelling, and they can be used in gasoline and oil applications. Nitrile rubbers are attacked by strong oxidizing agents, ketones, ethers, and esters.

Polyacrylic rubber.   Polyacrylic rubbers have good resistance to petroleum products, aliphatic hydrocarbons, and animal and vegetable fats and oils. They will swell in aromatic hydrocarbons, alcohols, and ketones. Polyacrylic rubbers are deteriorated by aqueous media, steam, glycols, and caustic environments. They have good flex resistance and low permeability to hydrogen, helium, and carbon dioxide, but they have fairly poor weathering characteristics since they are affected by water.

Butyl rubber.   Butyl rubber is quite resistant to many acid and alkali solutions and is very resistant to swelling by animal and vegetable oils. Vulcanized butyl rubber swells and deteriorates rapidly when exposed to aliphatic and aromatic solvents. It has low permeability to gases. Butyl is about equal to natural rubber in tear and abrasion resistance. It can be used in temperatures from -45 to $150°C$.

Chloroprene rubber (neoprene).   Neoprene is unaffected by aliphatic hydrocarbons, alcohols, glycols, fluorinated hydrocarbons, dilute mineral acids, concentrated caustics, and aqueous inorganic salt solutions. It is attacked by chlorinated hydrocarbons, organic esters, aromatic hydrocarbons, phenols, ketones, concentrated nitric and sulfuric acid, and strong oxidizing agents. Neoprene has excellent resistance to abrasive wear, sunlight, and weathering. It can be used in temperatures from −40 to 110°C.

Ethylene propylene (EPR and EPDM) rubber.   These elastomers provide resistance to water, acids, caustics, and phosphate-based hydraulic fluids. They offer mild resistance to aromatic hydrocarbons and are attacked by aliphatic and halogenated hydrocarbons. They have excellent weathering properties.

Fluoroelastomers (Viton).   The fluoroelastomers provide excellent resistance for aliphatic hydrocarbons, chlorinated solvents, animal, mineral, and vegetable oils, gasoline, jet fuels, dilute acids, alkaline media, and aqueous inorganic salt solutions. They have fair to poor resistance to oxygenated solvents, alcohols, aldehydes, ketones, esters, and ethers. Fluoroelastomers retain their physical properties over a temperature range from −68 to 205°C. They have low permeability rates with air and low water absorption.

Polyurethane diisocyanate elastomers.   These are commonly referred to as urethane or polyester rubbers. The solid polyurethanes are clear, flexible plastics with excellent resistance to most mineral and vegetable oils, greases, fuels, and aliphatic, aromatic, and chlorinated hydrocarbons. They are softened by alcohols, have limited use in weak acid solutions, and are unacceptable in concentrated-acid use. Caustic solutions degrade urethane rubbers. Urethanes have good weathering, abrasion resistance, load-bearing, and impact-resistance properties.

Research is constantly being pursued to develop new elastomers which will provide maximum chemical resistance. At this time, the most commonly used material for hazardous chemical protective garments is butyl rubber. Butyl rubber provides resistance to a large number of chemicals and maintains its physical properties over a wide range of temperature conditions.

RESPIRATORY PROTECTION

Providing respiratory protection to a response individual is by far the most important aspect of protective equipment. Toxic materials have three entry routes into the body: (1) inhalation, (2) skin absorption, and (3) ingestion. Inhalation presents the quickest and most direct route into the body and to the bloodstream. The level of protection which can be provided ranges from a single-use air-purifying respirator to a positive-pressure self-contained breathing apparatus (SCBA). Therefore, the proper selection of a respirator for an individual chemical spill hazard becomes a systematic evaluation.

The two basic respiratory hazards encountered are oxygen deficiency and contaminated atmospheres. Normal oxygen content in the atmosphere is 20.9% by volume. Oxygen content below 16% will not support combustion and is considered unsafe for human exposure. At low oxygen concentrations an individual can collapse immediately without warning, and death can result in minutes. While 16% oxygen (at sea level) is considered the lowest level for safe human exposure, current legislation requires that a work area have not less than 19.5% oxygen (at sea level). When exposure conditions are being assessed, it is important to remember that oxygen deficiency can occur in enclosed spaces where the oxygen is displaced by other gases or by vapors, or by fire and rust where the oxygen is consumed.

Air contaminants include particulate solids or liquids, gaseous material in the form of a true gas or vapor, or a combination of gas and particulate matter. The type of respiratory hazard posed to the response individual dictates the particular respirator to be used.

Particulate Hazards

Particulate hazards may be classified according to their chemical and physical properties and their effect on the body. Particle diameter in micrometers (μm) is one of the most important properties. Particles below 10 μm in diameter have a better opportunity to enter the respiratory tract, and particles in the range of 1 to 2 μm in size can reach the deep lung spaces.[15] A healthy lung will generally clean out the particles in the 5- to 10-μm range since these particles remain in the upper airways. With increased exposures or diseased systems, the efficiency of the lung is reduced.

The types of particulate hazards are classified as follows:

1. *Dust.*   Mechanically generated solid particulates (0.5 to 10 μm).

2. *Mist and fog.*   Liquid particulate matter (5 to 100 μm).

3. *Fumes.*   Solid condensation particles of small diameter (0.1 to 1.0 μm).

4. *Smoke.*   Chemically generated particulates (solid and liquid) of organic origins (0.01 to 0.3 μm).

Gaseous Contaminants

Gaseous contaminants can also be classified according to their chemical properties:

1. Inert gases (helium, argon, etc.), which do not metabolize in the body but displace air to produce an oxygen deficiency

2. Acid gases (SO_2, H_2S, HCL, etc.), which are acids or produce acid reactions with water

3. Alkaline gases (NH_3, etc.), which are alkalies or produce alkalies by reaction with water

4. Organic gases, which exist as true gases or vapors from organic liquids

5. Organometallic gases (metals attached to organic groups such as tetraethyl lead and the organic phosphates)

Exposure Levels

The degree of effect of both gaseous and particulate hazards depends mostly on the airborne concentration of contaminants and the length of exposure. As a result, a listing of threshold-limit values (TLVs) and immediately-dangerous-to-life-or-health (IDLH) values is published by the American Conference of Governmental Industrial Hygienists (ACGIH).[18] Categories of TLVs and IDLH values are specified as follows:

1. *TLV time-weighted average (TWA).*   The time-weighted average concentration for a normal 8-h workday or 40-h workweek to which nearly all workers can be repeatedly exposed without adverse effect.

2. *TLV short-term exposure limit (STEL).*   The maximum concentration to which workers can be exposed for a period up to 15 min without suffering from (*a*) intolerable irritation, (*b*) chronic or irreversible tissue change, or (*c*) narcosis of sufficient degree to increase accidents, impair self-rescue, or materially reduce work efficiency, provided that no more than four excursions per day are permitted with at least 60 min between exposure periods and provided also that the TLV-TWA limit is not exceeded. STEL should be considered a maximum allowable concentration that is not to be exceeded at any time during the 15-min excursion period.

3. *TLV ceiling (TLV-c)*. Concentration that should not be exceeded even instantaneously.

4. *Immediately dangerous to life or health (IDLH)*. Conditions that pose an immediate threat to life or health or conditions that pose an immediate threat of severe exposure to contaminants, such as radioactive materials, which are likely to have adverse cumulative or delayed effects on health.

The TLV values established by the ACGIH are intended for workplace environments where exposures to hazardous chemicals are or can be controlled and well defined. In an open-space spill environment, the parameters are not always as well defined as that. However, the use of TLVs, STELs, and IDLHs provides an important input into the selection of respiratory equipment.

Hazard Assessment

Proper assessment of the hazard involved with a particular spill is the first step in protecting response personnel. The initial survey at the spill site is generally performed by local officials, who determine a general hazard-nonhazard condition. The identity of the substance may or may not be known. Upon the arrival of the response team, air samples must be taken with proper sampling instruments to determine the identity and concentration of the spilled chemical. Samples should be taken in areas where the highest concentrations of the chemical would be expected. If a person is sent in to collect the samples, the highest level of protection available should be provided. After the samples have been collected and analyzed, decisions on hazard and level of protection can be confidently made. The next step is the selection of respiratory devices.

Respirator Types

Respiratory devices vary in design, application, and protective capability. After assessing the inhalation hazard, the user must understand the specific use and limitations of available equipment in order to ensure proper selection. Respiratory protective devices are tested and approved by the National Institute for Occupational Safety and Health–Mine Safety and Health Administration (NIOSH-MSHA) in Morgantown, West Virginia, for protection against a wide variety of inhalation hazards. Whenever a respiratory protective device is used, it is essential that the device be NIOSH-MSHA-approved. There are two basic types of respirators: air-purifying and atmosphere-supplying.

Air-Purifying Respirators

Air-purifying devices remove contaminants from the atmosphere and are to be used *only* in atmospheres containing at least 19.5% oxygen by volume. The common types of air-purifying respirators are mechanical filter respirators, chemical cartridge respirators, combinations of mechanical and chemical respirators, and gas masks.

Mechanical filter respirators provide protection against airborne particulate matter including dusts, metal fumes, mists, and smokes. They consist essentially of a soft face-piece of either half-mask or full-facepiece design. Depending on the design of the mask, one or two mechanical filter elements which contain fibrous material (usually resin-impregnated wool) can be attached. The filter removes the harmful particles by physically trapping them as air passes through it during inhalation; gaseous material passes through the filter. High-efficiency filters are used for dusts, fumes, and mists with TLVs less than 0.05 mg/m^3. These filters have very small pores and can be used for radionuclide filtration in the range of 0.3 μm. Two special filters have been developed for asbestos and radon daughter nuclides. NIOSH-MSHA certifies mechanical filter respirators under 30 CFR Part II, Subpart K.

Chemical cartridge respirators (Fig. 13-1) afford protection against light concentrations (10 to 1000 ppm by volume) of certain acid gases, alkaline gases, organic vapors, and mercury vapors by using various chemical filters to purify the inhaled air. In contrast to mechanical filters, the chemical cartridge respirators contain sorbents to remove harmful gases and vapors. These sorbents work on either an absorption or an adsorption principle and usually are either charcoal or silica gel. Since there is only a specified amount of sorbent in a cartridge, high concentrations of vapors will quickly saturate the material and cause a breakthrough. Concentrations for which individual cartridges are effective are provided by the manufacturer. NIOSH-MSHA approves chemical cartridge respirators under 30 CFR Part II, Subpart L.

Combination mechanical filter–chemical filter respirators utilize dust, mist, or fume filters with a chemical cartridge for dual or multiple exposure. Respirators sometimes need independently replaceable mechanical filters because the dust filter clogs before the chemical cartridge is exhausted.

Gas masks (Fig. 13-2) have been used effectively for protection against certain particulates, vapors, and gases. They provide simple operation, ease of maintenance, and economy. However, because they still are air-purifying devices, they should not be used in an oxygen-deficient atmosphere and generally should not be used in atmospheres with concentrations of toxic gases and vapors greater than 2% (20,000 ppm) by volume. If specific exposure concentrations are *suspected* of exceeding the specific limitations, only a self-contained breathing apparatus or a supplied-air respirator should be used. Various types of gas masks and canisters are approved by the U.S. Bureau of Mines under Schedule 14 for protection against specific gases. As shown in Table 2, each canister is specifically labeled and color-coded to indicate the type of protection afforded.

Fig. 13-1 Chemical cartridge respirator. [*Courtesy MSA.*]

Fig. 13-2   Gas mask with canister. [*Courtesy MSA.*]

When used within the limits of concentration and time for which they are designed, air-purifying respirators are effective tools in spill operations. It is extremely important to monitor the concentrations of contaminant in the wearer's breathing zone whenever an air-purifying respirator is used. Serious injury can result if the concentration limits are exceeded. Whenever the limits are exceeded, the response individual must use an atmosphere-supplying respirator.

TABLE 2   Color Code for Cartridges and Gas-Mask Canisters
(ANSI K13.1-1973)

Atmospheric contaminants to be protected against	Color assigned
Acid gases	White
Organic vapors	Black
Ammonia gas	Green
Carbon monoxide gas	Blue
Acid gases and organic vapors	Yellow
Acid gases, ammonia, and organic vapors	Brown
Acid gases, ammonia, carbon monoxide, and organic vapors	Red
Other vapors and gases not listed above	Olive
Radioactive materials (except tritium and noble gases)	Purple
Dusts, fumes, and mists (other than radioactive materials)	Orange

NOTES:

1. A purple stripe shall be used to identify radioactive materials in combination with any vapor or gas.
2. An orange stripe shall be used to identify dusts, fumes, and mists in combination with any vapor or gas.
3. When labels only are colored to conform with this table, the canister or cartridge body shall be gray, or a metal canister or cartridge body may be left in its natural metallic color.
4. The user shall refer to the wording of the label to determine the type and degree of protection that the canister or cartridge will afford.

   SOURCE: *Basic Elements of Respiratory Protection*, Mine Safety Appliances Co., Pittsburgh, 1976.

Atmosphere-Supplying Respirators

An atmosphere-supplying respirator provides the highest level of protection possible for a response individual. It isolates the individual's respiratory tract from any contaminated atmosphere. Supplied-atmosphere respirators fall into three main types: the hose-mask respirator, air-line respirators, and self-contained breathing apparatus (SCBA).

The hose-mask respirator has limited application and is not used for atmospheres immediately dangerous to health. This unit does not maintain positive pressure in the facepiece. The hose mask is either a lung-powered model with a maximum hose length of 22.9 m (75 ft) or a blower model with a hose length of 76.2 m (250 ft). These devices provide the lowest level of protection of any atmosphere-supplying respirators.

The second type, air-line respirators, comes in three main varieties: continuous-flow, demand-flow, and pressure demand-flow. Continuous-flow respirators are normally used when there is a constant supply of breathing air such as that provided by a compressor. They should be employed when only respiratory protection is needed. Demand-type air-line respirators with half masks or full facepieces deliver air only when the individual inhales with exhalation to the atmosphere. Such respirators are used when the air supply is restricted to compressed-air cylinders. Since negative pressure is created in the facepiece (as a result of inhalation), the facepiece must fit tightly around the face; otherwise contaminated air will be drawn in. To solve the negative-pressure problem presented in the demand mode, there is a pressure demand air-line respirator (Fig. 13-3) which provides positive pressure in the facepiece during both inhalation and exhalation.

NIOSH-MSHA approves air-line respirators under 30 CFR Part II, Subpart J. The most significant requirements for approval are a maximum hose length of 91.4 m (300 ft) and a maximum inlet pressure of 861.8 kPa (125 psig). With the longest hose length available, the air supply rate to the facepiece and the exhalation resistance must be within the limits established by NIOSH-MSHA for the particular respirator. The compressed air must meet the most recent requirements of the Compressed Gas Association's speci-

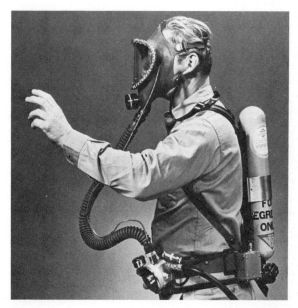

Fig. 13-3   Pressure demand air-line respirator. [*Courtesy MSA.*]

fication G7.1 (ANS1 Z86.1-1973) for Type 1, Class D gaseous air. The specification currently requires that the carbon monoxide level not exceed 20 ppm, the carbon dioxide level not exceed 1000 ppm, and condensed hydrocarbons not exceed 5 mg/m^3.

The third type of atmosphere-supplying respirator in this category is the self-contained breathing apparatus. Respirators which fall into this type are self-generating respirators, demand and pressure demand SCBAs, rebreathers, and emergency-egress SCBAs.

The self-generating apparatus has a nominal service life of 1 h. It uses a chemical canister which evolves oxygen and removes the exhaled carbon dioxide in accordance with breathing requirements. The canister normally contains potassium superoxide and evolves oxygen when contacted by the carbon dioxide and moisture in exhaled breath. It also retains the carbon dioxide and moisture, thus preventing fogging in the facepiece. This respirator depends only on replacement chemical canisters; no bottled oxygen supply is needed. The advantages of this system are simplicity of construction and use and lessened need for maintenance.

Demand and pressure demand respirators use high-pressure air supplies (1.65×10^4 kPa, or 2400 psig). Mine Safety Appliances Co., Scott Aviation Division of A-T-O Inc., and the Survivair Division of U.S. Divers Company are the only companies manufacturing these devices in the United States. These are open-circuit devices; that is, the air breathed comes from a compressed-air supply and is released to the atmosphere upon exhalation. In the demand mode a negative pressure is produced in the face mask, and inward leakage of contaminants is possible. In the pressure demand mode (Fig. 13-4), a slight positive pressure is maintained in the face mask to prevent contaminants from being drawn in. The ultimate in SCBA design, pressure demand respirators can be used in IDLH atmospheres.

These devices weigh 14.5 kg with a 30-min air supply. A new composite construction of aluminum and fiberglass reduces the bottle weight and affords a 45-min air supply. Depending on the individual's workload, the air supply can last from about 25 min for an extreme workload to 45 min for a light workload. The devices are back-mounted, and some units can be switched from a demand mode to a pressure demand mode. Because of the negative pressure created in the demand mode, in the future only pressure demand respirators will be manufactured.

The oxygen cylinder rebreathers in use today are constant-flow types with the rate of oxygen depletion being determined by the breathing rate. They automatically compensate for the varying breathing demands of the individual. The device consists of a small cylinder of 1.65×10^4 kPa (2400 psig) oxygen, reducing and regulating valves, a breathing bag, a facepiece, and a chemical container to remove carbon dioxide from the exhaled air. The principle of operation is fairly straightforward. High-pressure oxygen from the cylinder is reduced in pressure to a breathing level by means of a reduction and regulating valve.

In some units there is a constant flow plus a lung-controlled valve which supplies any needed additional flow. Other units have only an inhalation valve which delivers oxygen from the breathing bag to the wearer's face. Exhaled breath passes down another tube into a container which holds the carbon dioxide removal system [commonly soda lime (soda sorb), Baralyme, or lithium hydroxide] and then passes through a cooler. The purified exhaled air then flows into the breathing bag, where it is mixed with the incoming oxygen from the cylinder.

The rebreathing principle provides the most efficient use of the oxygen supply. Rebreathing units now manufactured are approved by NIOSH-MSHA for 45-min, 2-h, 3-h, and 4-h duration. For the long-term respiratory protection which may be needed in spill cleanup operations, the rebreather can be a useful tool. Moreover, rebreathers weigh much less than pressure demand or demand SCBAs (7.7 kg versus 14.5 kg).

Fig. 13-4 Pressure demand apparatus. [*Courtesy MSA.*]

Positive-pressure rebreathers are currently being approved by NIOSH-MSHA as demand closed-circuit systems. An approval schedule for pressure demand closed-circuit systems needs to be implemented by NIOSH-MSHA to ensure that these devices satisfy the pressure demand requirements. These devices will solve the problem of leakage which could occur in present rebreathers owing to negative pressure in the facepiece.

The final type of SCBA provides a small oxygen supply for emergency egress only. It is rated at 3 to 15 min and should never be used to enter a contaminated area. The emergency-egress bottle can be provided as ancillary equipment to the primary breathing supply if the main supply of air becomes nonfunctional. An automatic or a manual switch can be integrated easily into the system to connect the emergency air.

NIOSH-MSHA has approval schedules for the SCBA under 30 CFR Part II, Subpart 14. Except for special application equipment not covered by the approval schedules, all SCBAs should bear NIOSH-MSHA approval.

Selection of Respirators

Systematic selection of proper respiratory protective equipment should be made by personnel who are knowledgeable about inhalation hazards and respiratory equipment. NIOSH publishes *Criteria Documents*[19] to provide guidance in worker safety areas. Part of this publication is the Respirator Decision Logic, whose purpose is "to provide the necessary criteria for the selection of respirators and to assure technical accuracy and uniformity when selecting respirators for different substances having similar characteristics. The decision logic is a step-by-step elimination of inappropriate respirators until only the acceptable ones remain."

The Respirator Decision Logic uses extensively the concept of respiratory protection factors.[20] When both the contaminant ambient concentration and the TLV are known, the protection factor may be used to select a respirator so that the concentration inhaled by the wearer will not exceed the appropriate limit. Maximum-use concentrations for various types of respirators can be established for different chemicals. If the identity of the chemical is unknown, the concentration is unknown, or the appropriate exposure level is not defined, an SCBA must be used initially until the appropriate information is available. In ill-defined situations it is prudent to have the highest level of protection since the risk to the response individual is thus much lower.

Respiratory Program

For safe use of any respiratory protective device, it is extremely important that the individuals in a spill response team, both supervisors and workers, be properly instructed in the selection, use, and maintenance of respirators. A minimum training program should include the following steps:

1. Instruction in the types of hazards likely to be encountered and a discussion of what can happen if the proper device is not used

2. Discussion of the different types of respirators and how they are selected

3. Discussion of the respirators' capabilities and limitations

4. Instruction and training in actual use and close supervision to ensure that the respirators continue to be properly used

5. Classroom and field training to recognize and cope with emergency situations

Training will provide response personnel with an opportunity to handle each device, have it fitted properly, test its facepiece-to-face seal, wear it in normal air for a familiarity trial, and finally wear it in a test atmosphere (amyl acetate or irritant smoke).

Respiratory devices should never be worn when a satisfactory face seal cannot be obtained. Many conditions, such as facial hair or an unusually structured face, may prevent a satisfactory face seal. Beards should not be allowed for individuals required to wear respirators. A bearded wearer cannot obtain a good facepiece-to-face seal, lack of which can jeopardize the individual's safety and health.[21]

Respirators must be cleaned and disinfected after each use and stored in a clean, dry area. During cleaning, they should be inspected for defects. Defects should be immediately repaired, parts replaced, or new respirators obtained.

The effectiveness of the respiratory program should be checked routinely for compliance with the steps outlined above. If operational scenarios change, the program should be updated to reflect the changes. Proper selection, use, and care of the respiratory protective devices employed in hazardous material spills is essential to the health and safety of response individuals. Failure to choose the correct device or improper use and care can easily result in serious injury or death.

PORTABLE PERSONNEL MONITORS

In any response situation, there is a strong likelihood that personnel will encounter situations requiring entry into a hazardous environment. Hazards which response personnel are likely to encounter are oxygen deficiency, explosive or flammable atmospheres, toxic chemicals, and degradation of chemical protective material. With any of these hazards, response individuals must be able to be forewarned so that they can retreat to a safe environment, assess the new situation, and decide on further courses of action.

Ideally, a portable real-time instrument should be available to monitor each of the situations described above. At present, there is no single device which can monitor all possible hazards. The following sections discuss the capabilities of existing monitoring technology.

Oxygen Deficiency

Three basic types of detection systems for oxygen deficiency are available. One utilizes a sealed electrochemical cell. The oxygen diffuses through a Teflon membrane to the electrolyte, and an electrochemical reaction generates a current flow. The magnitude of the current is proportional to the amount of oxygen present. The oxygen content can be read directly.

A second type of oxygen detector, termed a galvanic cell, consists of a positive carbon electrode and a negative zinc electrode. The electrodes are immersed in an electrolyte, and when no oxygen is present, the carbon electrode is polarized and inhibits current flow. When oxygen is present, it diffuses to the carbon electrode and depolarizes it. This allows current to flow in proportion to the oxygen concentration.

At present, there is only one commercially available solid-state oxygen monitor. This device uses doped zirconium dioxide (ZrO_2) as the sensing element. The ZrO_2 gauge is specific for oxygen and has a fast response. A wide range of concentrations can be handled, but most units are used for combustion, exhaust gas control, or furnace atmosphere control. A disadvantage of the solid-state device is that the cell must be operated at high temperatures, thereby requiring a large amount of power. Also, the monitor is not yet portable.

The most commonly manufactured oxygen detector is the electrochemical cell. Most of the devices can measure a concentration of from 0 to 40% oxygen with an accuracy of $\pm 1\%$ of full scale and will sound an audible alarm when the concentration falls below 19.5% by volume.

Flammable-Vapor Monitors

Commercial vendors currently make two basic types of flammable-vapor detectors: catalytic oxidation monitors and metal oxide gas sensors (MOGS). Catalytic oxidation of principally organic vapors produces a temperature change in the catalyst support (platinum or Hopcalite) in a temperature-sensitive device. The change in temperature is converted into an electrical signal which activates the meter or alarm. These monitors are used extensively to measure concentrations of oxidizable organic vapors in the range of 100 to 10,000 ppm in air. They are nonspecific but can indicate to the response individual that an organic vapor is present at dangerous levels in the surrounding atmosphere.

Metal oxide sensors are based on the semiconductive properties of n-type metal oxides (SnO_2, TiO_2, ZnO). They are so sensitive that at levels well below those for flammability (a few hundred parts per million) the sensor can become saturated. The use of these detectors is limited for warning at higher concentrations. Although they can be employed to detect low explosive limits, they could prove to be better utilized in detecting toxic chemical limits.

Catalytic oxidation is a proven and effective means of detecting explosive limits [lower explosive limit–lower flammability limit (LEL-LFL)] for oxidizable vapors. The devices can and probably will be essential in providing information to the response individual who makes initial hazard-nonhazard assessments.

Toxic-Vapor Monitors

Different types of detectors are offered commercially to monitor the air for various toxic gases and vapors. Available detectors are largely designed to detect a particular substance such as CO, H_2S, NH_3, or Hg. Presently, the most widely applicable personnel monitoring devices are based on the MOGS technology or the electrolytic semiconductor technology. Other types of sensors are based on optical absorption, photoionization, flame ionization, and chemical ionization. Adaptations of mass spectroscopy, gas chromatography, and infrared spectrophotometry offer the advantage of greater compound specificity than many other devices, but they generally require more complex and larger hardware than is suitable for personnel monitoring. A discussion of the various toxic-vapor monitors can be obtained from manufacturers of analytical equipment.

Currently available toxic-vapor monitors which span a wide range of chemicals are colorimetric tubes such as those offered by National Draeger, Inc. (Fig. 13-5), Mine Safety Appliances Co., Bendix Corporation, Bacharach Instruments Co., and MDA Scientific, Inc. Although a number of systems have been developed, the most popular ones use an inert granular support on which a color-forming reagent is deposited. The coated granules are packed into a short glass tube which is sealed at both ends until it is to be used. When the tube is in use, the ends are opened and the air to be analyzed is pumped through at a known rate. Specific contaminants in the atmosphere react with the reagent color former, and a constantly lengthening stain progresses along the tube length. At present, colorimeter tubes are cumulative detectors. The outlook for real-time monitoring in this technology is promising, and the result will be a valuable tool for the response individual.

Material Degradation

Since there is no universal material which is resistant to all chemicals and flexible enough to serve as a protective garment, materials may degrade from chemical exposure. Usually,

Fig. 13-5 Colorimetric detector tubes and sampling pump manufactured by National Draeger, Inc.

such degradation will not be a quick, catastrophic failure. Therefore, provision of a real-time warning to an individual wearing a protective garment is important to prevent skin and respiratory contact with toxic chemicals.

Presently, there are no early-warning alarm systems which have been integrated into protective clothing. More development work needs to be done to provide this capability. Two possible alternatives which could provide the necessary alarms are using a sacrificial swatch or outer layer and integrating an electrical system into material during fabrication of a garment. On the assumption that an attack against clothing would have a slow reaction rate, it would be possible to apply an outer sacrificial layer or swatch which would show the effect of deterioration by a pronounced color change or other change in appearance. This change could be readily seen by the response individual and, in effect, would provide a timely alarm.

Soles of boots might be equipped with an intermediate layer of electrical wiring which would cause an alarm if the circuit were interrupted. For outer garments, a possible device would consist of two closely spaced systems of wiring used as a capacitor. Any shrinking or swelling of the elastomer matrix would be detected as a change in capacitance.[22]

SYSTEM INTEGRATION

As the scenario for a particular spill develops, the OSC makes decisions about cleanup operations. The OSC should have at his or her disposal the necessary protective equipment and a knowledge of the chemical or chemicals of concern and the behavior of the chemical or chemicals to assess properly the level of protection required for personnel entering a contaminated area. From this information and from input by knowledgeable sources of personnel safety, a proper combination of protective equipment items can be assembled into an ensemble.

The ultimate in system integration is the totally encapsulated suit (Figs. 13-6, 13-7, 13-8, 13-9, and 13-10). Such a suit will be necessary for exposure to chemicals which

Fig. 13-6 A 1-h life support system incorporating a water–wet ice body-cooling capability and a BioPak 60P. [*U.S. Coast Guard.*]

Fig. 13-7 A 1-h totally encapsulated suit and life support system prior to sealing the man in the ensemble. [*U.S. Coast Guard.*]

pose a serious skin and respiratory threat. All the individual protective items are designed into this ensemble to provide the response individual with a clean environment. Whenever an individual is enclosed in an encapsulated suit, a heat exchanger must be provided along with the ensemble. The most pressing need for heat removal is during hot summer days, when heat prostration can occur after as little as 15 min inside an ensemble. Since many spills occur in remote areas, it is important to have a heat sink which can be easily carried by the wearer of the suit. Transportability of the cooling unit, considerations of hose length, and large contaminated areas preclude the use of vortex tubes for cooling in remote areas.

Two technologies have emerged as acceptable for the removal of body heat. The first uses liquid air or liquid oxygen evaporation as the cooling medium. The gas from the evaporated liquid is blown through a ducting system inside the protective ensemble to provide both cooling and respiratory air. These units generally weigh from 16 to 19 kg. The second technology uses a water-to-wet-ice heat exchanger. The water passes through a contact heat exchanger wrapped around a block of ice and is pumped through a cooling vest on the individual. Including ice, such a unit weighs approximately 7 kg. When it is used in conjunction with a positive-pressure rebreather, the total weight ranges from 14 to 16 kg.

Both systems provide adequate breathing and cooling for the response individual. The use of liquid air is somewhat limited by availability of vendors of liquefied air, the cost

Fig. 13-8 A U.S. Coast Guard strike team member moving an over-packed drum to shore in a spill exercise. [*U.S. Coast Guard.*]

Fig. 13-9 A U.S. Coast Guard strike team member driving a forklift while wearing a totally encapsulated suit. [*U.S. Coast Guard.*]

Fig. 13-10   Reserve buoyancy in a totally encapsulated suit will help keep the wearer afloat if he or she accidentally falls into the water. [*U.S. Coast Guard.*]

of manufacturing liquid air, and logistics in remote areas. Water and ice are easily obtained and can be supplied at low cost.

CONCLUSION

The ability to give adequate protection to a spill response individual is well within existing technology. While some aspects of personnel protection will be better defined as experience in spill response grows, the most important decision that an OSC will make is what level of protection is necessary. Whenever the proper decision is not clear, a prudent approach is to keep personnel in the maximum level of protection, i.e., a totally encapsulated suit. It is far better to overprotect initially than to risk injury or death.

REFERENCES

1. *Exxon Elastomers/Chemical Resistance Handbook,* Exxon Chemical Company, Houston, 1974.

2. *Enjay Butyl Rubber Chemical Resistance Handbook,* SYN-66-1082, Exxon Chemical Company, Houston.

3. *Seals of Du Pont Viton—The Affordable Rubber,* A-84800, E. I du Pont de Nemours & Co., Wilmington, Del., 1973.

4. *Handbook of Laboratory Safety,* 2d ed., Chemical Rubber Co., Cleveland, 1970.

5. C. A. Harper, *Handbook of Plastics and Elastomers,* McGraw-Hill Book Company, New York, 1975.

6. H. E. Christensen et al., *The Toxic Substances List,* U.S. Department of Health, Education, and Welfare, 1972.

7. *Chemical Data Guide for Bulk Shipment by Water,* Publication No. 835, U.S. Coast Guard, 1973.

8. G. F. Bennett (ed.), *Proceedings of the 1978 National Conference on Control of Hazardous Material Spills,* Miami Beach, Fla., April 1978.

9. G. F. Bennett (ed.), *Proceedings of the 1976 National Conference on Control of Hazardous Material Spills,* New Orleans, April 1976.

10. *Chemical Hazards Response Information System (CHRIS),* AD 757-472-3-4, Arthur D. Little, Inc., Cambridge, Mass., 1972.

11. Lyman Fourt et al., *The Comfort and Function of Clothing,* Technical Report 69-74-CE, Army Contract DAAG 17-67-C-0139, 1969.
12. M. Windholz et al., *The Merck Index,* 9th ed., Merck & Co., Inc., Rahway, N.J., 1976.
13. *Fire Protection Guide on Hazardous Materials,* 6th ed., Nos. 325A, 325M, 49, 491M, and 704M, National Fire Protection Association, Boston, 1975.
14. Karel Verscheuren, *Handbook of Environmental Data on Organic Chemicals,* Van Nostrand Reinhold Company, New York, 1977.
15. G. D. Clayton and F. E. Clayton, *Patty's Industrial Hygiene and Toxicology,* 3d ed., vol. I, John Wiley & Sons, Inc., New York, 1978.
16. R. H. Perry and C. H. Chilton, *Chemical Engineers' Handbook,* 5th ed., McGraw-Hill Book Company, New York, 1973.
17. N. Irving Sax, *Dangerous Properties of Industrial Chemicals,* 5th ed., Van Nostrand Reinhold Company, New York, 1979.
18. *TLVs: Threshold Limit Values for Chemical Substances and Physical Agents in the Workroom Environment with Intended Changes for 1976,* American Conference of Governmental Industrial Hygienists, Cincinnati, 1976.
19. "Respirator Decision Logic," *Criteria Documents,* National Institute for Occupational Safety and Health, Cincinnati, 1978.
20. E. C. Hyatt, *Respirator Protection Factors,* Los Alamos Scientific Laboratory Informal Report No. LA-6084-MS, Los Alamos, N. Mex., 1976.
21. E. C. Hyatt et al., "Effect of Facial Hair on Respirator Performance," *American Industrial Hygiene Association Journal,* April 1973, pp. 135–142.
22. *Performance Requirements for a Spill Response Personnel Monitor,* draft final report for Coast Guard Contract DOT-CG-73211-A, Arthur D. Little, Inc., 1978.

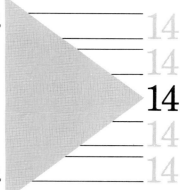

Ultimate Disposal

PART 1
Transportation / 14-2

PART 2
Landfill / 14-11

PART 3
Incineration on Land / 14-20

PART 4
Incineration at Sea / 14-37

PART 1
Transportation

Ronald J. Buchanan, Ph.D.*
Manager, Environmental Affairs
Conversion Systems, Inc.

Legal Mandates / 14-3
Problem Definition / 14-3
Objectives / 14-4
Personnel Organization / 14-4
Budget / 14-4
Manifest System / 14-5
System Mechanics / 14-7
Facility Regulations / 14-9
Program Evaluation / 14-9

Once spilled hazardous materials have been contained and cleaned up, a twofold problem still remains: (1) a "hazardous" waste results, and (2) this waste must be transported to a site for proper disposal. Although substantial difficulties may arise when attempting to transport and dispose of hazardous waste by cargo manifesting or locating appropriate disposal facilities, these problems are tractable and can be overcome by a rational management program.

This part of Chap. 14 is devoted to consideration of the handling of chemical waste, including waste from hazardous material spill incidents. It will present in concise fashion the program developed for the state of New Jersey, with particular emphasis on transportation. The rudiments of this program are applicable to any state and are organized as follows:

1. Legal mandates

2. Problem definition

3. Objectives

4. Personnel organization

5. Budget

6. Manifest system

*Former chief, Bureau of Hazardous and Chemical Wastes, New Jersey Department of Environmental Protection.

7. System mechanics
8. Facility regulations
9. Program evaluation

LEGAL MANDATES

In October 1976, the Congress amended the Solid Waste Disposal Act (Public Law 94-580) to include general provisions known as the Resource Conservation and Recovery Act (RCRA),[1] of which Subtitle C encompasses hazardous waste management. This law requires the U.S. Environmental Protection Agency (U.S. EPA) to promulgate criteria and standards as follows:

1. Identification and listing of hazardous waste
2. Standards applicable to generators
3. Standards applicable to transporters
4. Standards applicable to storage, treatment, and disposal facilities
5. Permits for treatment, storage, or disposal
6. Authorized state programs
7. Inspections
8. Federal enforcement
9. Retention of state authority

Subsequently, the U.S. EPA began to prepare such criteria and standards; to date it has either proposed or promulgated most of them.

Owing to the urgent problems of chemical waste handling, hazardous material spills, and illegal waste disposal facing New Jersey, the state's Solid Waste Management Administration decided to undertake an aggressive program to resolve these issues: (1) a legislative effort[2] for statewide hazardous waste management regulations encompassing generators, transporters, and disposal sites (which have been completed), (2) an increased enforcement effort, and (3) planning for future facility needs. The state's management program was developed in close coordination with the U.S. EPA and is consistent with federal regulations.

PROBLEM DEFINITION

Many northeastern states share a common problem of high chemical industry density, vast quantities of chemical wastes, and numerous hazardous materials spill incidents each year. These problems are particularly acute in New Jersey, which supports about 15,000 chemical-consuming firms that produce 5.4 million metric tons of chemical waste and experiences some 1800 separate hazardous material spills each year. Moreover, the U.S. EPA's 1981 ocean disposal ban is expected to intensify the problem by adding over 5 million metric tons of chemical waste to the state's land-based waste disposal burden. The total estimated chemical (hazardous) wastes generated within New Jersey are reported in Table 1-1.

TABLE 1-1   Total Potential Hazardous Waste Generation in New Jersey, by Industry

Standard industrial classification (SIC) category	Industry	Quantity (dry metric tons per year)
22	Textiles	1,886
281	Inorganic chemicals	47,710
282	Plastic materials and synthetics	6,410
283	Pharmaceuticals	17,920
2851	Paints and coatings*	6,605
286, 287, 289	Organic chemicals†	53,840
2911	Petroleum refining	25,790
30	Rubber products	531
3111	Leather tanning and finishing	885
333, 334	Metal smelting and refining	50,200
3471	Electroplating and metal finishing	3,462
355, 357	Special machinery	5,062
367	Electronic components	1,436
3691, 3691, 3692	Storage and primary batteries	447
Total		223,200

*Includes solvent reclamation operations and factory-applied coatings.
†Includes pesticides and explosives.
SOURCE: Roy F. Weston, Inc., report to the Solid Waste Management Administration, 1977.

OBJECTIVES

The immediate objectives of the Hazardous Waste Management Program are fivefold: (1) to identify waste generators, (2) to improve definition of the composition and quantities of such wastes, (3) to monitor waste movement, (4) to control waste handling and disposal, and (5) to determine the types of facilities necessary to accommodate these wastes. Of course, the more obvious accruement of improved environmental conditions is an overall long-term objective.

PERSONNEL ORGANIZATION

Estimates of Bureau of Hazardous and Chemical Wastes personnel and staffing divisions were developed on the basis of chemical waste generation data, frequency of hazardous material spills, number of registered chemical waste haulers (300), type and distribution of disposal facilities (27), and budget constraints. Specifically, four main divisions were formulated: the manifest system, the technical assistance element, the engineering review element, and the inspection-enforcement element. A separate planning element exists as an individual bureau within the Solid Waste Management Administration and acts as a consultant to the Hazardous Waste Management Program. The organization of the Bureau of Hazardous and Chemical Wastes is shown in Fig. 1-1.

BUDGET

The bureau's overall budget for staff and equipment was $1.2 million for the fiscal year 1980–1981. Annual costs to operate the manifest system cost were estimated to be

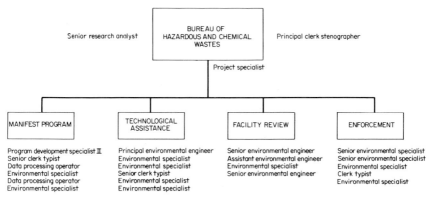

Fig. 1-1   Organization of the New Jersey Bureau of Hazardous and Chemical Wastes.

$90,000. The program was funded by a 75 percent federal share through a U.S. EPA grant and a 25 percent state matching allocation.

MANIFEST SYSTEM

A manifest, which is defined as an itemized list of a vehicle's cargo, is used primarily to identify the components of a particular consignment. Taking this consideration one step further, the state's manifest was designed not only to monitor but to control chemical waste movement. The manifest cycle accounts for the waste from its point of generation to the point of ultimate disposal. This system allows the state to receive notice of a waste shipment from two points: (1) from the generator when the waste is transferred from its control and (2) from the waste management facility when the waste is treated and/or disposed of. Such dual reporting will aid the Solid Waste Management Administration in controlling illicit disposal of wastes as well as in responding to spills and incidents during transport.

The manifest format contains five sections (see Fig. 1-2) including:

1. Waste generator's section: waste type, quantity, hauler, disposal facility

2. Two hauler sections signed respectively at pickup and consignment delivery to the disposal facility

3. Two disposal facility sections signed respectively upon waste receipt or rejection

The manifest cycle is shown in Fig. 1-3. As noted, the manifest is a bill of lading with five sections. The generator completes Sec. I at the time of waste collection, retains (in reverse order) Copy E, and submits Copy D to the administration. The hauler indicates acceptance of the waste in Sec. II at the time of collection and completes Sec. III after consignment delivery to the disposal facility. Upon arrival at the appropriate facility the operator indicates receipt of the waste in Sec. IV or rejection in Sec. V, remands Copy C to the hauler, retains Copy B, and forwards Copy A to the administration.* For wastes

---

*Since the federal manifest system is somewhat different from New Jersey's in allowing one final copy to be returned to the generator by the facility operator, the state has subsequently modified its system to add this component.

Form VHW-001 (10/80)

STATE OF NEW JERSEY
DEPARTMENT OF ENVIRONMENTAL PROTECTION

Please TYPE all information.

HAZARDOUS WASTE MANIFEST

PART A: SEND TO DISPOSER'S STATE   DOCUMENT NO.   NJ 0070702

GENERATOR NAME	PHONE (INCLUDE AREA CODE)	EPA ID NO.

ADDRESS (STREET - CITY - STATE -ZIP CODE)

TRANSPORTER NO. 1	PHONE (INCLUDE AREA CODE)	EPA ID NO.

ADDRESS (STREET - CITY - STATE - ZIP CODE)

TRANSPORTER NO. 2	PHONE (INCLUDE AREA CODE)	EPA ID NO.

ADDRESS (STREET - CITY - STATE - ZIP CODE)

TREATMENT, STORAGE OR DISPOSAL (TSD) FACILITY	PHONE (INCLUDE AREA CODE)	EPA ID NO.

SITE ADDRESS (STREET - CITY - STATE - ZIP CODE)

IF MORE THAN TWO TRANSPORTERS ARE TO BE UTILIZED, FILL OUT THE FOLLOWING AS APPROPRIATE
THIS FORM IS NO. _____ OF A TOTAL OF _____.   THE FIRST MANIFEST DOCUMENT NO. IS   NJ →

	PROPER US DOT SHIPPING NAME	US DOT HAZARD CLASS	UN NUMBER	FORM	NET QUANTITY	UNITS	CONTAINERS NO.	TYPE	EPA HAZ CODE	EPA WASTE TYPE
1.										
2.										
3.										
4.										
5.										
6.										

SPECIAL HANDLING INSTRUCTIONS INCLUDING CONTAINER EXEMPTION (i.e. IDENTIFICATION OF ADDITIONAL WASTES INCLUDED IN SHIPMENT OF A NONHAZARDOUS NATURE WHICH DO NOT HAVE TO BE MANIFESTED)

GENERATOR'S CERTIFICATION: This is to certify that the above named materials are properly classified, described, packaged, marked and labelled and are in proper condition for transportation according to the applicable regulations of the Department of Transportation, U.S. EPA and the State. The wastes described above were consigned to the Transporter named. The Treatment, Storage or Disposal Facility can and will accept the shipment of hazardous waste, and has a valid permit to do so. I certify that the foregoing is true and correct to the best of my knowledge.

GENERATOR'S SIGNATURE - ALSO PRINT SIGNATURE	TITLE	DATE SHIPPED	EXPECTED ARRIVAL DATE
		MO. DAY YR.	MO. DAY YR.

TRANSPORTER NO. 1 SIGNATURE AND CERTIFICATION OF RECEIPT OF SHIPMENT - ALSO PRINT SIGNATURE	TRANSPORTER NO. 1 SWA REGISTRATION NO.	DATE RECEIVED
		MO. DAY YR.

-------------------------------- *TEAR AT THIS PERFORATION* --------------------------------

GENERATOR EPA ID NO.

PART B: SEND TO DISPOSER'S STATE

TRANSPORTER NO. 1 SIGNATURE AND CERTIFICATION OF DELIVERY AND NON-TAMPERING WITH SHIPMENT-ALSO PRINT SIGNATURE		DATE DELIVERED
		MO. DAY YR.

TRANSPORTER NO. 2 SIGNATURE AND CERTIFICATION OF RECEIPT OF SHIPMENT-ALSO PRINT SIGNATURE	TRANSPORTER NO. 2 SWA REGISTRATION NO.	DATE RECEIVED
		MO. DAY YR.

TRANSPORTER NO. 2 SIGNATURE AND CERTIFICATION OF DELIVERY AND NON-TAMPERING WITH SHIPMENT-ALSO PRINT SIGNATURE		DATE DELIVERED
		MO. DAY YR.

TREATMENT STORAGE OR DISPOSAL FACILITY INDICATION OF ANY DIFFERENCES BETWEEN MANIFEST AND SHIPMENT OR LISTING OF REASONS FOR AND DISPOSITION OF REJECTED MATERIALS	HANDLING METHOD
	1 4
	2 5
TSD FACILITY EPA ID NO.	3 6

TREATMENT STORAGE OR DISPOSAL FACILITY SIGNATURE & CERTIFICATION OF RECEIPT OF SHIPMENT - ALSO PRINT SIGNATURE	TITLE	DATE RECEIVED
		MO. DAY YR.

In case of emergency or spill immediately call the State the Emergency occurred in and the N.J. Dept. of Environmental Protection
(609) 292-5560 (Day)   (609) 292-7172 (Night)

DOCUMENT NO.   NJ 0070702

S-004360

Fig. 1-2   Solid waste manifest.

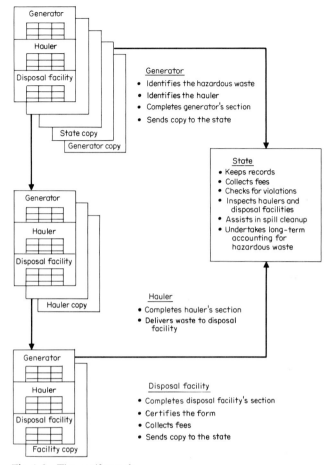

Fig. 1-3 The manifest cycle.

originating out of state or destined for an out-of-state disposal site, the facility is required to forward the appropriate manifest copy to the state.

SYSTEM MECHANICS

An in-house computer-based system was previously developed to license and monitor collector-haulers of solid waste in New Jersey. A manifest system which would allow the Solid Waste Management Administration to monitor all shipments of hazardous wastes disposed of in the state was subsequently requested. The system was designed jointly by the Solid Waste Management Administration and personnel from the Department of Transportation data center to provide the following:

1. An inventory and listing of authorized hazardous waste facilities
2. A report of total waste types by generator, facility, and county of origin

3. A report of manifests by generator and by facility

4. A listing of hazardous wastes and their codes

5. A listing of late manifests

6. Error reports

7. Collection of data for history reports

Manifest information from the generating company and the disposal facility are stored on a master tape, and weekly runs are made to match data. Missing manifests that are received by a specified time interval are noted as being late. Shipments transferred between haulers or rejected by the facility are reported, as are other discrepancies. To accomplish this, the system maintains the following files:

1. Authorized facility file

2. Waste code file

3. Manifest number file

4. Active manifest file

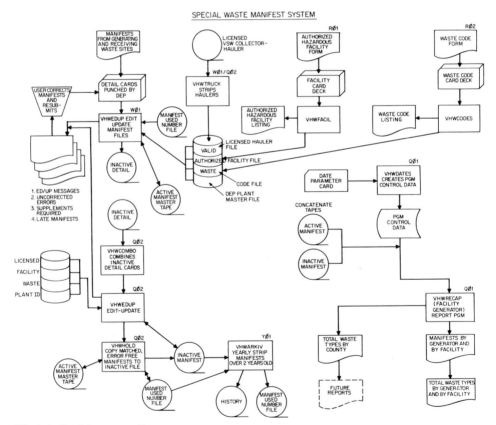

Fig. 1-4 Special waste manifest system.

5. Inactive manifest file

6. Manifest history file

Several programs have been written to update and maintain the system's files and produce necessary reports. The relationship between input forms, files, and programs is graphically illustrated in Fig. 1-4.

The computerized manifest system is interfaced with the existing data bank used by the Bureau of Air Pollution Control as well as the system under development for the Division of Water Resources. The gaps between the three links in the overall environmental control system (air, water, land) are being closed.

FACILITY REGULATIONS

Having adopted manifest regulations, the administration initiated preparation of a separate set of chemical and hazardous waste facility rules. These new rules addressed the issues of facility siting, design, operation, closure, and perpetual maintenance. After 8 months of staff work in conjunction with the Hazardous Waste Advisory Committee, facility rules were finalized and proposed in the *State Register*.[3] The Department of Environmental Protection promulgated these rules within 90 days after a public hearing.

The seven primary aspects of the regulations are unique, since certain areas of responsibility heretofore only implied have now been spelled out. These aspects are:

1. Precisely detailed application requirements

2. Exemption of on-site waste storage if short-term or low-volume as long as this exemption does *not* pose an environmental threat

3. Requirements for siting, design, operation, and closure for "secure" landfills

4. Perpetual maintenance provisions for long-term care and funding

5. Provisions for operational risk hazard analysis and guidelines

6. Emergency contingency plan requirements and guidelines

7. Staged approval process to expedite application review

Overall, the new facility rules are much more concise and in some instances stricter than the federal regulations under the RCRA. In fact, administrative personnel reviewed preliminary RCRA drafts to ensure, insofar as possible, that state regulations were consistent with those of the U.S. EPA.

PROGRAM EVALUATION

Program evaluation is based upon four tangible criteria: (1) the number of manifest forms received and processed (7000 per month), (2) the number of technical assistance problems handled (200 per month), (3) the number of facility engineering designs reviewed (4 per month), and (4) the number of inspections and enforcement actions undertaken (20 per month). To date excellent success has been realized from the program.

Data from this evaluation are routed back to the program to improve definition of the scope of the state's hazardous waste problem, reevaluate program objectives, determine staffing requirements, assess monetary constraints and requirements, plan for future disposal facility needs, and suggest regulatory modifications necessary to keep abreast of the

TABLE 1-2   Quarterly Manifest Summary Report, May 1–July 1, 1978 (Metric Tons)

	Acid solution	Alkaline solution	Oil and oil sludges	Solvents	Still bottoms	Other
Generated in New Jersey	13,710	5,827	4,693	8,718	3,494	23,925
Generated in Connecticut and disposed of in New Jersey	286	47	200	113	9	160
Generated in Delaware and disposed of in New Jersey	476		126	3		72
Generated in Massachusetts and disposed of in New Jersey				100		25
Generated in Maryland and disposed of in New Jersey	14	14	36	56		90
Generated in New York and disposed of in New Jersey	223	603	192	5,200	148	1,677
Generated in Pennsylvania and disposed of in New Jersey	2,375	148	71	597	77	2,750
Generated in Rhode Island and disposed of in New Jersey				180	20	

state's ever-changing situation. More important, however, information extracted from the manifest system (see summary in Table 1-2) has, for the first time, availed to the administration accurate data and important insight into the magnitude of the state (and regional) hazardous waste problem. The Bureau of Hazardous and Chemical Wastes has been able to identify and locate hazardous waste generators, the types and quantities of such wastes, the distribution of these wastes, and the destination of hazardous wastes within (and to a lesser extent outside) the state of New Jersey.

REFERENCES

1. *Resource Conservation and Recovery Act,* Pub. L. 94-580, 94th Cong., 2d Sess., 1976.
2. *Solid Waste Management Act,* as amended, Pub. L. 1975, New Jersey Department of Environmental Protection, 1976.
3. *Rules of the Bureau of Solid Waste Management,* as amended, N.J.A.C. 7:26–1 et seq., New Jersey Department of Environmental Protection, 1974.

PART 2
Landfill

A. Parker
Hazardous Materials Service
Harwell Laboratory

Introduction / 14-11
Behavior of Hazardous Materials in Landfill Sites / 14-12
Research on Hazardous Wastes in Landfill Sites in the United
    Kingdom / 14-12
Research on Behavior of Hazardous Materials in Landfill in Countries Other
    Than the United Kingdom / 14-14
Assessment of Potential Risks in Landfill / 14-15
Other Landfill Options / 14-16
Conclusions / 14-17
Appendix / 14-17

INTRODUCTION

Spills of hazardous chemicals can vary in size from a few milliliters to many thousands of liters and can encompass a very wide range of potentially hazardous chemicals. Obviously the nature and scale of the spillages will influence the method of response to such incidents. In many instances the spilled chemicals will be adsorbed on materials such as sawdust, soda ash, and activated charcoal. In other incidents it may be possible to remove the bulk of the spilled material by pumping, in the case of liquids, or by removal by road or rail transport if solid material has been spilled.

Normally, however, it is not possible to remove completely the spilled chemical. Therefore, those responsible for site decontamination are frequently faced with the problem of disposal of a basically inert material, i.e., soil which has been contaminated. The problem arises with the disposal of the contaminant rather than the adsorbent.

Disposal in a sanitary landfill is the most economical method, provided that this does not in itself cause environmental hazards. The most obvious of these "secondary" hazards is that of groundwater pollution. However, if volatile materials are involved, adverse effects on staff operating the landfill should be considered. Moreover, potential reactions of incompatible hazardous wastes with other material within the landfill should not be overlooked.

BEHAVIOR OF HAZARDOUS MATERIALS IN LANDFILL SITES

During the last decade, owing mainly to public concern, there has been much research into the behavior of hazardous wastes in landfill sites. If spillages of hazardous materials are to be landfilled either in a pure form or admixed with inert adsorbents, such research can obviously be used to predict the effects in a landfill. It may be able to answer questions such as the following:

1. Should the recovered material be landfilled at all, or would other disposal routes (e.g., incineration) be more applicable?
2. What type of site should be used? Is it necessary to use a containment site (Class 1),* or will a site which allows some escape of the spilled material be acceptable?
3. Should the recovered material be pretreated before landfill is considered? For instance, strongly acid wastes may require neutralization before disposal.
4. Should the recovered material be encapsulated in concrete or plastic or alternatively treated by a fixation technique such as the Chemfix† or Sealosafe‡ process prior to landfill?
5. To what extent will chemical reactions occur between the hazardous material and other substances in the landfill?
6. If codisposal with municipal waste is practiced, will this have a beneficial or an adverse effect?

Current research has enabled many of these questions to be answered, although it will not always be possible to give a definitive answer for all types of chemical spillages. Landfill research has been mainly concerned with the behavior of hazardous wastes which are generated on a large scale by industry and subsequently landfilled. Information on specific chemicals (which often are transported in smaller quantities by road, rail, and air) may be lacking. However, given the present state of knowledge, it is usually possible to make an intelligent prediction about their likely behavior. The range of chemicals that has been involved in actual spills is large, and many examples are given in the *Proceedings of the National Conferences on Control of Hazardous Material Spills* held in the United States in 1972, 1974, 1976, 1978, and 1980.[1,2,3,4,11] and sponsored by the U.S. Environmental Protection Agency (U.S. EPA) and other bodies.

RESEARCH ON HAZARDOUS WASTES IN LANDFILL SITES IN THE UNITED KINGDOM

In 1971 considerable concern was expressed about cyanide dumping on landfill sites in England. This, together with other events, led to the passing of the Deposit of Poisonous Wastes Act in 1972. It became apparent that if the Department of the Environment were to be able to offer guidance to waste disposal authorities on the selection and control of landfill sites, more fundamental information was required on the behavior of hazardous materials. Therefore a large-scale research program lasting for 3 years was initiated in 1973. A report incorporating results obtained up to March 1977 has been published.[5]

---

*A Class 1 site is one situated in strata of very low permeability such as clay.
†Trademark of Chemfix Inc., 1675 Airline Highway, Kenner, Louisiana 70663.
‡Trademark of Stablex Corp., 2 Radnor Corporate Center, Radnor, Pennsylvania 19087.

The investigation involved the study of 19 existing landfill sites in the United Kingdom. They were selected both for their varying geological characteristics and the variety of potentially hazardous wastes within them. Obviously, it was not possible to select sites containing every type of waste and covering every geological formation, but the selection was considered to be reasonably representative. The research showed many examples of attenuation of pollutants from landfill through the unsaturated zone to the groundwater. In general, pollutant plumes were limited in their extent. This finding was of very great practical importance since many of the sites had been operating for many years prior to this investigation. In addition, there have been very few recorded instances in the United Kingdom in which groundwater pollution from sanitary landfills or even old-fashioned refuse dumps has occurred. However, it is clear that sites should be selected carefully with particular attention being paid to hydrogeology. Obviously a landfill, particularly one being used for the deposition of hazardous wastes, should not be situated in highly permeable strata in close proximity to an aquifer.

This research has given valuable information on specific classes of chemicals which may be involved in spills and eventually find their way to landfill sites. Thus, with halosolvents, evaporation together with adsorption on refuse largely prevents the release of such materials to groundwater.[5] It is probable that these solvents remain associated with any mineral oil present in other wastes. Further information is given in Waste Management Paper No. 15 in the Appendix. If spilled solvents are applied to the landfill surface, evaporation could be of great importance, but this will depend on the method of deposition. Obviously with flammable solvents the potential risk of fire on the landfill should be recognized, and if solubility is marked, a containment site should be selected as a precautionary measure.

Cyanide spills have occurred, so it is possible that contaminated waste may require disposal, although in practice treatment with hypochlorite solution at the incident is more likely to be used. However, in the presence of domestic waste various processes operate which tend to destroy cyanide or convert it to harmless substances. Thus, cyanide is (1) converted to gaseous hydrogen cyanide, which can be lost by diffusion; (2) hydrolyzed to ammonium formate; (3) converted to thiocyanates by reaction with sulfur compounds; or (4) combined with cations to form insoluble complex cyanides; or (5) under aerobic conditions there is a possibility of biodegradation. Thus the codisposal of small amounts of cyanide with domestic waste is unlikely to cause problems provided the cyanide is well diluted by domestic waste; this was demonstrated by a pilot-scale experiment in which only 3 percent of added cyanide waste was found after 3 years of operation.

The chief problem with the landfilling of spilled metallic compounds, whether solids, liquids, or sludges, is that soluble cations or metallic complexes may appear in landfill leachates. However, there are attenuating mechanisms in the presence of domestic waste which are beneficial. Thus the near neutrality of the landfill environment will facilitate the formation of insoluble hydroxides and carbonates. In addition, under anaerobic conditions sulfur-containing materials in the waste will be reduced with the formation of sulfides, which will further help in the insolubilization of heavy metals. In practice, there is evidence that mercury is rendered virtually insoluble in landfill by this mechanism. Although precipitation is an important attenuating process, attention must be made to the valence of the material being deposited. As an example, groundwater pollution has been caused by deposition of hexavalent chromium, whereas if the metal is in the trivalent form, formation of an insoluble hydroxide can easily occur. So the landfilling of spilled tannery wastes, which contain significant quantities of trivalent chromium, presents no problem, but care should be taken when dealing with chromates and dichromates.

Phenols and phenolic-containing materials are commonly transported by road and rail and so may be involved in accidents involving spillage. Their behavior in a landfill is of particular interest since the acceptable limit for phenols in potable water is only 0.002 mg/L. This very low limit is necessary since objectionable taste and odor are produced when phenols are chlorinated during water purification. Many samples of waste as well as liquid from outside the boundaries of industrial landfill sites contain phenol, and although laboratory experiments have shown that phenols are biodegraded, field measurements show that the rate of degradation is slow. Also although phenols are adsorbed to some extent on domestic waste, this adsorption is reversible, so that this effect will not eliminate, although it will decrease, the downward rate of migration of phenol in the landfill. Research indicates that phenols can cause serious problems in landfills, and as their presence in leachates is extremely undesirable, deposition in containment sites is strongly recommended.

Polychlorinated biphenyls (PCBs) have been recognized as a potential environmental hazard because of their persistence in the food chain and their chemical stability. Although residues from spilled material are probably best incinerated, some research has been carried out on their behavior in landfill sites. It appears that PCBs remain relatively inert within the landfill and are not leached out in high concentrations by water.

Spillages of acids and alkalies, which happen fairly frequently, are usually dealt with by neutralization *in situ*. The resulting neutral salts can then be safely landfilled. This neutralization procedure is desirable since the landfilling of acids could result in the production of hazardous gases such as hydrogen sulfide. Also, if the buffering action of other waste is exceeded, lowering of the pH could result in the solubilization of metal species with the attendant risk of groundwater pollution.

Although not investigated in the research program described above, the behavior of pesticides in landfill is of interest. Some research has been carried out in the United States, and there is much published work on the behavior of pesticide residues in soil and water. For example, review articles by Edwards[6] and Matsumura[7] give many useful references, particularly on the degradation of pesticides and their loss from soils.

It appears that the landfilling of small quantities of pesticides is unlikely to cause serious problems, but the indiscriminate dumping of arsenical insecticides[8] should be avoided. If not, problems could arise, as in Minnesota in March 1975, when 11 people developed arsenic poisoning from grasshopper bait thought to have been buried between 1934 and 1936.

RESEARCH ON BEHAVIOR OF HAZARDOUS MATERIALS IN LANDFILL IN COUNTRIES OTHER THAN THE UNITED KINGDOM

Although results from recent research in the United Kingdom have been extensively quoted, it should be emphasized that much research is also being carried out in other countries, especially in the United States. The scope of the work of the U.S. EPA is outlined in the *Proceedings of the Hazardous Waste Research Symposium,* held in 1976 in Tucson, Arizona.[9]

Another important area of research is that carried out under the auspices of the North Atlantic Treaty Organization Committee on the Challenges of Modern Society (NATO/CCMS) pilot study on disposal of hazardous wastes. This work involves the United States, the United Kingdom, the Netherlands, Canada, Norway, France, Belgium, and West Germany. The research, which is still continuing, should give information relevant to the disposal of spilled hazardous materials since it can be divided into three main sectors:

1. Leachate generation, control, and attenuation

2. Site selection with special reference to hydrogeological and geological considerations

3. Technology of waste disposal including codisposal

ASSESSMENT OF POTENTIAL RISKS IN LANDFILL

When the cleanup of a spilled material is occurring, those responsible should be considering the method for disposal of some of the spilled material plus contaminated inert material (e.g., earth). Landfill is the most attractive option, but research is beginning to indicate its limitations. Unfortunately at present there is no exhaustive reference which will give an authoritative answer for all chemicals. However, in the United Kingdom the Department of the Environment has issued a series of technical memorandums covering the disposal of many classes of wastes including PCBs, mineral oils, cyanides, halogenated hydrocarbon solvents, metal-finishing wastes, mercury, tars, solvents, halogenated organics, wood preservatives, tannery products, asbestos, pharmaceuticals, arsenic-bearing wastes, and pesticides. A list of published memorandums is given in the Appendix. The department intends to publish additional memorandums on acids, cadmium, and biocides.

The guidelines for landfills given in these memorandums may be summarized as follows:

Waste	Landfill recommendations
PCBs	Small intermittent quantities which cannot readily be incinerated can be landfilled. As a guide, the disposal rate should not exceed about one small item such as a capacitor in every 7 metric tons of waste.
Cyanide	Solid cyanide waste containing more than 1000 g/m^3 cyanide should not be landfilled. Class 1 sites have a maximum permissible level of 1000 g/m^3. In addition, the overall content of the landfill should be limited to 10 g/m^3 averaged over any 1000 m^3. Class 2 and 3 sites should have permissible limits of 10 g/m^3 and 1 g/m^3 respectively.
Halogenated hydrocarbon solvents	Containment sites are recommended for metal-cleaning wastes. Occasional 0.2-m^3 drums can be accepted. The maximum deposit should not exceed a few metric tons per week. Recommendations are given as to site management.
Metal-finishing wastes	Landfill is suitable provided that attention is paid to site selection. Details of codisposal and segregation methods are given.
Mercury wastes	Landfill is suitable with dilute-and-disperse philosophy being adopted when possible. Mercury content should not normally exceed 2 percent total nonalkyl mercury or 0.01 percent alkyl mercury. During landfill, the objective should be not to increase the assumed national average of 2 mg mercury/kg fill by more than a further 2 mg mercury/kg.
Asphaltic and distillation wastes	The wide-ranging nature of wastes makes it difficult to define disposal routes, but landfill can generally be used. Containment sites are required for wastes containing water-soluble toxic components. Only minimal quantities of flammable liquids should be

Waste	Landfill recommendations
	landfilled. Acid tars may present problems. Useful information is given on landfill management.
Solvent wastes (excluding halogenated hydrocarbons)	Substantial quantities of dirty solvents and associated solid or semisolid wastes are currently deposited in landfill sites. These are usually discharged into lagoons or trenches in the landfill surface or simply emptied on the surface. Since substantial fire, inhalation, and contact hazards exist, waste disposal authorities are advised to accept only small quantities. The possibility of surface-water or groundwater contamination is recognized.
Tannery wastes	These do not present any significant problem during disposal to a sanitary landfill. Since they contain chromium and sulfides, they should be segregated from other industrial wastes, particularly those of an acidic nature.
Pharmaceutical wastes	Landfill appears to be generally acceptable, but wastes which would inhibit microbial decomposition should not be landfilled in harmful concentrations. Toxic components should be restricted to containment sites.
Arsenic-bearing wastes	Landfill is widely used, especially for wastes containing less than 1 percent arsenic. Criteria for disposal of many wastes are given in detail, consideration being given to potential water pollution.
Pesticides	Codisposal of limited quantities with other biodegradable industrial, commercial, or domestic wastes is likely to be satisfactory. Maximum daily loadings shall not exceed 20 ppm. Landfill is the normal disposal route, but special care must be taken during handling.

Obviously, while these memorandums or guidelines are very useful in giving recommendations for the landfill of many classes of hazardous materials, they do not cover all materials which might be involved in spills. Further information may be located in the 16 volumes of an EPA compendium on recommended methods of reduction, neutralization, recovery, or disposal of hazardous waste.[10] These volumes, which are edited by R. S. Ottinger et al., list 500 hazardous waste constituents.

In a spills incident, rapid information may be required by those responsible for decontamination. Fire services, which are so often involved, frequently need swift guidance as to the method of treatment and subsequent disposal of contaminated material. To meet this demand several countries have set up chemical emergency centers staffed by qualified chemists who provide a 24-h service. An interesting development is the installation of a central computer which can contain recommended disposal routes for spilled chemicals. Information can be retrieved by staff at emergency centers, or some emergency services can obtain information directly from the computer via a telephone link.

OTHER LANDFILL OPTIONS

During recent years there has been considerable interest in the fixation of hazardous materials prior to landfill; this option may therefore be of interest to those involved in spill cleanup. Two of the most commonly used methods are the Chemfix and Sealosafe processes. Liquids are treated so that they form essentially a calcium silicate lattice in which liquid remains trapped. The resultant slurry can be pumped into the landfill site, where it becomes solid. These processes are very effective for immobilizing heavy metals, but some reservations have been expressed about their ability to retain water-soluble organics during leaching. Other options for treatment of spilled materials prior to land-filling include placement in metal or plastic drums with the addition of concrete or wrapping in sealed plastic sacks. If these are buried within a landfill, their contents will only slowly be released.

CONCLUSIONS

The amount of hazardous wastes involved in spills is usually small compared with the large tonnage which arises from industrial processes and is deposited in landfill sites. If it has been demonstrated that at a particular site disposal of hazardous chemicals is a safe practice, any additional spilled waste should not cause a problem, provided that the site's capacity is not overloaded. However, the situation is more complex if the chemical is not normally landfilled as waste, but if the amount deposited is small compared with the volume of refuse (i.e., a large dilution applies), the risk should be minimal. Deposition of water-soluble toxics in a containment (Class 1) site should be obligatory, and this should apply to other substances whose behavior cannot be predicted. Moreover, for some toxic materials landfilling may present a hazard to workers on the site or to people nearby. Such materials should not be landfilled but rather destroyed by incineration or by chemical treatment.

Only in exceptional circumstances is land burial at the source of the spillage likely to be safe. Normally there will not be time to investigate thoroughly the local hydrogeology, and even if this is satisfactory, the site may be exhumed at a future date, thus causing an environmental hazard. This is far less likely to happen with a well-documented landfill site even though hazardous materials may have been deposited in it.

Finally, the economics of spill cleanup should be considered since the transport of large quantities of contaminated material to landfill sites can be very costly, especially if these are situated at a considerable distance from the incident. Efforts should be made to avoid unnecessary dilution of the spilled material so that the major proportion can be collected and returned to its source at high concentration. The minimum amount of adsorbent should be used so that this, together with the minimum amount of contaminated soil, remains to be transported to the landfill site. Local legislation may also require an estimate of the quantity of hazardous material present in both the contaminated adsorbent and the soil prior to landfill.

APPENDIX*

1. *Reclamation, Treatment and Disposal of Wastes—An Evaluation of Options,* Waste Management Paper No. 1, H. M. Stationery Office, London, 1976.

2. *Waste Disposal Surveys,* Waste Management Paper No. 2, H. M. Stationery Office, London, 1976.

*The Waste Management Papers listed in the Appendix have been prepared by the U.K. Department of the Environment in London and are available from Her Majesty's Stationery Office, 49 High Holborn, London WC IV 6HB, England.

3. *Guideline for the Preparation of a Waste Disposal Plan,* Waste Management Paper No. 3, H. M. Stationery Office, London, 1976.

4. *The Licensing of Waste Disposal Sites,* Waste Management Paper No. 4, H. M. Stationery Office, London, 1976.

5. *The Relationship between Waste Disposal Authorities and Private Industry,* Waste Management Paper No. 5, H. M. Stationery Office, London, 1976.

6. *Polychlorinated Biphenyl (PCB) Wastes—A Technical Memorandum on Reclamation, Treatment and Disposal Including a Code of Practice,* Waste Management Paper No. 6, H. M. Stationery Office, London, 1976.

7. *Mineral Oil Wastes—A Technical Memorandum of Arisings, Treatment and Disposal Including a Code of Practice,* Waste Management Paper No. 7, H. M. Stationery Office, London, 1976.

8. *Heat-Treatment Cyanide Wastes—A Technical Memorandum of Arisings, Treatment and Disposal Including a Code of Practice,* Waste Management Paper No. 8, H. M. Stationery Office, London, 1976.

9. *Halogenated Hydrocarbon Solvent Wastes from Cleaning Processes—A Technical Memorandum on Reclamation and Disposal Including a Code of Practice,* Waste Management Paper No. 9, H. M. Stationery Office, London, 1976.

10. *Local Authority Waste Disposal Statistics 1974/75,* Waste Management Paper No. 10, H. M. Stationery Office, London, 1976.

11. *Metal Finishing Wastes—A Technical Memorandum on Arisings, Treatment and Disposal Including a Code of Practice,* Waste Management Paper No. 11, H. M. Stationery Office, London, 1976.

12. *Mercury Bearing Wastes—A Technical Memorandum on Storage, Handling, Treatment and Recovery of Mercury Including a Code of Practice,* Waste Management Paper No. 12, H. M. Stationery Office, London, 1977.

13. *Tarry and Distillation Wastes and Other Chemical Based Residues—A Technical Memorandum on Arisings, Treatment and Disposal Including a Code of Practice,* Waste Management Paper No. 13, H. M. Stationery Office, London, 1977.

14. *Solvent Wastes (Excluding Halogenated Hydrocarbons)—A Technical Memorandum on Reclamation and Disposal Including a Code of Practice,* Waste Management Paper No. 14, H. M. Stationery Office, London, 1977.

15. *Halogenated Organic Wastes—A Technical Memorandum on Arisings, Treatment and Disposal Including a Code of Practice,* Waste Management Paper No. 15, H. M. Stationery Office, London, 1978.

16. *Wood Preserving Wastes—A Technical Memorandum on Arisings, Treatment and Disposal Including a Code of Practice,* Waste Management Paper No. 16, H. M. Stationery Office, London, 1978.

17. *Wastes from Tanning, Leather Dressing and Fellmongering—A Technical Memorandum on Recovery, Treatment and Disposal Including a Code of Practice,* Waste Management Paper No. 17, H. M. Stationery Office, London, 1978.

18. *Asbestos Waste—A Technical Memorandum on Arisings and Disposal Including a Code of Practice,* Waste Management Paper No. 18, H. M. Stationery Office, London, 1978.

19. *Wastes from the Manufacture of Pharmaceuticals, Toiletries and Cosmetics—A Technical Memorandum on Arisings, Treatment and Disposal Including a Code of Practice,* Waste Management Paper No. 19, H. M. Stationery Office, London, 1978.

20. *Arsenic-Bearing Wastes—A Technical Memorandum on Recovery, Treatment and Disposal Including a Code of Practice,* Waste Management Paper No. 20, H. M. Stationery Office, London, 1980.

21. *Pesticide Wastes—A Technical Memorandum on Arisings and Disposal Including a Code of Practice,* Waste Management Paper No. 21, H. M. Stationery Office, London, 1980.

REFERENCES

1. *Proceedings of the 1972 National Conference on Control of Hazardous Material Spills,* Houston, March 1972.

2. G. F. Bennett (ed.), *Proceedings of the 1974 National Conference on Control of Hazardous Material Spills,* San Francisco, August 1974.

3. G. F. Bennett (ed.), *Proceedings of the 1976 National Conference on Control of Hazardous Material Spills,* New Orleans, April 1976.

4. G. F. Bennett (ed.), *Proceedings of the 1978 National Conference on Control of Hazardous Material Spills,* Miami Beach, Fla., April 1978.

5. *Co-operative Programme of Research on the Behavior of Hazardous Wastes in Landfill Sites,* H. M. Stationery Office, London, 1978.

6. C. A. Edwards, "Pesticide Residues in Soil and Water," in C. A. Edwards (ed.), *Environmental Pollution by Pesticides,* Plenum Press, New York, 1973.

7. F. Matsumura, "Degradation of Pesticide Residues in the Environment," in C. A. Edwards (ed.), *Environmental Pollution by Pesticides,* Plenum Press, New York, 1973.

8. *Hazardous Waste Disposal Damage Reports,* SW-141, U.S. Environmental Protection Agency, 1975.

9. "Residual Management by Land Disposal," *Proceedings of the Hazardous Waste Research Symposium,* EPA-600/9-76-015, Tucson, Ariz., 1976.

10. R. S. Ottinger, J. L. Blumenthal, D. F. Dal Porto, G. Gruber, M. J. Santy, and C. C. Shih, *Recommended Methods of Reduction, Neutralization, Recovery or Disposal of Hazardous Waste,* vols. 1–14, EPA-600/2-73-053a–p, U.S. Environmental Protection Agency, 1973.

11. *Proceedings of the 1980 National Conference on Hazardous Material Spills,* Louisville, Ky., May 1980.

PART 3
Incineration on Land

Rudy G. Novak
Charles Pfrommer, Jr.
IT Enviroscience, Inc.

Introduction / 14-20
Identification, Characterization, and Categorization of Hazardous
    Spills / 14-22
Cleanup and Transportation / 14-24
Incineration Systems: Principles and Characteristics / 14-25
Preliminary Combustion Evaluation and Testing / 14-33
Monitoring and Follow-Up / 14-34
Glossary / 14-35

INTRODUCTION

Incineration is an effective method for disposing of many hazardous material spills. However, determination of whether incineration is appropriate for a particular spill and whether a specific incineration system is adequate depends on several factors.

There are various types of municipal and industrial incineration systems, each designed for optimal disposal performance on a specific feed material profile. These materials are usually wastes or by-products that are loosely controlled and have little, if any, apparent economic value. Hazardous material spills, on the other hand, often consist of relatively pure and sometimes concentrated product materials that may be recoverable or be recycled. Frequently the product must be destroyed because of product contamination and/or safety considerations. The basic factors to be considered in the incineration of hazardous materials are identified and reviewed in this part, together with equipment capabilities and operating conditions.

Incineration can take place under a variety of combustion conditions. Oxidation is the dominant process for efficient destruction, with pyrolysis occurring incidentally or transforming the material into a better physical form for oxidation. Incineration is therefore defined here as a high-temperature oxidation reaction between spilled hazardous materials and air under controlled conditions of retention time, temperature, and turbulence within a specified combustion chamber.

Although the desired end products of the incineration reaction are materials of greatly reduced toxicity, hazard, and reactivity (i.e., carbon dioxide, water, and low concentrations of hydrogen chloride and sulfur dioxide), less desirable combustion products can also be generated. These less desirable products may include carbonaceous (soot) and salt particulates, metal oxides, free halogens, carbon monoxide, oxides of nitrogen, or partly oxidized compounds.

If the combustion reaction proceeds primarily under pyrolytic or reducing conditions, a broad profile of decomposition products or partly oxidized organics, including oxygenated organochlorines such as phosgene, may be generated. These products are difficult, if not impossible, to remove from the flue gases by conventional gas-cleaning devices. The products that are removed require additional treatment or disposal. Therefore, it is imperative that the formation of these less desirable combustion products be prevented by selecting proper operating conditions.

Potential environmental and toxicological properties of all emissions and effluents (flue gases, scrubber-water purge streams, ash), along with regulatory stipulations and guidelines, must be considered for each spill profile when incineration is being contemplated as the method of disposal. When the characteristics of a spill profile are properly matched to the materials-handling and -destruction capabilities of an incineration system operated within documented guidelines, optimum destruction efficiencies can be expected. However, each disposal situation is specific, and confirmatory data must be obtained for the efficient destruction of any hazardous material involved.

Incineration of hazardous materials requires the following considerations: handling, blending, and feeding of the hazardous materials; combustion and combustion control; flue-gas cleaning and control systems; discharge and treatment of water purges from flue-gas cleaning equipment; ash disposal; cleanup of the container that transported the hazardous material; safety of the entire disposal operation; environmental health considerations within the total operation; environmental impact; and identifying, interpreting, and applying local, state, and federal regulations and guidelines.

Although incineration can be applied to a broad spectrum of hazardous materials, it may be a poor or totally inadequate choice for the disposal of certain materials. For example, hazardous materials that contain heavy metals (lead, mercury, arsenic, etc.) or that decompose violently (peroxides, perchlorates) are not candidates for incineration unless their undesirable characteristics are first modified by some form of pretreatment.

The hazardous spilled material must first be identified and characterized as to quantity, concentration, method of containerization, and significant physical, chemical, toxicological, and environmental properties. Regulations that apply to the safe and proper transportation and disposal of the spilled hazardous material should be investigated as described in Part 1 of this chapter. If incineration is to be used, the spilled material must be containerized and transported to the incineration site (unless a mobile incinerator can be used). The incineration system must be able to operate at the recommended combustion temperature and flue-gas retention time and to accept and efficiently destroy the spilled hazardous material (and any harmful by-products) with a minimum of pretreatment.

During the incineration operation, emissions, ash, and water purges should be monitored to provide documented results on the impact of the total disposal operation. Monitoring of ambient conditions may also be highly desirable. Adequate operating and analytical records should be retained.

The components of the spilled material must be correctly identified and compatible systems selected for their incineration to ensure that the basic obligations of safe handling and proper ultimate disposal are met in a satisfactory manner. The selection can be made from a broad range of commercial or custom-designed systems. The basic factors that should be considered in the incineration of hazardous material spills are identified and reviewed in this discussion.

IDENTIFICATION, CHARACTERIZATION, AND
CATEGORIZATION OF HAZARDOUS SPILLS

Identification

The content and characteristics of the hazardous spill must be identified and at least partly defined before any containment, transfer, cleanup, and disposal operations can be safely devised, evaluated, or initiated. Generic descriptions are of limited use and must ultimately be reinforced with specific product and compound information. Physical and chemical properties such as flammability, toxicity, stability, corrosiveness, oxidizing reactivity, vapor pressure, and environmental sensitivity are some of the significant characteristics that dictate safe and acceptable cleanup and disposal methods for hazardous waste spills. Since incineration almost always involves handling, recontainerization, transportation, and oxidation at high temperatures, identifying the proper spill parameters for further definition is vital for safe and efficient disposal.

Characterization

The definition of a spilled hazardous material should include at a minimum an assessment of the following characteristics:

1. **Safety.**   Human toxicity, environmental sensitivity, stability.
2. **Chemical components.**   Heavy metals, halogens, phosphorus, sulfur, salts, reactive groups.
3. **Physical characteristics.**   Melting point, viscosity, suspended solids, thermal stability.
4. **General characteristics.**   Active ingredients, water content, major components.

Shipping or manufacturer's records may adequately define these parameters, but sampling and analysis may be necessary when such information is not available. The degree of detail required will also depend on the nature and quantity of the material spilled and the type of incineration system or systems available. Sampling of the spilled material must be carefully planned and executed so that a representative sample can be obtained since several liquid and solid phases may be present in any type of containment device.

The relation of the above hazardous spill characteristics to incineration systems is shown in Table 3-1. The significance and impact of the relations given in the table should be evaluated when incineration of hazardous material is considered. For specific disposal requirements, some of the characteristics will prove to be vitally significant; in other cases, they may not apply. The significance of each characteristic is determined by the combined impact of a number of factors, such as quantity, concentration, toxicity, environmental impact, and safety considerations.

To evaluate these relations and their significance, expert consultation must be requested. This assistance may be obtained from the manufacturer of the hazardous material, regulatory agencies, certified disposal companies, or experienced hazardous waste incineration consulting firms.

Categorization

Hazardous material spills that are candidates for incineration may be placed in one of the following categories:

TABLE 3-1   Impact of Hazardous Spill Characteristics on Incineration

Hazardous spill characteristics and components	Primary impact
Human toxicity	Safety in transporting, receiving, handling, blending, and feeding; required destruction efficiency
Compound stability	Safety, detonation, explosion, and heat sensitivity, plus first item
Environmental sensitivity	Required destruction efficiency and emission control considerations, plus first item
Major components, active ingredients	Same as first three items
Physical form	Spill preconditioning and preparation; feed system; incinerator type
Thermal stability	Combustion temperatures; flue-gas retention times
Water content	Auxiliary-fuel and drying requirements
Melting point	Heat release rate; hearth type
Inorganic salts; organic and inorganic phosphorus	Emission particulate control considerations; ash-slagging properties; refractory life
Organic chlorine and sulfur	Wet flue-gas scrubbing system; neutralization; corrosivity
Organic bromide	Bromine gas emissions
Fluoride	Refractory degradation; corrosion
Heavy metals (As, Zn, Hg, Pb)	Emission control considerations; ash disposal; scrubber-water purge streams

1. Liquids
2. Sludges
3. Combustible solids
4. Contaminated noncombustible solids

Categorization encourages rapid and optimal matching of spill characteristics with available incineration types that are capable of accepting, handling, and destroying selected categories of materials. For instance, an incinerator may be designed to handle only one or two of the cited categories. An example is a grate-type municipal incineration system that is designed for combustible solids (refuse) and not for wet sludges that could flow and drop through the grate openings. Subcategories of the broad groupings may also be beneficial. The following guidelines are offered as an aid to categorization.

Liquids

Liquids can range from organic materials that support combustion to contaminated water that requires auxiliary fuel for combustion to be initiated and completed. Liquids are usually fed into an incinerator through a burner nozzle designed both to atomize the liquid into a very fine mist and to supply combustion air. If the liquid can support combustion and can be properly atomized, it may usually be substituted for fuels.

It may be more difficult to dispose of liquids with a low heat of combustion, such as contaminated water, because a dual system is required for handling both the liquid and the auxiliary fuel needed. Any water present in liquid spills must be evaporated before the organic components can be thermally destroyed. The rate at which a system can handle an aqueous stream may also be significantly lower than if the waste could support combustion by itself.

Sludges

Sludges are generally thick or viscous materials that are pumpable. They frequently have a high water or other liquid content and require drying before effective combustion can be initiated. An incinerator with a grate should generally not be used for the destruction of hazardous sludges.

Combustible Solids

Combustible solids include solid organic chemicals, contaminated plastic, wooden materials, tree branches, shrubbery, and seaweed. Also included are straw and other combustible sorbent materials purposely used to clean up a spill. The size of the solids will determine whether the type of feed mechanism on an existing incinerator is suitable.

Contaminated Noncombustible Solids

These solids frequently can account for the major quantity of materials to be disposed of following a spill of hazardous material. Such materials include contaminated sand, dirt, and rubble from the spill site. Since the sand and dirt may contain significant quantities of water, the incineration rate may be limited by the drying rate. Metallic drums initially containing the spilled material and those used to transport the spilled material to the incinerator should be included in this category.

CLEANUP AND TRANSPORTATION

Although other parts deal extensively with the safe and proper cleanup and transportation of spilled hazardous materials, the products used for the cleanup and the containers employed for transporting the spill to the incinerator can also affect the efficiency of the final disposal operation.

Liquids from spills are frequently contaminated with water, which reduces their heating value, or with soil and other solids from the spill site that may need to be removed so that the liquids can be incinerated without plugging pipelines or burner nozzles. The liquids will usually be transported to the incinerator in tank trucks or in drums. These containers need to be cleaned or disposed of after the hazardous material has been transferred to storage.

Sludges may be difficult for many incineration systems to handle because of limited feed mechanism capabilities or inadequate internal sludge-conveying mechanisms. They can be transported and handled as liquids if they can be pumped or as solids if they cannot.

The incinerator size and feed mechanism should be considered when determining the type of container to use for transporting solids. Fiber containers should be considered for this purpose because they can be burned in many systems, thus eliminating the need for cleaning the transport containers. However, some rotary-kiln incinerators can accept metal drums containing solids as a direct feed charge. The combustion characteristics (heat release rate) of the solids and the heat capacity of the incineration system will determine the amount of material that can be put in a container.

Although containers can also be used for noncombustible solids, the quantity of these solids, such as contaminated soil and other materials, may make their use impractical. If dump trucks are used to haul solids, some means of feeding the material into the incinerator must be available without creating additional hazards to personnel or contaminating equipment. The trucks or other vessels must be cleaned after they transport the contaminated materials.

INCINERATION SYSTEMS: PRINCIPLES AND CHARACTERISTICS

A large number of different incinerator types exist. Many are designed primarily for handling nonhazardous municipal-type refuse. Relatively few qualify as hazardous material waste disposal systems with integrated materials-handling, flue-gas-cleaning, and ash and scrubbing-water disposal facilities. Operating personnel must be properly trained and made responsible for maintaining control of the hazardous material until it is destroyed. The disposal system must be capable of operating under proper conditions and of achieving required destruction efficiencies for the hazardous material.

In general, open-pit incinerators, small stationary-hearth refuse incinerators, and similar units are not applicable because of their limited integrated-system capability, limited instrumentation for combustion control, and inadequate gas-cleaning systems. Large municipal incinerator facilities may also be unacceptable for disposal of hazardous material spills owing to generally lower operating temperatures, open-type grates, limited firing capabilities for liquid residues, and lack of secure and proper storage facilities. This section presents a brief overview of incineration principles, operating conditions, and types of incinerators available for the destruction of hazardous spills.

Combustion

Combustion is a complex process involving chemical and physical reactions, reaction kinetics, catalysis, combustion aerodynamics, and heat transfer. Overall combustion reaction equations describe a relatively simple process. Although they are perfectly accurate as an overall description, they do not cover the details of moving from reactants to products. The mechanism entails several localized chemical reactions that include pyrolysis, reduction, acidolysis, and oxidation. In these reactions, ions, electrons, free radicals, free atoms, and molecules form, combine, and/or decompose. The process is repeated over and over.

The complexity of combustion is further aggravated by the nature of the feed material, or "fuel," whose composition is usually imprecisely known and often subject to major fluctuations. For example, if the organic chlorine level of the material exceeds 30 percent by weight, free chlorine may be formed as a combustion product that is difficult to remove from flue gases. The generation of free chlorine can be minimized by diluting the material with other organics to keep the organic chlorine level below 30 percent and/or by injecting steam or water during the combustion process to direct equilibrium conditions away from free chlorine.

Liquid materials should be atomized through a burner nozzle into a preheated combustion chamber to reduce the reaction time, optimize destruction efficiencies, and maintain combustion control. The atomized spray must first be vaporized; thereafter it will burn according to its physical and chemical properties and the environment within the combustion chamber. The speed of the combustion reaction is generally determined by the rate of vaporization, which is approximately proportional to the diameter of the atomized drop.

When atomized drops are present in fairly high concentrations, the vaporizing drops tend to generate sufficient vapor for a flame to form around the group of drops. No reaction occurs to any significant degree within the cone of the spray; instead, the drops vaporize within the spray cone. The vapor thus formed diffuses to the boundaries of the spray cone, where the drops feed a surrounding sheath of flames that in turn feed back heat to maintain evaporation.

Sludges, solids, and organically contaminated noncombustible solids start combustion by heating, drying out, and pyrolyzing. Drying of solids and sludges can be hindered if agitation of some form is not used, since these materials tend to form an insulating crust. Water first evaporates out of the solids, which then thermally decompose into combustible volatiles that escape into the combustion space and leave a residue of char. The gas phase above the solids is the place where a large percentage of combustion occurs when materials of cellulosic origin are being incinerated. The amount of vapor-phase combustion that takes place also depends on the type of combustion facility.

The glowing char reacts with oxygen from the combustion air to form primarily carbon dioxide. This reaction usually proceeds relatively slowly as the result of the minimum surface exposure to atmospheric oxygen per unit mass of char. Some form of agitation of the mass is required for continuous exposure of fresh surface.

Operating Conditions

Appropriate operating conditions for the incineration of a spilled hazardous material depend on several factors relative to both the material and the design of the incineration system. Of primary concern are the operating temperature and the retention time of the gaseous products of combustion within the system.

For the four hazardous material spill categories cited previously, minimum operating flue-gas temperatures of 1000°C at a retention time of ¾ s should be utilized to achieve desired destruction efficiencies. Criteria based on Toxic Substances Control Act and Resource Conservation and Recovery Act regulations promulgated by the U.S. Environmental Protection Agency for PCBs could result in combustion standards even more stringent than cited here. Therefore, local, state, and federal regulations should be checked before a spilled hazardous material is to be incinerated.

Another significant parameter is the amount of excess air introduced during the combustion process. Excess air is the air in excess of that needed to supply the stoichiometric quantity of oxygen for the complete oxidation of all organic materials and fuels being incinerated. With fuel oils and other clean-burning liquids at elevated temperatures, 20 percent excess air may be adequate, but solids, sludges, and other liquids may require excess air levels of 50 to 100 percent or greater. Higher levels of excess air, in the range of 150 to 200 percent, may be necessary when there are poor air turbulence and lower temperatures, which are encountered in rotary kilns.

Thermally stable materials may require more stringent operating conditions. If information is lacking on a spilled material, a preliminary test burn should be conducted in an incineration system to verify that operating conditions are sufficient to destroy the material within regulatory guidelines.

The nature of the materials being incinerated and the regulatory codes determine the degree of flue-gas cleaning required. For example, a venturi scrubber operating at a 5- to 7.5-kPa water column pressure differential may adequately control particulate emissions from general municipal refuse, whereas a 15- to 17.5-kPa pressure differential may be necessary to control particulates generated from materials containing phosphorus or salts since these components tend to form fine submicrometer particulates.

Regulatory codes differ on both the controlled specific emission components, such as particulates and SO_2, and the allowable levels of each component. These codes must be considered when a spilled hazardous material is to be incinerated.

Incinerators

Many types of incinerators are manufactured and designed by a large number of equipment and design firms. Incineration installations are generally operated under guidelines derived from design considerations, emission criteria, operating experience, sampling tests, and equipment limitations. Since incineration is a specialty area, a glossary of some of the basic terminology has been prepared and is included at the end of this part.

Incinerators can be classified by their basic characteristics. One such classification is based on the movement of solid material charged into the system. If the charge remains stationary during the incineration process, the incinerator is a fixed-bed type. If the charge travels or is turned during incineration, the unit is classified as a moving-bed type. The types of incinerators in each of the two basic classifications are identified as follows:

1. Fixed-bed incinerators
    a. Open-pit
    b. Closed-chamber
       (1) Single-chamber
       (2) Multiple-chamber

2. Moving-bed incinerators
    a. Rotary-kiln
    b. Monohearth or multiple-hearth
    c. Moving-grate
    d. Fluid-bed

Flexibility to accept and efficiently destroy various hazardous materials depends on the incinerator type. The primary advantages and disadvantages of each type are given in Table 3-2. The types are described below.

Open-Pit Incinerator

The features that make this unit generally inadequate for the incineration of hazardous wastes are poor temperature and combustion control, absence of gas cleaning, and the required manual method of batch feeding. Theoretically, combustible and contaminated noncombustible solids may be fired in this unit. The open-pit incinerator may be unsafe for disposal of hazardous materials because operating personnel can be exposed to the flue gases. Liquid and sludge materials are generally not fired in this type of unit.

Single-Closed-Chamber Incinerator

This type basically consists of a refractory-lined combustion chamber fired by one or more liquid burner nozzles. Combustion air is usually added through the burner nozzles; secondary or tempering air can be introduced around the burner.

A wide range of liquid wastes can be incinerated, provided that the heating value is sufficient to maintain temperature for complete combustion. When a low-heat-value liquid is incinerated, it must be blended with a liquid of higher heat value or auxiliary fuel must be used. If the burner nozzle has two fuel ports, the auxiliary fuel and the low-heat-value liquid can be fired concurrently through the same nozzle instead of through two separate nozzles. The liquid material must be atomized through a burner nozzle by air, steam, or mechanical means, although air or steam atomization is preferred.

Since a significant amount of ash in the form of submicrometer inorganic salts or metal oxide particulates will usually be carried out of the combustion chamber with the flue gas, some degree of flue-gas cleaning will be necessary. However, a certain amount of heavier inorganic particulate will drop out within the incinerator, requiring an occasional shutdown and cleanout of the combustion chamber.

TABLE 3-2 Incinerator Capabilities

Incinerator type	Form of incineratable materials	Major advantages	Major disadvantages
Fixed-bed			
Open-pit	Combustible solids (principally municipal refuse)	Many available	Generally unacceptable for hazardous materials; most air pollution codes not met
Single-chamber	Liquids	Many available; simpler combustion control	Limited to liquids
Multiple-chamber	Combustible solids and liquids; limited capability for noncombustible solids and sludges	Low capital; often off-the-shelf unit operation; manual ash removal	No forced bed mixing or turnover; programmed batch incineration cycle
Moving-bed			
Rotary-kiln with secondary combustion	Approximately equal capability for all categories	Very flexible for accepting broad waste profile; potential feed capability for solids, sludges in drums, and other bulk containers; forced turnover and mixing of bed; continuous ash discharge	Limited availability; potentially higher refractory maintenance
Monohearth or multiple-hearth	All categories*†	Heat economy for multiple-hearth units; single-hearth units with secondary combustion capable of being used for combustible solids	Feed materials required to be of uniform size to be conveyable across hearth; potential for plugging of rabble arms; secondary combustion possibly necessary if feed materials pyrolize readily or if volatiles can be emitted from the top hearth
Moving-grate	Combustible solids (principally municipal refuse)	Underfire air provided to aid combustion	Possibility of damage to grate by plastics or melting materials and of unburned solids falling through grates
Fluid-bed	Sludges and liquids; limited capability for combustible and noncombustible solids	Secondary combustion chamber not required; fluctuations in feed rate or fuel value smoothed by bed heat sink; heat economy	Possibility of destruction of bed fluidization by melting salts or solids

*Solids must be small and relatively uniform in size.
†Secondary combustion may be required.

Multiple-Closed-Chamber Incinerator

This system consists of two separate but connected combustion chambers in series, with the flue gases generated in the first chamber entering the second chamber for exposure to direct flame contact and higher temperatures. It has potential for the incineration of combustible solids, liquids, contaminated noncombustible solids, and sludges since incomplete and pyrolytic products of combustion emanating from the first combustion chamber can be destroyed in the secondary combustion chamber, which generally has higher temperatures, increased retention time, and increased oxidizing capability. Flue-gas-cleaning requirements are the same as those for the single-chamber unit.

The fixed-bed multiple chamber has several limitations for incineration of nonliquid hazardous materials. Since the bed is not turned over or mixed efficiently, wet materials such as sludges or contaminated noncombustible solids may bake externally, leaving their centers incompletely combusted.

The multiple-chamber system is usually operated in a batch mode and follows a programmed incineration cycle. The cycle usually consists of heat-up, charging, combustion with additional charging, burn-down, cool-down, and ash removal. The program is sequenced and scheduled for optimum disposal of a specific feed profile. However, variable or difficult-to-oxidize spill material profiles, such as wet or sludgy thermally stable materials, may not be properly destroyed within the programmed cycle. The cyclical operating nature is somewhat disruptive to smooth functioning for both the combustion operation and the air pollution control equipment. Ash removal is usually a manual operation and can be hazardous.

Rotary-Kiln Incinerators

Rotary-kiln industrial incinerators are used as primary combustion chambers for a broad range of solid and liquid materials. This design offers an optimum amount of tumbling and turnover for sludges, combustible solids, and contaminated noncombustible solids. The kiln does not need internal moving parts since the solid material is conveyed by the rotating inclined action of the kiln walls. Ash removal is continuous, and the feed can be continuous or semicontinuous. Combustion of pyrolyzed and volatilized compounds is completed in a secondary combustion chamber located downstream of the kiln.

An artist's sketch and a schematic of two rotary-kiln incineration complexes are presented in Figs. 3-1 and 3-2. The figures show solids-receiving and -feeding facilities, liquids feeding, the rotary kiln and secondary combustion chambers, ash removal, and the associated flue-gas-cleaning equipment. Figure 3-3 is a diagram of a typical rotary-kiln incineration system.

Depending on its diameter and feed mechanism, a rotary kiln may accept solids or sludges in metal containers as large as 0.21-m^3 (55-gal) drums without jamming the drum, metal rims, or covers. The kiln can accept empty or full drums, sorbents, pallets, tree branches, wet sand or soil, and similar materials involved in a hazardous spill. Incineratable liquids capable of being atomized through a burner nozzle can be fired concurrently with the solid materials. It is highly desirable that the liquids be fired with the solid materials in the kiln in order to promote solid combustion and ultimate high-efficiency destruction.

Solid materials are fed into the kiln at a controlled rate through a feed chute or a ram-charging mechanism. They are heated primarily by radiant heat from the burning atomized liquid materials and by direct contact with the heated rotating-kiln walls. Radiation, however, is the primary heat transfer mechanism. Combustion and overfire air are usually added at the feed end of the kiln. This source of air is also important in the combustion aerodynamics that occurs within the kiln. Burning proceeds rapidly at the center of the kiln, with burnout occurring in about the last third of its length.

Fig. 3-1 Artist's sketch of a rotary-kiln incineration complex.

Although a degree of pyrolysis occurs within the bed of burning solids, the nonvolatile pyrolytic products are destroyed during the burnout and the volatile pyrolytic and evaporated materials are destroyed in the secondary combustion chamber. Appropriate emission controls are necessary when particulates and acid gases are being removed from the products of combustion.

To prevent the possibility of explosion or rapid temperature excursions, drummed material should not contain freestanding combustible liquids or even combustible liquids

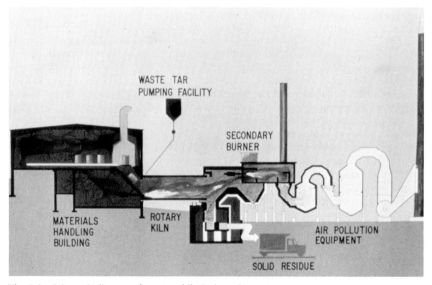

Fig. 3-2 Schematic diagram of a rotary-kiln incineration system.

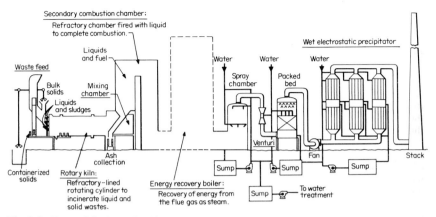

Fig. 3-3   Rotary-kiln thermal oxidation system.

of high vapor pressure that are bound within solids or sludges. A series of holes should be punched in the drums or other containers or their covers removed to provide adequate venting. Small quantities of the material should be tested for thermal stability and exothermic decomposition at elevated temperatures before significant quantities are charged into the high-temperature environment of the rotary-kiln primary combustion chamber.

Monohearth or Multiple-Hearth Incinerator

This type of incinerator utilizes mechanical agitation, provided either by the refractory-lined hearth or by the air-cooled rabble arms' being rotated slowly while the incineratable materials are fed along the periphery of the hearth. The waste materials heat, dry, ignite, and burn during their spiral travel across and down the series of drying, burning, burnout, and cooling hearths. The solid material is turned over to expose fresh surface during its travel and is conveyed in opposite directions on consecutive hearths.

With a single-hearth unit, the entire incineration process takes place on one hearth. In a multiple-hearth unit, the top hearths are utilized for drying, and consecutive lower hearths are used for combustion, burnout, and cooling. Combustion and overfire air is usually introduced tangentially on the burning hearth, with cooling air introduced in the cooling zone. The exit gases from a multiple-hearth unit leave below combustion temperatures as a result of the heat recovery and drying that take place on the upper hearths. Since a multiple-hearth unit has more hearth area than a single-hearth unit of equal diameter, its disposal capacity is greater.

Multiple-hearth incinerators are commonly used for disposal of dewatered activated sludges from wastewater treatment systems. If the sludge is a wet chemical sludge, some organics from the top hearths may be volatilized by steam distillation and carried off with the exhausted flue gases. If this occurs to any significant extent, secondary combustion of the flue gases is required. In a single-hearth unit the flue gases leave at combustion temperatures. As a result, the requirement for secondary combustion will depend on the total combustion efficiency as affected by the temperature, the retention time, the excess air, and the aerodynamics within the single-hearth incinerator.

Since large solids, wire, cloth, and sticky or slagging materials could build up on or plug the conveying rabble arms, they must be excluded or shredded to prevent the incinerator from being plugged. The incinerator usually requires preheating by liquid fuels, and it may require auxiliary fuel, fired through tangentially mounted burners located at one or more hearths, if the heat content of the solid material is not adequate to sustain combustion.

Moving-Grate Incinerator

Grates are basically cast-iron and steel conveyers that convey combustible solids (primarily refuse) through an incineration chamber while allowing combustion to proceed from the top of the solids. Combustion and grate cooling air is supplied by an air plenum below the grates.

Although incinerator grates come in a large variety of designs, most fall in the traveling- or reciprocating-grate category. Reciprocating grates provide mild agitation of the combustible solids in addition to conveying them through the chamber. The degree of burnout can be controlled by varying the throughput, grate speed, agitation, and temperature of the system.

Although municipal incinerators can achieve up to 98 to 99 percent destruction of the combustible materials, destruction efficiencies may be lower if the refuse is wet or if a nonagitating type of grate system is employed. Usually, very little supplementary fuel is fired into the combustion chamber.

Large quantities of industrial combustible materials are not usually incinerated alone in this type of unit because low-melting-point wastes will flow through and in between the grates before combustion is completed. These materials can also ignite and burn in an uncontrolled manner on and below the grates, possibly causing the air passages to become plugged by fused ash and melted wastes. In addition to creating poor combustion, this can cause the grates and drive mechanisms to become jammed or warped. If significant quantities of combustible industrial materials are incinerated in these units, it is anticipated that the resultant high temperatures, chemical corrosion, and plugging would make grate maintenance costs relatively high.

Fluid-Bed Incinerator

Fluid-bed designs have been applied industrially to incinerate liquids, sludges, and relatively homogeneous combustible and noncombustible solids that are produced within a narrow chemical profile. They offer the advantage of maintaining stable combustion over a broad aqueous-concentration range since the bed media provide a significant heat sink.

Intimate mixing of the combustion air with the waste occurs since the combustion air is also used for fluidization of the bed. As a result, combustion is efficient, with a minimum of excess air, thus minimizing the usage of auxiliary fuels. Secondary combustion usually is not necessary.

The most serious disadvantage of the fluid-bed incinerator is the potential fusion and subsequent defluidization of the bed resulting from low-melting-point eutectic mixtures that can be formed from the incineration of diverse materials. A low-melting-point ash generally presents serious technical problems in a fluid bed. Materials handling of solids into and out of the combustion chamber can also create difficult problems. Relatively small and uniform sizes of waste are required. Bulk items such as drums and fiber containers cannot be charged directly into the system, thereby limiting the flexibility of the unit to handle spilled material in containerized form. Size reduction and feeding of the charge into the pressurized fluid-bed chamber generally present a very difficult technical problem.

Emission Control

After the waste has been incinerated, the flue gases should pass through a gas-cleaning system to prevent accidental release of vaporized feed materials or thermal degradation products and to remove salt and oxide particulates and acid gases. However, gas-cleaning systems are not substitutes for good combustion or operating practices.

Particulates and acid gases are commonly removed by scrubbing in venturi or packed-tower scrubbers, although new systems such as wet electrostatic precipitators and high-energy air filters are now being utilized. Of specific interest is the removal efficiency needed for meeting particulate and other emission code requirements. With some combination of material profiles and codes, removal efficiencies of 50 to 70 percent are adequate. Other combinations may require removal efficiencies greater than 95 percent. The efficiency of any emission control system will depend on the nature of the components to be removed and on its design and type, energy input, and flue-gas preconditioning.

Packed scrubbing towers commonly use pH-controlled scrubbing water to remove acid gases, halogens, or other materials. Hydrogen chloride is an example of a common acid gas generated by the incineration of chlorinated organic materials.

However, scrubbing towers are usually not efficient for removing particulates that are generated from industrial wastes. Fabric filters, dry electrostatic precipitators, and cyclones can at best remove only particulate matter and not acid vapors or other gases. The use of these devices for cleaning incinerator flue gases is generally limited to removing particulates that are not highly deliquescent and are solids at the normal operating temperature of these devices. Condensed-liquid aerosols or deliquescent materials will build up, corrode the equipment, and be very difficult to remove from the collection equipment.

Ash Disposal

The ash generated by incineration is normally disposed of in landfills. It may contain various salts, heavy metals, and other materials that can be detrimental if care is not taken to use appropriate secured landfill sites. The classification of the landfill should be checked as to its compatibility with the projected ash profile, and approval should be obtained for disposal of the ash at that site. The ash should be analyzed to ascertain that the combustion process was complete and that no hazardous organic materials remain.

Effluent Disposal

Wet scrubbing systems for removing acid gases, salts, and other constituents from the combustion gases will generate wastewater that must be purged from the system. The streams may be high in salts, especially when alkalis are used to remove or neutralize dissolved acid gases. Generally these streams are recycled until inorganic concentrations are built up to a predetermined level. Since the purge stream could be large (i.e., 0.17 to 0.58 L/s or greater), approval for discharge to a sewer system must be obtained.

PRELIMINARY COMBUSTION EVALUATION AND TESTING

Combustion performance is evaluated in relation to experience with existing systems and similar materials and to applicable local, state, and federal regulations. Although regulations vary, the following flue-gas parameters are typical of those that may be of interest and for which maximum levels may be stipulated:

- Particulates
- Combustion efficiency
- Total hydrocarbons
- Carbon monoxide
- Nitrogen oxides

- Acid mist
- Free halogen
- Opacity
- Specific bed material components

A study of the hazardous material spill characteristics identified previously, along with their relationship to incineration (Table 3-1), will give clues as to which of these parameters may be difficult to meet. In addition, very preliminary laboratory combustion experiments conducted in a crucible and muffle furnace can safely identify some of the following characteristics and potential problems:

- Drying requirements
- Combustion stability
- Fast heat release
- Decomposition
- Detonation
- Potential emissions
- Ash-handling properties

If any significant questions as to the system's capability for destroying the spilled material remain, a test burn should be conducted on a small charge of the hazardous spilled material. The following is typical of the type of data that could be gathered during this activity:

1. **Burn characteristics.**   The optimum charging rates of the hazardous solids relative to the capacity, heat release rate of the solids, temperature, burnout, and auxiliary-fuel usage of the incineration system should be determined.

2. **Emissions.**   Emission levels of particulates, acid gases, and other flue-gas components may be of concern because of the applicable regulations, the charging rate, and the chemical nature of the spilled materials. A test burn offers an excellent opportunity to sample and analyze the flue gases in order to develop an anticipated-performance data base. Samples from the scrubbing-water purge stream should also be analyzed at this time.

3. **Ash characteristics.**   Ash-handling properties can be verified during a test burn by monitoring the physical characteristics of the ash during its residence in the incineration chamber. Additional data on ash softening and melting point can be determined in the laboratory from ash samples collected during the test.

4. **Monitoring testing.**   Emission, effluent, and ash-monitoring techniques to be used during the full-scale burn can be tested and evaluated during the test burn.

This preliminary combustion evaluation and testing activity is highly recommended for identification of potential problems and for development of appropriate solutions at minimum risk and environmental exposure.

MONITORING AND FOLLOW-UP

During the full-scale incineration of the spilled hazardous material, analytical data should be developed through monitoring, and the following records should be kept:

1. Dates and amount of spilled hazardous material received

2. Dates and amount incinerated and amount remaining

3. Destruction efficiencies

4. Analytical data for verification of destruction efficiencies, emission levels, scrubber purge, and ash quality

5. Volume or weight measurements of purge streams, ash, and flue gases

6. Incinerator system operating data

The monitoring readouts should be as nearly continuous and immediate as possible. If a significant time lapse in receiving analytical data is anticipated, sufficient basic data must be developed during earlier tests to ensure regulatory compliance and acceptable destruction efficiencies. Some of the monitoring activities can be complex and may require special development or adaptation for flue-gas or effluent analysis.

After the spilled hazardous materials have been incinerated, equipment components that come into contact with the untreated spilled material should be cleaned and checked to prevent operating personnel from being inadvertently exposed to the materials. These components include drums, dump trucks, tanks, storage bins, conveyers, shredders, and other equipment that could have come into contact with the material. Ash that is generated during the incineration should be disposed of at an appropriate secure landfill.

Incineration is a complex and specialized technique that involves many problem areas, such as control, safety, emissions, effluents, and other functions and variables relating to operation of the system. Therefore, disposal of spilled hazardous materials by incineration should not be attempted without expert and experienced consultation from recognized professionals in this field.

GLOSSARY

Afterburner   Burner located downstream and away from the primary combustion chamber so that the combustion gases from the primary chamber pass through its flame to complete the oxidation of smoke and other incompletely burned materials.

Burner   Device used to feed liquid or gaseous fuels or wastes into the combustion chamber in proportion with air in order to produce stable flame.

Combustion Air   Air introduced into the incineration system to supply oxygen for the combustion process. It can include both underfire and overfire air as a general term.

Combustion Chamber, Primary   Chamber within an incinerator where initial ignition, pyrolytic decomposition, and burning of fuels and wastes occur.

Combustion Chamber, Secondary   Chamber containing the afterburner, utilized to complete the combustion of gaseous materials and particulates emitted from the primary combustion chamber.

Excess Air   Air remaining after a fuel has been completely burned or air supplied in addition to the theoretical quantity.

Fly Ash   Particulate matter suspended in the gaseous products of combustion.

Hearth   Solid surface upon which wet or liquid materials are placed for drying and/or burning.

Heat of Combustion   Amount of heat liberated by the combustion of a unit quantity of fuel or waste.

Overfire Air   Air introduced above, adjacent to, or beyond the fuel bed.

Secondary Air   Term used synonymously with overfire air.

Underfire Air   Supplied air that passes through the fuel bed to support combustion.

GENERAL REFERENCES

1. R. E. Bastian and W. R. Seeman, "The Design and Operation of a Chemical Waste Incinerator for the Eastman Kodak Company," paper presented at the Eighth Biennial National Waste Processing Conference and Exhibit, sponsored by the American Society of Mechanical Engineers, Chicago, May 7–10, 1978.

2. Frank L. Cross, Jr., *Handbook on Incineration,* Technomic Publishing Co., Inc., Westport, Conn., 1972, pp. 63, 64.

3. W. H. Elliott, Jr. and V. B. McCormick, "Incineration of Hazardous Substances," paper presented at the Seventieth Annual Meeting of the Air Pollution Control Association, Toronto, June 20–24, 1977.

4. *A Guide for Incineration of Chemical Plant Wastes,* Technical Guide SW-3, Manufacturing Chemists Association, Washington, adopted 1974.

5. *I.I.A. Incinerator Standards,* Incinerator Institute of America, New York, 1968.

6. *Materials for Construction for Shipboard Waste Incinerators,* National Academy of Sciences, National Materials Advisory Board, Washington, 1977, pp. 35–89.

7. R. G. Novak et al., "How Sludge Characteristics Affect Incinerator Design," *Chemical Engineering,* vol. 84, no. 10, 1977, pp. 131–136.

8. W. L. O'Connell, "How to Attack Air-Pollution Control Problems," *Chemical Engineering: Deskbook Issue,* vol. 83, no. 22, 1976, pp. 97–106.

9. W. R. Seeman, "Planning a Solid Waste Management Processing Center for a Petrochemical Complex," paper presented at the Seventy-ninth National Meeting of the American Institute of Chemical Engineers and the Eighth Petrochemical and Refining Exposition, Houston, Mar. 16–20, 1975.

10. T. T. Shen, M. Chen, and J. Lauber, "Incineration of Toxic Chemical Wastes," *Pollution Engineering,* vol. 10, no. 10, 1978, pp. 45–50.

PART 4
Incineration at Sea

E. E. Finnecy
Hazardous Materials Service
Harwell Laboratory

Introduction / 14-37
Rationale of Ocean Incineration / 14-38
History of Ocean Incineration / 14-38
Incinerator Ships / 14-40
Environmental Impact of Ocean Incineration / 14-41
Legislation / 14-45
Application of Ocean Incineration to Wastes Resulting from Spills / 14-48
Conclusions / 14-49

INTRODUCTION

Incineration of wastes at sea is considered to be one of the methods for the ultimate disposal of some hazardous wastes with the least environmental impact. Its application to the treatment of residues resulting from the cleanup of hazardous material spills has been rare, if it has been used at all. The reasons for this are clear and involve problems with the logistics of the disposal operation, the availability of suitable ships equipped with suitable incinerators, the cost of the disposal, and legislative requirements.

Nevertheless, more restrictive legislation on the ultimate disposal of certain hazardous wastes at land-based installations may encourage a greater demand for ocean incineration, which could lead to more ships offering the service. This could, in turn, lead to a reduction of charges as a result of increased competition. Thus measures such as the U.S. Resource Conservation and Recovery Act,[1] the United Kingdom's disposal site licensing arrangements under the Control of Pollution Act, 1974,[2] the Directive on Toxic and Dangerous Wastes promulgated by the Council of the European Communities,[3] and other national and international legislation may all combine to produce conditions favorable to an expansion in the use of ocean incineration.

A further problem militating against a wider use of incineration at sea is that although the environmental impact of the process has been extensively studied for certain types of organochlorine-containing wastes, there has been little, if any, authoritative work on the effects of burning other wastes at sea.

Despite these difficulties in the application of ocean incineration to the disposal of wastes resulting from spill cleanup, the technique is of interest as one component in the armory of available methods. Future developments could make it easier to apply.

RATIONALE OF OCEAN INCINERATION

All spilled material that can be recovered should be reclaimed. Not only is this sound sense, it is also the official view in those countries that have formed a discernible waste management policy. It also seems clear that spilled materials that can be washed away or disposed of to landfill sites with acceptable environmental consequences will be so treated.

It seems reasonable to consider incineration at sea for materials for which, for one reason or another, the options mentioned above are not available. In particular, it is appropriate to consider ocean incineration when the choice of disposal methods lies between some form of chemical treatment, incineration on land, and, perhaps, some form of encapsulation or solidification.

In coming to a conclusion the manager will have to consider the following factors:

1. Nature of the waste
2. Object of treatment
3. Technical suitability of the options
4. Availability and accessibility of treatment methods
5. Legislative requirements
6. Environmental impact of available and accessible treatment methods
7. Economic aspects

Incineration at sea, like its land-based equivalent, is clearly most readily applicable to flammable materials. A major difference between the two operations lies in the treatment required for the effluent gases. It is claimed that ocean incineration offers these advantages:

1. No off-gas-cleaning plant is needed.
2. No liquid scrubber effluent requiring treatment and disposal is produced.
3. The mechanical simplicity of the incinerator system allows the more easy achievement of high combustion temperatures necessary for the complete destruction of some of the more refractory organic materials.

The ability to operate without effluent-gas-cleaning equipment is based on one of the following assumptions:

1. Any gases released during combustion will be diluted by dispersion and dilution in the atmosphere to very low concentrations before any possible contact with human beings can occur.
2. Any such gases will be absorbed in and diluted by the vast volume of the sea. Since seawater is slightly alkaline and contains significant quantities of bicarbonates, it will be effectively buffered against pH changes caused by the absorption of acid gases.

Similar arguments apply to particulate emissions from incineration at sea; any airborne particulates produced will either be present in very low concentrations in air by the time they reach inhabited land or will have settled on and be dispersed in the sea.

HISTORY OF OCEAN INCINERATION

The first commercial use of incineration at sea for the disposal of industrial wastes was in 1969, when the West German company Stahl-und Blech-Bau GmbH (SBB) of Gelsenkirchen offered this service to industry using the *Matthias I,* a small converted coastal tanker, which could load 500 metric tons of waste on each trip. The vessel was equipped with an incinerator, specially designed by SBB, with a combustion efficiency greater than 99.5 percent. This efficiency was acceptable to the Netherlands government, under whose flag the ship operated.

Matthias I incinerated 6000 metric tons of waste in 1969, 12,000 metric tons in 1970, and 27,000 metric tons in 1971. It was essentially an experimental vessel, and in 1972 SBB brought the larger *Matthias II* into service. This vessel was capable of loading 1000 metric tons of waste on each voyage. A second company began operations in 1972, when Ocean Combustion Services of Rotterdam, a Dutch subsidiary of the West German shipping company Hansa Lines, converted a chemical tanker to the seagoing incinerator ship *M/T Vulcanus.* This vessel was larger than *Matthias I* or *Matthias II,* with a tank capacity for about 3500 m³ of waste.*

In 1974, *Vulcanus* took part in a series of trial burns of waste from the Shell Chemical Company's Deer Park, Texas, plant. The burns took place in the Gulf of Mexico and were conducted under various permits issued by the U.S. Environmental Protection Agency (U.S. EPA). The environmental effects of these burns were extensively studied by various teams of research workers, and several reports were issued. In 1976, *Vulcanus* burned approximately 10,400 metric tons of the herbicide Agent Orange on behalf of the United States government. These burns, carried out west of Johnston Atoll in the Pacific Ocean, were also the subject of intensive scrutiny for their environmental impact, as was a further burn of Shell waste in 1977 in the Gulf of Mexico. Other burns took place in the North Sea.

The largest ship so far used for ocean incineration, SBB's *Matthias III,* went into service in 1976. This vessel is a modified tanker of 19,300 deadweight tons and was designed to dispose of 15,000 metric tons of liquid waste and 1500 tons of solid waste on each voyage.

Matthias III differs from the other vessels not only in its ability to accept solid waste but in its ability to burn waste while steaming. (The earlier vessels could burn only while holding a fixed position.) SBB planned that *Matthias III* would operate a regular service sailing from Rotterdam to the eastern seaboard of the United States, returning to the Mediterranean and thence to Rotterdam. Wastes collected at its various stopping points would be burned during the Atlantic crossing.

Certain problems were experienced during the early burns on board *Matthias III,* and the ship was withdrawn from service for modification. However, SBB ran into financial difficulties, and its two ships were acquired by the Dutch parent company MENEBA.

A new incinerator ship, the *Vesta,* underwent trials in the North Sea during the spring of 1979. This vessel is operated by Firma Lehnkering AG of Duisburg, West Germany. Information about this ship is very scanty, but it is small (handling about 1000 metric tons of liquid waste per voyage), and it is operated mainly on behalf of two major West German chemical companies.

Several other companies have, from time to time, expressed interest in ocean incineration, and some have announced that they intend to offer the service. At least one company, the Dutch Bos-Kalis organization, made a quite extensive design study, which has been reported by Van Gulik.[4] The U.S. Maritime Administration has also made an extensive study of the requirements for a chemical waste incinerator ship[5] and its environmental impact. More recently the United States company Global Marine Development, announced[6] that it was preparing proposals for a number of companies to incinerate their waste in the Gulf of Mexico and the Atlantic Ocean beginning in mid-1981.

---

*In October 1980, it was announced that a United States firm, Waste Management of Northbrook, Illinois, had purchased Ocean Combustion Services and its assets, including the *Vulcanus.*

TABLE 4-1   Specifications for *M/T Vulcanus* Incinerator Vessel

Overall length..................	102 m
Breadth......................	14.4 m
Maximum draft	7.4 m
Deadweight	4768 metric tons
Speed	10–13 knots
Tank capacity.................	3503 m³ (15 tanks, 115–574 m³)
Tank construction	All tanks, pumps, pipes, etc., of uncoated low-carbon steel
Loading equipment	Not available, but capable of being placed on board if required
Hose connections	102, 152, and 203 mm in diameter
Waste type	Must be liquid and pumpable by equipment on board; solids in pieces up to 50 mm in largest dimension may be acceptable; must not be corrosive to mild steel
Incinerator capacity	20–25 metric tons/h

INCINERATOR SHIPS

Fairly detailed information is available on *Vulcanus*[7] and is summarized in Table 4-1. *Vulcanus* is a double-hulled tanker with a minimum clearance between the tanks and hull of 1.10 m. The vessel met the requirements of the Intergovernmental Maritime Consultative Organization (IMCO) that were in force when it entered service. Brief details of its incinerators are given in Table 4-2.

Before waste is burned, the incinerators are preheated with fuel oil to 1200°C. Wastes are fed to the incinerators by pumps connected to one or more tanks. The feed rate is controlled to maintain combustion zone temperatures above 1200°C; fuel oil can be added if the waste gives insufficient heat release to achieve such a temperature. Alarm systems prevent operation of the incinerators if combustion conditions are inadequate for complete destruction of the wastes.

The control panel on *Vulcanus* has meters recording temperatures within the incinerator, a time clock with date, control lamps showing when burners and pumps are switched on, and a display from a navigation system (originally a Decca Navigator Mark 21). This panel is photographed every 15 min by an automatic camera in a sealed box; governmental officials seal the box at the start of each voyage and inspect it at the end.

Few details have been published about *Matthias II* beyond those already reported. The incinerator is fitted toward the bow of the ship and is of a diameter approaching the

TABLE 4-2   Description of Incinerators Fitted to *M/T Vulcanus*

Maximum outside diameter	5.5 m
Maximum inside diameter	4.8 m
Overall height	10.45 m
Air supply (by fan)	90,000 m³/h
Volume of combustion chamber	88 m³
Dwell time	0.5–1.5 s
Number of incinerators	2
Number of burners/incinerators	3

beam of the vessel. Like *Vulcanus,* this vessel can accept only pumpable liquid wastes. Data on *Matthias III* can be found in References 8 and 9.

ENVIRONMENTAL IMPACT OF OCEAN INCINERATION

The environmental impact of the incineration of chlorinated hydrocarbon wastes at sea has been extensively studied.[5,7,10,11,12,13] In 1973 French workers from the Centre d'Études et de Recherches de Biologie et d'Océanographie Médicale (CERBOM) were requested by Incimer (the French agents for SBB) to study the consequences of incinerating various liquid chlorinated wastes at sea. They were attempting to predict, from data collected in the North Sea, the effects that this operation could have were it to be conducted in the Mediterranean Sea. The studies they were able to make in the time available to them were, they admitted, incomplete, but they nevertheless felt that certain conclusions could legitimately be drawn:[13]

1. The process does not seem to bring about changes in the sea's biomass.

2. The absorption, by the sea, of the hydrochloric acid emitted in the plume does not seem to affect the productivity of the sea. However, they did conclude that if the plume entered the sea in large quantities, signs existed to suggest that there was some effect. For example, in these conditions they observed discoloration of *Diogenes* species.

3. No food chain concentration was observed for chlorinated hydrocarbons or for mercury and lead, toxic metals which could be associated with haloorganic wastes. Other toxic materials such as cadmium and benzopyrenes were not studied, except in the negative sense that no disturbances of the links in the nutrition chain were observed.

The CERBOM workers made the general observation that they had found no harmful effects from the ocean incineration of the wastes burned during the period of their study. They were of the opinion that their evidence was quite favorable to incineration at sea.

There were, admittedly, gaps in this work, which focused mainly on the effect on marine life of the absorption of the combustion products by the sea. No evidence was offered in the report of the impact of the emission while the plume was airborne. No chemical analysis was given for the wastes burned, nor was any analysis reported for the combustion gases produced during combustion. Laboratory experiments were performed when the combustion products were found to consist of carbon monoxide and dioxide, water vapor, traces of undefined hydrocarbons, and hydrogen chloride. The last gas was released at a rate equivalent to 123 L/kg of waste burned (corresponding to about 20 percent weight per weight of chlorine in the waste).

Later in 1974, *M/T Vulcanus* carried out two research burns of chlorinated hydrocarbon waste from the Shell Chemical Company's Deer Park, Texas, plant. During these burns about 12,600 metric tons of waste were destroyed. The operations were extensively studied, and detailed reports exist from the Tereco Corporation, which was responsible for the sea-level monitoring of the impact of the emissions,[10] and from the U.S. Maritime Administration, which was responsible for the overall management of the experimental aspects of the operation on behalf of the U.S. EPA.[5]

The major conclusion from the Tereco Corporation workers was that no significant difference was detected between the pH and chlorinity values of the test and control areas even though the tests were sufficiently discriminatory to show up differences in sampling techniques and to detect differences between the day and night carbon dioxide content of the water. They concluded that ocean incineration was, on balance, the best method then available for the relatively harmless disposal of organochlorine wastes.

The more extensive report from the Maritime Administration[5] listed more detailed conclusions:

1. The design and operation of *Vulcanus* was adequate for the destruction of the wastes burned.

2. The ship's design did not initially include provision for stack emission monitoring, wind speed monitoring, or excess airflow recording.

3. The feed rates did not exceed the permitted limit of 25 metric tons/h, and flame temperatures complied with the 1200°C minimum and 1350°C averages required.

4. Stack gases were monitored for oxygen, carbon monoxide, carbon dioxide, chlorine, hydrogen chloride, and unburned organochlorine compounds. The findings indicated a combustion efficiency in excess of 99.9 percent.

5. The plume trailed from the stack at an angle of about 20° to the horizontal, reaching a maximum altitude of 850 m above mean sea level. The plume fanned out to a horizontal width of 1200 m at a distance 2400 m downwind from the stack.

6. Maximum hydrogen chloride concentrations measured in the plume occurred between 100 and 240 m above sea level and up to 400 m downwind. The maximum value measured was 3 ppm (the time-weighted average threshold-limit value for hydrogen chloride is 5 ppm for occupational exposure).

7. The addition of ammonia to the plume gave visual confirmation of its extent.

8. Monitoring of the sea surface showed a maximum hydrogen chloride concentration of 7 ppm about 6 m above sea level.

9. Marine monitoring surveys indicated no measurable increase in toxic metals or organochlorine compounds in the water or marine life. No adverse effects were observed on migratory birds.

10. Results of the project indicated that at-sea incineration of the wastes in question was compatible with the intent of the U.S. Marine Protection, Research, and Sanctuaries Act and that it was an alternative that could be considered for these wastes.

The U.S. EPA in its response to the environmental-impact statement prepared by the Maritime Administration (presented in Reference 5, Volume I, pages IX–4, IX–15) had certain reservations. Among the most important was the need for some safety shutdown procedure if a mishap should lead to a drop in combustion efficiency, leading in turn, perhaps, to the emission of unacceptable quantities of pollutants.

Details of the wastes burned during the two burns are shown in Tables 4-3 and 4-4.

In 1976 *M/T Vulcanus* performed three burns totaling about 10,400 metric tons of the United States military herbicide Agent Orange, a formulation containing 2,4-D (2,4-dichlorophenoxyacetic acid) and 2,4,5-T (2,4,5-trichlorophenoxyacetic acid) contaminated by dioxin (TCDD, or 2,3,7,8-tetrachlorodibenzo-*p*-dioxin), the material whose release gave rise to the catastrophe at the Italian town of Seveso. The first burn was conducted under an EPA research permit, while the two subsequent operations were under EPA special permits. The operation was extensively monitored, and reports were produced.[14,15] The permit conditions required extensive safety measures, and the burns were observed by EPA officials and representatives of the U.S. Air Force. Monitoring included on-line measurement of carbon monoxide, carbon dioxide, oxygen, and chlorinated hydrocarbon emissions from both the *Vulcanus* incinerators. Combustion efficiencies exceeded 99.9 percent during all burns, and all the permit conditions were met. The special emergency safety equipment was not required.

In March 1977, a further batch of 4100 metric tons of waste from the Shell Chemical

TABLE 4-3   Major Components of Waste Feeds in the *Vulcanus* Research Burns

Component	Research Burn I	Research Burn II
	Concentration (percent by weight)	
1,2,3-Trichloropropane	27	28
Tetrachloropropyl ether	6	6
1,2-Dichloroethane	11	10
1,1,2-Trichloroethane	13	13
Dichlorobutanes and heavier	11	10
Dichloropropanes and lighter	20	22
Allyl chloride	3	3
Dichlorohydrins	9	8
Specific gravity	1.30	1.29
Total quantity burned (metric tons)	4200	8400

SOURCE: U.S. Department of Commerce, Maritime Administration, *Final Environmental Impact Statement—Maritime Administration—Chemical Waste Incinerator Ship Project,* vols. I and II, U.S. National Technical Information Service Reports Nos. PB253978–253979, 1975–1976.

Company's Deer Park plant was incinerated in the Gulf of Mexico. This burn was done under an EPA special permit and was again extensively observed, a report being produced by TRW Inc.[11] These wastes also had a chlorine content of 63% by weight. The incineration process was monitored by an industrial field sampling team. Waste destruction efficiencies and combustion efficiencies were determined by five methods, each with independent means of sampling, analysis, and calculation. Efficiencies of at least 99.9 percent were observed at waste feed rates of 22 metric tons/h. Flame temperatures averaged

TABLE 4-4   Elemental Analyses of Waste Feeds in *Vulcanus* Research Burns

Major elements	Research Burn I	Research Burn II
	Concentration (percent by weight)	
Carbon	29	29.3
Hydrogen	4	4.1
Oxygen	4	3.7
Chlorine	63	63.5
Metallic elements	Concentration (ppm)	
Copper	0.51	1.1
Chromium	0.33	0.1
Nickel	0.25	0.3
Zinc	0.14	0.3
Lead	0.05	0.06
Cadmium	0.0014	0.001
Arsenic	<0.01	<0.01
Mercury	<0.001	<0.002

SOURCE: U.S. Department of Commerce, Maritime Administration, *Final Environmental Impact Statement—Maritime Administration—Chemical Waste Incinerator Ship Project,* vols. I and II, U.S. National Technical Information Service Reports Nos. PB253978–253979, 1975–1976.

1535°C, and dwell times were calculated as 0.9 s. An automatic waste shutoff system was incorporated to stop the waste flow should flame temperatures fall below 1200°C. Optical pyrometers were used to monitor the flame temperature directly, and thermocouples were used to monitor the wall temperature directly and the flame temperature indirectly, a statistical correlation between the two temperatures having been obtained.

Stack emission analysis disclosed traces of certain organochlorine compounds known to be ingredients in the waste. For example, between 0.1 and 0.6 mg/m^3 of trichloropropane were found, and small quantities of chlorinated and alkyl-substituted benzenes were found in some samples.

The TRW investigators experienced some trouble with the carbon monoxide, carbon dioxide, and nitrogen oxide on-line analyzers. This was believed to be due to the particularly hostile environment to which they were exposed. Improvements in these analyzers were seen to be needed but held to be entirely feasible within current technology.

TRW Inc. produced a further report in which it made a detailed comparison between the environmental impact of at-sea incineration and its land-based counterpart.[12] The main conclusions from this review are:

1. The major uncertainties about at-sea incineration relate to:
   a. The size distribution and composition of particulate emissions.
   b. The need for more data to describe the behavior of the exhaust gas plume.
   c. The extent to which the hydrogen chloride could add to acid rain problems.

2. More fail-safe and contingency plans have been developed for at-sea incineration than for its land-based counterpart.

3. Malfunction is likely to have more serious acute effects from incinerators on land than from those at sea.

4. The scrubber effluent from land-based incinerators can contribute total dissolved solids to water supplies.

5. Land-based units offer the possibility of hydrochloric acid recovery.

6. Ocean incinerators can dispose of organochlorine wastes at a faster rate than land units.

In the face of the evidence presented, the U.S. EPA gave guarded approval to the incineration of certain organochlorine wastes at sea[16] but pointed out that this approval applied only to wastes similar to those tested. The U.S. Maritime Administration commissioned a cost and feasibility study from Global Marine of ship conversion for future incinerator vessels, believing that United States wastes could support four such ships.[6,14]

Insofar as conclusions can be drawn from the published evidence, the following seem reasonable:

1. Ocean incineration of chlorinated hydrocarbon wastes can be done in such a manner as to ensure combustion and destruction efficiencies in excess of 99.9 percent for certain wastes containing up to 63% organically bound chlorine.

2. This operation can be conducted without putting the crew of the ship at risk from hazardous emissions or from hazard during the loading and feeding of the wastes to the incinerator.

3. The environmental impact of the uncontrolled emission of gaseous combustion products is minimal when carried out at a carefully chosen site.

4. Adequate means of monitoring the burn to ensure compliance with permit conditions exist.

LEGISLATION

Legislative control of the disposal of wastes at sea is exercised through a net of international conventions and national laws. Those measures controlling the construction and operation of ships will not be covered here; neither will general health and safety-at-work law be considered insofar as it applies to ships.

The first important international convention controlling the pollution of the sea by waste disposal from ships was the Convention for Prevention of Pollution of the Sea by Oil of 1954, amended in 1962, 1969, and 1971. It was followed by a series of conventions concerned with wastes other than oil, though all include sections on oily wastes. The first of these was the Convention for the Prevention of Marine Pollution by Dumping from Ships and Aircraft (Oslo, 1972), which was concerned with a defined area of the northeast Atlantic Ocean and the North Sea. It was followed by the Convention for the Prevention of Marine Pollution by Dumping of Wastes and Other Matter (London, 1972), covering all seas and oceans. This is the most important single piece of international legislation in the area of sea disposal and is now binding on the signatory powers (these now comprise more than 100 states including almost all industrialized nations).

Three further conventions applying to defined sea areas have been enacted:

1. Helsinki (or Baltic) Convention, relating to the Baltic Sea
2. Barcelona Convention, relating to the Mediterranean Sea
3. Kuwait Convention, relating to the Persian Gulf area

In general terms, these conventions add restrictions to those contained in the London Convention and set up authorities to control the disposal of waste at sea.

The London Convention applies to vessels of the signatory powers wherever they load the wastes or to vessels of any flag loading waste for sea disposal from ports in any of the signatory states. All wastes destined for dumping at sea must be licensed by an appointed authority within one of the convention states, and the disposal must be carried out at a predetermined spot using an established procedure. The appointed authorities can stipulate the packaging methods to be used for waste consigned to sea disposal. The convention also defines an inspection procedure and gives wide powers to official inspectors from any duly appointed controlling authority.

In general, each batch of waste must be individually licensed, and a charge can be made for the license. The authority can demand a chemical analysis of the waste, can have this done, and make a charge for analysis.

Certain substances are subject to more stringent control. Thus the following materials are prohibited from dumping at sea:

1. Organohalogen compounds.
2. Mercury and mercury compounds.
3. Cadmium and cadmium compounds.
4. Persistent synthetic materials which may float and remain in suspension in the sea (e.g., plastics).
5. Crude oil, fuel oil, heavy diesel oil, lubricating oils, hydraulic fluids, and any mixtures containing them taken on board for the purpose of dumping.
6. High-level radioactive waste.
7. Biological and chemical warfare materials.

8. The above prohibitions do not apply to materials that are readily rendered harmless by natural processes in the sea, provided that these do not render edible marine organisms unpalatable and do not endanger the health of humans or domestic animals.

9. The above prohibitions do not apply to wastes or other materials (e.g., sewage sludges and dredging spoil) contaminated by the materials referred to in Pars. 1 to 5.

The following substances are subject to strict control:

1. Wastes containing significant quantities of arsenic, lead, copper, zinc, and their compounds, organosilicon compounds, cyanides, fluorides, and pesticides and their by-products not on the prohibited list.

2. Containers and bulky items which may sink and interfere with fishing.

3. If large quantities of acids and alkalies are to be dumped, their content of beryllium, chromium, nickel, vanadium, and their compounds should be strictly controlled.

4. Radioactive materials not prohibited elsewhere.

All other wastes or materials to be dumped into the sea are subject to general licensing requirements set out in Annex III of the London Convention.

Ratification of the convention required that each nation introduce national legislation that would give legal power to the convention for that nation and its ships. Some national legislation is listed in Table 4-5.

By April 1976 the following countries had ratified the London Convention; i.e., they had introduced domestic legislation that was at least as restrictive as the convention:

Canada	New Zealand
Denmark (excluding the Faroe Islands)	Norway
Dominican Republic	Philippines
Guatemala	Spain
Haiti	Sweden
Hungary	U.S.S.R.
Ireland	United Kingdom
Jordan	United States
Mexico	

In addition, the following countries had lodged instruments of accession with the appropriate bodies and thus had, in effect, ratified the treaty:

Afghanistan	Panama
Cuba	Tunisia
Kenya	United Arab Emirates
Nigeria	Zaire

The London Convention thus entered the realm of international law and became binding on the contracting parties. However, it is difficult to apply the terms of the London, Oslo, Barcelona, and Helsinki conventions to incineration at sea, since the material to be burned

TABLE 4-5   Selected National Legislation Relating to Dumping Wastes at Sea

Country	Law	Responsible authority
Canada	Ocean Dumping Control Act, 1975	Ministry of the Environment
Denmark	Statute No. 312, June 1975; the "Dumping Act," Art. 4, Saltwater Fishing Act, 1965	Directorate of the Environment
European Economic Community (EEC)	Directive of May 4, 1976, on pollution caused by certain dangerous substances discharged into the aquatic environment of the Community; proposed directive (January 12, 1976) on dumping wastes at sea	Operated by appropriate authorities in each member state of the EEC
France	Law No. 76-600, July 7, 1976 (specific reference to incineration at sea)	Ministry of the Environment
Japan	Marine Pollution Prevention Law, 1970, as amended in 1970 and 1973; implementation orders of 1971, 1972, and 1973	Ministry of Transport
Netherlands	Marine Waters Pollution Act	Ministry of Transport
United Kingdom	Dumping at Sea Act, 1974	Ministry of Agriculture, Fisheries, and Food
United States	Marine Protection, Research, and Sanctuaries Act, 1972, as amended in 1974, 1975, and 1976	U.S. Environmental Protection Agency
West Germany	Water Conservation Law, as amended in March 1974	German Hydrographic Institute

is clearly not being dumped. (The terms of these conventions do apply to incinerator residues and possibly to gaseous emissions. If these eventually enter the sea a license may be required.)

Several governments have taken steps to bring incineration at sea under control. Thus in the United States permits have been required from the U.S. EPA. Similarly, permits have been necessary in the United Kingdom, the Netherlands, and other countries. IMCO was instructed, in its role as Secretariat to the London Convention, to formalize the position of ocean incineration. As a result of this initiative, new rules controlling incineration at sea entered into effect on March 12, 1979.[17]

Included in the new regulations are requirements relating to approval of the incineration system by appropriate national authorities and provisions required for operation of the incinerators, recording devices, and inspection and notification of licenses. An IMCO spokesman has stated[17] that existing techniques and monitoring devices will allow incinerator ships to operate. The new regulations define a marine incineration facility as a vessel, platform, or other artificial structure operating for the purpose of incineration at sea. Activities related to the normal operations of ships or platforms are excluded.

Among the operational requirements, the new regulations demand that the combustion efficiency of the incinerator must be 99.95 $\pm$ 0.05 percent and that incineration must take place at a temperature in excess of 1250°C unless tests prove that adequate combustion efficiency can be achieved at lower temperatures.

APPLICATION OF OCEAN INCINERATION TO
WASTES RESULTING FROM SPILLS

Typical characteristics of spilled material as presented for disposal will include the following:

1. The material may be contaminated to a greater or lesser extent by other materials involved in the spill. For example, spilled material resulting from a road accident involving multiple loads will usually be contaminated.
2. The material may be contaminated with materials used to fight any fire that occurred at the time of the spill.
3. The material may be contaminated by gasoline, diesel oil, lubricating oil, etc.
4. The material may be contaminated with materials used as absorbents or with materials used to construct barriers to prevent spread of the spill.
5. The material may be contaminated with soil or other material onto which the spill occurred, including vegetation.
6. The material may be of unknown or ill-defined composition.
7. The material may be in containers that have suffered varying degrees of damage.
8. The quantity of material for disposal can range from a few kilograms or less to thousands of metric tons.

Among the licensing requirements that will have to be met before a material can be accepted for sea disposal are:

1. The waste will have to be chemically characterized sufficiently to determine that it does not fall into the prohibited classes.
2. The waste will have to be sufficiently characterized for a reasonable environmental-impact assessment to be made.
3. The waste will have to be presented in a form acceptable on the vessel conducting the incineration.
4. The waste will have to be in a form such that it can be burned in the ship's incinerator.
5. Licensing arrangements for the disposal of any solid residue after combustion will have to be made.

It may be difficult to organize the incineration of the waste at sea for other, operational reasons:

1. Because only two ships offer the service, it may be difficult to arrange for disposal. In addition, obtaining a license is likely to be time-consuming if the waste is of unknown composition.
2. The spill may occur at a place remote from a port visited by the incineration vessel.
3. The waste may have to be packaged and stored for a period before it can be loaded on the ship.
4. The costs are likely to be high.

It is difficult to be precise about the cost of ocean incineration. Among the components of the price will be:

1. The cost of packaging the spilled material

2. The cost of transporting it to a safe storage place and any special surveillance needed during transport or storage

3. The cost of obtaining chemical analyses of the waste

4. Any charge for the license, including possibly the cost of preparing an environmental-impact assessment

5. The cost of loading the waste onto the ship, and any special dues or fees required

6. Possible insurance costs

7. The charge made by the company operating the ship

Incineration charges were about US$50 per metric ton f.o.b. for the incineration of Shell Chemical wastes in 1974.[18] It should be noted that this cost was for a complete load for the *Vulcanus*. Such large quantities will probably be rare when spilled materials are concerned. It is likely that the total costs of incinerating wastes originating from spill cleanup will be considerably higher than this. Cost inflation since 1974 will also contribute to raising prices.

CONCLUSIONS

It has been shown that with a properly designed and operated shipborne incinerator certain organochlorine waste materials can be burned at sea with minimal environmental impact, and the legislative situation has been formalized by new rules from IMCO. However, little effort has yet been devoted to examining the impact of ocean incineration of materials other than organochlorine wastes. Before incineration at sea of significant amounts of materials other than these can occur, the environmental impact of such disposal will have to be studied, using, for example, the U.S. EPA research permit system to legalize the disposals.

A wider adoption of incineration at sea for the disposal of spilled material residues would seem to require that certain conditions are met:

1. More ships offering the service, perhaps several small vessels operating regularly out of several ports, would have advantages.

2. An infrastructure for the collection, transport, and storage of residues prior to loading.

3. Convenience for delivery to the vessel.

Thus, ocean incineration would appear to be a soundly based disposal method for many waste materials, but its application to spill residues poses institutional and organizational difficulties. As presently practiced, it seems unlikely that it will be widely used for this specialized disposal operation. However, changes in the number of vessels suitably equipped and in their manner of operation could alter this situation.

REFERENCES

1. *Resource Conservation and Recovery Act,* Pub. L. 94-580, 94th Cong., 2d Sess., 1976.

2. U.K. Parliament, *Control of Pollution Act, 1974,* H. M. Stationery Office, London, 1974.

3. Council of the European Communities, "Directive on Toxic and Dangerous Wastes," Mar. 20, 1978, 78/319/EEC, *Official Journal,* L84, Mar. 31, 1978.

4. B. van Gulik, "Design of a Sea-Borne Incinerator," *Proceedings of the Second Annual Conference on Pollution,* Manchester, Mar. 15–16, 1972, British Institute of Management, Manchester Branch.

5. U.S. Department of Commerce, Maritime Administration, *Final Environmental Impact Statement—Maritime Administration—Chemical Waste Incinerator Ship Project,* vols. I and II, U.S. National Technical Information Service Reports Nos. PB253978–253979, 1975–1976.

6. *Chemical Engineering,* vol. 85, no. 22, Oct. 30, 1978, p. 60.

7. T. A. Wastler et al., *Disposal of Organochlorine Wastes by Incineration at Sea,* U.S. National Technical Information Service Report No. PB246243, U.S. Environmental Protection Agency, July 1975.

8. J. Gallay, "Incineration of Hazardous Waste Water Residues at Sea," paper presented at the National Conference on Management and Disposal of Industrial Residues, Washington, Feb. 3–5, 1975.

9. A. Wright, "Maritime Incineration—A Clean Answer to Chlorinated Wastes," *Chemical Age,* vol. 115, no. 3027, July 22, 1977, pp. 11, 15.

10. Tereco Corporation, *Sea Level Monitoring of the Incineration of Organic Chloride Waste by M/T "Vulcanus" in the Northern Gulf of Mexico: Shell Waste Burn No. 2,* prepared for the U.S. Environmental Protection Agency, U.S. National Technical Information Service Report No. PB 253265, January 1975.

11. TRW Inc., *At-Sea Incineration of Organo-Chlorine Wastes on Board the M/T "Vulcanus,"* prepared for the U.S. Environmental Protection Agency, U.S. National Technical Information Service Report No. PB272110, September 1977.

12. TRW Inc., *Environmental Assessment: At-Sea and Land-Based Incineration of Organo-Chlorine Wastes,* prepared for the U.S. Environmental Protection Agency, U.S. National Technical Information Service Report No. PB283642, April 1978.

13. M. Aubert et al., *Effect on the Marine Environment of the Combustion at Sea of Some Industrial Waste,* CERBOM, Parc de la Côte, Avenue Jean Lorrain, Nice, France, 1974.

14. "Seagoing Furnace Destroys Toxics," *EPA Journal,* vol. 4, no. 8, 1978, pp. 16–17.

15. TRW Systems Inc., *At-Sea Incineration of Herbicide Orange on Board the M/T "Vulcanus,"* final report prepared for the U.S. Environmental Protection Agency, U.S. National Technical Information Service Report No. PB281690, April 1977.

16. L. J. Ricci, "Offshore Incineration Gets Limited U.S. Backing," *Chemical Engineering,* vol. 83, no. 1, Jan. 5, 1976, pp. 86, 88.

17. "Ocean Dumping Convention Amended to Provide Rules on Incineration," *International Environment Reporter,* vol. 1, no. 11, Nov. 10, 1978, pp. 362–363.

18. "Concern over Ocean Incineration," *Offshore Engineering,* vol. 5, May 1977, p. 30.

The Future

Dennis M. Stainken
Allied Corporation *

Spill Impact / 15-2
Risk Analysis / 15-3
Prevention / 15-3
Spill Response / 15-4
Cleanup / 15-5
Ultimate Disposal / 15-5
Equipment Development / 15-6
Superfund / 15-6

Future developments in hazardous material control and spill amelioration will, it is hoped, address key problems identified in earlier chapters of the *Handbook* and enhance the technologies currently in use. Critical areas for further development will be found in impact evaluation, assessment of hazard and risk, spill prevention, response, cleanup, and ultimate disposal. This chapter presents a summary of accomplishments and unsolved problems in several of these areas and predicts future trends in spill control management, hazardous material disposal, and waste disposal site technology.

SPILL IMPACT

The impact of hazardous material spills on several key areas (municipal wastewater treatment plants, the environment, and economics) was discussed in Chap. 4. Prior to 1972, it was common practice to discharge materials to municipal wastewater treatment plants with the expectation that dilution and biodegradation would occur. The planned disposal of some hazardous materials worked for compounds which were biodegradable, but most of the materials are relatively difficult to assimilate in biological systems. Generally, a municipal wastewater treatment plant (POTW) should not be used as an ultimate sink for hazardous wastes. However, spills and illegal discharges do result in hazardous materials' entering POTWs, and these materials can damage a POTW physically, destroy or hinder the biological process, or pass through the plant and exit in the effluent. Activities necessary for managing hazardous material spills could be classed in three categories: preventive, countermeasure, and operational. In the first stages, a spill must be detected or anticipated. An inventory of hazardous materials being used by those discharging to a POTW treatment system should be maintained to narrow the search for a spilling entity. Risk estimations can be made from inventories to formulate countermeasures for spills into a POTW. If necessary, restrictive ordinances, fines, penalties, and flow routing can be initiated. Once a spill reaches a POTW, a variety of physical and biological procedures can be used to minimize its impact.

---

*The views reflected in this chapter are those of tthe author and do not reflect the opinion or policies of Allied Corporation.

Spills of hazardous materials into the aquatic environment have three major effects on an ecosystem: (1) they destroy all or part of the indigenous biota, (2) they impair ecosystem function, and (3) they place the ecosystem in disequilibrium. Aquatic ecosystems essentially provide transport, biodegradation, assimilation, and recycling functions which are impaired or lost when impacted by spills. To prevent environmental damage, materials should be screened by an environmental hazard evaluation process before being manufactured, transported, or used.

To decrease potential spill damage, spill contingency plans should include knowledge of local ecosystems to predict and prevent effects and to establish monitoring sites. Local ecosystems should be evaluated to determine their sensitivity or resilience to stress.

Industries impacting sensitive areas should employ greater safeguards to ensure spill prevention, especially when stress-persistent nonbiodegradable hazardous materials are involved. An area of future research which should be expanded is the restoration of perturbed ecosystems. Damage may result from the spill or the cleanup technology, and the critical issue is whether or not the impacted ecosystem returns to "normal."

A series of legislative directives and mandates governs spill control. The paucity of data on both the number of hazardous substance discharges and the costs of discharge events makes it difficult to conduct an economic-impact analysis. There has been little work addressing the identification and quantification of the major parameters affecting the economic impact of hazardous material spills. Providing additional data on cleanup costs should improve current economic analyses.

RISK ANALYSIS

In Chap. 5 the problems of assessing hazard and risk were discussed. The principal hazards associated with spills, the ways in which these hazards could occur, the means of quantifying various types of hazards, and the methods of evaluating the risk that certain hazards might occur were presented. Factors governing the severity of a spill were the material's intrinsic properties, the dispersive energy, the quantity spilled, environmental factors, and population density in the vicinity. The risk of a material's spreading and potential impact can be evaluated on the basis of the following data: volatilization, evaporation rate, flash point, flammability, toxicity, capability to detonate or form fireballs, etc.

Hazard quantification can use several types of indices based on available data. However, the question of how best to quantify the hazard and the risk is still open. This is an area that needs refinement of methodology. How to calculate, predict, and quantify damage from a spill is another area which needs further research to minimize the effects of hazardous material releases.

PREVENTION

Spill prevention was emphasized in Chap. 6. In plant operations, spill prevention should be approached as accident prevention. The probable causes of spills should be identified and eliminated, thereby reducing the odds of a spill's occurring. The requirements for a prevention program should include records of past spills and personnel involved (in the plant, consultant firms, contractors, insurance agencies, federal and state environmental agencies).

Elements to be considered in the design of a prevention program should include personnel identification and training, equipment, the environment (security, weather, geography), facility inspections, records, and product knowledge of personnel. Surveys should be conducted inside and outside the facility to identify potential problem areas, and these problems should be corrected by facility changes or personnel training. Additionally, the plant should be operated to avoid sudden upset. Damage from hazardous material spills should diminish as plant facilities improve and implement adequate spill prevention and contingency plans.

The prevention of hazardous material spills during transportation, whether by truck, train, or vessel, is the primary goal of the U.S. Department of Transportation and the U.S. Coast Guard. These agencies provide strict safety regulations for the transport of hazardous materials and the design of carriers and packaging for such materials. The critical components of spill prevention involve personnel training and/or certification and equipment inspection and maintenance. It is hoped that spills may be reduced or minimized as improved operating procedures, packaging, and carriers are designed and implemented.

SPILL RESPONSE

Chapter 7 addressed spill response planning concepts. Spill response plans are generally designed to prevent damage and reduce material loss. Training programs are an effective tool in spill prevention, but they are only part of an overall spill prevention program. A typical prevention program will identify potential hazardous materials and possible sources of spills. This survey includes shipping, storage, volume handled, and physical and chemical properties of the material. On the basis of these data, spill containment and response planning can then be implemented.

Many industries maintain cooperative or trade organization facilities for addressing spills. A critical aspect of a response plan is to have adequate spill cleanup capabilities or response teams available. Successful spill response planning should address at least the manufacturing, storage, usage, and transportation aspects of hazardous materials. The critical aspect of spill response planning is the implementation of all parts of the plan.

Problems and techniques for managing hazardous material spills into watercourses were discussed in Chap. 8. The success of a response to a hazardous material spill into water often depends on the quality of the sampling, analysis, and detection program implemented to support it. Because many hazardous materials dissolve or sink, sophisticated technology is needed to discover and monitor their presence. Critical aspects of response are the correct implementation of sampling and sample preservation, chemical analyses, remote sensing, and aerial reconnaissance. Safety procedures should be concomitantly followed to ensure worker safety. Future developments in instrumentation technology should result in more rapid and sensitive chemical analyses and remote sensing. It will be critical to furnish on-site rapid identification and quantification of spilled materials and equally important to provide constant availability of these services.

Aquatic toxicity data are often necessary to evaluate the hazards of a spill and cleanup procedures. Bioassays, one of the most commonly used techniques, can be employed to evaluate the toxicity of a mitigative technique and thus determine the acceptability of the "cleaned" effluent and the zone of influence of a spill. They can utilize a variety of organisms from laboratory stocks, and techniques are relatively consistent.

Toxicity can be defined by death or behavioral effects. Methods of testing toxicity in spill situations include portable bioassays, in situ and static bioassays, luminescent-bacteria-toxicity-monitoring systems, and CAM (cholinesterase antagonist monitor) devices. The primary difficulty has been, and will continue to be, interpretation of the environmental significance of the data generated.

During a spill it is critical to be able to describe and predict the movement and dispersion of spilled hazardous materials. Water quality models (Chap. 8) can furnish such descriptions in terms of spatial and temporal distributions of material in the water body. While a number of models exist, they are complex. Simpler models which may be refined to site-specific situations are available. However, it will be necessary to validate these models with appropriate refinement before they are used in many spill situations.

CLEANUP

Spill cleanup technology and concepts were addressed in Chap. 9. After a spill has occurred, the major question to be answered is "What are the appropriate ameliorative techniques?" Depending on the physical-chemical properties of the spilled materials, various approaches can be used to treat spills. Physical techniques may include neutralization, sedimentation, coagulation-precipitation, filtration, ion exchange, chemical oxidation, and carbon adsorption. Collection followed by disposal or biodegradation treatment measures may then be employed. All aspects of available spill cleanup technologies are areas for future development. Many possible methods of treating large spills of hazardous materials, especially those in major water bodies, have only been tested in the laboratory. These techniques must be expanded, refined, and developed to address spills in the field.

Biological spill cleanup techniques also need much study. Biological countermeasures are attractive because bacteria are almost always present and presumably will metabolize the spilled material to carbon dioxide, water, etc. Theoretically, biological countermeasures may be applied *in situ* to a spill or by pumping the material to a portable biological treatment system. However, these approaches still need substantial research and development. There are basic problems in the production, storage, and deployment of appropriate bacteria. Once these problems are resolved, it will also be necessary to avoid introducing undesirable bacteria and causing secondary problems (e.g., oxygen depletion due to aeorobic biodegradation).

ULTIMATE DISPOSAL

Ultimate disposal of hazardous material wastes will continue to be a major area of concern. Problems in transportation and considerations in handling waste from hazardous material spill incidents were discussed in Chap. 14, and the state of New Jersey's program was presented as a model. Aspects of the program included identifying and defining the composition and quantities of wastes and the means for controlling the handling and disposal of wastes. The major element in this program was the use of a cargo manifest system to ensure adherence to prescribed regulations. The emergence of a nationwide program as a result of the Resource Conservation and Recovery Act will bring uniformity, at least to the United States.

The ultimate disposal of hazardous wastes by landfilling was also discussed in Chap. 14. Aspects of problems in separating wastes from absorbents and soils were presented. Contaminated soils can often be put into sanitary landfills if they do not cause groundwater, toxicity, or reactivity problems. However, there are several problems, including the sheer bulk of the soils, etc., which must often be removed and transported for great distances. Additional questions which must often be addressed are: "Should the wastes be incinerated?" "What type of site should be used?" "Should the material be pretreated or encapsulated before landfilling?"

Research into these problems have revealed that leaching and some attenuation can occur, depending on site geology, hydrology, and type of waste. Many classes of compounds evaporate, biodegrade, or are absorbed and immobilized at the site. The major problems in landfilling hazardous wastes appear to be site selection and administering proper landfill techniques. In general, landfilling at a spill site may be unwise because of the many unknowns.

The most promising technology for the disposal of hazardous wastes is incineration, at sea or on land. Although technical problems exist and capital costs are high, the optimum mode of safe disposal of hazardous wastes will be incineration.

EQUIPMENT DEVELOPMENT

Future trends in the management of hazardous material spills will encompass several areas. As spill treatment technology evolves and matures and as contingency plans are implemented, it is expected that standardization of spill treatment equipment and its increased availability will evolve. Although it is doubtful that a single piece of equipment will be suitable for all spill situations, it is probable that designs and applications of the equipment (e.g., mobile carbon treatment systems, incineration systems, and laboratories) will become increasingly similar. As technology for spill control develops, implementation of improved regulations governing spill cleanup may generate a further need for the development of unique types of equipment. Increased availability of spill treatment equipment will be equally necessary to ensure timely deployment of spill containment and ameliorative devices. The need for this technology and equipment deployment will be rapidly identified as refined spill contingency plans are compiled and reviewed.

Spill prevention and contingency planning will continue to improve as experience is gained from case studies, model programs, and refined regulations. Adequate contingency planning should also ensure that major toxicity questions are resolved (i.e., data should be obtained before a spill), applicable programs are implemented, and necessary equipment and material are on site or available. Cooperative agreements between similar industries, state and local agencies, and management personnel should increase to maximize the deployment of available resources when necessary and to ensure the implementation of a safe, economical, and technically sound approach to spill control.

Further refinements in spill control technology still need to be developed. Critical areas for development will remain the same: better spill prevention and detection devices, more rapid and specific chemical identification and quantification procedures and devices, and better spill control and containment equipment and techniques. A particular area for future development is the establishment of more effective and environmentally sound cleanup and restorative technologies.

An equally critical area for future development is the assessment of the biotic impact of spill events. Ecosystem research is complicated and necessarily long-term. However, many geographic areas, plant sites, etc., would benefit from carefully planned and conducted baseline studies. What constitutes biotic damage, assessment of the longevity of spill effects, and development of habitat restoration techniques are difficult academic questions which will need answers to ensure safe, efficient spill management. The development of a better toxicity data base is equally necessary to ensure that proper health and safety procedures will be employed.

SUPERFUND

Future developments in the management of hazardous substances discharges in the United States will be seriously influenced by the Superfund law (Public Law 96-510). This law (the Comprehensive Environmental Response, Compensation, and Liability Act of 1980) provides significantly broadened legislative authority for response to spills and other releases of hazardous polluting chemicals. It establishes a $1.6 billion trust fund for the mitigation of hazardous substances releases from all sources, including uncontrolled waste disposal sites, and for the amelioration of pollution not only in navigable waters but also in groundwater, soils, sediments, and the atmosphere.

As a consequence of this legislation, the future will bring new and improved technology for multimedia cleanup, removal, and remedial action at spills and mismanaged hazardous waste sites. Procedures and guidelines will be specified and equipment developed for the identification, detection, containment, stabilization, and removal of hazardous chemical releases and the repair and restoration of damaged natural resources.

Methodology will be developed for on-site treatment of hazardous wastes, leachates, and contaminated soils, sediments, and groundwater. Increased attention will be placed on *in situ* remedial action because of significant problems associated with the transportation of hazardous substances off site. Alternative means will be devised for disposing of or otherwise destroying pollutants on site. New approaches will be developed for the destruction of refractory compounds, particularly in soils and sediments. Accelerated efforts will also be made on methodology for returning contaminated environments to their predamaged condition.

As a result of Superfund, the future should yield more meaningful rules and regulations governing chemical discharges. Improved contingency plans, training programs, and public education should further enhance the prevention, control, and treatment of hazardous substances releases and thereby reduce the risk of damage from such releases to human health and welfare or the environment.

Appendix

Conversion Factors

Conversion Factors

To convert from	To	Multiply by
cubic meter (m³)	gallon (gal)	2.64 E + 02 (U.S.) 2.20 E + 02 (British)
cubic meter (m³)	barrel (55 gal U.S.)	4.76 E + 00
cubic meter/second (m³/s)	gallon/minute (gpm)	4.40 E + 00 (U.S.) 3.67 E + 00 (British)
cubic meter/second (m³/s)	millions of gallons/day (mgd)	2.28 E + 01 (U.S.) 1.90 E + 01 (British)
degree Celsius (°C)	degree Fahrenheit (°F)	$(1.8 \times °C) + 32$
hectare (h)	acres	2.471 E + 00
joule (J)	British thermal units (Btu)	9.55 E − 02
kilogram (kg)	pounds (lb)	2.20 E + 00
pascal (Pa)	psi	1.45 E − 04
pascal (Pa)	kilopascal	1.0 E − 03

Index

Absorption, **4**-13
Acrylonitrile, **7**-45
Activated carbon, **8**-20, **9**-9, **9**-33, **9**-54, **12**-8, **12**-14, **12**-28
  column, field construction, **9**-10
  dioxin, **11**-37
  isotherms, **12**-17
  PCB, **12**-8
  pesticides, **11**-8, **12**-28
  phenol, **12**-14
  powdered, **9**-10
  reactivator, **9**-36
Activated sludge, **9**-45
ADR (*see* European Agreement Concerning the International Carriage of Dangerous Goods by Road)
Adsorption, **4**-13
Adsorption isotherms, **12**-17
Aerial reconnaissance, **8**-11
Aeronautics Act, **2**-12
AFFF (aqueous film-forming foams), **10**-26
Agent Orange, **14**-39
Agriculture, U.S. Department of, **7**-4
Air monitoring, **11**-13
Air movement, induced, **10**-22
Air-purifying respirators, **13**-8
Air waybill, **1**-24
Alarms, **6**-31, **8**-10
All-Union Scientific Research Institute for Water Protection (U.S.S.R.), **2**-24

Allegheny County Sanitation Authority (ALCOSAN), **4**-4
Alliance, Ohio, pesticide fire, **11**-4
Allied Chemical Corporation, emergency assistance by, **3**-24
American Bureau of Shipping, **6**-26
American Cyanamid Co., emergency assistance by, **3**-24
American National Standards Institute (ANSI), **10**-40, **10**-53
Ammonia, **9**-5, **9**-51, **10**-14, **10**-30, **10**-46
  dispersion, **10**-42
  evaporation, **10**-45
  pipelines, **10**-41
  properties: chemical, **10**-36
    physiological, **10**-36
    reactivity, **10**-36
  releases, **10**-41, **10**-49
  rollover in storage tanks, **10**-42
  storage, **10**-40
  toxicity, **5**-9
  vapor concentration, **10**-46
Ammonium hydroxide, **10**-46
Analysis:
  continuous monitoring, **4**-4
  field kits, **8**-6
  gases, **10**-13
  risk, **5**-19, **14**-9, **15**-2
  TCDD, **11**-24
  water, **8**-2

Analytical support, **9**-29
Animals, dead, **11**-25
Aquatic ecosystems, **4**-15
  management, **4**-19
  rehabilitation, **4**-16
  vulnerability, **4**-16
Aquatic toxicity, **7**-44, **7**-45
Aquatic toxicology, **8**-14
Aqueous film-forming foams (AFFF), **10**-26
Arctic Waters Pollution Prevention Act, **2**-12
Asphyxiation from liquid gases, **10**-68, **10**-69
Association of American Railroads (AAR), **6**-46
Association of Special Libraries and Information Bureaux (Aslib), **3**-40
Atmosphere-supplying respirators, **13**-11
Atmospheric dispersion, **10**-13
Atmospheric turbulence, **10**-13
Atomic Energy Control Act, **2**-12
Auditing, **7**-31, **7**-44
Australia, Hazchem use, **3**-32

Bacteria, **9**-40
  luminescent, monitoring, **8**-17
  storage of, **9**-47
Barcelona Convention, **14**-45
Barges, **6**-26
Barriers, **9**-3
  inflatable plastic, **9**-45
  migration, **9**-31
Battelle Memorial Institute, Pacific Northwest Laboratories, **3**-9
Belgium:
  governmental information systems, **3**-42
  laws, **2**-15
Bioassay, **8**-14, **15**-4
Biodegradation, **9**-57
  phenol, **9**-45, **9**-58
Biological accumulation, **4**-16
Biological countermeasures, **9**-40
Biological monitoring, **12**-29
Biological treatment, **9**-44
Biological zone of influence, **8**-14
Biomonitoring, **8**-16
Boiling-liquid–escaping-vapor explosion (BLEVE), **7**-65, **10**-69
Booms, **6**-21, **7**-17, **9**-3, **9**-21
Box 21 Rescue Squad (Dayton, Ohio), **7**-23
British Railways Board, **7**-35
Brooklyn Center, Minn., fire, **11**-6

Bulk Chemicals Code (IMCO), **1**-6, **7**-32
Bulk storage, **7**-44
Bulk Terminals in Chicago, Ill., silicon tetrachloride incident, **11**-11
Bulk transportation of hazardous substances, **3**-36
Bureau of Explosives (AAR), **6**-47
Bursting disk in dioxin accident, **11**-22
Butadiene, fire fighting, **10**-77

Cabot Corporation, **11**-12
CAM (continuous aqueous monitor), **8**-11, **8**-16
Canada:
  governmental information systems, **3**-45
  laws, **2**-10
Canada Shipping Act, **2**-11
Canada–United States Marine Pollution Contingency Plan, **7**-11
Carbamate detection, **8**-11, **8**-16
Carbon (see Activated carbon)
Carbon dioxide measurement, **10**-5
Cargo loading, **6**-31
Cartridges and gas-mask canisters, **13**-10
Case histories (spill response summaries), **9**-5
Catastrophe Control Service (Federal Republic of Germany), **7**-40
Cavtat, **7**-32
Cedar Bluff, Ala., Pyranol incident, **12**-6
CEFIC (European Council of Chemical Manufacturers Federation), **3**-34
Centre d'Études et de Recherches de Biologie et d'Océanographie Médicale (CERBOM), **14**-41
Centre Opérationnel de la Direction de la Sécurité Civile (CODISC), **3**-42
CEQ (Council on Environmental Quality), **7**-4
CERBOM (Centre d'Etudes et de Recherches de Biologie et d'Océanographie Médicale), **14**-41
CERCLA (see Comprehensive Environmental Response, Compensation, and Liability Act)
Chemical Hazards Response Information System (see CHRIS)
Chemical Industries Association (CIA), **3**-32, **7**-31, **7**-59, **10**-40
Chemical Industry Scheme for Assistance in Freight Emergencies (Chemsafe), **3**-36, **7**-35, **7**-57
Chemical Waste Law (Netherlands), **2**-17

Chemical Transportation Emergency Center (*see* CHEMTREC)
Chemicals:
  data sheet, **7**-33
  hazard potential, **I**-6
  hazards response information system (*see* CHRIS)
  major spills, **5**-21
  production, **I**-4
  reactor, **5**-14
  sewer, **7**-46
  waste haulers, **14**-4
  waste transportation, **2**-17, **3**-36, **14**-11, **14**-24
Chemsafe (Chemical Industry Scheme for Assistance in Freight Emergencies), **3**-36, **7**-35, **7**-57
CHEMTREC (Chemical Transportation Emergency Center), **I**-9, **3**-11, **6**-51, **7**-16, **7**-83, **7**-89, **10**-57
Chicago, Ill., Bulk Terminals incident, **11**-11
Chloracne, **11**-18
CHLOREP (Chlorine Emergency Plan), **3**-23, **10**-57
Chlorinated hydrocarbon incineration, **14**-41
Chlorine, **7**-60, **10**-31, **10**-52
  contingency plan, **3**-23, **10**-55, **10**-64
  dispersion, **10**-61
  disposal, **10**-63
  emergency plan, **3**-23, **10**-57
  flash, **10**-61
  kit, **7**-17
  leak control, **10**-57
  leaks, **10**-56
  neutralization, **10**-61, **10**-64
  placards, **10**-55
  plume, **10**-62
  properties, **10**-53
  releases, **10**-53
  respirators, **10**-55
  toxicity, **5**-9
Chlorine Institute, **3**-23, **10**-53
CHRIS (Chemical Hazards Response Information System), **I**-9, **3**-2, **3**-11, **7**-81, **10**-3
Chromium, **9**-56
CIA (*see* Chemical Industries Association)
Civil Aviation Authority, **7**-38
Civil Defense Service (Federal Republic of Germany), **7**-40
Classification of hazards, **1**-22, **3**-26

Clean Water Act (1977), **2**-2, **3**-7, **4**-23, **6**-34, **7**-9, **7**-13, **7**-50, **7**-77
Cleanup, **7**-7, **7**-79, **9**-1, **14**-24, **15**-5
  analytical support, **9**-29
  biological, **9**-40
  contractors, **6**-3, **6**-22, **7**-82
  cost, **4**-30, **11**-6
  dioxin, **11**-40
  equipment, **3**-5, **6**-20
  manual, **9**-14, **9**-29
  materials, **6**-18
  pesticides, **11**-5, **12**-19
  technological development, **9**-24
Clothing, protective, **13**-4
  pesticides, **12**-20
  (*See also* Personnel, protection)
Clouds:
  flammable, **5**-8
  hydrochloric acid, **11**-11
  toxic, **5**-8
Coagulation, **9**-52
Coagulation-sedimentation, **9**-33
Coastal areas, **7**-77
CODISC (Centre Opérationnel de la Direction de la Sécurité Civile), **3**-42
Codisposal with municipal waste, **14**-12
Combustion, **14**-25
  evaluation, **14**-33
  gas treatment, **14**-21
  solids, **14**-24
Command post, **7**-92
Commerce, U.S. Department of, **7**-4
Commission of the European Communities, **3**-40, **11**-37
Commissione Bonifica, **11**-37
Common carriers, transportation by, **4**-29
Communications, **7**-61, **12**-29
Comprehensive Environmental Response, Compensation, and Liability Act (CERCLA), **2**-1, **7**-4, **9**-32, **9**-38, **15**-6
Compressed-air quality, **13**-11
Computers:
  data bank, **3**-40
  information **3**-3, **3**-9, **3**-42
CONCAWE (Conservation of Clean Air and Water in Europe), **7**-38
Concentration techniques, **9**-32
Confinement, **10**-75
Conservation of Clean Air and Water in Europe (CONCAWE) **7**-38
Consultants, **6**-4
Consumer Product Safety Act, **2**-8
Containment, **6**-20, **7**-13, **7**-43, **9**-3, **9**-29
Containment ponds, **6**-15

Contingency plans, I-7, 3-7, 3-48, 6-21, 11-7, 14-9
  chlorine, 3-23, 10-55, 10-64
  evaluation checklist, 7-94
  industrial, 3-24
  international, 7-11
  local, 7-7, 7-20
  pesticides, 3-23
  radiological, 3-27
  U.S. Coast Guard, 7-81
Continuous aqueous monitor (CAM), 8-11, 8-16
Continuous monitoring, 4-4
Contractors, cleanup, 6-3, 6-22, 7-82
Control of Pollution Act, 2-20, 14-37
Convention for the Prevention of Marine Pollution by Dumping from Ships and Aircraft, 1-13, 14-45
Convention for the Prevention of Marine Pollution by Dumping of Wastes and Other Matter, 14-45
Convention for Prevention of Pollution of the Sea by Oil, 14-45
Convention on Standards of Training, Certification and Watchkeeping for Searfarers, 7-73
Cooperatives, 7-11, 7-47, 15-4
Corrosives, 7-35
Cost recovery, 7-13
Council on Environmental Quality (CEQ), 7-4
Council of European Communities, 3-27, 14-37
Countermeasures, 4-10, 7-13, 11-12
  biological, 9-40
Covers, film and foam, 10-25
Crop-spraying aircraft, 3-32
Cryogenic cooling, 10-25
Cryogenic gases, 10-66
Cryogenic liquids, 5-7
Cyanides, 9-57
  disposal, 14-13
Cyclic colorimeter, 8-11

DABAWAS, 3-42
Dams:
  porous peat moss, 9-15
  underflow, 6-20
Dangerous Goods Code (IMCO), 1-6, 3-26
Data banks, environmental information, 3-40
Dayton, Ohio, 7-23
Dead animals, 11-25

Deaths and injuries, 6-35
  employee, 4-34
  ratio, 5-18
Decision making for spill prevention, 6-22
Declaration of Inspection, 6-29
Decontamination, 7-70, 11-7, 11-23
  dioxin, 11-37
  ship, 7-75
  trailer, 9-28
Deepwater Port Act, 2-3, 4-25
Defense, U.S. Department of, 7-4
Deflagration, 10-69
Denmark, laws, 2-15
Deposit of Poisonous Wastes Act, 14-12
Deposition isopleths, 11-31
Dermatological symptoms, 11-25
Detection, 4-3
  combustible gases, 10-6, 10-75
  flame, 10-75
  low-temperature, 10-74
  organophosphates, 8-11, 8-16
  process, 10-74
  remote, 8-9
  tubes, 10-4
  water, 8-2
Detector tubes, 10-4
Detonation, 5-12
Developmental equipment, 15-6
Dikes, 6-11, 10-75, 11-12
  instability detection, 9-26
  portable foam, 9-31
Dioxin (TCDD), 5-15, 11-18, 14-42
  cleanup, 11-40
  decontamination, 11-37
  incineration, 14-32
  irradiation, 11-36
  microbial degradation, 11-36
  monitoring, 11-24
  photolysis, 11-35
  toxicology, 11-33
Directive on Toxic and Dangerous Wastes, 2-14, 14-37
Disaster relief agencies, 7-88
Disaster Research Center, 7-21
Discharge:
  frequency, 4-28
  volume, 4-29
Discovery, 7-12
Dispersion, 7-31, 11-12, 11-18
  ammonia, 10-42
  chlorine, 10-61
  coefficients, 10-15
  liquid gases, 10-71
  models, 8-22, 10-13, 11-17

Dispersion, models (*Cont.*):
   dioxin, **11**-29
   water, **9**-45
Disposal, **I**-12, **7**-13, **5**-5
   chlorine, **10**-63
   flammables, **14**-13
   incinerator: ash, **14**-32
      scrubbing solutions, **14**-33
   metallic compounds, **14**-3
   PCB, **9**-34, **12**-3, **14**-14
   pesticide residues, **11**-5, **12**-22
   phenol, **14**-13
   radioactive materials, **1**-17
   sorbents, **6**-19
   (*See also* Ultimate disposal)
Distribution Emergency Response System,
   3-24
Documentation, **7**-13, **9**-27
Dow Chemical Company, emergency assistance by, **3**-24
Drainage, **6**-6, **7**-47, **11**-3
Dredging, **9**-15, **9**-33
Drinking-water contamination, **11**-5, **11**-25,
   12-13, **12**-18
Drums: leaks, **9**-4, **12**-20
   overpacks, **9**-4
   storage, **6**-13
Dump sites, abandoned, **12**-8
Dumping wastes at sea, national legislation,
   14-47
du Pont de Nemours & Co., E. I., emergen-
   cy assistance by, **3**-24

ECDIN (Environmental Chemicals Data
   Information Network), **3**-40
Economic decisions for spill prevention, **6**-
   23
Economic-impact analysis, **15**-3
Ecosystem damage, **4**-33
   (*See also* Aquatic ecosystems)
EERU (*see* Environmental Emergency Re-
   sponse Unit)
Elastomers, chemical resistance, **13**-4
Emergencies, environmental (*see* Environ-
   mental emergencies)
Emergency, transportation (*see* Transporta-
   tion, emergency)
Emergency action center, **2**-15, **11**-12
Emergency action code, **3**-31
Emergency chlorine kit, **10**-58
Emergency international response systems,
   3-43
Emergency port task forces, **7**-77

Emergency shutdown systems, **10**-75
Emergency telephone numbers, **7**-89
Employee injury, **4**-34
Encapsulation, **14**-12
Energy, U.S. Department of, **3**-23, **7**-4
Environment, Department of the (Canada),
   3-45, **7**-35, **7**-59, **14**-12
Environmental Chemicals Data Informa-
   tion Network (ECDIN), **3**-40
Environmental Contaminants Act, **2**-12
Environmental damage costs, **4**-33
Environmental effects, **4**-15
Environmental emergencies:
   federal action, **2**-3
   judicial orders, **2**-7
   preparedness systems, **2**-18
   program, **3**-49
Environmental Emergency Response Unit
   (EERU), **7**-8, **8**-6, **9**-26, **9**-36
Environmental impact, **4**-15, **7**-2
Environmental Protection Agency (EPA)
   (*see* U.S. Environmental Protection
   Agency)
Environmental response team (ERT), **7**-8,
   7-83
Equipment:
   cleanup, **3**-5, **6**-20
   damaged, cost of, **4**-32
   development, **15**-6
   needs, **7**-47
   response, **7**-61
   sources, **7**-47
   spacing, **10**-76
   spill team, **7**-24
ERT (environmental response team), **7**-8,
   7-83
Escape-type respirator, **10**-56
Ethylene, fire fighting, **10**-77
Ethylene oxide, **7**-67
European Agreement Concerning the International Carriage of Dangerous
   Goods By Road (ADR), **2**-15, **3**-26, **3**-
   29, **3**-34, **7**-56
European Council of Chemical Manufacturers Federation (CEFIC), **3**-34
European Economic Community, **2**-14
European governmental response plans, **7**-
   29
European information systems (*see* Infor-
   mation systems, governmental, Eu-
   rope)
Evacuation, **7**-31, **7**-82, **7**-92, **11**-12
   distances, **10**-16
Evaluation, **7**-12

Evaluation checklist, 7-94
Evaluation sheet, 7-36
Evaporation, 10-13
Explosion, 10-65, 13-2
  mortality index, 5-16
  unconfined vapor cloud, I-7, 5-11
Explosives, 5-12
Explosives Act, 2-12
Exposure level, 13-7
Exposure radius, 10-17

Farming, land, of oily wastes, 9-35
Fault-tree risk analysis, 5-19
Federal Disaster Relief Act, 2-1
Federal Emergency Management Agency
  (FEMA), 2-6, 7-4, 9-27
Federal government agencies (*see names of
  specific agencies, for example*; Agricul-
  ture, U.S. Department of; Transpor-
  tation, U.S. Department of)
Federal Insecticide, Fungicide, and Roden-
  ticide Act, 2-8
Federal Office of the Environment Plan-
  ning and Information System (UM-
  PLIS), 3-42
Federal Railroad Administration, 6-45
Federal Railroad Safety Act, 6-44
FEMA (Federal Emergency Management
  Agency), 2-6, 7-4, 9-27
Field bioassays, *in situ*, 8-20
Field-constructed carbon columns, 9-10
Field kits, 8-6
Field laboratories, 8-8
Field treatment systems, 12-14
Films, 9-36
  surfactant, 10-26
Filtration, mixed-media, 9-33
Finland, laws, 2-15
Fire, 7-34, 13-2, 10-65
  ammonia, 10-35
  control systems: dry chemical, 10-76
    high-expansion foam, 10-77
    inerting, 10-78
    low-expansion foam, 10-77
    mortality index, 5-16
    ventilation, 10-78
    water, 10-78
  hazard prediction, 10-73
  hydrocarbon, 10-26
  liquid gases, 10-69
  pesticides, 11-2
  protective materials, 10-76
  ships, 7-75

Fire departments, I-7, 3-42, 4-3, 7-15, 7-50,
  10-59, 11-12
  authority, 7-86
  preplanning, 7-86, 11-7
  training, 7-93
Fire-fighting foams, 10-26
Fireballs, 5-11
Fish and Wildlife Service, 7-5
Fisheries Act, 2-11
Fishery Conservation and Management
  Act, 4-26
Fixation technique, 14-12
Flame detection, 10-75
Flammability limits, 5-8
  liquid gas, 10-67
Flammability measurement, 10-3
Flammables, 6-5
  disposal, 14-13
  liquid, 7-35
  liquid gases, 10-65
  vapors, 10-21
    monitoring, 13-15
Flash vaporization, gas, 5-4
Flash-off, 10-14
Flashing, liquid gas, 10-71
Flixborough, England, 4-33, 5-3
Floating materials, recovery, 9-20
Floor drains, 6-16
Fluorochemical surfactant foams, 10-26
Flushing, 4-3, 7-37, 7-51
Foam diking system, portable, 9-31
Foam plugs, 9-4
Foams, 7-17, 9-31, 11-7, 11-13
  aqueous film-forming (AFFF), 10-26
  fire-fighting, 10-26
  fluorochemical, 10-26
  Light Water ATC, 10-30
  mitigation times, 10-29
  new systems, 10-30
  protein-based, 10-27
  quick-setting, 6-20
  suppression capability, 10-28
  surfactant, 10-27
  Type L, 10-30
  Universal, 10-30
Framework Directive on Wastes (EEC), 2-
  14
France:
  governmental information systems, 3-42
  laws, 2-16

Gas Carrier Code (IMCO), 7-32
Gas chromatograph, 10-10

Gas chromatography–mass spectrography (GC-MS), 11-24
Gas-mask canisters, 13-10
Gases:
  analysis, 10-13
  chromatograph, 10-10
  chromatography–mass spectrometry, 11-24
  control, 9-13
  flammable, 10-65
    detection, 10-6, 10-75
  flash vaporization, 5-4
  heavier-than-air, 10-73
  liquid (see Liquid gases)
  monitoring (see Monitoring, gases)
  OSHA limits, 10-11
  plume, 10-13
  releases, 7-35
  spills, 5-3, 10-14
  toxicity, 5-8
  (See also Ammonia; Chlorine; Liquefied natural gas; Liquefied petroleum gases)
Gauging devices, 6-31
GC-MS (gas chromatography–mass spectrometry), 11-24
Gel formation, 10-23
Geography, plant, 6-6
Germany, Federal Republic of, 3-42
  laws, 2-16
  response plans, 7-40
GESAMP (Joint Group of Experts on the Scientific Aspects of Marine Pollution), 1-4
Givaudin Research Co., 11-19, 11-22
Governmental information systems (see Information systems, governmental)
Governmental plans:
  European, 7-28
  U.S., 7-3
Great Lakes, 7-11, 7-77
Greece, laws, 2-17
Groundwater, 7-39, 14-13
  recovery and treatment, 9-3, 9-17
Grouting, soil, 9-31
Guthion, 12-21

HACS (Hazard Assessment Computer System), 3-2, 7-81
Hague Protocol, 1-23
Halosolvents, disposal, 14-15
Harmful quantities, determination of, 4-24

Harwell, U.K., National Response Center, 1-10, 3-31, 3-36, 7-35, 7-57
Hazard(s), 13-2
  assessment, 5-1, 7-82, 13-8
    computer system (HACS), 3-2, 7-81
    handbook, 3-3
  calculation of area, 10-17
  classes, 1-22, 3-26
  control systems: liquid gases, 10-74
    vapors, 10-21
  detection systems, 10-74
    manufacturers of, 10-78
  gaseous, 13-7
  identification, 3-26
  information, labeling systems, 3-27
  level, PCB, 12-13
  particulate, 13-7
  prediction, fire, 10-73
  quantification, 5-16
Hazard Assessment Computer System (HACS), 3-2, 7-81
Hazardous Chemical Data Manual, 3-3
Hazardous Emergency Leak Procedure (HELP), 3-24
Hazardous materials:
  categorization, 14-22
  characterization, 14-22
  destruction, 14-21
  identification, 14-22
  survey, 7-87
Hazardous Materials System (HAZMATS), 3-45, 7-92
Hazardous Materials Transportation Act, 2-8, 6-44
Hazardous substance(s):
  control, 4-24
  labeling road tankers, 3-27, 7-37
Hazardous waste sites, uncontrolled releases, 9-24
Hazards profile, 1-8
Hazchem, 3-31
  signs, 7-57
  use: Australia, 3-32
    Netherlands, 3-32
    Sweden, 3-32
Hazfile, 3-40
HAZMATS (Hazardous Materials System), 3-45, 7-92
Health and Human Services, U.S. Department of, 7-4
Health and Safety Commission (U.K.), 2-18
Health and Safety Executive (U.K.), 7-31
Health and Safety at Work Act (U.K.), 7-31
Health screening program, 7-22

Heavier-than-air gas, **10**-73
Heavier-than-water liquids, **12**-26
Heavy metal detection, **8**-11
Heavy metal precipitation, **9**-53
HELP (Hazardous Emergency Leak Procedure), **3**-24
Helsinki (Baltic) Convention, **14**-45
Highway authority (U.K.), **7**-37
Holes, plugging, **7**-17
Home Office handbook (U.K.), **7**-35
Hot-tap, pipe, **11**-13
Hydrocarbons:
  fires, **10**-26
  volatile, **10**-26
  (*See also* Liquefied natural gas; Liquefied petroleum gases)
Hydrochloric acid cloud, **11**-11
Hydrogen cyanide, toxicity, **5**-9
Hydrogen sulfide, **10**-14
  toxicity, **5**-9
Hydrogeology, **14**-13

IATA (*see* International Air Transport Association)
ICAO (International Civil Aviation Organization), **1**-25
ICMESA factory, dioxin incident at, **11**-18
Identification, **7**-67
  chemical analysis, **7**-82
  hazard, **3**-26
  hazardous materials, **14**-22
  United Nations identification number, **3**-27, **7**-35, **7**-56, **7**-72
IDLH (immediately-dangerous-to-life-or-health) values, **13**-7
Illinois Environmental Protection Agency, **11**-12
IMCO (*see* Intergovernmental Maritime Consultative Organization)
Immediately-dangerous-to-life-or-health (IDLH) values, **13**-7
Impact, **15**-1
  biological treatment process, **4**-7
  economic, **15**-3
  environmental, **4**-15, **7**-2
  municipal facilities, **4**-1
Incineration:
  dioxin, **14**-32
  emission control, **14**-3
  glossary, **14**-35
  land, **14**-20
  liquids, **14**-23
  mobile, **9**-34

Incineration (*Cont.*):
  monitoring, **14**-34
  operation, **14**-21
  organocholine, **14**-37
  sea (*see* Ocean incineration)
  sludge, **14**-24
  systems, **14**-25
  types, **14**-27
  (*See also* Combustion)
Incinerator disposal: ash, **14**-32
  scrubbing solutions, **14**-33
Incinerators, types of, **14**-27
Induced air movement, **10**-22
Industrial contingency plans, **3**-24
Industrial response, **7**-48
  pollution control officer, **7**-54
Industrial response plans, **7**-43, **7**-56
Industrial team response, **7**-60
Information systems:
  computerized, **3**-3, **3**-9, **3**-42
  governmental: Canada, **3**-45
    Europe, **3**-25, **7**-56
      Belgium, **3**-42
      France, **3**-42
      German Federal Republic, **3**-42
      Netherlands, **7**-39
      Switzerland, **3**-42
      United Kingdom, **3**-25
    United States of America, **3**-2
      Coast Guard, **3**-2
      EPA, **3**-9
    National Fire Protection Association, **3**-32
Information transfer, **9**-36
Infrared analyzer, **10**-10
Initiation of action, **7**-12
Inspection, **6**-7
Institut voor Chemie-Ingenieurstechnik, **3**-42
Institut für Wasserforschung, **3**-42
Insurance agencies, **6**-5
Interagency Radiological Assistance Plan (IRAP), **3**-23, **7**-4
Intergovernmental Maritime Consultative Organization (IMCO), **1**-2, **1**-13, **6**-29, **7**-72, **14**-47
  Bulk Chemicals Code, **1**-6, **7**-32
  Dangerous Goods Code, **1**-6, **3**-26
  Gas Carrier Code, **7**-32
Interior, U.S. Department of the, **7**-5
International Air Transport Association (IATA), **1**-21
  Cargo Traffic Conference Resolution 618, **1**-22

International Air Transport Association (IATA) (*Cont.*):
  Restricted Articles Regulations, 1-24, 3-26
International Chamber of Shipping (ICS), 7-73
International Civil Aviation Organization (ICAO), 1-25
International contingency plans, 7-11
International Convention for the Prevention of Pollution From Ships (MARPOL), 1-5
International Convention Relating to Intervention on the High Seas in Cases of Oil Pollution Casualties, 1-11
International Convention for the Safety of Life at Sea (SOLAS), 1-3
International Maritime Dangerous Goods Code, 1-3, 7-73
International Register of Potentially Toxic Chemicals (IRPTC), 3-42
International Regulations Concerning the Carriage of Dangerous Groups by Rail (RID), 3-26, 3-29, 7-56
Interstate Commerce Commission (ICC), 6-44
Intervention on the High Seas Act, 2-6
Inventories, 4-5
Investigation, 4-5
Ion exchange, 9-54
IRAP (Interagency Radiological Assistance Plan), 3-23, 7-4
Ireland, laws, 2-17
IRPTC (International Register of Potentially Toxic Chemicals), 3-42
Irradiation, dioxin, 11-36
Istituto Superiore di Sanità, 11-19
Italian Parliamentary Commission, 11-19

Jacksonville, Fla., 7-23
Jefferson County, Ky., 7-23
Jettisoning, 7-74
Joint Committee on Fire Brigade Operations (U.K.), 3-32
Joint Group of Experts on the Scientific Aspects of Marine Pollution (GESAMP), 1-4
Justice, U.S. Department of, 7-5

Kepone, 4-18
Kingston, Tenn., 12-5
Kuwait Convention, 14-45

Labels (*see* Placards)
Labor, U.S. Department of, 7-5
Laboratories: field, 8-8
  mobile, 8-8, 9-14, 9-30
  off-site, 8-8
Land farming of oily wastes, 9-35
Land incineration, 14-20
Land response plans, 7-64
Landfill, 14-9
Landfill risk, 14-15
Laws:
  Carriage of Dangerous Goods, 2-16
  disposal of waste at sea, 14-45
  environmental protection, 2-5
  national: Belgium, 2-15
    Canada, 2-10
    Denmark, 2-15
    Finland, 2-15
    France, 2-16
    German Democratic Republic, 2-16
    German Federal Republic, 2-16
    Greece, 2-17
    Ireland, 2-17
    Luxembourg, 2-17
    Netherlands, 2-17
    Norway, 2-18
    Sweden, 2-18
    U.S.S.R., 2-20
    United Kingdom, 2-18
    United States, 2-2
  Nuisance, 2-17
  Pollution of Surface Waters, 2-17
  Protection, Use, and Care of Waters, 2-16
  Toxic Wastes, 2-15
  Waste Disposal, 2-17
  Waste Disposal and Waste Recovery, 2-16
Leachate recovery, 9-19
Leaching, 12-16, 12-28
Leaks, 6-15
  chlorine, 10-56
Legal decisions in spill prevention, 6-22
LEL (lower explosion limit), 10-3
Level alarms, 6-31
LFL (lower flammability limit), 10-23, 10-69
Light Water ATC (foam system), 10-30
Lime, 12-8
Liquefied natural gas (LNG), 1-14, 10-26, 10-66
  firefighting, 10-77
Liquefied petroleum gases (LPG), 5-4, 7-38, 7-75, 10-66
Liquid gases, 10-65

Liquid gases (*Cont.*):
  boil-off, **10**-72
  control, **10**-70
  dispersion, **10**-71
  flammability limits, **10**-67
  flammable, **10**-65
  flashing of spilled, **10**-71
  hazards, **10**-66
  properties, **10**-66
  pumps, **10**-70
  spill control, **10**-70
  toxicity, **10**-68
Liquids: flammable, **7**-35
  heavier-than-water, **12**-26
  incineration, **14**-23
  phase modification, **10**-23
  spills, **5**-3
  volatile, **10**-13
LNG (*see* Liquefied natural gas)
Loading, **6**-52
Local contingency plans, **7**-7, **7**-20
Local response plans, **7**-15
Logic diagram, **7**-53
Logical decisions in spill prevention, **6**-23
Lombardy Regional Council, **11**-23, **11**-37
London Dumping Convention, **4**-13
London Fire Brigade, **3**-32
Low-temperature detection, **10**-74
Lower explosion limit (LEL), **10**-3
Lower flammability limit (LFL), **10**-23, **10**-69
LPG (*see* Liquefied petroleum gas)
Luxembourg, laws, **2**-17

Major accidents and natural disasters, responsibility for, in U.K., **7**-30
Manifests, **7**-74, **14**-2
Manual, spill control, **9**-14, **9**-29
Marine Environment Protection Committee (MEPC), **1**-3
Marine Pollution Control Unit (U.K.), **7**-32
Marine Protection, Research, and Sanctuaries Act, **1**-17, **2**-7, **2**-12, **14**-47
Maritime operations, spill prevention in, **6**-25
Maritime Safety Committee, **1**-3
MARPOL (International Convention for the Prevention of Pollution from Ships), **1**-5
Maryland Water Resources Administration, **12**-13
Mask, canister-type, **10**-56

Materials Transportation Bureau (MTB), **6**-47
Matthias I, II, and *III*, **14**-39
Medical assistance, **7**-82
Medical surveillance, **7**-83, **11**-9
Medicoepidemiological Commission, **11**-33
Metallic compounds, disposal, **14**-3
Meteorological conditions, **10**-15
Methanol, biological degradation, **9**-45
Methyl methacrylate, **7**-45
Microbial decomposition, **9**-35
Microbial degradation, **12**-14
  dioxin, **11**-36
Microinvertebrates, monitoring, **8**-18
Microorganisms, **9**-41
Milan, Italy, **11**-18
Ministry of Agriculture, Fisheries, and Food (U.K.), **7**-30
Mitigation, **7**-13, **8**-14
Mitigation times, foams, **10**-29
Mobile incineration, **9**-34
Mobile laboratories, **8**-8, **9**-14, **9**-30
Mobile physical-chemical treatment unit, **9**-12, **9**-32
Mobile spill alarm system, **8**-10
Mobile stream diversion system, **9**-36
Models, water quality, **8**-22
Monitoring, **4**-3, **9**-14
  air, **11**-13
  biological, **12**-29
  dioxin, **11**-24
  flammable vapors, **13**-15
  gases, **10**-3
    meter calibration, **10**-11
    oxygen deficiency, **13**-15
    toxic vapors, **13**-16
  incineration, **14**-34
  robot monitors, **8**-10
  toxicity (*see* Toxicity, monitoring)
Montgomery County (Ohio) Combined General Health District, **7**-23
Mortality index, **5**-16
MTB (Materials Transportation Bureau), **6**-47

NACA (National Agricultural Chemicals Association), **3**-23
NATES (National Analysis of Trends in Emergency Systems), **3**-49
National Agricultural Chemicals Association (NACA), **3**-23
  pesticide safety team network, **3**-23

National Analysis of Trends in Emergency Systems (NATES), 3-49

National Arrangements for Dealing with Incidents Involving Radioactivity (NAIR), 7-30, 7-37

National Association of Waste Disposal Contractors of the United Kingdom, 3-30, 7-37

National Chemical Emergency Center (see Harwell, U.K.)

National Emergency Equipment Locator System (NEELS), 3-45, 7-91

National Fire Protection Association (NFPA), 7-23
information system, 3-32

National Institute for Occupational Safety and Health (NIOSH), 9-27, 13-8

National Oceanic and Atmospheric Administration (NOAA), 7-4

National Oil and Hazardous Substance Pollution Contingency Plan, 2-4, 3-2, 7-3, 7-18, 7-77, 9-27, 9-40

National Pollution Discharge Elimination System (NPDES), 4-24, 7-9

National Response Center, 3-3, 4-25, 7-5

National response team (NRT), 7-6, 7-79, 7-83, 9-27

National Strike Force, 2-5, 7-8, 7-79

National Transportation Safety Board (NTSB), I-8

National Water Council, 7-39

NATO/CCMS (North Atlantic Committee on the Challenges of Modern Society), 14-14

Natural Resources, Department of, 3-48

Navigation safety, 6-25, 6-33

NEELS (National Emergency Equipment Locator System), 3-45, 7-91

Netherlands, the:
governmental information systems, 7-39
Hazchem use, 3-32
response plans, 7-39
specialist information, 3-42

Neutralization, 4-10, 9-50
chlorine, 10-61, 10-64
in situ, 9-15
pesticide, 12-30

New Jersey, 14-1

NFPA (see National Fire Protection Association)

NIOSH (National Institute for Occupational Safety and Health), 9-27, 13-8

NOAA (National Oceanic and Atmospheric Administration), 7-4

North Atlantic Committee on the Challenges of Modern Society (NATO/CCMS), 14-14

Norway, laws, 2-18

Notification, 6-22, 7-12, 7-53
level, U.K., 5-18
telephone numbers, 7-6

NPDES (National Pollution Discharge Elimination System), 4-24, 7-9

NRC (see Nuclear Regulatory Commission)

NRT (see National response team)

NTSB (National Transportation Safety Board), I-8

Nuclear Regulatory Commission (NRC), 3-24, 7-18, 7-50, 7-79, 7-84, 7-89, 9-27

Number, United Nations identification, 3-27

Occupational Safety and Health Act, 2-8

Occupational Safety and Health Administration (OSHA), 7-5

Ocean Dumping Act (see Marine Protection, Research, and Sanctuaries Act)

Ocean incineration, 1-19, 14-37
cost, 14-48
environmental impact, 14-41
incinerator ships, 14-37
licensing, 14-48

Odor, 12-27

Off-site laboratories, 8-8

Ohio Environmental Protection Agency (OEPA), 11-5

Ohio River Valley Water Sanitation Commission (ORSANCO), 4-4, 8-10

OHM-TADS (see Oil and Hazardous Materials Technical Assistance Data System)

Oil Companies International Marine Forum (OCIMF), 7-73

Oil and Hazardous Materials Spills Branch (OHMSB), U.S. EPA, 9-36

Oil and Hazardous Materials Technical Assistance Data System (OHM-TADS), I-9, 3-9, 7-92, 12-2

Oil Pollution Prevention regulation, 7-11

Oil spills, magnitude of problem, I-2

Oleum, 10-30

On-scene coordinator (OSC), 2-5, 3-3, 3-9, 5-45, 6-34, 7-5, 7-8, 7-77, 7-81, 9-36, 11-4, 13-2, 13-17

Operating personnel, 6-28

Organocholine, incineration, 14-37

Organometallic compounds, 7-34

Organophosphate:
  detection, **8**-11, **8**-16
  pesticide, **12**-22
ORSANCO (Ohio River Valley Water Sanitation Commission), **4**-4, **8**-10
OSC (*see* On-scene coordinator)
OSHA (*see* Occupational Safety and Health Administration)
Oslo Convention, **1**-13
Outer Continental Shelf Lands Act, **2**-5, **4**-26
Overpressurization, **10**-69
Oxidation-reduction, **4**-11, **9**-56
Oxygen:
  deficiency, **13**-2
  depletion, **5**-10

Packaging and Labelling of Dangerous Substances Regulations, **7**-37
PCB (polychlorinated biphenyl), **4**-18, **8**-8, **9**-5, **12**-2, **14**-26
  disposal, **9**-34, **12**-3, **14**-14
  transportation accident, **12**-25
Peat-moss dams, porous, **9**-15
Permits for dumping matter at sea, **1**-15
Personnel:
  monitors, **13**-14
  operating, **6**-28
  protection, **7**-8, **11**-8, **11**-37
    clothing, **12**-20, **13**-4
  (*See also* Safety; Safety equipment; Training)
Pest Control Products Act, **2**-12
Pesticides, **12**-18
  cleanup, **11**-5, **12**-19
  detection, **8**-11, **8**-16
  disposal, **9**-33, **14**-16
  fires, **11**-2
  information, **3**-23
  neutralization, **12**-30
  organophosphates, **12**-22
  residues, disposal, **11**-5, **12**-22
  toxicology, **11**-21
Phenol, **12**-12
  biodegradation, **9**-45, **9**-58
  disposal, **14**-13
Phosgene, toxicity, **5**-9
Phosphorus, **7**-34
Photolysis, dioxin, **11**-35
Physical-chemical treatment systems, **9**-9, **9**-50
Physical-chemical treatment unit, mobile, **9**-12, **9**-32

PIAT (public information assistance team), **7**-9
Pipe alleys, **6**-12
Pipeline:
  ammonia, **10**-41
  spills, **7**-38
Pipelines Act, **7**-38
PIRS (Pollution Incident Reporting System), **3**-7, **4**-29
Placards, **2**-18, **3**-27, **6**-47, **7**-87
  chlorine, **10**-55
Plant(s):
  design, **11**-7
  geography, **6**-6
  location, **11**-7
  operations, **6**-2
Plant supervisor, response, **7**-50
Plug flow, **12**-30
Plugs, foam, **9**-4
Plume:
  chlorine, **10**-62
  gases, **10**-13
  trajectory, **11**-30
  water, **8**-6
Poisons:
  classified list, **2**-16
  information service, **3**-40
  pesticides, **11**-6
Police Department, Chicago, **11**-12
Pollution charges, **4**-37
Pollution Incident Reporting System (PIRS), **3**-7, **4**-29
Pollution report (POLREP), **7**-14
Pollution of the sea, **1**-7, **2**-26
POLREP (pollution report), **7**-14
Polychlorinated biphenyl (*see* PCB)
Polyelectrolytes, **9**-52
Polyneuropathy, **11**-18
Ponds, containment, **6**-15
Ports and Tanker Safety Act, **7**-79
Ports and Waterways Safety Act, **1**-7, **2**-8
Precipitation, **4**-11
  heavy metals, **9**-53
Preparedness systems for pollution emergencies, **2**-18
Prevention, **I**-10, **4**-8, **6**-1, **7**-43, **7**-72, **9**-26, **15**-2
  cost of, **4**-36
  liquid gases spills, **10**-70
  maritime operations spills, **6**-25
  rail transport spills, **6**-44
  transportation spills, **6**-52
Process disruption, cost, **4**-31

Product(s):
  knowledge, **6**-8
  loss, 6-3
    value of, **4**-30
  separators, **6**-15
Proposals for Dangerous Substances (Conveyance by Road), **2**-19
Propylene tanker disaster, **5**-4
Protection of personnel (*see* Personnel, protection)
Protein-based foams, **10**-27
Pseudomonads, **9**-47
Public information assistance team (PIAT), **7**-9
Pumps, liquid gas, **10**-70

Quantification of hazards, **5**-16
Quick-setting foams, **6**-20

Radioactive materials, **3**-23
  disposal, **1**-17
Railroads, **4**-32, **6**-43, **7**-35
RAM model for calculating hazard area, **10**-18
RCRA (*see* Resource Conservation and Recovery Act)
Reconnaissance, aerial, **8**-11
Records, **6**-3
Recovery, **9**-3
Reference books, **7**-92
Refrigerants, **10**-25
Regional contingency plan, **3**-2
Regional response team (RRT), **2**-5, **7**-5, **7**-79
Regions, U.S., **7**-10
Regulations, **4**-22
Regulatory controls, **4**-36
Rehabilitation:
  of aquatic ecosystems, **4**-16
  in TCDD accident, **11**-23
Remote detection, **8**-9
Remote sensing, **8**-9
Removability, **4**-24
Reportable quantities, **4**-25
Reporting, **3**-48
Research and development:
  U.S. Coast Guard, **7**-80
  U.S. EPA, **9**-24
Resource Conservation and Recovery Act (RCRA), **14**-3, **14**-26, **14**-37
Respirator Decision Logic, **13**-13

Respirators:
  air-purifying, **13**-8
  atmosphere-supplying, **13**-11
  chlorine, **10**-55
  escape-type, **10**-56
  protective, **13**-6
  selection of, **13**-13
  training program for respiratory protection, **13**-14
Response, **6**-34, **15**-4
  capability, **7**-17
  checklist, **7**-62
  industrial, **7**-48
    pollution control officer, **7**-54
  industrial team, **7**-60
  methods, **3**-3, **7**-82
  phases, **7**-12
  plans, **7**-1, **9**-27
    fire service, **7**-86
    governmental: European, **7**-29
      U.S., **7**-3
    industrial, **7**-43, **7**-56
    land, **7**-64
    local, **7**-15
    transportation, **7**-72
    U.S. Coast Guard, **7**-77
  plant supervisor, **7**-50
  systems, **3**-45
Restoration, **9**-35, **15**-6
Restricted Articles Regulations, **1**-24
Restricted Articles Shipping Certification (IATA), **1**-23
Retention basins, **6**-15
Revolving fund, **2**-5
RID (*see* International Regulations Concerning the Carriage of Dangerous Goods by Rail)
Risk, **6**-43
  analysis, **5**-19, **14**-9, **15**-2
  assessment, **5**-1
  estimation, **4**-5
  landfill, **14**-15
River basin organizations (U.S.S.R.), **2**-24
Road Haulage Association, **7**-37
Road Traffic Act, **2**-17, **2**-20
Robot monitors, **8**-10
RRT (*see* Regional response team)
Rule 36 of the Texas Railroad Commission, **10**-15

Safe Drinking Water Act, **2**-2, **2**-5
Safe Working Practices aboard Merchant Ships (U.K.), **7**-75

Safety, **I**-11
blowoff areas, **6**-16
cleanup site, **9**-27
in decontamination at scene of accident, **7**-70
in marine emergencies, **7**-32
during maritime transportation, **7**-72
in mobile spill response units, **9**-14
of personnel, **7**-13, **7**-88
(*See also* Personnel, protection)
in pesticide fires, **11**-7
of the public, **7**-88
in sampling, analysis, and detection, **8**-2
Safety equipment:
personnel, **13**-2
totally encapsulated suit, **13**-19
Safety and Reliability Directorate of the U.K. Atomic Energy Authority, **5**-19
Sampling and analysis, **9**-29
preservation techniques, **8**-3
water, **8**-2
San Carlos, Spain, fireball, **5**-11
SCBA (*see* Self-contained breathing apparatus)
Scientific support coordinator (SSC), **7**-9
Sea:
dumping wastes at, national legislation, **14**-47
incineration at (*see* Ocean incineration)
pollution of the, **1**-7, **2**-26
spills, **7**-32
SEACHEM (Spillages Emergency Action for Chemicals), **7**-32
Security, **6**-5
Sediment treatment, **9**-33
Sedimentation, **9**-51
Self-contained breathing apparatus (SCBA), **7**-74, **10**-56, **11**-8, **11**-12, **13**-12
Sensing, remote, **8**-9
Separation techniques, **9**-32
Seveso, Italy, **5**-3, **5**-15, **7**-29, **11**-8
Shell Chemical Company, **14**-41
Shipper's certification, **1**-24
Shipper's liability, **1**-24
Shipping papers, **6**-47
Ships:
decontamination, **7**-75
fire, **7**-75
incinerator, **14**-37
vessel design, **6**-25
Shoreline contamination, **7**-34
Silicon tetrachloride, **10**-31, **11**-11
Sinking chemicals, recovery, **9**-15

SKIM (*see* Spill Cleanup Inventory System)
Skimmers, **9**-21
Sludge, activated, **9**-45
Sludge incineration, **14**-24
Sodium dodecyl benzene sulfonate, **7**-45
Soil:
contamination, **11**-24
grouting, **9**-31
recovery and treatment, **9**-21, **9**-33
SOLAS (International Convention for the Safety of Life at Sea), **1**-3
Solid spills, **5**-3
Solid Waste Disposal Act, **14**-3
Solids, combustion of, **14**-24
Sorbents, **6**-19, **9**-21
Sorption, **10**-23
Source control, **9**-2
SPCC (*see* Spill prevention control and countermeasure)
Spill alarm system, mobile, **8**-10
Spill area prediction, **10**-73
Spill Cleanup Inventory System (SKIM), **3**-5, **7**-6, **7**-91
Spill Control Association of America, **6**-4
Spill prevention control and countermeasure (SPCC), **I**-2, **I**-10, **6**-21, **7**-9
Spill response summaries, **9**-5
Spill team equipment, **7**-24
Spillages Emergency Action for Chemicals (SEACHEM), **7**-32
Spiller's responsibility, **7**-18
SSC (scientific support coordinator), **7**-9
Standard federal regions, **7**-10
Standard Transportation Commodity Code (STCC), **6**-47
State, U.S. Department of, **7**-5
STCC (Standard Transportation Commodity Code), **6**-47
Storage drums, **6**-13
Storage tanks, **6**-11
Straw, **12**-29
Stream, leaks into, **9**-4
Stream bypass, **9**-9, **12**-29
Stream diversion system, mobile, **9**-36
Styrene, **7**-45
Sulfur dioxide, toxicity, **5**-9
Sulfur trioxide, **10**-30
Superfund (*see* Comprehensive Environmental Response, Compensation, and Liability Act)
Suppression capability, foams, **10**-28
Surface cooling, **10**-24
Surfactant foams, **10**-27
Surveillance, **4**-3

Sweden
  Hazchem use, 3-32
  laws, 2-18
Switzerland, 11-22
  governmental information systems, 3-42

Tank(s): liquid gas, 10-70
  portable, 9-14
  storage, 6-11
Tank cars, 7-35, 7-46
Tank trucks, 6-5, 6-31, 7-66
Tankermen, 6-28
Tankers, liquefied gas, 10-70
TAT (technical assistance team), 7-18
TCDD (see Dioxin)
TCP (trichlorophenol), 11-18
TEAP (Transportation Emergency Assistance Program), 7-89, 10-57
TEC (U.S. Army Technical Escort Center), 11-14
Technical assistance team (TAT), 7-18
Technical decisions for spill prevention, 6-24
Technical Information Center, Oil and Hazardous Materials Spills Branch, U.S. EPA, 9-36
Technicoscientific Commission, Rome, 11-39
Technology development:
  for cleanup, 9-24
  for restoration, 15-6
Telephone numbers, emergency, 7-89
TEM model for calculating hazard area, 10-18
2,3,7,8-Tetrachlorodibenzo-p-dioxin (TCDD) (see Dioxin)
Tetraethyl lead, 7-32
Texas Railroad Commission, 10-15
Threshold-limit value (TLV), 10-4, 10-21, 13-7
TNT equivalent, 5-12
Toluene, 7-45, 12-26
Topping off, 6-29
Torrey Canyon, I-2, 7-35
Toxic and Hazardous Materials Group (U.K.), 7-59
Toxic Substances Control Act, 2-7, 14-26
Toxic vapors, 7-31
Toxicity, 13-2
  aquatic, 7-44, 7-45
  gases, 5-9
  liquid gas, 10-68
  measuring devices, portable, 8-16

Toxicity (Cont.):
  monitoring: fish, 8-18
    luminescent bacteria, 8-17
    microinvertebrates, 8-18
  tests, 8-14
Toxicology, 12-24
  aquatic, 8-14
  dioxin, 11-33
  pesticides, 11-21
Trace contaminants, definition, 1-18
Trade, Department of (U.K.), 7-32
Traffic control, 6-8
Trailer, decontamination, 9-28
Training, 3-48, 6-22, 6-44, 11-7
  fire service, 7-93
  maritime, 6-29, 7-75
  respiratory protection, 13-14
Trajectory, plume, 11-30
Trans-Alaska Pipeline Act, 2-3
Transport emergency cards (Tremcards), 3-34, 7-58
Transportation, 7-82
  accidents, I-7
    ammonia, 10-42
    PCB, 12-25
    pesticide, 12-18
    phenol, 12-12
  ammonia, 10-41
  bulk, 3-36
  chemical wastes, 2-17, 3-36, 14-11, 14-24
  chlorine, 10-52
  common carriers, 4-29
  emergency: assistance, industrial, 3-24
    cards, 3-34, 7-58
    reporting procedures, 3-24
  maritime placards, 3-27
  poisons, 2-16
  rail, 4-32, 6-43, 7-35
  response plans, 7-72
  ship (see Ships)
  spills, 6-52, 7-31
  vehicle inspection, 6-8
  vessel design, 6-25
Transportation, U.S. Department of, 4-29, 6-47, 6-51, 7-5, 10-41
Transportation of Dangerous Goods Act, 2-10, 3-49
Transportation Emergency Assistance Program (TEAP), 7-89, 10-57
Transportation of Explosives Act, 6-44
Treatment, 9-3, 7-49
  biological, 9-44
  chemical and physical systems, 9-9, 9-50
  field system, 12-14

Treatment (*Cont.*):
  mobile, **8**-14
Tremcards (transport emergency cards), **3**-34, **7**-58
Trichlorobenzene, **12**-5
Trichlorophenol (TCP), **11**-18
Trion, Ga., Pyranol incident, **12**-6

UKHIS (United Kingdom Hazard Information System), **3**-31, **7**-67
Ullage gas, **10**-71
Ultimate disposal, **9**-33, **14**-1
  (*See also* Disposal; Incineration)
Unconfined vapor cloud explosions, **I**-7, **5**-11
Uncontrolled waste disposal sites, **15**-7
Underflow dam, **6**-20
Union Carbide Corp., emergency assistance by, **3**-24
U.S.S.R. laws, **2**-20
United Kingdom (U.K.):
  Advisory Committee on Major Hazards, **5**-8
  Atomic Energy Authority, **3**-36
  governmental information systems, **3**-25
  Hazard Information System (UKHIS), **3**-31, **7**-67
  laws, **2**-18
  response plans, **7**-29
  responsibility for major accidents and natural disasters in, **7**-30
United Nations (UN):
  Committee of Experts on the Transport of Dangerous Gases, **1**-21, **3**-26
  Conference on the Law of the Sea, **2**-22
  environment program, **3**-42
  identification number, **3**-27, **7**-35, **7**-56, **7**-72
U.S. Army:
  Chemical Systems Laboratory, **8**-6
  Corps of Engineers, **2**-6
  Technical Escort Center, **11**-14
U.S. Coast Guard (USCG), **6**-25, **7**-5, **9**-28, **10**-21, **10**-41, **11**-12, **12**-24
  contingency planning, **7**-81
  information system, **3**-2
  Marine Environmental Protection Division, Pollution Response Branch, **3**-5
  research and development, **7**-80
  response plans, **7**-77
  strike team, **1**-10
U.S. Environmental Protection Agency (U.S. EPA), **I**-2

U.S. Environmental Protection Agency (U.S. EPA) (*Cont.*):
  authority, **2**-3
    shared with U.S. Coast Guard, **7**-77
  criteria and standards for solid waste disposal, **14**-3
  definition of a hazardous spill, **7**-43
  emergency response team, **7**-8, **7**-83
  field analysis kit, **8**-6
  fire department training programs, **7**-23
  hazardous materials list, **7**-81
  incineration on land, **14**-26
  incineration at sea, **14**-39, **14**-44
  information system, **3**-9
  manual for field response personnel, **9**-15
  Oil and Hazardous Materials Spills Branch, Edison, N.J., **9**-25
  on-scene coordinator assignment and appointment, **3**-9, **6**-34
  PCB information, **12**-2
  pesticide fires, **11**-2
  physical-chemical treatment unit, mobile, **9**-12
  regional contingency plans, **7**-91
  research and development, **9**-24
  responsibility, **7**-5
    (*See also* authority, *above*)
  temporary leak patching during transport, **9**-4
  vapor hazard control, **10**-24
U.S. Geological Survey (USGS), **7**-5
U.S. government agencies (*see names of specific agencies, for example:* Agriculture, U.S. Department of; Transportation, U.S. Department of)
U.S. governmental information systems, **3**-2, **3**-9
U.S. governmental response plans, **7**-3
U.S. laws, **2**-2
  (*See also names of specific laws*)
U.S. Maritime Administration, **14**-39
Universal Gelling Agent, **10**-24
Unloading, **6**-5, **6**-9
Upper flammable limit (UFL), **10**-69
USCG (*see* U.S. Coast Guard)
USGS (U.S. Geological Survey), **7**-5

Valves and fittings, leaks from, **6**-15
Vaporization from pools, **5**-6
Vapors, **4**-32
  clouds, **5**-7
  control, **9**-31

Vápors, flammable (*Cont.*):
  emissions, **6**-5
  exhausts, **6**-16
  flammable, **10**-21
    monitoring, **13**-15
  liquid gases, **10**-65
  phenol, **12**-12
  suppression, **10**-26
  toxic, **7**-31
  unconfined vapor cloud explosions, **I**-7,
    5-11
Vegetation contamination, **11**-26
Volatile chemical vapor suppression, **9**-31
Volatile liquids, **10**-13
Vulcanus, **14**-39

Warsaw Convention, **1**-23

Wastewater treatment, **7**-39, **7**-51, **9**-58
  pesticide impact on, **11**-3
Water analysis, **8**-2
  field kit, **9**-29
Water Pollution Act (Ireland), **2**-17
Water quality models, **8**-22
Water-reactive chemicals, **10**-30
Water Resources Act (U.K.), **2**-19
Water supplies, **4**-16, **4**-35
  protection, **7**-38
  treatment plant, danger to, **11**-4
  (*See also* Drinking-water contamination)
Water treatment technology, **9**-50
  manual for, **9**-29
Weather, **6**-5, **10**-15
Weirs, **6**-20
Whitehouse, Fla., oil spill, **12**-8
Wipe tests of building surfaces, **11**-25

About the Editors

Gary F. Bennett

Dr. Bennett is a professor of biochemical engineering at the University of Toledo and a consultant to several organizations in the field of hazardous materials, including the City of Toledo Fire Department and the U.S. Environmental Protection Agency. He has been program director for and editor of several national conferences on hazardous materials spills and the problems of uncontrolled hazardous waste sites as well as currently serving as editor of the *Journal of Hazardous Materials*. He is the recipient of several awards by both the Toledo section and the national organization of the American Institute of Chemical Engineers. Dr. Bennett received his bachelor of science degree from Queens University in Kingston, Ontario, and his master of science and doctor of philosophy degrees from the University of Michigan in Ann Arbor.

Frank S. Feates

Dr. Feates is a director in the Air, Noise, and Waste Directorate of the Department of the Environment in London, England. Previously he was with the U.K. Atomic Energy Authority in Harwell as head of the Environmental Group and commercial manager of biological and chemical programs related to industrial research. In this capacity he established and headed the Chemical Emergency Response Centre to provide information for and respond to chemical spills in the United Kingdom. He was the initial editor of the *Journal of Hazardous Materials* and has written extensively himself on hazardous spills. Dr. Feates is a graduate of the University of London.

Ira Wilder

Mr. Wilder is chief of the Oil and Hazardous Materials Spills Branch of the U.S. Environmental Protection Agency in Edison, New Jersey. He has been involved for more than 12 years in the initiation, planning, execution, and management of research and development programs related to the prevention, control, and abatement of multimedia pollution from spills of hazardous chemicals and mismanaged hazardous waste disposal sites. He has numerous publications in the field and serves on the editorial board of the *Journal of Hazardous Materials*. Prior to joining the EPA, Mr. Wilder worked at the U.S. Naval Applied Science Laboratory and managed research programs involving the storage, handling, safety, and fire protection of petroleum products and hazardous materials. He received his bachelor of science degree in chemical engineering from the City College of New York.